ANNUAL REVIEW OF ENTOMOLOGY

ANNUAL REVIEW OF ENTOMOLOGY

VOLUME 36, 1991

THOMAS E. MITTLER, *Editor*

University of California, Berkeley

FRANK J. RADOVSKY, *Editor*

North Carolina State Museum of Natural Sciences, Raleigh

VINCENT H. RESH, *Editor*

University of California, Berkeley

ANNUAL REVIEWS INC.   4139 EL CAMINO WAY   P.O. BOX 10139   PALO ALTO, CALIFORNIA 94303-0897

R̲ ANNUAL REVIEWS INC.
Palo Alto, California, USA

International Standard Serial Number: 0066–4170
International Standard Book Number: 0–8243–0136-6
Library of Congress Catalog Card Number: A56-5750

∞ The paper used in this publication meets the minimum requirements of American National Standard for Information Sciences—Permanence of Paper for Printed
Library Materials, ANSI Z39.48-1984.

Typesetting by Kachina Typesetting Inc., Tempe, Arizona; John Olson, President
Typesetting Coordinator, Janis Hoffman

PRINTED AND BOUND IN THE UNITED STATES OF AMERICA

PREFACE

With this volume of the *Annual Review of Entomology,* my colleagues Frank J. Radovsky and Vincent H. Resh join me as the designated Editors of the *Review.* This change in their status (from Associate Editors) reflects an overdue recognition of the fact that we have shared equally, for several years, the numerous duties for which the scientific editors of an annual review are responsible. In close collaboration with a professional Production Editor, we have striven to ensure that the *Annual Review of Entomology* provides our readership with entomological reviews of the highest quality (and at the lowest cost a nonprofit organization such as Annual Reviews Inc. can offer).

As members of an Editorial Committee consisting of eight eminent entomologists from the USA and Canada, we also participate in the vital process of selecting a broad spectrum of topics for inclusion in successive volumes of the *Review* and in the nomination of potential authors to write those articles.

On average, we publish 25 chapters each year in the *Annual Review of Entomology.* Although editors of journals that publish considerably more manuscripts may envy our relatively light and seasonal work load, even our invited articles frequently pose complex and demanding challenges. Consistently, the publication process involves a series of most interesting and rewarding activities. As we follow successive articles from their conception to their birth two years later, we become progressively more attached to the *Review.* No wonder editors remain devoted to this task for so many years (and agree to serve a succession of five-year terms at the invitation of Annual Reviews Inc.).

Above all, we thank the authors of Volume 36 for their valuable contributions. The conscientious efforts of our Production Editor, Amanda Suver, and the high quality of the work done by our compositors and printers also deserve special recognition.

T. E. Mittler,
For the Editors

OTHER REVIEWS OF ENTOMOLOGICAL INTEREST

From the *Annual Review of Ecology and Systematics,* Volume 21 (1990)

Biogeography of Nocturnal Insectivores: Historical Events and Ecological Filters, William E. Duellman and Eric R. Pianka

Function and Phylogeny of Spider Webs, William G. Eberhard

Host Specialization in Phytophagous Insects, John Jaenike

Multivariate Analysis in Ecology and Systematics: Panacea or Pandora's Box?, Frances C. James and Charles E. McCulloch

Plant-Pollinator Interactions in Tropical Rain Forests, Kamaljit S. Bawa

From the *Annual Review of Genetics,* Volume 24 (1990)

Molecular Mechanisms Regulating Drosophila *P Element Transposition,* Donald C. Rio

Genetics of Early Neurogenesis in Drosophila melanogaster, José A. Campos-Ortega and Elisabeth Knust

Genetics of Circadian Rhythms, Jeffrey C. Hall

From the *Annual Review of Microbiology,* Volume 44 (1990)

The Generation of Genetic Diversity in Malaria Parasites, Glenn A. McConkey, Andrew P. Waters, and Thomas F. McCutchan

Sexual Differentiation in Malaria Parasites, Pietro Alano and Richard Carter

From the *Annual Review of Neuroscience,* Volume 14 (1991)

Genetic and Molecular Bases of Neurogenesis in Drosophila melanogaster, José A. Campos-Ortega and Yuh Nung Jan

From the *Annual Review of Phytopathology,* Volume 28 (1990)

Epidemiology of Barley Yellow Dwarf: A Study in Ecological Complexity, Michael E. Irwin and J. Michael Thresh

Protease Inhibitors in Plants: Genes for Improving Defenses against Insects and Pathogens, Clarence A. Ryan

Annual Review of Entomology
Volume 36, 1991

CONTENTS

GENE AMPLIFICATION AND INSECTICIDE RESISTANCE, *Alan L.
Devonshire and Linda M. Field*   1

INDUCTION OF DEFENSES IN TREES, *Erkki Haukioja*   25

SELF-SELECTION OF OPTIMAL DIETS BY INSECTS, *G. P. Waldbauer
and S. Friedman*   43

EVOLUTION OF OVIPOSITION BEHAVIOR AND HOST PREFERENCE IN
LEPIDOPTERA, *John N. Thompson and Olle Pellmyr*   65

AVERMECTINS, A NOVEL CLASS OF COMPOUNDS: Implications for
Use in Arthropod Pest Control, *Joan A. Lasota and Richard
A. Dybas*   91

TRAP CROPPING IN PEST MANAGEMENT, *Heikki M. T. Hokkanen*   119

THE SENSORY PHYSIOLOGY OF HOST-SEEKING BEHAVIOR IN
MOSQUITOES, *M. F. Bowen*   139

PROSPECTS FOR GENE TRANSFORMATION IN INSECTS, *Alfred M.
Handler and David A. O'Brochta*   159

BIOSYSTEMATICS OF THE CHEWING LICE OF POCKET GOPHERS,
Ronald A. Hellenthal and Roger D. Price   185

THE FUNCTION AND EVOLUTION OF INSECT STORAGE HEXAMERS,
William H. Telfer and Joseph G. Kunkel   205

MANAGEMENT OF DIABROTICITE ROOTWORMS IN CORN, *Eli Levine
and Hassan Oloumi-Sadeghi*   229

BIOLOGICAL CONTROL OF CASSAVA PESTS IN AFRICA, *H. R. Herren
and P. Neuenschwander*   257

SAMPLING AND ANALYSIS OF INSECT POPULATIONS, *Eizi Kuno*   285

ARTHROPOD BEHAVIOR AND THE EFFICACY OF PLANT PROTECTANTS,
Fred Gould   305

SENSILLA OF IMMATURE INSECTS, *Russell Y. Zacharuk and Vonnie
D. Shields*   331

TRANSMISSION OF RETROVIRUSES BY ARTHROPODS, *L. D. Foil and C. J. Issel*   355

ECOLOGY AND MANAGEMENT OF TURFGRASS INSECTS, *Daniel A. Potter and S. Kristine Braman*   383

BEHAVIORAL ECOLOGY OF PHEROMONE-MEDIATED COMMUNICATION IN MOTHS AND ITS IMPORTANCE IN THE USE OF PHEROMONE TRAPS, *Jeremy N. McNeil*   407

WHITEFLY BIOLOGY, *David N. Byrne and Thomas S. Bellows, Jr.*   431

AEDES ALBOPICTUS IN THE AMERICAS, *Karamjit S. Rai*   459

ENVIRONMENTAL IMPACTS OF CLASSICAL BIOLOGICAL CONTROL, *Francis G. Howarth*   485

MATERNAL EFFECTS IN INSECT LIFE HISTORIES, *Timothy A. Mousseau and Hugh Dingle*   511

BIONOMICS OF LEAF-MINING INSECTS, *H. A. Hespenheide*   535

VEGETATIONAL DIVERSITY AND ARTHROPOD POPULATION RESPONSE, *D. A. Andow*   561

LYME BORRELIOSIS: Relation of Its Causative Agent to Its Vectors and Hosts in North America and Europe, *R. S. Lane, J. Piesman, and W. Burgdorfer*   587

ECOLOGICAL AND EVOLUTIONARY SIGNIFICANCE OF PHORESY IN THE ASTIGMATA, *M. A. Houck and B. M. OConnor*   611

INSECT HERBIVORY ON EUCALYPTUS, *C. P. Ohmart and P. B. Edwards*   637

OFF-HOST PHYSIOLOGICAL ECOLOGY OF IXODID TICKS, *Glen R. Needham and Pete D. Teel*   659

INDEXES

Subject Index   683

Cumulative Index of Contributing Authors, Volumes 27–36   692

Cumulative Index of Chapter Titles, Volumes 27–36   695

ANNUAL REVIEWS INC. is a nonprofit scientific publisher established to promote the advancement of the sciences. Beginning in 1932 with the *Annual Review of Biochemistry,* the Company has pursued as its principal function the publication of high quality, reasonably priced *Annual Review* volumes. The volumes are organized by Editors and Editorial Committees who invite qualified authors to contribute critical articles reviewing significant developments within each major discipline. The Editor-in-Chief invites those interested in serving as future Editorial Committee members to communicate directly with him. Annual Reviews Inc. is administered by a Board of Directors, whose members serve without compensation.

For the convenience of readers, a detachable order form/envelope is bound into the back of this volume.

Annu. Rev. Entomol. 1991. 36:1–23

GENE AMPLIFICATION AND INSECTICIDE RESISTANCE

Alan L. Devonshire and Linda M. Field

Insecticides and Fungicides Department, AFRC Institute of Arable Crops
Research, Rothamsted Experimental Station, Harpenden, Hertfordshire, AL5 2JQ,
United Kingdom

KEY WORDS: evolution of resistance, biochemistry of resistance, DNA methylation,
              Myzus persicae, Culex pipiens quinquefasciatus

PERSPECTIVES AND OVERVIEW

Of the two topics drawn together in this review, insecticide resistance will be
familiar to many readers because it has long been recognized as one of the
major challenges facing applied entomologists. Consequently, it has been
dealt with extensively over the years in various reviews (37, 41, 63, 67),
including several in the *Annual Review of Entomology,* which have recently
covered the economics of resistance (47), its ecological genetics (82), and
methods for its detection based on conventional and molecular techniques
(12). Fifteen years have passed since Plapp's (74) review of "The Biochemical Genetics of Resistance," which updated Oppenoorth's (66) contribution
with an identical title published in 1965. Since then, the term "biochemical
genetics" has taken on new connotations. The original objective was to use
classical genetic techniques of interbreeding to isolate major resistance genes
in a susceptible background, map them in relation to known marker genes,
and study their inheritance patterns (degree of dominance). Now, biochemical
geneticists have at their disposal the sophisticated techniques of molecular
biology that can identify changes at the DNA level. It is therefore timely to
reappraise progress in this area and in particular to focus on advances in the
understanding of mechanisms underlying the massive overproduction by

1

0066-4170/91/0101-0001$02.00

some resistant insects of esterases that detoxify insecticides. In two species, enhanced enzyme production has been firmly established to result from gene amplification, and this mechanism is the possible cause of resistance in many more cases.

This review relies heavily on the examples of gene amplification in the aphid *Myzus persicae* (35) and the mosquito *Culex pipiens quinquefasciatus* (60). However, these examples are set against the wider perspective of similar genetic events in other organisms and in cell cultures that develop resistance to cytotoxic agents.

Gene duplication has been recognized for many years as a major force in evolution (51); the redundant copy of the gene thereby generated can accumulate mutations that would otherwise be forbidden because they destroy the vital function of the original gene. This duplication and divergence constitutes one of the two broad classes of repetitive genes, that is those showing variant repetition (50); they are best exemplified by the least diverged gene families such as those for globins, actins, and immunoglobulins. Membrane ion channels, which show considerable homology to each other, appear to have acquired their functional specificity for particular ions through duplication and divergence (40) and are close to the subject of this review because they can be the targets of insecticides. They often comprise heterooligomers or in some cases (e.g. the sodium channel) the duplicated sequences have diverged but remained together as a coherent transcription unit producing a single large polypeptide that spans the lipid bilayer many times in a way similar to the oligomeric channels.

The other broad class of duplicated genes are those showing dosage repetition (50) or amplification. In this case, many identical gene copies are present in each cell to satisfy a very high demand for particular products that could not be achieved by a single copy gene even with a very active promoter. The best studied examples are the genes encoding the RNA components of the protein synthesizing machinery (50). In eukaryotes, they typically occur in one or several blocks of tandemly repeated ribosomal DNA sequences that alternate with nontranscribed spacer regions, the copy number per haploid genome ranging from less than 100 to more than 10,000 (50). These blocks are usually chromosomal but in many unicellular eukaryotes (as well as in oocytes) they also occur on extrachromosomal elements. In insects, the chorion genes responsible for the synthesis of egg shell proteins become developmentally amplified in ovarian follicle cells by selective over-replication of these loci (17).

Besides meeting the normal metabolic requirements of the cell, gene amplification can also play a major role in both the genesis of cancer cells (17) and their ability to defend themselves against the cytotoxic drugs used for cancer therapy (43, 92). A number of oncogenes have been shown to be amplified in cell lines and in tumors; increased expression at the mRNA and protein levels apparently confers a growth advantage on the cell. Such amplifications are often associated with chromosomal translocation of the oncogene, which can lead to rearrangements that generate abnormal proteins (17). When cancer cells are challenged by cytotoxic drugs, resistance often builds up by the selection of cells carrying amplified genes that protect the cell against the drug (see below). Although such selection in cell cultures is thought to act on pre-existing amplifications, the disruption of nucleotide synthesis by these drugs can induce DNA changes likely to engender amplification events (87).

The similarities between drug-resistant cells, which had already been shown unequivocally by molecular biological techniques to arise from gene amplification, and the massive quantitative changes in the esterase protein in insecticide-resistant *M. persicae* led to the original hypothesis of Devonshire & Sawicki (23) that gene amplification could be the basis of insecticide resistance.

RESISTANCE TO TOXIC SUBSTANCES BASED ON GENE AMPLIFICATION

Many studies of cultured mammalian cells have shown that these cells develop resistance to cytotoxic substances by amplification of the gene encoding the target protein of the toxin (for reviews, see 43, 92). The most-studied examples are amplification of the dihydrofolate reductase (DHFR) gene in cells treated with the anticancer drug methotrexate (MTX) and of the multifunctional CAD protein (carbamyl-P synthetase, aspartate transcarbamylase, dihydro-orotase) gene in response to treatment with N-(phosphonacetyl)-L-aspartate (PALA). The levels of enzyme resulting from gene amplification can be very high, as in a Chinese hamster ovary (CHO) cell line selected with MTX that had DHFR as 30% of the cytoplasmic protein with a corresponding increase in DHFR mRNA (58). In general, the overproduced enzyme in resistant cells has the same kinetic properties as in susceptible cells, but some amplified DHFR genes are altered to produce a protein with reduced affinity for MTX (36). In addition to these key studies on DHFR and CAD genes, treatment of cells with other anticancer drugs has led to the selection of specific resistance resulting from amplification of the target genes.

Mammalian cell cultures also develop resistance to toxins by amplification of genes responsible for detoxication mechanisms; for example, treatment of mouse cells with cadmium can select amplified genes encoding low–molecular weight metallothioneins that sequester toxic metals, and multidrug resistance in cells (hamster, mouse, and human) selected with a broad spectrum of drugs results from elevated levels of a P-glycoprotein encoded by an amplified multigene family (80). Because the P-glycoprotein is homologous to a bacterial transport protein, it probably functions as an energy-dependent export pump to remove toxins nonspecifically. A third gene amplification causing increased detoxication and likely to confer cross-resistance is that of the glutathione S-transferase gene in cells resistant to the chemotherapeutic alkylating agents, bifunctional nitrogen mustards (49).

Gene amplification can also confer resistance on nonmammalian cells. Plant cells selected for resistance to the herbicides L-phosphinothricin and glyphosate had amplified genes for the target enzymes, glutamine synthetase and 5-enolpyruvylshikimate-3-phosphate synthetase, respectively (26, 88). Cultured mosquito cells selected with MTX contained 14- to 20-fold more DHFR (28), and recent evidence suggests enrichment of DNA likely to contain the DHFR coding sequence (89). Similarly, many microorganisms have developed resistance to toxins by gene amplification. Amplification of a metallothionein gene in the yeast *Saccharomyces cerevisiae,* selected with copper, could be enhanced by treatment with chemical carcinogens (1), suggesting that, as in cell cultures, compounds that interfere with DNA replication can increase the rate of gene amplification.

In contrast to the extensively studied gene amplification in cell cultures and microorganisms, there have been very few reports of the same phenomenon in intact higher organisms. The major examples are in human cancer tissue, in which multidrug resistance associated with amplified P-glycoprotein genes has been found in the tumors of patients treated with chemotherapeutic agents (91).

Two very interesting reports discuss in situ gene amplification in nontumorigenic mammalian tissues—a mean two- to three-fold increase in metallothionein (MT1) gene copy number in the livers of rats injected with cadmium (48) and a 100-fold amplification of a catalytically deficient allele of the *CHE* gene that encodes human serum cholinesterase in blood samples of a family exposed to organophosphorus (OP) insecticides (76, 90).

It was not clear whether the former induced change in MT1 copy number resulted from selective replication of DNA in liver cells already carrying extra MT1 genes or from rapid de novo gene amplification, possibly resulting from cadmium interference with DNA replication (48). Although evidence clearly pointed to gene amplification, its toxicological significance is difficult to assess because the transcriptional competence of the amplified genes was nullified by posttranscriptional events, possibly the specific degradation of MT1 mRNA, resulting in no observable increase in metallothionein.

The *CHE* gene amplification occurred in a man whose parents had been exposed to OP insecticides during the period when he was conceived. This amplification was suggested to result from the selection of a de novo amplification event very early in his development, probably at spermatogenesis because of the continued active cell division that makes amplification and selection more likely. The finding that both the man and his father possessed the same characteristic *CHE* DNA restriction fragments (although the father's were not amplified) supported the spermatogenesis hypothesis (76). Serum cholinesterase is thought to protect against OP poisoning by acting as a sink when it binds the insecticide, as is well established for the esterases of resistant aphids (21). However, the man's 100-fold increase in *CHE* gene copy number did not give rise to an elevated serum cholinesterase; thus this amplification exhibited no apparent selective advantage in this respect. Investigators proposed that any advantage might have been conferred early in his life when the genes were developmentally expressed because of evidence for transient expression in many embryonic cell types (90).

BIOCHEMISTRY OF INSECTICIDE RESISTANCE

Insects colonize many niches, and their biochemical processes are well adapted to cope with the variety of toxic chemicals produced by host plants sometimes specifically as a defense against their feeding. Many of these natural toxins are lipophilic and are made more polar by the insects' metabolisms so that they are readily removed via the water-based excretory system. Three main classes of enzyme contribute to this process. The cytochrome P-450 mediated mixed-function oxidases introduce oxygen, often as hydroxyl groups, into aromatic rings (81), glutathione *S*-transferases catalyze the conjugation of molecules that have an electrophilic center with the thiol group of the tripeptide glutathione (15), and the hydrolases cleave esters and amides with a consequent increase in polarity as their components are released (41). These primary metabolic changes are often complemented by secondary processes involving conjugation of the introduced functional groups with sugars, amino acids, or anions such as sulphate to increase polarity even further. The role of such processes in protecting insects is exemplified by the fact that mixed-function oxidase activity is dramatically higher in polyphagous lepidopterans, which can encounter a broader spectrum of plant-derived toxicants, than in mono- or oligophagous species (81). Similarly, dietary manipulation of *Spodoptera frugiperda* can lead to marked changes in glutathione *S*-transferase activity with concomitant effects on tolerance to OPs (15), apparently involving a quantitative rather than a qualitative change in the enzyme(s).

Heritable resistance to insecticides can arise from either a qualitative change in the toxicant's biochemical target, rendering it less sensitive, or a quantitative change in activity of the pre-existing metabolic defense enzymes (67). The class of insecticide-degrading enzyme that becomes more active in a particular insect seems to depend on both feeding behavior and the chemical involved. Thus, plant eaters such as lepidopteran larvae and some coleopterans, which encounter a whole spectrum of compounds within the leaf, tend to develop resistance by increases in the versatile mixed-function oxidases (81), especially when the insecticide is a pyrethroid or carbamate, whereas acarine and homopteran species that feed more selectively seem predisposed to acquire increased hydrolase activity, especially when the toxicant is an OP.

Increases in these enzymes can be determined either with the insecticides themselves as substrates (usually by using radiometric assays) or, perhaps more commonly, by using model substrates. In almost all cases where resistance has been linked with such increased metabolism, only enzyme activity as such was determined. Activity could increase by qualitative changes generating an enzyme with a higher catalytic efficiency, as in the mutant aliesterases of houseflies (67) or by generating increased amounts of an enzyme, as in aphids (21) and mosquitoes (59). Which of these mechanisms occurs in a particular instance often cannot be determined readily, especially when the insects contain several isoenzymes that can mask the altered Michaelis constant (K_m) of one mutant isoenzyme. Indeed to determine the apparent K_m for such a mixture is inappropriate because this parameter has meaning only in the context of a single enzyme and a failure to detect a change in the K_m in such circumstances cannot be taken as unequivocal evidence that changes are quantitative rather than qualitative. Nevertheless, many of the reported instances of increased enzyme activity linked with resistance have probably arisen from the production of more enzyme, owing to changes in gene dosage or regulation rather than mutation of the structural genes.

AMPLIFIED RESISTANCE GENES IN INSECTS

Myzus persicae *Esterases*

Increased esterase activity measured by the hydrolysis of naphthyl esters as model substrates is commonly associated with resistance, but usually the toxicological consequence of this association for insecticide hydrolysis is not established, and neither is the nature of the greater enzyme activity. However, in *M. persicae*, only one of several esterases is responsible for the marked increase in total esterase activity (4, 18), and this enzyme detoxifies insecticides both by hydrolysis (19) and by sequestering them when its catalytic center is phosphorylated or carbamylated (21).

Many clonal cultures were established from glasshouse and field populations, from which susceptible and six different levels of resistance were identified by comparing esterase levels and LD_{50}s with those of standard susceptible aphids (23). The amount of esterase protein increased approximately two-fold between each of the progressively more resistant clones; the most resistant contained approximately 10 picomoles per insect, equivalent to 1% of its protein. The maintenance of the seven constitutive levels of esterase production in the clones led to the hypothesis that they arose from gene amplification analogous to the well-established phenomenon in mammalian cell cultures (43, 92), the regular doubling in the aphid esterase suggesting a succession of gene duplications. Although all clones examined at that time fell into one of the seven categories when esterase activity distribution curves were determined, the doubling now appears to reflect the threshold for unequivocal recognition of a clone with significantly higher esterase content, and need not have implications for the mechanism by which the higher levels of amplification are generated. Indeed, studies on large populations from the field, using an immunoassay to determine esterase content, show broad distributions (30) in which susceptible aphids are readily discernible, but the resistant fraction does not contain clear components corresponding to the distribution curves of the six standard resistant clones comprising the geometric series.

In addition to the dramatic quantitative variation in insecticide-degrading esterase, the enzyme can also occur in two very closely related forms, E4 and FE4 (22), with only one being overproduced in each resistant aphid. They can be distinguished electrophoretically (20), although they have only a slightly different M_r (65,000 for E4 compared with 66,000 for FE4) and they also have slightly different (approximately two-fold) catalytic efficiencies with naphthyl esters or OPs, but not carbamates, as substrate (20, 22). These minimal qualitative differences have little toxicological impact and do not detract from the underlying importance of the massive overproduction of esterase protein. However, the genetic basis of this difference is of interest since it appears to depend on the karyotype of the aphid; those with a normal karyotype overproduce FE4 whereas aphids having a commonly occurring A1,3 translocation (10) have more of the smaller E4 protein (31).

In the geometric series originally defined by Devonshire & Sawicki (23), variants V1, V2, V4, and V8 were of normal karotype and only the more resistant variants V16, V32, and V64 had the translocation, but this boundary is not rigid. The very resistant nontranslocated aphids commonly found on peaches in southern Europe have FE4 levels corresponding to the E4 content of translocated V16 aphids (22), and, conversely, translocated Japanese clones (95) show only low levels of resistance (3- to 15-fold). It is not clear whether the latter exemplify a special situation, discussed later, in which susceptible and slightly resistant aphids retaining their parents' translocations arise from the spontaneous loss of resistance within some clonal populations of the most resistant variants.

More recent molecular studies have confirmed the original gene amplification hypothesis and elucidated the nature of the difference between E4 and

FE4. Poly A^+ RNA extracted from resistant aphids contained much more esterase mRNA than that from susceptible aphids as judged by in vitro translation and immunoprecipitation with an E4 antiserum (24). The polypeptides synthesized in vitro from the RNA of translocated and normal karyotype aphids showed the same difference (1000) in M_r as the native E4 and FE4 (although both were approximately 9000-M_r smaller since they were not glycosylated), indicating different primary structures.

The large difference in E4 message content between susceptible and resistant aphids enabled the preliminary identification of E4 clones in a cDNA library prepared from size-fractionated mRNA of the most resistant variant. Differential hybridization, initially of colony blots and then of plasmid DNA, to ^{32}P-labelled poly A^+ RNA of susceptible and resistant aphids identified three putative E4 clones from a library of 1200 recombinants, and two were subsequently confirmed as containing E4 sequences (both approximately 1 kb from a message of approximately 2 kb) by using hybrid-arrested translation (35). These were used as probes in studying the nucleic acids extracted from susceptible and resistant aphids.

Probing of RNA and DNA from various aphid clones has shown increases in E4-related sequences in line with their esterase content, establishing unequivocally that gene amplification is the underlying mechanism (35). Southern blots of DNA showed that the extent of amplification in the most resistant aphids is approximately 64-fold and that translocated and normal karyotype aphids have markedly different restriction patterns with EcoRI. All resistant aphids studied to date have one of only two qualitative restriction patterns correlating with the form of esterase overproduced. Thus, in normal-karyotype aphids overproducing FE4, the amplified sequences are on a 4-kb EcoRI fragment, whereas in translocated aphids they are associated with an 8-kb fragment. The homogeneity of the EcoRI fragments within, and conservation between, aphid clones is consistent with esterase gene copies being grouped together in tandem arrays rather than scattered throughout the genome. Although this observation is supported by restriction analysis with a variety of other enzymes that also showed that the amplified unit (amplicon) exceeds 25 kb (32), further work is required to establish this fact unequivocally.

The 4-kb and 8-kb genomic EcoRI restriction fragments have now been cloned and analyzed using restriction mapping (32) and partial sequencing (L. M. Field, unpublished information). The 5' end of the E4 was identified by homology with the protein sequence of the N terminus (for which at least the first 40 amino acids are the same for both esterases) and both genes appear to be identical in this region. Probing with cDNA made by priming poly A^+ RNA with oligo dT and intentionally limited to approximately 500 bases identified the position of the 3' ends. This established that the FE4 gene has a maximum length of 3.2 kb and that the E4 gene is of similar size but differs in the last 250 bases.

Culex *Species Esterases*

Increased naphthyl esterase activity is also commonly associated with insecticide resistance in the *Culex pipiens* complex and other culicine mosquitoes, although the picture here is somewhat more complicated than in aphids due to the involvement of a larger number of esterases. The association was first recognized in *C. pipiens pipiens* in 1970 (103), and it was later shown that only one of several alleles of the *Est-2* locus (Est-$2^{0.64}$) was more active in insects resistant to chlorpyrifos (73); this locus was later renamed *Est-B* (71). However, chlorpyrifos resistance was then found to be closely linked with increased activity of *Est-3* [equivalent to *Est-A* (71)], which appeared to constitute the resistance gene. Similar increases in esterase activity have been identified in resistant strains of *C. tarsalis* (75), and in *C. pipiens quinquefasciatus, C. pipiens pallens,* and *C. pipiens molestus* (55). Much of the early work, and the relationships and nomenclature of the esterases, has been reviewed by Villani et al (98) and Wood et al (102). Although, as in aphids, early indications of a doubling in esterase activity between strains pointed to a series of gene duplications (70, 98), only later were such increases shown to arise from changes in the quantity of enzyme (59).

The early work revealed changes in only one esterase corresponding to each of the putative A and B loci, but many examples have since shown increases in activity of various esterases A and B [so called on the basis of their preferential hydrolysis of α- or β-naphthyl acetate, and unrelated to the long-established nomenclature of Aldridge (2)]. Resistance in French populations of *C. pipiens pipiens* has been associated mainly with increased activity of only esterase A1 but the A2 and B2 forms now seem to be present in populations from southern France (52). Similarly, Californian *C. pipiens quinquefasciatus* showed higher esterase B1 activity in the 1970s (70) but had increased esterase A2 and B2 activity in 1984–85; 90% of these wild-type insects had both B1 and A2/B2 (78). Resistant *C. pipiens quinquefasciatus* from Africa and Asia also have more A2 and B2 (52). Parathion-resistant *C. tarsalis* from California has increased activity of the A3 and B3 forms of the esterases (75), whilst *C. tritaeniorhynchus* resistant to a variety of OPs and carbamates contains elevated levels of various A and B forms tentatively identified as esterases A1, A2, B1, B2, and B3 (96).

The esterases A and B seem to comprise two gene families that have homology within, but not between, the groups. Thus, an antiserum to esterase A1 cross-reacts with other esterases A, but not with esterases B, whilst a B1 antiserum only reacts with members of the B group (59). These immunological studies also demonstrated quantitative increases in esterase protein; the Californian Tem-R strain of *C. pipiens quinquefasciatus* has at least 500 times more esterase B1, equivalent to 6–12% of its total protein, whilst esterase A1 is overproduced by 70-fold (1–3% of total protein) in the French resistant strain (S54) of *C. pipiens pipiens* (59). The distinctiveness of the A and B loci had previously been established by crossing studies that clearly identified two tightly linked loci (16, 71, 78). This linkage also appears to extend to a structural association between the esterases A and B because their members commonly occur together showing elevated activity, i.e. A2 with B2, A3 with B3, etc. However, the evidence from crossing and immunological studies does not support the suggestion (57) that the esterases are different polymeric forms of one protein subunit produced by a single gene. Although the esterases A are indeed homodimers of 60,000 M_r polypeptides, they appear to be distinct from the polypeptides of esterases B, which are monomers of M_r 67,000 (38). When analyzed on polyacrylamide gradient gels, esterase A1 from the S54 strain of *C. pipiens pipiens* sometimes appears as two bands of M_r 118,000 and 134,000 or as a single broad band of M_r 120,000 (38); the molecular basis of this variability is not yet known.

The way in which the mosquito esterases confer cross resistance to various insecticides, and in particular the relative contributions of the esterases A and B, is yet to be established. In French *C. pipiens pipiens,* chlorpyrifos resistance was conferred by the A locus (73), whereas the high resistance of the Tem-R strain of *C. pipiens quinquefasciatus* was associated with the B1 locus (71). As in *M. persicae* (21), the massive overproduction of any esterase protein by resistant insects may result in the detoxication of insecticidal esters first by sequestration and then by hydrolysis when the inhibited esterase reactivates.

Alongside these electrophoretic and biochemical studies of mosquitoes, molecular biological work has confirmed that gene amplification is again the basis of elevated esterase (60). In vitro translation of polyA$^+$ RNA from susceptible and resistant larvae, coupled with immunoprecipitation by the B1 antiserum, established that the Tem-R strain has more of a band corresponding to the 67,000 M_r native protein (i.e. there appears to be no glycosylation). Larval RNA from Tem-R insects was used to prepare a cDNA library in the expression phage λgt11 from which was isolated one recombinant expressing a 135,000 M_r β-galactosidase fusion protein containing a 19,000 M_r peptide recognized by the esterase B1 antiserum. The cDNA insert (0.7 kb) was purified and confirmed as an esterase B1 sequence by hybrid selection of mRNA coupled with immunoprecipitation. Probing of *Eco*RI digests of genomic DNA from susceptible and resistant adults revealed an abundant 2.1-kb fragment in Tem-R and only weak hybridization to a 2.8-kb fragment in susceptible insects. Slot blots of serially diluted DNA showed that the amplification was > 250-fold.

The B1 probe has now been used to examine DNA of *Culex* strains collected recently from the field for comparison with the Tem-R strain that has been maintained in the laboratory for approximately 15 years (77). The probe recognizes sequences encoding esterases B1, B2, and B3 and shows diversity in the restriction patterns according to the form of esterase over-produced. This result is analogous to the finding that 4-kb and 8-kb *Eco*RI fragments are diagnostic of the FE4 and E4 esterases, respectively, in *M. persicae* (35). Two field strains of the *C. pipiens* complex with elevated esterase B1 (though not homozygous) yielded only a 2.1-kb fragment as in TemR, whereas a strain overproducing esterase B2 gave a single fragment of 9 kb. Elevated esterase B3 has so far only been found in *C. tarsalis,* a species that does not interbreed with any of the pipiens complex, and this study revealed two amplified fragments of 3.6 kb and 3.9 kb. The DNA from strains without increased esterase B (including one, S54, with elevated esterase A1) had only weak hybridization to the B1 probe, but produced different restriction patterns, indicating heterogeneity of nonamplified flanking sequences (77).

Sequence data have now been obtained for the B1 gene (61, 72), and the deduced N-terminal amino acid sequence, up to 30 residues past the putative active-site serine, shows considerable homology with acetylcholinesterases from *Drosophila* and *Torpedo,* human butyrylcholinesterase, and esterases of *Heliothis,* rabbit liver, and *Drosophila.* Preliminary analysis of the B2 and B3 genes confirms they they differ from the B1 gene (72). The B1 amplicons in Tem-R comprise a 25-kb core, with just one copy of the gene, flanked by variable regions containing repetitive DNA sequences, which might play a role in the amplification process (61).

Other Insects

Many other examples show that increased esterase activity is associated with insecticide resistance, but the biochemical and molecular mechanisms responsible have not yet been established. Duplication of esterase genes is common in various *Drosophila* species [reviewed by Oakeshott et al (65)] but appears to be a mechanism for generating enzyme diversity (variant repetition) rather than increasing enzyme production. The genes occur as small tandem duplications at loci that vary between species, probably reflecting chromosomal rearrangement and not independent evolutionary events. The tandem *Est6* and *EstP* genes of *D. melanogaster* show extensive (60–70%) homology at the DNA and protein levels but have differential expression during development, suggesting distinct physiological functions. Differences in activity of the *Est6* enzyme between species of the melanogaster group, and between their tissues, appear to involve changes in transcription due to *cis*-acting promoter polymorphisms rather than changes in copy number (65).

Although no evidence has indicated that insecticides have influenced the evolution of qualitative or quantitative changes in esterase genes in *Drosophila*, the commonly occurring duplication of the metallothionein gene (see below) might have been selected for by the application of copper-containing fungicides in orchards and vineyards (54). So far in the strains studied, only a simple duplication to generate two copies of the metallothionein gene has been detected, yet this appears to have practical implications because larvae with the duplicated gene produce more metallothionein mRNA and tolerate higher concentrations of copper and cadmium.

One of the other ubiquitous mechanisms of resistance to insecticides, increased mixed-function oxidase activity, is being characterized at the molecular level in *Drosophila* and houseflies. In the former, cytochrome P-450 has been classified into two subsets, P-450A, which is constitutive, and P-450B, which is expressed at high levels only in resistant strains (93). Increased expression involves *trans*-acting genes on chromosomes II and III both mapping to major insecticide resistance loci. By using a P-450B-specific cDNA probe, the resistant strain was shown to produce at least 20 times more mRNA of a significantly smaller size than that of the susceptible strain, in line with the differences in M_r between the A and B proteins. Furthermore, southern blots identified structural variation in the DNA between strains but no substantial differences in band intensity, appearing to rule out gene amplification as the underlying mechanism (100).

Variation in cytochrome P-450 and its toxicological consequences have been studied most intensively in houseflies, in relation to both insecticide resistance and induction by many endogenous and exogenous compounds (81). At least six forms of P-450 appear to be involved, showing different isoenzyme compositions between susceptible and resistant strains. A P-450 cDNA clone was isolated by using resistant houseflies (Rutgers strain) with intrinsically higher P-450 content that had been treated with phenobarbital to increase this content even further (29). By using the cDNA as a probe, the level of a specific P-450 mRNA was shown to increase when resistant flies were treated with phenobarbital. At that time, the researchers did not know whether insecticide resistance resulted from mutation(s) that increased the expression of some P-450 genes or modified the catalytic activity of some P-450 proteins (29). However, it has now been established (R. Feyereisen, personal communication) that mRNA levels corresponding to the cDNA are elevated in the original Rutgers strain compared to several other resistant and susceptible strains, but, as in *Drosophila*, with no massive gene amplification.

As a corollary, the overproduction of P-450 by a dioxin-resistant mouse cell line did not result from gene amplification, but from increased transcription. This overproduction was associated with hypomethylation of the variant

P-450 gene compared with the wild type gene, although a cause-effect relationship was not established (45).

Circumstantial evidence supports that gene amplification might be responsible for increases in the third group of enzymes conferring insecticide resistance, the glutathione S-transferases. In a series of 12 housefly strains, activity toward 3,4-dichloronitrobenzene (DCNB) varied over a 30-fold range correlating with resistance, the enzyme(s) apparently accounting for up to 6% of total soluble protein in the most resistant strain. This massive overproduction, together with the regression of activity in strains bred without selection, led Oppenoorth (67) to draw a parallel with established examples of gene amplification. Similarly, OP resistance in the predatory mite *Phytoseiulus persimilis* appeared to be controlled by several autosomally inherited factors that were codominant, yet the only biochemical mechanism identified was increased glutathione S-transferase activity (39). Furthermore, shifts in LC_{50} during selection occurred as a succession of steps, each showing a two-fold increase. These data indicated that gene amplifiction might be the underlying mechanism.

PROPERTIES OF AMPLIFIED DNA

Location and Stability

The amplified genes of microorganisms are generally present as simple tandem repeats of identical units within small defined segments of DNA. However, the amplicons of mammalian cultured cells are generally large, highly repeated, and undergo continuous rearrangement (43, 92); they may be integrated into chromosomes in sufficient numbers to be detectable as homogeneously staining regions (HSRs) or be located on extrachromosomal elements, double minute chromosomes (DMs) (43, 92), or their submicroscopic precursors, episomes (83). The location affects stability; amplified genes on chromosomes are stable for long periods without selection, but those on acentromeric DMs segregate unequally at cell division and are therefore unstable (43, 92). Initially, MTX tends to select unstable resistance on DMs, but higher concentrations select for integrated DHFR genes (46). CAD gene amplification tends to be stable, although some unstable lines containing DMs have been reported (56). Even apparently stably amplified genes on HSRs may be lost, over several years, in the absence of drug selection (5), contrasting with the loss of DMs, which can occur in only 20–30 cell doublings (43). There are also examples of spontaneous loss of multidrug resistance in mammalian cells without loss of amplified genes but with greatly decreased P-glycoprotein mRNA, suggesting transcriptional control (13).

No evidence indicates that amplified genes in resistant insects are associated with DMs; the duplicated MT1 genes of *Drosophila* are on chromosome

III (68) and the amplified esterase genes of *Culex* are grouped on a single chromosome, probably II (62). Insecticide-resistant aphids with amplified esterase genes have no visible HSRs or DMs (R. L. Blackman, personal communication) and the amplified genes are retained in the absence of selection, suggesting a chromosomal location. Even though resistance can be unstable in *M. persicae*, this does not result from a loss of the amplified genes but from their transcriptional control (see below).

Size and Structure

Generally, amplicons are much larger than the gene under selection (43). For example, size estimates of the DHFR amplicon range from approximately 150 kb to nearer 3000 kb. It seems that initially DHFR amplification involves large and variable amounts of DNA but that nonselected sequences are subsequently lost (53). Giulotto et al (42) studying single-step mutants with an average of 2.6 copies of the CAD gene found only 3 novel joints (see below) and calculated that this indicates a 10,000-kb amplicon. The techniques of pulsed field electrophoresis now provide the opportunity to characterize these very large amplified units, and even to isolate them for cloning and studying in yeast artificial chromosomes.

Amplified genes and their flanking regions within the amplicon have been studied largely using restriction analysis with cDNA probes for the genes of interest. In many cases, only quantitative changes have been identified; for example, amplified CAD (99) and DHFR genes (64) were present on fragments corresponding to the nonamplified fragments in susceptible cells, suggesting that structure of the genes or their immediate flanking DNA does not change during amplification. However, some cell lines do have new restriction fragments associated with amplification. Any extra copy of a DNA sequence in a simple tandem arrangement will generate a small new sequence between the copies, i.e. a novel joint (92), but detection of this sequence depends on the restriction enzymes used and the location of the sequence detected by the probe. Tyler-Smith & Alderson (97) found the same four restriction fragments with a DHFR cDNA probe in wild-type mouse cells and three MTX-resistant lines, but all three resistant lines had an extra band. Probing with subportions of the cDNA showed that the new fragments were at the ends of only some copies of the DHFR gene and appeared to have been generated early in the amplification process and subsequently amplified. Ethidium bromide staining of DNA co-amplified with these genes revealed distinct patterns of amplified sequences, suggesting that recombination had linked some DHFR genes to new flanking sequences (11). Gene amplification can thus involve major DNA rearrangements in cell cultures.

The duplicated MT1 genes in *Drosophila* chromosomes occur as simple tandem repeats, sometimes containing transposable element-like insertions (54). In one selected strain, the two 2.2-kb transcription units have 0.3 kb and

1.4 kb flanking DNA with a 6 base-pair novel joint forming part of a larger inverted repeat (68). Restriction analysis of DNA from translocated resistant aphids revealed an E4-related 15-kb fragment that is not present in the susceptible strain (35) and might represent a novel joint. In mosquito, the amplified esterase DNA is on a fragment that is a different size from that of the unamplified gene (60, 77), suggesting a rearrangement.

Mechanisms Generating Amplification

The mechanism of gene amplification in drug-resistant cells has been difficult to study because it does not occur synchronously and can only be examined many generations after the primary event (91), but various models have been proposed (43, 92). An early model involved gene duplication by random unequal crossing-over between sister chromatids and then further misalignment producing expanded arrays of tandem repeats with spacers. However this mechanism cannot explain the frequent appearance of DMs before HSRs. In an alternative model, saltatory (unscheduled) replication occurs more than once at the same origin within a cell cycle, generating multiple unattached DNA strands that are either released (DMs) or integrated by end-to-end ligation and recombination (HSRs). This repeated replication could explain the sudden appearance of many gene copies, the presence of HSRs and DMs, and the initiation of chromosome breaks or translocations. It also predicts that if DNA replication were temporarily inhibited, reinitiation of already replicated sequences would be likely and treatments that interfere with DNA replication (e.g. hydroxyurea, UV light, or MTX) do increase the frequency of DHFR gene amplification (1, 43, 87). Schimke (86) suggested that there is a fundamental problem in the faithful replication of complex chromosomes in cell culture, which results in multiple initiations (followed by occasional recombination to cause often observed chromosome abnormalities such as gene amplification, increased sister-chromatid exchange, polyploidy, breakage, translocation, and inversions). The latter are often involved in gene amplification and models for the generation of tandem arrays of inverted duplications have been proposed (44, 69). However, amplified genes can also be arranged in a head-to-tail tandem fashion with no inverted sequences (e.g. 79).

The mechanisms of gene amplification in insects have not yet been elucidated. The MT1 gene in *Drosophila* does not show extensive homology in its flanking sequences, suggesting that homologous recombination (unequal crossing over) is not involved in the duplication of this gene (68). Pasteur et al (72) concluded that the extensive amplification of esterase B1 genes in mosquitoes was unlikely to be achieved in a single step, and the same argument could apply to the esterase genes in *M. persicae* in which the observed intermediate copy numbers (35) make amplification of E4 and FE4 genes likely to proceed stepwise. The apparent divergence of the aphid esterase genes (E4 and FE4) might be a consequence of the A1,3 translocation event; evidence indicates a link between amplification and both translocations and the presence of transposable elements. Translocation might play an indirect causal role by moving a block of sequences away from unfavorable sites (92), rather than simply being a by-product of the instability of amplified DNA. Since amplified esterase genes are found in resistant aphids of both karyotypes, the A1,3 translocation cannot be an essential part of the process; it may play a role for E4, but amplification of FE4 genes must occur by a different mechanism or involve an as yet undetected translocation. In *Drosophila,* copies of transposed P-elements can be retained at their original site, thus leading to their duplication (27). The generation of many copies of esterase genes in resistant mosquitoes and aphids could be mediated by similar mobile genetic elements, and insecticide selection of insects with amplified esterase genes might then play a role in the dispersion and spreading of transposable elements in natural populations. In support of this hypothesis, mosquitoes have two repetitive DNA sequences dispersed in the genome that are co-amplified with the esterase gene in insecticide-resistant insects (61).

ORIGINS AND INHERITANCE OF AMPLIFIED INSECTICIDE RESISTANCE GENES

Selection

Insecticide resistance is generally believed to arise from selection acting on random variation, i.e. it is pre-adaptive, but Wood (101) suggested that insecticides might act both by selection and by increasing mutation rates. Evidently, the rate of gene amplification in cell cultures is accelerated by treatments that interfere with DNA replication, including exposure to the cytotoxic selection agents themselves (43, 87). Indeed, choriocarcinoma cells treated with MTX exhibit DMs but not multiple DHFR genes, showing amplification in the absence of selection (84). Insecticides may also play a more fundamental role in the development of resistance if they too can increase the rate at which gene amplification occurs. Although some insecticides have been shown to increase DNA and chromosome abnormalities (e.g. 14), which in turn can be associated with gene amplification mechanisms, most of the evidence for such effects comes from experiments using artificial and extreme conditions and any practical implications are yet to be established.

Inheritance

Early experiments on the co-inheritance of resistance and elevated esterase levels in both mosquitoes and aphids were done before gene amplification was implicated. In *C. pipiens quinquefasciatus*, the highly active esterase B1 was indissociable from resistance and inherited as a monofactorial dominant character (70). *M. persicae* is normally parthenogenetic in laboratory culture, and technical problems complicate breeding through the sexual phase; the few successful crosses between a susceptible and a moderately resistant clone of normal karyotype were difficult to interpret in Mendelian terms (8, 9), possibly because of recombination within the amplified sequences. However, a very resistant clone heterozygous for the chromosome translocation also appeared to be heterozygous for the elevated E4 character that was inherited monofactorially (10). Thus in both mosquitoes and aphids, resistance and high esterase can be co-inherited as a single factor. Both can have a range of esterase levels that, in aphids at least, is known to result from different numbers of esterase genes (35). The finding that amplified B1 genes in *C. pipiens quinquefasciatus* are inherited as a single coherent unit led Tabashnik to model the evolution of insecticide resistance by gene amplification as if it were a single locus comprising a series of alleles of increasing potency (94). He used the data of Pasteur et al (70) and concluded that gene amplification–based resistance could develop more rapidly than other forms of resistance. However, for the model to be robust and have predictive value, the amplified genes must be stable and form a sufficiently cohesive unit to be continually inherited as a single entity. Whilst this occurred over a few controlled crosses in the laboratory, over many generations, recombination within the amplified region may well occur or expression of the genes may change (see below), both of which could profoundly upset the predicted outcome.

Instability

Research has long indicated that the inheritance of insecticide resistance, now known to involve gene amplification, can be unstable in the absence of selection; indeed, instability has been taken to indicate that resistance arises by amplification (39, 67). However, one must always consider that instability might result from an initial low-level contamination with susceptible insects or heterozygosity of the resistance gene, associated with faster reproduction of susceptible insects in the absence of insecticides. Instability has been reported in both mosquitoes (70, 72) and aphids, and studies of the latter have well established that contamination is not responsible (31, 85).

The molecular basis of instability has been examined in *M. persicae* in which changes within asexual clones can be studied without the complication of possible meiotic recombination. Recombination between homologous chromosomes (endomeiosis) has been proposed as a mechanism for generating diversity in aphids and other parthenogenetic insects, but Blackman (7) argued that this is unlikely in aphids. Resistance is usually stable for long periods but is lost in some clones in a variegated way, i.e. a small proportion of offspring lose resistance and high esterase content in a single generation (31, 85). Such reversion has only been confirmed in aphids with the chromosome translocation that is retained during the process (31). Unlike normal susceptible aphids, revertants show intraclonal variation in esterase level on which insecticide selection can act, leading to recovery of resistance (31). At a molecular level, reversion involves loss of elevated esterase and its mRNA but not of the amplified genes, suggesting that transcriptional control is responsible (33).

In vertebrates, switches in gene activity are often associated with changes in the presence of 5-methylcytosine (5 mC) in CpG doublets within or close to the gene (25). Most studies have exploited the different abilities of the isoschizomers *Hpa*II and *Msp*I to cleave DNA at CCGG sites containing 5 mC; *Msp*I, unlike *Hpa*II, can cut when the internal cytosine is methylated, and thus restriction patterns are different for the two enzymes if 5 mC is present in CpG doublets (25). This approach has shown that in the majority of cases, methylation is associated with gene inactivation and demethylation with an increase in transcriptional activity (25). Furthermore, clusters of CpGs have been identified in the promotor regions (6), which suggests that methylation affects transcription by interfering with DNA-protein interactions. So far, little evidence shows that methylation plays a role in controlling gene expression in insects (25), but experiments with *Msp*I and *Hpa*II have shown that changes in 5 mC correlate with altered expression of amplified esterase genes in aphids and, surprisingly, that esterase genes are methylated in resistant aphids expressing the genes, but not in revertants where the genes are inactive (33). DNA methylation has been associated with gene amplification in some other systems such as asparagine synthetase genes in cell cultures, although in this system increased expression was associated with demethylation of sites in the 5' region (3). Recent work on the role of 5 mC in aphids indicates a cluster of potential methylation sites at the 5' end of the esterase genes that might influence their expression (32). If 5 mC does play a causal role, it might be possible to alter esterase gene expression and hence resistance status using treatments that interfere with DNA methylation; although this might help elucidate the process, one would likely encounter considerable difficulties in registering such a product for field use.

Whatever the mechanism of reversion in aphids, it has major implications. It is important to identify revertant aphids because they constitute a hidden resistance potential within populations (31). This identification is possible using a combination of enzyme immunoassay and DNA probing on single insects (34). Recent work has shown that unexpressed, nonmethylated, amplified E4 genes are present in British field populations as well as in glasshouses (L. M. Field, unpublished information). Unmethylated amplified FE4 genes might also occur in aphids with low esterase, although resistance by amplification of FE4 genes has previously been thought to be stable.

CONCLUDING REMARKS

The role of gene amplification in insecticide resistance, although widely invoked, has so far only been conclusively demonstrated in two species. In both cases, an esterase is involved, and in aphids both gene copy number and gene transcription can vary. Almost all studies of gene amplification in response to environmental stress and of the role of DNA methylation in gene expression have so far utilized cultured cells or microorganisms. The identification of these processes in insects provides excellent opportunities for understanding these important evolutionary events in intact higher eukaryotes.

ACKNOWLEDGMENTS

We thank authors who made typescripts available to us prior to publication and Mrs. Jenny Large for typing the multiple copies of the manuscript that evolved during the writing of this review.

Literature Cited

1. Aladjem, M. I., Koltin, Y., Lavi, S. 1988. Enhancement of copper resistance and *CupI* amplification in carcinogen-treated cells. *Mol. Gen. Genet.* 211:88–94
2. Aldridge, W. N. 1953. Serum esterases. *Biochem. J.* 53:110–7
3. Andrulis, I. L., Barrett, M. T. 1989. DNA methylation patterns associated with asparagine synthetase expression in asparagine-overproducing and -auxotrophic cells. *Mol. Cell. Biol.* 9:2922–27
4. Beranek, A. P. 1974. Esterase variation and organophosphate resistance in populations of *Aphis fabae* and *Myzus persicae*. *Entomol. Exp. Appl.* 17:129–42
5. Biedler, J. L., Melera, P. W., Spengler, B. A. 1980. Specifically altered metaphase chromosomes in antifolate-resistant Chinese hamster cells that overproduce dihydrofolate reductase. *Cancer Genet. Cytogenet.* 2:47–60
6. Bird, A. P. 1986. CpG-rich islands and the function of DNA methylation. *Nature* 321:209–13
7. Blackman, R. L. 1979. Stability and variation in aphid clonal lineages. *Biol. J. Linn. Soc.* 11:259–77
8. Blackman, R. L., Devonshire, A. L. 1978. Further studies on the genetics of the carboxylesterase regulatory system involved in resistance to organophosphorus insecticides in *Myzus pertsicae* (Sulzer). *Pestic. Sci.* 9:517–21
9. Blackman, R. L., Devonshire, A. L.,

Sawicki, R. M. 1977. Co-inheritance of increased carboxylesterase activity and resistance to organophosphorus insecticides in *Myzus persicae* (Sulzer). *Pestic. Sci.* 8:163–66
10. Blackman, R. L., Takada, H., Kawakami, K. 1978. Chromosomal rearrangement involved in the insecticide resistance of *Myzus persicae*. *Nature* 271:450–52
11. Bostock, C. J., Tyler-Smith, C. 1981. Gene amplification in methotrexate-resistant mouse cells II. Rearrangement and amplification of non-dihydrofolate reductase gene sequences accompanying chromosomal changes. *J. Mol. Biol.* 153:219–36
12. Brown, T. M., Brogdon, W. G. 1987. Improved detection of insecticide resistance through conventional and molecular techniques. *Annu. Rev. Entomol.* 32:145–62
13. Capranico, G., De Isabella, P., Castelli, C., Supino, R., Parmiani, G., Zunino, F. 1989. P-glycoprotein gene amplification and expression in multidrug-resistant murine P388 and B16 cell lines. *Br. J. Cancer* 59:682–85
14. Chen, H. H., Hsueh, J. L., Sirianni, S. R., Huang, C. C. 1981. Induction of sister-chromatid exchanges and cell cycle delay in cultured mammalian cells treated with eight organophosphorus pesticides. *Mutat. Res.* 88:307–16
15. Clark, A. G. 1989. The comparative enzymology of the glutathione *S-*

transferases from non-vertebrate organisms. *Comp. Biochem. Physiol.* 92B: 419–66

16. Curtis, C. F., Pasteur, N. 1981. Organophosphate resistance in vector populations of the complex of *Culex pipiens* L. (Diptera:Culicidae). *Bull. Entomol. Res.* 71:153–61

17. Delidakas, C., Swimmer, C., Kafatos, F. C. 1989. Gene amplification: an example of genome rearrangement. *Curr. Opin. Cell Biol.* 1:488–96

18. Devonshire, A. L. 1975. Studies of the carboxylesterases of *Myzus persicae* resistant and susceptible to organophosphorus insecticides. In *Proc. 8th British Insecticide and Fungicide Conf.* 1:67–73. London: British Crop Protection Council

19. Devonshire, A. L. 1977. The properties of a carboxylesterase from the peach-potato aphid, *Myzus persicae* (Sulz.), and its role in conferring insecticide resistance. *Biochem. J.* 167:675–83

20. Devonshire, A. L. 1989. Insecticide resistance in *Myzus persicae:* from field to gene and back again. *Pestic. Sci.* 26:375–82

21. Devonshire, A. L., Moores, G. D. 1982. A carboxylesterase with broad substrate specificity causes organophosphorus, carbamate and pyrethroid resistance in peach-potato aphids *(Myzus persicae)*. *Pestic. Biochem. Physiol.* 18:235–46

22. Devonshire, A. L., Moores, G. D., Chiang, C. 1983. The biochemistry of insecticide resistance in the peach-potato aphid, *Myzus persicae*. In *IUPAC Pesticide Chemistry, Human Welfare and the Environment*. ed. J. Miyamoto, P. C. Kearney, S. Matsunaka, D. H. Hutson, S. D. Murphy pp. 191–96. New York: Pergamon

23. Devonshire, A. L., Sawicki, R. M. 1979. Insecticide-resistant *Myzus persicae* as an example of evolution by gene duplication. *Nature* 280:140–41

24. Devonshire, A. L., Searle, L. M., Moores, G. D. 1986. Quantitative and qualitative variation in the mRNA for carboxylesterases in insecticide-susceptible and resistant *Myzus persicae* (Sulz). *Insect Biochem.* 16:659–65

25. Doerfler, W. 1983. DNA methylation and gene activity. *Annu. Rev. Biochem.* 52:93–124

26. Donn, G., Tischer, E., Smith, J. A., Goodman, H. M. 1984. Herbicide-resistant alfalfa cells: an example of gene amplification in plants. *J. Mol. Appl. Genet.* 2:621–35

27. Engels, W. R. 1983. The P family of transposable elements in *Drosophila. Annu. Rev. Genet.* 17:315–44

28. Fallon, A. M. 1984. Methotrexate resistance in cultured mosquito cells. *Insect Biochem.* 14:697–704

29. Feyereisen, R., Koener, J. F., Farnsworth, D. E., Nebert, D. W. 1989. Isolation and sequence of a cDNA encoding a cytochrome P-450 from an insecticide-resistant strain of the house fly, *Musca domestica. Proc. Natl. Acad. Sci. USA* 86:1465–69

30. ffrench-Constant, R. H., Devonshire, A. L. 1988. Monitoring frequencies of insecticide resistance in *Myzus persicae* (Sulzer) (Hemiptera:Aphididae) in England during 1985–86 by immunoassay. *Bull. Entomol. Res.* 78:163–71

31. ffrench-Constant, R. H., Devonshire, A. L., White, R. P. 1988. Spontaneous loss and reselection of resistance in extremely resistant *Myzus persicae* (Sulzer). *Pestic. Biochem. Physiol.* 30:1–10

32. Field, L. M. 1989. *The molecular genetic basis of insecticide resistance in the peach-potato aphid* Myzus persicae *(Sulzer)*. PhD thesis. London: Counc. Natl. Acad. Awards. 152 pp.

33. Field, L. M., Devonshire, A. L., ffrench-Constant, R. H., Forde, B. G. 1989. Changes in DNA methylation are associated with loss of insecticide resistance in the peach-potato aphid *Myzus persicae* (Sulz.) *FEBS Letts.* 243:323–27

34. Field, L. M., Devonshire, A. L., ffrench-Constant, R. H., Forde, B. G. 1989. The combined use of immunoassay and a DNA diagnostic technique to identify insecticide-resistant genotypes in the peach-potato aphid, *Myzus persicae* (Sulz). *Pestic. Biochem. Physiol.* 34:174–78

35. Field, L. M., Devonshire, A. L., Forde, B. G. 1988. Molecular evidence that insecticide resistance in peach-potato aphids (*Myzus persicae* Sulz.) results from amplification of an esterase gene. *Biochem. J.* 251:309–12

36. Flintoff, W. F., Weber, M. K., Nagainis, C. R., Essani, A. K., Robertson, D., Salser, W. 1982. Overproduction of dihydrofolate reductase and gene amplification in methotrexate-resistant Chinese hamster ovary cells. *Mol. Cell. Biol.* 2:275–85

37. Ford, M. G., Holloman, D. W., Khambay, B. P. S., Sawicki, R. M., eds. 1987. *Combating Resistance to Xenobiotics*. Chichester: Ellis Horwood. 320 pp.

38. Fournier, D., Bride, J. M., Mouchès, C., Raymond, M., Magnin, M., et al.

1987. Biochemical characterization of the esterases A1 and B1 associated with organophosphate resistance in *Culex pipiens* L. complex. *Pestic. Biochem. Physiol.* 27:211–17

39. Fournier, D., Pralavorio, M., Cuany, A., Bergé, J. B. 1988. Genetic analysis of methidathion resistance in *Phytoseiulus persimilis* (Acari: Phytoseiidae). *J. Econ. Entomol.* 81:1008–13

40. Franciolini, F., Petris, A. 1989. Evolution of ionic channels in biological membranes. *Mol. Biol. Evol.* 6:503–13

41. Georghiou, G. P., Saito, T., eds. 1983. *Pest Resistance to Pesticides*. New York: Plenum. 809 pp.

42. Giulotto, E., Saito, I., Stark, G. R. 1986. Structure of DNA formed in the first step of CAD gene amplification. *EMBO J.* 5:2115–21

43. Hamlin, J. L., Milbrandt, J. D., Heintz, N. H., Azizkhan, J. C. 1984. DNA sequence amplification in mammalian cells. *Int. Rev. Cytol.* 90:31–82

44. Hyrien, O., Debatisse, M., Buttin, G., Robert de Saint Vincent, B. 1988. The multicopy appearance of a large inverted duplication and the sequence at the inversion joint suggests a new model for gene amplification. *EMBO J.* 7:407–17

45. Jones, P. B., Miller, A. G., Israel, D. I., Galeazzi, D. R., Whitlock, J. P. 1984. Biochemical and genetic analysis of variant mouse hepatoma cells which overtranscribe the cytochrome P_1-450 gene in response to 2,3,7,8-tetrachlorodibenzo-*p*-dioxin. *J. Biol. Chem.* 259:12357–63

46. Kaufman, R. J., Schimke, R. T. 1981. Amplification and loss of dihydrofolate reductase genes in a Chinese hamster ovary cell line. *Mol. Cell. Biol.* 1:1069-76

47. Knight, A. L., Norton, G. W. 1989. Economics of agricultural pesticide resistance in arthropods. *Annu. Rev. Entomol.* 34:293–314

48. Koropatnick, J. 1988. Amplification of metallothionein-1 genes in mouse liver cells *in situ:* extra copies are transcriptionally active (42737). *Proc. Soc. Exp. Biol. Med.* 188:287–300

49. Lewis, A. D., Hickson, I. D., Robson, C. N., Harris, A. L., Hayes, J. D., et al. 1988. Amplification and increased expression of alpha class glutathione *S*-transferase-encoding genes associated with resistance to nitrogen mustards. *Proc. Natl. Acad. Sci. USA* 85:8511–15

50. Long, E. O., Dawid, I. B. 1980. Repeated genes in eukaryotes. *Annu. Rev. Biochem.* 49:727–64

51. MacIntyre, R. J. 1976. Evolution and ecological value of duplicate genes. *Annu. Rev. Ecol. Syst.* 7:421–68

52. Magnin, M., Marboutin, E., Pasteur, N. 1988. Insecticide resistance in *Culex quinquefasciatus* (Diptera: Culicidae) in West Africa. *J. Med. Entomol.* 25:99–104

53. Mariani, B. D., Schimke, R. T. 1984. Gene amplification in a single cell cycle in Chinese hamster ovary cells. *J. Biol. Chem.* 259:1901–10

54. Maroni, G., Wise, J., Young, J. E., Otto, E. 1987. Metallothionein gene duplications and metal tolerance in natural populations of *Drosophila melanogaster*. *Genetics* 117:739–44

55. Maruyama, Y., Yasutomi, K., Ogita, Z. I. 1984. Electrophoretic analysis of esterase isozymes in organophosphate-resistant mosquitoes *(Culex pipiens)*. *Insect Biochem.* 14:181–88

56. Meinkoth, J., Killary, A. M., Fournier, R. E. K., Wahl, G. M. 1987. Unstable and stable CAD gene amplification: Importance of flanking sequences and nuclear environment in gene amplification. *Mol. Cell. Biol.* 7:1415–25

57. Merryweather, A. T., Crampton, J. M., Townson, H. 1990. Purification and properties of an esterase from organophosphate-resistant strain of the mosquito *Culex quinquefasciatus*. *Biochem. J.* 266:83–90

58. Milbrandt, J. D., Heintz, N. H., White, W. C., Rothman, S. M., Hamlin, J. L. 1981. Methotrexate-resistant Chinese hamster ovary cells have amplified a 135-kilobase-pair region that includes the dihydrofolate reductase gene. *Proc. Natl. Acad. Sci. USA* 78:6043–47

59. Mouchès, C., Magnin, M., Bergé, J. B., de Silvestri, M., Beyssat, V., et al. 1987. Overproduction of detoxifying esterases in organophosphate-resistant *Culex* mosquitoes and their presence in other insects. *Proc. Natl. Acad. Sci. USA* 84:2113–16

60. Mouchès, C., Pasteur, N., Bergé, J. B., Hyrien, O., Raymond, M., et al. 1986. Amplification of an esterase gene is responsible for insecticide resistance in a California *Culex* mosquito. *Science* 233:778–80

61. Mouchès, C., Pauplin, Y., Agarwal, M., Lemieux, L., Herzog, M., et al. 1990. Characterization of amplification core and esterase B1 gene responsible for insecticide resistance in *Culex*. *Proc. Natl. Acad. Sci. USA* 87:2574–78

62. Nance, E., Heyse, D., Britton-Davidian, J., Pasteur, N. 1990. Chromosomal organization of the amplified

esterase B1 gene in organophosphate-resistant *Culex pipiens quinquefasciatus* Say (Diptera, Culicidae). *Genome.* 33:148–52

63. National Research Council. 1986. *Pesticide Resistance, Strategies and Tactics for Management.* Washington DC: Natl. Acad. Press. 471 pp.

64. Nunberg, J. H., Kaufman, R. J., Chang, A. C. Y., Cohen, S. N., Schimke, R. T. 1980. Structure and genomic organisation of the mouse dihydrofolate reductase gene. *Cell* 19:355–64

65. Oakeshott, J. G., Healy, M. J., Game, A. Y. 1990. Regulatory evolution of β-carboxyl esterases in Drosophila. In *Ecological and Evolutionary Genetics of Drosophila,* ed. J. S. F. Barker, T. Slarmer, R. McIntyre. New York: Plenum. In press

66. Oppenoorth, F. J. 1965. Biochemical genetics of insecticide resistance. *Annu. Rev. Entomol.* 10:185–206

67. Oppenoorth, F. J. 1985. Biochemistry and genetics of insecticide resistance. In *Insect Control,* Vol. 12., eds. G. S. Kerkut, L. I. Gilbert, pp. 731–73. New York: Pergamon

68. Otto, E., Young, J. E., Maroni, G. 1986. Structure and expression of a tandem duplication of the *Drosophila* metallothionein gene. *Proc. Natl. Acad. Sci. USA* 83:6025–29

69. Passananti, C., Davies, B., Ford, M., Fried, M. 1987. Structure of an inverted duplication formed as a first step in a gene amplification event: implications for a model of gene amplification. *EMBO J.* 6:1697–1703

70. Pasteur, N., Georghiou, G. P., Iseki, A. 1984. Variation in organophosphate resistance and esterase activity in *Culex quinquefasciatus* Say from California. *Génét. Sél. Evol.* 16:271–84

71. Pasteur, N., Iseki, A., Georghiou, G. P. 1981. Genetic and biochemical studies of the highly active esterases A and B associated with organophosphate resistance in mosquitoes of the *Culex pipiens* complex. *Biochem. Genet.* 19:909–19

72. Pasteur, N., Raymond, M., Pauplin, Y., Nancé, E., Heyse, D. et al. 1990. Role of gene amplification in insecticide resistance. In *Pesticides and Alternatives: Inovative Chemical and Biological Approaches to Pest Control.* ed. J. E. Casida. Amsterdam: Elsevier. In press

73. Pasteur, N., Sinègre, G. 1975. Esterase polymorphism and sensitivity to Dursban organophosphorus insecticide in *Culex pipiens pipiens* populations. *Biochem. Genet.* 13:789–803

74. Plapp, F. W. 1976. Biochemical genet-

ics of insecticide resistance. *Annu. Rev. Entomol.* 21:179–98

75. Prabhaker, N., Georghiou, G. P., Pasteur, N. 1987. Genetic association between highly active esterases and organophosphate resistance in *Culex tarsalis.* *J. Am. Mosq. Control Assoc.* 3:473–75

76. Prody, C. A., Dreyfus, P., Zamir, R., Zakut, H., Soreq, H. 1989. *De novo* amplification within a "silent" human cholinesterase gene in a family subjected to prolonged exposure to organophosphorous insecticides. *Proc. Natl. Acad. Sci. USA* 86:690–94

77. Raymond, M., Beyssat-Arnaouty, V., Sivasubramanian, N., Mouchès, C., Georghiou, G. P., Pasteur, N. 1989. Amplification of various esterase B's responsible for organophosphate resistance in *Culex* mosquitoes. *Biochem. Genet.* 27:417–23

78. Raymond, M., Pasteur, N., Georghiou, G. P., Mellon, R. B., Wirth, M. C., Hawley, M. K. 1987. Detoxifying esterases new to California in organophosphate-resistant *Culex quinquefasciatus* (Diptera: Culicidae). *J. Med. Entomol.* 24:24–27

79. Requena, J. M., Jópez, M. C., Jimenez-Ruiz, A., de la Torre, J. C., Alonso, C. 1988. A head-to-tail organization of hsp70 genes in *Trypanosoma cruzi.* *Nucleic Acids Res.* 16:1393–406

80. Riordan, J. R., Deuchars, K., Kartner, N., Alon, N., Trent, J., Ling, V. 1985. Amplification of P-glycoprotein genes in multi-drug resistant mammalian cell lines. *Nature* 316:817–19

81. Ronis, M. J. J., Hodgson, E. 1989. Cytochrome P-450 monooxygenases in insects. *Xenobiotica* 19:1077–92

82. Roush, R. T., McKenzie, J. A. 1987. Ecological genetics of insecticide and acaricide resistance. *Annu. Rev. Entomol.* 32:361–80

83. Ruiz, J. C., Choi, K., Von Hoff, D. D., Roninson, I. B., Wahl, G. M. 1989. Autonomously replicating episomes contain *mdr*1 genes in a multidrug-resistant cell line. *Mol. Cell. Biol.* 9:109–15

84. Sakai, K. -I., Wake, N., Fugino, T., Yasuda, T., Katon, H., et al. 1989. Methotrexate-resistant mechanisms in human choriocarcinoma cells. *Gynecol. Oncol.* 34:7–11

85. Sawicki, R. M., Devonshire, A. L., Payne, R. W., Petzing, S. M. 1980. Stability of insecticide resistance in the peach-potato aphid, *Myzus persicae* (Sulzer). *Pestic. Sci.* 11:33–42

86. Schimke, R. T. 1986. Methotrexate re-

sistance and gene amplification. *Cancer* 57:1912–17

87. Schimke, R. T., Hoy, C., Rice, G., Sherwood, S. W., Schumacher, R. I. 1988. Enhancement of gene amplification by perturbation of DNA synthesis in cultured mammalian cells. *Cancer Cells* 6:317–23

88. Shah, D. M., Horsch, R. B., Klee, H. J., Kishore, G. M., Winter, J. A., et al. 1986. Engineering herbicide tolerance in transgenic plants. *Science* 233:478–81

89. Shotkoski, F. A., Fallon, A. M. 1990. Genetic changes in methotrexate-resistant mosquito cells. In press

90. Soreq, H., Zakut, H. 1990. Amplification of butyrlcholinesterase and acetyl-cholinesterase genes in normal and tumor tissues: Putative relationship to organophosphorous poisoning. *Pharmac. Res.* 7:1–7

91. Stark, G. R., Giulotto, E., Saito, I., 1987. DNA amplification in tumors and drug-resistant cells. In *New Avenues in Developmental Cancer Chemotherapy*, pp. 495–502. New York: Academic

92. Stark, G. R., Wahl, G. M. 1984. Gene amplification. *Annu. Rev. Biochem.* 53:447–91

93. Sundseth, S. S., Kennel, S. J., Waters, L. C. 1989. Monoclonal antibodies to resistance-related forms of cytochrome P450 in *Drosophila melanogaster*. *Pestic. Biochem. Physiol.* 33:176–88

94. Tabashnik, B. E. 1990. Implications of gene amplification for evolution and management of insecticide resistance. *J. Econ. Entomol.* In press

95. Takada, H. 1979. Esterase variation in Japanese populations of *Myzus persicae* (Sulzer) (Homoptera: Aphididae), with special reference to resistance to organophosphorous insecticides. *Appl. Entomol. Zool.* 14:245–55

96. Takahashi, M., Yasutomi, K. 1987. Insecticidal resistance of *Culex tri-*

taeniorhynchus (Diptera:Culicidae) in Japan: genetics and mechanisms of resistance to organophosphorus insecticides. *J. Med. Entomol.* 24:595–603

97. Tyler-Smith, C., Alderson, T. 1981. Gene amplification in methotrexate-resistant mouse cells. I. DNA rearrangement accompanies dihydrofolate reductase gene amplification in a T-cell lymphoma. *J. Mol. Biol.* 153:203–18

98. Villani, F., White, G. B., Curtis, C. F., Miles, S. J., 1983. Inheritance and activity of some esterases associated with organo-phosphate resistance in mosquitoes of the complex of *Culex pipiens* L. (Diptera:Culicidae). *Bull. Entomol. Res.* 73:153–70

99. Wahl, G. M., Padgett, R. A., Stark, G. R. 1979. Gene amplification causes overproduction of the first three enzymes of UMP synthesis in *N*-(Phosphonacetyl)-L-asparate-resistant hamster cells. *J. Biol. Chem.* 254:8679–89

100. Waters, L. C., Ch'ang, L. Y., Kennel, S. J. 1990. Studies on the expression of insecticide resistance-associated cytochrome P450 in *Drosophila* using cloned DNA. *Proc. 7th Int. Congr. Pestic. Chem. Hamburg.* (Abstr.)

101. Wood, R. J. 1981. Insecticide resistance: genes and mechanisms. In *Genetic Consequences of Man-Made Change.* ed. J. A. Bishop, L. M. Cook, pp. 53–96. New York: Academic

102. Wood, R. J., Pasteur, N., Sinègre, G. 1984. Carbamate and organophosphate resistance in *Culex pipiens* L. (Diptera:Culicidae) in southern France and the significance of *Est-3A*. *Bull. Entomol. Res.* 74:677–87

103. Yasutomi, K., 1970. Studies on organophosphate-resistance and esterase activity in the mosquitoes of the *Culex pipiens* group. *Jpn. J. Sanit. Zool.* 21:41–45

Annu. Rev. Entomol. 1990. 36:25–42

INDUCTION OF DEFENSES IN TREES

Erkki Haukioja

Laboratory of Ecological Zoology, Department of Biology, University of Turku, SF-20500 Turku, Finland

KEY WORDS: insect population dynamics, insect-induced resistance, insect-induced amelioration

INTRODUCTION

Like all biological traits, plant defenses evolve by natural selection if the benefits of defense are higher than the costs, and if genetic variance in the defensive traits is available for natural selection to work on. Induced, rather than constitutive, plant defenses are assumed to develop if defenses are costly, if the need for defenses, and therefore also the benefits derived from them, are intermittent, and if the plants can anticipate and respond to high levels of consumption. In addition, the plant must be able to survive its first encounter with herbivores.

Induced defenses belong to a class of induced responses. All these responses need not be defensive, even if they hamper insect performance. I therefore prefer to use the more neutral term "resistance", rather than "defense" (38, 58). If herbivore-induced responses are true defenses, molded by natural selection, they are probably predictable, assuming we can measure the relevant parameters. Herbivore-induced responses may also represent something other than true defenses, for example, incidental by-products of recovery after damage. In such cases, general predictions based on evolutionary reasoning alone may not predict occurrence of induced responses and environmental correlates very well. This lack of accuracy is also true if induced responses are in fact defenses but are primarily targeted against something

25

other than herbivores. Surprisingly, insect damage may also make the plant more accessible to subsequent herbivory (40). In such cases, the induced response cannot be a defense against the herbivore.

Totally apart from their evolutionary origin and physiological mechanisms, herbivore-induced plant responses may affect the insect. Induced alterations in tree quality differ from other food-related qualitative factors in herbivore population dynamics in that they may function in a density-dependent way and may therefore affect the stability of insect populations (38). The consequences of induced responses for insect populations depend critically on insect lifespan relative to the duration of the qualitative alteration of the tree. The terms *rapid induced resistance* (RIR) and *delayed induced resistance* (DIR) have been applied to cases in which insects encounter the consequences of damage brought about by the same or an earlier generation, respectively (44). The same response may be experienced as DIR by a short-lived insect but as RIR by a long-lived one; for insect species with univoltine generations, responses within a single season are rapid ones, while multiannual effects function as DIR. RIR functions as a stabilizing and DIR as a destabilizing agent for the insect population (44, 58).

Entomologically, the most interesting question concerning induced resistance in plants is whether it exists, and if so, whether it affects insect population dynamics. Accordingly, many tests—especially concerning induced resistance in trees—have been conducted during the last 15 years to find out how insects are affected by previous consumption. The key finding has been that the effects on insects vary from strongly negative to slightly positive (12, 29, 40, 44). In this article, I review insect-induced responses in trees. To avoid duplication with some recent reviews (36, 44, 58, 76, 82), I concentrate in particular on the reasons for the variation in results. I limit my discussion to induced responses occurring in tree foliage.

ARE HERBIVORE-INDUCED RESPONSES IMPORTANT FOR HERBIVORES?

After herbivory, repair mechanisms necessitate chemical changes in plant tissue. Only some of the changes may be relevant to the herbivory, and which alterations, if any, indicate defensive responses is unclear. Relevant changes may take place in compounds not measured or in physical traits, and herbivores may be able to counteract presumably adverse chemical responses. Chemical analyses have nevertheless been used extensively to determine induced resistance in foliage, although by themselves the analyses seldom reveal whether the determined compounds are important for herbivores. Biotests with insects therefore provide the most realistic way to assess whether induced changes in plants really affect herbivores. At the same time, biotests provide data about the efficacy of herbivore-induced alterations in plant quality for insect performance.

I discuss separately studies in which previous consumption has been found to elicit adverse, beneficial, or no effects at all in the performance of individual insects. Misclassifying results of single studies happens when true differences are not statistically significant or if significant treatment effects are found because of erroneous practices. Therefore, I deal with sources of errors in each case.

Adverse Effects

A growing number of studies show that damage to foliage may induce alterations in leaves then present or in those that develop later, with adverse effects on insects (see 12, 44). There are examples in which the original damage was artificial and in which it was natural, and in which leaf quality was tested either chemically and/or by biotests (3, 42, 43, 49, 51, 65, 83). Impairment was recorded immediately (within a few hours after the damage) and/or was carried over to the following years (3, 18, 45).

Criticism of the suggested importance of induced plant resistance has focused on experimental design, and in particular on the statistical procedures applied (29, 44). In the first tests of induced resistance, significance testing was often performed with the wrong error variance. Using trees instead of insects as the error variance term usually eliminates this danger but is much more laborious because of the large number of individual trees needed.

No Effect

In a number of tests of induced plant resistance, no significant effect was found in insect performance (e.g. 35, 66, 74). These studies indicate either that the treatments do not elicit induced resistance in trees, that they do but the insects could overcome them, or that the design of the experiments was flawed.

Many studies do in fact involve methodological flaws so that experimental procedures may have masked true induced responses. A largely ignored source of error is the large between-tree and particularly within-tree variation, which leads to high pretreatment variance in the quality of foliage (31; J. Suomela & A. Nilson, unpublished data). This reduces the probability of detecting treatment effects, although it also hints that induced responses in foliage are not strong compared to variation due to other reasons. More specific sources of error result, for instance, from the fact that, when natural herbivory is measured, the trees to be defoliated are usually chosen by the insects rather than the researcher. Thus, the defoliated trees are likely to have represented a biased sample of the tree population, and may have been different, presumably more suitable, from the start, compared to the un-defoliated trees used as controls (67, 70). The precise manner in which the foliage has been damaged varies in different tests, and may affect the results (44). For instance, artificial defoliation may not be such a powerful cue as natural defoliation (34, 35, 43, 71). This may result from the absence of some stimulus (such as saliva; see 35) in the simulated damage, or equally from the generally crude form of the artificial damage. Manual defoliation usually removes large chunks, obviously is less discriminating with regard to leaf parts, and, unlike natural herbivory, takes place once or a few times only.

Biological reasons may potentially explain the nonsignificant responses. Plants may differ so that some species respond, others do not, or so that only some individuals respond, or so that the response is only under specific conditions. Surprisingly few data concern variation in induced resistance among tree species. Interspecific variation seems to be large in DIR (6, 71). Edwards et al (21) analyzed RIR in a number of British trees and found pronounced variability, but because they used only one tree for each species, one cannot distinguish between interspecific variation and variation between individual trees. Genetic variability in plant materials may be an important cause for the lack of demonstrated significant induced resistance (12). If so, this is easy to eliminate by means of careful experimental controls.

Plants may elicit induced resistance only under some conditions. If this expression relates to obvious traits such as tree age, the pattern is relatively easy to detect: e.g. pines induce defensive responses when the trees are young (55, 60) and birches when the leaves are young (46, 68). But if there is temporal, and especially annual, variation in induced responses, the repeatability of the tests is impaired. At present, there are no published results based on multiple induction treatments in which all experimental and control trees were simultaneously randomized as to treatments and used exactly similar defoliation procedures each time. Such experiments are urgently needed to determine whether variable results in different tests depend on variation in methods or represent a true biological phenomenon.

The lack of an induced defensive response in trees may not result from plant intricacy alone; insects too may vary so that some species do respond and others are not sensitive. In addition to differences in their detoxification systems, insects may also vary in their capacity to handle low quality foods because of behavioral adjustments. For example, crowded geometrid larvae managed relatively better on low quality than on high quality diets; for solitary larvae either no difference was observed or the larvae did better on a high quality diet (48). In the case of RIR, some insect species were affected adversely while others were not (35), or effects varied only in magnitude (32). In DIR, different species of birch herbivores were uniformly affected in the year following complete defoliation (32), indicating that the chemical alterations in plant quality were nonspecific and had a similar effect even on species that were seasonally separated (23, 32, 56, 68).

Ameliorative Effects

In some studies, insects are reported to perform better in trees that are defoliated than in nondefoliated controls (e.g. 49, 59, 74, 80, 94). The differences are slight, and most studies involve statistical problems similar to those in early studies reporting induced resistance. Also, in some studies, the positive effects of previous consumption can be explained at least partially by the nonrandom choice of experimental trees (80, 94): the trees originally chosen by insects were better food even after defoliation. This result may merely show that the tree had no induced resistance, or that the response was so weak that it could not reverse the effects of the original choice by insects.

However, in birch, browsing or damage to apical buds, but not to leaf laminas alone, consistently made the foliage in the next growth season less resistant (although not necessarily significantly) than the foliage of control trees (16, 40, 49, 67; K. Danell & E. Haukioja, unpublished data). These observations have two interesting implications. First, they show that simulations of herbivory may produce either lower or higher quality foliage in the same species, and obviously in the same individual, depending on the type of damage. This difference points to the significance of the manner of the herbivory simulation. Second, the observations make possible a scenario whereby under certain conditions insects may create positive feedback loops via the quality of their host trees (40, 49).

INDUCED PLANT RESPONSES AND REGULATION OF HERBIVORE POPULATIONS

Zoological interest in induced plant defenses was stimulated by their potential significance in the regulation of herbivore populations (3, 41). Induced defenses, unlike other food-related factors (such as spatial and temporal variation in plant quality and quantity), are density-dependent, at least on an off-on basis. However, opinion regarding the importance of induced resistance is polarized (64, 79). Often a disproportionate emphasis is placed on single regulative factors.

Although the adverse effects of previous consumption on traits that affect the performance of individual insects have been well documented in several studies, the magnitude of these effects is often so small that impact on the insect population remains uncertain. On the other hand, laboratory experiments usually totally ignore the possibility that induced compounds might entice predators (see 4, 17). Nevertheless, one can try to evaluate the effect on insect populations by calculating the probable outcomes, preferably on a per generation basis (44). The insect growth rate is usually the first trait to respond when food quality changes; larval growth rates often decline by 5–10% when the insect is reared on the foliage of trees previously damaged (44). In the case of minor changes in growth rates, insects in bioassays are often able to compensate by prolongation of the growth period. When insect growth rates are strongly retarded, final weights and fecundities change concomitantly. Per-generation fecundities in insects on previously damaged trees typically range from 5–10% (in RIR) up to more than 70% (DIR) below the corresponding values on control trees (12, 44). The latter approaches the impact of parasitoids.

The importance of slight decreases in foliage quality, such as those found in RIR, on insect population density is largely an open question. Even minor induced changes in plant quality may be important because they compel the larvae to move more (20), making them vulnerable to other regulative factors (95). If there are no compensatory density-dependent factors, the average density of the insects affected decreases. Field verification is lacking.

One should realize that demonstrations of even strong RIR in no way explain the occurrence of insect outbreaks—RIR is basically a stabilizing factor in insect population dynamics. DIR offers a potential explanation for outbreaks and more regular cycles (37). DIR may have much stronger effects on insect performance than RIR but, again, no hard data record its effects on insect populations. Circumstantial support for the importance of DIR is found, e.g. in the observation that the two defoliating insects with the most regular cyclic fluctuation in densities, the larch budmoth, *Zeiraphera diniana* (3), and the autumnal moth, *Epirrita autumnata* (45), are also the two species that show the most convincing evidence of strong DIR in host trees. In these cases, the time lags of induced resistance are also of sufficient length, three to four years, to contribute to the observed ten-year cycle.

Insect species whose densities show less regular outbreaks offer little proof of effective DIR after peak defoliation. However, even if the decrease phase of an outbreak is not caused by DIR, induced amelioration in foliage quality in the increase phase is still possible (40). If insect damage at low or moderate densities is able to improve the quality of the following leaf generation, it simultaneously creates a positive feedback loop. It has been hypothesized that insects feeding on apical parts of shoots, especially on buds, are the most likely to elicit such an effect, most probably in mature, slowly growing stands (40). According to this scenario, insect-induced amelioration in foliage quality would have a potential similar to that of mass attacks by bark-beetles, which can break the defensive barrier of resins in coniferous trees (75). Although proven cases of insect-induced alterations in foliage quality in outbreak species are few, obviously more options are open for induced plant responses to modify the population dynamics of forest insects than has been previously assumed.

CHEMICAL BASIS OF INDUCED RESPONSES

General theories of plant defenses (26, 78) share the basic assumption that secondary compounds are responsible for plant resistance, including induced defenses, and that these, rather than nutritive factors, represent the critical chemical factors explaining variance in herbivore performance. Although certain secondary compounds are highly toxic for insects, the general applicability of both the above assumptions is unclear. The first difficulty in evaluating them is to decide which secondary compounds to measure. In practical work the implicit assumption has been that the induction of second-ary compounds, and the defensive commitments of plants, can be monitored by analyzing the contents of certain broad classes of chemical compounds, such as total phenols. Many studies of RIR and DIR in woody plants have in fact demonstrated accumulation of total phenols or specific phenol groups (62, 72, 81, 83, 87, 90). However, although foliage phenol content is often negatively correlated with herbivore performance (51, 81), the causal rela-tionships are seldom clear (61). A test in which phenol production was blocked by phenylalanine ammonia lyase—but pretreatment phenol levels remained the same—did not confirm the importance of total phenols for insects (33, 34).

The validity of approaches concentrating on the determination of broad chemical groups of secondary compounds is unclear, and work with the chemical basis of winter resistance of trees against hares has shown that different species of trees are protected by specific groups of chemicals (9, 77). The most detailed analysis of the chemical basis of induced defenses of woody plants against insects concerns aspen leaves, which within 24 hours after artificial damage accumulate salicortin and tremulacin (11). These chemicals are converted to 6-HCH (6-hydroxy-2-cyclohexenone) by leaf esterases. Within the insect gut, 6-HCH is converted to various toxic com-pounds. All these compounds in artificial diets reduced the pupal weights of aspen tortrix (11). Clausen et al (10) suggested that, in aspen, RIR and DIR may be interconnected so that DIR involves an increase in the concentrations of the same compounds, especially tremulacin, that are responsible for the RIR.

The relative importance of allelochemicals vs nutritive factors on insect performance is hard to determine because the concentrations of some puta-tively important secondary constitutents correlate negatively with nutritive compounds essential for herbivores: the foliage in growth seasons following defoliation is usually richer in phenols and in fiber, and poorer in nitrogen, amino acids, and sugars (3, 41, 91). Therefore, to exclude the importance of chemical changes in nutritive compounds is difficult. Most tests of the importance of nutritive factors merely screen the nitrogen content, which is important at low concentrations (28) but not necessarily after that. On the other hand, the older literature in forest entomology has granted a pivotal role to sugars (84). In particular, the amount of total hexoses correlates well with insect performance (54), and may contribute to DIR as well (E. Haukioja & T. Jensen, unpublished data). A shortage of vital nutrients as a defense has not been regarded as a viable tactic for a tree because insects may compensate for the low quality of forage by feeding longer and therefore consuming more (63). Whether this happens in the field is unclear. If insect mortality increases on low-quality diets, the poor nutrient content may be an active defensive tactic (50).

Browsing-induced amelioration in foliage quality has been shown in birches to correlate with higher quantities of several nutrients and chlorophyll, and lower quantities of phenols (16; K. Danell & E. Haukioja, unpublished data).

WHY DO PLANTS SHOW INDUCED RESISTANCE?

Informally, zoologists often view plant responses as defenses on the basis of the outcome, i.e. a negative effect on the intruder. However, induced responses may decrease insect performance without any defensive commitments by the tree. For instance, high-quality tissue may be in short supply because it was consumed by herbivores in the first encounter, or passive deterioration in foliage quality may have resulted from the drying of nearby leaf tissue. Refoliated foliage may differ from mature leaves simply because of its age (24, 52, 68). Correspondingly, the defensive vs incidental nature of herbivore-induced responses in plants is fertile soil for debate and depends largely on the criteria we use to define defense (39, 82).

The most direct tests of induced plant defenses as evolutionary products against herbivory should demonstrate that the three general preconditions of natural selection are satisfied: i.e. the trait has both variation and a hereditary basis and fitness varies among individuals with different traits (22). Even if these conditions are fulfilled, they do not prove that the trait evolved as a defense, merely that natural selection is currently molding it.

In practical tests with long-lived organisms such as trees, only variation in induced responses is easy to quantify; the heredity of the responses can be studied but this study is laborious, and fitness differences are almost impossible to measure. In practice, we have to try to measure the benefits and costs of suggested defenses to find out whether the former exceed the latter. The benefits of induced defenses have been evaluated in the field in only a few cases, and even then as reduction in consumption. For RIR, some field tests indicate that induced responses reduce further consumption near the damage point within a consumed leaf (30) and in other leaves (86), but such results have not been found in all studies (35). The DIR may also decrease consumption by increasing mortality of feeders (3, 69). However, the relevant point for the plant is not whether consumption diminishes but whether the reduction in consumption is so great that it cancels the costs of the induced defense. The costs of induced defenses in woody plants are even less well known than the benefits. The primary difficulty is that herbivore-induced responses in plants normally follow tissue damage, and we have to be able to distinguish between the detrimental effects of the damage per se and the costs of the induced responses triggered by the damage. No data elucidate this point. Certainly, energetic costs can be calculated for the production of induced compounds such as phenols, but the repercussions of these costs in terms of lifetime reproductive output, survival, or growth of the tree are not known. When induced defenses are based upon a low content of necessary nutritive constituents, the costs may be high if the critical compound limits vital functions in the plant, as in the effect of nitrogen on photosynthesis (27). On the other hand, the costs may be minimal if low foliage quality is achieved, e.g. by rapidly transferring sugars from leaves to less vulnerable organs.

Induced responses indicate high phenotypic plasticity, and therefore variability, but the key condition for their evolution is genetic variability in the strength of induced responses. To my knowledge, this variability has not been quantified in trees, although different provenances may differ in induced responses (42).

Because direct testing of the defensive nature of induced responses is difficult, several indirect approaches have been used. These include the search for triggering mechanisms so specific that logically they have to be targeted against the intruder and manipulation of tree resource levels thought to be important for the induced response.

In approaches that try to demonstrate sophisticated triggering methods, the results are conflicting. Hartley & Lawton (35) found that insect feeding, or the application of insect saliva, elicited a stronger response in phenolic profiles in leaves than artificial damage, but they found no consistent effect on insects. Although the difference may have resulted from microorganisms in the saliva, the experiment confirms that tree reactions vary depending on the type of damage. Clausen et al (11) suggested that after foliar damage, an induced accumulation of phenolic glycosides in the leaves takes place via transportation from nearby internodes. In addition, glycosides were transformed enzymatically into various toxic compounds. Both these findings indicate a function in leaves after damage but not prior to it. Similarly, Baldwin (2) showed that damage-induced nicotine in tobacco leaves is conveyed from the roots. Possible communication among individual trees also represents such a complex triggering system for induced responses that it may indicate a true defensive response. Reported cases of aerial communication are questionable (29). The only nonpseudoreplicated case in which defoliation of nearby trees influenced the suitability of other trees did not preassume aerial communication (51): for instance, mycorrhizal-mediated root connections offer a potential channel for chemical information. Haukioja et al (51) found that minute nutrient applications to soil, including those in the form of insect frass, augmented the detrimental effects of defoliation on moth larvae. But the effects were the same when similar amounts of nutrients in commercial fertilizers were applied.

Approaches intended to determine whether defoliation-induced deterioration in foliage quality simply follows loss of nutrients (65, 87) have produced detailed hypotheses on conditions in which DIR is limited by either carbon or soil-derived nutrients (5, 8, 88). Direct testing of hypotheses concerning the relative role of carbon vs nutrients is difficult as long as we do not know whether the poor performance of insects is based, for instance, on an excess of carbon-based or nitrogen-based secondary compounds, or on a shortage of mineral nutrients or of carbon-based nutritive constituents such as sugars. So far, therefore, only some alternatives can be excluded. In a study to determine whether DIR is modified by nutrient availability, Haukioja & Neuvonen (43) found no reduction in DIR when birch trees were strongly fertilized at the time of defoliation. This indicates that the poor quality of foliage was not caused by a shortage of mineral nutrients. The result has been confirmed in a three-year experiment in which prolonged fertilizing did not alleviate DIR. More interestingly, shading also did not interact with defoliation to affect insect performance, indicating that a simple reduction in carbon availability had no effect either (F. S. Chapin, E. Haukioja, S. Neuvonen, & J. Suomela, unpublished data).

Theories based on the evolution of plant defenses superimposed by the constraints of the carbon-nutrient balance of the plant (13, 88) have predicted that defense expressions would be more flexible in rapidly growing species. On these grounds, Bryant et al (7) predicted that induced responses would be more common in rapidly than in slowly growing species. They found corroborative evidence in some African tree and bush species (6, 7). However, Neuvonen et al (71), comparing four species of subarctic trees, arrived at just the opposite result. Clearly, we do not have a general theory that successfully predicts the occurrence of induced defenses in various ecosystems.

Some studies have demonstrated mitigated induced responses in plants growing in crowded conditions, which indicates existence of active, resource-demanding induced responses (1, 57).

To conclude, the purely defensive nature of the induced accumulation of secondary compounds, or induced decreases of nutritive factors, is difficult to demonstrate. This is particularly true if the defensive function has evolved secondarily to the end products of other processes (88). We cannot exclude nondefensive explanations, based on passive changes in foliage quality, especially underlying DIR. However, it is equally clear that nondefensive theories do not adequately explain numerous details found in experimental results with respect to both RIR and DIR.

WHY ARE PLANT RESPONSES HARD TO PREDICT?

Inconsistencies in tests that try to corroborate general predictions about induced defenses in woody plants may have several causes. First, the more general the predictions, the more difficult it is to measure the relevant evolutive parameters, such as costs and benefits, in long-lived trees. Second, some of the assumptions behind the predictions may be true only partially or not at all. For instance, a number of predictions are based on the assumption that induced defenses are crucial for fluctuations in herbivore density (41, 79). But although herbivore-induced resistance can obviously modify insect population dynamics, it may not have evolved for defense against insects and need not function in the manner predicted on the basis of the defensive role. When induced responses do not represent defenses in an evolutionary sense, predictions made on the basis of defensive benefits will probably be erroneous. A further complication is that if induced amelioration, rather than or in addition to induced defenses, plays a pivotal role in the fluctuating densities of insect populations (40), we may be trying to explain by means of defenses something that is based on a failure of defenses. Third, the relevant predictions may not be possible without a detailed knowledge of the mechanisms available to trees.

A number of very basic features in trees differ completely from corresponding solutions in animals and may constrain tree responses: most trees are composed of modular structures; their carbon intake is at least partially sink-regulated; and they lack a rapidly responding holistic control system (15, 39, 49). These characteristics may affect, for example, variation within and between trees, thereby affecting experimental procedures. Furthermore, the most basic characteristics of trees may constrain the generality of certain traditional assumptions applied in theorizing about plant-herbivore interactions. For instance, if photosynthesis is strongly sink-regulated, reasoning based on the sharing of a fixed amount of carbon among alternative uses (25) loses its strength.

Studies of induced responses contain several seemingly odd details, which are hard to understand in terms of a totally integrated holistic individual but which make sense when trees are viewed as modular systems whose regulation is based on decentralized local hormonal control (39). When individual leaves in trees are damaged, the immediate induced responses are first seen in those leaves that are phyllotactically nearest to the damaged ones, not necessarily in those that are physically closest to the point of the damage (92). Thus the structure of the plant—in this case the vascular system—determines the detailed pattern of plant behavior. Another nonobvious feature of induced responses in trees is that the response may be local and not affect the whole tree: when individual branches or ramets within a birch are defoliated, these, and not the whole tree, respond (90), although other parts may also be affected (43). Tuomi et al (90) argued that because the branches, not the whole tree, respond to defoliation chemically (increased phenol content, lower nitrogen) as well as developmentally (89), the relevant explanation is a local shortage of resources. However, the key point need not be a shortage of resources as such but the allocation of resources within an individual tree. The integration of the tree is based on the ability of meristems to draw resources from the common pool according to the number and depth of the sinks that they can create (39). The defoliation of a branch may simply prevent its meristems from competing with those in nondefoliated parts, although resources per se would not be in short supply.

The proposal has been made that RIR helps to spread the damage over the canopy (19, 53) without necessarily decreasing the amount of consumption incurred by the tree. This spreading may be beneficial to the plant because compensation for dispersed damage may be easier (93) and/or because RIR repels herbivores from the most valuable young leaves (19). However, understanding the value of leaves demands a detailed knowledge of their role in relation to their position. In birch, the growing long-shoot leaves are the only young ones after the spring burst of foliage. The proposal has been made that they are strongly defended because the growth of the long shoots is the means by which the tree competes for light (20). The youngest leaves in fact elicit the strongest induced resistance (19, 35). The assertion that this helps the tree to compete for light, however, depends critically on the assumption that birch growth, i.e. elongation of the long shoots, is affected by the leaves in the elongating shoots. Nevertheless, removing even half of the growing leaves in long shoots did not reduce the growth of the shoot (49). Thus, repelling herbivores from young long-shoot leaves does not necessarily increase the ability of the tree to compete for light because the resources for shoot elongation were collected during the previous season. Instead, young long-shoot leaves seem to feed their own axillary bud, affecting the growth of the shoot starting from the bud the next season (49). After bud formation is completed, the young leaves may no longer be more valuable than mature leaves (49), and, in the field, long-shoot leaves experience a higher rate of consumption than mature leaves (73).

The most basic structural and functional solutions must make possible the sessile life of a tree. The decentralized hormonal control makes plants able to recover even after severe damage because by activation of previously suppressed meristems, the flow of nutrients and water is directed to sites where growth can proceed best (39). Simultaneously, this local control may dramatically increase internal variation and may be responsible for partial local responses within a tree. Trees are compartmentalized into at least semi-independent functional units, reducing the probability of diseases spreading over the whole individual (85). However, the decentralized regulative system also leads to concomitant threats. For example, induced amelioration may result from the modular and compartmentalized structure of trees, and damage to local centers of hormone production may modify behavior of previously suppressed meristems. In birch, damage to the apical meristems of branches (i.e. to terminal buds) was enough to ameliorate foliage quality in the next growth season (49). If insects cause equivalent effects by preferentially consuming apical buds [as they do in the field (49)], they may diminish the negative consequences of defoliation and may even modify plant suitability (14, 40) with potentially interesting possibilities for the behavior of insect populations (40, 49).

CONCLUSIONS

The consumption of foliage triggers induced responses in trees; these responses may either lessen or increase the performance of individual insects. Potentially the responses may create complex competitive situations, but we do not know their role in the population dynamics of insects, nor their relative importance compared with other agents regulating insect populations. To find out whether induced plant responses are important for the population dynamics of insects, we need factorial experiments in which these responses are studied simultaneously with other potential regulators of density. Such tests are laborious and expensive to conduct, and, before undertaking them, we have to know the potential sources of variability in the tests. The variability of the results is due in part to methodological weaknesses but also to the biological properties of trees. An important source of variability may be the large within-tree variation in foliage quality, which increases the error variance and may severely hamper attempts to obtain statistically sound proof of tree responses. Another important factor may be the way in which the damage is simulated; seemingly minor variations in defoliation treatment may actually lead to opposite outcomes.

From an evolutionary point of view, trees represent an interesting case; they are architecturally complex, compartmentalized modular organisms in which induced responses may be local. For instance, the defoliation of an individual branch in a birch makes the branch respond largely autonomously from other parts of the same tree. This kind of behavior obviously depends on the decentralized hormonal control of trees. The modular, locally regulated structure and function of trees may also explain why certain types of damage may induce better foliage quality, obviously as a consequence of damage-induced hormonal imbalance in the tree.

Currently, results conflict as to whether induced defensive responses are true defenses. To determine the costs and benefits of putative induced defenses we need long-term studies concerning the success of individual trees whose induced responses are known to differ, and data about the persistence of differences within trees. No evidence supports the notion of induced plant responses specifically targeted against certain insect species, but the specificity of some responses is too high to be merely a by-product of the recovery.

In the chemical analysis of the induced responses of woody plants, only the first steps have been taken, and detailed chemical analyses of specific secondary compounds are needed. In addition, a largely neglected possibility is that reduced foliage quality, especially in delayed induced resistance, may be caused in part by reduced amounts of sugars, rather than by increased concentrations of allelochemicals alone.

ACKNOWLEDGMENTS

I thank Sinikka Hanhimäki, Pekka Niemelä, Seppo Neuvonen, and Janne Suomela for comments on the manuscript, Ellen Valle for correcting the language, and the Academy of Finland for providing grant support of the research.

Literature Cited

1. Baldwin, I. T. 1988. Damage-induced alkaloids in tobacco: pot-bound plants are not inducible. *J. Chem. Ecol.* 14:1113–20
2. Baldwin, I. T. 1990. Mechanism of damage-induced alkaloid production in wild tobacco. *J. Chem. Ecol.* 15:1661–80
2a. Barbosa, P., Schultz, J. C., eds. 1987. *Insect Outbreaks.* San Diego: Academic
3. Benz, G. 1974. Negative Rückkopplung durch Raum- und Nahrungskonkurrenz sowie zyklische Veränderung der Nahrungsgrundlage als Regelprinzip in der Populationsdynamik des Grauen Lärchenwicklers, *Zeiraphera diniana* (Guenée) (Lep., Tortricidae). *Z. Angew. Entomol.* 76:196–228
4. Bergelson, J. M., Lawton, J. H. 1988. Does foliage damage influence predation on the insect herbivores of birch? *Ecology* 69:434–45

5. Bryant, J. P., Chapin, F. S. III, Reichardt, P., Clausen, T. 1985. Adaptation to resource availability as a determinant of chemical defense strategies in woody plants. *Recent Adv. Phytochem.* 19:219–37
6. Bryant, J. P., Heitkonig, I., Kuropat, P., Owen-Smith, N. 1990. Effects of severe defoliation upon the long-term resistance to insect attack and chemistry of leaves of six southern African savanna woody species. *Am. Nat.* In press
7. Bryant, J. P., Kuropat, P. J., Cooper, S. M., Frisby, K., Owen-Smith, N. 1989. Resource availability hypothesis of plant antiherbivore defence tested in a South African savanna ecosystem. *Nature* 340:227–29
8. Bryant, J. P., Tuomi, J., Niemelä, P. 1988. Environmental constraints of constitutive and long-term inducible defenses in woody plants. In *Chemical*

Mediation of Coevolution, ed. C. K. Spencer, pp. 376–89. New York: Academic

9. Bryant, J. P., Wieland, G. D., Reichardt, O. B., Lewis, V. E., McCarthy, M. C. 1983. Pinosylvin methyl ether deters snowshoe hare feeding on green alder. *Science* 222:1023–25

10. Clausen, T. P., Bryant, J. P., Reichardt, P. B., Werner, R. A. 1990. Long-term and Short-term induction in quaking aspen: related phenomena? See Ref. 76, In press

11. Clausen, T. P., Reichardt, P. B., Bryant, J. P., Werner, R. A., Post, K., Frisby, K. 1989. Chemical model for short-term induction in quaking aspen *(Populus tremuloides)* foliage against herbivores. *J. Chem. Ecol.* 15:2335–46

12. Coleman, J. S., Jones, C. G. 1990. A phytocentric perspective of phytochemical induction by herbivores. See Ref. 76, In press

13. Coley, P. D., Bryant, J. P., Chapin, F. S. 1985. Resource availability and plant antiherbivore defense. *Science* 230:895–99

14. Craig, T. P., Price, P. W., Itami, J. K. 1986. Resource regulation by a stem-galling sawfly on the arroyo willow. *Ecology* 67:419–25

15. Crawley, M. J. 1983. *Herbivory: The Dynamics of Animal-Plant Interactions.* Oxford: Blackwell. 437 pp.

16. Danell, K., Huss-Danell, K. 1985. Feeding by insects and hares on birches earlier affected by moose browsing. *Oikos* 44:75–81

17. Dicke, M., Sabelis, M. W. 1989. Does it pay plants to advertize for body-guards? In *Causes and Consequences of Variation in Growth Rate and Productivity of Higher Plants,* ed. H. Lambers. The Hague: SPB Academic. In press

18. Edwards, P. J., Wratten, S. D. 1983. Wound induced defences in plants and their consequences for patterns of insect grazing. *Oecologia* 59:88–93

19. Edwards, P. J., Wratten, S. D. 1987. Ecological significance of wound-induced changes in plant chemistry. In *Insects—Plants,* ed. V. Labeyrie, G. Farbres, D. Lachaise, pp. 213–18. Dordrect: Junk

20. Edwards, P. J., Wratten, S. D., Gibberd, R. 1990. The impact of inducible phytochemicals on food selection by insect herbivores and its consequences for the distribution of grazing damage. See Ref. 76, In press

21. Edwards, P. J., Wratten, S. D., Greenwood, S. 1986. Palatability of British trees to insects: constitutive and induced defences. *Oecologia* 69:316–19

22. Endler, J. A. 1986. *Natural Selection in the Wild.* New Jersey: Princeton. 336 pp.

23. Faeth, S. H. 1986. Indirect interactions between temporally separated herbivores mediated by the host plant. *Ecology* 67:479–94

24. Faeth, S. H. 1987. Community structure and folivorous insect outbreaks: the role of vertical and horizontal interactions. See Ref. 2a, pp. 135–71

25. Fagerström, T., Larsson, S., Tenow, O. 1987. On optimal defence in plants. *Funct. Ecol.* 1:73–81

26. Feeny, P. 1976. Plant apparency and chemical defence. *Recent Adv. Phytochem.* 10:1–40

27. Field, C., Mooney, H. A. 1986. The photosynthesis-nitrogen relationship in wild plants. In *On the Economy of Plant Form and Function,* ed. T. J. Givnish, pp. 25–56. New York: Cambridge

28. Fox, L. R., Macauley, B. 1977. Insect grazing on Eucalyptus in response to variation in leaf tannins and nitrogen. *Oecologia* 26:145–62

29. Fowler, S. V., Lawton, J. H. 1985. Rapidly induced defences and talking trees: The devil's advocate position. *Am. Nat.* 126:181–95

30. Gibberd, R., Edwards, P. J., Wratten, S. D. 1988. Wound-induced changes in the acceptability of tree-foliage to Lepidoptera: within leaf-effects. *Oikos* 51:43–47

31. Gill, D. E. 1986. Individual plants as genetic mosaics: ecological organisms versus evolutionary individuals. In *Plant Ecology,* ed. M. J. Crawley, pp. 321–43. Trowbridge: Blackwell

32. Hanhimäki, S. 1989. Induced resistance in mountain birch: defence against leaf-chewing insect guild and herbivore competition. *Oecologia* 81:242–48

33. Hartley, S. E. 1988. The inhibition of phenolic biosynthesis in damaged and undamaged foliage and its effects on insect herbivores. *Oecologia* 76:275–83

34. Hartley, S. E., Lawton, J. H. 1987. Effects of different types of damage on the chemistry of birch foliage, and the responses of birch feeding insects. *Oecologia* 74:432–37

35. Hartley, S. E., Lawton, J. H. 1990. Biochemical basis and significance of rapidly induced changes in birch foliage. See Ref. 76, In press

36. Harvell, C. D. 1990. The ecology and evolution of inducible defenses. *Q. Rev. Biol.* In press

37. Haukioja, E. 1980. On the role of plant defences in the fluctuation of herbivore populations. *Oikos* 35:202–13
38. Haukioja, E. 1982. Inducible defences of white birch to a geometrid defoliator, *Epirrita autumnata. Proc. 5th Int. Symp. Insect-Plant Relationships,* pp. 199–203. Wageningen: Pudoc
39. Haukioja, E. 1990. Toxic and nutritive substances as plant defence mechanisms against invertebrate pests. In *Pests, Pathogens and Plant Communities,* ed. J. Burdon, S. R. Leather, pp. 219–31. Oxford: Blackwell.
40. Haukioja, E. 1990. Positive and negative feedbacks in insect-plant interactions. In *Population Dynamics of Forest Insects,* ed. A. D. Watt, S. R. Leather, N. A. C. Kidd, M. Hunter, pp. 113–22. Andover: Intercept.
41. Haukioja, E., Hakala, T. 1975. Herbivore cycles and periodic outbreaks. Formulation of a general hypothesis. *Rep. Kevo Subarctic Res. Stat.* 12:1–9
42. Haukioja, E., Hanhimäki, S. 1985. Rapid wound-induced resistance in white birch *(Betula pubescens)* foliage to the geometrid *Epirrita autumnata,* a comparison of trees and moths within and outside the outbreak range of the moth. *Oecologia* 65:223–28
43. Haukioja, E., Neuvonen, S. 1985. Induced long-term resistance of birch foliage against defoliators, defensive or incidental? *Ecology* 66:1303–8
44. Haukioja, E., Neuvonen, S. 1987. Insect population dynamics and induction of plant resistance: the testing of hypotheses. See Ref. 2a, pp. 411–32
45. Haukioja, E., Neuvonen, S., Hanhimäki, S., Niemelä, P. 1988. The autumnal moth *Epirrita autumnata* in Fennoscandia. In *Dynamics of Forest Insect Populations,* ed. A. A. Berryman, pp. 163–78. New York: Plenum
46. Haukioja, E., Niemelä, P. 1979. Birch leaves as a resource for herbivores: seasonal occurrence of increased resistance in foliage after mechanical damage to adjacent leaves. *Oecologia* 39:151–59
47. Haukioja, E., Niemelä, P., Sirén, S. 1985. Foliage phenols and nitrogen in relation to growth, insect damage, and ability to recover after defoliation, in the mountain birch *Betula pubescens* ssp. *tortuosa. Oecologia* 65:214–22
48. Haukioja, E., Pakarinen, E., Niemelä, P., Iso-Iivari, L. 1988. Crowding-triggered phenotypic responses alleviate consequences of crowding in *Epirrita autumnata* (Lep., Geometridae). *Oecologia* 75:549–58
49. Haukioja, E., Ruohomäki, K., Senn, J.,

Suomela, J., Walls, M. 1990. Consequences of herbivory in the mountain birch *(Betula pubescens* ssp. *tortuosa):* importance of the functional organization of the tree. *Oecologia.* 82:238–47
50. Haukioja, E., Ruohomäki, K., Suomela, J., Vuorisalo, T. 1990. Nutritive quality as a defense against herbivores. *For. Ecol. Manage.* In press
51. Haukioja, E., Suomela, J., Neuvonen, S. 1985. Long-term inducible resistance in birch foliage, triggering cues and efficacy on a defoliator. *Oecologia* 65:363–69
52. Hunter, M. D. 1987. Opposing effects of spring defoliation on late season caterpillars. *Ecol. Entomol.* 12:373–82
53. Janzen, D. H. 1979. New Horizons in the biology of plant defenses. See Ref. 80a, pp. 331–50
54. Jensen, T. S. 1988. Variability of Norway spruce *(Picea abies* L.) needles; performance of spruce sawflies *(Gilpinia hercyniae* Htg.). *Oecologia* 77:313–20
55. Karban, R. 1987. Herbivory dependent on plant age: hypothesis based on acquired resistance. *Oikos* 48:336–37
56. Karban, R., Adamchak, R., Schnathorst, W. C. 1987. Induced resistance and interspecific competition between spider mites and a vascular wilt fungus. *Science* 235:678–80
57. Karban, R., Brody, A. K., Schnathorst, W. C. 1989. Crowding and a plant's ability to defend itself against herbivores and diseases. *Am. Nat.* 134:749–60
58. Karban, R., Myers, J. H. 1989. Induced plant responses to herbivory. *Annu. Rev. Ecol. Syst.* 20:331–48
59. Landsberg, J., Morse, J., Khanna, P. 1990. Tree dieback and insect dynamics in remnants of native woodlands on farms. *Proc. Ecol. Soc. Aust.* In press
60. Leather, S. R., Watt, A. D., Forrest, G. I. 1987. Insect-induced chemical changes in young lodgepole pine *(Pinus contorta):* the effect of previous defoliation on oviposition, growth and survival of the pine beauty moth, *Panolis flammea. Ecol. Entomol.* 12:275–81
61. Martin, J. S., Martin, M. M., Bernays, E. A. 1987. Failure of tannic acid to inhibit digestion or reduce digestibility of plant protein in gut fluids of insect herbivores: Implications for theories of plant defense. *J. Chem. Ecol.* 13:605–21
62. Mattson, W. J., Palmer, S. R. 1987. Changes in levels of foliar minerals and phenolics in trembling aspen, *Populus tremuloides,* in response to artificial defoliation. In *Mechanisms of Woody Plant Defenses against Insects: Search*

for Pattern, ed. W. J. Mattson, J. Levieux, C. Bernard-Dagan, pp. 155–68. New York: Springer

63. Moran, N., Hamilton, W. D. 1980. Low nutritive quality as defence against herbivores. *J. Theor. Biol.* 86:247–54

64. Myers, J. H. 1988. Can a general hypothesis explain population cycles of forest Lepidoptera? *Adv. Ecol. Res.* 18:179–242

65. Myers, J. H., Williams, K. S. 1984. Does tent caterpillar attack reduce the food quality of red alder foliage? *Oecologia* 62:74–79

66. Myers, J. H., Williams, K. S. 1987. Lack of short or long term inducible defenses in the red alder-western tent caterpillar system. *Oikos* 48:73–78

67. Neuvonen, S., Danell, K. 1987. Does browsing modify the quality of birch foliage for *Epirrita autumanta* larvae? *Oikos* 49:156–60

68. Neuvonen, S., Hanhimäki, S., Suomela, J., Haukioja, E. 1988. Early season damage to birch foliage affects the performance of a late season herbivore. *J. Appl. Entomol.* 105:182–89

69. Neuvonen, S., Haukioja, E. 1984. Low nutritive quality as defence against herbivores: induced responses in birch. *Oecologia* 63:71–74

70. Neuvonen, S., Haukioja, E. 1985. How to study induced plant resistance? *Oecologia* 66:456–57

71. Neuvonen, S., Haukioja, E., Molarius, A. 1987. Delayed induced resistance against a leaf-chewing insect in four deciduous tree species. *Oecologia* 74:363–69

72. Niemelä, P., Aro, E.-M., Haukioja, E. 1979. Birch leaves as a Resource for herbivores. Damage-induced increase in leaf phenolics with trypsin-inhibiting effects. *Rep. Kevo Subarctic Res. Stat.* 15:37–40

73. Niemelä, P. Haukioja, E. 1982. Seasonal patterns in species richness of herbivores: Macrolepidopteran larvae of Finnish deciduous trees. *Ecol. Entomol.* 7:169–75

74. Niemelä, P., Tuomi, J., Mannila, R., Ojala, P. 1984. The effect of previous damage on the quality of Scots pine foliage as food for Diprionid sawflies. *Z. Angew. Entomol.* 98:33–43

75. Raffa, K. F., Berryman, A. A. 1983. The role of host plant resistance in the colonization of behaviour and ecology of bark beetles (Coleoptera: Scolytidae): *Ecol. Monogr.* 53:27–49

76. Raupp, M. J., Tallamy, D. W., eds. 1990. *Phytochemical Induction by Herbivores,* New York: Wiley. In press

77. Reichardt, P. B., Bryant, J. P., Clausen, T. P., Wieland, G. D. 1984. Defense of winter-dormant Alaska paper birch against snowshoe hares. *Oecologia* 65:58–59

78. Rhoades, D. F. 1979. Evolution of plant chemical defense against herbivores. See Ref. 80a, pp. 3–54

79. Rhoades, D. F. 1985. Offensive-defensive interactions between herbivores and plants, their relevance in herbivore population dynamics and ecological theory. *Am. Nat.* 125:205–38

80. Roland, J., Myers, J. H. 1987. Improved insect performance from host-plant defoliation: winter moth on oak and apple. *Ecol. Entomol.* 12:409–14

80a. Rosenthal, G. A., Janzen, D. H., eds. 1979. *Herbivores. Their Interaction with Secondary Plant Metabolites.* New York: Academic

81. Rossiter, M. C., Schultz, J. C., Baldwin, I. T. 1988. Relationships among defoliation, red oak phenolics, and gypsy moth growth and reproduction. *Ecology* 69:267–77

82. Schultz, J. C. 1988. Plant responses induced by herbivores. *Trends Ecol. Evol.* 3:45–49

83. Schultz, J. C., Baldwin, I. T. 1982. Oak leaf quality declines in response to defoliation by Gypsy moth larvae. *Science* 217:149–51

84. Schwenke, W. 1968. Neue Hinweise auf eine Abhängigkeit der Vermehrung blatt- und nadelfressender Forstinsekten vom Zuckergehalt ihrer Nahrung. *Z. Angew. Entomol.* 61:365–69

85. Shigo, A. L. 1984. Compartmentalization: a conceptual framework for understanding how trees grow and defend themselves. *Annu. Rev. Phytopathol.* 22:189–214

86. Silkstonme, B. E. 1987. The consequences of leaf damage for subsequent insect grazing on birch (*Betula* spp.): A field experiment. *Oecologia* 74:149–52

87. Tuomi, J., Niemelä, P., Haukioja, E., Sirén, S., Neuvonen, S. 1984. Nutrient stress: An explanation for plant antiherbivore responses to defoliation. *Oecologia* 61:208–10

88. Tuomi, J., Niemelä, P., Fagerström, T. 1990. Carbon allocation, phenotypic plasticity and induced defences. See Ref. 76, In press

89. Tuomi, J., Niemelä, P., Jussila, I., Vuorisalo, T., Jormalainen, V. 1989. Delayed budbreak: a defensive response of mountain birch to early-season defoliation? *Oikos* 54:87–91

90. Tuomi, J., Niemelä, P., Rousi, M., Sirén, S., Vuorisalo, T. 1988. Induced

accumulation of foliage phenols in mountain birch: branch response to defoliation? *Am. Nat.* 132:602–8

91. Valentine, H. T., Wallner, W. E., Wargo, P. M. 1983. Nutritional changes in host foliage during and after defoliation, and their relation to the weight of gypsy moth pupae. *Oecologia* 57:298–302

92. Watson, M. A. 1986. Integrated physiological units in plants. *Trends Ecol. Evol.* 1:119–23

93. Watson, M. A., Casper, B. B. 1984. Morphogenetic constraints on patterns of carbon distribution in plants. *Annu. Rev. Ecol. Syst.* 15:233–58

94. Williams, K. S., Myers, J. H. 1984. Previous herbivore attack of red alder may improve food quality for fall webworm larvae. *Oecologia* 63:166–70

95. Wratten, S. D., Edwards, P. J., Winder, L. 1988. Insect herbivory in relation to dynamic changes in host plant quality. *Biol. J. Linn. Soc.* 35:339–50

Annu. Rev. Entomol. 1991. 36:43–63

SELF-SELECTION OF OPTIMAL DIETS BY INSECTS

G. P. Waldbauer and S. Friedman

Department of Entomology, University of Illinois, 320 Morrill Hall, 505 South Goodwin Avenue, Urbana, Illinois 61801

KEY WORDS: nutrition, nutritional ecology, insect-plant interactions, feeding behavior, food selection

INTRODUCTION

Since 1915, when the first study of dietary self-selection (32) was published, the length of the list of animals known to self-select has grown at an accelerating pace. Cafeteria experiments with chemically defined diets showed that some of these animals, when offered the separate, purified nutrient components of their usual diet, eat the nutrients in a balance that more or less resynthesizes the original diet and that is often superior to it. Other animals eat two or more natural foods in proportions that yield a more favorable balance of nutrients than will any one of these foods alone. This behavior is usually called *dietary self-selection* (29, 93), although the authors of some papers use terms other than self-selection to designate this phenomenon, e.g. optimization of the nutrient mix (125), diet mixing (44), nutritional wisdom (130), and diet balancing (21, 47). In the ecological literature, information on self-selection can be found in some papers on optimal foraging.

We use the term self-selection by insects in the same sense as has been used for over 60 years in the literature on mammals. Unlike diet switching, self-selection is a continuous regulation of food intake that involves frequent shifts between foods. We propose that the two basic criteria of self-selection are: (*a*) that the animal's choice of foods or nutrients is nonrandom, and (*b*) that it benefits from self-selecting. Self-selection has been extensively studied

43

0066-4170/91/0101-0043$02.00

in mammals, especially laboratory rats, in several other vertebrates, in a few invertebrates, and in a protozoan. Until 1973, it had not been reported for insects, although many insects obviously have the opportunity or need to select from a variety of natural foods.

Self-selection is included in the recently proposed definition of diet switching by insects (110). We suggest that diet switching be redefined to exclude self-selection and to include only long-term shifts in dietary intake that occur between life stages or generations. Lepidoptera, for instance, are often folivores as larvae and nectarivores as adults. This switch differs from self-selection because it is permanent and mediated by metamorphic changes in anatomy, physiology, and behavior.

Diet switching between instars has a different temporal organization than self-selection, which occurs within instars. First instars of the grasshopper *Dociostaurus maroccanus* eat only tender leaves of the grass *Poa bulbosa,* but later instars eat many grasses and some forbs (117). Young corn earworm larvae *(Heliothis zea),* more sensitive to allelochemics than older larvae, tend to feed on the anthers of cotton blossoms, which apparently have a lower concentration of allelochemics than do other parts of the blossom (84). Gypsy moths *(Lymantria dispar),* when fed oak foliage during early instars and pine foliage during late instars, are superior in development rate, survival, and fecundity to larvae fed only oak foliage (6, 7).

Perspectives and Significance

Self-selection is ubiquitous; known examples range across kingdoms (Protista and Animalia), across invertebrate phyla (Mollusca and Arthropoda), and across vertebrate classes (Chondrichthyes, Aves, and Mammalia). It probably evolved before the Metazoa arose and may well be common to and be mediated by similar mechanisms in most if not all animals. Thus, work on insects will contribute to understanding the fundamental basis of self-selection by all animals.

Dietary self-selection by insects, a newly discovered dimension of insect feeding behavior, is central to understanding how insects interact with their host plants and other foods and offers heretofore untried experimental approaches to exploring the behavioral and physiological bases of feeding by insects. The interaction of insectan and other herbivores with plants is the foundation of almost all terrestrial ecosystems. With the possible exception of nematodes, insects are usually the most abundant herbivores in terrestrial ecosystems in terms of numbers of species, numbers of individuals, and biomass. Since integrated pest management (IPM) is essentially applied ecology, a knowledge of self-selection will enhance our ability to design IPM programs for the control of damage by pests of plants.

Self-Selection and Optimal Foraging

Although dietary self-selection has received relatively little attention from ecologists (but see 112), a knowledge of self-selection has important consequences for optimal foraging theory, the focus of many theoretical ecologists for the past two decades (28, 57, 112). Optimal foraging theory seeks to understand the foraging behavior of animals by mathematical modeling. Many models include assumptions that the animal increases its fitness by adopting feeding patterns that maximize the intake per unit time of some critical dietary currency, often assumed, especially in older models, to be energy (82, 112). As pointed out by Pyke (82) and Stephens & Krebs (112), many other optimal foraging models, especially more recent ones, recognize that this maximization may have constraints i.e. that self-selection may occur (82, 112). For example, the maximization of energy intake by moose (14, 38, 112) is constrained by their need to eat low-energy aquatic plants in order to obtain sufficient sodium. One study (28) stated that typical predictions of maximization models are ". . . that large foods should be preferred to small, . . . patchy foods to spaced-out foods; calorie-rich foods to calorie-poor foods . . .." Work on self-selection, however, indicates that the situation is rarely simple, that most animals are adapted to eat so as to achieve a favorable balance of nutrients rather than to maximize the intake rate of calories or some single nutrient (11, 12, 24, 93, 98, 99, 108, 120, 122).

Investigators (28, 112) have argued that optimal foraging models for predators can justifiably assume the maximization of one dietary currency because animals are likely to be a well-balanced diet for other animals, thus obviating the need to diversify food intake in order to achieve nutritional balance. The finding that wolf spiders, *Pardosa ramulosa,* eat three prey species in nonrandom proportions (44), and that nestling European bee eaters, *Merops apiaster,* grow better on a mixed diet of insect species than on only one species (60), argues that these assumptions are not always justified.

Many optimal foraging theorists recognize that the diets of herbivores tend to be more complex than the diets of predators—herbivores may have to eat a variety of plant species or plant parts to obtain all of their required nutrients in sufficient quantity or to minimize their intake of the various toxins that are so widely distributed among plants. Herbivorous insects do tend to eat a mixed intake of plant species or of different tissues from the same species (8–10, 25, 40, 48, 58, 63, 66, 79, 96, 103). This general rule may have some exceptions. For example, an adult lepidopteran that comes out of the pupal stage with sufficient protein, lipid, and micronutrients to produce its quota of eggs may seek only to fuel its flight muscles by maximizing its caloric intake of nectar, although many adult Lepidoptera also feed from carrion, feces, urine puddles, or pollen. Indeed, a great deal of optimal foraging theory is applied to nectarivores (28, 112).

Self-Selection by Vertebrates and Other Nonarthropods

Some early studies demonstrated dietary self-selection by vertebrates offered a choice of natural foods (127). Pigs grew exceptionally rapidly when given a free choice of grains and other feeds (32); chickens grew better when they were offered a variety of natural foods rather than a predetermined diet (78); and newly weaned human infants thrived when they made their own choices from a selection of foods (29).

Curt P. Richter, in his studies with rats, was the first to use separate nutrients in self-selection experiments. His work still ranks among the best in the field of self-selection. When Richter et al (93) offered laboratory rats a choice of casein, sucrose, olive oil, mineral solutions, and natural vitamin sources, self-selecting rats grew as well as did a control group on the standard diet and utilized their food more efficiently for growth, gaining slightly more weight than the controls while reducing their intakes of solids and calories by 36.4% and 18.7%, respectively. He also showed that when adrenalectomized rats, which normally die from excessive salt loss, were given free access to sodium chloride, they increased their intake of it and remained alive and free of overt symptoms of insufficiency (87, 92). Parathyroidectomized rats, which have convulsions and die because they lose calcium, increased their intake of calcium lactate by a factor of 3.9, thus avoiding convulsions and maintaining normal growth. They resumed the lower rate of calcium lactate ingestion after the reimplantation of parathyroids (91). Finally, pancreatectomized rats, which exhibit the symptoms of diabetes mellitus, avoided these symptoms when offered a free choice by eating less sucrose and more fat than did intact rats (94). Conversely, intact rats, made hypoglycemic by insulin injections, increased their glucose intake (88).

Most rats self-select well and thrive, but a few die because they select poorly, often because they fail to ingest sufficient protein (80, 104). Some workers seemingly failed to demonstrate self-selection (12, 77, 98, 128), but these discrepancies have been explained (98), and many other workers have since demonstrated self-selection by rats and various other animals.

Because we are not aware of a recent review, we cite a few examples from the recent literature on self-selection by vertebrates. The notable papers on laboratory rats include References 2, 13, 22, 71, 115. Dormice (Glis glis) underwent cyclic periods of weight gain accompanied by large increases in the self-selected carbohydrate intake, usually followed by increased lipid intake (99). In a natural habitat, moose (Alces alces) selected nonrandomly and usually consistently when offered a cafeteria of various aquatic plant species, generally favoring plants with a high sodium content (38). They eat a balance of aquatic plants and foliage from woody plants, the former largely for their high sodium content and the latter largely for their caloric content (14, 112). European bee-eater nestlings grew better on a diet of mixed insect species than they did on a diet of only one species, possibly explaining why the parent birds do not take only the energy-maximizing prey types (60). Herbivorous parrotfish *(Sparisoma radians)* grew best on a mixed diet of plants and usually ate a nonrandom balance of plant species independent of the plants' relative abundance (64).

Except for insects, little is known about self-selection by invertebrates. Plate limpets, *Acmaea scutum* (Phylum Mollusca) ate a tender alga and a tough alga in a 60:40 ratio, thus minimizing the excessive tooth wear associated with eating the tough alga (59). When offered two species of prey, the protozoan *Stentor coeruleus* ate a mixed diet and reproduced faster on mixtures of algae and protozoa than on a diet of only one of these (83).

THE CRITERIA FOR SELF-SELECTION

Nonrandomness of Choice

Nonrandomness of choice, one of the two basic criteria of self-selection, can be demonstrated in at least two ways. First, if two or more foods are presented (e.g. to stored-grain insects) as a homogeneous mixture of small, dry particles, the self-selected intake ratio can be compared with the ratio of these foods in the mixture (120). Many foods cannot be presented in this way. Two agar–based diets, for example, are best presented as separate blocks in an arena. The two blocks presented to a self-selector are nutritionally incomplete but complementary; one may lack sugar but have protein, while the other may lack protein but have sugar, although both blocks are otherwise nutritionally complete (122). Nonrandomness is then demonstrated by comparing these arenas with control arenas containing two identical and nutritionally complete blocks arbitrarily designated A and B. Self-selector and control arenas are paired by always placing the A block and the same incomplete block in the same relative positions in their respective arenas. An insect offered two identical control blocks of adequate diet tends to eat from only one of the blocks (122). Insects offered two identical blocks of inadequate diet move more often and thus tend to divide their feeding evenly between the two blocks (101). [Note that the mean intake ratio (A block:B block) will be 50:50 in either case.]

Self-selecting animals not only make nonrandom choices, but the individuals of a uniform group also tend to eat at least the major nutrients in more or less consistent proportions. The self-selected protein:carbohydrate ratios of 60 rats ranged only from 17:83 to 30:70, although possible ratios ranged from 7:93 to 78:22 (115). Insects also tend to be consistent self-selectors. Some are known to have optimal protein:carbohydrate ratios (51, 53, 68, 122), and the concentration of protein in the diet is critical for at least some species (30). Thus, it is not surprising that the self-selected protein:

carbohydrate ratios of 25 *H. zea* larvae ranged only from 62:38 to 90:10, although ratios of from 0:100 to 100:0 were possible (121). Furthermore, 11 groups of 10 cockroach nymphs self-selected protein:carbohydrate ratios that ranged only from 12:88 to 21:79, although all ratios (0:100 to 100:0) were again possible (24).

Selection for minor nutrients may be less consistent. The only data available for insects (100) show that the range of ratios was broad when *H. zea* larvae self-selected in order to obtain B-vitamins. The intake ratios of self-selecting larvae (diet lacking only choline chloride:diet lacking only most of the other B-vitamins) ranged from 96:4 to 22:78, while all ratios were possible. Nevertheless, most of these larvae ate enough of both diets to survive. Perhaps the range of ratios would have been narrowed if the two complementary diets had been differently flavored so as to facilitate the association of each one with the satisfaction of a particular nutrient need.

The Benefits of Self-Selection

The second criterion of self-selection is met if self-selectors fare as well or better on the self-selected mix than on other mixtures of the same ingredients or on a diet lacking a self-selected nutrient. We anticipate that the improvement is usually quantitative; the insect self-selects a more or less optimal balance of nutrients. The qualitative nutritional requirements of insects vary little among species, although insects, as a group, feed on a wide variety of organic substances (34, 35, 51). The optimal balance of nutrients often varies with the species, age, or sex of the individual, usually falling within a narrow range for individuals of the same species, age, and sex (54). Diets may vary in nutritional value for a given species, depending upon the content of vitamins or sterols (19, 37), the amino acid balance (53, 69), the proportion of total nitrogen (73), and/or the level of digestible carbohydrate (53).

Self-selection is obviously beneficial if survival, weight gain, development rate, or fecundity is enhanced because, most likely, the self-selected intake is more efficiently utilized for growth and metabolism than are other intakes. The benefit may, however, not be obvious if the insect increases its food utilization efficiency without affecting the above-mentioned parameters. From a strictly physiological point of view, increased efficiency per se may seem to offer little advantage because the animal has the alternative of simply eating more rather than increasing efficiency. An ecological perspective, however, leaves no doubt that increased efficiency has survival value for insects (124). For cockroaches (24) and other insects that must forage for scattered resources, increased efficiency and the concomitant reduction in the amount of food that must be eaten results in less time spent foraging and, thus, less exposure to predators. Increased efficiency also enhances an insect's ability to compete for food when it is scarce.

Statistical Considerations

The conventional t-test shows a significant difference between controls and self-selectors if the mean intake ratio of the latter differs widely from the usual 50:50 mean intake ratio of the controls. This difference occurred with *H. zea* larvae, which had mean self-selector (protein:carbohydrate) and control intake ratios of 79:21 and 47:53, respectively, (122) and with cockroach nymphs, which had corresponding ratios of 16:84 and 51:49 (24). However, if the self-selectors, although choosing nonrandomly, have a mean intake ratio close to 50:50, the t-test will not show a difference. Control groups tend to have mean ratios of about 50:50 usually because about half of the insects feed mostly or solely from the A block, while the rest feed mostly or solely from the B block. Self-selectors will also have 50:50 mean intake ratios if each insect takes about half of its intake from one of the incomplete diets and the other half from the complementary diet (100). The statistical test of choice is, therefore, a modified Levene's test (118). The amounts eaten by each insect, self-selector or control, from each of its two diet blocks are expressed as percentages of the total it ate from both blocks. The difference between these two percentages, the modified Levene's number, is a measure of feeding variance. A large difference shows that the insect ate mainly from one block; a smaller difference shows that its feeding from the two blocks was more nearly equal. The modified Levene's numbers of self-selectors and controls can be compared by a paired or two sample t-test, depending on the organization of control and experimental arenas. An arcsine transformation may be employed with very large or very small percentages.

Dietary Self-Selection and Food Utilization Efficiencies

The physiological basis of changes in efficiency can be explored by determining the quantitative parameters of dry weight food utilization used by nutritional ecologists (106, 119): RCR (relative consumption rate), RGR (relative growth rate), AD (approximate digestibility), ECD [efficiency of conversion of digested (assimilated) food to biomass], and ECI (efficiency of conversion of ingested food to biomass). These parameters are related such that ECI = AD × ECD and RGR = RCR × ECI. AD measures the portion of consumed food that passes through the gut wall into the hemolymph and is thus available for metabolism and growth; it is affected by factors such as the food's rate of passage through the gut and its content of indigestible matter. ECD measures the proportion of assimilated food that is converted to body mass; it is affected by metabolic rate, a deficiency of vitamins (49), and/or other imbalances. These parameters often suggest the underlying causes of differences in utilization efficiencies. When, for example, the sucrose content of their diet was increased, the ECI of *H. zea* larvae declined, despite a significant increase in AD, because of a large concomitant decrease in ECD (122). These larvae assimilated more sucrose than they could use and catabolized the excess, which was insensibly passed out of the body as water and carbon dioxide (52, 122). Cockroach nymphs that self-selected ate significantly less than did the corresponding controls, although relative growth rates did not differ between these two treatments (24). Like the rats of Richter et al (93), these nymphs grew as much as the controls while eating less food. This increase in efficiency was caused by significant increases in both AD and ECD resulting from the selection of the optimal protein:carbohydrate ratio.

H. zea larvae that feed in maize ears may not benefit from increased efficiency because they are protected from most predators and are surrounded by more food than they can eat. Nevertheless, when fed maize kernels, they maximized their intake of germ within feasible limits, thus increasing ECD and thereby eating less while gaining as much weight as did larvae that were fed only endosperm (25). Increasing efficiency may seem superfluous on maize, but the behavior that results in increasing germ intake may have evolved on other plants. *H. zea* larvae are polyphagous and feed mainly on fruits, some of which (e.g. the fruits of *Phaseolus vulgaris*) are so small that a larva must eat several of them during its lifetime. Thus, the larvae can be expected to feed so as to reduce the number of pods that must be attacked. Thereby, as do cockroaches, they enhance their ability to compete for food, decrease foraging time, and thus minimize exposure to predators.

THE POTENTIAL FOR SELF-SELECTION BY INSECTS

Although very few cases of self-selection by insects or other arthropods have been documented, many, if not most, insects have at least the opportunity to select a favorable balance from their natural foods. Predators may take several prey species (81). Omnivorous cockroaches (97) and mole crickets (42) eat both animal and vegetable foods. Anautogenous female mosquitoes take blood and nectar for protein and energy, respectively (50). Some basically herbivorous species are facultative predators (73), and most predaceous Hemiptera also suck plant juices (45). Strictly herbivorous insects also have opportunities to self-select. Some feed on more than one plant species, as do several grasshoppers (66), but even insects that feed on only one plant have a choice of tissues that vary in nutrient content. Seed and fruit feeders, such as confused flour beetle larvae *(Tribolium confusum)* (27) and *H. zea* larvae (75) have a choice of several tissues or structures. Tobacco hornworm larvae *(Manduca sexta)* sometimes attack fruits, although they eat mostly foliage (75). Many adult insects, including syrphid flies (103) and *Heliconius* butterflies (40, 41), eat both pollen and nectar. Some adult Neuroptera eat insects, honeydew, and pollen (113, 114). The possibility that honey bees *(Apis mellifera)* self-select is interesting from a sociobiological perspective.

Genetic variability within a population of colonies for the tendency to collect pollen vs nectar was shown by the successful selection of high- and low-pollen hoarding lines (46). This variability is apparently based on local adaptation to different environments (20, 46). Recent evidence also shows a genetic variability in the tendency to collect pollen vs nectar among workers belonging to the different "subfamilies" that arise within a colony because of polyandry (95). Genetically determined foraging specialization on nectar or pollen may increase the efficiency with which a colony adjusts its proportional intake of nectar and pollen in response to changing environmental conditions (95).

Plant species and plant parts often vary in nutrient composition. This variation provides ample latitude for herbivores to adjust their nutrient balance by varying the mix of species and/or parts they eat. Plant parts and species may vary greatly in amino acid balance and total nitrogen content (73, 105, 109). Old leaves are usually low in total nitrogen, while young leaves, seeds, and other parts of fruits tend to have more nitrogen (70). The parts of seeds may vary in nutrient content. While wheat endosperm is low in vitamins and has only 8–13% protein and about 70% starch, wheat germ is rich in vitamins and has about 25% protein and 20% sugars (67). Furthermore, germ and endosperm proteins may differ in composition and nutritional quality, as in maize kernels (126). Other nutrients, e.g. trace elements (116), may also vary with plant species, structure, or tissue.

DIETARY SELF-SELECTION BY INSECTS AND OTHER ARTHROPODS

Selection from Natural Foods

The first demonstration of self-selection by an arthropod is the report that *T. confusum* larvae, when fed a 1:1:1 mixture of small particles of the three fractions of the wheat kernel—germ, bran, and endosperm—did not feed randomly but selected a mix of 81% germ, 17% endosperm, and 2% bran (120). The self-selected mix supported better growth than any one fraction alone or wheat ground to a flour so fine (17) that the larvae could not feed selectively. This may have been because this mix, unlike a single fraction or whole wheat flour, provided a protein: carbohydrate ratio (57:43) close to the optimal 50:50 ratio (68). These larvae, given more of each wheat fraction than they could eat, easily increased their germ intake—possibly beyond what is possible for larvae fed unfractionated kernels that have a germ:endosperm: bran ratio of only 3:82:15. It had already been shown that *T. confusum* larvae, when fed whole kernels of wheat with an injury in the area of the germ, ate all of the germ and only a little of the surrounding endosperm (36).

When *H. zea* larvae were offered whole maize kernels, they fed preferentially on the germ but also ate some of the endosperm (25). Eating endosperm might in part be a "search for the germ" (25), a view supported by the results of an experiment in which a suboptimal defined diet was layered over an optimal defined diet in small plastic cups. *H. zea* larvae placed in these cups burrowed through the suboptimal layer with little or no feeding and then fed upon the optimal layer (25). Thus, larvae fed whole kernels seem to maximize their intake of germ within feasible limits, thus increasing utilization efficiency and decreasing the amount of food that they must eat (25). Larvae that ate cut portions of kernels that included the germ and some endosperm utilized their food more efficiently than did larvae that ate cut portions that included only endosperm, although these latter larvae did survive to complete development.

Locusts and grasshoppers eat a mix of fresh and dry leaves even when fresh leaves are abundant (63). The survival, growth, and fecundity of *Melanoplus differentialis* were improved by eating both turgid and wilted sunflower leaves (62). Migratory locust nymphs *(Schistocerca gregaria)*, when given a choice of fresh and lyophilized wheat leaves, ate both and had higher RGRs and ECDs than did nymphs that were given only fresh leaves (63). They selected a balance of fresh and dry leaves that regulated their water balance (96) and optimized their dry matter intake (63). Self-selection may have been nonrandom because, during a two-hour opportunity to self-select, nymphs transferred from a diet of only fresh leaves took 75% of their meals from dry leaves, while those transferred from a diet of only dry leaves took only 25% of their meals from dry leaves (63). The mean duration of the meals is not specified.

In addition to these findings, some reports, usually not identified as studies of dietary self-selection, indicate that other insects or spiders may self-select. These reports establish only one of the two criteria of self-selection—either the nonrandomness of choice or the acquisition of a benefit. Most spiders require more than one prey species to survive (85). Wolf spiders ate three prey species in nonrandom proportions (44). The investigator argued that this mixed diet provided the ten essential amino acids in a balance that was assumed to be optimal because it is similar to the balance in the spider's own body (44). However, survival, growth, fecundity, and the efficiency of food utilization were not measured, and amino acids were measured only in the preys' hemolymphs rather than in all of the tissues that the spiders ingest (55). Thus, the acquisition of a benefit was not demonstrated.

The larvae of several species of Noctuidae eat a mix of different plant species, a behavior correlated with fast development and increased survival (74). Grasshoppers may also eat a mixed diet of plant species (117). When fed a mix of three plants, *Euthystira brachyptera* grew better and was more fecund than when fed only one species (58). *Melanoplus bilituratus, M.*

differentialis, and *M. sanguinipes* benefited similarly when they ate a mix of plant species rather than any one plant species (8–10, 48, 66, 79). The crops of field-collected *M. bivittatus* and *M. femurrubrum* contained both forbs and grasses (4). The food utilization efficiencies of these grasshoppers on single and mixed plant diets were measured, but, as in the other studies mentioned in this paragraph, measurements of the separate intakes of the different plants in the mixed diet were not made (5). Hence, it is not known if selection by any of these noctuids, locusts, or grasshoppers is nonrandom, or if individuals of the same species, age, and sex choose more or less the same optimal mix. Other reports establish neither criterion of self-selection, but note only that some mites and insects more or less continuously ingest a mixture of different foods (72, 110). These studies frequently assume, probably often correctly, that these foods complement each other in nutrient content (110).

Selection from Defined Diets

The use of defined diets makes possible chemical manipulations and standardizations not possible with natural foods. Last instar *H. zea* larvae (121) moved back and forth between diets (26) to self-select a mean 79:21 protein:carbohydrate ratio (122) when offered two diets that were identical and nutritionally complete, except that one lacked casein (the only protein) and the other lacked sucrose (the only digestible carbohydrate). Control larvae offered two identical and adequate diets (50:50 casein:sucrose) chose one of them apparently at random, ate only or almost only from it (122), and seldom moved between diets (26). An 80:20 protein:carbohydrate ratio was nutritionally superior to 100:0, 50:50, and 20:80 ratios when incorporated in otherwise equivalent diets and fed to *H. zea* larvae throughout the last instar (122). On the 80:20 diet, ECI and ECD were higher than on diets that had more sugar or altogether lacked sugar, although diets high in sugar content gave higher ADs.

Groups of 10 brown-banded cockroach nymphs *(Supella longipalpa),* allowed to self-select from the beginning of the third instar to the adult molt, selected a mean 16:84 protein:carbohydrate ratio when given a choice between two defined diets that were nutritionally complete except that one lacked casein (the only protein) and the other lacked glucose (the only digestible carbohydrate) (24). Groups of 10 control nymphs, offered blocks of two identical and nutritionally complete diets (50:50 casein:sucrose), fed randomly, dividing their intake almost evenly between the two blocks. Self-selectors were superior to controls in all food utilization parameters. In contrast to *H. zea, S. longipalpa* forced to feed throughout their entire nymphal lives on a single diet with a 20:80 casein:glucose ratio (approximating the 16:84 ratio) had a higher mortality rate than nymphs given a 50:50 diet or the opportunity to self-select (24). A 20:80 diet was nutritionally unsuitable as the only choice, perhaps because it does not match closely enough the varying needs of these insects throughout nymphal life. In support of this idea, measurements of casein and glucose consumption made throughout the first and last nymphal stadia showed that in each case carbohydrate intake was very high at first but decreased during the stadium, finally equalling the level of protein intake, which remained low and more or less constant through each of these stadia. Time-lapse photography showed that *H. zea*'s self-selected protein:carbohydrate ratio apparently also varied during the last larval stadium.

H. zea's and *S. longipalpa*'s mean self-selected protein:carbohydrate ratios, 79:21 and 16:84, respectively, are almost opposites, a difference that reflects the differences in their growth rates and life styles. On its defined diet, *S. longipalpa* requires 256 days to complete the nymphal stage, while *H. zea* completes the larval stage in only 17 days on its defined diet (24, 122). *S. longipalpa* nymphs are active because they must search for scattered food, find shelter, and avoid predators (24). The sedentary *H. zea* larva may spend its life within one ear of maize, where it is protected from most predators and food is abundant (25). Thus, *H. zea* larvae must eat a large proportion of protein to permit rapid tissue building, but being sedentary, use relatively little sugar to meet the energy requirement for activity. Conversely, the slow-growing *S. longipalpa* nymphs need not accumulate protein rapidly but must eat a large proportion of sugar to meet the higher activity requirements and the proportionally greater maintenance requirements that result from its slow growth rate.

A mean intake ratio measured over an entire stadium or life stage reveals the overall nutrient balance required during that period. However, measuring self-selection during several intervals of the first and last instars of *S. longipalpa* demonstrates that it is a dynamic process that is sensitive to the insect's changing nutritional needs as they vary from week to week (24). More recent work (I. Ahmad, G. P. Waldbauer, & S. Friedman, unpublished data) shows that the protein:carbohydrate ratio self-selected by *M. sexta* larvae varies from day to day, at least during the fifth stadium.

Self-selection from defined diets by the oligophagous and mobile fifth instar nymphs of the acridid *Locusta migratoria* and by the polyphagous and relatively sedentary last instar larvae of the noctuid *Spodoptera littoralis* was recently demonstrated (108). These insects were fed during a conditioning period of 4, 8, or 12 h on either a nutritionally complete diet, one lacking protein but including carbohydrate, or one lacking carbohydrate but including protein. At the end of the conditioning period, the insects were transferred to arenas in which they could choose between two incomplete but complementary diets for 9 h, one lacking only protein and one lacking only carbohydrate. Both species selected for the nutrient missing from the conditioning diet, even when the conditioning period had been only 4 h long (108).

Adult female black blow flies, *Phormia regina,* do not develop eggs if they are fed only a sucrose solution and starve if they are fed only an aqueous yeast extract as a protein source (15). Females with access to both sucrose and yeast exhibited a distinctive and nonrandom feeding pattern during the first vitellogenic cycle. The volume of sucrose solution consumed, consistently greater than yeast consumption, gradually increased until it peaked on the fifth day. A peak of yeast intake on the first day or two was followed by a second peak, which corresponded with the period of vitellogenesis on the third to fifth days. This second peak was followed by a period of low yeast intake that continued until after oviposition, when vitellogenesis began again (15). *P. regina* and many other Diptera (76) must exhibit similar feeding patterns in nature as they exploit honeydew or nectar for carbohydrate and feed on pollen, feces, or carrion for protein.

BEHAVIORAL AND PHYSIOLOGICAL CONTROL OF SELF-SELECTION

Behavioral Hypotheses

An animal can gain information about its food by only two mechanisms: (*a*) via peripheral sensilla; this involves mainly chemosensory stimuli in insects; and (*b*) via a metabolic route, as when a toxin or lack of a nutrient causes a metabolic upset, which may be perceived by internal sense organs. Accordingly, two behavioral hypotheses for the control of self-selection have been proposed, one based largely on sensory stimuli and one based primarily on metabolic feedback.

Richter (89, 90) proposed a sensory hypothesis: that the animal has innate nutritional wisdom based in specific hungers for specific nutrients and in the ability to recognize sensorially each of the nutrients that it requires. [We later referred to this as the "hypothesis of the innate flavor template" (26, 124).] However, this is not likely to be a general explanation for dietary self-selection (98). Because animals require over 30 different nutrients, the hypothesis requires that as many different recognition templates be hard-wired in the brain. Furthermore, no innate appetites can be demonstrated for most nutrients, except that rats have an innate hunger and recognition mechanism for sodium chloride (98) and chickens for water (22, 98).

A metabolic feedback mechanism, probably coupled with learning, is more flexible and, being the more parsimonious concept, seems more likely to have evolved. This hypothesis has been variously stated (3, 22, 98), but until recently little evidence supported it. Konrad Lorenz (65), in considering Richter's self-selection work, offered a behavioral hypothesis—that animals eat some of each available food, form an engram of how it feels afterwards, and then eat the foods that produce positive feedback. We restated his hypothesis and named it the "malaise hypothesis." It proposes that an animal feeds on a toxic or nutrient-deficient food until the toxin or deficiency causes a metabolic upset (malaise) that induces searching behavior. The wandering animal may then encounter a different food that provides positive feedback because it lacks the toxin or has the nutrient in question (23, 24, 124).

Sensory stimuli probably play an important but secondary role. Insects may learn to associate sensory stimuli with the presence of a required nutrient; they may feed on an artificial diet only if it has a food-like flavor or odor, or they may feed on an alternate diet more readily if it is sensorially distinguishable from the other diet. Indeed, insects do little feeding on agar gels that lack nutrients (123), and some do not feed on gels that contain flavorless nutrients. Thiamine-deficient rats did not self-select for a diet with thiamine until it was flavored with anise (98).

Less than 30 years ago, most investigators thought that animals cannot learn to associate the sensory stimuli from a nutrient-deficient or toxic food with the delayed consequences of eating it (107, 130). Traditional learning theory held that the unconditioned stimulus (e.g. malaise due to a deficiency) must follow immediately after the conditioned stimulus (i.e. a sensory stimulus from the deficient food). Abundant evidence now shows that delayed learning mechanisms of this type do occur (98, 107). In an early study, rats were conditioned to reject a novel food if they were X-irradiated and thus made ill some time after eating it (86). The converse was also shown. Thiamine-deficient rats were conditioned to drink a saccharin-flavored, non-nutritive solution containing no thiamine by injecting them with thiamine after they drank this solution (39). Although few reports document food aversion learning by insects, it is known to occur (16, 31, 61), and the induction of feeding preferences has been demonstrated (56).

Recent experiments support the malaise hypothesis. Preliminary work (102) showed that last instar *H. zea* larvae are not stimulated to bite by mannitol, although it is as nutritious for them as is sucrose, and that sorbose, although not nutritious for these larvae, does stimulate them to bite, although much more weakly than does sucrose (102). These two carbohydrates were then used alone or in combination to separate the effects of phagostimulation and metabolic feedback from nutrients on self-selection by last instar *H. zea* larvae. When offered a choice of two diets, one containing protein but lacking a digestible carbohydrate and another lacking protein but containing mannitol as the only digestible carbohydrate, the larvae did not self-select to add the presumably tasteless mannitol to their intake of the protein diet. However, when given a choice between two diets that were both sapid because they both contained protein—but only one of which contained a digestible carbohydrate (again the tasteless mannitol)—the larvae greatly preferred the mannitol-containing diet to the one that lacked a carbohydrate. These results show that the mannitol was perceived via the metabolic route, but that both metabolic feedback and chemosensory stimulation are required to elicit self-selection. Furthermore, although *H. zea* larvae did not self-select for diets containing either the tasteless mannitol or the non-nutritious sorbose as the only carbohydrate, they did self-select to obtain similar diets containing both sorbose and mannitol (101). Although metabolic feedback is primary, both it and phagostimulation were required to elicit self-selection.

The Physiological Basis of Self-Selection

Broadly speaking, understanding the physiological basis of self-selection requires answers to four questions: 1. What nutrient imbalances or deficiencies cause foods to be suboptimal? 2. How does this cause a metabolic upset? 3. How is the upset perceived as a malaise; i.e. how does it affect the central nervous system (CNS)? And 4. how does the CNS bring about changes in proportional food consumption, i.e. self-selection?

Work with *H. zea* larvae begins to answer the first two questions. First, both a diet lacking sugar and one containing too much sugar were suboptimal, being utilized for growth less efficiently than an apparently optimal diet (122). Second, utilization parameters also indicate how a nutritional imbalance might cause a metabolic upset. When *H. zea* larvae were fed a diet with too much sugar, ECD declined significantly while AD increased (122). As mentioned above, the most likely explanation is that these insects, unable to prevent the assimilation of excess sugar, had to catabolize the excess and eliminate it as water and carbon dioxide.

In 1912, in a brilliant paper (49) that established the existence of "accessory growth factors" (vitamins), Hopkins showed that the absence of these factors causes a great waste of assimilated food. In other words, an animal cannot efficiently utilize protein for growth (i.e. ECD markedly decreases) if vitamins are absent or in short supply (43). Similarly, *H. zea* larvae on a defined diet have higher ECDs when some sucrose is present than when it is absent (122). The ingestion of sucrose, a good source of energy, apparently spares these insects from metabolizing protein for energy. The ability to utilize protein for growth efficiently, which is reflected by ECD, may be limited by an insufficiency or absence of any other required or helpful nutrient. Thus, we propose the hypothesis that an insect may, on the metabolic level, perceive a shortage of vitamins or of any other required or helpful nutrient as a change in the concentration of some metabolite associated with the utilization of protein.

An answer to the third question, that of the link between metabolic upset and the CNS, is suggested by work on mammals. In rats, brain and plasma serotonin and tryptophan increase after the consumption of carbohydrate or the injection of insulin (33). Drugs that suppress serotonin production in rats lead to decreased protein consumption, while those that activate serotonin receptors or cause an increase in serotonin production suppress carbohydrate intake or increase protein intake (129). That serotonin is involved in self-selection is, thus, likely but not entirely established (18).

Although insects are known to have serotonin in the CNS (111), little is known about the control of appetite in insects by serotonin. *H. zea* larvae, fed a diet high in sucrose but lacking protein, had higher levels of brain serotonin than did larvae fed a protein-containing diet (23). When an inhibitor of serotonin production was added to both diets offered to self-selecting larvae, brain serotonin levels were reduced and carbohydrate consumption increased by 85% (23). Adding free tryptophan, a serotonin precursor, to the diets did not result in a significant decrease in carbohydrate intake, but protein consumption increased significantly (23). These findings suggest that serotonin is involved in a feedback loop that regulates the intake of at least protein and carbohydrate.

A possible answer to the fourth question, i.e. how the CNS brings about changes in food consumption, is suggested by recent experiments with *L. migratoria* (1), in which the level of dietary protein appears to affect the sensitivity of chemoreceptors on the maxillary palpi. This may, in turn, produce changes in food intake. Decreases in the sensitivity of these receptors seem to be correlated with increases in blood osmolality and amino acid concentration as they vary with the protein content of the diet (1).

CONCLUSIONS

The discovery that insects self-select from defined diets reveals a new dimension of the insect-food interaction and offers a new experimental approach to exploring the behavioral and physiological bases of feeding. This discovery may have other wide-ranging consequences, notably for understanding insect-host plant interactions, which are fundamental to comprehending the organization of ecosystems and to devising integrated pest management systems. Furthermore, since dietary self-selection is widespread, if not universal, among insects and other animals, work with insects contributes to our understanding of how all animals interact with their foods.

ACKNOWLEDGMENTS

M. R. Berenbaum, G. E. Robinson, and J. H. Willis reviewed the manuscript. G. E. Robinson contributed information on honey bees. The authors were partially supported by grant 88-37153-3433 from the USDA Competitive Research Grants Program while preparing this article.

Literature Cited

1. Abisgold, J. D., Simpson, S. J. 1988. The effect of dietary protein levels and haemolymph composition on the sensitivity of the maxillary palp chemoreceptors of locusts. *J. Exp. Biol.* 135:215–29

2. Ackroff, K., Schwartz, D., Collier, G. 1986. Macronutrient selection by foraging rats. *Physiol. Behav.* 38:71–80

3. Ashley, D. V. M. 1985. Factors affecting the selection of protein and carbohydrate from a dietary choice. *Nutr. Res.* 5:555–71

4. Bailey, C. G., Mukerji, M. K. 1976. Feeding habits and food preferences of *Melanoplus bivittatus* and *M. femurrubrum* (Orthoptera: Acrididae). *Can. Entomol.* 108:1207–12

5. Bailey, C. J., Mukerji, M. K. 1976. Consumption and utilization of various host plants by *Melanoplus bivittatus* (Say) and *M. femurrubrum* (DeGeer) (Orthoptera: Acrididae). *Can. J. Zool.* 54:1044–50

6. Barbosa, P., Martinat, P., Waldvogel, M. 1986. Development, fecundity and survival of the herbivore *Lymantria dispar* and the number of plant species in its diet. *Ecol. Entomol.* 11:1–6

7. Barbosa, P., Waldvogel, M., Martinat, P., Douglass, L. W. 1983. Developmental and reproductive performance of the gypsy moth, *Lymantria dispar* (L.) (Lepidoptera: Lymantriidae), on selected hosts common to mid-Atlantic and southern forests. *Environ. Entomol.* 12:1858–62

8. Barnes, O. L. 1955. Effect of food plants on the lesser migratory grasshopper. *J. Econ. Entomol.* 48:119–24

9. Barnes, O. L. 1963. Food-plant tests with the differential grasshopper. *J. Econ. Entomol.* 56:396–99

10. Barnes, O. L. 1965. Further tests of the effect of food plants on the migratory grasshopper. *J. Econ. Entomol.* 58:475–79

11. Bazely, D. R. 1989. Carnivorous herbivores: mineral nutrition and the balanced diet. *Trends Ecol. Evol.* 4:155–56

12. Barnett, S. A. 1975. *The Rat, a Study in Behavior.* Chicago: Univ. Chicago Press. 318 pp.

13. Bellush, L. L., Rowland, N. E. 1986. Dietary self-selection in diabetic rats: an overview. *Brain Res. Bull.* 17:653–61

14. Belovsky, G. E. 1978. Diet optimization in a generalist herbivore: the moose. *Theor. Popul. Biol.* 14:105–34

15. Belzer, W. R. 1978. Patterns of selective protein ingestion by the blowfly *Phormia regina. Physiol. Entomol.* 3:169–75

16. Berenbaum, M. R., Miliczky, E. 1984. Mantids and milkweed bugs: Efficacy of aposematic coloration against invertebrate predators. *Am. Midl. Nat.* 111:64–68

17. Bhattacharya, A. K., Waldbauer, G. P. 1970. Use of the faecal uric acid method in measuring the utilization of food by *Tribolium confusum. J. Insect Physiol.* 16:1983–90

18. Blundell, J. E. 1984. Serotonin and appetite. *Neuropharmacology* 23:1537–51

19. Brust, M., Fraenkel, G. 1955. The nutritional requirements of the larvae of a blowfly, *Phormia regina* (Meig.). *Physiol. Zool.* 28:186–204

20. Calderone, N. W., Page, R. E. Jr. 1988. Genotypic variability in age polyethism and task specialization in the honey bee, *Apis mellifera* (Hymenoptera: Apidae). *Behav. Ecol. Sociobiol.* 22:17–25

21. Castonguay, T. W., Applegate, E. A., Upton, D. E., Stern, J. S. 1983. Hunger and appetite: old concepts/new distinctions. *Nutr. Rev.* 41:101–10

22. Castonguay, T. W., Collier, G. H. 1986. Diet balancing: some limitations. *Nutr. Behav.* 3:43–55

23. Cohen, R. W., Friedman, S., Waldbauer, G. P. 1988. Physiological control of nutrient self-selection in *Heliothis zea* larvae: the role of serotonin. *J. Insect Physiol.* 34:935–40

24. Cohen, R. W., Heydon, S. L., Waldbauer, G. P., Friedman, S. 1987. Nutrient self-selection by the omnivorous cockroach *Supella longipalpa. J. Insect Physiol.* 33:77–82

25. Cohen, R. W., Waldbauer, G. P. Friedman, S. 1988. Natural diets and self-selection: *Heliothis zea* larvae and maize. *Entomol. Exp. Appl.* 46:161–71

26. Cohen, R. W., Waldbauer, G. P. Friedman, S., Schiff, N. M. 1987. Nutrient self-selection by *Heliothis zea* larvae: a time-lapse film study. *Entomol. Exp. Appl.* 44:65–73

27. Cotton, R. T. 1956. *Pests of Stored Grain and Grain Products.* Minneapolis: Burgess. 306 pp. Revised ed.

28. Crawley, M. J. 1983. *Herbivory, the Dynamics of Animal-Plant Interactions, Studies in Ecology,* Vol. 10. Berkeley, CA: Univ. Calif. Press. 437 pp.

29. Davis, C. M. 1928. Self-selection of diets by newly weaned infants: an ex-

perimental study. *Am. J. Dis. Child.* 36:651–89

30. Davis, G. R. F. 1972. Refining diets for optimal performance. In *Insect and Mite Nutrition*, ed. J. G. Rodriguez, pp. 172–81. Amsterdam: North Holland. 702 pp.

31. Dethier, V. G. 1980. Food-aversion learning in two polyphagous caterpillars, *Diacrisia virginica* and *Estigmene congrua*. *Physiol. Entomol.* 5:321–25

32. Evvard, J. M. 1915. Is the appetite of swine a reliable indication of physiological needs? *Proc. Iowa Acad. Sci.* 22:375–403 + 5 plates

33. Fernstrom, J. D., Wurtman, R. J. 1971. Brain serotonin content: increase following ingestion of carbohydrate diet. *Science* 174:1023–25

34. Fraenkel, G. 1959. The raison d'etre of secondary plant substances. *Science* 129:1466–70

35. Fraenkel, G. 1969. Evaluation of our thoughts on secondary plant substances. *Entomol. Exp. Appl.* 12:473–86

36. Fraenkel, G., Blewett, M. 1943. The natural foods and the food requirements of several species of stored products insects. *Trans. R. Entomol. Soc. Lond.* 93:457–90

37. Fraenkel, G., Blewett, M. 1946. Linoleic acid, vitamin E and other fat-soluble substances in the nutrition of certain insects, *Ephestia kuehniella*, *E. elutella*, *E. cautella* and *Plodia interpunctella* (Lep.) *J. Exp. Biol.* 22:172–90

38. Fraser, D., Chavez, E. R., Paloheimo, J. E. 1984. Aquatic feeding by moose: selection of plant species and feeding areas in relation to plant chemical composition and characteristics of lakes. *Can. J. Zool.* 62:80–87

39. Garcia, J., Ervin, F. R., Yorke, C. H., Koelling, R. A. 1967. Conditioning with delayed vitamin injections. *Science* 155:716–18

40. Gilbert, L. E. 1972. Pollen feeding and reproductive biology of *Heliconius* butterflies. *Proc. Natl. Acad. Sci USA* 69:1403–7

41. Gilbert, L. E. 1975. Ecological consequences of a coevolved mutualism between butterflies and plants. In *Coevolution of Animals and Plants*. ed. L. E. Gilbert, P. H. Raven, pp. 210–40. Austin, Texas: Univ. Texas Press. 263 pp.

42. Godan, D. 1964. Untersuchungen über den Einfluss tierischer Nahrung auf die Vermehrung der Maulwurfsgrille (*Gryllotalpa gryllotalpa* L.). *Z. Angew. Entomol.* 51:207–23

43. Gordon, H. T. 1959. Minimal nutrition-

al requirements of the German roach, *Blattella germanica* L. *Ann. NY Acad. Sci.* 77:290–351

44. Greenstone, M. H. 1979. Spider feeding behaviour optimizes dietary essential amino acid composition. *Nature* 282:501–3

45. Hagen, K. S. 1987. Nutritional ecology of terrestrial insect predators. See Ref. 109a, pp. 533–77

46. Hellmich, R. L. II, Kulincevic, J. M., Rothenbuler, W. C. 1985. Selection for high and low pollen-hoarding honey bees. *J. Hered.* 76:155–58

47. Hill, W., Castonguay, T. W., Collier, G. H. 1980. Taste or diet balancing? *Physiol. Behav.* 24:765–67

48. Hodge, C. 1933. Growth and nutrition of *Melanoplus differentialis* Thomas (Orthoptera: Acrididae). I. Growth on a satisfactory mixed diet and on diets of single food plants. *Physiol. Zool.* 6:306–28

49. Hopkins, F. G. 1912. Feeding experiments illustrating the importance of accessory factors in normal dietaries. *J. Physiol. Lond.* 44:425–60

50. Horsfall, W. R. 1955. *Mosquitoes: Their Bionomics and Relation to Disease.* New York: Ronald. 723 pp.

51. House, H. L. 1962. Insect nutrition. *Annu. Rev. Biochem.* 31:653–72

52. House, H. L. 1965. Insect nutrition. In *The Physiology of Insecta*, ed. M. Rockstein, 2:769–813. New York: Academic. 905 pp.

53. House, H. L. 1970. Choice of food by larvae of the fly, *Agria affinis*, related to dietary proportions of nutrients. *J. Insect Physiol.* 16:2041–50

54. House, H. L. 1974. Nutrition. In *The Physiology of Insecta*, ed. M. Rockstein, 5:1–62. New York: Academic. 2nd ed.

55. Humphreys, W. F. 1980. [Comment on M. H. Greenstone's paper] Spider feeding behaviour optimises dietary essential amino acid composition. *Nature* 284:578

56. Jermy, T., Hanson, T. E., Dethier, V. G. 1968. Induction of specific food preference in lepidopterous larvae. *Entomol. Exp. Appl.* 11:211–30

57. Kamil, A. C., Krebs, J. R., Pulliam, H. R., eds. 1987. *Foraging Behavior.* New York: Plenum. 676 pp.

57a. Kamil, A. C., Sargent, T. D., eds. 1981. *Foraging Behavior, Ecological, Ethological, and Psychological Approaches.* New York: Garland. 534 pp.

58. Kaufmann, T. 1965. Biological studies on some Bavarian Acridoidea (Orthoptera), with special reference to their

feeding habits. *Ann. Entomol. Soc. Amer.* 58:791–801

59. Kitting, C. L. 1980. Herbivore-plant interactions of individual limpets maintaining a mixed diet of intertidal marine algae. *Ecol. Monogr.* 50:527–50

60. Krebs, J. R., Avery, M. I. 1984. Diet and nestling growth in the European bee eater. *Oecologia* 64:363–68

61. Lee, J. C., Bernays, E. A. 1988. Declining acceptability of a food plant for the polyphagous grasshopper *Schistocera americana:* the role of food aversion learning. *Physiol. Entomol.* 13:291–301

62. Lewis, A. C. 1984. Plant quality and grasshopper feeding: effects of sunflower condition on preference and performance in *Melanoplus differentialis. Ecology* 65:836–43

63. Lewis, A. C., Bernays, E. A. 1985. Feeding behavior: selection of both wet and dry food for increased growth in *Schistocerca gregaria* nymphs. *Entomol. Exp. Appl.* 37:105–12

64. Lobel, P. S., Ogden, J. C. 1981. Foraging by the herbivorous parrotfish *Sparisoma radians. Marine Biol.* 64:173–83

65. Lorenz, K. 1965. *Evolution and the Modification of Behavior.* Chicago: Univ. Chicago Press. 121 pp.

66. MacFarlane, J. H., Thorsteinson, A. J. 1980. Development and survival of the twostriped grasshopper, *Melanoplus bivittatus* (Say) (Orthoptera: Acrididae), on various single and multiple plant diets. *Acrida* 9:63–76

67. MacMasters, M. M., Hinton, J. J. C., Bradbury, D. 1971. Microscopic structure and composition of the wheat kernel. In *Wheat Chemistry and Technology.* ed. Y. Pomeranz, pp. 51–113. St. Paul, MN: Amer. Assoc. Cereal Chemists. 821 pp.

68. Magis, N. 1963. Recherche de l'optimum protidique et de l'optimum glucidique dans la nutrition des *Tribolium. Bull. Soc. R. Sci. Liege* 32:737–50

69. Manoukas, A. G. 1981. Effect of excess levels of individual amino acids upon survival, growth and pupal yield of *Dacus oleae* (Gmel.) larvae. *Z. Angew. Entomol.* 91:309–15

70. Mattson, W. J. Jr. 1980. Herbivory in relation to plant nitrogen content. *Annu. Rev. Ecol. Syst.* 11:119–61

71. Maxwell, G. M., Fourie, F., Bates, D. J. 1988. The effect of restricted and unrestricted "cafeteria" diets upon the energy exchange and body composition of weanling rats. *Nutr. Rep. Int.* 37:629–37

72. McMurtry, J. A., Rodriguez, J. G.

1987. Nutritional ecology of phytoseiid mites. See Ref. 109a, pp. 609–44

73. McNeill, S., Southwood, T. R. E. 1978. The role of nitrogen in the development of insect/plant relationships. In *Biochemical Aspects of Plant and Animal Coevolution.* ed. J. B. Harborne, pp. 77–98. London: Academic. 435 pp.

74. Merzheevskaya, O. I. 1988. *Larvae of Owlet Moths (Noctuidae), Biology, Morphology and Classification.* Washington: Smithsonian Inst. and Nat. Sci. Found. 419 pp. (Translated from the Russian)

75. Metcalf, C. L., Flint, W. P., Metcalf, R. L. 1962. *Destructive and Useful Insects.* New York: McGraw-Hill. 1087 pp.

76. Oldroyd, H. 1964. *The Natural History of Flies.* New York: Norton. 324 pp.

77. Overman, S. R. 1976. Dietary self-selection by animals. *Psychol. Bull.* 83:218–35

78. Pearl, R., Fairchild, T. E. 1921. Studies on the physiology of reproduction in the domestic fowl. XIX. On the influence of free choice of food materials on winter egg production and body weight. *Am. J. Hyg.* 1:253–77

79. Pickford, R. 1962. Development, survival and reproduction of *Melanoplus bilituratus* (Wlk) (Orthoptera: Acrididae) reared on various food plants. *Can. Entomol.* 94:859–69

80. Pilgrim, F. J., Patton, R. A. 1947. Patterns of self-selection of purified dietary components by the rat. *J. Comp. Physiol. Psych.* 40:343–48

81. Price, P. W. 1975. *Insect Ecology.* New York: Wiley. 514 pp.

82. Pyke, G. H. 1984. Optimal foraging theory: a critical review. *Annu. Rev. Ecol. Syst.* 15:523–75

83. Rapport, D. J. 1981. Foraging behavior of *Stentor coeruleus:* A microeconomic interpretation. See Ref. 57a, pp. 77–92

84. Reese, J. C. 1981. Insect dietetics: complexities of plant-insect interactions. In *Current Topics in Insect Endocrinology and Nutrition.* ed. G. Bhaskaran, S. Friedman, J. G. Rodriguez, pp. 317–35. New York: Plenum. 362 pp.

85. Reichert, S. E., Harp, J. M. 1987. Nutritional ecology of spiders. See Ref. 109a, pp. 645–72

86. Revusky, S. H., Bedarf, E. W. 1967. Association of illness with prior ingestion of novel foods. *Science* 155:219–20

87. Richter, C. P. 1936. Increased salt appetite in adrenalectomized rats. *Am. J. Physiol.* 115:155–61

88. Richter, C. P. 1942. Increased dextrose

appetite of normal rats treated with insulin. *Am. J. Physiol.* 135:781–87

89. Richter, C. P. 1942–43. Total self-regulatory functions in animals and human beings. *Harvey Lect. Ser.* 38:63–103

90. Richter, C. P. 1955. Self-regulatory functions during gestation and lactation. In *Gestation, Transactions of the Second Conference.* ed. C. A. Villee, pp. 11–93. New York: Josiah Macy, Jr. Foundation. 262 pp.

91. Richter, C. P., Eckert, J. F. 1937. Increased calcium appetite of parathyroidectomized rats. *Endocrinology* 21:50–54

92. Richter, C. P., Eckert, J. F. 1938. Mineral metabolism of adrenalectomized rats studied by the appetite method. *Endocrinology* 22:214–24

93. Richter, C. P., Holt, L. E. Jr., Barelare, B. Jr. 1938. Nutritional requirements for normal growth and reproduction in rats studied by the self-selection method. *Am. J. Physiol.* 122:734–44

94. Richter, C. P., Schmidt, E. C. H. Jr. 1941. Increased fat and decreased carbohydrate appetite of pancreatectomized rats. *Endocrinology* 28:179–92

95. Robinson, G. E., Page, R. E. Jr. 1989. Genetic determination of nectar foraging, pollen foraging, and nest-site scouting in honey bee colonies. *Behav. Ecol. Sociobiol.* 24:317–23

96. Roessingh, P., Bernays, E. A., Lewis, A. C. 1985. Physiological factors influencing preference for wet and dry food in *Schistocerca gregaria* nymphs. *Entomol. Exp. Appl* 37:89–94

97. Roth, L. M., Willis, E. R. 1960. The biotic associations of cockroaches. *Smithson. Misc. Coll.* 141:1–470

98. Rozin, P. 1976. The selection of foods by rats, humans, and other animals. *Adv. Study Behav.* 6:21–76

99. Schaefer, A., Piquard, F., Haberey, P. 1976. Food self-selection during spontaneous body weight variations in the dormouse (*Glis glis* L.). *Comp. Biochem. Physiol.* 55A:115–18

100. Schiff, N. M., Waldbauer, G. P., Friedman, S. 1988. Dietary self-selection for vitamins and lipid by larvae of the corn earworm, *Heliothis zea. Entomol. Exp. Appl.* 46:249–56

101. Schiff, N. M., Waldbauer, G. P., Friedman, S. 1989. Dietary self-selection by *Heliothis zea* larvae: roles of metabolic feedback and chemosensory stimuli. *Entomol. Exp. Appl.* 52:261–70.

102. Schiff, N. M., Waldbauer, G. P., Friedman, S. 1989. Response of last instar *Heliothis zea* larvae to carbohydrates: stimulation of biting, nutritional value. *Entomol. Exp. Appl.* 52:29–38

103. Schneider, F. 1969. Bionomics and physiology of aphidophagous Syrphidae. *Annu. Rev. Entomol.* 14:103–24

104. Scott, E. M. 1946. Self selection of diet. I. Selection of purified components. *J. Nutr.* 31:397–406

105. Scriber, J. M. 1977. Limiting effects of low leaf-water content on the nitrogen utilization, energy budget, and larval growth of *Hyalophora cecropia. Oecologia* 28:269–87

106. Scriber, J. M., Slansky, F. Jr. 1981. The nutritional ecology of immature insects. *Annu. Rev. Entomol.* 26:183–211

107. Shettleworth, S. J. 1984. Learning and behavioral ecology. In *Behavioral Ecology, an Evolutionary Approach.* ed. J. B. Krebs, N. B. Davies, pp. 170–94. Sunderland, MA: Sinauer. 493 pp.

108. Simpson, S. J., Simmonds, M. S. J., Blaney, W. M. 1988. A comparison of dietary selection behaviour in larval *Locusta migratoria* and *Spodoptera littoralis. Physiol. Entomol.* 13:225–38

109. Slansky, F. Jr., Feeny, P. 1977. Stabilization of the rate of nitrogen accumulation by larvae of the cabbage butterfly on wild and cultivated food plants. *Ecol. Monogr* 47:277–86

109a. Slansky, F. Jr., Rodriguez, J. G., eds. 1987. *Nutritional Ecology of Insects, Mites, Spiders, and Related Invertebrates.* New York: Wiley. 1016 pp.

110. Slansky, F., Jr., Rodriguez, J. G. 1987. Nutritional ecology of insects, mites, spiders, and related invertebrates: an overview. See Ref. 109a, pp. 1–69

111. Sloley, B. D., Downer, R. G. H., Gillott, C. 1986. Levels of tryptophan, 5-hydroxytryptamine, and dopamine in some tissues of the cockroach *Periplaneta americana. Can. J. Zool.* 64:2669–73

112. Stephens, D. W., Krebs, J. R. 1986. *Foraging Theory.* Princeton, NJ: Princeton Univ. Press. 247 pp.

113. Tauber, C. A., Tauber, M. J. 1973. Diversification and secondary intergradation of two *Chrysopa carnea* strains (Neuroptera: Chrysopidae). *Can. Entomol.* 105:1153–67

114. Tauber, M. J., Tauber, C. A. 1974. Dietary influence on reproduction in both sexes of five predaceous species (Neuroptera). *Can. Entomol.* 106:921–25

115. Theall, C. L., Wurtman, J. J., Wurtman, R. J. 1984. Self-selection and regulation of protein:carbohydrate ratio in foods adult rats eat. *J. Nutr.* 114:711–18

116. Tinker, P. B. 1981. Levels, distribution and chemical forms of trace elements in food plants. *Phil. Trans. R. Soc. Lond. Ser. B.* 294:41–55

117. Uvarov, B. P. 1977. *Grasshoppers and Locusts,* Vol. 2. London: Centre for Overseas Pest Research. 613 pp.

118. Van Valen, L. 1978. The statistics of variation. *Evol. Theory* 4:33–43

119. Waldbauer, G. P. 1968. The consumption and utilization of food by insects. *Advan. Insect Physiol.* 5:229–88

120. Waldbauer, G. P., Bhattacharya, A. K. 1973. Self-selection of an optimum diet from a mixture of wheat fractions by the larvae of *Tribolium confusum. J. Insect Physiol.* 19:407–18

121. Waldbauer, G. P., Cohen, R. W., Friedman, S. 1984. An improved procedure for laboratory rearing of the corn earworm, *Heliothis zea* (Lepidoptera: Noctuidae). *Great Lakes Entomol.* 17:113–18

122. Waldbauer, G. P., Cohen, R. W., Friedman, S. 1984. Self-selection of an optimal nutrient mix from defined diets by larvae of the corn earworm, *Heliothis zea* (Boddie). *Physiol. Zool.* 57:590–97

123. Waldbauer, G. P., Fraenkel, G. S. 1961. Feeding on normally rejected plants by maxillectomized larvae of the tobacco hornworm, *Protoparce sexta*

(Lepidoptera: Sphingidae). *Ann. Entomol. Soc. Amer.* 54:477–85

124. Waldbauer, G. P., Friedman, S. 1988. Dietary self-selection by insects. In *Endocrinological Frontiers in Physiological Insect Ecology.* ed. F. Sehnal, A. Zabza, D. L. Denlinger, 1:403–22. Wroclaw, Poland: Wroclaw Tech. Univ. Press 600 pp.

125. Westoby, M. 1974. An analysis of diet selection by large generalist herbivores. *Am. Nat.* 108:290–304

126. Wilson, C. M. 1983. Seed protein fractions of maize, sorghum, and related cereals. In *Seed Proteins, Biochemistry, Genetics, Nutritive Value.* ed. W. Gottschalk, H. P. Müller, pp. 271–307. The Hague: Junk. 531 pp.

127. Woods, R. 1949. A history of self-selection of diet. I. *Borden's Rev. Nutr. Res.* 10:1–9

128. Woods, R. 1949. Self-selection of diet. II. Experimental investigation. *Borden's Rev. Nutr. Res.* 10:10–16

129. Wurtman, J. J., Wurtman, R. J. 1979. Drugs that increase central serotoninergic transmission diminish elective carbohydrate consumption by rats. *Life Sci.* 24:895–904

130. Zahorik, D. M., Houpt, K. A. 1981. Species differences in feeding strategies, food hazards, and the ability to learn food aversions. See Ref. 57a, pp. 289–310

Annu. Rev. Entomol. 1991. 36:65–89
Copyright © 1990 by Annual Reviews Inc. All rights reserved

EVOLUTION OF OVIPOSITION BEHAVIOR AND HOST PREFERENCE IN LEPIDOPTERA

John N. Thompson and Olle Pellmyr[1]

Departments of Botany and Zoology, Washington State University, Pullman, Washington 99164

KEY WORDS: host selection, hosts shifts, host specificity, genetics of behavior, sensory cues

---

Therefore the study of butterflies—creatures selected as the types of airiness and frivolity—instead of being despised, will some day be valued as one of the most important branches of Biological science.

H. W. Bates, 1876 (7)

OVERVIEW AND PERSPECTIVES

Oviposition behavior has been at the center of many of the major debates on the ecology and evolution of interactions between insects and plants: the causes of host specificity, the origins of host shifts, the potential for sympatric speciation, the modes of coevolution, and the pattern of attack on host plants within local populations. The selectivity of ovipositing females may often provide the initial basis for divergence of insect populations onto different plant species, and it may drive the evolution of some plant defenses.

Ever since Ehrlich's & Raven's (37) paper on coevolution of butterflies and plants, Lepidoptera has probably been the taxon in which the greatest number of species has been studied for some aspects of oviposition behavior (Coleoptera and Diptera perhaps tying for second place). Consequently, one can

[1]Current address: Department of Biological Sciences, University of Cincinnati, Cincinnati, Ohio 45221

0066-4170/91/0101-0065$02.00

infer patterns in the evolution of oviposition behavior within the Lepidoptera from a wealth of studies, and this order has continued to be among the major test groups for ideas on coevolution. Multiyear observational studies and detailed, experimental studies of oviposition behavior in both butterflies and moths have increased considerably since Gilbert's & Singer's (51) review on butterfly ecology and Chew's & Robbins' (18) review of egg laying in butterflies. In writing this review, we have tried to avoid overlap with the papers already reviewed in detail by these authors.

We consider here the behavioral, genetic, and ecological determinants of oviposition behavior as they influence preference for plants and plant parts in both butterflies and moths. Our concern is with how oviposition behavior contributes to the evolution of preference and specificity for (a) plant species, (b) individual plants within populations, and (c) plant parts. The review has two obvious biases. Although butterflies constitute only about five percent of lepidopteran species (89), they have received the lion's share of attention for obvious reasons. Unlike most moths, they are relatively large, showy, day-flying, and often possible to follow in the field, at least for short periods of time. Our review, then, is biased somewhat toward butterflies as a result of the past pragmatism of researchers. The other bias is that the review focuses almost exclusively on phytophagous lepidopterans; very little is known in a quantitative way about oviposition behavior in nonphytophagous species, and even less about larviposition in the few known cases of ovovivipary (59).

PREFERENCE AND SPECIFICITY FOR PLANT SPECIES

Most work on oviposition behavior is based upon the idea that females, when confronted with an array of potential hosts, will exhibit a hierarchy in their preferences (25, 130, 153–155, 167, 168). When a number of potential host plants are available, a female will lay most eggs on her most preferred plant species (or habitat or plant part), fewer eggs on her next preferred plant, and so on. Using this criterion, specificity is the number of plant species on which a female will oviposit when offered all plants in a simultaneous choice trial. [Jaenike (64) recently reviewed the factors influencing host specificity in phytophagous insects.]

Most studies of preference and specificity in ovipositing Lepidoptera have used some type of simultaneous choice trial. Singer (130, 131, 133), however, has argued that insects encounter plants one at a time rather than simultaneously, and therefore has used a different protocol for evaluating preference and specificity. He proposed the length of time over which a female refuses all hosts except one as the index of specificity. In both views of preference and specificity, a host shift involves either a change in the preference hierarchy or the use of a host lower in the hierarchy in the absence of the more preferred host.

A common problem with studies of host preference is failure to design experiments in a way that allows a clear interpretation of the pattern of oviposition preference among ovipositing females within a population. A common experimental design is to place a group of females (sometimes as many as 80) in a single cage together with the test plants, then count the final number of eggs laid after some length of time. This design masks any variation in oviposition preference that may occur among females. The final distribution of eggs may be a composite of females that differ in their degree of specificity or even in how they rank the host plants. Consequently, no female in the population may actually distribute her eggs in the way indicated by the composite distribution. Moreover, competition among the females for oviposition sites may lead to a more uniform distribution of eggs than would occur if females were tested individually. Studies in which females have been tested individually (e.g. 88, 112, 130, 140, 142, 144, 146, 153, 167) have shown variation among females within populations. These results highlight the need to test females individually when conducting trials on oviposition preference.

Genetics of Oviposition Behavior

Only a handful of lepidopteran species have so far been studied in any detail for the genetic bases of oviposition behaviors. The few available studies have focused mostly on preference for plant species. Significant differences in host preference have been found among families within populations of *Colias eurytheme* (Pieridae) (142), *Euphydryas editha* (Nymphalidae) (134), and *Papilio zelicaon* and *Papilio oregonius* (Papilionidae) (153). In *C. eurytheme*, females differed strongly in their preference for either alfalfa or vetch. Full-sib analysis and mother-offspring analysis showed high heritability ($h^2=0.54-0.71$) of preference. Maternal effects could potentially have contributed to some of the similarity in preference among sibs.

Singer et al (134) also found a high heritability ($h^2=0.9$) in oviposition preference for different plant species in *E. editha,* but they could not eliminate nongenetic maternal effects as a major cause of the high correlation between females and their female offspring. The study was complicated somewhat by comparing the results for wild-caught female parents in the field with laboratory-reared offspring in a greenhouse. If anything, this complication might be expected to lower the estimate of heritability rather than artifically raise it because estimates of heritabilities are environment-specific. Results of Ng (88) and Singer et al (134) also suggest that oviposition preference and larval performance may be correlated within populations and may vary among individuals such that females prefer the plant species on which their larvae have the greatest chance of surviving during their first 10 days of growth. Whether this correlation results from a genetic correlation between these traits, pleiotropic effects, or maternal effects has not yet been determined.

For *Papilio,* variation in oviposition preference was found both in *P. zelicaon,* which was locally oligophagous, and *P. oregonius,* which was monophagous in the field (153). Interspecific crosses showed that the differences in preference between these two species were inherited primarily through one or more loci on the X chromosome, which may be modified by additional autosomal loci (154). Interspecific hybrid females tended toward their paternal species in their preference for host plants. (Since females are the heterogametic sex in Lepidoptera, females receive only one X chromosome, and that comes from their father.)

In the only other study to use genetic crosses to analyze the genetics of host preference, Schneider & Roush (124) showed that populations of *Heliothis virescens* (Noctuidae) from Mississippi and the Virgin Islands, which varied in their mean relative preference for two plant species, produced hybrid offspring that were somewhat intermediate in preference regardless of the direction of the cross.

Chemistry of Host Choice

Recent studies of the chemical cues used by lepidopterans in choosing host plants indicated some of the ways in which preference hierarchies among potential hosts can become established (113). In some species, the concentrations of particular chemical compounds determines the relative preference of females for potential host plants or their surrogates (e.g. agar disks, impregnated paper disks), whereas in other species the relative proportions of compounds determine the female's response. In the buckeye butterfly, *Junonia coenia* (Nymphalidae), which feeds on plants rich in iridoid glycosides, females oviposit preferentially on agar disks with the highest concentrations of these compounds when presented with disks ranging in concentration from 0 to 1% (98). In contrast, *Heliothis subflexa* (Noctuidae) responded to extract leaf washings, but no dosage response was apparent, at least in tests using olfactometers (79). *H. virescens* females can choose between cotton and groundcherry even when olfaction and vision are prevented, indicating that contact chemoreception and mechanoreception alone allow discrimination in this species (105).

In *Hadena bicruris* (Noctuidae), which lays its eggs in the white flowers of *Silene latifolia,* search behavior and orientation to the flowers are induced by a dose-dependent response toward the floral volatiles (13). After floral visits, a moth may approach nonfragrant white objects, but she alights only in the presence of olfactory stimuli. Other floral odors experienced during initial feeding may also lead to oviposition behavior, but these other odors are overridden by the host odor.

Swallowtail butterflies have been studied intensively for the combinations of plant compounds that elicit oviposition (40). For the species studied so far (*Papilio polyxenes* in North America, and *Papilio xuthus* and *Papilio pro-*

tenor in Japan), the oviposition response of females after contact with im- pregnated filter paper depends upon the relative proportions of several plant compounds (42, 60, 90, 91, 93). Individual compounds elicit either weak oviposition responses or no response at all. Weak oviposition responses can also be elicited by combinations of subsets of these compounds, but strong responses require all the compounds in certain proportions. Moreover, *P. polyxenes* females appear to use a combination of contact stimulants and volatiles in choosing hosts. Using artificial plants made of wire branches and sponge leaves, Feeny et al (43) found that females laid more eggs on model plants treated with a combination of contact stimulants and volatiles from carrot leaves than they did on plants treated only with contact stimulants.

As in swallowtail butterflies, pine beauty moths, *Panolis flammea* (Noc- tuidae), in free flight cages prefer particular ratios of plant compounds, in this case $\beta:\alpha$ pinene, when choosing plants for oviposition (73). Yellow peach moths, *Conogethes punctiferalis* (Pyralidae), are attracted to and oviposit on ripe host fruits on the basis of fruit volatiles. Attraction is much increased when the volatiles of particular mold fungi are present, and otherwise unac- ceptable substrates can be rendered acceptable by mold infection (61). In codling moths, *Cydia pomonella* (Olethreutinae), females are induced by particular isomers of α-farnesene from the fruit to oviposit on or near it; larvae that hatch a few centimeters away from an apple use the same compounds to find the fruit (165).

A combination of chemical stimulants and deterrents determines use of a plant species by ovipositing *Pieris* (Pieridae) females. Water-soluble com- pounds other than glucosinolates appear to be primarily responsible for host choice in *Pieris rapae* [sinigrin was a poor stimulant in Renwick's & Radke's (116) bioassay]. Specific cardenolides have recently been identified in cruci- fers as chemical deterrents to *Pieris brassicae* and *P. rapae* oviposition (117, 120). *Erysimum cheiranthoides* contains both oviposition attractants and deterrents for oviposition by *P. rapae* (115), but the deterrents appear to outweigh the attractants.

In addition to the plant compounds involved in species recognition, other chemicals can provide information to ovipositing females on the relative quality of a particular species or individual. An often-discussed factor is nitrogen, but its effects on oviposition are still poorly understood. Nitrogen fertilization of plants variously affected oviposition in *P. rapae* in different experiments. Some experiments, using potted plants in small plots, have shown that females oviposit preferentially on plants augmented with nitrogen fertilizer (86, 175). More recently, however, Letourneau & Fox (74) found that nitrogen addition had a positive effect on egg densities in small-scale trials using potted plants but had no effect on egg densities in larger-scale trials with field-grown plants. Changes in nitrogen levels are often accom- panied by changes in other chemical constituents of plants (e.g. sugars, secondary compounds), which complicates the interpretation of the effects of nitrogen per se. By sampling neighboring pairs of ragwort *(Senecio jacobaea)* plants—one of which had an egg mass, the other of which did not—van der Meijden et al (162) found that cinnabar moths, *Tyria jacobaeae* (Arctiidae), selected plants with a high concentration of both organic nitrogen and sugars. Plants rich in only nitrogen or sugars, or poor in both, were less likely to receive eggs. The effects of alkaloids, which are positively correlated with nitrogen and negatively correlated with soluble carbohydrate levels, were tested in cage trials but showed no effect on oviposition. The varying results among studies undoubtedly result in part from the different ways in which nitrogen levels interact with other plant compounds under various ecological conditions.

Ecological Determinants of Host Plant Choice

A recent forum in *Ecology* (141) on the factors responsible for the number of hosts used by phytophagous insects highlighted the diversity of hypotheses on the evolution of host specificity. Allelochemicals, nutritional chemistry, plant morphology, natural enemies, feeding modes (e.g. parasitism vs grazing) and, in some instances, mutualists all received support as potentially major determinants of host preference and specificity. Much of the debate was over whether one factor (e.g. allelochemicals, predation) was of overriding importance in determining the choice of plant species by ovipositing females. The real problem, however, as pointed out by several of the contributors to the forum, is not to find which of these factors is the single most important determinant of host choice because the answer will vary from species to species. Rather, the important ecological and evolutionary problem is to understand patterns in how these factors interact.

Studies of the phylogeny of host choice in Lepidoptera have shown convincingly that no single factor governs the evolution of host specificity. As pointed out by a number of researchers (e.g. 11, 37, 78), butterflies have broad patterns of host choice that relate clearly to plant chemistry and taxonomic affinities, such as the use of crucifers by *Pieris*, legumes by *Colias*, or umbellifers by the *Papilio machaon* complex. But other studies of butterflies and moths have shown that host shifts or lack of use of some plant species does not always follow from the pattern of distribution of particular plant compounds (137, 150). Moreover, the patterns of host use within some lepidopteran families do not mimic the taxonomic affinities of their hosts. In the rich lepidopteran fauna of Santa Rosa National Park (66), Janzen (65) recorded sphingid species from 16 plant families, and saturniid species in the park use even more plant families. Janzen (65) notes that the connecting theme in host use by these moths is that all the families used by sphingids are traditionally viewed as containing "toxic small molecules," and all those used by the saturniids have phenol-rich or aromatic foliage.

Similarly, studies of Microlepidoptera indicate a large number of host shifts onto quite different plant families within moth families. The family Prodoxidae, which includes the yucca moths, contains genera that oviposit on Rosaceae, Umbelliferae, Saxifragaceae, Agavaceae, and probably Myrtaceae, i.e. families in five different angiosperm orders (32).

In a review of host associations within the paraphyletic group Microlepidoptera, Powell (102) found that several moth families, such as the Gelechiidae, Gracillariidae, and Tortricidae, have radiated onto at least 70 plant families. A relationship between host use and insect diversity emerges when one uses Powell's data for families and subfamilies, updates them to fit the most recent family concepts (89), excludes the largely scavenging Tineidae, and adds new records for four families. Family or subfamily size (measured as number of species with available host data) is very highly correlated ($r^2=0.85$) with number of host families (O. Pellmyr, unpublished data). Similarly high coefficients were obtained for other families of Lepidoptera within particular geographic regions. This may suggest that the frequency of cross-familial host shifts is similar among Lepidoptera, regardless of their phylogenetic affinities and degree of specialization. The number of species within a lepidopteran taxon evidently has to be considered, together with other factors, in attempting to explain the diversity of hosts used by that taxon.

Ovipositional "mistakes"—ovipositions onto plant species outside the normal range of acceptable hosts—provide an opportunity for probing into the critical stimuli that determine how females pick and choose among plants (18, 132). Such mistakes may be the raw material for host shifts. They may mark a broadening of the number of plant species used by an insect population, favoring females that save time in searching for hosts by adding this species to those they use. Or, these mistakes may mark the beginnings of a complete shift onto a new plant species. Alternatively, they may simply serve to select against females that are less specific than others in their choice of host plants (45). These mistakes, however, have not been studied in any systematic way for any lepidopteran species.

Oviposition Preference and Larval Performance

Again, where the food of the young depends on where the mother places her eggs, as in the case of the caterpillars of the cabbage-butterfly, we may suppose that the parent stock of the species deposited her eggs sometimes on one kind and sometimes on another of congenerous plants (as some species now do), and if the cabbage suited the caterpillars better than any other plant, the caterpillars of those butterflies, which had chosen the cabbage, would be most plentifully reared, and would produce butterflies more apt to lay their eggs on the cabbage than on the other congenerous plants.

C. Darwin, 1844 (28)

A major working hypothesis on the evolution of oviposition behavior is that females will choose plant species that maximize larval survival and growth. Most studies on preference-performance correlations in Lepidoptera have focused on population-level correlations: that is, mean oviposition preference as compared with survival and growth of larvae averaged across all females tested (3, 21, 109, 110, 131, 132, 136, 172). As Via (164) first pointed out, however, the relevant evolutionary question is rather how variation in preference and performance is distributed among individuals. Recent studies of *E. editha* by Singer et al (134) showed that both preference and performance (in this case, mean weight of 10-day old offspring) vary within a population of this species, with some individuals preferring and growing faster on *Collinsia parviflora* than on *Plantago lanceolata*. Moreover, individual females tend to oviposit on the plant species on which their larvae can reach the higher weight during the first ten days after oviposition. To date, this is the only lepidopteran species for which a preference-performance correlation has been studied at the level of individual females.

Recent work on *P. zelicaon* and *P. oregonius* showed that oviposition preference is genetically independent of larval performance in these species, at least with respect to the physical linkage of loci (157). Unlike the results for oviposition preference, interspecific crosses have provided no evidence that the loci governing survivorship on different hosts, growth rates, and pupal masses are X-linked. Consequently, host shifts in these butterflies must involve loci on two or more chromosomes.

Most studies of Lepidoptera analyzing the preference/performance hypothesis have focused on survival, growth rate, and pupal mass in the absence of natural enemies of eggs, larvae, and pupae. Nonetheless, ovipositing females in some lepidopteran species clearly choose hosts specifically because these plants harbor ant species that protect the larvae from predators or parasitoids (3, 99). In some species, such a *Ogyris amaryllis* in Australia, ovipositing females actually choose a plant species of lower nutritional value rather than other available plants because that species is attended by particular ant species (4). Of course, not all species tended in some way by ants show this behavior. Each of the five species of *Maculinea* (Lycaenidae) in Europe is tended in later instars by a different species of *Myrmica* ant. Ovipositing females lay eggs regardless of the presence of ants, and the early instars develop on the plants in the absence of ants (146).

Another means of achieving enemy-free space was proposed for *Greya subalba* (Prodoxidae) (152). Females oviposit into developing seeds of an umbelliferous plant, ovipositing in only a few of the seeds in an umbellet. This pattern of oviposition results in a broad distribution of larvae within and among plants. In a population of *Lomatium dissectum* studied for three consecutive years, virtually all plants had between 10 and 65% of their seeds attacked by *G. subalba* (152). Females of a braconid wasp, *Agathis thomp-*

soni, carefully search umbellets for *G. subalba* larvae and are the only known major enemies of the larvae while they are within the schizocarps (151). Selection may have favored females that distribute their eggs among umbellets in a way that maximizes unpredictability of larval dispersion to a searching parasitoid.

A lack of correlation between oviposition preference and larval performance can also result from at least five other factors (155). First, the preferred plant may be rare. *E. editha* in coastal California prefer to oviposit, and their larvae fare better, on *Scrophularia californica* than on *Diplacus aurantiacus. Diplacus,* however, is the much more abundant host and receives more eggs than *Scrophularia* (172, 174). Second, a plant commonly chosen for oviposition but poor for larval growth may be a recent addition to a habitat, and selection may not have had sufficient time to favor females that avoid that plant species. For example, *Pieris napi* in Colorado oviposits on seven cruciferous hosts, including two recently introduced species that are lethal to larvae (17, 118). Third, a host plant may be favorable for larval growth under some conditions, but it may sometimes grow in a habitat unfavorable for flight of ovipositing females or for larval growth (e.g. 107). Fourth, females may oviposit preferentially on plants that allow their offspring to sequester particular secondary plant compounds for defense, even if those plants result in slower growth of larvae. Sequestration of plant compounds has been found in several lepidopteran species (12, 39, 52, 75), but preference for such plants at the expense of larval growth rate has not been shown. Fifth, in species whose larvae feed as grazers and move among several plants during development, rather than feed as parasites on an individual plant, selection may not consistently favor females that oviposit on a certain plant species (unless that species is particularly favorable for the survival of eggs and early instar larvae or is clumped, allowing for larvae to move from one plant to another).

MOVEMENT AMONG PLANTS

The specificity of ovipositing females in selecting plants makes lepidopterans potentially powerful agents of natural selection on plant species. Work over the past 20 years has shown that females of some species reject many potential host individuals as they search for sites to lay their eggs. Females discriminate among plant species, among genotypes within plant species (e.g. 8), among plants in different microhabitats, among plants of different sizes and physiological conditions, and among plant parts.

Prealighting Versus Postalighting Behavior

The precise behavioral sequence used by females when choosing plants for oviposition varies among species. Most research, however, seems to be conducted under the overall paradigm that females obtain cues and make choices at three different levels of resolution when searching for a potential host: choice of habitat in which to search for host plants, choice of specific plants on which to land within a habitat, and the decision to oviposit or not after landing on a plant (18, 33, 95, 114). In choosing a habitat or a plant individual on which to alight, females in some lepidopteran species have been shown to use the amount of sunlight hitting plants or some correlate of the amount of light (e.g. 54), leaf shape (48, 106, 139), specific wavelength contribution to reflected light (122), plant volatiles (43, 104), or plant volatiles combined at short range with some component of plant reflectance in the visible spectrum (13, 57, 176). Once a female lands on a plant, she may still reject it; physical (e.g. 105, 138) and chemical factors may affect her decision whether to oviposit or not. In some species, females drum the plant surface with some or all tarsi and, only then, do they either oviposit or fly off (41, 62). Scherer & Kolb (122) showed that drumming and egg-laying in *P. brassicae* are elicited by specific wavelengths, representing either wavelength-specific behavior or possibly color vision. Morphology of the petals affects oviposition in *H. bicruris;* females also apparently identify the sex of the unisexual flower by partially inserting the proboscis into the tubular corolla, suggesting use of another physical cue in choosing female flowers. Chemotactile or olfactory stimuli may also influence oviposition in *H. bicruris* at the postalighting stage of host choice (13). This case demonstrates well how the process of oviposition behavior may be influenced by numerous cues.

Whether each step of the behavorial sequence assesses the same aspects of plant quality or different aspects is unknown. Rausher (111) and Papaj & Rausher (95), however, suggested that information obtained by females prior to landing upon a plant may be quite different from that obtained by females after landing. Cues used prior to alighting may act mostly to maximize oviposition rate and the overall chance of larval survival. Postalighting cues may be used primarily to assess the suitability for larval growth of a particular plant relative to other plants of the same species in the population. In both the pipevine swallowtail *(Battus philenor)* (95, 112) and the zebra swallowtail *(Eurytides marcellus)* (27), the particular plants chosen by ovipositing females sustain a higher larval survival than plants rejected by those females. In contrast, Wilkund (166) found that *Leptidea sinapis* (Pieridae) females land on plant species in proportion to the plants' density, indicating that females land on plants at random and plant recognition occurs (through drumming) only after a female has alighted. Thus, both species recognition and quality of the individual plant are determined at the postalighting stage.

The tendency to alight on nonhost plants or to reject plants after alighting may vary among populations. In a population of *P. rapae* in Britain, females in flight cages were less likely to land on nonhosts or to reject plants after landing than were females taken from a population in Australia and flown under the same conditions in Britain (69). This difference may reflect differences in the behavioral sequence used by females during oviposition or in their learning ability.

Learning may produce a correlation between prealighting choices and postalighting choices. [Learning in phytophagous insects was recently reviewed in detail by Papaj & Prokopy (94).] After naive females of *B. philenor* land upon a number of plants in an experimental enclosure, they learn to land more often on large young plants than on large old plants. (Changes in landings on small plants are less pronounced over time.) These large young plants are more likely to elicit an oviposition response after landing than are old large plants (95). *B. philenor* females can also learn leaf shapes to use as cues in choosing host plants (106). Learning has been shown to influence aspects of subsequent oviposition in some butterfly species (94) but has so far not been demonstrated in any moths. *P. rapae* can learn to associate some colored papers and leaf disks with sinigrin, and some combinations of sinigrin and visual stimuli may elicit oviposition to varying degrees in the laboratory (159, 160). Whether such learning has any effect in the field is unknown. A study of the closely related *P. brassicae* (122) purported to show absence of associative learning of visual cues associated with food, but the experimental design was insufficient in that only a single, constantly rewarding stimulus was provided during the conditioning phase. Except for *B. philenor*, modification of host specificity through learning is still poorly understood because few studies have tested for this effect. *C. eurytheme* (145), *E. editha* (142), *P. zelicaon*, and *P. oregonius* (153, 154) do not appear to modify their specificity to particular plant species with experience, at least under the experimental conditions in which these species have been tested.

Egg-Load Assessment

When resource competition or cannibalism is likely, mechanisms in females may evolve to recognize and avoid conspecific egg loads or larvae on hosts. Such conditions may arise when host plant species or the parts fed upon by larvae are small (127, 149). Visual recognition and avoidance of conspecific or confamilial eggs is present in *Heliconius* (Heliconiidae) (10, 173), pierids (127, 170), *Danaus* (Danaidae) (121), and *Battus* (108). Chemical recognition of conspecifics is present in a grain-feeding pyralid (*Ephestia;* 20), a foliage-feeding noctuid (*Trichoplusia;* 114), a seed-eating noctuid (*Hadena;* 14), and two seed-eating prodoxids (*Tegeticula;* 2, 70); at least the latter three are likely to experience resource competition, and *Hadena* is also cannibalistic. Both visual and chemical cues are employed in *P. brassicae* (121). The chemical constituents are not known for any lepidopteran oviposition-deterring pheromone, but they probably are substances of relatively low volatility (9). In *P. brassicae,* deterrence persisted for seven weeks under laboratory conditions (125). A coevolutionary consequence of egg-load assessment is the evolution of egg-mimicking structures in some *Heliconius* hosts (173).

Effect of Plant Spatial Distribution

All possible patterns of egg distribution relative to plant density have been observed. At one extreme, females of some species sometimes lay more eggs on plants in patches than on isolated plants. For example, *Euphydryas anicia* females make sharper turns and fly shorter distances between landings when encountering host plants within patches than on isolated host plants. In one year of a three-year study, this resulted in an aggregation of females and eggs within host patches (92). Harassment by males, however, served somewhat as a counterbalance to this aggregation in other years because it caused females to fly farther and sometimes to leave a patch altogether. In *L. sinapis* (Pieridae), females land on plants at random and the distribution of their eggs among potential host plant species corresponds to the relative densities of these plants within a community (166). Finally, at the other extreme from *E. anicia, Pieris* spp. often lay eggs on relatively isolated plants (24, 119, 126). *P. rapae* appears to spread the risk among plants, laying few eggs within any one patch regardless of plant density (26, 68, 119). Other interpretations of why some species appear to oviposit on relatively isolated plants include (*a*) a greater likelihood of encountering isolated plants if females regularly move between nectar plants and oviposition sites, and (*b*) a potentially higher probability of encountering isolated plants if females always pick the nearest plant from a random point (18).

In natural communities, plant size can be inversely correlated with plant density, resulting in a different pattern of oviposition than would be expected from plant density alone. For *Depressaria pastinacella* (Oecophoridae), isolated plants in natural populations are more likely to be attacked than plants within patches (147, 156). This apparent preference for isolated plants seems actually to result from differences in relative sizes of isolated plants as compared with those within patches. Isolated plants are often larger and have floral buds available for oviposition for a longer period of time than do plants within patches. If isolated plants are compared with plants of the same size growing within patches, the plants in the patches have a higher probability of attack. Similarly, *Anthocharis cardamines* (Pieridae) females are more likely to oviposit on large plants because these plants have floral buds available for a longer time than small plants (22). *Cactoblastis cactorum* (Pyralidae) primarily attacks plants near previously attacked plants and shows some preference for larger individuals (87). The results for these species indicate how plant size structure in natural populations can change the pattern of egg distribution from what might occur in agricultural fields in which plants are of nearly uniform size.

Selection may favor different behaviors in lepidopterans that graze on several plants during development than in those that feed parasitically on a single host individual (148, 155). Grazing favors females capable of finding patches in which larvae can move among plants, and it favors larvae capable of finding new host individuals. Although insects whose larvae feed parasitically on plants are often aggregated on large hosts, some grazing Lepidoptera show quite different patterns of distribution. In Sweden, for example, *P. napi* and *Pontia daplidice* (Pieridae) prefer to oviposit on rosettes or on seedlings of their host plants when the hosts are abundant (44). The smaller plants, nearer to the warm ground, support faster growth rates than do larger plants.

The availability of nectar for adult butterflies can greatly affect how females distribute their eggs among potential host plants. Nonetheless, as Murphy (83) noted, few studies have investigated the distribution of adult resources as part of the explanation for nonrandom distributions of eggs among potential host plants. Nectar and pollen sources have been shown to influence the pattern of movement of adult females (e.g. 33, 35, 45, 50, 84, 128). Murphy (83) found that *Penstemon newberryi* shrubs near nectar plants were more likely to receive eggs from *Euphydryas chalcedona* than were plants away from nectar plants. The same pattern was observed in a different *E. chalcedona* population for oviposition on *D. aurantiacus.* On the other hand, *A. cardamines* females minimize feeding time by exploiting a broad spectrum of plants for nectar in the larval host-plant habitat (170). The availability of nectar may sometimes result in oviposition on hosts that are relatively unsuitable for larval growth. Courtney (21) found that, among several cruciferous hosts, *A. cardamines* laid most of their eggs on plants that also provided the females with nectar, even though these plants were not very good for larval growth. In contrast, Odendaal et al (92) found no correlation between the distribution of females and that of nectar plants in *E. anicia;* and, in *L. sinapis,* females actually search for nectar and host plants in different habitats (166).

The combination of available mating sites, nectar plants, and hosts for oviposition is likely to determine the overall metapopulation structure of most butterfly populations. The only species, however, that has been studied intensively to show how these resources influence metapopulation structure is *E. editha* (36, 56). Adult individuals of *E. editha* are often fairly sedentary, although the extent to which individuals disperse varies among populations (38, 50). Local populations of *E. editha* fluctuate widely between years in numbers of adults, and extinction of these local breeding groups is fairly common (36). Recolonization of patches depends upon movement of ovipositing females into those areas. By transplanting 100 postdiapause larvae into each of 38 vacant habitats, Harrison (55) found that the chance of a gravid female establishing a new local population that lasts two years is 6.25%.

Long-distance dispersal before or during oviposition occurs in some lepidopteran species and may have consequences for gene flow. Few quantitative studies have been attempted; the wing-tagging of monarchs is the most extensive (15, 161). On a more limited scale, gravid females of the incurvariid *Perthida glyphopa*, with a wingspan of less than 5 mm, readily migrate to individuals of its *Eucalyptus* host that are isolated by 1200 m from the tree's nearest conspecifics (77). The spruce budworm, *Choristoneura fumiferana* s.l. (Tortricidae), deposits part of her eggs before dispersal flight, then departs on longer flights, especially when moth density is high (53).

EGG PLACEMENT

Selection of Plant Parts

Closely related species or genera of Lepidoptera sometimes differ in the plant parts they choose for oviposition, but the evolutionary origins for these differences have seldom been studied in detail. Heliconiine butterflies, for example, include genera that specialize on old leaf tissue and others that specialize on new shoots (10), and *Heliconius*, a genus of new shoot specialists, appears to be of relatively recent evolutionary origin (49). *Depressaria* includes species in which females ovipost almost exclusively on flowers or floral buds and other sibling species that oviposit both on floral buds and leaves (149). The *Depressaria* species that are more flexible in their choice of oviposition sites are those that feed on plants that are relatively small and flower more irregularly than related potential host plants. Few communitywide analyses have been made of the plant parts used by lepidopterans, but Janzen (67) found that more than 95% of the lepidopteran species in Santa Rosa National Park eat green leaves, and about 37% feed exposed on leaf surfaces.

The selection of plant parts by ovipositing females can influence how an interaction evolves between a particular insect species and its host plant. Attack on some plant parts (e.g. old leaves) may have very little effect on plant fitness, whereas attack on other parts (e.g. meristems, flowers) may greatly affect plant survival or reproduction. The evolution of mutualism between yucca moths and yuccas is a graphic illustration of how selection of plant parts by ovipositing females and the behavior of those females can influence how an interaction evolves. Seed parasites in the genus *Tegeticula* have become the major pollinators of their yucca hosts through behaviors exhibited by ovipositing females. Females of some *Tegeticula* (Prodoxidae) species actively deposit pollen on the stigmas of a flower after oviposition. However this behavior evolved, it effectively guarantees that the offspring of this female will have developing seeds on which to feed (6). Genera closely related to *Tegeticula* are internal parasites of other parts of yucca stalks (30) or various parts of other plants (32, 152). Some of these species, such as the floral-stalk miners of monocarpic yuccas, may be commensalistic with their hosts, whereas other species, such as the nonpollinating seed-parasitic *G. subalba,* are antagonistic and can decrease seed output by an average of 25% or more in a population of its *Lomatium* hosts (152).

Even within the general category of a plant part (e.g. leaves), females can be highly selective in where they lay their eggs. For example, eggs of leaf-mining moths are sometimes aggregated among leaves within host plants; leaf age, damage by other herbivores, and position on a plant all influence choice by ovipositing females (5, 16, 129). Some butterflies are more likely to lay their eggs on the underside of leaves rather than the upperside of leaves (82, 158, 171) or those receiving particularly high levels of sunlight (54).

Short-term studies of only a few years are probably insufficient for evaluating the selection pressures and constraints that influence the choice of oviposition sites within plants. For example, gypsy moth (*Lymantria dispar;* Lymantriidae) females in Japan usually lay their eggs in tree canopies in regions without snow, but lay them either mostly below the snow or both above and below the typical snow depths in regions with snow. In an eight-year study of mortality of egg masses in five regions, Higashiura (58) found that survival of egg masses below the snow was approximately two to four times higher than above the snow during four years in which predation by birds, which is concentrated during the snowy season, was high. During other years, when avian predation was low, survival above and below the snow was approximately the same. In the absence of avian predation, oviposition in the tree canopy is advantageous because newly hatched larvae must move up the tree to migrate by ballooning. Avian predation on egg masses and sites favorable for newly hatched larvae may therefore be conflicting selection pressures on site selection by ovipositing females. Higashiura argues that avian predation induces unpredictable strong selection pressure against exposed egg masses during many years, causing the maintenance of a behavioral polymorphism in snowy regions such that some females lay their egg batch above the snow whereas others lay thir eggs below the snow. Whether this is truly a genetic polymorphism, however, is not known.

A special case of egg placement is oviposition away from the host, notably by dropping eggs during flight. This behavior has evolved a number of times, and appears to be related to cases of superabundance of the host (169) or polyphagy. Under such circumstances, the reduction of search time in laying each egg may compensate for failures of some first-instar larvae to reach a host. Egg-dropping may also be correlated with production of small and/or very numerous eggs (up to 18,000 per female in some hepialids; 19). Oviposition away from the host is also common among species that overwinter as eggs and feed on herbaceous hosts. Wiklund (169) proposed that such behavior might in some cases result from selection for enemy-free space.

Although some closely related taxa show major differences in use of plant parts, there is a basic pattern of similarity of use at the family level (102). At least among Microlepidoptera, this pattern reflects the functional morphology of the female terminalia; although highly variable within the order, it is quite homogeneous within families (85). For example, a piercing-cutting ovipositor constrains the female to place eggs inside penetrable tissue, and, unless an empty space can be reached, often only one egg at a time can be laid. This is a time-consuming mode of oviposition. A piercing-cutting ovipositor evolved very early in the Lepidoptera, probably as part of the *Bauplan* of Glossata (71). It is conserved in two groups (suborder Dacnonypha and superfamily Incurvarioidea), and the cutting ovipositor has been modified only once among the approximately 400 species of Incurvarioidea. Outside these groups, however, the ovipositor has been heavily modified at least three other times, facilitating external deposition of eggs. All these evolutionary events coincide with remarkable radiation, particularly in Ditrysia. These radiations no doubt coincided also with many other changes, but the patterns observed suggest that the plasticity in patterns of egg laying found primarily in butterflies is predominantly behavioral in origin, and that morphological factors may constrain such plasticity, at least in primitive Lepidoptera.

Single-Egg Versus Batch Laying

A variety of hypotheses has been proposed on the ecological conditions that favor laying of eggs in clusters rather than singly, and Chew & Robbins (18) reviewed this literature through the early 1980s. Some recent mathematical models have suggested that clutch size should generally decrease as the number of females ovipositing in a patch increases (96, 97, 135) or, under some conditions, as larval competition increases (63).

Clutch size varies among some closely related species, among populations within species, and sometimes even among females within a population. This variation indicates that clutch size is not under severe phylogenetic constraints in at least some lepidopteran taxa. In two closely related species of *Colotis* (Pieridae), one species lays eggs singly near the main trunk, whereas the other species lays eggs in clusters of about 30 on the outside of leaves (72). *P. brassicae* and *P. rapae* use the same host plant but lay their eggs in batches or singly, respectively; this may be linked to ecological specialization on high and low density stands of the host (29). In a transcontinental experiment, *P. rapae* females from a population in Britain were more likely to lay more than one egg after alighting than were females from a population in Australia flown under the same experimental conditions (69). *Aporia crataegi* (Pieridae) females in Morocco vary their average clutch size with host species, and two other butterflies appear to vary clutch size depending on host plant density (23). In *B. philenor*, where clutch size apparently varies between hosts (143), individuals in at least some populations adjust clutch size for host quality (100).

Very little is known about moths with regard to variation in clutch size. In cage trials, *Diatraea saccharalis* (Pyralidae) clutch size varied with plant part, but multiple females were flown in each cage (138). So, one cannot determine whether individual females varied their clutch sizes as they encountered different plant parts or whether females differing in clutch sizes preferred different plant parts. Powell (101) and Powell & Common (103) present voluminous data primarily for caged females of numerous Tortricinae; batch size and architecture is rather stable at the tribal level. Individuals of species that lay batches often lay smaller batches later in their lives. Detailed studies of select taxa among tribes focusing on individual variation could provide interesting phylogenetic and ecological data on the factors that determine these patterns.

Consequence of Brachyptery

Brachyptery, or even aptery, in females has evolved one or more times within at least ten families of Lepidoptera (O. Pellmyr, unpublished data). The evolution of flightlessness sets special constraints on oviposition, particularly the inability to move between habitats and among potential host plants (e.g. 46).

Among flightless species, eggs are typically laid in clusters. Patterns of oviposition behavior in lineages that include some species with sedentary females suggest that loss of flight ability seldom evolved before the evolution of a clustered distribution of eggs. Among bombycoids such as lymantriids, the transition from winged to wingless condition is gradual; many species have winged females that rarely or never fly. The often-noxious larvae of many of these species are gregarious during most or all instars. In *L. dispar*, females vary between subspecies in whether they fly before oviposition, but in all cases they lay a single batch (58). Several closely related families, such as lasiocampids, endromids, and thaumetopoeids, also have winged females that seldom fly; in many of these instances, the larvae are gregarious as well. In geometrids, brachyptery has evolved at least four times, at least three of them in association with adult activity at very low ambient temperature. Tree-feeding taxa, such as *Operophthera* (Geometridae), may distribute their eggs over one individual tree; taxa feeding on smaller hosts may move between a handful of host individuals. Genera that include primitively winged species, such as *Lycia* and *Phigalia* (Geometridae), show no evidence of different egg distribution patterns. In epipyropids, the larvae of which are parasitic on Homoptera, at least one species is brachypterous; all species cluster their eggs on the host plant of the host insect, or even on the female's cocoon, often laying thousands of eggs (31). Thus, the evidence does not suggest that clustering evolved after brachyptery.

Flightlessness may not have been causal in the evolution of batch oviposition, but it usually prevents a female from choosing among host plants. It is tempting to hypothesize that brachyptery in fact is likely to evolve only when female host choice has little effect on fitness. In some brachypterous species, other means of dispersal have evolved. Larvae can disperse through ballooning during early instars (*Lymantria:* 58, 76; *Orgyia:* 80; *Alsophila:* 46). In the brachypterous *Orgyia pseudotsugata* (Lymantriidae), distances drifted typically were less than 150 m, while in *L. dispar,* over 75% dispersed less than 75 m. This type of dispersal is undirected, however, and the larva has very limited ability to choose among hosts.

If brachyptery has significant effects on population structure—through batch oviposition and restricted female movement—it could contribute to the evolution of specialized host strains (81). In the gynogenetic and rarely sexual *Alsophila pometaria* (Geometridae), Mitter et al (81) and Futuyma et al (46) indeed found host strains that differed in frequency between neighboring maple and oak stands. Differential survival between genotypes was largely attributable to phenological factors and behavioral differences (47). No differences in frequency were found in mixed-species stands. Because adults often oviposit on the tree on which they fed as larvae (123), Futuyma et al (46) proposed that the lack of differentiation observed in mixed stands resulted from larval dispersal.

FUTURE STUDIES

The past 15 years have resulted in major advances in our understanding of the sequence of lepidopteran oviposition behaviors leading to the distribution of a female's eggs among host plants. Among these advances are specific experimental protocols for studying oviposition preference and specificity, significant progress on the behavioral genetics of oviposition preference relative to larval performance, elucidation of how chemical attractants and deterrents influence acceptance of potential hosts, indications of some of the differences between prealighting and postalighting behaviors, and a richer understanding of how the distribution of resources affects the movement of females as they search for plants. Nonetheless, detailed studies on each of these components of oviposition behavior have been made on only a few species. So, determining general patterns from the results is still difficult.

All these components of behavior have yet to be studied for any one species. The genera that come closest to having had some species studied for most of these components are *Euphydryas, Papilio,* and *Pieris.* These are all strong-flying butterflies. Fragmentary pieces of information from other families suggest that perhaps the variability of many butterflies is unusual rather than the norm. Studies are badly needed on nocturnal taxa. An unexplored possibility in this regard would be to work at high latitudes in the summer, where otherwise nocturnal moths execute their oviposition behaviors under the midnight sun. With the advent of information for Lepidoptera other than butterflies, it should also become increasingly possible to develop a better phylogenetic perspective on oviposition behavior within the order.

ACKNOWLEDGMENTS

We thank Hal Hansel, John Jaenike, Michael Singer, and Wayne Wehling for helpful comments on the manuscript. This work was supported by NSF grants BSR-8705394 and BSR-8817337 and USDA Competitive Grant WNR-8900340.

Literature Cited

1. Ahmad, S., ed. 1983. *Herbivorous Insects: Host-seeking Behavior and Mechanisms.* New York: Academic
2. Aker, C. L., Udovic, D. 1981. Oviposition and pollination behavior of the yucca moth, *Tegeticula maculata* (Lepidoptera: Prodoxidae), and its relation to the reproductive biology of *Yucca whipplei* (Agavaceae). *Oecologia* 49:96–101
3. Atsatt, P. R. 1981a. Ant-dependent food plant selection by the mistletoe butterfly *Ogyris amaryllis* (Lycaenidae). *Oecologia* 48:60–63
4. Atsatt, P. R. 1981b. Lycaenid butterflies and ants: selection for enemy-free space. *Am. Nat.* 188:638–54
5. Auerbach, M., Simberloff, D. 1989. Oviposition site preference and larval mortality in a leaf-mining moth. *Ecol. Entomol.* 14:131–40
6. Baker, H. G. 1986. Yuccas and yucca moths—a historical commentary. *Ann. Mo. Bot. Gard.* 73:556–64
7. Bates, H. W. 1876. *A Naturalist on the River Amazons,* p. 348. London: Murray. 4th ed.
8. Beach, R. M., Todd, J. W. 1988. Oviposition preference of the soybean looper (Lepidoptera: Noctuidae) among four soybean genotypes differing in larval resistance. *J. Econ. Entomol.* 81:344–48
9. Behan, M., Schoonhoven, L. M. 1978. Chemoreception of an oviposition deterrent associated with eggs in *Pieris brassicae. Entomol. Exp. Appl.* 24:163–79
10. Benson, W. W. 1978. Resource partitioning in passion vine butterflies. *Evolution* 32:493–518
11. Berenbaum, M. 1983. Coumarins and caterpillars: a case for coevolution. *Evolution* 37;163–79
12. Boppré, M. 1986. Insects pharmacophagously utilizing defensive plant chemicals (Pyrrolizidine alkaloids). *Naturwissenschaften* 73:17–26
13. Brantjes, N. B. M. 1976a. Riddles around the pollination of *Melandrium album* (Mill.) Garcke (Caryophyllaceae) during the oviposition by *Hadena bicruris* Hufn. (Noctuidae, Lepidoptera). I + II. *Proc. K. Ned. Akad. Wet. Ser. C.* 79:1–12, 127–41
14. Brantjes, N. B. M. 1976b. Prevention of superparasitation [sic] of *Melandrium* flowers (Caryophyllaceae) by *Hadena* (Lepidoptera). *Oecologia* 24:1–6
15. Brower, L. P. 1985. New perspectives on the migration biology of the monarch butterfly, *Danaus plexippus* L. In *Migration: Mechanisms and Adaptive Significance,* ed. M. A. Rankin, pp. 748–85 Austin, TX: Univ. of Texas Press
16. Bultman, T. L., Faeth, S. H. 1986. Selective oviposition by a leaf-miner in response to temporal variation in abscission. *Oecologia* 64:117–20
17. Chew, F. S. 1977. Coevolution of pierid

butterflies and their cruciferous food plants. II. The distribution of eggs on potential food plants. *Evolution* 31:568–79

18. Chew, F. S., Robbins, R. K. 1984. Egg-laying in butterflies. See Ref. 163, pp. 65–79
19. Common, I. F. B. 1970. Lepidoptera. In *The Insects of Australia,* sponsor CSIRO, pp. 765–866, Carlton: Melbourne Univ. Press
20. Corbet, S. A. 1973. Oviposition pheromone in larval mandibular glands of *Ephestia kuehniella. Nature* 243:537–38
21. Courtney, S. P. 1981. Coevolution of pierid butterflies and their cruciferous foodplants. III. *Anthocharis cardamines* (L.). Survival, development and oviposition on different hostplants. *Oecologia* 51:91–96
22. Courtney, S. P. 1982. Coevolution of pierid butterflies and their cruciferous foodplants. IV. Crucifer apparency and *Anthocharis cardamines* (L.) oviposition. *Oecologia* 52:258–65
23. Courtney, S. P. 1984. The evolution of egg clustering by butterflies and other insects. *Am. Nat.* 123:276–81
24. Courtney, S. P. 1988. Oviposition on peripheral hosts by dispersing *Pieris napi* (L.) (Pieridae). *J. Res. Lepid.* 26:58–63
25. Courtney, S. P., Chen, G. K., Gardner, A. 1989. A general model for individual host selection. *Oikos* 55:55–65
26. Cromartie, W. J. 1975. The effect of stand size and vegetational background on the colonization of cruciferous plants by herbivorous insects. *J. Appl. Ecol.* 12:517–33
27. Damman, H., Feeny, P. 1988. Mechanisms and consequences of selective oviposition by the zebra swallowtail butterfly. *Anim. Behav.* 36:563–73
28. Darwin, C. 1909. Essay of 1844. In *The Foundations of the Origin of Species,* ed. F. Darwin. Cambridge: Cambridge Univ. Press
29. Davies, C. R., Gilbert, N. 1985. A comparative study of the egg-laying behaviour and larval development of *Pieris rapae* L. and *P. brassicae* L. on the same host plants. *Oecologia* 67:278–81
30. Davis, D. R. 1967. A revision of the moths of the subfamily Prodoxinae (Lepidoptera: Incurvariidae). *Bull. US Nat. Mus.* 255:1–170
31. Davis, D. R. 1987. Epipyropidae. In *Immature Insects,* ed. F. W. Stehr, pp. 456–60, Dubuque, Iowa: Kendall-Hunt
32. Davis, D. R., Pellmyr, O., Thompson, J. N. 1990. Biology and systematics of *Greya* Busck and *Tetragma* n.gen.

(Lepidoptera: Prodoxidae). *Smithson. Contrib. Zool.* In press
33. Douwes, P. 1968. Host selection and host finding in the egg-laying female *Cidaria albulata* L. (Lep. Geometridae). *Opusc. Entomol.* 33:233–79
34. Douwes, P. 1975. Distribution of a population of the butterfly *Heodes virgaureae. Oikos* 26:332–40
35. Ehrlich, P. R., Gilbert, L. E. 1973. Population structure and dynamics of the tropical butterfly *Heliconius ethilla. Biotropica* 5:69–82
36. Ehrlich, P. R., Murphy, D. D., Singer, M. C., Sherwood, C. B., White, R. R., Brown, I. L. 1980. Extinction, reduction, stability and increase: the responses of checkerspot butterfly *(Euphydryas)* populations to California drought. *Oecologia* 46:101–5
37. Ehrlich, P. R., Raven, P. H. 1964. Butterflies and plants: a study in coevolution. *Evolution* 18:586–608
38. Ehrlich, P. R., White, R. R., Singer, M. C., McKechnie, S. W., Gilbert, L. E. 1975. Checkerspot butterflies: a historical perspective. *Science* 188:221–28
39. Eisner, T., Meinwald, J. 1987. Alkaloid-derived pheromones and sexual selection in Lepidoptera. In *Pheromone Biochemistry,* ed. G. D. Prestwich, G. J. Blomquist, pp. 251–96. Orlando: Academic
40. Feeny, P. 1990. Chemical constraints on the evolution of swallowtail butterflies. In *Herbivory: Temperate and Tropical Perspectives.* ed. P. W. Price, T. M. Lewinsohn, W. W. Benson, G. W. Fernandes. New York: Wiley. In press
41. Feeny, P., Rosenberry, L., Carter, M. 1983. Chemical aspects of oviposition behavior in butterflies. See Ref. 1, pp. 27–76
42. Feeny, P., Sachdev, K., Rosenberry, L., Carter, M. 1988. Luteolin 7-O-(6"-O-Malonyl)-b-D-glucoside and Trans-chlorogenic acid: oviposition stimulants for the black swallowtail butterfly. *Phytochemistry* 27:3439–48
43. Feeny, P., Städler, E., Åhman, I., Carter, M. 1989. Effects of plant odor on oviposition by the black swallowtail butterfly, *Papilio polyxenes* (Lepidoptera: Papilionidae). *J. Insect Behav.* 2:803–27
44. Forsberg, J. 1987. Size discrimination among conspecific hostplants in two pierid butterflies: *Pieris napi* L. and *Pontia daplidice* L. *Oecologia* 72:52–57
45. Futuyma, D. F. 1983. Selective factors in the evolution of host choice by phytophagous insects. See Ref. 1, pp. 227–44
46. Futuyma, D. F., Leipertz, S. L., Mitter,

C. 1981. Selective factors affecting clonal variation in the fall cankerworm *Alsophila pometaria* (Lepidoptera: Geometridae). *Heredity* 47:161–72

47. Futuyma, D. F., Cort, R. P., van Noordwijk, I. 1984. Adaptation to host plants in the fall cankerworm (*Alsophila pometaria*) and its bearing on the evolution of host affiliation in phytophagous insects. *Am. Nat.* 123:287–96

48. Gilbert, L. E. 1975. Ecological consequences of coevolved mutualism between butterflies and plants. In *Coevolution of Animals and Plants,* ed. L. E. Gilbert and P. H. Raven, pp. 210–40. Austin, TX: Univ. Texas Press

49. Gilbert, L. E. 1984. The biology of butterfly communities. See Ref. 163, pp. 41–54

50. Gilbert, L. E., Singer, M. C. 1973. Dispersal and gene flow in a butterfly species. *Am. Nat.* 107:58–72

51. Gilbert, L. E., Singer, M. C. 1975. Butterfly ecology. *Annu. Rev. Ecol. Syst.* 6:365–97

52. Grant, A. J., O'Connell, R. J., Eisner, T. 1989. Pheromone-mediated sexual selection in the moth *Utetheisa ornatrix:* olfactory receptor neurons responsive to a male-produced pheromone. *J. Insect Behav.* 2:371–85

53. Greenbank, D. O. 1973. The dispersal process of spruce budworm moths. *Maritimes Forest Research Centre Information Report M-X-39.* 25 pp.

54. Grossmueller, D. W., Lederhouse, R. C. 1985. Oviposition site selection: an aid to rapid growth and development in the tiger swallowtail butterfly, *Papilio glaucus. Oecologia* 66:68–73

55. Harrison, S. 1989. Long-distance dispersal and colonization in the Bay checkerspot butterfly, *Euphydryas editha bayensis. Ecology* 70:1236–43

56. Harrison, S., Murphy, D. D., Ehrlich, P. R. 1988. Distribution of the Bay checkerspot butterfly, *Euphydryas editha bayensis:* evidence for a metapopulation model. *Am. Nat.* 132:360–82

57. Haynes, K. F., Baker, T. C. 1989. An analysis of anemotactic flight in female moths stimulated by host odour and comparison with the males' response to sex pheromone. *Physiol. Entomol.* 14:279–89

58. Higashiura, Y. 1989. Survival of eggs in the gypsy moth *Lymantria dispar.* II. Oviposition site selection in changing environments. *J. Anim. Ecol.* 58:413–26

59. Hinton, H. E. 1981. *Biology of Insect Eggs,* Vol. 2. Oxford: Pergamon

60. Honda, K. 1986. Flavanone glycosides as oviposition stimulants in a papilionid butterfly, *Papilio protenor. J. Chem. Ecol.* 12:1999–2010

61. Honda, H., Ishiwatari, T., Matsumoto, Y. 1988. Fungal volatiles as oviposition attractants for the yellow peach moth, *Conogethes punctiferalis* (Guenée) (Lepidoptera: Pyralidae). *J. Insect Physiol.* 34:205–11

62. Ilse, D. 1937. New observations on responses to colour in egg-laying butterflies. *Nature* 140:544–45

63. Ives, A. R. 1989. The optimal clutch size of insects when many females oviposit per patch. *Am. Nat.* 133:671–87

64. Jaenike, J. 1990. Host specialization in phytophagous insects. *Annu. Rev. Ecol. Syst.* 21: In press

65. Janzen, D. H. 1984. Two ways to be a tropical big moth: Santa Rosa saturniids and sphingids. *Oxford Surv. Evol. Biol.* 1:85–140

66. Janzen, D. H. 1987. Insect diversity of a Costa Rican dry forest: why keep it, and how? *Biol. J. Linn. Soc.* 30:343–56

67. Janzen, D. H. 1988. Ecological characterization of a Costa Rican dry forest caterpillar fauna. *Biotropica* 20:120–35

68. Jones, R. E. 1977. Movement patterns and egg distribution in cabbage butterflies. *J. Anim. Ecol.* 46:195–212

69. Jones, R. E. 1987. Behavioral evolution in the cabbage butterfly (*Pieris rapae*). *Oecologia* 72:69–76

70. Kingsolver, R. W. 1984. *Population biology of a mutualistic association: Yucca glauca and Tegeticula yuccasella.* PhD thesis, Lawrence: Univ. Kansas

71. Kristensen, N. P. 1984. Studies on the morphology and systematics of primitive Lepidoptera (Insecta). *Steenstrupia* 10:141–91

72. Larsen, T. B. 1988. Differing oviposition and larval feeding strategies in two *Colotis* butterflies sharing the same food plant. *J. Lepid. Soc.* 42:57–58

73. Leather, S. R. 1987. Pine monoterpenes stimulate oviposition in the pine beauty moth, *Panolis flammea. Entomol. Exp. Appl.* 43:295–97

74. Letourneau, D. K., Fox, L. R. 1989. Effects of experimental design and nitrogen on cabbage butterfly oviposition. *Oecologia* 80:211–14

75. Löfstedt, C., Vickers, N. J., Roelofs, W. L., Baker, T. C. 1989. Diet related courtship success in the Oriental fruit moth, *Grapholita molesta* (Tortricidae). *Oikos* 55:402–8

76. Mason, C. J., McManus, M. L. 1981. Larval dispersal of the gypsy moth. In *The Gypsy Moth: Research toward Inte-*

grated Pest Management, ed. C. C. Doane, M. L. McManus, pp. 161–202. Washington, DC: USDA

77. Mazanec, Z., Justin, M. J. 1986. Oviposition behaviour and dispersal of *Perthida glyphopa* Common (Lepidoptera: Incurvariidae). *J. Aust. Ent. Soc.* 25:149–59

78. Miller, J. S. 1987. Host-plant relationships in the Papilionidae (Lepidoptera): parallel cladogenesis or colonization? *Cladistics* 3:105–20

79. Mitchell, E. R., Heath, R. R. 1987. *Heliothis subflexa* (GN.) (Lepidoptera: Noctuidae): demonstration of oviposition stimulant from groundcherry using novel bioassay. *J. Chem. Ecol.* 13:1849–58

80. Mitchell, R. G. 1979. Dispersal of early instars of the Douglas-Fir Tussock moth. *Ann. Entomol. Soc. Amer.* 72:291–97

81. Mitter, C., Futuyma, D. J., Schneider, J. C., Hare, J. D. 1979. Genetic variation and host plant relations in a parthenogenetic moth. *Evolution* 33:777–90

82. Moore, G. J. 1986. Host plant discrimination in tropical satyrine butterflies. *Oecologia* 70:592–95

83. Murphy, D. 1983. Nectar sources as constraints on the distribution of egg masses by the checkerspot butterfly, *Euphydryas chalcedona* (Lepidoptera: Nymphalidae). *Env. Entomol.* 12:463–66

84. Murphy, D. D., Menninger, M. S., Ehrlich, P. R. 1984. Nectar source distribution as a determinant of oviposition host species in *Euphydryas chalcedona*. *Oecologia* 62:269–71

85. Mutuura, A. 1972. Morphology of the female terminalia in Lepidoptera, and its taxonomic significance. *Can. Entomol.* 104:1055–71

86. Myers, J. H. 1985. Effect of physiological condition of the host plant on the ovipositional choice of the cabbage white butterfly, *Pieris rapae*. *J. Anim. Ecol.* 54:193–204

87. Myers, J. H., Monro, J., Murray, N. 1981. Egg clumping, host plant selection and population regulation in *Cactoblastis cactorum* (Lepidoptera). *Oecologia* 51:7–13

88. Ng, D. 1988. A novel level of interactions in plant-insect systems. *Nature* 334:611–13

89. Nielsen, E. S. 1989. Phylogeny of major lepidopteran groups. In *The Hierarchy of Life*, ed. B. Fernholm, K. Bremer, H. Jörnvall, pp. 281–321. Amsterdam: Elsevier

90. Nishida, R., Fukami, H. 1989. Oviposi-

tion stimulants of an Aristolochiaceae-feeding swallowtail butterfly, *Atrophaneura alcinous*. *J. Chem. Ecol.* 15:2565–75

91. Nishida, R., Ohsugi, T., Kokubo, S., Fukami, H. 1987. Oviposition stimulants of a *Citrus*-feeding swallowtail butterfly, *Papilio xuthus* L. *Experientia* 43:342–44

92. Odendaal, F. J., Turchin, P., Stermitz, F. R. 1989. Influence of host-plant density and male harassment on the distribution of female *Euphydryas anicia* (Nymphalidae). *Oecologia* 78:283–88

93. Ohsugi, T., Nishida, R., Fukami, H. 1985. Oviposition stimulant of *Papilio xuthus*, a *Citrus*-feeding swallowtail butterfly. *Agric. Biol. Chem.* 49:1897–1900

94. Papaj, D. R., Prokopy, R. J. 1989. Ecological and evolutionary aspects of learning in phytophagous insects. *Annu. Rev. Entomol.* 34:315–50

95. Papaj, D. R., Rausher, M. D. 1987. Components of conspecific discrimination behavior in the butterfly *Battus philenor*. *Ecology* 68:245–53

96. Parker, G. A., Begon, M. 1986. Optimal egg size and clutch size: effect of environment and maternal phenotype. *Am. Nat.* 128:573–95

97. Parker, G. A., Courtney, S. P. 1984. Models of clutch size in insect oviposition. *Theor. Popul. Biol.* 26:27–48

98. Pereyra, P. C., Bowers, M. D. 1988. Iridoid glycosides as oviposition stimulants for the buckeye butterfly, *Junonia coenia* (Nymphalidae). *J. Chem. Ecol.* 14:917–28

99. Pierce, N. E., Elgar, M. A. 1985. The influence of ants on host plant selection by *Jalmenus evagoras*, a myrmecophilus lycaenid butterfly. *Behav. Ecol. Sociobiol.* 16:209–22

100. Pilson, D., Rausher, M. D. 1988. Clutch size adjustment by a swallowtail butterfly. *Nature* 333:361–63

101. Powell, J. A. 1964. Biological and taxonomic studies on tortricine moths, with reference to the species in California. *Univ. Calif. Publ. Entomol.* 32. 317 pp.

102. Powell, J. A. 1980. Evolution of larval food preferences in Microlepidoptera. *Annu. Rev. Entomol.* 25:133–59

103. Powell, J. A., Common, I. F. B. 1985. Oviposition patterns and egg characteristics of Australian tortricine moths (Lepidoptera: Tortricidae). *Aust. J. Zool.* 33:179–216

104. Ramaswamy, S. B. 1988. Host finding by moths: sensory modalities and behaviours. *J. Insect Physiol.* 34:235–49

105. Ramaswamy, S. B., Ma, W. K., Baker, G. T. 1987. Sensory cues and receptors for oviposition by *Heliothis virescens*. *Entomol. Exp. Appl.* 43:159–68
106. Rausher, M. D. 1978. Search image for leaf shape in a butterfly. *Science* 200:1071–73
107. Rausher, M. D. 1979a. Larval habitat suitability and oviposition preference in three related butterflies. *Ecology* 60:503–11
108. Rausher, M. D. 1979b. Egg recognition: its advantage to a butterfly. *Anim. Behav.* 27:1034–40
109. Rausher, M. D. 1981. Host plant selection by *Battus philenor* butterflies: the roles of predation, nutrition, and plant chemistry. *Ecol. Monogr.* 51:1–20
110. Rausher, M. D. 1982. Population differentiation in *Euphydryas editha* butterflies: larval adaptation to different hosts. *Evolution* 36:581–90
111. Rausher, M. D. 1983. Ecology of host-selection behavior in phytophagous insects. In *Variable Plants and Herbivores in Natural and Managed Systems*. ed. R. F. Denno, M. C. McClure, pp. 223–57. New York: Academic
112. Rausher, M. D., Papaj, D. R. 1983. Host plant selection by *Battus philenor* butterflies: evidence for individual differences in foraging behavior. *Anim. Behav.* 31:341–47
113. Renwick, J. A. A. 1989. Chemical ecology of oviposition in phytophagous insects. *Experientia* 45:223–28
114. Renwick, J. A. A., Radke, C. D. 1980. An oviposition deterrent associated with frass from feeding larvae of the cabbage looper, *Trichoplusia ni* (Lepidoptera: Noctuidae). *Environ. Entomol.* 9:318–20
115. Renwick, J. A. A., Radke, C. D. 1987. Chemical stimulants and deterrents regulating acceptance or rejection of crucifers by cabbage butterflies. *J. Chem. Ecol.* 13:1771–75
116. Renwick, J. A. A., Radke, C. D. 1988. Sensory cues in host selection for oviposition by the cabbage butterfly, *Pieris rapae*. *J. Insect Physiol.* 34:251–57
117. Renwick, J. A. A., Radke, C. D., Sachdev-Gupta, K. 1989. Chemical constituents of *Erysimum cheiranthoides* deterring oviposition by the cabbage butterfly, *Pieris rapae*. *J. Chem. Ecol.* 15:2161–68
118. Rodman, J. E., Chew, F. S. 1980. Phytochemical correlates of herbivory in a community of native and naturalized Cruciferae. *Biochem. Syst. Ecol.* 8:43–50
119. Root, R. B., Kareiva, P. M. 1984. The search for resources by cabbage butterflies *(Pieris rapae)*: ecological consequences and adaptive significance of Markovian movement in a patchy environment. *Ecology* 65:147–65
120. Rothschild, M., Alborn, H., Stenhagen, G., Schoonhoven, L. M. 1988. A stophanthidine glycoside in the Siberian wallflower: a contact deterrent for the large white butterfly. *Phytochemistry* 27:101–8
121. Rothschild, M., Schoonhoven, L. M. 1977. Assessment of egg load by *Pieris brassicae* (Lepidoptera: Pieridae). *Nature* 266:352–55
122. Scherer, C., Kolb, G. 1987. Behavioral experiments on the visual processing of color stimuli in *Pieris brassicae* L. (Lepidoptera). *J. Comp. Physiol. A* 160:645–56
123. Schneider, J. C. 1980. The role of parthenogenesis and female aptery in microgeographic, ecological adaptation in the fall cankerworm, *Alsophila pometaria* Harris (Lepidoptera: Geometridae). *Ecology* 61:1082–90
124. Schneider, J. C. Roush, R. T. 1986. Genetic differences in oviposition preference between two populations of *Heliothis virescens*. In *Evolutionary Genetics of Invertebrate Behavior*. ed. M. D. Huettel, pp. 163–71. New York: Plenum
125. Schoonhoven, L. M., Sparnaay, T., van Wissen, W., Meerman, J. 1981. Seven-week persistence of an oviposition-deterrent pheromone. *J. Chem. Ecol.* 7:583–88
126. Shapiro, A. M. 1975. Ecological and behavioral aspects of coexistence in six crucifer-feeding pierid butterflies in the central Sierra Nevada. *Am. Midl. Nat.* 93:424–33
127. Shapiro, A. M. 1981. The pierid red-egg syndrome. *Am. Nat.* 117:276–94
128. Sharp, M. A., Parks, D. R., Ehrlich, P. R. 1974. Plant resources and butterfly habitat selection. *Ecology* 55:870–75
129. Simberloff, D., Stiling, P. D. 1987. Larval dispersion and survivorship in a leaf-mining moth. *Ecology* 68:1647–57
130. Singer, M. C. 1982. Quantification of host preference by manipulation of oviposition behavior in the butterfly, *Euphydryas editha*. *Oecologia* 52:224–29
131. Singer, M. C. 1983. Determinants of multiple host use by a phytophagous insect population. *Evolution* 37:389–403
132. Singer, M. C. 1984. Butterfly-hostplant relationships: host quality, adult choice and larval success. See Ref. 163, pp. 82–88

133. Singer, M. C. 1986. The definition and measurement of oviposition preference in plant-feeding insects. In *Insect-Plant Interactions*, ed. T. A. Miller, J. A. Miller, pp. 65–94. New York: Springer-Verlag

134. Singer, M. C., Ng, D., Thomas, C. D. 1988. Heritability of oviposition preference and its relationship to offspring performance within a single insect population. *Evolution* 42:977–85

135. Skinner, S. W. 1985. Clutch size as an optimal foraging problem for insect parasitoids. *Behav. Ecol. Sociobiol.* 17:231–38

136. Smiley, J. T. 1978. Plant chemistry and the evolution of host specificity: new evidence from *Heliconius* and *Passiflora. Science* 201:745–47

137. Smiley, J. T. 1985. Are chemical barriers necessary for evolution of butterfly-plant associations? *Oecologia* 65:580–83

138. Sosa, O. Jr. 1988. Pubescence in sugarcane as a plant resistance character affecting oviposition and mobility by the sugarcane borer (Lepidoptera: Pyralidae). *J. Econ. Entomol.* 81:663–67

139. Stanton, M. L. 1982. Searching in a patchy environment: foodplant selection by *Colias p. eriphyle* butterflies. *Ecology* 63:839–53

140. Stanton, M. L., Cook, R. E. 1983. Sources of intraspecific variation in the hostplant seeking behavior of *Colias* butterflies. *Oecologia* 60:365–70

141. Strong, D. R. 1988. Insect host range. *Ecology* 69:885

142. Tabashnik, B. E., Wheelock, H., Rainbolt, J. D., Watt, W. B. 1981. Individual variation in oviposition preference in the butterfly, *Colias eurytheme. Oecologia* 50:225–30

143. Tatar, M. 1989. Swallowtail clutch size reconsidered. *Oikos* 55:135–36

144. Thomas, C. D., Ng, D., Singer, M. C., Mallet, J. L. B., Parmesan, C., Billington, H. L. 1987. Incorporation of a European weed into the diet of a North American herbivore. *Evolution* 41:892–901

145. Thomas, C. D., Singer, M. C. 1987. Variation in host preference affects movement patterns within a butterfly population. *Ecology* 68:1262–67

146. Thomas, J. A., Elmes, G. W., Wardlaw, J. C., Woyciechowski, M. 1989. Host specificity among *Maculinea* butterflies in *Myrmica* ant nests. *Oecologia* 79:452–57

147. Thompson, J. N. 1978. Within-patch structure and dynamics in *Pastinaca sativa* and resource availability to a specialized herbivore. *Ecology* 59:443–48

148. Thompson, J. N. 1982. *Interaction and Coevolution.* New York: Wiley

149. Thompson, J. N. 1983. Selection pressures on phytophagous insects on small host plants. *Oikos* 40:438–44

150. Thompson, J. N. 1986a. Patterns in coevolution. In *Coevolution and Systematics*, ed. A. R. Stone, D. J. Hawksworth, pp. 119–43. Oxford: Oxford Univ. Press

151. Thompson, J. N. 1986b. Oviposition behaviour and searching efficiency in a natural population of a braconid parasitoid. *J. Anim. Ecol.* 55:351–60

152. Thompson, J. N. 1987. Variance in number of eggs per patch: oviposition behaviour and population dispersion in a seed parasitic moth. *Ecol. Entomol.* 12:311–20

153. Thompson, J. N. 1988a. Variation in preference and specificity in monophagous and oligophagous swallowtail butterflies. *Evolution* 42:118–28

154. Thompson, J. N. 1988b. Evolutionary genetics of oviposition preference in swallowtail butterflies. *Evolution* 42:1223–34

155. Thompson, J. N. 1988c. Evolutionary ecology of the relationship between oviposition preference and performance of offspring in phytophagous insects. *Entomol. Exp. Appl.* 47:3–14

156. Thompson, J. N., Price, P. W. 1977. Plant plasticity, phenology, and herbivore dispersion: wild parsnip and the parsnip webworm. *Ecology* 58:1112–19

157. Thompson, J. N., Wehling, W., Podolsky, R. 1990. Evolutionary genetics of host use in swallowtail butterflies. *Nature* 344:158–60

158. Tiritilli, M. E., Thompson, J. N. 1988. Variation in swallowtail/plant interactions: host selection and the shapes of survivorship curves. *Oikos* 53:153–60

159. Traynier, R. M. M. 1984. Associative learning in the ovipositional behaviour of the cabbage butterfly, *Pieris rapae. Physiol. Entomol.* 9:465–72

160. Traynier, R. M. M. 1986. Visual learning in assays of sinigrin solution as an oviposition releaser for the cabbage butterfly, *Pieris rapae. Entomol. Exp. Appl.* 40:25–33

161. Urquhart, F. A., Urquhart, N. R. 1977. Overwintering areas and migratory routes of the monarch butterfly (*Danaus plexippus,* Lepidoptera: Danaidae) in North America with special reference to the western population. *Can. Entomol.* 109:1583–89

162. van der Meijden, E., van Zoelen, A. M., Soldaat, L. L. 1989. Oviposition by the cinnabar moth, *Tyria jacobaeae*, in relation to nitrogen, sugars and alkaloids of ragwort, *Senecio jacobaea*. *Oikos* 54:337–44

163. Vane-Wright, R., Ackery, P. R., eds. 1984. *The Biology of Butterflies*. New York: Academic

164. Via, S. 1986. Genetic covariance between oviposition preference and larval performance in an insect herbivore. *Evolution* 40:778–85

165. Wearing, C. H., Hutchins, R. F. N. 1973. α-Farnesene, a naturally occurring oviposition stimulant for the codling moth, *Laspeyresia pomonella*. *J. Insect Physiol.* 19:1251–56

166. Wiklund, C. 1977. Oviposition, feeding and spatial separation of breeding habitats in a population of *Leptidea sinapis* (Lepidoptera). *Oikos* 28:56–68

167. Wiklund, C. 1981. Generalist vs. specialist oviposition behaviour in *Papilio machaon* (Lepidoptera) and functional aspects on the hierarchy of oviposition preferences. *Oikos* 36:163–70

168. Wiklund, C. 1982. Generalist versus specialist utilization of host plants among butterflies. In *Proceedings of the 5th International Symposium on Insect-Plant Relationships*, pp. 181–91. Wageningen: Cent. Agric. Pub. Doc.

169. Wiklund, C. 1984. Egg-laying patterns in butterflies in relation to their phenology and the visual apparency and abundance of their host plants. *Oecologia* 63:23–29

170. Wiklund, C., Åhrberg, C. 1978. Host plants, nectar source plants, and habitat selection of males and females of *Anthocaris cardamines* (Lepidoptera). *Oikos* 31:169–83

171. Williams, E. H. 1981. Thermal influences on oviposition in the montane butterfly *Euphydryas gilletti*. *Oecologia* 50:342–46

172. Williams, K. S. 1983. The coevolution of *Euphydryas chalcedona* butterflies and their larval host plants. III. Oviposition behavior and host plant quality. *Oecologia* 56:336–40

173. Williams, K. S., Gilbert, L. E. 1981. Insects as selective agents on plant vegetative morphology: egg mimicry reduces egg laying by butterflies. *Science* 212:467–69

174. Williams, K. S., Lincoln, D. E., Ehrlich, P. R. 1983. The coevolution of *Euphydryas chalcedona* butterflies and their larval host plants. II. Maternal and host plant effects on larval growth, development, and food-use efficiency. *Oecologia* 56:330–35

175. Wolfson, J. L. 1980. Oviposition response of *Pieris rapae* to environmentally induced variation in *Brassica nigra*. *Entomol. Exp. Appl.* 27:223–32

176. Youngman, R. R., Baker, T. C. 1989. Host odor mediated response of female navel orangeworm moths (Lepidoptera: Pyralidae) to black and white sticky traps. *J. Econ. Entomol.* 82:1339–43

Annu. Rev. Entomol. 1991. 36:91–117

AVERMECTINS, A NOVEL CLASS OF COMPOUNDS: Implications for Use in Arthropod Pest Control

Joan A. Lasota and Richard A. Dybas

Merck Sharp & Dohme Research Laboratories, Agricultural Research and Development, Three Bridges, New Jersey 08887

KEY WORDS: abamectin, avermectin use in crop protection, novel acaricide, miticide, insecticide class, avermectin use in IPM, ivermectin

DISCOVERY

Avermectins represent a novel class of macrocyclic lactones that have demonstrated nematicidal, acaricidal, and insecticidal activity. They are a mixture of natural products produced by a soil actinomycete, *Streptomyces avermitilis* MA-4680 (NRRL 8165), which was isolated in culture at the Kitasato Institute from a soil sample collected at Kawana Ito City, Shzuoka Prefecture, Japan. The discovery of the avermectins from this organism in 1976 has greatly influenced the arsenal of chemicals available for control of household and agricultural arthropod pests as well as parasites of mammals. The avermectins were identified in screens for natural products with antihelminthic activity for use in animal health. Discovery of the antihelminthic activity of the avermectins, first described in 1979 (17, 38, 74), can be attributed to several significant factors in the design of the screen. Stapley & Woodruff (112) present a review of the early phases of the avermectin program and note the following factors that likely contributed to the successful discovery of these natural products: (*a*) soil microorganisms from a variety of locations that possessed unique morphological characteristics were chosen for incorporation in the screens to enable investigators to discover new materials; (*b*) fermentation broths were tested for activity in vivo in a host animal, allowing the

91

simultaneous identification of efficacy and toxicity; and (c) the screens were designed to test for activity against more than one host parasite, thereby increasing the potential for finding an effective product with biological activities as well as for revealing toxicity to the host.

In the screening procedure, cultures were fermented and whole broth tests were conducted for six days via diet incorporation (0.0002% of the diet) on mice infected with *Nematospiroides dubius,* a gastrointestinal nematode. The in vivo tests demonstrated good antihelminthic activity (associated with the actinomycete mycelia) against nematodes at concentrations that did not produce adverse toxicological effects to the rodent. The isolation, characterization, and structure determination of the avermectin natural products have been described (17a, 42).

Burg et al (17) described the characteristics of the *S. avermitilis* culture and the avermectins produced by fermentation. When studied taxonomically, *S. avermitilis* was found to contain sporophores forming compact spirals that open as the culture ages. Spherical or oval spores form chains of 15 spores or more, and electron microscopy revealed a smooth spore surface. The *S. avermitilis* spore mass is brownish gray in color. In addition to these characteristics, *S. avermitilis* was identified as a new, distinctive species based on its cultural and carbon utilization patterns and its production of melanoid pigments. Miller et al (74) provide a description of the isolation and chromatographic properties of the producing culture.

STRUCTURE AND CHEMISTRY

The avermectins are a family of macrocyclic lactones that consist primarily of four major (A_1a, A_2a, B_1a, B_2a) components and four homologous minor (A_1b, A_2b, B_1b, B_2b) components. Avermectins are designated as A_1, A_2, B_1, and B_2, referring to mixtures of the homologous pairs containing at least 80% of the *a* component and no more than 20% of the *b* component. The difference between the A and B series is a methoxy group at the C_5 position of the cyclohexene moiety in the A series and an hydroxyl group in the B series. A double bond links carbons 22 and 23 (C_{22}, C_{23}) in the 1 series; this bond is reduced in the 2 series and is a hydroxyl group at C_{23}. The *a* series has a secondary butyl substitution at C_{25}, whereas the *b* series has an isopropyl group in that position (Figure 1).

Mass spectroscopy and carbon-13 NMR were used to elucidate the related avermectin structures (6). X-ray crystallography was used to further confirm the structure and to identify the relative and absolute stereochemistry (111). Avermectins characteristically possess a rigid 16-membered lactone ring system, a spiroketal that forms two six-membered rings, and a cyclohexene-diol or methoxy-cyclohexenol *cis*-fused to a five-membered cyclic ether. In

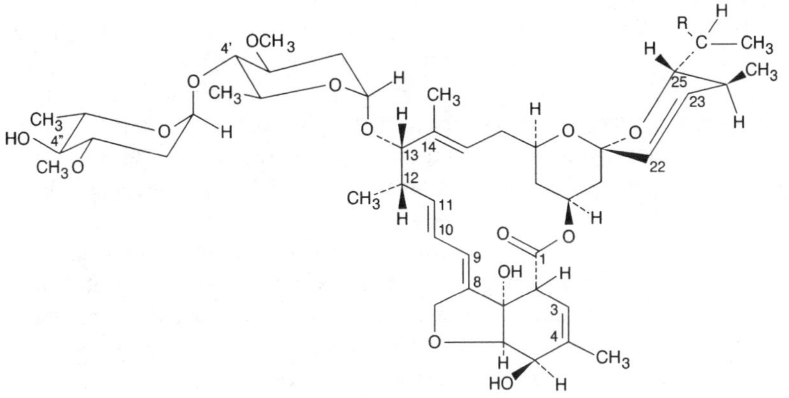

a-Component R = $C_2 H_5 \geq$ 80%
b-Component R = $CH_3 \leq$ 20%

AVERMECTIN B1
ABAMECTIN

Figure 1 Chemical structure of abamectin.

addition, the avermectin components possess a disaccharide substituent that contains two identical α-L-oleandrose units coupled to the C-13 by an oxygen bond. Dybas (30) and Fisher & Mrozik (41, 42) further review the chemistry of avermectins.

Comparative reviews of the chemistry of the avermectins and milbemycins, related members of the class of 16-membered ring macrolides, are presented in several comprehensive studies (27, 41). The milbemycins, a complex of 13 closely related fermentation products of *Streptomyces hygroscopicus aureolacrimosus,* are structurally related to the avermectins; however, the former lack the disaccharide substitute at C-13 (116).

MECHANISM OF ACTION

The differential susceptibility of invertebrates and vertebrates provides an acceptable therapeutic ratio for avermectin use in human and veterinary medicine and agriculture. The differential distribution of GABAergic (gamma-aminobutyric acid) neurons is believed to be at least partially responsible for the selective toxicity of avermectins to specific invertebrates as compared to mammals; in the latter, the GABAergic neurons are restricted to the central nervous system. Wright (129) and Turner & Schaeffer (122) reviewed the mode of action of avermectins. Although biochemical, electrophysical, and pharmacological studies on the mechanism of action of avermectins have involved the use of several of the different avermectin components, little evidence has shown qualitative differences between them (41, 124), suggesting that the members of the entire avermectin family act in a similar fashion. However, despite years of research, the exact mechanism of action of the avermectins has been difficult to specifically define, possibly because of activity at different sites, differing sensitivities to avermectins (based on target species), and poor avermectin solubility in aqueous solutions (122).

Fritz et al (43), in one of the earliest reports on avermectin's mechanism of action, used the lobster stretcher muscle as a model system whereby GABA acts as a neurotransmitter between nerve and muscle cells. In this model, avermectin increased the muscle permeability to chloride ions, consequently reducing excitatory potentials and input resistance, believed to be caused by a reduction in membrane permeability. This inhibition and subsequent reversal caused by the action of picrotoxin, a chloride ion channel blocker, and the similar actions of bicuculline, a GABA receptor antagonist, demonstrate the activity of avermectin as a chloride channel agonist. This mode of action was subsequently confirmed (5, 73).

Work by Kass et al (63, 64) using *Ascaris lumbricoides* as a model system yielded results that demonstrated avermectin's function as a GABA agonist that stimulates GABA release from presynaptic inhibitory membranes. In response to the presence of avermectin, chloride channels remain open, resulting in the simultaneous flow of chloride and sodium ions. Because of the stimulatory effect of avermectin that keeps the chloride channels open, a negative charge is maintained at the motor neuron, essentially blocking signals for inhibitory or excitatory action.

In addition to the evidence that action of avermectin is related to GABA-gated chloride ions, physiological evidence suggests an increase in membrane permeability to chloride ions, which functions as a response separate from the GABA-mediated chloride channels (110). The affinity of avermectin for retinol binding proteins has also been proposed as a possible mechanism of action for avermectin (106); however, additional research is required to further substantiate this hypothesis. Avermectins act on GABA-gated chloride channels in vertebrates. However, in invertebrates, the exact mechanism by which chloride channels are activated is unknown. Activation of chloride channels occurs at avermectin levels 10^{-3} below those that activate GABA-gated channels.

SPECTRUM OF ACTIVITY

The eight individual components found to comprise the avermectin complex demonstrate varying degrees of potency against a spectrum of nematode species (38). The B-series components have proven to be more biologically active than the A series. Compounds of the B_1 and B_2 groups vary with regard to their relative toxicities; the differences are believed to be related to target species and method of application (i.e. orally vs parenterally) (21). Although avermectins are highly active against nematodes, they demonstrate no activity against cestodes or trematodes (124). This difference is attributed to the lack of GABA-mediated nerve synapses or neuromuscular junctions in the platy-helminths. Wang & Pong (124) noted no significant antibacterial, anti-protozoal, or antifungal activity exhibited by the avermectins. Avermectins do, however, demonstrate high activity against molluscs and arthropods. Activity against the latter group is explored in greater detail below. The insecticidal activity of the avermectins was first demonstrated in laboratory assays against the confused flour beetle, *Tribolium confusum* (85). Subsequent research elucidated activity against a wide spectrum of insects and mites (60, 97). For arthropod pests of importance in crop production, B_1a was found to be more active than the B_2a or A_2a components, with the exception of the corn rootworm *(Diabrotica undecimpunctata)*, for which avermectin B_2a has the greatest activity. On the basis of its high intrinsic toxicity to arthropods compared to the other natural avermectins and numerous synthetic variants, avermectin B_1 was subsequently selected for development as a crop protection agent.

Synthetic chemical studies following the discovery of the avermectins led to the synthesis of ivermectin, a semisynthetic avermectin registered for veterinary and medical use. Ivermectin is the common name for 22,23-dihydroavermectin, a semisynthetic derivative of B_1 produced from the natural products by reduction of the bond between C_{22} and C_{23} (20, 21, 35, 42). Ivermectin is produced commercially for veterinary and medical purposes and is marketed under the trade names CARDOMEC, EQVALAN, IVOMEC, HEARTGARD 30, MECTIZAN, and ZIMECTRIN.[1]

Ivermectin has demonstrated broad spectrum activity against a variety of helminthic species (20, 21, 37, 38, 81) at a fraction of the concentrations of other antihelminthic compounds. It has demonstrated activity against genera of the following superfamilies: Trichostrongyloidea, Strongyloidea, Metastrongloidea, Rhabditoidea, Ascaridoidea, Oxyuroidea, Spiruriidea, Filaroidea, and Trichuroidea and is efficacious against both mature and immature developmental stages of roundworms and against nematodes in a wide variety of domestic animals.

Ivermectin is currently used as a human drug (MECTIZAN®) for control of *Onchocerca volvulus,* a microfilaria that is transmitted by *Simulium* black flies and causes onchocerciasis, the disease known as African river blindness.

---

[1]All tradenames in this review are registered to Merck and Co., Inc.

The most current review of ivermectin and its applications is provided by Green et al (49a).

Abamectin is the common name assigned to avermectin B_1, which has been developed as a pesticide for control of agricultural and household arthropod pests. AFFIRM, AVID, AGRIMEC, AGRI-MEK, VERTIMEC and ZEPHYR are the trade names assigned to abamectin. The spectrum of activity of abamectin and its application are discussed in subsequent sections.

ENVIRONMENTAL FATE AND TOXICITY

This section explores the environmental fate and toxicological effects of abamectin with regard to the product's use in agriculture. Wislocki et al (127) present a thorough review of the environmental considerations involving abamectin use in crop protection, including the compound's effects on non-target organisms. Abamectin is environmentally acceptable because it is used at low rates, is rapidly lost from the environment and does not bioaccumulate in biological systems.

Abamectin rapidly breaks down in the presence of ultraviolet light, degrading oxidatively and photooxidatively. In water, upon exposure to sunlight, avermectin B_1a was found to have a half-life of less than 12 hours (127). The rapid photodegradation of abamectin also occurs in organic solvents (83). Abamectin rapidly degrades to multiple products in soil exposed to sunlight, resulting in a half-life of 21 hours (127). Avermectin B_1a and B_1b degrade rapidly on thin film surfaces irrespective of the presence of light. However, the presence of light accelerates degradation, resulting in a half-life of four to six hours (69).

In soil-binding studies, results of which are used to determine the bioavailability of a material to biological systems, abamectin was found to bind tightly to soil particles, making it an immobile pesticide lacking bioavailability in the environment (127).

Soil metabolism studies on sandy loam, clay, and construction-grade soil yielded half-lives of 20 to 47 days, and abamectin degraded to a minimum of 13 products (16). Microbial metabolism under aerobic conditions yielded 8a-hydroxy avermectin B_1a as the major degradate. The bioavailability of residues to plants has been determined in a rotational crop study utilizing various soil types including those with a wide range of organic matter content (80). Following abamectin application at rates considerably higher than the maximum allowable abamectin-use rate for a commercial situation, researchers found no indication of bioconcentration of abamectin residues in the plants (80).

Results of soil dissipation field studies indicate that following multiple weekly applications, abamectin soil residues are rapidly depleted (P. C.

Tway, unpublished data). Crop residue data for celery, cotton foliage, citrus fruits, and tomato fruits also demonstrated rapid abamectin depletion, even after multiple weekly applications. Metabolism studies have further substantiated rapid degradation on these crops (16, 59, 70a, 79).

Field studies designed to assess the fate of abamectin in the aquatic environment by simulating pesticide drift and run off confirmed rapid degradation of abamectin in water and on soil (127). Wislocki et al (127) reported no accumulation or persistence of abamectin in bluegill sunfish, probably because of the large size of the abamectin molecule, which prevents it from being truly lipophilic. Rat (70) and goat (71) metabolism studies yielded similar results, indicating that abamectin does not bioconcentrate in individual animal tissues. As a consequence, abamectin is not expected to bioaccumulate in the food chain following its use as a crop protection agent.

Studies to define the effect of abamectin on nontarget organisms (127) have identified the mysid shrimp as the most sensitive invertebrate; with a 96-h LC_{50} of 0.022 ppb. Pink shrimp, daphnids, blue crabs, and Eastern oysters all displayed considerably less sensitivity to abamectin (127). The photolytic degradation products of abamectin in water and the major soil metabolite were all found to be less toxic to *Daphnia* than the abamectin parent compound (48-h $LC_{50} = 0.34$ ppb), indicating that abamectin degrades to metabolites with lower toxicity.

Results of an acute fish toxicity test demonstrated little variability (LC_{50}s of 3.2 to 42 ppb) between the five species tested, which contrasts with the high variability (four orders of magnitude difference) seen with the invertebrates (127). The results of chronic studies conducted on mysid shrimp, daphnia, and rainbow trout indicated that chronic exposure did not significantly increase the toxicity of abamectin to aquatic organisms. Abamectin is an acutely toxic compound, but its bioaccumulation is low in aquatic organisms.

Results of testing abamectin for its effects on nontarget terrestrial organisms (127) indicate no adverse affects on nitrogen-fixing bacteria or earthworms *(Eisenia foetida)* at levels expected in the environment.

Honeybees were found to be very sensitive to abamectin following direct contact, contact via their food, and through exposure to treated foliage. However, because of the rapid depletion of abamectin residues, exposure to foliage 24 to 48 hours following treatment did not adversely affect the bees.

The LD_{50} of abamectin for the mallard duck in response to technical abamectin is 84.6 mg/kg and the LC_{50} value (eight-day dietary) is 908 ppm. The bobwhite quail was found to be less sensitive to abamectin with an LD_{50} greater than 2,000 mg/kg and an LC_{50} value of 1,417 ppm. Thus, abamectin is not selectively toxic to birds.

Therefore, because of its low use rate, rapid degradation in the environment, strong binding to soil, and lack of bioaccumulation, abamectin will not be present in the environment at levels that would have an adverse impact on the environment following its use in agricultural crop production.

ARTHROPOD CONTROL AND NONCROP APPLICATIONS

Red Imported Fire Ant

Abamectin, under the tradename AFFIRM®, represents the first commercial registration for this compound. It claims to control red imported fire ant (RIFA), *Solenopsis invicta*. The ants' presence in the southern United States warrants control because of the painful sting of RIFA workers to humans and animals, direct and indirect crop damage, and damage to farm machinery resulting from extensive mound construction. AFFIRM is a bait toxicant, formulated as 0.011% w/v abamectin in soybean oil on pregelled defatted corn grits.

Abamectin has been identified as a highly potent inhibitor of red imported fire ant queen reproduction (68). Destruction or retardation of the queen's reproductive capabilities is an important element in red imported fire ant control because worker ants cannot rear replacement queens. Tests in which abamectin in soybean oil bait was fed to queen-right colonies revealed that abamectin concentrations of 0.025% or greater resulted in death or sterility of the queen (as well as in the mortality of a large population of workers and brood). Bait with 0.0025% abamectin caused queen sterility and complete brood mortality (by four weeks); however, little worker mortality resulted. Only slight reduction in brood production and no worker mortality occurred with a concentration of 0.00025%.

The histological effects of abamectin toxicity to the reproductive system of RIFA queens have been described (44) and include extreme hypertrophy of the squamous epithelium, possibly resulting from the utilization of the metabolic products for the production of larger and more numerous epithelial cells in lieu of use of these products for normal egg production. Pycnosis (abnormal clumping of chromatin in the nurse cells) is believed to interfere with the movement of nourishment to developing eggs from the nurse cells. Additional abnormalities include minimal egg production, production of abnormally small eggs, and the absence of yolk within the eggs.

In studies involving ant worker toxicity, abamectin demonstrated a rate-related delayed response. This delayed effect is critical for the success of a bait toxicant because workers must distribute the toxicant throughout the fire ant colony prior to successful worker reduction. In laboratory studies (125), 0.1% abamectin resulted in 3% mortality one day after treatment, although 75% worker mortality occurred ten days following treatment. In other laboratory tests (126), abamectin (in soybean oil) at a rate of 0.025% resulted in 71–100% mortality of ant workers 16 days after treatment. In field evaluations, 0.07% abamectin in soybean oil (1 lb of bait/acre or 0.5 g/ha) provided 87% reduction in the population index at a six-week evaluation and 85% control at 12 weeks (68, 126). During both evaluation times, at least 99% of the remaining colonies had no brood.

In individual fire ant–mound trials, 0.0055% and 0.011% abamectin (in soybean oil and impregnated on pregelled defatted corn grits) applied in three, seven, or nine tablespoons per mound reduced the occurrence of worker brood 30 days post-treatment (8). Lemke & Kissan (66) reported greater than 75% reduction in active fire ant colonies 16 weeks post-treatment following individual mound treatments with abamectin (AFFIRM W002). At a concentration of 0.011% (one, four, and seven tablespoons per mound) AFFIRM caused 55–82% mortality 12 to 21 weeks following individual mound treatment. Although most rapid kill was observed with the seven-tablespoon rate, by 21 weeks post-treatment, the four- and seven-tablespoon rates provided equivalent control (50).

Thus, abamectin has demonstrated high activity against RIFA and has potential for use in baits as an ingestion toxicant, resulting in sterilization of queens and gradual destruction of worker populations in mounds, which are not readily recolonized after treatment.

Cockroaches

Cochran (24) published one of the few studies on the physiological effects of abamectin on cockroaches. The acute and reproductive effects of this compound on the German cockroach, *Blattella germanica,* was evaluated via the oral route to determine the potential of this material for use in bait formulations. These studies were conducted using various cockroach life stages, including male and female nymphs and newly emerged adult females. The insects were fed pulverized dog food treated with various abamectin concentrations (0.03–30.00 ppm).

Following 10 days of abamectin feeding by adult females, greater than 85% mortality was observed with the 6.5 ppm concentration, and, 20 days following exposure, 60% mortality was seen at rates as low as 3.0 ppm. Thirty days continual exposure resulted in 40% mortality at 0.3 ppm.

In terms of reproductive effects, only half of the adult females treated with 3.0-ppm abamectin mated. Oothecal egg hatch was greatly reduced in females ingesting 0.3 ppm or higher abamectin concentrations. In surviving females that failed to reproduce, ovarial development was stunted, often resulting in misshapen eggs. Individual oocyte length measurements demonstrated a dose-dependent retardation of oocyte development. Feeding inhibition, believed to result from intoxication (versus unpalatability or repellency), was observed at concentrations of 6.5 ppm and higher.

The effects of abamectin ingestion on the F_1 progeny of the surviving females was also examined. Abamectin was found to affect the rate of progeny development. Longer oothecal maturation and oothecal carrying periods were observed in progeny of females fed 0.3- and 3.0-ppm concentrations of abamectin.

Large nymphal male German cockroaches were used to evaluate the effects of abamectin on the male reproductive system. Feeding inhibition was seen with the 0.3-ppm and greater abamectin concentrations. Subsequently enhanced mortality was observed with all treatments (0.03–30.0 ppm); however, seven-day mortality was only observed in the case of the 30.0-ppm treatment. When male abamectin survivors were mated with untreated females, normal sized oothecae, which yielded normal percent egg hatch, were produced.

Small nymphs (instars 1–2) were found to be the stage most susceptible to abamectin; high mortality was observed at 3.0-, 10.0-, and 30.0-ppm concentrations after 13 days. Feeding inhibition was a symptom of abamectin toxicity with all abamectin rates (0.03 to 30.0 ppm). Medium-sized nymphs (instars 3–4) were also killed by the abamectin concentrations tested, although the effects were less dramatic. Abamectin did not affect the subsequent reproduction of these nymphs. A study on mature nymphs given feeding choices confirmed the palatability of abamectin and its lack of repellency to German cockroaches.

Abamectin's potential as an effective treatment for cockroach control has been demonstrated in indoor urban situations (130). Abamectin at 0.02% (200 ppm) and 0.1% (1000 ppm) concentrations (applied via a self-pressurized crack and crevice spray device) was tested in single-family dwellings exhibiting high German cockroach populations. The 0.02% concentration provided a minimum of 85% reduction of cockroaches throughout the eight-week duration of the study, and the 0.1% rate gave 96% control after two weeks and 94% control four and eight weeks after treatment.

Thus, abamectin successfully controls German cockroaches in laboratory and field (urban) situations. It affects nymphal and adult mortality and has adverse reproductive effects at sublethal concentrations. Since the 1985 reports, abamectin has been critically tested in various bait station models and against an array of cockroach species in many locations (J. M. Gillespie, unpublished data) and has been shown to be highly effective for cockroach control. The compound's registration as a bait station is expected in 1990.

AGRICULTURAL AND HORTICULTURAL APPLICATIONS

Mites

The most widespread worldwide commercial use of abamectin has been as an acaricide. Abamectin has repeatedly demonstrated excellent initial and residual control of immature and adult motile mites on a variety of ornamentals, food crops, and cotton, performing in this capacity at concentrations magnitudes lower than conventional acaricides (31, 33, 35, 97). Abamectin has demonstrated broad spectrum activity against mites in the families Tetranychidae, Eriophyidae, and Tarsonemidae, with LC_{90} values in the range of 0.02–0.24 ppm.

Although few studies have been conducted on eriophyid gall mites, abamectin at 0.025 and 0.063 ppm was shown (under laboratory conditions) to penetrate galls formed on the plant stem by *Eriophyes discoridis* and cause 97% and 84% mite mortality, respectively (39). In leaf dip bioassays, the LC_{90} value of abamectin was found to be 0.24 ppm (compared with 35 ppm for fenvalerate), and in a leaf residual bioassay, 1 ppm demonstrated high activity for ten days. In contrast to these effects of abamectin on the gall mite population, the LC_{90} for the predaceous mite *(Phytoseius finitimus)* population was 100 ppm (17 ppm for fenvalerate). This selective toxicity suggests abamectin's potential for use in integrated management systems.

The efficacy of abamectin against the eriophyid citrus rust mite, *Phyllocoptruta oleivora,* an economically important pest on citrus in the United States, Brazil, and other citrus-growing areas, has been documented (12, 35, 72). Feeding damage by this pest results in fruit russetting and reduction in fruit size, juice quality, and internal solids. A LC_{90} value of 0.02 ppm has been established for adult mites (72). Maximum kill was achieved 48 to 72 hours following treatment. Abamectin did not exhibit ovicidal activity even at 25 ppm, the highest rate tested.

Although abamectin reduced initial *P. oleivora* populations below economic threshold levels in preliminary field trials with application rates of 13.5–54.0 g ai/ha, lengthy residual activity did not occur. The use of emulsified paraffinic oil at a rate of 0.25% or one gallon per acre (9.35 L/ha) (nonacaricidal concentrations) in combination with abamectin significantly improved the residual control of citrus rust mite populations, demonstrating up to 16 weeks of activity. Abamectin 0.15 emulsifiable concentrate (EC) in combination with 0.25% paraffinic crop oil is recommended for control of citrus rust mite on orange, grapefruit, lemon, and tangerine without phytotoxic effects. Additional laboratory and field studies have confirmed the enhanced contact and residual activity of abamectin when used in combination with oil (31) and additional successes with this combination are to be expected.

Although paraffinic oil is recognized as a vehicle of transport for abamectin movement into foliar tissue, resulting in enhanced abamectin penetrability, it must be recognized that cuticular waxiness (as a characteristic of host species) and leaf toughness (as a characteristic of foliage age) may also affect the ability and rate of abamectin penetration. Light (wavelength or intensity) and surfactants have been investigated as factors that may influence abamectin's efficacy in the field (75). Quantitative differences in abamectin's foliar penetration and initial and residual mite activity on lima bean, peach, and azalea were seen under similar light regimes, these differences possibly resulting from a combination of host species and leaf age.

Differences in translaminar acaricidal activity between French bean, cotton, and chrysanthemum were attributed to differences in the leaves' cuticular waxes (chrysanthemum is the most waxy and French bean the least) (131). The addition of oil to the abamectin formulation enhanced the compound's translaminar activity, resulting in reduced LC_{50} values for *Tetranychus urticae* and *Aphis fabae* for the abamectin plus oil combination. However, LC_{50} values for *T. urticae* were 30.1 and 33.4 ppm (dorsal or ventral leaf surfaces), whereas 50% mortality of aphid populations was only achieved with concentrations of 450 ppm or greater. Although differences in intrinsic toxicity of abamectin to various arthropod species (97) is probably the most important factor contributing to differential abamectin toxicity, greater susceptibility of mites versus aphids may result from the mites' feeding on leaf tissue (versus selective phloem feeding by the aphid), allowing contact with more toxicant.

Contrary to abamectin's high toxicity ($LC_{90} = 0.02$ ppm) and long residual activity against citrus rust mite, lower intrinsic toxicity ($LC_{90} = 0.24$ ppm) (35) and inconsistent field residual activity (two to six weeks) were found for citrus red mite, *Panonychus citri* (33).

In laboratory studies, abamectin was reportedly highly toxic to the tomato russet mite, *Aculops lycopersici*, with an LC_{90} value of 0.0096 ppm against adults (105). Abamectin was found to be much less toxic ($LC_{90} = 12.58$ ppm) to *Homeopronematus anconai*, a tydeid predatory mite, which supports the integration of abamectin into pest management systems without forgoing predator safety.

Twenty-one days of residual control of the tarsonemid broad mite, *Polyphagotarsonemus latus*, on ornamentals was achieved with abamectin at 1.8% w/v used at a rate of 25 ml/100 liters of water (53). In laboratory trials, abamectin was shown to be highly toxic ($LC_{90} = 0.05$ ppm) to adult *P. latus* on citrus fruit (35). Although not ovicidal, abamectin at a rate of 0.1 ppm provided 100% control of eclosing immatures. In field trials, a combination of abamectin (14 to 28 g ai/acre) and 0.25% emulsified paraffinic oil provided greater than 90% initial mite control and residual fruit protection for a minimum of four weeks (R. D. Brown, unpublished data).

The activity of abamectin against tetranychid spider mites has been extensively investigated because of the economic importance of this mite family on horticultural and agricultural crops (22, 26, 33, 98). A high intrinsic contact activity ($LC_{90} = 0.02$ to 0.15 ppm) has been reported for spider mites (35, 97). All motile mite stages are highly sensitive to abamectin, although no true ovicidal activity has been seen. Abamectin's ability to penetrate the egg chorion just prior to hatching has somewhat reduced egg hatch. This reduction is evidenced by paralysis of newly eclosed immatures and inhibition of eclosion (47). El-Banhawy & Anderson (38a) further demonstrated the influence of egg age and temperature on *T. urticae* egg susceptibility to abamectin. Abamectin is a highly successful commercial acaricide for control of spider mites on ornamentals, displaying long residual activity under greenhouse conditions (47, 49, 86).

Abamectin's long residual activity against spider mites resulting from its foliar uptake has been documented (35, 132). Although cotton foliar surface residues were shown to decompose by 82 and 98% (16) after two and eight days, rapid penetration of abamectin into cotton foliage within several hours of application was shown. By eight days, 1.3% of tritiated abamectin had been absorbed into the cotton foliage. In field-treated laboratory bioassays, the LT_{50} for abamectin was 21 to 28 days when applied against *T. urticae*. Control for 14 days and 35% mortality was seen for seven days in outdoor conditions (132). At a concentration of 30 ppm, abamectin provided 100% control of adult female mites for 28 days under greenhouse and outdoor conditions.

Excellent spider mite control on cotton under field conditions has been documented (33). Abamectin applied at 22.4 g ai/ha provided 92 to 96% residual mite control for 46 days. A similar study reported 90 and 96% initial mite control 13 and 18 days postapplication, and 85 to 95% residual control at 41 and 49 days post-treatment when abamectin was applied at rates of 9 to 10 g ai/ha. At 16.8 g ai/ha abamectin provided 100% residual control after 21 days. In a comparative field trial of abamectin at 11.2 and 22.4 g ai/ha with and without paraffinic oil, abamectin alone provided 92 and 98% control for 3 to 14 days, whereas abamectin at 11.2 g ai/ha in combination with oil provided 98% mite control three days post-treatment. However, after 14 days, only 69% control was observed. Additional studies have demonstrated 80% or greater mite control by 7 to 10 days after treatment with abamectin applied at the recommended rates of 11.2 to 22.4 g ai/ha and greater than 90% control by 14 days post-treatment. Abamectin's lack of dramatic initial mite reduction seven days after application results from its lack of ovicidal activity and the need for newly hatched immatures to accumulate lethal dosages of the compound, a process usually requiring an additional three or four days. Effective tetranychid mite control on commercially grown cotton with 11.2 g ai/ha of abamectin has resulted in reduced crop damage, which ultimately translates to increased yields (29).

Field trials have demonstrated at least 80% *T. urticae* control on tomatoes with abamectin applied at 11.2 to 22.4 g ai/ha (33). Abamectin applied at 40 g ai/ha successfully controlled the twospotted spider mite on canteloupe for four weeks (92).

Abamectin applied to pears at a rate of 14 to 28 g ai/ha in combination with

0.25% parraffinic oil reduced twospotted spider mite, European red mite *(Panonychus ulmi),* and pear rust mite *(Epitrimerus pyri)* populations by at least 85% seven days after application. Residual activity, sufficient to suppress populations below threshold levels, was maintained for a minimum of four weeks (33). On strawberry, abamectin has been shown to successfully control motile mites when two back-to-back applications of abamectin 11.2 or 22.4 g ai/ha are timed seven days apart. An average of 68 and 79% control over a 35-day time period was seen with this spray schedule versus 52% control from a single application of 11.2 g ai/ha for the same evaluation periods (D. M. Dunbar, unpublished data). Thus, two properly timed abamectin applications successfully disrupt the *T. urticae* life cycle on strawberry. Abamectin (7.4 g ai/400 liters of water) (0.00625 lb ai/100 gallons) applied to almonds provided 85% initial control of *Tetranychus pacificus* six days following treatment; 48-day residual activity was demonstrated with a combination of abamectin and narrow-range spray oil (33). The predatory mite, *Metaseiulus occidentalis,* and six-spotted thrips, *Scolothrips sexmaculatus,* were not adversely affected by the combination.

Leafminers

The reservoir of abamectin that remains within the mesophyll layer of treated leaves makes the chemical easily accessible for ingestion by mining larvae. Numerous cases of successful leafminer control on ornamentals and vegetables have been documented (33, 47, 113).

Laboratory bioassays (leaf dip technique) to evaluate abamectin toxicity to *Liriomyza trifolii* were conducted using leafminer-infested cowpea leaves (65). The LC_{50} values for larvae and pupae were found to be 0.377 ppm and 0.202 ppm, respectively, and the LC_{90} values were 1.473 ppm and 0.724 ppm. Abamectin was shown to cause an increase in aberrant puparial forms. Although abamectin is not ovicidal, its translaminar activity results in control of hatching larvae, which ultimately prevents mine formation (47). Topical assays to assess toxicity to adult females resulted in an LD_{50} of 0.404 ppm (89). An LC_{50} value of 0.386 ppm was reported from contact assays using third-instar larvae. In laboratory choice tests, adult female *L. trifolii* demonstrated feeding and oviposition repellency to abamectin-treated chrysanthemums for up to seven days post-treatment. Laboratory studies to evaluate leafminer toxicity to a chemical with a unique mode of action, such as abamectin, are important steps in recognizing the need for establishing baseline toxicity data, a valuable tool for use in resistance-monitoring programs.

Additional laboratory studies on the effects of abamectin residues on adult *L. trifolii* found female mortality to be induced by ingestion and contact toxicity, whereas male mortality was exclusively related to contact toxicity (108). Although female repellency was not seen, feeding, leaf-puncturing (as evidenced by reduced stippling), and oviposition were reduced.

In greenhouse studies involving *L. trifolii* control on ornamentals, weekly abamectin applications, particularly during continuous fly infestations, resulted in excellent leafminer control on chrysanthemum, gerbera, and gypsophila (19, 48, 52, 87, 96, 123), although Robb & Parrella (102) found no differences in control between 7- and 14-day applications. In terms of phytotoxicity, Price (95) reported no noticeable foliage or flower damage to 35 collective cultivars of chrysanthemum, gerbera, or snapdragon. Parrella (88) also reported no abamectin phytotoxicity on chrysanthemum.

In field trials to evaluate abamectin efficacy against *L. trifolii* leafminers that demonstrated 30-fold permethrin resistance, abamectin effectively terminated mine development and produced marketable chrysanthemums. Thus, cross-resistance with permethrin was not seen. Abamectin was shown to not adversely affect the *L. trifolii* parasitoids, *Diglyphus intermedius* or *Ganespidium hunteri*, on chrysanthemums (52), supporting the use of this compound in integrated pest management (IPM) programs (62).

Abamectin's performance against agromyzid leafminers on vegetable crops has been comparable to that seen on ornamentals (33). On field-sprayed tomatoes, abamectin efficacy was evidenced by reduced leaf stippling and oviposition and low adult mortality; however, higher (80%) adult mortality occurred 24 h following exposure to abamectin-dipped foliage (107). Larval mortality occurred soon after hatch or just prior to complete eclosion, thus preventing mine formation. In trials on field-grown peppers, abamectin was shown to be efficacious against *L. trifolii* without detrimentally affecting the hymenopterous leafminer parasites, *Chrysonotomyia* spp., *Chrysocharis* spp., or *Discorygma* spp. (22a).

IPM programs for celery in which abamectin is a key component of control recommendations have been proposed (119, 120). Although abamectin was shown to effectively control *L. trifolii* populations, it did not significantly affect the species composition of the leafminer parasite complex, including species in the eulophid and pteromalid families. In addition, the seasonal percent parasitism, adult parasite mortality, and immature parasite emergence and survival from treated foliage were not adversely affected (119).

Psyllids

Abamectin rates in the range of 14 to 28 g ai/ha applied in postbloom sprays significantly reduced pear psylla, *Psylla pyricola*, nymphal populations (33). In combination with 0.25% paraffinic oil or a minimum of one gallon per acre (9.35 liters/ha), increased residual activity is seen for up to four weeks. Abamectin control of psylla populations and decreased fruit damage by the sooty mold fungus that develops on psylla honeydew were achieved (18). Since dormant abamectin sprays at 11.2 g ai/ha failed to reduce psylla populations below the economic threshold level of 0.5 nymphs per leaf, abamectin is only recommended after petal fall for control of psylla nymphs.

Lepidoptera

Abamectin was shown to have a wide range of activity against lepidopteran pest species (31, 33, 97, 113, 128). LC_{90} values derived from foliar bioassays using neonate larvae ranged from 0.02 ppm for tobacco hornworm *(Manduca sexta)* to 1.5 ppm for corn earworm *(Heliothis zea)* and 6.0 ppm for southern armyworm *(Spodoptera eridania)*. A 400-fold difference in LC_{50} values was seen between *Heliothis virescens* and *S. eridania* (7). Bull (15) also reported significant differences in abamectin susceptibility between *H. virescens, H. zea,* and *S. eridania,* and Christie & Wright (23) showed differences in abamectin toxicity to various larval instars of *Spodoptera littoralis* and *Heliothis armigera*. A greater abamectin toxicity in ingestion versus residual contact bioassays has been indicated (7, 15, 25, 133).

In diet-incorporation bioassays against neonate *Spodoptera exigua,* the LC_{50} value (ng ai/ml diet) for abamectin was 755.4 (121). Although abamectin residues were shown to have immediate antifeedant effects, this action was no longer evident seven days after treatment. Moar & Trumble (76) reported an LD_{50} of 0.814 μg ai/ml for *S. exigua* in a diet-incorporation bioassay.

At field rates of 0.071 and 0.71 g/ha applied to spruce budworm *(Choristoneura occidentalis)* eggs, Robertson et al (103) reported inhibited larval development. Following treatment of LC_{50} concentrations to sixth-instar larvae, the egg fertility of mated adult survivors was significantly reduced. Infertility possibly resulted from sterility rather than inhibition of hatching. For sixth-instar gypsy moth *(Lymantria dispar)* larvae, the LD_{50} value for abamectin (24 h post-application) was found to be 0.124 μg/g and 0.024 μg/g for third-instar larvae (28). Although abamectin at 1.0 ppm affected egg hatch little, this rate resulted in death of newly emerging larvae. Differences in abamectin susceptibility to various larval instars was also found.

In topical bioassays involving male and female pink bollworm *(Pectinophora gossypiella)* adults, abamectin demonstrated LC_{50} values in the range of 0.032 to 0.044 μg/insect (9, 10). Mating was prevented for up to 72 h with the LD_{10} or LD_{50} dose of abamectin and oviposition was prevented with the LD_{10} dose. Egg hatch was also severely retarded following adult treatments (9). Further effects of abamectin on adult pink bollworms including neurobiological effects have been reported (3, 4) and have been reviewed extensively by Strong & Brown (113). Because of abamectin's rapid photolytic degradation and its relatively weak contact toxicity against lepidopteran larvae, particularly noctuids, the value of this compound on cotton is limited (33).

Tomato pinworm *(Keiferia lycopersicella)* neonate larvae were found to be highly susceptible to abamectin in laboratory foliar residual bioassays, with an LC_{95} of 0.031 ppm (D. J. Schuster & J. L. Taylor, unpublished data); however, less susceptibility was seen in internal-foliage feeding (LC_{95} = 15.1 ppm). For commercial field applications, abamectin is recommended at rates of 11.2 to 22.4 g ai/ha for control of this pest. Trumble (118) reported 93–96% reduction in tomato fruit damage with abamectin applied weekly at these rates. However, in another study, only 70–75% less damage occurred after five weekly applications.

In studies to evaluate the toxicity of abamectin to the soybean looper, *Pseudoplusia includens*, Beach & Todd (11) found topical LD_{50} values at 48 h to range from 0.024 to 0.203 μg/larva. Larvae reared on soybean foliage were found to be more susceptible than those reared on diet. Adult moths fed on an abamectin/sucrose solution showed reduced survival, fertility and fecundity. The adult LC_{50} at 72 h was found to be 6.1 mg/L.

Laboratory evaluations of abamectin toxicity to the lesser peachtree borer, *Synanthedon pictipes*, and the peachtree borer, *Synanthedon exitiosa*, were conducted for eggs, larvae, and adults (C. E. Yonce, J. W. Snow, J. L. Taylor, unpublished data). Except in one test, borer eggs were not affected by rates of 10 or 20 ppm. No acute toxicity was seen with treated larvae at rates up to 20 ppm, although pupation was retarded. This rate reduced oviposition when adults were treated prior to mating. Egg viability was reduced following treatment of females prior to mating, of gravid females, and of males that were subsequently mated with untreated females.

The effects of abamectin on codling moth *(Cydia pomonella)* eggs demonstrated toxicity only at concentrations greater than 100 ppm (101). Toxicity to seven-day-old larvae following topical application occurred at concentrations greater than approximately 5 ppm. In diet-incorporation bioassays, 100% mortality of neonate and ten-day-old larvae occurred with 0.05 and 0.32 ppm, respectively. Although diet assays with 0.025 ppm were not acutely toxic, such treatment of neonate and ten-day-old larvae resulted in reduced egg production. Adults of the European grape berry moth *(Lobesia botrana)* were found to be less susceptible than larvae to abamectin-treated grape bunches (58). Larvae subjected to grape bunches that had been dipped in a solution of 10^{-3}% ai abamectin incurred 98% mortality. Fifty-six percent mortality was evidenced with 8×10^{-6}% ai. In field trials on tobacco, differential abamectin susceptibility was seen between tobacco hornworm, controlled at 5.6 g ai/ha, and tobacco budworm, requiring 24.4 g ai/ha for control (61).

The diamondback moth, *Plutella xylostella*, is highly susceptible to abamectin, making this compound ideal for control of this pest, particularly where resistance has developed to commercially used insecticides (67, 114, 115) and where resistance management strategies must be employed. Abamectin's translaminar and surface residue activity against diamondback moth has been reported (2). The leaf tissue activity (reported as LC_{50} at 96 h) against second-instar larvae ranged from 0.011 μg/ml on Chinese cabbage to 2.79 μg/ml on cabbage; for brussels sprouts it was 1.59 μg/ml (0.99 in combination with safflower oil at 2500 μg/ml). These differences in susceptibility might be a function of differences in plant cuticular waxes. There may also be good translaminar activity against first instars that feed selectively on specific foliar tissue layers in a fashion similar to leafminers. Residual bioassays indicated no cross-resistance with known resistance to malathion, DDT, and cypermethrin (1). In studies to determine the sublethal effects of abamectin to *P. xylostella*, use of 1.8 ng/ml reduced weight gain of fourth-instar larvae. Second-instar larvae feeding on plants treated at this low rate had a 50% reduction (compared to control) in oviposition when adults.

Miscellaneous Insects

Although many species of insect pests do not represent major targets for commercial abamectin use, mention of pertinent publications that deal with the effects of abamectin is warranted. The lethal and sublethal effects of abamectin on tephritid fruitflies have been investigated (5). Additionally, abamectin toxicity to thrips (51, 78, 84), diaspid scale insects (91, 93), aphids (117), chrysomelid and curculionid beetles (94, 100), and vespids (90) have been reported.

In tests of abamectin as a fabric protectant (13, 14), 100%-wool test cloth, treated with acetone solutions of abamectin (0.00005–0.10% by weight of the cloth) was used to evaluate the effects of this compound on larvae of the black carpet beetle *(Attagenus unicolor)*, furniture carpet beetle *(Anthrenus flavipes)*, and the webbing clothes moth *(Tineola bisselliella)*. These limited studies suggest a niche for abamectin as a fabric protectant.

EFFECTS ON BENEFICIAL ARTHROPODS

The effects of abamectin on beneficial arthropods has been intensively investigated because of the importance in many agricultural ecosystems of naturally occurring and introduced beneficial populations and the potential for integration of these organisms into pest management systems. Selective acaricides that demonstrate effective pest control, do not adversely affect beneficials, and allow population expansion of beneficials are deemed essential in mite-management practices. Survival of beneficial mites may be further enhanced by the preservation of refuges and development of resistant predatory strains (46). Although an abamectin reservoir within the mesophyll layer of leaf tissue is accessible to phytophagous mites and leafminers, parasitic and predatory arthropods continue to proliferate because of the short-lived surface residues.

Differential toxicities of abamectin to the phytoseid predatory mite, *M. occidentalis,* and to two phytophagous spider mites, *T. urticae* and *P. ulmi,* have been reported (46). Predatory mite larvae exposed to abamectin concentrations ranging from 0.001 ppm to 0.1 ppm matured and produced viable eggs; however, activity, survival, and developmental rates of the progeny were reduced when larvae were exposed to 0.1 ppm.

In laboratory studies on the differential toxicity of abamectin to *T. urticae* and *M. occidentalis* on bean and almond foliage (55), no differences were seen with abamectin at rates of 0.025, 0.05, 1.0, and 5.0 ppm alone or in combination with oil in either species. However, predatory females produced more progeny (although larval development was not different) on foliage treated with abamectin plus oil. Field-treated and aged almond foliage did not affect *M. occidentalis* 96 h post-treatment, although after the same time period the foliage was extremely toxic to *T. urticae* adults. This suggests that a reservoir of abamectin within the mesophyll leaf layer was accessible to *T. urticae* but had sufficiently dissipated from the foliage surface so as not to be harmful to the predator. On field-treated bean foliage, abamectin plus oil (3.0 ppm) provided 33 days residual activity against *T. urticae,* whereas at 48 to 96 h post-treatment, *M. occidentalis* could survive. Following immediate short exposures to abamectin- (3 ppm) treated foliage, *M. occidentalis* exposed for 300 s exhibited increased mortality. Reduced fecundity was seen in females exposed for 30 s. This study demonstrates the need for proper abamectin-application timing to encourage survival of a residual predator population on treated foliage.

A more sophisticated mechanism for enhancing the survival of predaceous mites involves genetic improvement of the population through selection for resistance to abamectin. Hoy & Ouyang (57) selected a heterogeneous *M. occidentalis* population 20 times and achieved a 3.8-fold increase in adult female survival and an increase in egg production.

In laboratory bioassays on the effects of field-applied foliar residues on three beneficial arthropods *(Aphytis melinus,* an aphelinid parasitoid of California red scale; *Cryptoloemus montrouzieri,* a coccinellid mealybug predator; and *Euseuis stipulatus,* a phytoseid predator of mites and citrus thrips) recommended rates of abamectin did not cause significant residual mortality to *A. melinus* or *C. montrouzieri* (77). Since abamectin residues dissipated to insignificant levels six days post-treatment, *E. stipulatus* populations were not adversely affected, supporting abamectin's use in IPM systems.

Abamectin at a concentration of 1.0 ppm showed slight contact toxicity to immature *Amblyseius gossipi* predatory mites. Toxicity increased when the immatures were fed abamectin-treated prey. Resultant adult females oviposited at significantly reduced rates (99). Ovicidal effects of abamectin on *T.*

urticae and *A. gossipi* have also been investigated (40). The secondary effects of abamectin on *Phidippus audax* spiders, following ingestion of abamectin-treated *H. zea* larvae, include reduced weight gain within 48 h. Eighty percent of the spiders that preyed upon larvae subjected to diet treated with 10 ppm abamectin were torpid or dead. Percent mortality increased to 100% when abamectin was increased to 40 ppm.

RESISTANCE

Pesticide resistance has become a commercially significant problem, especially to acaricides in which short mite-generation times and an enormous reproductive potential contribute to rapid genetic selection. To assess the potential for resistance development in mites, laboratory and greenhouse selection studies were conducted using *T. urticae* and *T. pacificus* populations collected from greenhouse, strawberry, and almond colonies (56). After 4, 6, or 15 selections, LC_{50} values for *T. urticae* populations were 0.12 to 0.24 ppm—not significantly different from the LC_{50} values (0.05 to 0.15 ppm) prior to selection (56). Similarly, no response to selection was achieved with the *T. pacificus* colonies that were selected 2 and 15 times. Unselected and selected *T. pacificus* colonies had LC_{50} values of 0.04 and 0.12 ppm and 0.06 and 0.08 ppm, respectively. Although the magnitude of selection pressure that would yield resistant colonies is uncertain, the frequency of resistance alleles was shown to be low in selected mite populations. Additionally, because one or more of the test colonies in this study demonstrated resistance to propargite, cyhexatin, or fenbutatin oxide, this study has established that cross-resistance of abamectin to these compounds is unlikely.

Laboratory bioassays were conducted on *T. urticae* colonies collected from commercial cotton plantings. Although the mites exhibited varying degrees of resistance to dicofol and propargite, standard cotton acaricides (54), abamectin was highly toxic to all field isolates; LC_{50} values were in the range of 0.0017–0.0049 ppm for contact toxicity and 0.014–0.077 ppm for residual toxicity. These results confirm a high degree of susceptibility of this mite to abamectin and a lack of cross-resistance to industry standards.

Roush & Wright (104) report a lack of abamectin cross-resistance in houseflies *(Musca domestica)* to various insecticides (diazinon, dieldrin, DDT, permethrin) of known resistance. However, conflicting results (109) indicate polygenic cross-resistance (5.9- and 25-fold) to abamectin in two pyrethroid-resistant housefly strains. Results of field successes in which abamectin was applied to populations with demonstrated resistance to the standard compounds and of laboratory resistance studies indicate a niche for abamectin in agroecosystems where resistance to the standard insecticides or acaricides has occurred or where resistance could occur.

4''-EPI-METHYLAMINO-4'' DEOXYAVERMECTIN B₁

In 1988, a derivative of abamectin, 4''-epi-methylamino-4'' deoxyavermectin B$_1$ (EMA), a second-generation avermectin insecticide, was approved for field development by Merck Sharp & Dohme Research Laboratories (32). EMA was obtained by introducing an amino substituent at the 4'' position in the terminal dissaccharide unit of abamectin (82). This significant breakthrough in avermectin chemistry helped broaden the activity spectrum of this class of compounds by identifying an extremely potent compound with high toxicity against lepidopteran species at very low concentrations. Whereas abamectin demonstrates wide differential toxicity to lepidopterans, EMA has exhibited high intrinsic activity against a wide range of species in this order (34).

In laboratory foliar ingestion bioassays, EMA had LC$_{90}$ values in the range of 0.002 to 0.014 ppm against neonate lepidopteran larvae (34). In contrast to abamectin, which demonstrated considerably less intrinsic toxicity against economically important species such as *Spodoptera frugiperda, S. exigua, H. zea, H. virescens,* and *Trichoplusia ni,* the LC$_{90}$ values for EMA were 0.01, 0.005, 0.002, 0.003, and 0.014 ppm, respectively (33, 34). This represents a 1,166-fold increase in potency over abamectin in the case of *S. eridania* and about a 1,500-fold increase in the case of *S. exigua* (121). Although EMA is less toxic to *T. urticae* (LC$_{90}$ = 0.29 ppm) than abamectin (LC$_{90}$ = 0.03 ppm), EMA was also found to be toxic to the Colorado potato beetle, *Leptinotarsa decemlineata* (LC$_{90}$ = 0.032 ppm) and Mexican bean beetle, *Epilachna varivestis* (LC$_{90}$ = 0.20 ppm). As is abamectin, EMA is considerably less toxic to *A. fabae,* the bean aphid (LC$_{90}$ = 19.9 ppm).

EMA is significantly more toxic than the commercially used carbamate or pyrethroid standards. For example, EMA is 1,720, 884, and 268 times more toxic to *S. eridania* on a ppm basis than the commercial standards methomyl, thiodicarb, and fenvalerate, respectively (36). Against *H. virescens,* EMA was found to be 3,300, 1,666, and 500 times more potent than methomyl, thiodicarb, and fenvalerate (36).

Preliminary field trials have further substantiated the excellent performance of EMA against economically damaging pests on a variety of crops including cole crops, leafy vegetables, and sweet corn (34). At field-applied rates in the range 5.6–22.4 g ai/ha, an 18 g ai/L emulsifiable concentrate formulation of EMA (MK-243) resulted in reduction in pest populations and increase in marketable yield. Although all rates provided acceptable control and reduced yield reduction, MK-243 applied at rates of 11.2 to 22.4 g ai/ha provided the most consistent results.

ACKNOWLEDGEMENTS

The authors thank Connie Graf for her dedicated clerical assistance during the preparation of this manuscript.

Literature Cited

1. Abro, G. H., Dybas, R. A., Green, A. S. J., Wright, D. J. 1988. Toxicity of avermectin B1 against a susceptible laboratory strain and an insecticide-resistant strain of *Plutella xylostella* (Lepidoptera: Plutellidae). *J. Econ. Entomol.* 81:1575–80

2. Abro, G. H., Dybas, R. A., Green, A. S. J., Wright, D. J. 1989. Translaminar and residual activity of avermectin B1 against *Plutella xylostella* (Lepidoptera: Plutellidae). *J. Econ. Entomol.* 82:385–88

3. Agee, H. R. 1985. Neurobiology of the bollworm moth: Effects of abamectin on neural activity and on sensory systems. *J. Agric. Entomol.* 2:325–36

4. Agee, H. R. 1985. Neurobiology of the bollworm moth: Response of neurons in the central nervous system to abamectin. *J. Agric. Entomol.* 2:337–44

5. Albrecht, C. P., Sherman, M. 1987. Lethal and sublethal effects of avermectin B1 on three fruit fly species (Diptera: Tephritidae). *J. Econ. Entomol.* 80:344–47

6. Albers-Schonberg, G., Arison, B. H., Chabala, J. C., Douglas, A. W., Eskola, P., et al. 1981. Avermectins. Structure determination. *J. Am. Chem. Soc.* 103:4216–21

7. Anderson, T. E., Babu, J. R., Dybas, R. A., Mehta, H. 1986. Avermectin B1: Ingestion and contact toxicity against *Spodoptera eridania* and *Heliothis virescens* (Lepidoptera: Noctuidae) and potentiation by oil and piperonyl butoxide. *J. Econ. Entomol.* 79:197–201

8. Apperson, C. S., Powell, E. E., Brown, M. 1985. Efficacy of individual mound treatments of MK-936 and Amdro against the red imported fire ant (Hymenoptera: Formicidae). *J. Ga. Entomol. Soc.* 19:508–16

9. Bariola, L. A. 1984. Pink bollworm, *Pectinophora gossypiella*, (Lepidoptera: Gelechiidae) effects of low concentrations of selected insecticides on mating fecundity in the laboratory. *J. Econ. Entomol.* 77:1278–82

10. Bariola, L. A., Lingren, P. D. 1984. Comparative toxicities of selected insecticides against pink bollworm (Lepidoptera: Gelechiidae) moths. *J. Econ. Entomol.* 77:207–10

11. Beach, R. M., Todd, J. W. 1985. Toxicity of avermectin to larval and adult soybean looper (Lepidoptera: Noctuidae) and influence on larval feeding and adult fertility and fecundity. *J. Econ. Entomol.* 78:1125–28

12. Bianchi, R. A., Undurraga, J. M., Dybas, R. A. 1984. Effect of avermectin B1 (MK-936) in the control of citrus rust mites at Santa Gertrudis, S. P. *Proc. IX Brazilian Entomol. Cong. 1984.* Londrina, Parana: Entomol. Soc. Brazil

13. Bry, R. E. 1990. Avermectin B1a as a long-term protectant of woolen fabric. *J. Entomol. Sci.* In press

14. Bry, R. E., Lang, J. H. 1984. Avermectin B1a: Effectiveness against three species of fabric insects. *J. Ga. Entomol. Soc.* 19:523–32

15. Bull, D. L. 1986. Toxicity and pharmacodynamics of avermectin in the tobacco budworm, corn earworm, and fall armyworm (Noctuidae: Lepidoptera). *J. Agric. Food Chem.* 34:74–78

16. Bull, D. L., Ivie, G. W., MacConnell, J. G., Gruber, V. F., Ku, C. C., et al. 1984. Fate of avermectin B1a in soil and plants. *J. Agric. Food Chem.* 32:94–102

17. Burg, R. W., Miller, B. M., Baker, E. E., Birnbaum, J., Currie, S. A., et al. 1979. Avermectins, a new family of potent antihelminthic agents: Producing organism and fermentation. *Antimicrob. Agents Chemother.* 15:361–67

17a. Burg, R. W., Stapley, E. O. 1989. Isolation and characterization of the producing organism. See Ref. 20a, pp. 24–32

18. Burts, E. 1985. SN72129 and avermectin B1, two new pesticides for control of pear psylla, *Psylla pyricola* (Homoptera: Psyllidae). *J. Econ. Entomol.* 78:1327–30

19. Calabretta, C., Nucifora, A., Calabro, M. 1987. Present state of the control of *Liriomyza trifolii* (Burgess) (Diptera: Agromyzidae) with chemical, biotechnical and biological methods on protected gerbera crops in Sicily. *Dif. Delle Plante* 10:87–95

20. Campbell, W. C. 1981. An introduction to the avermectins. *New Z. Vet. J.* 29:174–78

20a. Campbell, W. C., ed. 1989. *Ivermectin and Abamectin.* New York: Springer-Verlag. 363 pp.

21. Campbell, W. C., Fisher, M. H., Stapley, E. O., Albers-Schonberg, G., Jacob, T. A. 1983. Ivermectin: A potent new antiparasitic agent. *Science* 221:823–28

22. Canargo, de M., de Lilia, K. P. C., de Hermano, V. 1987. Toxicity of abamectin to the spider mite *Tetranychus urticae* (Koch, 1936). *An. Soc. Entomol. Brasil* 16:223–27

22a. Chandler, L. D. 1985. Response of

Liriomyza trifolii (Burgess) to selected insecticides with notes on Hymenopterous parasites. *Southwest. Entomol.* 10:228–35

23. Christie, P. T., Wright, D. J. 1990. Activity of avermectin B1 against larval stages of *Spodoptera littoralis* Boisd. and *Heliothis armigera* Hubn. (Lepidoptera: Noctuidae): Possible mechanisms determining differential toxicity. *Pest. Sci.* In press

24. Cochran, D. G. 1985. Mortality and reproductive effects of avermectin B1 fed to German cockroaches, *Blattella germanica*. *Entomol. Exp. Appl.* 37:83–88

25. Corbitt, T. S., Green A. S. J., Wright, D. J. 1989. Relative potency of avermectin B1 against larval stages of *Spodoptera littoralis* and *Heliothis virescens*. *Crop Prot.* 8:127–32

26. Daiber, K. C. 1985. Tomato pests in South Africa. *Hort. Sci.* 2:1–5

27. Davies, G. H., Green, R. H. 1986. Avermectins and milbemycins. *Nat. Prod. Rep. England* 3:87–121

28. Deecher, D. C., Brezner, J., Tanenbaum, S. W. 1987. Avermectin B1a and milbemycin D as contact toxicants for gypsy moth (Lepidoptera: Lymantriidae) larvae and eggs. *J. Econ. Entomol.* 80:1284–87

29. Dunbar, D. M., Dybas, R. A., Norton, J. A. 1989. Abamectin 0.15 EC (ZEPHYR): Miticide/insecticide for spider mite control on San Joaquin Valley cotton. *Proc. 1989 Beltwide Cotton Insect Res. Control Con.* pp. 286–88. Memphis: Natl. Cotton Counc. Am.

30. Dybas, R. A. 1983. Avermectins: Their chemistry and pesticidal activities. In *IUPAC Pesticide Chemistry, Human Welfare and the Environment*, ed. J. Miyamota, P. C. Kearney, pp. 83–90. New York: Pergamon

31. Dybas, R. A. 1985. Synergistic avermectin combination for treating plant pests. *US Patent 4,560,677*

32. Dybas, R. A. 1988. Discovery of novel second generation avermectin insecticides for crop protection. Presented at Int. Entomol. Congr., 18th, Vancouver

33. Dybas, R. A. 1989. Abamectin use in crop protection. See Ref. 20a, pp. 287–310

34. Dybas, R. A., Babu, J. R. 1988. 4''-epi-methylamino-4'-deoxyavermectin B1 (MK-243): A novel insecticide for crop protection. *Proc. 1988 Brighton Crop Protection Con. Pests and Diseases,* 1:57–64. Surrey, UK: Brit. Crop Protect. Counc.

35. Dybas, R. A., Green, A. S. J. 1984. Avermectins: their chemistry and pesticidal activity. *Proc. 1984 Br. Crop Prot. Con. Pests Diseases,* 9B-3:947–54. Surrey, UK: Brit. Crop Protect. Counc.

36. Dybas, R. A., Hilton, N. J., Babu, J. R., Preiser, F. A., Dolce, G. J. 1989. Novel second-generation avermectin insecticides and miticides for crop protection. In *Novel Microbial Products For Medicine and Agriculture,* ed. A. L. Demain, G. A. Somkuti, J. C. Hunter-Cevera, H. W. Rossmoore, pp. 203–12. New York: Elsevier

37. Egerton, J. R. 1980. 22, 23-dihydroavermectin B1, a new broadspectrum antiparasitic agent. *B. Vet. J.* 136:88–97

38. Egerton, J. R., Ostlind, D. A., Blair, L. S., Eary, C. H., Suhayda, D., et al. 1979. Avermectins, a new family of potent anthelmintic agents: Efficacy of the B1a component. *Antimicrob. Agents Chemother.* 15:372–78

38a. El-Banhawy, E. M., Anderson, T. E. 1985. Effects of avermectin B1 and fenvalerate on the survival, reproduction and egg viability of the two-spotted spider mite, *Tetranychus urticae* (Acari: Tetranychidae). *Int. J. Acarol.* 11:11–16

39. El-Banhawy, E. M., El-Bagoury, M. E. 1985. Toxicity of avermectin and fenvalerate to the eriophyid gall mite *Eriophyes discoridis* and the predacious mite *Phytoseius finitimus* (Acari: Eriophyidae; Phytoseidae). *Int. J. Acarol.* 11:237–40

40. El-Banhawy, E. M., Reda, A. S. 1988. Ovicidal effects of certain pesticides on the twospotted spider mite, *Tetranychus urticae,* and the predacious mite, *Amblyseius gossipi* (Acari: Tetranychidae: Phytoseiidae). *Insect Sci. Appl.* 9:369–72

41. Fisher, M. H., Mrozik, H. 1984. The avermectin family of macrolide-like antibiotics. In *Macrolide Antibiotics,* ed. S. Omura, pp. 553–606. New York: Academic

42. Fisher, M. H., Mrozik, H. 1989. Chemistry. See Ref. 20a, pp. 1–23

43. Fritz, L. C., Wang, C. C., Gorio, A. 1979. Avermectin B1a irreversibly blocks postsynaptic potentials at the lobster neuromuscular junction by reducing muscle membrane resistance. *Proc. Nat. Acad. Sci. USA* 76:2062–66

44. Glancy, B. M., Lofgren, C. S., Williams, D. F. 1982. Avermectin B1a: Effects on the ovaries of red imported fire ant, *Solenopsis invicta,* (Hymenoptera: Formicidae) queens. *J. Med. Entomol.* 19:743–47

45. Grafius, E., Hayden, J. 1988. Insecticide and growth regulator effects on the leafminer, *Liriomyza trifolii* (Diptera: Agromyzidae), in celery and observations on parasitism. *Great Lakes Entomol.* 21:49–54

46. Grafton-Cardwell, E. E., Hoy, M. A., 1983. Comparative toxicity of avermectin B1 to the predator *Metaseiulus occidentalis* (Nesbitt) (Acari: Tetranychidae). *J. Econ. Entomol.* 76:1216–20

47. Green, A. S. J., Dybas, R. A. 1984. Avermectin B1: Control of mites on ornamentals. *Proc. 1984 B. Crop Prot. Con. Pests Diseases,* 11A-7:1129–33. Surrey, UK: Brit. Crop Protect. Counc.

48. Green, A. S. J., Heijne, B., Schreurs, J., Dybas, R. A. 1985. Serpentine leafminer *(Liriomyza trifolii)* (Burgess) control with abamectin (MK-936) in Dutch ornamentals, a review of the processes involved in the evolution of the use directions, and a summary of the results of phytotoxicity evaluations. *Meded. Fac. Landbouwwet. Rijksuniv. Gent* 50/2b:603–22

49. Green, A. S. J., Heijne, B., Schreurs, J., Dybas, R. A. 1985. Twospotted spider mite (*Tetranychus urticae* Koch) control with abamectin (MK-936) on roses. *Meded. Fac. Landbouwwet. Rijksuniv. Gent* 50/2b:623–32

49a. Greene, B. M., Brown, K. R., Taylor, H. R. 1989. Use of ivermectin in humans. See Ref. 20a, pp. 311–23

50. Greenblatt, J. A., Norton, J. A., Dybas, R. A., Harlan, D. P. 1986. Control of the red imported fire ant with abamectin (AFFIRM®), a novel insecticide, in individual mound trials. *J. Agric. Entomol.* 3:233–41

51. Grout, T. G., Morse, J. G. 1986. Insect growth regulators: Promising effects on citrus thrips (Thysanoptera: Thripidae). *Can. Entomol.* 118:389–92

52. Hara, A. 1986. Effects of certain insecticides on *Liriomyza trifolii* (Burgess) (Diptera: Agromyzidae) and its parasitoids on chrysanthemums in Hawaii. *Proc. Hawaii. Entomol. Soc.* 26:65–70

53. Heungens, A., Degheele. 1986. Control of the broad mite, *Polyphagotarsonemus latus* (Banks), with acaricides on *Psophocarpus tetragonolobus* and *Ricinus communis. Parasitica* 42:3–10

54. Hilton, N. J., Dybas, R. A. 1990. Abamectin (ZEPHYR®): A novel avermectin miticide/insecticide for spider mite control on cotton. *Proc. 1989 Beltwide Cotton Insect Res. Control Con.* pp. 284–85. Memphis: Natl. Cotton Counc. Am.

55. Hoy, M. A., Cave, F. E. 1985. Labora-

tory evaluation of avermectin as a selective acaricide for use with *Metaseiulus occidentalis* (Nesbitt) (Acarina: Phytoseiidae). *Exper. Appl. Acarol.* 1:139–52

56. Hoy, M. A., Conley, J. 1987. Selection for abamectin resistance in *Tetranychus urticae* and *T. pacificus* (Acari: Tetranychidae). *J. Econ. Entomol.* 80:221–25

57. Hoy, M., Ouyang, Y. 1989. Selection of the western predatory mite, *Metaseiulus occidentalis* (Acari: Phytoseiidae), for resistance to abamectin. *J. Econ. Entomol.* 82:35–40

58. Ishaaya, I., Gurevitz, E., Ascher, K. R. S. 1983. Synthetic pyrethroids and avermectin for controlling the grapevine pests, *Lobesia botrana, Cryptoblabes gnidiella* and *Drosophila melanogaster. Phytoparasitology* 11:161–66

59. Iwata, I., MacConnell, J. G., Flor, J. E., Putter, I., Dinhoff, T. M. 1985. Residues of avermectin B1a on and in citrus fruits and foliage. *J. Agric. Food Chem.* 33:467–71

60. James, P. S., Picton, J., Riek, R. F. 1980. Insecticidal activity of the avermectins. *Vet. Rec.* 106:59

61. Johnson, A. W. 1985. Abamectin for tobacco insect control. *Tab. Int.* 187:135–38

62. Jones, V. P., Parrella, M. P., Hodel, D. R. 1986. Biological control of leafminers in greenhouse chrysanthemums. *Calif. Agric.* 40:10–12

63. Kass, I. S., Stretton, A. O. W., Wang, C. C. 1984. The effects of avermectin and drugs related to acetylcholine and 4-aminobutyric acid on neurotransmission in *Ascaris suum. Mol. Biochem. Parasit.* 13:213–25

64. Kass, I. S., Wang, C. C., Walrond, J. P., Stretton, A. O. W. 1980. Avermectin B1a, a paralyzing anthelmintic that affects interneurons and inhibitory motorneurons in *Ascaris suum. Proc. Natl. Acad. Sci. USA* 77:6211–15

65. Leibee, G. L. 1988. Toxicity of abamectin to *Liriomyza trifolii* (Burgess) (Diptera: Agromyzidae). *J. Econ. Entomol.* 81:738–40

66. Lemke, L. A., Kissan, J. B. 1987. Evaluation of various insecticides and home remedies for control of individual red imported fire ant colonies. *J. Entomol. Sci.* 22:275–81

67. Liu, M. Y., Tzeng, Y. J., Sun, C. N. 1981. Diamondback moth resistance to several synthetic pyrethroids. *J. Econ. Entomol.* 74:393–96

68. Lofgren, C. S., Williams, C. F. 1982. Avermectin B1a: Highly potent inhibitor of reproduction by queens of the red

imported fire ant (Hymenoptera: Formicidae). *J. Econ. Entomol.* 75:798–803

69. MacConnell, J. G., Demchak, R. J., Preiser, F. A., Dybas, R. A. 1989. A study of the relative stability, toxicity and penetrability of abamectin and its 8,9-oxide. *J. Agric. Food Chem.* 37:1498–1501

70. Maynard, M. S., Halley, B., Green-Erwin, M., Alvaro, R., Gruber, V. F., et al. 1990. The distribution, elimination, and levels of the total residue and unchanged parent compound from rats administered avermectin B1a. *J. Agric. Food Chem.* In press

70a. Maynard, M. S., Iwata, Y., Wislocki, P. G., Ku, C. C., Jacob, T. A. 1989. Fate of avermectin B1a on citrus fruits. 1. Distribution and magnitude of the avermectin B1a and 14C residue on citrus fruits from a field study. *J. Agric. Food Chem.* 37:178–83

71. Maynard, M. S., Wislocki, P. G., Ku, C. C. 1989. Fate of avermectin B1a in lactating goats. *J. Agric. Food Chem.* 37:1491–97

72. McCoy, C. W., Bullock, R. C., Dybas, R. A. 1982. Avermectin B1: A novel miticide active against citrus mites. *Proc. Fla. State Hort. Soc.* 95:51–56

73. Mellin, T. N., Busch, R. D., Wang, C. C. 1983. Postsynaptic inhibition of invertebrate neuromuscular transmission by avermectin B1a. *Neuropharmacology* 22:89–96

74. Miller, T. W., Chaiet, L., Cole, D. J., Flor, J. E., Goegelman, R. T., et al. 1979. Avermectins, a new family of potent anthelmintic agents: isolation and chromatographic properties. *Antimicrob. Agents Chemother.* 15:368–71

75. Mizell, R. F. III, Schiffhauer, D. E., Taylor, J. L. 1986. Mortality of *Tetranychus urticae* Koch (Acari: Tetranychidae) from abamectin residues: Effects of host plant, light, and surfactants. *J. Entomol. Sci.* 21:329–37

76. Moar, W. J., Trumble, J. T. 1987. Toxicity, joint action, and mean time of mortality of Dipel 2X, avermectin B1, neem and thuringiensin against beet armyworms (Lepidoptera: Noctuidae). *J. Econ. Entomol.* 80:588–92

77. Morse, J. G., Bellows, T. S., Gaston, I. K., Iwata, Y. 1987. Residual toxicity of acaricides to three beneficial species on California citrus. *J. Econ. Entomol.* 80:953–60

78. Morse, J. G., Brawner, O. L. 1986. Toxicity of pesticides to *Scirtothrips citri* (Thysanoptera: Thripidae) and implications to resistance management. *J. Econ. Entomol.* 79:565–70

79. Moye, H. A., Malagodi, M. H., Yoh, J., Deyrup, C. L., Chang, S. D. C., et al. 1990. Avermectin B1a metabolism in celery: A residue study. *J. Agric. Food Chem.* 38:290–97

80. Moye, H. A., Malagodi, M. H., Yoh, J., Leibee, G. L., Ky, C. C., Wislocki, P. G. 1987. Residues of avermectin B1a in rotational crops and soils following soil treatment with [14C] avermectin B1a. *J. Agric. Sci. Chem.* 35:859–64

81. Mrozik, H. 1985. Chemistry and biological activities of avermectin derivatives in biotechnology and its application to agriculture. *B. Crop Prot. Counc. Monogr.* 32:133–43

82. Mrozik, H., Eskola, P., Linn, B., Lusi, A., Shih, T., et al. 1989. Discovery of novel avermectins with unprecedented insecticidal activity. *Experientia* 45:315–16

83. Mrozik, H., Eskola, P., Reynolds, G. F., Arison, B. H., Smith, G. M., Fisher, M. H. 1988. Photoisomers of avermectins. *J. Org. Chem.* 53:1820–23

84. Oetting, R. D. 1987. Laboratory evaluation of the toxicity and residual activity of abamectin to *Ecinothrips americanus*. *J. Agric. Entomol.* 4:321–26

85. Ostlind, D. A., Cifelli, S., Lang, R. 1979. Insecticidal activity of the antiparasitic avermectins. *Vet. Rec.* 105:168

86. Overman, A. J., Price, J. F. 1984. Application of avermectin and cyromazine via drip irrigation and fenamiphos by soil incorporation for control of insect and nematode pests in chrysanthemums. *Proc. Fla. State Hort. Soc.* 97:304–6

87. Pandolfo, F. M. 1985. Studies on the chemical control of *Liriomyza trifolii* on gerbera. *Inf. Fitopatol.* 35:49–59

88. Parrella, M. P. 1983. Evaluations of selected insecticides for control of permethrin-resistant *Liriomyza trifolii* (Diptera: Agromyzidae) on chrysanthemum. *J. Econ. Entomol.* 76:1460–64

89. Parrella, M. P., Robb, K. L., Virzi, J. K., Dybas, R. A. 1988. Analysis of the impact of abamectin on *Liriomyza trifolii* (Burgess) (Diptera: Agromyzidae). *Can. Entomol.* 120:831–37

90. Parrish, M. D., Roberts, R. B. 1984. Toxicity of avermectin B1 to larval yellowjackets, *Vespula maculifrons* (Hymenoptera: Vespidae). *J. Econ. Entomol.* 77:769–72

91. Peleg, B. A. 1987. Effect of avermectin B1 on the development and fecundity of the California red scale, *Aonidiella aurantii*, and on two of its natural enemies: *Chilocorus bipustulatus* and *Aphytis holoxanthus*. *Phytoparasit.* 15:115–23

92. Perring, T. M. 1987. Seasonal abun-

dance, spray timing and acaricidal control of spider mites on cantaloupe. *J. Agric. Entomol.* 4:12–20

93. Pfeiffer, D. G. 1985. Toxicity of avermectin B1 to San Jose scale (Homoptera: Diaspididae) crawlers, and effects on orchard mites by crawler sprays compared with full-season applications. *J. Econ. Entomol.* 78:1421–24

94. Pienkowski, R. L., Mehring, P. R. 1983. Influence of avermectin B1 and carbofuran on feeding by alfalfa weevil, *Hypera postica,* larvae (Coleoptera: Curculionidae). *J. Econ. Entomol.* 76:1167–69

95. Price, J. F. 1983. Field evaluations of new pesticides for control of leafminers, mites and aphids on flower crops. *Proc. Fla. State Hort. Soc.* 96:287–91

96. Price, J. F., van de Vrie, M. 1989. Behaviour of *Liriomyza trifolii* leafminer flies and responses to host plant and insecticides. *Entomologie* 77:(Abst. #11.20)

97. Putter, I., MacConnell, J. G., Preiser, F. A., Haidri, A. A., Ristich, S. S., Dybas, R. 1981. Avermectins: Novel insecticides, acaricides and nematicides from a soil microorganism. *Experientia* 37:963–64

98. Ramalho, F. S., de Jesus, F. M., Menezes, J. N. 1986. Evaluation of acaricides against twospotted spider mite (*Tetranychus urticae* Koch) on cotton. *Anais da Sociedade Entomologica do Brasil.* 15:247–56

99. Reda, A. S., El-Banhawy, E. M. 1988. Effect of avermectin and dicofol on the immatures of the predacious mite *Amblyseius gossipi* with a special reference to the secondary poisoning effect on the adult female (Acari: Phytoseiidae). *Entomophaga* 33:349–55

100. Reed, D. K., Reed, G. L. 1986. Activity of avermectin B1 against the striped cucumber beetle (Coleoptera: Chrysomelidae). *J. Econ. Entomol.* 79:943–47

101. Reed, D. K., Tromley, N. J., Reed, G. L. 1985. Activity of avermectin B1 against codling moth (Lepidoptera: Olethreutidae). *J. Econ. Entomol.* 78:1067–71

102. Deleted in proof

103. Robertson, J. L., Richmond, C. E., Preisler, H. K. 1985. Effects of avermectin B1 on the western spruce budworm (Lepidoptera: Tortricidae). *J. Econ. Entomol.* 78:1129–32

104. Roush, R. T., Wright, J. E. 1986. Abamectin: Toxicity to house flies (Diptera: Muscidae) resistant to synthetic organic insecticides. *J. Econ. Entomol.* 79:562–64

105. Royalty, R. N., Perring, T. M. 1987. Comparative toxicity of acaricides to *Aculops lycopersici* and *Homeopronematus anconai* (Acari: Eriophyidae, Tydeidae). *J. Econ. Entomol.* 80:348–51

106. Sami, B. P., Vaid, A. 1988. Specific interaction of ivermectin with retinol-binding protein from filarial parasites. *J. Biochem.* 249:929–32

107. Schuster, D. J., Everett, P. H. 1983. Response of *Liriomyza trifolii* (Diptera: Agromyzidae) to insecticides on tomatoes. *J. Econ. Entomol.* 76:1170–76

108. Schuster, D. J., Taylor, J. L. 1988. Longevity and oviposition of adult *Liriomyza trifolii* (Diptera: Agromyzidae) exposed to abamectin in the laboratory. *J. Econ. Entomol.* 81:106–9

109. Scott, J. G. 1989. Cross-resistance to the biological insecticide abamectin in pyrethroid-resistant house flies. *Pestic. Biochem. Physiol.* 34:27–31

110. Scott, R., Duce, I. 1985. Effects of 22,23 dihydroavermectin B1a on locust *Schistocerca gregaria* muscles may involve several sites of action. *Pestic. Sci.* 16:599–604

111. Springer, J. P., Arison, B. H., Hirshfield, J. M., Hoogsteen, K. 1981. The absolute stereochemistry and conformation of avermectin B2a aglycone and avermectin B1a. *J. Am. Chem. Soc.* 103:4221–24

112. Stapley, E. O., Woodruff, H. B. 1982. Avermectin, antiparasitic lactones produced by *Streptomyces avermitilis* isolated from soil in Japan. *Trends in Antibiotic Research.* pp. 154–70. Tokyo: Jpn. Antibiotic Res. Assoc.

113. Strong, L., Brown, T. A. Avermectins in insect control and biology: A review. *Bull. Entomol. Res.* 77:357–89

114. Sudderuddin, K. I., Pooi-Fong, K. 1978. Insecticide resistance in *Plutella xylostella* collected from the Cameron highlands of Malaysia. *FAO Plant Prot. Bull.* 26:53–57

115. Sun, C. N., Chi, H., Feng, H. T. 1978. Diamondback moth resistance to diazinon and methomyl in Taiwan. *J. Econ. Entomol.* 71:551–54

116. Takiguchi, Y., Mishima, H., Okuda, M., Terao, M., Aoki, A., Fukuda, R. 1980. Milbemycins, a new family of macrolide antibiotics: Fermentation, isolation and physico-chemical properties. *J. Antibiot.* 33:1120–27

117. Tedders, W. L., Payne, J. A., Taylor, J. L. 1984. Laboratory evaluation of aver-

mectin B1 against the foliar aphids of pecan. *J. Ga. Entomol. Soc.* 19:344–50

118. Trumble, J. T. 1983. Suppression of lepidopterous larvae and leafminers on tomato. *Insectic. Acaric. Tests* 8:150

119. Trumble, J. T. 1985. Integrated pest management of *Liriomyza trifolii:* Influence of avermectin, cyromazine, and methomyl on leafminer ecology in celery, *Apium graveolens. Agric. Ecosys. Environ.* 12:181–88

120. Trumble, J. T. 1985. Planning ahead for leafminer control. *Calif. Agric.* 39:8–9

121. Trumble, J. T., Moar, W. J., Babu, J. R., Dybas, R. A. 1987. Laboratory bioassay of the acute and antifeedant effects of avermectin B1. *J. Agric. Entomol.* 4:21–28

122. Turner, M. J., Schaeffer, J. M. 1989. Mode of action of ivermectin. See Ref. 20a, pp. 73–88

123. van de Vrie, M., Price, J. F., de Jong, J. 1986. Behaviour of *Liriomyza trifolii* leafminer flies and responses to host plants and insecticides. *Meded. Fac. Landbouwwet. Rijksuniv. Gent* 51:879–84

124. Wang, C. C., Pong, S. S. 1982. Actions of avermectin B1a on GABA nerves. In *Membranes and Genetic Disease,* pp. 373–95. New York: Liss

125. Williams, D. F. 1983. The development of toxic baits for the control of the red imported fire ant. *Fla. Entomol.* 66:162–71

126. Williams, D. F. 1985. Laboratory and field evaluation of avermectin against the red imported fire ant. *Southwest. Entomol.* (Suppl.) 7:27–33

127. Wislocki, P. G., Grosso, L. S., Dybas, R. A. 1989. Environmental aspects of abamectin use in crop protection. See Ref 20a, pp. 182–200

128. Wolfenbarger, D. A., Johnson, A. W., Herzog, G. A., Tappan, W. B. 1985. Activity of avermectin in the laboratory and the field against the boll weevil and *Heliothis* spp. on cotton and flue-cured tobacco. *Southwest. Entomol.* 7:17–26

129. Wright, C. G., Dupree, H. E. 1985. Acephate and avermectins for German cockroach control (Dictyoptera, Blattellidae). *J. Entomol. Sci.* 20:20–23

130. Wright, D. J. 1985. Biological activity and mode of action of avermectins. *Proc. Neurotox. 85th Bath, United Kingdom.* Chichester, UK: Horwood

131. Wright, D. J., Loy, A., Green, A. S. J., Dybas, R. A. 1985. The translaminar activity of abamectin (MK-936) against mites and aphids. *Meded. Fac. Landbouwwet. Rijksuniv. Gent* 50/2b:595–601

132. Wright, D. J., Roberts, I. T. J., Androher, A., Green, A. S. J., Dybas, R. A. 1985. The residual activity of abamectin (MK-936) against *Tetranychus urticae* (Koch) on cotton. *Meded. Fac. Landbouwwet. Rijksuniv. Gent* 50/2b:633–37

133. Wright, J. E., Jenkins, J. N., Villavaso, E. J. 1985. Evaluation of avermectin B1 (MK-936) against *Heliothis* spp. in the laboratory and in field plots and against the boll weevil in field plots. *Southwest. Entomol.* (Suppl.) 7:11–16

Annu. Rev. Entomol. 1991. 36:119–38

TRAP CROPPING IN PEST MANAGEMENT

Heikki M. T. Hokkanen

Institute of Plant Protection, Agricultural Research Centre of Finland, SF-31600 Jokioinen, Finland

KEY WORDS: insect pest control, integrated pest management, insect behavior, agricultural ecology, agroecosystem diversity

PERSPECTIVES

Trap crops are plant stands that are grown to attract insects or other organisms like nematodes to protect target crops from pest attack. Protection may be achieved either by preventing the pests from reaching the crop or by concentrating them in a certain part of the field where they can economically be destroyed. For example, in soybeans, one can attract 70–85% of the stink bug population to a trap crop that covers only 1–10% of the total crop area (56).

The principle of trap cropping rests on the fact that virtually all pests show a distinct preference to certain plant species, cultivars, or a certain crop stage. Manipulations of the stands in time and space so that attractive host plants are offered at the critical time in the pest's and/or the crop's phenology lead to the concentration of the pests at the desired site, the trap crop. Techniques of manipulation range from establishing an early or a late trap crop of the same cultivar as the main crop to planting a completely different plant species. To enhance the attractiveness of trap crops, specific chemical compounds, such as insect pheromones (31), plant kairomones (57), or insect-food supplements (29a), may be used.

Farmers are motivated to utilize trap cropping because of the difficulties in coping with the pest situation in other ways. Sometimes the cost of chemical pesticides and the number of treatments required is so high that more economical ways have to be developed (80); additionally, the pests have often evolved resistance to the commonly used pesticides (69, 81), which requires some alternative control strategies. In some cases, no effective

119

0066-4170/91/0101-0119$02.00

chemical pesticides are availalbe, as for the treatment of cauliflower just before harvest (38).

Further motivations to use trap cropping are the economic and environmental benefits often associated with this strategy. Savings in insecticide costs are usually substantial and outweigh the costs of the operations (80); also, the marketable yields are often clearly higher than in conventional farming (38, 69, 77). One major benefit of trap cropping is that the main crop seldom needs to be treated with insecticides, and thus the natural control of pests may remain fully operational in most of the field. Also, the overall use of pesticides is clearly less than in conventional farming, making the strategy environmentally attractive.

Trap cropping relates to many recent developments in ecology and agriculture. Particularly the concepts of intercropping (e.g. 6, 8, 33), crop diversification (e.g. 5, 13, 23, 35), and integrated pest management (e.g. 29, 40, 50), as well as the general awakening of agricultural ecology have clear connections to the theory and practice of trap cropping. In fact, progress in these and other fields, including insect behavior, allelochemicals, and biotechnology, coupled with advances in technology and crop science, may open up new avenues for the rational and profitable use of trap cropping as an important part of the integrated production of crops.

Besides its potential role in improving the environmental soundness and overall performance of conventional agriculture, trap cropping techniques may have special importance to subsistence farming in the developing countries. Additionally, the increasing sector of organic farming also could exploit this strategy of pest control.

Apart from its direct use to control pests, trap cropping has often been utilized for many other purposes. In fact, until the 1970s, it was mostly used for studying the ecology of certain pest species, as for example the dispersal, hibernation, winter survival, and spring emergence of the cotton boll weevil *Anthonomus grandis* (31). For species that are difficult to find at a certain stage, like the cutworms *Euxoa* spp., this technique has been valuable (16). Agronomists have also used it to facilitate pesticide efficacy testings (36), and sometimes trap crops can be effective in obtaining pest survey information (18).

Yet another function of trap crops is their use for attracting natural enemies of pest insects to the fields and concentrating them there to enhance naturally occurring biological control. This type of trap cropping has been practiced in the Soviet Union particularly for the protection of cabbage plants (1, 2, 84), but also of other crops (9), and has been used to a certain extent elsewhere (e.g. 27, 52, 53, 61).

Brief reviews of trap cropping practices have earlier been included in general reviews of cultural control methods (e.g. 29, 35, 55, 74, 77), but to my knowledge no comprehensive review has been attempted before now.

Although trap cropping has successfully been employed against parasitic plants, nematodes, and insects (29), this review concentrates only on its direct use against insect pests.

OVERVIEW OF TRAP CROPPING

The principle of insect control by trap cropping has been known for centuries and is still exploited in many traditional farming systems either explicitly or implicitly. Pest control ideas based on this principle have emerged throughout this century, as documented in the entomological literature, but seldom have any of them led to practical uses. For example, in 1860, Curtis (24) recommended controlling the parsnip webworms *Depressaria depressella* and *Depressaria daucella* on carrots by setting parsnip plants 2–3 m apart among the carrots. The moths prefer parsnip and lay the eggs on these trap plants, where the larvae are easily destroyed. Similarly, controlling the cotton boll weevil through trap cropping had already been suggested by 1900, only a few years after the detection of the pest in the United States (31). Widespread, successful applications of trap cropping in practice, however, are a phenomenon of the last two decades (cf. 72).

Successful Uses

To date, practical applications of trap cropping in modern agriculture have been very few. Only about 11 pest species (seven species groups) are, or have been, successfully controlled in four crop ecosystems where trap crops have played a major role (Table 1): cotton, soybeans, potatoes, and cauliflower. Of these, the cotton and soybean trap cropping systems clearly have the greatest importance worldwide.

In addition to agricultural uses, the trap cropping principle has been exploited in controlling forest pests. Trap trees, or only logs, have long been set up in Europe to catch certain pest species. For example, control of the spruce bark beetle, *Ips typographus*, was practiced in this manner over 200 years ago (12). Currently, many of these traditional systems have been replaced with synthetic trap crops—pheromone traps—which are used to mass-trap forest pests (12).

Potential Uses

Suggestions for uses of trap cropping in agricultural pest management are abundant (Table 2). Altogether, at least 35–40 important pest species or species complexes (soybean stink bugs, corn rootworms) apparently could be controlled with the help of trap cropping. All major insect pest orders and many different production ecosystems are represented in these examples, suggesting that in principle trap cropping could be used much more than it actually is.

Table 1   Examples of trap crop systems successfully applied in agricultural practice

Controlled pests	Main crop	Trap crop	Location	Reference
Lygus bugs *(Lygus hesperus, Lygus elisus)*	cotton	alfalfa	California	76, 77
Cotton boll weevil *(Anthonomus grandis)*	cotton cotton	cotton cotton	USA Nicaragua	17, 31 80
Stink bugs *(Nezara viridula,* *Euschistus* spp., *Acrosternum hilare,* *Piezodorus guildinii)*	soybeans soybeans soybeans	soybeans soybeans soybeans, cowpea	SE USA Brazil Nigeria	56, 60, 83 49 42
Mexican bean beetle *(Epilachna varivestris)*	soybeans	snap beans	SE USA	63
Bean leaf beetle *(Cerotoma trifurcata)*	soybeans	soybeans	SE USA	60
Colorado beetle *(Leptinotarsa decemlineata)*	potato potato	potato potato	USSR Bulgaria	20, 28 15
Blossom beetle *(Meligethes aeneus)*	rape, cauliflower	rape, marigold	Finland	37, 38
Pine shoot beetle *(Tomicus piniperda)*	pine trees	pine logs	Great Britain	14
Spruce bark beetle *Ips typographus*	spruce	spruce trees, logs	Europe	12

The majority of proposed trap crop systems involve lepidopteran pests (about 10 spp.), while coleopteran, dipteran, and hemipteran pests are represented equally (about 5 spp. each). Oddly enough, the trap crop systems successfully adapted to practical use, however, do not as yet involve any lepidopterans, but rather some six to seven species each of hemipteran and coleopteran pests (Table 1).

CASE STUDES

Cotton

Trap cropping for the control of cotton pests has been practiced over wider areas than for any other purpose. Large-scale programs for the control of the cotton boll weevil in the southern United States and for lygus bugs in California were developed parallel to each other at the end of the 1960s.

Table 2 Examples of proposed trap crop systems

Pest to be controlled	Main crop	Trap crop	Location	Reference
Hemiptera	alfalfa	alfalfa	Wisconsin, USA	71
Cotton lygus	cotton	*Cissus*	Uganda	79
(*Lygus voessleri*)		*adenocaulis*		
Tarnished plant bug	pine	*Senecio*	Finland	39
(*Lygus rugulipennis*)	seedlings	*vulgaris*		
Brown plant hopper	rice	rice	Philippines	69
(*Nilaparvata lugens*)				
Green leafhopper	rice	rice	Philippines	70
(*Nephotettix virescens*				
+ RTV[a])				
Cotton whitefly	tomato	cucumber	Jordan	4
Bemisia tabaci	seedlings			
+ TYLCV[b])	tomato	cucumber	Iraq	3
	seedlings			
	tomato	cucumber	Israel	22
	seedlings			
Onion thrips	cotton	onion,	Egypt	32
(*Thrips tabaci*)	seedlings	garlic		
Corn rootworms	corn	corn	USA	36, 87
(*Diabrotica* spp.)				
Cucumber beetles	cucurbits	cucurbits	USA	57, 58
(*Diabrotica* spp.)				
Mustard beetle	turnip	white	Finland	41
(*Phaedon cochleariae*)		mustard		
Wireworms	strawberry	wheat	Switzerland	48
(*Agriotes* spp.)				
Citrus wood borer	citrus	*Cordia*	Brazil	59
(*Cratosomus*		*verbenacea*		
flavofasciatus)				
Shoot flies	corn	sorghum	India	67, 68
(*Atherigona* spp.)				
Melon fly	squash,	corn	Hawaii, USA	29
(*Dacus cucurbitae*)	melon			
Fruit fly	cucurbits	citrus,	Cap Verde	75
(*Dacus frontalis*)		corn		
Serpentine leafminer	chrysan-	bean	Canada	34
(*Liriomyza trifolii*)	themum			
Carrot rust fly	carrot	carrot	Finland	47
(*Psila rosae*)				
Corn stalk borer	corn	sorghum	India	66
(*Chilo partellus*)				
African cane borer	sugar cane	corn	Ivory Coast	21
(*Eldana saccharina*)				
Corn earworm	tomato	corn	North Carolina, USA	46
(*Heliothis zea*)	tomato	corn	Arkansas, USA	85
	strawberry	corn	California, USA	86
	cotton	corn	Arkansas, USA	54
	cotton	corn	Venezuela	64

Table 2 (*Continued*)

Controlled pests	Main crop	Trap crop	Location	Reference
Cutworms: *Euxoa messoria, E. tessellata*	rye	tobacco	Ontario, Canada	16
Fall armyworm (*Spodoptera frugiperda*)	sorghum	corn	Honduras	18
Cotton worm (*Spodoptera litura*)	tobacco	castor	Gujarat, India	19
Legume pod-borer (*Maruca testulalis*)	cowpea	*Crotalaria* spp.	Nigeria	43
European corn borer	corn	corn	North Caro- lina, USA	7
(*Ostrinia nubilalis*)	corn	potato	North Caro- lina, USA	46
	corn	corn	France	25
	cotton	corn	China	30
Anise swallowtail (*Papilio zelicaon*)	citrus	sweet fennel	California, USA	73
Turnip sawfly (*Athalia rosae*)	winter rape	rape	Hungary	65

[a] RTV = rice tungro virus.
[b] TYLCV = tomato yellow leaf curl virus.

BOLL WEEVIL   The cotton boll weevil is a serious problem in cotton production in North and Central America. It is claimed to be the most costly insect in the history of American agriculture (31), and it has been estimated that about one third of all insecticides used for agricultural purposes are applied for its control (62). In Nicaragua, weevil damage and control costs amounted in the 1983–84 season to over $40 million, 16% of the national agricultural export earnings (80).

An early fruiting trap crop of about 5% of the total cotton area has been used in the US, and a between-season trap crop is used in Nicaragua. Boll weevil trap cropping was greatly enhanced after the discovery and synthesis of the boll weevil pheromone grandlure in the beginning of the 1970s. An improved synthesis of grandlure now allows the establishment of a synthetic trap crop, a plot at the edge of the main cotton field that is 2–10% of the total area, baited with the pheromone and highly attractive to the immigrating weevils (10). The insects are then killed with insecticide sprays on the trap crop.

Trap cropping is now an integral part of cotton boll weevil management in the southern United States and plays an essential role in an attempt to eradicate the pest completely (26). Overall, the grandlure baited traps and trap crops offer an economically attractive and environmentally favorable means of reducing insecticides for control of the boll weevil throughout its entire range (31).

LYGUS BUGS   At the end of the 1960s, lygus bugs, a key pest of cotton in California, were shown to prefer lushly growing alfalfa over cotton, and strips of this crop interspersed in cotton fields virtually eliminated the need to spray the main crop for lygus control (72, 77, 78). Alfalfa is also excellent habitat for many important natural-enemy species, and the practice of interplanting increases the naturally occurring biological control of pests like *Heliothis zea* and spider mites in adjacent crops (76).

Strips of alfalfa about six meters wide were interplanted in cotton fields at 100 to 150 m intervals. Most alfalfa was sprayed four to six times with insecticides, and some fields were harvested for seed. The system required frequent irrigation to keep alfalfa attractive to *Lygus*, and crop rotation difficulties together with changes in federal regulations concerning cotton production soon caused interplanting cotton with alfalfa to be abandoned (77).

In Uganda, encouraging trials using *Cissus adenocaulis* as a trap crop for the cotton lygus, *Lygus voessleri*, have been reported (79). A single *Cissus* screen between the source of infestation and the cotton appeared effective enough for a good result. No further information regarding any practical implementation of these results is available.

Soybean

Several species of stink bugs, the Mexican bean beetle, and the bean leaf beetle are all major pests of soybeans. Their phenology in the soybean crop is closely bound to crop phenology, making trap cropping a suitable strategy for control (51). Early planted, early maturing soybean varieties near the main soybean plantings have been used for successful trap cropping of the bean leaf beetle and/or the stink bugs in the United States (56, 60, 83), Brazil (49), and Nigeria (42).

An area for the trap cropping of about 1–10% of the main crop proved sufficient to attract up to 85% of the pest populations (56). Appropriately timed insecticide applications for trap cropping control these pests and prevent their spread to the main crop, while conserving natural enemy populations throughout the remainder of the ecosystem. Also, the spread of the bean pod mottle virus, which is vectored by the bean leaf beetle, is retarded (35).

For the control of the Mexican bean beetle, 6–12 border rows of snap beans, planted approximately two weeks prior to the soybean, have been used with success (63). The beetle prefers snap beans over all other species of *Phaseolus* and other genera of beans (11). A trap crop area of about 0.2 ha proved adequate in these experiments regardless of the size of the main crop (1–30 ha) (63).

Potato

In Belorussia, integrated control of potato pests has been practiced since 1957 (28). Early-planted trap crops of potatoes are used to concentrate the attack of the Colorado beetle, and the beetles are destroyed on the trap crop periodically. In one experiment, the beetles were destroyed in the trap crop before oviposition, or when their numbers reached 5–10 per plant. Such crops over an area of 2-5 ha provided protection for 200–500 ha of main crop, reducing the numbers of beetles and delaying the infestation by two weeks. As a result, just one spray of the main crop was necessary instead of two (20). A trap crop of about 5% of the main crop area, planted 10–15 days earlier, is now recommended in Bulgaria for Colorado beetle control in potato production (15).

Cauliflower

In Finland, the rape blossom beetle, normally a pest on oilseed crucifers, became a serious pest on cauliflower in the beginning of the 1980s, often destroying up to one third of the whole harvest (37, 38). Farmers had no means of controlling the beetles on harvest ripe cauliflower, and the increased rape cultivation and the associated swarms of *Meligethes* forced farmers to abandon growing cauliflower at some localities.

A trap cropping system using several species of trap plants (chinese cabbage, oilseed and turnip rape, sunflower, and marigold), often in a mixture, was developed and successfully used to prevent the spread of the beetles to the cauliflower plants (37, 38). Because the beetles are highly mobile, the trap plants had to be grown in several strips, forming a set of barricades in the anticipated direction of infestation (Figure 1). About 15 strips, 5–20 m wide and with a total length up to 5 km (3–5 ha), were used for protection of 42–45 ha of cauliflower. Two to four insecticide sprays were needed to prevent the spread of the beetles to the crop plants, mainly because the heavy infestation quickly overloaded the trap crops unless they were periodically emptied. Despite the rather intensive management requirement of such trap cropping, the results were extremely positive, with an approximately 20% increase in marketable yield, and consequently good economic profits (38). This trap cropping system is practiced in Finland in areas where the blossom beetles are a problem to cauliflower cultivation.

ECONOMICS OF TRAP CROPPING

Several studies give detailed economic calculations on the profitability of trap cropping, and all of them indicate great benefits in terms of economic returns. On the average, a 10–30% overall increase in net profits, mainly resulting from reduced insecticide use, reduced pest attack, or both, has been reported. With the alfalfa-cotton interplanting system, a net gain of $40 in profits per hectare was reported, while cotton profits at that time averaged about $125–150 per ha (77). Area-wide suppression of the cotton boll weevil with trap cropping in Nicaragua reduced insecticide costs an average of 43%, representing an overall return of approximately 120% on investment. In normal-rainfall areas, however, the returns were about 250%, whereas the programs showed slight net losses in the dry, more marginal production areas (80). In Nicaragua, cotton yields were not affected by trap cropping, but, in the US, substantial yield increases have been reported (17) along with up to 50–75% savings in insecticide costs (10, 17).

Economic calculations of trap cropping practices in soybeans growing in Delaware show that "any field 9.2 ha or larger is potentially more economically treated with a trap crop than the alternative chemical control" (63). In Louisiana, at a farm isolated by a distance of about 10–15 km from other soybean fields, trap cropping protected the area of 1400–1600 ha of soybeans so effectively that, during the three study years, insecticide treatments were necessary only on 36, 97, and zero ha, respectively, presenting 90–100% savings in insecticide costs (60).

In the Philippines, trap-cropping control of the green leafhopper and the associated tungro virus in rice resulted in a 12% higher economic return than the chemical treatment, or 29% higher than the untreated controls (70). The rice yield was slightly lower in the trap crop treatments than in the chemical control, but the savings in insecticide and labor costs were substantial—enough to afford net benefits of 12%. Similar results have been reported for brown planthopper control in rice, except that the yields in the trap crop treatment were significantly higher than in the control (69).

The highest economic benefits have been reported from the protection of cauliflower with trap crops in Finland. There, the overall profit increase was about 20% during the three study years. Because this ratio depends on the size of the field to be protected (c.f. 38, 63), a more reliable figure is the return on investment, which for cauliflower was 616, 197, and 417% in the three study years, respectively. In fact, the calculated return per ha for the trap crops was two to three times higher than the return from the main crop (38).

The economics of using trap crops may not always be as encouraging as reviewed above. As pointed out by Saxena (69), "the concept of a trap crop is meaningful only when [rice] fields are likely to be invaded with a high pest population." Also, the value of the crop and the economic injury levels of the key pests affect the economic success of using trap crops. High-value crops with low tolerance to pests, for which the crop plant is a nonpreferred host, appear as the most likely candidates to be successfully protected by trap cropping (37). Examples of such systems are cauliflower/rape blossom beetle (37, 38) and strawberry/corn earworm (86).

If the trap crop is an early-sown part of the field containing the same crop plant as the main crop, it can be harvested with the main crop and the yield losses due to the trap cropping practice are minimal. If the trap crop species is different from the main crop but can be used for some additional purpose, such as human consumption, animal feed, or green manure, or if it serves as a nursery for natural enemies useful in adjacent crops, the economics of its use can still be more favorable than the plain calculations concerning the trap crop/main crop may show initially. Also, the potential environmental benefits obtainable through trap cropping should be considered.

STRATEGIES AND TECHNIQUES

Strategies of Trap Cropping

The trap crop must be more attractive to the pest than the main crop, at least at some critical time, but preferably over long periods. Differences in attractiveness can be achieved in two ways: (*a*) by the use of a more preferred plant species than, or a more preferred cultivar of, the same species as the main crop, both grown at the same time (Figure 1, *top*), and (*b*) by the use of the main crop plant, timed to be at the most attractive stage at the critical time for pest control, when the main crop is not yet attractive (Figure 1, *bottom*). Strategy *a* often has to be used in cases where a very mobile pest arrives in the fields at a vulnerable crop stage, and strategy *b* when the pests arrive in the fields before the crop is vulnerable or even attractive. For example, for the control of the rape blossom beetle on cauliflower, strategy *a* was used, but for the control of the same pest on turnip or oilseed rape, strategy *b* must be applied (37). Many of the examples given in Tables 1 and 2 have exploited strategy *a*, but note that most of the real successes rely on strategy *b*: cotton boll weevil, soybean pests, Colorado beetle, blossom beetle on rape, and the rice pests brown plant hopper and green leafhopper. Sometimes a dual strategy, *c*, may give the best results; this has been used for example with soybean pests.

Management of Trap Crops

Management of trap crops includes, for example, the choice of the trap species, determination of the size and location of the stands, their establishment and growing operations, possible manipulations to increase their attractiveness, and the pesticide treatments. Detailed knowledge of the quantitative and qualitative properties of kairomones in plant species and cultivars and their roles in insect attraction, arrest, compulsive feeding, and oviposition is essential for a scientifically designed trap crop system (57). Such knowledge can, for example, be used to choose cultivars with high levels of kairomones to be used in trap crops.

Knowledge of the behavior of the target insect is also necessary in the physical design of the size and arrangement of trap crops. For highly mobile pests, an extensive network of trap crops may be required (cf. Figure 1). For

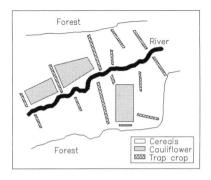

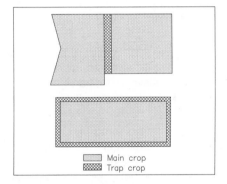

Figure 1   (Top) Placement of trap crop strips in cauliflower protection (see 38). An example of trap cropping strategy *a*: trap crop species different than the main crop; pests arrive at vulnerable crop stage. The strips are drawn excessively wide for clarity.   *(Bottom)* Generalized placement of trap crop strips in trap cropping strategy *b*: early plantings of the same species as the main crop; pests arrive early to the fields. Main crop not yet attractive at the time of pest arrival *(upper field)* or main crop already vulnerable at pest arrival *(lower field)*. Modeled after data in References 25, 38.

some pests and situations, the trap crop must be within the main crop (e.g. wireworms), but for some others it may work better when located further away (e.g. rape blossom beetle). Information about the overwintering or hibernation sites and the direction of infestation may also be a key to successful design of trap cropping.

The size of the trap crop may vary greatly, but in general about 10% of the crop area is used. The required trap size is, however, not likely to be a function of the size of the main crop, but rather of the number of pests (intensity of infestation) expected (37). Also, the mobility of the target pest affects the trap size, as does the need to protect the crop from attacks from just one or from many different directions. Therefore, no specific recommendations as to the size can be given; the overall impression from the literature is, however, that the trap area is usually too small to be effective rather than too large.

A successful establishment and subsequent management of the trap crop stand is of prime importance to the functioning of the system. This is often more difficult than anticipated, and, consequently, poor results are reported when inadequate management is the main reason for the failure. Difficulties in management have been of major concern in trap cropping for *Lygus* control in California cotton (77), boll weevil control in Nicaragua (80), and blossom beetle control in cauliflower (38). For example, the timing of trap crop flowering for blossom beetle control is very difficult because the development of the trap crop greatly depends on the rainfall and temperature patterns during the summer, and this varies greatly from year to year in Finland (38). Several sowing dates and crop mixtures have been used to secure the right timing.

Sometimes permanent trap crops might be used to reduce the operational costs. Possible examples include sweet fennel stands in citrus groves for the anise swallowtail (73), perennial corn for trap crop systems utilizing corn, and perennial flowers such as *Sonchus vulgaris* for the rape blossom beetle in cauliflower.

The success of trap cropping has sometimes been greatly improved through various other manipulations, particularly through the use of sex and aggregation pheromones. In cotton, the addition of grandlure pheromone baits in the trap crop facilitated the final breakthrough in the widespread adoption of this method of boll weevil control (10, 26). Pheromones have also been used in conjunction with trap trees in the control of *Ips* bark beetles (12).

Increase in the effectiveness of trap cropping has sometimes been obtained through innovations specific to that particular system. In cotton, early season square removal, for example through the use of a chemical, provides for a greater difference in phenology between the trap crop and the main crop, thus increasing the attraction power of the trap crop (45). Manipulation of the red imported fire ant as an efficient predator of the boll weevil in the trap crop may also be of value in some places (44).

An antifeedant treatment of trap potatoes was shown to increase the attractiveness of the trap to the Colorado beetle, and it also made the plants unsuitable for the development of the larvae (82). A similar natural phenomenon was observed with the legume pod borer, *Maruca testulalis,* which was more attracted to and laid more eggs on *Crotalaria juncea* than on any other test plant, including the crop plant cowpea. *C. juncea,* however, is not suitable for larval development of the borer, and thus makes a good candidate for a unique trap crop for *M. testulalis* (43). One might assume, however, that in such systems natural selection quickly would sort out strains of the pest that do not react to the trap-crop plant.

Integrating with Other Controls

Trap cropping is normally used as part of an integrated control program involving chemicals and often other cultural controls. Trap cropping is usually also motivated by the enhanced natural control in the major part of the crop. Benefits can sometimes be augmented, for example, through a skillful combination with biopesticides, other biological control agents, resistant plant varieties, or crop rotation schemes.

In experiments for the control of *Spodoptera litura* in a tobacco nursery, a trap crop of castor was tested together with neem seed extract, nucleopolyhedrosis virus preparation, the chitin inhibitor diflubenzuron, and chlorpyrifos (19). The damage by *S. litura* in each of the treatments was significantly reduced from that in the control, but there were no differences between the treatments. Economic calculations led to the conclusion that "the utilization of biocontrol agents and antifeedant in the management of *S. litura* with trap crop on all sides of the nursery is ecologically sound and economically beneficial to the farmers" (19).

In the Ivory Coast, successful trap cropping experiments for the control of the African sugar cane borer *Eldana saccharina* have been conducted using maize as a trap crop (21). Augmentative releases of the egg parasitoids *Trichogrammatoidea eldanae* and *Telenomus* near *T. dignus* were also used against the borer. The good results were further improved in the trap cropping system because the parasitoids reproduced more rapidly when infested maize (trap crop) was available.

An interesting combination of trap cropping and crop rotation for the control of the carrot fly, *Psila rosae*, was proposed by Kettunen et al (47). Crop rotation is a recommended cultural control method for the pest because the flies tend to remain close to the eclosion site. Young female flies could be effectively captured and retained in a small trap crop of carrots at the site of the previous year's crop while the main crop was shifted to a new location (47).

CONSTRAINTS AND RISKS

Various constraints and risks reduce the willingness of scientists to develop trap cropping methods as well as the willingness of farmers to adopt the developed methods. These problems have prevented trap cropping, despite its benefits, from being widely used.

Ecological Constraints

Many authors have generalized, often contradictorily, that only certain types of pests are amenable to control with trap cropping. Most commonly, this statement has referred to polyphagous vs oligophagous pests, but no clear pattern appears to emerge from these considerations. The only valid generalization seems to be that the pests to be controlled with the help of trap crops must show a clear preference for a particular host plant species, cultivar, or stage, which is phenologically different from the main crop at the time of critical appearence of the pest.

Other important pest properties determining the possibility of utilizing the trap crop principle include the dispersal mechanisms and the mobility of the pest. More or less passively dispersed pests, such as certain aphids, can hardly be attracted to a trap crop. Highly mobile pests also appear more difficult than relatively stationary pests to be economically controlled with trap cropping.

The ecological responses of pests to trap crops vary, and, in some cases despite good temporary aggregation, no clear benefits have been obtained. Corn rootworms aggregate on trap corn, sometimes producing good trap cropping results (36). However, these pests often show large annual fluctuations in numbers and damage irrespective of the cropping system, and researchers have suggested (87) that some density-dependent factor inherent in this particular trap cropping system causes the fluctuations.

Aggregation in the trap crop may affect the pest behavior or activity in a way not anticipated. For example, the trap crop may attract relatively more pests to the field than would have arrived there otherwise. It also may change, positively or negatively, the distribution or activity of important natural enemies in the field, which should be considered in an evaluation of the success or failure of the trap cropping. Aggregation of natural enemies in the trap crop, and their subsequent destruction through insecticide sprays, may completely negate the benefits of using the traps for pest control in some situations.

Further ecological questions pertinent to the trap-cropping concept deal with the factors that we want to—or rather, that we can—influence through the traps: pest population growth, extent of colonization, timing of colonization, and natural mortality. Finding answers to these and related questions sets the possibilities and limits for this method of pest control in each particular case.

Rational Constraints

Lack of detailed knowledge of the ecology and behavior of the target pest often is a limiting factor in using cultural controls such as trap cropping. If no encouraging, economically attractive examples from the neighborhood are available, most farmers will not venture unfamiliar techniques, even when the advisory service assures them of the benefits of such systems. Trap cropping also appears—often quite rightly—to many as a rather complicated, even difficult practice, sometimes quite different from the other activities taking place in the farm. Lack of suitable trap-crop varieties or special equipment and time constraints in running the operations simultaneously with other farm activities may simply make the practice impossible.

One significant logical aspect in the acceptance and utilization of trap cropping is the predictability of pest attacks. Trap crops are useless if pests are scarce; therefore a curative treatment with chemical pesticides appeals more to the farmer, at least when occasional pests are concerned. Reliable forecasting methods for pest attacks should greatly improve the rational use of trap cropping. To be useful in this context, however, the forecast should be known a long time in advance to facilitate the use of trap crops.

Another rational constraint to this method is the usual situation of several phenologically different important pest species occurring in the same field. Economically feasible trap cropping may also depend on the size of the operation. As a consequence, to set up trap crops to the extent necessary for successful control in relatively small-scale farming is often impossible (cf. 37, 38, 63, 80). Some of these problems may be avoided through the use of the different strategies of trap cropping; for example, little risk is involved in the use of the same crop cultivar as the main crop (strategy b). To overcome some of the rational constraints, large scale demonstration plots or projects should be designed to provide convincing examples as well as reliable data for extension purposes.

Agronomic Risks

Trap cropping sometimes involves risks that lessen its attractiveness to the farmer. Particularly, the potential of trap crops turning into pest nursery crops, which later create a hazard to all the crops nearby, must be considered. The risk is real, if for some reason (e.g. machine failure, unavailability of suitable pesticides, or time constraints) the spraying of pests from the trap crop is late or inefficient (cf. 56, 83). Occasionally, the trap crop may also create other pest problems, even when the main pest can be kept under control. For example, in cotton boll weevil management, early-planted trap crops or sprayings may initiate earlier infestations of other destructive pests, such as *Heliothis* spp. or *Spodoptera* spp., by providing a host plant earlier than normal and/or destroying beneficial insects in the trap crops, allowing rapid build-up of these pests and subsequent spread to the main plantings (31).

Successful establishment and maintenance of the quality of the trap crop are essential components of the system. These features are sometimes at risk because of agronomic properties of the trap plants or other factors, as seen for example with the difficulties in maintaining the quality of alfalfa trap crops in *Lygus* control (77).

Environmental Risks

Almost all the studies on trap cropping refer to the environmental benefits that can be obtained through this practice. These include a considerable reduction in overall pesticide use, a decreased pesticide load on the main crop (important if it is an edible crop), the protection of most of the natural enemies present in the main crop, and the possibility of placing the trap crops in such a way that the leakage or drift of pesticides to the surrounding environment is minimized (29).

Environmental risks in this context are seldom considered. They may not be many, but they include the possibility of increased evolution of pesticide resistance in the pest and the destruction of some natural enemies present in the trap crop, as well as the potential fertilizer, water, and soil-erosion impacts of the operation.

The evolution of pesticide resistance may be enhanced if virtually all the pest individuals aggregate on the trap crop, and the ones that survive spraying there constitute the majority of the remaining reproducing population. Some evidence to that effect comes from studies in Nicaragua, where the cotton boll weevils in general are significantly more resistant to methyl parathion than the weevil populations in the United States or El Salvador (81). Weevil larvae infesting cotton squares in the trap crops were found to be more resistant to insecticides than those infesting the main crop (81).

Some highly mobile natural enemies may be attracted to and to aggregate on the trap crop just as their host or prey is. They are likely to be destroyed by the insecticide sprays, and this may radically affect the balance between the enemy and the host/prey. Detailed studies on this topic are lacking, but in some situations negative effects on natural enemies are suspected (31).

CONCLUSIONS

Trap cropping is a useful strategy in the control of several pests in various agroecosystems. It offers significant economic and environmental benefits, and its use can be successfully integrated with other cultural, biological, and chemical control methods. At present, trap cropping is still practiced to a limited extent because of various ecological, rational, technical, and economic constraints. Many of these obstacles can be removed through ecological research; technical development in fields such as the detection, synthesis, and dispensing of allelochemicals; and with proper demonstration projects. The significance of trap cropping as an alternative to the conventional reliance on chemical pesticides may increase in the near future with the tightening policy on pesticides in most countries.

ACKNOWLEDGMENTS

I thank professors M. Markkula of the Agricultural Research Centre of Finland, J.C. van Lenteren of the Agricultural University, Wageningen, The Netherlands, and D. Pimentel of Cornell University, Ithaca, New York, for their interest and critical advice on the topic, as well as for their other support. The National Research Council for Agriculture and Forestry of the Academy of Finland has partly financed my research on trap cropping.

Literature Cited

1. Adashkevich, B. P. 1974a. Relations between aphids, their parasites and hyperparasites in vegetable field agrobiocenosis. See Ref. 49a, pp. 3–10
2. Adashkevich, B. P. 1974b. Number dynamics of predatory insects on nectariferous plants. See Ref. 49a, pp. 11–18
3. Al-Hitty, A., Sharif, H. L. 1987. Studies on host plant preference of *Bemisia tabaci* (Genn.) on some crops and effect of using host trap on the spread of Tomato Yellow Leaf Curl Virus to tomato in the plastic house. *Arab. J. Plant Prot.* 5:19–23 (In Arabic, English abstract)
4. Al-Musa, A. 1982. Incidence, economic importance, and control of tomato yellow leaf curl in Jordan. *Plant Dis.* 66:561–63
5. Altieri, M. A. 1983. Vegetational designs for insect-habitat management. *Environ. Manage.* 7:3–7
6. Andow, D. 1983. The effect of agricultural diversity on insect populations. In *Environmentally Sound Agriculture, Proceedings of the Fourth IFOAM International Scientific Conference.* ed. W. Lockeretz, pp. 91–116. New York: Praeger Science
7. Anderson, T. E., Kennedy, G. G., Stinner, R. E. 1984. Distribution of the European corn borer, *Ostrinia nubilalis* (Hübner) as related to oviposition preference of the spring colonizing generation in eastern North Carolina. *Environ. Entomol.* 13:248–51
8. Andrews, D. J., Kassam, A. H. 1976. The importance of multiple cropping in increasing world food supplies. In *Multiple Cropping,* ed. R. I. Papendick, P. A. Sanchez, G. P. Triplett, pp. 1–10. New York: ASA Special Publications 27
9. Antsiferova, T. A. 1971. Efficient honey plants for beekeeping and for biologic protection of plants. In *Proceedings of the XXIIIrd International Apicultural Congress, Moscow, USSR 1971,* pp. 543–546. Bucharest, Romania: Apimondia
10. Attractants lure boll weevils to cotton insecticides. *World Crops* 36:109
11. Augustine, M. G., Fisk, F. W., Davidson, R. H., LaPidus, J. B., Cleary, R.

W. 1964. Host-plant selection by the Mexican bean beetle, *Epilachna varivestis. Ann. Entomol. Soc. Am.* 57:127–34
12. Bakke, A., Riege, L. 1982. The pheromone of the spruce bark beetle *Ips typographus* and its potential use in the suppression of beetle populations. See Ref. 52a, pp. 3–15
13. Baliddawa, C. W. 1985. Plant species diversity and crop pest control. *Insect Sci. Appl.* 6:479–87
14. Bevan, D. 1974. Control of forest insects: there is a porpoise close behind us. In *Biology in Pest and Disease Control,* ed. D. Price-Jones, M. E., Solomon, pp. 302–12. Oxford: Blackwell
15. Bozhkov, D. K. 1985. Use of some aspects of the biology of the Colorado beetle *(Leptinotarsa decemlineata)* for its control. *Ekologiya Bulgaria* 16:75–77 (In Bulgarian, English summary)
16. Bucher, G. E., Cheng, H. H. 1970. Use of trap plants for attracting cutworm larvae. *Can. Entomol.* 102:797–98
17. Burris, E., Clower, D. F., Jones, J. E., Anthony, S. L. 1983. Controlling boll weevils with trap cropping, resistant cotton. *La. Agric.* 26(3):22–24
18. Castro, M., Pitre, H., Meckenstock, D. 1988. Potential for using maize as trap crop for the fall armyworm, *Spodoptera frugiperda* (Lepidoptera: Noctuidae), where sorghum and maize are intercropped on subsistence farms. *Fla. Entomol.* 71:273–78
19. Chari, M. S., Bharboda, T. M., Patel, S. N. 1985. Studies on integrated management of *Spodoptera litura* Fb. in tobacco nursery. *Tob. Res.* 11:93–98
20. Chausov, E. G. 1976. The trustworthiness of forecasts. *Zashch. Rast.* 12:42–43 (In Russian)
21. Cochereau, P. 1980. The sugar cane African borer *Eldana saccharina* Walker (Lepidoptera, Pyralidae, Gelenchiinae): studies on populations and parasites. *Entomol. Newsl. Int. Soc. Sugarcane Technol.* 8:6–7
22. Cohen, S., Berlinger, M. J. 1986. Transmission and cultural control of whitefly-borne viruses. *Agric. Ecosyst. Environ.* 17:89–97

23. Cromartie, W. J. 1981. The environmental control of insects using crop diversity. See Ref. 60a, 2:223–51
24. Curtis, J. 1860. *Farm Insects*. Glasgow, Scotland: Blackie & Son
25. Derridj, S., Lefer, H., Augendre, M., Durand, Y. 1988. Use of strips of *Zea mays* L. to trap European corn borer *(Ostrinia nubilalis* Hbn.) oviposition in maize fields. *Crop Prot.* 7:177–82
26. Dickerson, W. A. 1986. Grandlure: use in boll weevil control and eradication programs in the United States. *Fla. Entomol.* 69:147–53
27. Dong, C. X., Xu, C. E. 1984. Spiders in cotton fields and their protection and utilization. *China Cotton* 3:45–47 (In Chinese). *Rev. Appl. Entomol.* A 75(2), Ref. 811
28. Dorozhkin, N. A., Bel'skaya, S. I., Meleshkevich, A. A. 1975. Combined protection of potatoes. *Zashch. Rast.* 7:6–8 (In Russian)
29. Flint, M. L., van den Bosch, R. 1981. *Introduction to Integrated Pest Management*, p. 153. New York/London: Plenum. 240 pp.
29a. Hagen, K. S., Sawall, E. F. Jr., Tassan, R. L. 1970. The use of food sprays to increase effectiveness of entomophagous insects. *Proc. Tall Timbers Conf. Ecol. Anim. Control Habitat Manage.* 2:59–81
30. Hao, A. X. 1985. A preliminary discussion on the control of cotton insect pests. *China Cotton* 2:47–48 (In Chinese). *Rev. Appl. Entomol.* A 75(2), Ref. 813
31. Hardee, D. D. 1982. Mass trapping and trap cropping of the boll weevil, *Anthonomus grandis* Boheman. See Ref. 52a, pp. 65–71
32. Hassanein, M. H., Khalil, F. M., Eisa, M. A. 1970. Contribution to the study of *Thrips tabaci* Lind in Upper Egypt. *Bull. Soc. Entomol. Egypte* 54:133–40
33. Helenius, J. 1989. *Intercropping, Insect Populations and Pest Damage: Case Study and Conceptual Model*. PhD thesis. Univ. Helsinki, Dept. Agricultural and Forest Zoology Report, 14:1–49
34. Herbert, H. J., Smith, R. F., McRae, K. B. 1984. Evaluation of non-insecticidal methods to reduce damage to chrysanthemums by the leafminer *Liriomyza trifolii* (Diptera: Agromyzidae). *Can. Entomol.* 116:1259–66
35. Herzog, D. C., Funderburk, J. E. 1984. Ecological bases for habitat management and pest cultural control. In *Insects on Plants. Community Patterns and Mechanisms*, ed. D. R. Strong, J. H. Lawton, R. Southwood, pp. 217–49. Oxford/London: Blackwell Science

36. Hill, R. E., Mayo, Z. B. 1974. Trap corn to control corn rootworms. *J. Econ. Entomol.* 67:748–50
37. Hokkanen, H. M. T. 1989. Biological and agrotechnical control of the rape blossom beetle *Meligethes aeneus* (Coleoptera, Nitidulidae). *Acta Entomol. Fennica* 53:25–29
38. Hokkanen, H., Granlund, H., Husberg, G.-B., Markkula, M. 1986. Trap crops used successfully to control *Meligethes aeneus* (Col., Nitidulidae), the rape blossom beetle. *Ann. Entomol. Fennici* 52:115–20
39. Holopainen, J. K. 1989. Host plant preference of the tarnished plant bug *Lygus rugulipennis* Popp. (Het., Miridae). *J. Appl. Entomol.* 107:78–82
40. Horn, D. J. 1988. *Ecological Approach to Pest Management*. London: Elsevier Applied Science. 285 pp
41. Hukkinen, Y. 1931. Über das Auftreten und die Bekämpfung des Meerrettichblattkäfers *(Phaedon cochleariae* Fabr.) in Finnland. *Verhandlungen der Deutschen Gesellschaft für angewandte Entomologie E. V. auf der achten Mitgliederversammlung zu Rostock 1930*, pp. 76–84. Berlin: Paul Parey
42. Jackai, L. E. N. 1984. Using trap plants in the control of insect pests of tropical legumes. In *Proceedings of the International Workshop on Integrated Pest Control Grain Legumes, Goiania, Goias, Brasil, 1983*, pp. 101–12. Brasilia, Brasil: Dep. Difus. Technol., EMBRAPA. 417 pp
43. Jackai, L. E. N., Singh, S. R. 1981. Studies on some behavioral aspects of *Maruca testulalis* on selected species of *Crotalaria*, and *Vigna unguiculata*. *Trop. Grain Legume Bull.* 22:3–6
44. Jones, D., Sterling, W. L. 1979. Manipulation of red imported fire ants in a trap crop for boll weevil suppression. *Environ. Entomol.* 8:1073–77
45. Kennedy, C. W., Smith, W. C. Jr., Jones, J. E. 1986. Effect of early season square removal on three leaf types of cotton. *Crop Sci.* 26:139–45
46. Kennedy, G. G., Margolies, D. C. 1985. Mobile arthropod pests: management in diversified agroecosystems. *Bull. Entomol. Soc. Amer.* 31(3):21–27
47. Kettunen, S., Havukkala, I., Holopainen, J. K., Knuuttila, T. 1988. Nonchemical control of carrot rust fly in Finland. *Ann. Agric. Fenn.* 27:99–105
48. Klingler, J., Doane, J. F. 1974. Verminderung des Drahtwurmschadens an Erdbeersetzlingen durch Fangpflanzen. *Schweiz. Z. Obst- Weinbau* 110(27): 739–42

49. Kobayashi, T., Cosenza, G. W. 1987. Integrated control of soybean stink bugs in the Cerrados. *Jpn. Agric. Res. Q.* 20:229–36

49a. Kogan, A., ed. 1974. *Entomophages and Microorganisms in Plant Protection.* Kishinev, USSR: TsK KP Moldavii. 94 pp. (In Russian)

50. Kogan, M., ed. 1986. *Ecological Theory and Integrated Pest Management Practice.* New York: Wiley & Sons. 362 pp

51. Kogan, M., Turnipseed, S.G. 1987. Ecology and management of soybean arthropods. *Annu. Rev. Entomol.* 32:507–38

52. Kowalska, T. 1986. Action des plantes environnantes sur les populations d'insectes entomophages, en culture de chou tardif. *Colloq. INRA* 36:165–69

52a. Kydonieus, A. F., Beroza, M., eds. 1982. *Insect Suppresion with Controlled Release Pheromone Systems,* Vol. 2. Boca Raton, FL: CRC

53. Leius, K. 1967. Influence of wild flowers on parasitism of tent caterpillar and codling moth. *Can. Entomol.* 99:444–46

54. Lincoln, C., Isely, D. 1947. Corn as a trap crop for the cotton bollworm. *J. Econ. Entomol.* 40:437–38

55. Mayse, M. A. 1983. Culture control in crop fields: a habitat management technique. *Environ. Manage.* 7:15–22

56. McPherson, R. M., Newsom, L. D. 1984. Trap crops for control of stink bugs in soybean. *J. Ga. Entomol. Soc.* 19:470–80

57. Metcalf, R. L. 1985. Plant kairomones and insect pest control. *Ill. Nat. Hist. Surv. Bull.* 33:175–98

58. Metcalf, R. L., Rhodes, A. M., Ferguson, J. E., Metcalf, E. R. 1979. Bitter *Cucurbita* spp. as attractants for diabroticite beetles *Rep. Cucurbit Genet. Coop.* 2:38–39

59. Nascimento, A. S., Mesquita, A. L. M., Sampaio, H. V. 1986. Planta-armadilha no controle de coleobroca em citros. *Inf. Agropecu.* 12(140):13–18

60. Newsom, L. D., Herzog, D. C. 1977. Trap crops for control of soybean pests. *La. Agric.* 20(3):14–15

60a. Pimentel, D., ed. 1981. *Handbook of Pest Management in Agriculture,* Vols. 1, 2. Boca Raton, FL: CRC

61. Powell, W. 1986. Enhancing parasitoid activity in crops. In *Insect Parasitoids,* ed. J. K. Waage, D. J. Greathead, pp. 319–40. London: Academic

62. Rainwater, C. F. 1962. Where we stand on boll weevil control and research. *Proceedings of the Boll Weevil Research Symposium, State College, Missisippi, March 21, 1962.* See Ref. 52a, p. 66

63. Rust, R. W. 1977. Evaluation of trap crop procedures for control of Mexican bean beetle in soybeans and lima beans. *J. Econ. Entomol.* 70:630–32

64. Salazar, V. J., Martinez, N. 1982. Control del complejo *Heliothis* en siembras comerciales de algodon mediante el uso de maiz como cultivo trampa. *Agron. Trop.* 32:187–94

65. Saringer, G., Kacso, A. 1984. Pesticide-free method of control for the pseudocaterpillars of the turnip sawfly. *Nemzetközi Mezogard. Sz.* 28(4):71–74 (In Hungarian)

66. Sarup, P., Panwar, V. P. S., Marwaha, K. K., Siddiqui, K. H. 1977. Exploitation of polyphagy for the control of *Chilo partellus* (Swinhoe) infesting maize crop. *J. Entomol. Res.* 1:184–92

67. Sarup, P., Siddiqui, K. H., Marwaha, K. K. 1986. Possible limitations in exploring the trap of the preferred host sorghum for the control of shoot fly complex *(Atherigona soccata* Rondani and *A. naqvii* Steyskal) in spring sown maize. *J. Entomol. Res.* 10:119–24

68. Sarup, P., Siddiqui, K. H., Marwaha, K. K., Panwar, V. P. S. 1984. Changing pest complex of maize as exemplified by the shootfly *(Atherigona* spp.) preference for hosts in spring season. *J. Entomol. Res.* 8:115–19

69. Saxena, R. C. 1982. Colonization of rice fields by *Nilaparvata lugens* (Stål) and its control using a trap crop. *Crop Prot.* 1:191–98

70. Saxena, R. C., Justo, H. D., Palanginan, E. L. 1988. Trap crop for *Nephotetti virescens* (Homoptera: Cicadellidae) and tungro management in rice. *J. Econ. Entomol.* 81:1485–88

71. Scholl, J. M., Medler, J. T. 1947. Trap strips to control insects affecting alfalfa seed production. *J. Econ. Entomol.* 40:448–54

72. Sevacherian, V., Stern, V. M. 1974. Host plant preferences of Lygus bugs in alfalfa-interplanted cotton fields. *Environ. Entomol.* 3:761–66

73. Shapiro, A. M., Masuda, K. K. 1980. The opportunistic origin of a new citrus pest. *Calif. Agric.* 34(6):4–5

74. Speight, M. R. 1983. The potential of ecosystem management for pest control. *Agric. Ecosyst. Environ.* 10:183–99

75. Steffens, R. J. 1983. Ecology and approach to integrated control of *Dacus frontalis* on the Cap Verde Islands. *Fruit Flies of Economic Importance. Proceedings of the CEC/IOBC International Symposium, Athens, Greece, 1982,* ed.

R. Cavalloro, pp. 632–38. Rotterdam: A. A. Balkema. *Rev. Appl. Entomol. A* 71(9) Ref. 6407

76. Stern, V. M. 1969. Interplanting alfalfa in cotton to control lygus bugs and other insect pests. *Proc. Tall Timbers Conf. Ecol. Anim. Control Habitat Manage.* 1:55–69

77. Stern, V. M. 1981. Environmental control of insects using trap crops, sanitation, prevention, and harvesting. See Ref 60a, 1:199–209

78. Stern, V. M., Mueller, A., Sevacherian, V., Way, M. 1969. Lygus bug control in cotton through alfalfa interplanting. *Calif. Agric.* 23:8–10

79. Stride, G. O. 1969. Investigations into the use of a trap crop to protect cotton from attack by *Lygus vosseleri* (Heteroptera: Miridae). *J. Entomol. Soc. South. Afr.* 32:469–77

80. Swezey, S. L., Daxl, R. G. 1988. Areawide suppression of boll weevil (Coleoptera: Curculionidae) populations in Nicaragua. *Crop Prot.* 7:168–76

81. Swezey, S. L., Salamanca, M. L. 1987. Susceptibility of the boll weevil (Coleoptera: Curculionidae) to methyl parathion in Nicaragua. *J. Econ. Entomol.* 80:358–61

82. Szentesi, A. 1981. Antifeedant-treated potato plants as egg-laying traps for the Colorado beetle *Leptinotarsa decemlineata* Say, Col., Chrysomelidae). *Acta Phytopathol. Acad. Sci. Hung.* 16:203–9

83. Todd, J. W., Schumann, F. W. 1988. Combination of insecticide applications with trap crops of early maturing soybean and southern peas for population management of *Nezara viridula* in soybean (Hemiptera: Pentatomidae). *J. Entomol. Sci.* 23:192–99

84. Vorotintseva, A. F. 1974. Sowing of nectariferous plants as method of enriching garden biocenosis with beneficial insects. See Ref. 49a, pp. 26–30

85. Whitcomb, W. H. 1960. Sweet corn as a trap crop to protect early tomatoes from the tomato fruitworm. *Arkansas Farm Res.* 9(1):10

86. Wisenborn, W. D., Trumble, J. T., Voth, V. 1988. Corn earworm outbreaks in strawberries. *Calif. Agric.* 42(5):25–26

87. Witkowski, J. F., Owens, J. C. 1979. Corn rootworm behavior in response to trap corn. *Iowa State J. Res.* 53:317–24

Annu. Rev. Entomol. 1991. 36:139–58

THE SENSORY PHYSIOLOGY OF HOST-SEEKING BEHAVIOR IN MOSQUITOES

M. F. Bowen

Insect Neurobiology Program, SRI International, Menlo Park, California 94025

KEY WORDS: hematophagy, orientation, attractants, olfaction, chemoreception

INTRODUCTION

Mosquitoes depend on receptors for a variety of sensory modalities, including vision, hearing, mechanoreception, and chemoreception, to transduce environmental information into biologically useful signals. In all likelihood, each modality plays a role in the complex process of identifying and locating appropriate blood-meal hosts. Analysis of blood-feeding behavior in the field, as well as in the laboratory, is a challenging endeavor, not only because blood feeding is a composite of behaviors and the sensory input is complex (3, 29, 37), but also because of the variety of feeding strategies exhibited by mosquitoes in their natural habitats (27) and the variations in internal and external states that impinge upon the expression of the behavior (48, 49). The inter- and intraspecific variations in behavioral response to a host add yet another layer of complexity to the analysis (37, 84).

Given such circumstances, the realization that the process of finding and taking a blood meal can be broken down into a series of discrete stimulus-response behaviors (29, 37) has provided a useful and, indeed, essential operational premise in the effort to understand how and when a mosquito takes a blood meal.

One such stimulus-response behavior is host-seeking. Host-seeking is distinct from other behaviors in the blood-feeding repertoire such as landing,

139

0066-4170/91/0101-0139$02.00

probing, and biting. This distinction is not merely academic: each behavior is mediated by different stimuli detected by distinct sets of receptors, and the behaviors can be uncoupled from each other experimentally (65, 66). Furthermore, in the natural sequence of events leading up to blood feeding, the female must locate a host before it can take a blood meal.

The females' tendency to engage in host-seeking changes in concert with variations in physiological state such as age, reproductive status, and diapause (48, 49, 67). Olfactory receptors in the mosquito undergo alterations in sensitivity that are correlated with these changes in host-seeking behavior (10, 17, 18). Although the mosquito uses both visual and olfactory cues to orient toward a host (3), the olfactory system plays a prominent role in modulating the response. This review summarizes our current knowledge of the receptors responsible for detecting host attractants in the female mosquito and the physiological factors that influence olfactory responsiveness and host-seeking behavior.

In light of the extensive literature on host location in mosquitoes, this review does not attempt to be comprehensive. Emphasis is placed on recent literature that relates to mechanisms of olfactory-mediated host attraction. Further information and earlier literature can be found in more general reviews of blood-feeding and host location (27, 29, 37, 63, 90, 91, 94) and in those covering the endocrinology of host-seeking in mosquitoes (48, 49).

BEHAVIORAL ASPECTS OF HOST SEEKING

Host-seeking has been operationally defined as the in-flight orientation of the avid female toward a potential blood-meal host. The term *host-seeking* may be too general and teleological a term (45) to be of much use in the description of neurophysiological correlates of mosquito host attraction. This definition would benefit from more precise descriptions of behavioral responses of mosquitoes under different odor and temperature conditions, clarification of the range of effectiveness of various host attractants, and knowledge of the identity and characteristics of the receptors that detect odors associated with the various behavioral responses. A meaningful analysis of odor-mediated host-seeking behavior requires knowledge of (*a*) the specific chemical components of host odors and the amount released by the odor source per unit time, (*b*) the configuration of the stimulus in time and space, and (*c*) the contribution of each host odor to the individual responses that make up the composite behavior of host-seeking.

The Chemical Identity and Concentration of Behaviorally Relevant Host Volatiles

CO_2 and lactic acid are the best-described host emanations in terms of chemical identity and amounts released into the host airstream (1, 32, 85).

Not surprisingly these attractants have received the most attention in terms of sensory physiology and behavior. This emphasis does not imply that lactic acid and CO_2 are the only host attractants for mosquitoes because the intact host is still the most effective stimulus in eliciting host-seeking behavior (32, 85), and unidentified chemicals in addition to lactic acid are attractive to mosquitoes (2, 83, 85). 1-Octen-3-ol, for example, was first identified as a host attractant for tsetse (36) but has recently been shown to attract certain species of mosquitoes in the field (46, 92). Current techniques for the capture and chemical analysis of gas-phase emanations such as gas chromatography and atomic mass spectrophotometry should unravel the chemical identity and vapor-phase concentrations of additional host attractants. A knowledge of the emitted levels of identified volatiles is essential because different concentrations of olfactory stimulants may have very different effects on mosquito behavior (32, 85).

Lactic acid has been shown to elicit oriented flight behavior in mosquitoes under laboratory conditions (1, 85). L-lactic acid is more attractive than the D-form under certain conditions, although this observation has not been reconciled with the observation that D- and L-lactic acid are equally effective in eliciting an electrophysiological response from the lactic acid–excited cell (20). Lactic acid is reportedly repellent at high source concentrations (reviewed in 85). However, the low vapor pressure of lactic acid at room or skin temperature limits the amount of the chemical in the vapor phase, so repellent effects may result from the complex chemistry of lactic acid and the presence of secondary compounds rather than lactic acid per se (53). The stimulus measurement relevant to host-seeking behavior is the concentration of the chemical in the vapor phase emanating from the host. For lactic acid, this can be calculated from the evaporation rate estimated from 11 human subjects by Smith et al (85) as a lactic acid flux rate of $15.0 \times 10^{-11} \pm 4.3$ mol/s (range $= 9.2 \times 10^{-11}$ to 24.8×10^{-11} mol/s), which is well within the stimulus dynamic range of the olfactory receptors (see section on sensory aspects of host-seeking).

Tests of lactic acid as an attractant in the field have not been successful (89), although recent field trials suggest that the compound is somewhat attractive for certain species (46). The ability (or inability) of lactic acid to elicit host-seeking behavior must be evaluated in light of two considerations: (a) the probability that a "multi-component chemical stimulus" (32) is required, possibly involving as yet unidentified host odors as well as CO_2 (32, 46, 85, 92); and (b) the effect of the physiological state of the female mosquito on odor reception (10, 17, 18).

CO_2 is an established host attractant for mosquitoes both in the laboratory and in the field (32). Gillies (32) thoroughly reviewed the effect of CO_2 on mosquito behavior. He concludes that CO_2 both activates (induces take-off and sustained flight) and orients mosquitoes. Gillies (32) further points out that CO_2 elicits orientation in mosquitoes under laboratory conditions only when presented intermittently and that, in the field, CO_2 is probably experienced by the mosquito in this way because an odor released from a point source exists as a filamentous plume rather than a broad, homogeneous concentration gradient (70). Since both activation and orientation are part of the process of attraction, the attractive effects of other odors on host-seeking behavior cannot always be elicited unless CO_2 and the host odor are presented simultaneously. Lactic acid, for example, is ineffective in stimulating host-seeking unless CO_2 is also present (46, 85).

The concentration of CO_2 in atmospheric air is 0.03–0.04% and in human breath 4.5%. Excretion from the total human skin surface is about 0.3 to 1.5% of that expired from the lungs (see 32). Local atmospheric levels can vary considerably, depending on time of day and density of vegetation, so the CO_2 differential between atmospheric levels and biologically relevant objectives is considerable (32). Mosquitoes are electrophysiologically sensitive to changes in CO_2 levels as low as 0.01% (43). Unnaturally high CO_2 levels can have anomalous effects on behavior and physiology (71). Because CO_2 induces and maintains flight, mosquitoes may be reluctant to terminate flight and land under such conditions, particularly in the absence of other odors (32).

The Configuration of the Stimulus in Time and Space

Odor released from a point source exists in time and space as a discrete yet discontinuous plume carried downwind, usually in a turbulent airflow that frequently changes direction (70). Because of the filamentous nature and irregular path of the plume, odor is experienced downwind as a series of intermittent pulses (6). Insect olfactory receptors are well equipped to respond to such stimulation. For example, moth olfactory receptors can temporally reflect rapid, intermittent odor pulses (42, 80) and mosquitoes have fast on-off responses to lactic acid and CO_2 (23, 43). This phenomenon is also reflected at more central processing levels (e.g. 14). As demonstrated in moths and mosquitoes, the continuous presence of an odor stimulus can elicit anomalous behavioral responses that may not be representative of the response to a physiological odor source (4, 5, 32). For example, mosquitoes will not display sustained upwind flight in a wind tunnel unless CO_2 is presented in pulses (32, 74). Furthermore, the continuous presentation of stimulus can lead to adaptation in some olfactory receptors (5, 42).

Orientation and Approach to the Host

Both field and laboratory observations employing various devices such as traps, wind tunnels, and olfactometers have yielded a plethora of data regarding the role of odors in mediating host-seeking behavior (29, 37). The data are difficult to compare, perhaps because of variations in (a) species involved, (b) physiological state of the experimental subjects, (c) techniques employed, and (d) conditions under which the experiments were carried out. The most informative techniques, from the standpoint of the neurophysiological correlates of behavior, involve the visual analysis of flight behavior of individual insects under various odor conditions either in a wind tunnel or in the field (e.g. 15, 30). The videotaped data can be readily analyzed according to the protocol of Marsh et al (54). Flight characteristics such as duration, net ground velocity (distance along wind line per unit time), number of turns, turning frequency, turning severity (degrees per turn), angular velocity, and interreversal span (distance between reversal points measured perpendicular to the midline of the plume) can all be obtained from the video recordings.

Such studies are not yet widely available for mosquitoes but have been immensely informative in the analysis of moth pheromone orientation (4, 5). When pheromone is detected, male moths fly upwind, using visual cues to control speed and direction (optomotor anemotaxis), and initiate an internal program of self-steered counter-turning. The behavior is thus a product of both an internal program (idiothetic control) and external stimuli (allothetic control).

Researchers have good reasons to suspect that mosquitoes orient to hosts in a way similar to moths because most mosquitoes fly upwind to the source of host odors (see 92). The pattern and mechanism of upwind orientation in the mosquito may not be exactly identical to that in the Lepidoptera because the mosquito-host relationship is one of predator and prey, whereas mate-finding in the Lepidoptera is an intraspecific interaction of mutual advantage to both odor-emitter and recipient. For example, the hematophagous tsetse does not use self-steered counter-turning, although it does employ optomotor anemotaxis (15). Preliminary studies in our laboratory suggest that the mosquito also employs an irregular flight pattern similar to that seen in the tsetse. Any adaptive advantage to a regular versus an irregular flight path in predatory insects is yet to be determined. Also, the issue of how mosquitoes can orient upwind to odor sources in the dark remains unresolved. Mosquitoes can use visual cues in upwind orientation, as first demonstrated by Kennedy (44) in the *Aedes aegypti*. Gillett (31) has suggested that mosquitoes do not need visual cues to fly upwind but can do so by flying close to the ground and making periodic dips to detect wind shear. In addition, species are different with respect to the details of upwind flight behavior and the conditions under which it can be elicited (26, 46, 74, 89, 92). More studies on flight patterns of mosquitoes in response to host volatiles, both in the laboratory and in the field, are needed.

SENSORY ASPECTS OF HOST-SEEKING

Morphology and Ultrastructure of Olfactory Sensilla

Numerous studies have focused on the structure of the antennal sensilla in mosquitoes (39, 55–59, 61, 62). Similar morphology has been found in culicine and anopheline mosquitoes (39, 55, 56, 58, 61). Sensilla are cuticular extensions that house the sensory cell dendrites. The cell body is located at the sensillar base; the dendrites reside within the cuticular extension; and the axonal afferent nerves from many different sensilla are bundled together into two flagellar nerves that extend proximally before eventually synapsing with interneurons in the antennal lobes of the deutocerebrum (12, 38). In *A. aegypti,* for example, eight morphological types can be distinguished (58); (*a*) sensilla chaetica containing one neuron (mechanoreception), (*b*) sensilla ampullacea containing three neurons (probably thermoreception), (*c*) sensilla coeloconica or pit pegs containing three neurons (temperature reception), (*d*) sensilla basiconica or grooved pegs containing three to five neurons (olfaction), and (*e*) four types of sensilla trichodea (long pointed, short pointed, blunt-tipped type I and blunt-tipped type II), all of which house one or two olfactory neurons. The sensillar types are not uniformly distributed over the antenna and the total number of each type varies from six sensilla coeloconica per antenna to 507 blunt-tipped type I sensilla trichodea per antenna (58). Of a total of approximately 2,058 neurons per flagellar nerve, 93% are associated with known olfactory receptors (58). The thin-walled bulb-shaped organs, or pegs, on the palps each contain three neurons (56), at least one of which is sensitive to changes in CO_2 (43).

Electrophysiology

Ultrastructural studies have been necessary to identify specific sensilla and to determine the number of neurons associated with each. Electrophysiological studies of a single sensillum provide positive functional identification of the olfactory neurons housed within. Detailed, quantitative studies of olfactory electrophysiology of host attractants in mosquitoes have been limited to *A. aegypti* and *Culex pipiens.*

   Single-unit extracellular recordings have identified three types of neurons that detect host-derived stimuli: lactic acid, CO_2, and temperature. Each is associated with a morphologically distinct sensillar type. The response characteristics of these sensory receptors are typical of receptor cells in many other organisms (76).

LACTIC ACID–SENSITIVE CELLS   The sensilla basiconica or grooved pegs (57) contain receptors that are sensitive to lactic acid (23). One type responds to lactic acid with an increase in spike frequency and a second type exhibits a decrease in spike frequency. The dynamic range (range of stimulus intensities over which a receptor or population of receptors can respond without saturation) is the same for both the excited and inhibited cells. Individual cells display variable dynamic ranges (range fractionation) within an effective total range of between 2.7 and 40.0×10^{-11} mol/s (20). Lactic acid flux from a human hand falls well within this range (21, 85). In contrast to Kellogg (43), Davis & Sokolove (23) found that humidity responses were of insufficient sensitivity for the cells to be considered humidity receptors, i.e. 20 impulses/s for a 50% change in humidity; the cells merely depended on humidity for proper functioning (see also 86).

The grooved peg neurons are unresponsive to chemicals other than those closely related to lactic acid. The optimal stimulus configuration for these cells is a 3-carbon, α-hydroxy, monocarboxylic acid. Although the requirement of the α-side group is not rigidly specific (20), the cells display the highest sensitivity to lactic acid. The receptor does not discriminate between the l and d isomers of lactic acid (20).

The sensilla basiconica usually contain three but may have as many as five neurons (57) so that receptors sensitive to chemicals other than lactic acid may also be contained within the grooved pegs.

CO_2-SENSITIVE CELLS   The behavioral synergism between CO_2 and other host odors must occur centrally rather than at the peripheral level because most known odor receptors reside on the antennae whereas the receptors for CO_2 are located on the palps (43). The club-shaped pegs each house three neurons, one of which detects changes in CO_2 (43). The cells exhibit phasic-tonic responses to fluctuations in CO_2 and logarithmic sensitivity to stimulus. Changes in CO_2 levels as low as 0.01% can be detected. The receptors are apparently saturated at 4.0% CO_2, the level present in human breath. Although another cell found in this sensillum responds to organic solvents such as acetone, n-heptane, and amyl acetate, the published data do not indicate whether this response was physiological or pharmacological because the stimulus intensities were not given (43). The three neurons innervating each peg consist of two morphologically distinct types of cells (56). Which cell type is CO_2-sensitive and the quantitative response characteristics of the other neuron are unknown.

THERMORECEPTORS   Mosquitoes have acutely sensitive thermoreceptors located in the sensilla coeloconica (22). These consist of two types of cells: one displays an increase in spike frequency in response to sudden increases in temperature; the other increases its firing rate in response to decreases in temperature (22). The spontaneous firing rates of the cells depend on ambient temperature and the changes in spike frequency observed upon a step change in temperature depend on the starting temperature. The maximum response is observed at an ambient temperature of between 25° and 28°C. Maximum phasic sensitivity is observed in response to temperature changes of $\pm$ 0.2°C, but the cells can respond to changes as low as 0.05°C. Warm, moist convection currents arising from a host are important host-seeking cues, and currents having local thermal differentials of as much as 0.05°C exist at distances greater than two meters away from a 2- to 3-kg rabbit (E. E. Davis, unpublished observations). Such temperature changes are well within the range of detection of the mosquito thermoreceptors.

Neural Coding Characteristics

With respect to the mechanisms for encoding a specific behavior pattern, the terms *odor generalist* and *odor specialist* refer to the degree of specificity of a receptor for a given stimulus (42). The terms *labelled line* and *across fiber pattern* refer to the interpretation (i.e. perception) of the sensory signal by the central nervous system and its behavioral consequence (42). The characteristics of the mosquito olfactory receptors that have been examined so far indicate that these neurons are odor specialists, i.e. they respond to a relatively narrow range of chemical stimuli. Input from one type of odor specialist, CO_2, is adequate for orientation. In this context, the CO_2 receptor could be considered a labelled line. The input from several such odor specialists (receptors for lactic acid, CO_2, and temperature, and, possibly, receptors for as-yet-unidentified host odors) is necessary to evoke the complete response leading to the location and identification of an intact host, i.e. an across fiber pattern of odor specialist cells.

In the case of the lactic acid–sensitive cells, receptor sensitivity has been shown to be directly related to host-seeking behavior. The tendency of a given population of females to exhibit host-seeking behavior changes in conjunction with physiological events such as vitellogenesis and diapause (see section on sensory and endocrinological aspects of host-seeking). Actively host-seeking females invariably have receptors that are highly sensitive to lactic acid, but only low-sensitivity receptors are found on non-host-responsive females (10, 17, 18). When host-seeking behavior is inhibited, the dynamic range of the lactic acid–excited cells shifts to stimulus intensities that exceed the expected maximum lactic acid emission from a human hand (21). The response to stimulus of the lactic acid–excited cell is inhibited by the repellent DEET (N,N-diethyl-m-toluamide) (19). The lactic acid–inhibited cell does not undergo such changes in sensitivity (21).

Based on the net response-firing patterns of both types of neurons in host-responsive and non-host-responsive females in response to lactic acid alone and in combination with DEET, a model of the sensory control of host-seeking behavior has been proposed that considers the response characteristics of both the lactic acid–excited and –inhibited cells (21). The model assumes that host-seeking behavior is directly related to the total activity in this set of neurons. The role of the inhibited cell is to actively enforce host-seeking inhibition by decreasing the net afferent output to below the spontaneous firing rate when the sensitivity of the lactic acid–excited cells shifts to a lower sensitivity state (21). Such a net decrease is observed during egg maturation (see 18), during diapause (10), and in the presence of the repellent DEET (19), and each situation is characterized by the absence of host-seeking.

SENSORY AND ENDOCRINOLOGICAL ASPECTS OF HOST-SEEKING

Young and Nulliparous Females

Most female mosquitoes do not become host-responsive until several days after pupal-adult emergence (see 18). The appearance of host-seeking behavior in A. aegypti coincides with a progression in lactic acid–receptor sensitivity, which suggests a developmental process. Newly emerged (0–24 h post-emergence) females are nonresponsive and possess only silent (nonspiking) neurons. Older females (24–96 h post-emergence) that are not host-responsive possess more spiking neurons, most of which are nonresponsive or nonspecific; some neurons show specificity for lactic acid but have low sensitivity. Females of any age that exhibit host-seeking behavior (18–24 hours post emergence and older) possess neurons that are highly sensitive to lactic acid (18). The appearance of this behavior in A. aegypti roughly coincides with juvenile hormone-dependent previtellogenic ovarian development, but the correlation is incidental: juvenile hormone deprivation by allatectomy (removal of the corpora allata) within one hour of adult emergence fails to prevent the appearance of host-seeking even though ovarian follicles remain teneral (9). Lactic acid sensitivity is likewise unaffected by allatectomy (9). In A. aegypti, host-seeking behavior must be either independent of juvenile hormone or have a sensitivity threshold and/or sensitive period very different from that of the ovaries (28).

In C. pipiens, the development of host-seeking behavior after emergence may be juvenile hormone dependent. Movement toward a host is not observed in females that have been allatectomized at one hour post-emergence (R. Meola & J. Readio, personal communication). This effect was assessed using paired cages, of which one contained a host and the other contained experimental or control groups of mosquitoes. Only 1 of 92 allatectomized females moved to the host cage in overnight trials. Allatectomized females in which the corpora allata had been re-implanted moved readily to the host cage in overnight trials, and 5, 10, and 50 ng of the juvenile hormone analogue methoprene also restored the behavior in a dose-dependent manner. The movement of unoperated females in the absence of a host was similar to that of allatectomized females in the presence of a host. The effect of juvenile hormone-deprivation on host attractant-receptor sensitivity in *Culex* has not been explored. The finding that allatectomy affects receptor sensitivity would also support the notion that juvenile hormone initiates host-seeking behavior in *Culex;* however, a lack of a juvenile hormone–deprivation effect on receptor functioning would not preclude a role for juvenile hormone in host-seeking behavior. The behavior can also be modulated at levels other than the periphery, as during distension inhibition (see section on blood-fed and gravid females) and during the circadian cycle (see section on the circadian system).

Blood-Fed and Gravid Females

The immediate effect of a blood meal above a threshold volume (2.5 μl in young *A. aegypti* females) is inhibition of host-seeking resulting from the activation of stretch receptors that reside in the anterior part of the abdomen (reviewed in 48, 49). Whether this distension inhibition is mediated directly by nervous signals or by hormonal intermediaries is unknown. Whatever the mechanism of distension inhibition in mosquitoes, the inhibition is not effected peripherally: olfactory receptor sensitivity is unaffected by distension; lactic acid receptor sensitivity remains high for about 18–24 h after the blood meal (17).

After the blood meal is digested and distension is alleviated, humoral events related to oocyte maturation in mated females inhibit host-seeking until after oviposition, a phenomenon referred to as *oocyte-induced behavioral inhibition* (reviewed in 48, 49). In blood-fed females that go on to develop eggs, the ovaries release an initiating factor 6–12 hours after blood feeding. The ovarian factor stimulates the fat body to produce a hemolymph-borne substance that either directly or indirectly renders the peripheral olfactory receptors less sensitive to lactic acid (51).

The identities of the ovarian and fat body factors are unknown, although some evidence suggests that the ovarian initiating factor released within 12 h after a blood meal is an ecdysteroid (11). Large doses of ecdysteroids can inhibit host-seeking (7), but this effect is pharmacological and nonspecific (47).

The mechanism by which olfactory receptor sensitivity is shifted is unknown. The morphological arrangement of insect receptor neurons within the sensillum suggests several routes by which receptor function could be controlled via hemolymph-borne signals (8). The cell body lies at the base of the sensillum, which is filled with extrahemolymphatic fluid called receptor lymph. The dendrites reside within this lymphatic compartment, which has a composition apparently related to receptor function: high K^+ concentration (24) and stimulus-binding and degradative proteins (95). The receptor lymph and the receptor cells themselves are exposed to hemolymphatic signals such as hormones, and alterations in either could result in changes in receptor functioning (8). Although direct humoral control of receptor function has not been demonstrated, anatomical (60, 75), electrophysiological (8, 17, 51), and behavioral (8, 17, 51) experiments all suggest that such control can and does occur. The possibility that the antennal receptors are subject to efferent neural control cannot be ruled out, although to date such innervation has not been demonstrated.

HOST-SEEKING IN MOSQUITOES THAT UNDERGO ADULT DIAPAUSE

Many temperate-zone mosquitoes overwinter in a state of dormancy called diapause, during which reproduction and development are suspended. Diapause is induced by endocrine changes implemented through the reception of seasonal photoperiodic and thermal cues that presage the onset of inimical climatic conditions. The occurrence and physiology of embryonic, larval, and adult diapause in mosquitoes are the topics of a recent and comprehensive review by Mitchell (67), and adult diapause in *Culex* mosquitoes has also been recently reviewed (28).

The effect of diapause on host-seeking behavior has been examined only in species that undergo adult reproductive diapause. Diapausing females (adult male mosquitoes do not survive the winter) are inseminated, possess teneral ovarian follicles, and have greatly hypertrophied fat bodies resulting from increased lipid deposition (28, 67, 96). Depending on species and strain, diapausing females can show one of two patterns of blood-feeding behavior during diapause (reviewed in 96). Some groups may take occasional blood meals during diapause without developing eggs, a phenomenon known as *gonotrophic dissociation*. Other mosquitoes do not blood-feed or develop eggs during diapause, a situation referred to as *gonotrophic concordance*. Concordant mosquitoes depend entirely on plant juices to build up the lipid reserves necessary for successful overwintering (67, 96).

The best-described case of gonotrophic concordance is that of *C. pipiens*. This species does not host-seek during diapause (10, 66) but can be induced to bite and take a blood meal if placed in close proximity to a host (66). When blood feeding is induced in diapausing females, the meals taken are sub-threshold or incompletely digested and prematurely excreted, and thus not used for fat body development (68, 69). Analysis of the peripheral receptors corroborates these observations. The lactic acid–sensitive cells in diapausing females consist of low-sensitivity, nonspecific, and nonresponsive neurons whereas high-sensitivity receptors are present on both nondiapausing and postdiapausing females (10). Highly sensitive receptors are not found on diapausing animals (10), a condition that is reminiscent of that in teneral and gravid *A. aegypti* females (17, 18). The question arises, then, whether the diapause condition represents an interrupted state of imaginal development (akin to that in teneral *A. aegypti*) or whether the sensory system is mature, albeit inhibited (as in the gravid *A. aegypti*). Nonresponsive, nonspecific, and silent neurons may reflect undifferentiated receptors, so estimation of their numbers can give some idea of the stage of development of the nervous system. Preliminary data, showing the presence of primarily undifferentiated neurons in diapausing females, suggest that the peripheral sensory system in diapausing females is in a state of interrupted or delayed development (M. F. Bowen, in preparation). Although the data for the lactic acid–sensitive cells are inconclusive, the analysis of other receptor-cell groups (specifically, the neurons located in the sensilla trichodea that are sensitive to oviposition site-related volatiles) suggests that diapause interrupts peripheral sensory development: the number of nonresponsive neurons is higher in diapausing females than in nondiapausing or gravid females. These observations suggest that the peripheral sensory system undergoes developmental processes in early imaginal life and that this development can be influenced by physiological state so as to affect adult behavior.

Gonotrophic dissociation was first demonstrated in *Anopheles labranchiae atroparvus,* a species that remains sequestered in domestic shelters during the fall and winter and continues to take blood meals during this time without developing eggs (see 96). The phenomenon has since been confirmed in other anophelines (96). As Washino (96) points out, gonotrophic dissociation may occur only intermittently throughout the geographic range of a given species and the incidence of blood-feeding can vary considerably between populations. Because fat body development was roughly equivalent in dissociative and concordant *Anopheles freeborni* females and the survival rates of the two types of diapausing females did not differ, Washino (96) concluded that the selective advantage of blood-feeding during diapause was related to the relative availability of blood-meal hosts as compared to plant nutrient sources.

With respect to host-seeking behavior and its relationship to gonotrophic dissociation, several points require clarification. 1. It is not clear if females that exhibit gonotrophic dissociation can host-seek strictly speaking. 2. If host-seeking is indeed expressed by such females, the fact that this behavior can persist independently of physiological changes attendant on the diapause state suggests that the behavior in anophelines, unlike that in *Culex,* is not strictly controlled by diapause-inductive processes but is somewhat labile. The physiological conditions that result in the expression or nonexpression of blood-feeding behavior in these diapausing mosquitoes are unknown. 3. The sensory physiology of dissociation females has not been examined so one cannot evaluate the role of the peripheral sensory system in gonotrophic dissociation at this time.

Washino's (96) statement that "the physiological mechanism of diapause expressed as gonotrophic concordance or dissociation still is unclear and requires considerable clarification" remains relevant. The endocrinology of adult diapause in mosquitoes is not well understood. The failure of the follicles to develop is generally believed to result from low juvenile hormone titers in the adult (reviewed in 28, 67). Mitchell (67) points out that because the corpora allata do not affect lipid and glycogen synthesis, the involvement in diapause of the medial brain neurosecretory cells is a distinct possibility because these cells have been shown to control this metabolic pathway in other species. The analysis of diapause in mosquitoes, as well as in other insects, is blocked by a lack of basic information on hormone titers during development and complicated by the fact that the environmental cues that induce diapause are perceived in stages prior to that in which diapause is expressed. In adult diapause in mosquitoes, the larva and pupa are the stages sensitive to diapause-inducing stimuli (reviewed in 67). Although covert, the juvenile-stage events that result in adult diapause should be amenable to analysis using current techniques. The internal and external larval milieu may have profound effects on adult behavior as well as sensillar morphology and receptor functioning (see below). The effect of rearing conditions on adult behavior in nondiapausing as well as diapausing mosquitoes is largely unexplored and deserves more attention.

OTHER PHYSIOLOGICAL FACTORS IMPINGING ON THE EXPRESSION OF HOST-SEEKING

Nutritional State

Both larval and adult nutrition can affect the expression of host-seeking behavior in adult females. Rearing A. aegypti larvae on a suboptimal diet gives rise to adults that are not only smaller in size but also less likely to engage in host-seeking behavior (50). Providing sugar to such adults fails to increase host responsiveness (50). Because Culex larvae reared on a suboptimal diet can metamorphose into adults with impaired flight capacity (16), the ability to fly, rather than host-seeking per se, may be affected by larval nutritional conditions in A. aegypti.

Sugar-feeding in the adult can affect host responsiveness immediately after sugar ingestion (40) as well as during vitellogenesis: Sugar-deprived A. aegypti females are more likely to exhibit host-seeking behavior after a blood meal regardless of whether they develop eggs (reviewed in 48, 49). Such increased host-seeking is believed to result from the absence of oocyte-induced inhibition (reviewed in 48, 49). One would expect the peripheral host-attractant receptors to display high sensitivity in sugar-deprived, host-responsive females.

Aging

Aging alters host-seeking behavior in several ways; the effects depend upon whether the female is chronologically or gonotrophically old (reviewed in 48, 49). First, the threshold blood volume for distension inhibition is lower in chronologically old females. Second, the recovery of host-seeking after distension inhibition occurs more rapidly in gonotrophically aged females than in chronologically aged females, possibly because of faster blood digestion. Third, the onset of oocyte-induced inhibition of host-seeking behavior is delayed and the incidence is diminished as both gonotrophic and physiologic age increase.

The effect of aging on peripheral receptor sensitivity in mosquitoes has not been systematically examined. Age-related changes in receptor function occur in other Diptera. In blowflies, for example, the numbers of responsive salt and sugar receptors as well as the sensitivity of the remaining operative cells decrease with age in blowflies (77, 87, 88). The effect of chronologic as well as gonotrophic age on olfactory receptor functioning in mosquitoes deserves more attention, particularly since older populations in the field are of considerable epidemiological significance (48).

The Circadian System

Mosquitoes exhibit daily periods of activity and inactivity that are the external manifestations of endogenous circadian oscillators (41). Spontaneous flight activity has received the most attention, but host-seeking behavior is also expressed in a circadian pattern in many species (64, 78, 93). To express host-seeking behavior, the mosquito must be willing to fly, but evidence indicates that the two behaviors are not tightly coupled temporally and peaks of flight activity can precede and/or lag behind host-seeking (64, 79). Host-seeking behavior in *Culex* is expressed during the dark phase in both the laboratory (10) and in the field (64). The sensitivity of the lactic acid receptors does not vary throughout the day (M. F. Bowen, in preparation). High sensitivity to lactic acid can be observed in the light phase (when females are not host-responsive) as well as in the dark phase (when females are host-responsive) of the light-dark cycle. Thus, the control of the daily expression of host-seeking behavior in this species does not reside at the level of the peripheral sensory receptors.

Morphological Variation

Interspecific variations in sensillar number and density in mosquitoes have been described (55, 59, 61, 62). The significance of these variations is not clear, but the evolution of autogeny in the Culicidae offers one possible explanation (59). *Wyeomyia*, for example, is a genus that has both autogenous and anautogenous representatives. This group has retained piercing mouthparts, so differences in mouthpart morphology must not cause the loss of the blood-feeding habit in autogenous members of this group (25). The density of grooved peg sensilla in an autogenous, non-blood-feeding strain of *Wyeomyia smithii* is significantly lower than that in the anautogenous *Wyeomyia aporonema* (62). If the grooved pegs in this genus house host-attractant sensitive receptors, as in other mosquitoes, then the loss of sufficient afferent input from this set of sensory cells might account for the absence of blood-feeding behavior. This hypothesis would be greatly strengthened if intraspecific differences in sensillar number between blood-feeding and non-blood-feeding strains could be demonstrated.

In mosquitoes, autogeny is associated with precocious sexual receptivity and advanced ovarian maturation. Both of these phenomena are juvenile hormone dependent (33–35, 52), and studies in roaches suggest that this morphogenic hormone can also affect sensillar number. Elevated or prolonged preimaginal juvenile hormone levels reduce sensillar numbers and affect response to pheromone in adult roaches (81, 82). Experimental evidence from a non-blood-feeding strain of *W. smithii* that is autogenous for successive gonotrophic cycles (73) suggest that an early release of juvenile hormone in the pupal stage is responsible for precocious sexual maturation (72). One can speculate that juvenile hormone may also reduce sensillar numbers, thus limiting the number of receptors available for the detection of host attractants and rendering the strain less inclined to engage in host-seeking. This hypothesis remains to be tested.

OVERVIEW AND PROSPECTS FOR FUTURE RESEARCH

It has become apparent in recent years that the mosquito sensory system is not merely a passive conduit of information from the environment to the central nervous system. Besides the classic sensory functions of transduction and high-gain amplification of specific external chemical signals, olfactory neurons apparently contribute to the control of behavioral output by modulating sensory input. This rather surprising phenomenon comes about through a combination of the specific neural coding characteristics of the peripheral olfactory neurons and their responsiveness to systemic signals that act to change receptor sensitivity.

Sensory physiology does not provide a complete answer to the question of how behavior is controlled in mosquitoes, but, as Hocking (37) pointed out, "behavioral work is always more difficult in the absence of an adequate knowledge of sensory physiology." Our current understanding of mosquito behavior would greatly benefit from comparative studies of species of mosquitoes other than *Culex* and *Aedes*. Identification of the endogenous factors that act on peripheral receptor sensitivity would facilitate the study of the mechanism of receptor sensitivity modulation, not only by endogenous signals, but also by exogenous factors such as repellents. The effects of the larval environment, senescence, learning, and other host attractants on host-seeking and receptor functioning need to be incorporated into current models of host-seeking behavior. The mosquito has been treated much like a black box in the input-output analysis of behavior, and knowledge of central processing and projection patterns of sensory afferents would greatly expand our comprehension of mosquito host-seeking as it has for behavior in other taxa (13, 14, 38). Finally, more functional terminology that describes what the mosquito "is actually doing" rather than its "presumed state of mind" (45) would greatly facilitate the analysis of host attraction in mosquitoes.

ACKNOWLEDGMENTS

The author thanks Edward E. Davis, Marc J. Klowden, Roger W. Meola, Carl J. Mitchell, and Janice Readio for their suggestions on an earlier version of this review and Roger Meola and Janice Readio for providing unpublished data. Research on the sensory physiology of mosquitoes has been supported by National Institutes of Health grants AI-23336 to the author and AI-21267 to Edward E. Davis.

Literature Cited

1. Acree, F., Turner, R. B., Gouck, H. K., Beroza, M., Smith, N. 1968. L-lactic acid: A mosquito attractant isolated from humans. *Science* 161:1346–47
2. Ahmadi, A., McClelland, G. A. H. 1985. Mosquito-mediated attraction of female mosquitoes to a host. *Physiol. Entomol.* 10:251–55
3. Allen, S., Day, J. F., Edman, J. D. 1987. Visual ecology of biting flies. *Annu. Rev. Entomol.* 32:297–316
4. Baker, T. C. 1989. Pheromones and flight behavior. In *Insect Flight*, ed. G. J. Goldsworthy, C. H. Wheeler, pp. 231–55. Boca Raton, FL: CRC
5. Baker, T. C. 1989. Sex pheromone communication in the Lepidoptera: New research progress. *Experientia* 45:248–62
6. Baker, T. C., Haynes, K. F. 1989. Field and laboratory electroantennographic

measurements of pheromone plume structure correlated with oriental fruit moth behavior. *Physiol. Entomol.* 14:1–12
7. Beach, R. 1979. Mosquitoes: Biting behavior inhibited by ecdysone. *Science* 205:829–31
8. Blaney, W. M., Schoonhoven, L. M., Simmonds, M. S. J. 1986. Sensitivity variations in insect chemoreceptors: a review. *Experientia* 42:13–19
9. Bowen, M. F., Davis, E. E. 1989. The effects of allatectomy and juvenile hormone replacement on the development of host-seeking behavior and lactic acid receptor sensitivity in the mosquito *Aedes aegypti*. *Med. Vet. Entomol.* 3:53–60
10. Bowen, M. F., Davis, E. E., Haggart, D. A. 1988. A behavioural and sensory analysis of host-seeking behaviour in the

diapausing mosquito *Culex pipiens*. *J. Insect Physiol.* 34:805–13

11. Bowen, M. F., Loess-Perez, S. 1989. A re-examination of the role of ecdysteroids in the development of host-seeking inhibition in blood-fed *Aedes aegypti* mosquitoes. In *Host-Regulated Developmental Mechanisms in Vector Arthropods*. ed. D. Borovsky, A. Spielman, pp. 286–91. Vero Beach, FL: Univ. Fla-IFAS

12. Childress, S. A., McIver, S. B. 1984. Morphology of the deutocerebrum of female *Aedes aegypti* (Diptera: Culicidae). *Can. J. Zool.* 62:1320–29

13. Christensen, T. A., Hildebrand, J. G. 1987. Functions, organization, and physiology of the olfactory pathways in the lepidopteran brain. In *Arthropod Brain: Its Evolution, Development, Structure and Functions*, ed. A. P. Gupta, pp. 457–84. New York: Wiley

14. Christensen, T. A., Hildebrand, J. G. 1988. Frequency coding by central olfactory neurons in the sphinx moth *Manduca sexta. Chem. Senses* 13:123–30

15. Colvin, J., Brady, J., Gibson, G. 1989. Visually-guided, upwind turning behavior of free-flying tsetse flies in odor-laden wind: a wind tunnel study. *Physiol. Entomol.* 14:31–40

16. Dadd, R. H., Kleinjan, J. E. 1979. Essential fatty acid for the mosquito *Culex pipiens:* arachidonic acid. *J. Insect Physiol.* 25:495–502

17. Davis, E. E. 1984. Regulation of sensitivity in the peripheral chemoreceptor systems for host-seeking behaviour by a haemolymph-borne factor in *Aedes aegypti. J. Insect Physiol.* 30:179–83

18. Davis, E. E. 1984. Development of lactic acid-receptor sensitivity and host-seeking behaviour in newly emerged female *Aedes aegypti* mosquitoes. *J. Insect Physiol.* 30:211–15

19. Davis, E. E. 1985. Insect repellents: Concepts of their mode of action relative to potential sensory mechanisms in mosquitoes (Diptera: Culicidae). *J. Med. Entomol.* 22:237–43

20. Davis, E. E. 1988. Structure-response relationship of the lactic acid-excited neurons in the antennal grooved-peg sensilla of the mosquito *Aedes aegypti. J. Insect Physiol.* 34:443–50

21. Davis, E. E., Haggart, D. A., Bowen, M. F. 1987. Receptors mediating host-seeking behaviour in mosquitoes and their regulation by endogenous hormones. *Insect Sci. Appl.* 8:637–41

22. Davis, E. E., Sokolove, P. G. 1975. Temperature responses of antennal receptors of the mosquito, *Aedes aegypti. J Comp. Physiol.* 96:223–36

23. Davis, E. E., Sokolove, P. G. 1976. Lactic acid-sensitive receptors on the antennae of the mosquito, *Aedes aegypti. J. Comp. Physiol.* 105:43–54

24. De Kramer, J. J., Hemberger, J. 1987. The neurobiology of pheromone recognition. See Ref. 76a, pp. 433–72

25. Downes, J. A. 1971. The ecology of blood-sucking Diptera: An evolutionary perspective. In *Ecology and Physiology of Parasites*, ed. A. M. Fallis, pp. 232–58. Toronto: Univ. Toronto Press

26. Edman, J. D. 1979. Orientation of some Florida mosquitoes (Diptera: Culicidae) toward small vertebrates and carbon dioxide in the field. *J. Med. Entomol.* 15:292–96

27. Edman, J. D., Spielman, A. 1988. Blood feeding by vectors: physiology, ecology, behavior, and vertebrate defense. In *Epidemiology of Arthropod-Borne Viral Diseases*, ed. T. Monath, 1:153–89. Boca Raton, FL:CRC

28. Eldridge, B. F. 1987. Diapause and related phenomena in *Culex* mosquitoes: their relation to arbovirus disease ecology. In *Current Topics in Vector Research*, ed. K. F. Harris, pp. 1–28. New York: Springer-Verlag

29. Friend, W. G., Smith, J. J. B. 1977. Factors affecting feeding by bloodsucking insects. *Annu. Rev. Entomol.* 22:309–31

30. Gibson, G., Brady, J. 1988. Flight behaviour of tsetse flies in host odour plumes: the initial response to leaving or entering odour. *Physiol. Entomol.* 13:29–42

31. Gillett, J. D. 1979. Out for blood: flight orientation up-wind in the absence of visual cues. *Mosq. News* 39:221–29

32. Gillies, M. T. 1980. The role of carbon dioxide in host-finding by mosquitoes (Diptera: Culicidae). *Bull. Entomol. Res.* 70:525–32

33. Gwadz, R. W., Lounibos, L. P., Craig, G. B. 1971. Precocious sexual receptivity induced by a juvenile hormone analogue in females of the yellow fever mosquito, *Aedes aegypti. Gen. Comp. Endocrinol.* 16:47–51

34. Gwadz, R. W., Spielman, A. 1973. Corpus allatum control of ovarian development in *Aedes aegypti. J. Insect Physiol.* 19:1441–48

35. Hagedorn, H. H., Turner, S., Hagedorn, E. A., Pontecorvo, D., Greenbaum, P., et al. 1977. Post emergence growth of the ovarian follicles of *Aedes aegypti. J. Insect Physiol.* 23:203–6

36. Hall, D. R., Beevor, P. S., Cork, A., Nesbitt, B. F., Vale, G. A. 1984. 1-Octen-3-ol: A potent olfactory stimulant and attractant for tsetse isolated from cattle odours. *Insect Sci. Appl.* 5:335–39

37. Hocking, B. 1971. Blood-sucking behavior of terrestrial arthropods. *Annu. Rev. Entomol.* 16:1–26

38. Homberg, U., Christensen, T. A., Hildebrand, J. G. 1989. Structure and function of the deutocerebrum in insects. *Annu. Rev. Entomol.* 34:477–501

39. Ismail, I. A. H. 1964. Comparative study of sense organs in the antennae of culicine and anopheline female mosquitoes. *Acta Trop.* 21:155–68

40. Jones J. C., Madhukar, B. V. 1976. Effects of sucrose on blood avidity in mosquitoes. *J. Insect Physiol.* 22:357–60

41. Jones, M. D. R. 1976. Persistence in continuous light of a circadian rhythm in the mosquito *Culex pipiens fatigans* Wied. *Nature* 261:491–92

42. Kaissling, K.-E. 1987. *Wright Lectures on Insect Olfaction*, ed. K. Colbow. Burnaby, BC: Simon Fraser Univ. 189 pp.

43. Kellogg, F. E. 1970. Water vapour and carbon dioxide receptors in *Aedes aegypti*. *J. Insect Physiol.* 6:99–108

44. Kennedy, J. S. 1940. The visual responses of flying mosquitoes. *Proc. Zool. Soc. Lond. Ser. A* 109:221–42

45. Kennedy, J. S. 1986. Some current issues in orientation to odour sources. See Ref. 75a, pp. 11–25

46. Kline, D. L., Takken, W., Wood, J. R. Carlson, D. A. 1990. Field studies on the potential of butanone, carbon dioxide, honey extract, 1-octen-3-ol, L-lactic acid and phenols as attractants for mosquitoes. *Med. Vet. Entomol.* 4: In press

47. Klowden, M. J. 1982. Nonspecific effects of large doses of 20-hydroxyecdysone on the behavior of *Aedes aegypti*. *Mosq. News* 42:184–89

48. Klowden, M. J. 1988. Factors influencing multiple host contacts by mosquitoes during a single gonotrophic cycle. *Entomol. Soc. Am. Misc. Publ.* 68:29–36

49. Klowden, M. J. 1990. The endogenous regulation of mosquito reproductive behavior. *Experientia*. In press

50. Klowden, M. J., Blackmer, J. L., Chambers, G. M. 1988. Effects of larval nutrition on the host seeking behavior of adult *Aedes aegypti* mosquitoes. *J. Am. Mosq. Control Assoc.* 4:73–75

51. Klowden, M. J., Davis, E. E., Bowen, M. F. 1987. Role of the fat body in the control of host-seeking behavior in the mosquito *Aedes aegypti*. *J. Insect Physiol.* 33:643–46

52. Lea, A. O. 1968. Mating of virgin *Aedes aegypti* without insemination. *J. Insect Physiol.* 14:305–8

53. Lockwood, L. B., Yoder, D. E., Zienty, M. 1965. Lactic acid chemistry and metabolism of L- and D-lactic acids. *Ann. NY Acad. Sci.* 119:854–67

54. Marsh, D., Kennedy, J. S., Ludlow, A. R. 1978. An analysis of anemotactic zigzagging flight in male moths stimulated by pheromone. *Physiol. Entomol.* 3:221–40

55. McIver, S. B. 1970. Comparative study of antennal sense organs of female culicine mosquitoes. *Can. Entomol.* 102:1258–67

56. McIver, S. B. 1972. Fine structure of pegs on the palps of female culicine mosquitoes. *Can. J. Zool.* 50:571–82

57. McIver, S. B. 1974. Fine structure of antennal grooved-pegs of the mosquito, *Aedes aegypti*. *Cell Tiss. Res.* 153:327–37

58. McIver, S. B. 1978. Structure of sensilla trichodea of female *Aedes aegypti* with comments on innervation of antennal sensilla. *J. Insect Physiol.* 24:383–90

59. McIver, S. B. 1987. Sensilla of haematophagous insects sensitive to vertebrate host-associated stimuli. *Insect Sci. Appl.* 8:627–35

60. McIver, S. B., Baker, K. 1988. Putative neurosecretory cells in the antenna of *Aedes aegypti* (Diptera: Culicidae). Presented at Annu. Meet. Entomol. Soc. Am., Louisville, Kentucky

61. McIver, S. B., Charlton, C. C. 1970. Studies on the sense organs on the palps of selected culicine mosquitoes. *Can. J. Zool.* 48:293–95

62. McIver, S. B., Hudson, A. 1972. Sensilla on the antennae and palps of selected *Wyeomyia* mosquitoes. *J. Med. Entomol.* 9:37–45

63. Meola, R., Readio, J. 1988. Juvenile hormone regulation of biting behavior and egg development in mosquitoes. In *Advances in Disease Vector Research*, ed. K. F. Harris, 5:1–24. New York: Springer-Verlag

64. Meyer, R. P., Reisen, W. K., Eberle, M. E., Milby, M. M. 1986. The nightly host-seeking rhythms of several culicine mosquitoes (Diptera: Culicidae) in the southern San Joaquin valley of California. In *Proc. Papers 54th Annu. Conf. Calif. Mosq. Vector Control Assoc.* March 16–19, p. 136. Sacramento: CVCVA (Abstr.)

65. Mitchell, C. J. 1981. Diapause termination, gonoactivity, and differentiation of host-seeking behavior from blood-feeding behavior in hibernating *Culex tarsalis* (Diptera: Culicidae). *J. Med. Entomol.* 18:386–94

66. Mitchell, C. J. 1983. Differentiation of host-seeking behavior from blood-feeding behavior in overwintering *Culex pipiens* (Diptera: Culicidae) and observations on gonotrophic dissociation. *J. Med. Entomol.* 20:157–63

67. Mitchell, C. J. 1988. Occurrence, biology, and physiology of diapausing in overwintering mosquitoes. In *The Arboviruses: Epidemiology and Ecology*, ed. T. P. Monath, 1:191–217. Boca Raton, FL:CRC

68. Mitchell, C. J., Briegel, H. 1989. Inability of diapausing *Culex pipiens* (Diptera: Culicidae) to use blood for producing lipid reserves for overwintering survival. *J. Med. Entomol.* 26:318–26

69. Mitchell, C. J., Briegel, H. 1989. Fate of the blood meal in force-fed, diapausing *Culex pipiens* (Diptera: Culicidae). *J. Med. Entomol.* 26:332–41

70. Murlis, J. 1986. The structure of odour plumes. See Ref. 75a, pp. 27–38

71. Nicolas, G. N., Sillans, D. 1989. Immediate and latent effects of carbon dioxide on insects. *Annu. Rev. Entomol.* 34:97–116

72. O'Meara, G. F., Lounibos, L. P. 1981. Reproductive maturation in the pitcher-plant mosquito., *Wyeomyia smithii*. *Physiol. Entomol.* 6:437–43

73. O'Meara, G. F., Lounibos, L. P., Brust, R. A. 1981. Repeated egg clutches without blood in the pitcher-plant mosquito. *Ann. Entomol. Soc. Am.* 74:68–72

74. Omer, S. M. 1979. Responses of females of *Anopheles arabiensis* and *Culex pipiens fatigans* to air currents, carbon dioxide and human hands in a flight tunnel. *Entomol. Exp. Appl.* 26:142–51

75. Pass G., Sperk G., Aricola H., Baumann E., Penzlin H. 1988. The antennal heart of the American cockroach. a neurohormonal release site containing octopamine. In *Neurobiology of Invertebrates: Transmitters, Modulators and Receptors*, ed. J. Salanki, K. S-Rozsa, 36:341–50. *Symposia Biologica Hungarica*. Budapest, Hungary: Akademiai Kiado

75a. Payne, T. P., Birch, M. C., Kennedy, C. É. J., eds. 1986. *Mechanisms in Insect Olfaction*. Oxford: Clarendon

76. Perkel, D. H., Bullock, T. H. 1968. Neural coding. *Neurosci. Res. Prog. Bull.* 6:221–348

76a. Prestwich, G. D., Blomquist, G. J., eds. 1987. *Pheromone Biochemistry*. Orlando, FL: Academic

77. Rees, C. J. C. 1970. Age-dependency of response in an insect chemoreceptor sensillum. *Nature* 227:740–42

78. Reisen, W. K., Aslamkhan, M. 1978. Biting rhythms of some Pakistan mosquitoes (Diptera: Culicidae). *Bull. Entomol. Res.* 68:313–30

79. Rozendaal, J. A. 1989. Biting and resting behavior of *Anopheles darlingi* in the Suriname rainforest. *J. Am. Mosq. Control Assoc.* 5:351–58

80. Rumbo, E. R., Kaissling, K.-E. 1989. Temporal resolution of odour pulses by three types of pheromone receptor cells in *Antheraea polyphemus*. *J. Comp. Physiol.* A165:281–91

81. Schafer, R. 1977. The nature and development of sex attractant specificity in cockroaches of genus *Periplaneta* III. Normal intra- and interspecific behavioral responses and responses of insects with juvenile hormone-altered antennae. *J. Exp. Zool.* 199:73–84

82. Schafer, R., Sanchez, T. V. 1976. The nature and development of sex attractant specificity in cockroaches of the genus *Periplaneta* II. Juvenile hormone regulates sexual dimorphism in the distribution of antennal olfactory receptors. *J. Exp. Zool.* 198:323–36

83. Schreck, C. E., Smith, N., Carlson, D. A., Price, G. D., Haile, D. Godwin, D. R. 1981. A material isolated from human hands that attracts female mosquitoes. *J. Chem Ecol.* 8:429–38

84. Service, M. 1985. *Anopheles gambiae:* Africa's principal malaria vector, 1902–1984. *Bull. Entomol. Soc. Am.* 31:8–12

85. Smith, C. N., Smith, N., Gouck, H. K., Weidhaas, D. E., Gilbert, I. H., et al. 1970. L-lactic acid as a factor in the attraction of *Aedes aegypti* (Diptera: Culicidae) to human hosts. *Ann. Entomol. Soc. Am.* 63:760–70

86. Städler, E., Schoni, R., Kozlowski, M. W. 1987. Relative air humidity influences the function of the tarsal chemoreceptor cells of the cherry fruit fly (*Rhagoletis cerasi*). *Physiol. Entomol.* 12:339–46

87. Stoffolano, J. G. 1973. Effect of age and diapause on the mean impulse frequency and failure to generate impulses in labellar chemoreceptor sensilla of *Phormia regina*. *J. Gerontol.* 28:35–39

88. Stoffolano, J. G., Damon, R. A.,

Desch, C. E. 1978. The effect of age, sex, and anatomical position on peripheral responses of taste receptors in blowfies, genus *Phormia* and *Protophormia*. *Exp. Gerontol.* 13:115–24

89. Stryker, R. G., Young, W. W. 1970. Effectiveness of carbon dioxide and L(+)lactic acid in mosquito light traps with and without light. *Mosq. News* 30:388–93

90. Sutcliffe, J. F. 1986. Black fly host location: a review. *Can. J. Zool.* 64:1041–53

91. Sutcliffe, J. F. 1987. Distance orientation of biting flies to their hosts. *Insect Sci. Appl.* 8:611–16

92. Takken, W., Kline, D. L. 1989. Carbon dioxide and 1-octen-3-ol as mosquito attractants. *J. Am. Mosq. Control Assoc.* 5:311–16

93. Taylor, D. M., Bennett, G. F., Lewis, D. J. 1979. Observations on the host-seeking activity of some Culicidae in the Tantramar marshes, New Brunswick. *J. Med. Entomol.* 15:134–37

94. Visser, J. H. 1988. Host-plant finding by insects: orientation, sensory input and search patterns. *J. Insect Physiol.* 34:259–68

95. Vogt, R. G. 1987. The molecular basis of pheromone reception: Its influence on behavior. See Ref. 76a, pp. 385–431

96. Washino, R. K. 1977. The physiological ecology of gonotrophic dissociation and related phenomena in mosquitoes. *J. Med. Entomol.* 13:381–88

Annu. Rev. Entomol. 1991. 36:159–83

PROSPECTS FOR GENE TRANSFORMATION IN INSECTS[1]

Alfred M. Handler

Insect Attractants, Behavior, and Basic Biology Research Laboratory, Agricultural Research Service, US Department of Agriculture, Gainesville, Florida 32604

David A. O'Brochta

Center for Agricultural Biotechnology, Department of Entomology, University of Maryland, College Park, Maryland 20742

KEY WORDS: germline transformation, gene vectors, P-element mobility, transposons

PERSPECTIVES AND OVERVIEW

The ability to manipulate genetic material in vitro and integrate it into a host genome has proven to be one of the more powerful methods of genetic analysis, as well as a means to manipulate an organism's biology. In insects, the use of gene transformation is equally significant in its potential to facilitate an understanding of insect genetics, biochemistry, development, and behavior. A more complete understanding of insect biology would in turn certainly enhance current methods, and promote development of new methods, to manage populations of both beneficial and pest species. Despite the benefits to be derived from gene-transfer, the routine and efficient introduction of exogenous DNA into insect genomes is limited to the genus *Drosophila*. Although DNA has been integrated into the genomes of three mosquito species (76, 80, 86), this integration has apparently resulted from rare random integration events, and the utility of this method is uncertain. The inability to achieve routine gene transfer in insects therefore makes the nature

[1]The US Government has the right to retain a nonexclusive royalty-free license in and to any copyright covering this paper.

of this review more prospective than retrospective. Current methods allow the in vivo analysis of manipulated insect genes, such as by extrachromosomal transient expression (somatic transformation) (70, 72) or integration of genes from nondrosophilid species into *Drosophila* (83). While these methods are useful for specific applications in some insects, they are inherently limited in terms of the types of genes that can be analyzed and the scope of analyses possible (42).

The primary method of gene transformation in *Drosophila* has utilized gene vectors derived from the P transposable element that has inherent mobility properties. Initially, the transposition of P observed in drosophilids distantly related to *D. melanogaster* (8), which do not normally contain P, suggested that P might be phylogenetically unrestricted and therefore useful as a gene vector in other insects. The development of P vectors allowing the dominant selection of neomycin-resistant transformed insects (111) permitted the testing of this notion. However, testing of P gene vectors in several nondrosophilid species, using both germline and somatic transformation tests, has thus far revealed a lack of P-vector function in these insects.

Given that P-element–based gene vectors in their present state are either nonfunctional or very inefficient in nondrosophilid insects, attempts to modify the P-element system or to develop new gene vectors or methodologies for gene transfer must be considered. Modification of the P-element requires methods that will allow a further understanding of normal P function in *Drosophila,* and a testing of P modifications in heterologous systems. The development of new gene vectors will require the isolation of new transposable elements and an evaluation of their ability to serve as vectors. In some organisms, DNA with mobility properties has been revealed using hybrid dysgenic-induced mutations and restriction fragment length polymorphisms, but these methods are most efficient in species that have been well characterized genetically and are easily subject to molecular studies. In addition, some, if not most, of the transposons identified with these methods will not fit the criteria for gene vector development, or, as with P, their use may be restricted to a few related species. Clearly, most current approaches to gene vector development are not trivial and may prove especially difficult for use in particular species.

Another general approach to gene-transfer may utilize random integration of DNA into chromosomes by nonhomologous recombination. This is routinely used in mammalian systems (37), and some success has been reported in various insect species. However, the relatively low frequency of random integration in insects would suggest that for this technique to be efficient, methods for mass introduction of DNA into germ cells and mass selection for integration will be required.

Beyond development of a means to integrate exogenous DNA, it has become clear that improvements in DNA delivery systems and selection schemes for integration are required. Indeed, while the P-element system has reportedly failed in nondrosophilids, low survival rates and inconsistencies in the selection schemes used actually make a critical evaluation impossible. In addition, a lack of recovery of transformants could result from limitations in the amounts or stability of the DNA delivered. A proper evaluation of these points is essential before a critical consideration of vector systems can be made.

APPLICATIONS OF GENE TRANSFORMATION

Basic Applications

The use of gene transformation in a wide variety of organisms has already demonstrated the vast amount of biological information that might be gained with such techniques (37, 47, 117). Of particular importance to insects are the various aspects of genetic analysis that have been impeded because of difficulties in cytogenetics, individual mating, rearing, and long generation times, among others. Transformation techniques coupled with in vitro DNA manipulation should prove invaluable to the identification and isolation of insect genes and to defining their structure-function relationships. While some of this information may be derived from transfection in insect cell lines or transient extrachromosomal expression, clearly the interactive influences of specific genetic components during developmental processes require whole animal germline transformation.

TRANSPOSON TAGGING   Genes are generally identified in most organisms by mutational analysis. In the absence of localized mutant alleles, genes can be isolated from genomic or cDNA libraries as a function of their homology to previously isolated genes or their transcriptional activity. Many insects are not amenable to mutational analysis, and the range of genes that can be isolated as described is somewhat limited. Transposon-based gene vectors can overcome these limitations because many show little site specificity when they integrate into a genome, and as a result integration can often disrupt a normal gene function, leading to a mutant phenotype (see 64). If the disrupted gene is of interest based upon its mutant phenotype, it can be subsequently localized and isolated using the transposon DNA with which it is tagged for chromosome in situ hybridization or to probe an appropriate library. Transposon tagging has been used to isolate dozens of genes from D. melanogaster (56) and similar applications using mammalian retroviral gene vectors have been used to identify and isolate genes involved in mammalian development (108).

MUTANT RESCUE   An efficient gene transformation system could also be used to isolate genes without relying on tagging methods, as well as to confirm the identity of putative clones. This may be achieved by integrating cloned DNA containing a putative wild-type gene into the genome of a mutant animal. Rescue or complementation of the mutant phenotype would confirm the presence of the functional wild-type gene within the vector. More precise localization of the gene within the cloned DNA can then be accomplished by systematically deleting DNA followed by successive rounds of gene transformation. This method has been used effectively to identify and isolate genes from *D. melanogaster*, but only after their position had been established within the order of tens or hundreds of kilobases because the number of germline transformants that can be generated is limited. For *Drosophila*, this limitation is directly related to the number of embryos that can be injected. If efficient methods for introducing vector DNA into large numbers of insects existed in conjunction with an efficient system for selecting for vector integration, then entire genomic libraries could be introduced by gene transformation methods (95). This method would permit gene isolation by mutant rescue to be implemented with virtually any gene without prior localization as is currently necessary.

ENHANCER TRAPS   The identification and isolation of genes using mutagenesis and tagging methods has been very fruitful. Yet these methods may exclude genes showing pleiotropy or a lethal phenotype. These limitations have been overcome in *Drosophila* by a method called *enhancer trapping*, which relies on the ability of relatively strong regulatory sequences within the genome to promote the activity of vector-encoded sequences inserted nearby. This method was originally developed to identify bacterial transcription units and to determine their orientation (13). Recently, a similar scheme was developed in *D. melanogaster* using the *Escherichia coli* β-galactosidase gene under the promoter control of the weak P-element promoter (2, 92, 118). The activity of this promoter is low enough so that β-galactosidase activity does not reach levels detectable by whole-tissue histochemical staining. If, however, the element transposes into a site adjacent to a transcriptional enhancer, the activity of the fusion gene will be increased and β-galactosidase activity will be detected. The pattern of expression revealed histochemically allows one to determine if the enhancer identified is associated with a gene involved in a developmental process of interest. This strategy has revealed a number of developmentally regulated loci and will undoubtedly find routine use in the future.

GENE REPLACEMENT   Most gene transfer methods in eukaryotes result in a gene insertion into a nonhomologous site within the genome without affecting the resident homologous allele. This complicates some analyses when the expression of the resident allele interferes with the function or assessment of the introduced gene. Such a problem is minimized when the resident gene is a recessive null mutant allele whose activity may be superseded by an introduced functional allele. Most insects, however, have few mutant strains or null alleles. This limitation may be overcome by the actual replacement of the resident gene with an in vitro modified copy. In addition, gene insertion into an unusual genomic site may result in a position effect that alters normal gene expression. For enhancer traps, this effect is desirable; however, often the alteration confuses the analysis of a particular gene or gene-construct. Gene replacement would ameliorate such position effects by allowing integration into the normal genetic milieu.

Gene replacement is routinely accomplished in yeast (47) and has been demonstrated in plants (94), mammalian cell lines (114), and slime molds (21). These events depend on homologous recombination between DNA within the vector and the chromosome, and are quite rare. The successes mentioned have resulted largely from powerful selection schemes that can be imposed on large numbers of cells in vitro.

Field Applications

Since the one insect shown to be amenable to gene transformation, *D. melanogaster,* is of limited agricultural or medical importance, the potential field applications of gene-transfer remain somewhat more prospective. Nevertheless, at least the basic biological information to be gained as it relates to chemical resistance mechanisms, sex determination, hybrid sterility, and hormone action and metabolism will enhance current insect management programs and will help develop new ones. One particular biological control method that might rapidly take advantage of gene-transfer techniques is the sterile-male release program (62), which could be greatly enhanced with efficient means of genetically sexing and sterilizing males. Classical genetic techniques, generally using a selectable gene linked to a male-specific Y chromosome, have undergone considerable study. Although some success has been achieved using chemical resistance to select males in mosquitoes (105), other genetic-sexing strains have proven highly susceptible to breakdown resulting from chromosome instability, recombination, and mutant reversions. (30, 48). Molecular techniques could minimize these difficulties by creating relatively small chimeric genes that have a selectable gene-product coding region linked to a sex-specific regulatory region. The genetic damage due to transformation would be negligible compared to chromosome rearrangements and mutation induction required by classical techniques. Implementation of such schemes may be easier than for other field applications using transformed insects because most strategies would allow the containment of genetically altered breeding parental insects, limiting release to sterile-male progeny.

STRATEGIES FOR TRANSFORMING ORGANISMS

The production of transgenic plants and animals has relied on a variety of methods. The simplest methods are perhaps the least understood from a mechanistic standpoint and involve the direct introduction of DNA, usually as a plasmid, into a plant or animal cell nucleus. The DNA, once present in the nucleus, randomly integrates into the host's genome by an illegitimate recombination event. This method is used to transform insect and noninsect cell lines and to create transgenic mice, plants, nematodes, and slime molds. The frequency of transformation using this method can be quite high. For example 20–30% of the mouse oocytes injected with DNA will result in a transgenic mouse (37). However, a disadvantage of this method is that often the DNA integrates as multiple copies oriented in a head to tail fashion (15), which can complicate subsequent analysis of the transformants. The random integration of DNA into insect genomes has been observed, although more data are necessary to properly evaluate frequencies and possible mechanisms (34, 76, 80, 86).

Improvements in the integration frequency can be achieved by physically linking the DNA to be integrated to vector DNA with inherent mobility properties. Vectors currently used to transform the germline of organisms have usually been derived from either retroviruses such as amphotropic murine leukemia viruses (37), transposons such as the P-element from *Drosophila* (27), or infectious agents such as the Ti plasmid from *Agrobacterium tumefaciens* (117). A variety of nonretroviral vectors have been developed for mammalian and insect cell lines but these are not useful for whole animal transformation.

Gene Transformation in Drosophila

Early attempts were made to genetically transform *D. melanogaster* by soaking mutant embryos in solutions of wild-type genomic DNA (31, 32). A variety of somatic mosaics resulted, but these had no clear inheritance of completely reverted phenotypes. The conclusion was that genetic transformation had occurred as a result of episomal transmission and not chromosomal integration. Subsequent attempts to revert *vermillion* mutant lines by injection of wild-type DNA met with some success (34, 69), although this result was not repeated and integration was never verified biochemically. The efficient and routine integration of DNA into a *Drosophila* host genome awaited the development of gene vectors created from modified P-element transposons (102, 103), and more recently, the *hobo* element (6). As in other organisms, these gene vectors have taken advantage of the inherent mobility properties of transposable elements. Although we know that transposable elements comprise a significant proportion of the *Drosophila* genome, not all transposons are conducive to gene vector activity. Early consideration was given to retroviruslike elements such as *copia,* but their relatively low mobility made them poor candidates for gene vectors (96).

An appreciation of how the P gene vectors function in *Drosophila,* possible restrictions on their function in other insects, and how this system may serve as a model for new gene vectors requires an understanding of P biology [see Engels (27) for an extensive review]. Furthermore, since *Drosophila* is the only insect subject to routine gene transformation, a review of the methodology can illuminate some of the special problems and factors to be considered in transforming insects.

P-ELEMENT VECTORS   The P-element is a highly mobile transposon discovered by virtue of the genetic defects that resulted from an induction of its mobility in interstrain crosses of *D. melanogaster*. These collective defects, including elevated mutation rates, chromosome rearrangements, sterility, and male recombination, are known as hybrid dysgenesis (57). The ability of an autonomous 2.9-kilobase (kb) P-element to transpose depends upon the integrity of its 31–base pair (bp) terminal inverted repeats and an internal transcription unit composed of four open reading frames encoding an 87-kilodalton (kd) protein (54, 91). This protein, or *transposase*, is required for both P excision and transposition, although its precise mechanistic role is unknown. At least two levels of control regulate P-element movement, both of which directly affect transposase function. First, P-elements can only be mobilized when present in a poorly defined cellular state known as M cytotype (28, 55). This maternally inherited cellular property is usually found in strains lacking autonomous P-elements. Strains containing autonomous P-elements usually develop a cytotype (P) that represses P movement. The maternally inherited cellular factors that determine cytotype are unknown, though mutated P-element genes may play a role. Second, P-element movement is limited to the germline. Although P-element transcription occurs in somatic and germ cells, complete transcript processing (including splicing of all three introns) required for functional transposase production is germline specific (66). Somatic cells splice only the first two introns, resulting in a truncated nonfunctional transposase product (97).

The development of P-elements into gene vectors was facilitated by their ability to be structurally modified such that foreign DNA can be inserted without destroying their ability to be mobilized (102, 103). If the transposase gene is defective or deleted, mobility can be promoted by transposase provided in trans. Thus in practice, the P gene vector consists of plasmid DNA containing a marker gene surrounded by P terminal sequences but lacking the transposase gene. The vector is mobilized in germline tissue by the presence of an intact transposase (helper) gene located on a separate plasmid, which promotes P integration into a chromosomal site. Normally, only the P termini and intervening DNA (usually including a selectable marker gene) are integrated. The transposase helper has structurally altered (or deleted) terminal sequences that prevent its integration and subsequent inheritance. Without a source of transposase in succeeding generations, the vector DNA is stably inherited. For unknown reasons, part of the integration process involves creation of an 8-bp duplication of the chromosomal target site (91). Transformation markers have routinely used the wild type genes for *rosy* (103) and *white* eye color (60), *alcohol dehydrogenase* (35), and more recently the bacterial neomycin-resistance gene, neomycin phosophotransferase (111).

The efficiency of P transformation depends on the host used, the size of the vector, concentration of DNA, and the way embryos are handled. Approximately 80% of the embryos can be expected to hatch and 40–50% will reach adulthood following careful injection (95). A variable percentage, ranging from 10–90%, of these adults can be sterile because of genetic background effects or as a result of damage or abnormalities induced by injection. Of the resulting fertile adults, 10–60% can be expected to yield at least one transformed progeny, though very large P vectors such as those containing cosmid inserts can transform with efficiencies 10 times lower (95). This evaluation in *Drosophila* indicates that, using a P vector of moderate size, one should anticipate, as a conservative estimate, recovering at least one transformant following the injection of 100–200 eggs.

Autonomous P-elements have been found only in strains of *D. melanogaster* and *D. willistoni*, though nonmobile P sequences exist in many species of the subgenus *Sophophora* (10, 17, 18, 65). Interestingly, these sequences are not found in some species most closely related to *D. melanogaster* such as *D. simulans* and *D. mauritiana*. Although the distribution of P is discontinuous, P has been mobilized in all drosophilids tested, as shown by gene transformation or embryonic mobility assays (90). P-mediated transformation was reported for *D. simulans* (19, 104) and a distantly related drosophilid, *D. hawaiiensis* (8), at frequencies expected for *D. melanogaster*. However, for *D. hawaiiensis*, only autonomous P elements, and not the altered P vector, were found to integrate, suggesting that in this species P activity is limited.

HOBO VECTORS   P-element vectors have been used almost exclusively for *Drosophila* transformation; however, a new transformation vector has been developed recently utilizing the *hobo* transposon. Although *hobo* was originally discovered because of its association with a glue protein gene (75), it was subsequently shown to transpose following certain interstrain crosses, a phenomenon closely paralleling P-M hybrid dysgenesis. While *hobo* shares several structural characteristics with P, notably a 3.0-kb length, 12-bp terminal inverted repeats, and production of an 8-bp duplication of integration target sequences, it has no sequence similarity with P, and the transposons do not cause cross-mobilization (5).

The *hobo* transformation vector, H[(ry^+)harl] (6) is analogous to the P vector Carnegie 20 (102) in that it has deleted internal sequences and a *rosy*$^+$ gene marker. Successful transformation was achieved in both "*hobo* dysgenic" embryos (having resident *hobo* elements), injected with vector alone, and in nondysgenic embryos, injected with vector and a complete *hobo* element helper plasmid (6). The efficiency of transformation in both cases was similar to that of P.

TRANSFORMATION METHODOLOGY   The generation of transgenic *Drosophila* has become, in the course of just a few years, a routine laboratory procedure. Specific details about the physical operations involved in producing transformants such as preparation of DNA, eggs, needles, and injection procedures have been thoroughly described by others (see 102, 109 for detailed protocols). In brief however, the general procedure involves collecting freshly laid eggs and dechorionating them either manually or chemically in diluted bleach. The eggs are attached to a glass coverslip with double-sided tape and desiccated briefly until they are slightly flaccid. The eggs are then covered with halocarbon oil and those still in the preblastoderm stage are injected at their posterior end with buffer containing the P-element vector and a transposase-producing helper plasmid. Larvae that hatch are placed on regular diet and adults (G_0) are backcrossed in individual matings. The resulting G_1 progeny are selected or screened for the presence of the marker gene integrated with the P-element, which can be verified by Southern analysis and chromosomal in situ hybridization.

Gene Transformation in Nondrosophilids

Before the initial attempts to transform *Drosophila*, successful somatic gene transformation was first reported in *Ephestia*, in which mutant larvae injected with wild-type DNA gave rise to adults with wild-type wing scales (14). Similar experiments were repeated in *Ephestia* using red eye mutants and in *Bombyx* using *white* mutants treated with wild-type DNA (88). In both experiments, a wild-type adult was recovered, but while the phenotype was inherited, transmission was non-Mendelian. Although one explanation of the unusual inheritance considered "replicating instabilities," it is more likely that extrachromosomal inheritance occurred.

Interest was regained in the possibility of transforming insects with P transformation in *D. melanogaster*, especially after the demonstration of P transformation in distantly related drosophilids (8) and the development of vectors (i.e. pUChsneo) containing the dominant-selectable marker gene neomycin phosophotransferase (NPT) (111). These factors indicated, respectively, that P mobility might not be phylogenetically restricted and that resistance to the neomycin-analog, Geneticin® (G418), could be used in many insect species as a selection for gene integration. The discovery that somatic restrictions on P could be alleviated by deletion of the third intron of the transposase gene (66, 97) also offered the possibility that the resulting P helper, pUChsπΔ2–3, could produce functional transposase in heterologous systems.

Attempts to transform insects using the pUChsπΔ2–3 helper and pUChsneo vector have been reported for several mosquito species, the Mediterranean fruitfly, *Ceratitis capitata,* and *Locusta migratoria.* In a process similar to the *Drosophila* protocol, these experiments involved injecting plasmids into preblastoderm embryos (G_0 generation), mating the surviving adults, and subjecting their G_1 progeny (or subsequent generations) to G418 selection. For some experiments, putative transformants were subjected to Southern blot analysis to verify and define the nature of the integration event.

MOSQUITOES   Thus far, the only nondrosophilids in which gene integration has been reported are mosquitoes, but apparently none of these integrations were P mediated. In *Anopheles gambiae,* one chromosomal integration event was detected resulting from 310 surviving G_0 adults mated *inter se* (80). However, Southern blot analysis indicated that this integration resulted from a single recombination event between the host genome and *white* gene sequences on pUChsneo. Similar experiments in *Aedes triseriatus* yielded two patterns of integration from 57 surviving G_0 adults backcrossed to uninjected mosquitoes (76). Curiously, neither integration was consistent with a P-mediated transposition or a simple recombination event as occurred in *A. gambiae.* Furthermore, no subsequent neomycin-resistant survivors showed evidence of chromosomal integration and the precise nature of the integrations originally observed remains obscure. In identical experiments with *Aedes aegypti,* from 71 surviving G_0 adults, one integration event was reported, but as in previous experiments, Southern analysis failed to demonstrate a P-mediated integration (86).

TEPHRITIDS AND LOCUSTS   Efforts to transform tephritid fruitflies, including the Caribbean fruitfly, *Anastrepha suspensa,* by our laboratory and *C. capitata* by others have failed to recover either P-mediated or random chromosomal integrations. In two experiments on the medfly, a total of over 20,000 embryos have been injected and nearly 900 G_0 adults mated (77, 99). In both experiments, G418-resistant G_1 larvae were selected, but resistance was variable or lost in succeeding generations and in no case did Southern blotting show P integration. In one of these experiments, the researchers (77) inferred that G418-resistance either was inherent in the subpopulations tested or resulted from extrachromosomal expression of the vector-encoded resistance gene. The general lack of persistence of plasmid DNA in *Drosophila* (72) would suggest that the former explanation is more plausible. Similar efforts to transform *L. migratoria* have also failed to reveal a P integration (116).

LIMITATIONS   The failure to recover germline transformants in heterologous systems by direct application of the *D. melanogaster* P-transformation system could result from a number of limitations including: (*a*) an inability to transcribe or translate the transposase gene, (*b*) mobility constraints due to vector size or construction, (*c*) an inefficient or ineffective selection system, (*d*) a lack of appropriate genomic target sites, or (*e*) a lack of requisite host-encoded cofactors or the presence of repressors. Unfortunately, one cannot critically test these various possibilities independent of one another using germline transformation experiments, and yet one must understand the limitations of the system before restrictions can be ameliorated. A common difficulty encountered in the transformation experiments described above was the relatively high frequency of G418-resistant lines recovered that did not have NPT gene integration. This frequency may have resulted from variability in sensitivity to G418 or leakiness in the resistance-phenotype, yet these possibilities are difficult to have controls for. Indeed, the variability in the selection raises the possibility that transformants may have been created but failed to express the resistance phenotype and were lost. Thus, the P-vector system has probably not been fairly evaluated by these experiments. Clearly, effective selection schemes are essential for transformation, and various methods exist other than G418-resistance. Screening for integration is not insurmountable even if it requires direct DNA hybridization. Of more immediate concern is whether the gene vector is capable of transposition.

EVALUATION OF P MOBILITY   Evaluating the mobility properties of a transposon-based gene vector in heterologous systems depends on first determining the mobility of the transposon independent of other vector DNA, especially selectable markers. Methods should be direct, rapid, quantitative, and preferably allow a systematic testing of modifications. In this respect, we have made efforts to analyze P-element function using excision assays that directly monitor P mobility in the soma of insect embryos. The rationale of this approach is based upon the observation that P-element insertions often result in gene inactivation, and that reversion of these mutations frequently results from the excision of the transposon (39). Since both P-element excision and transposition depend upon normal transposase function, restoration of gene function disrupted by a P insertion can be used to assess P mobility and transposase activity.

Rio et al (97) took advantage of these characteristics of P movement to develop a rapid assay that assesses P function in cell lines. Plasmids (pISP and pISP-2) were constructed that allow the detection of P mobility as a result of gene function restoration following P-element excision from a plasmid-encoded gene. Specifically, a small nonautonomous P-element sequence surrounded by *white* gene DNA was inserted into the *lacZα* peptide coding region of pUC8. In this configuration, *lacZα*, which is required in appropriate bacterial hosts for β-galactosidase activity, is nonfunctional. In the presence of functional transposase and any other required host factors, the P-element can be mobilized, resulting in a restoration *lacZα* function. When these plasmids are harvested after incubation in cells and transformed into appropriate bacteria, staining for β-galactosidase activity (blue coloration on X-gal media) acts as an indicator for excision of P from *lacZα*.

We have since modified the in vitro excision assay so that P functionality can be assessed in the insect embryonic soma, enabling us to directly address the question of P mobility in nondrosophilids (see 90 for detailed protocols). In general, the pISP indicator plasmid was injected into preblastoderm embryos in the presence of an endogenous (chromosomal) or exogenous (plasmid) source of transposase. After an overnight incubation, plasmids were recovered and subsequently transformed into bacteria. This assay tests the expression of the transposase gene into a functional gene product, as well as determining more generally whether the embryonic milieu is supportive of P mobility. A limitation of this assay is that the pISP indicator plasmid only permits detection of those excisions restoring *lacZα* function. Yet these may actually represent only a minority of all excisions because of events that fail to restore the *lacZα* reading frame. We therefore modified the indicator plasmid by inserting the *E. coli* S12 ribosomal protein gene into the P sequence. S12 acts dominantly to confer sensitivity to streptomycin in streptomycin-resistant bacteria (20) and, thus, the new indicator plasmid (pπstrep[s]) permits all P excisions to be monitored by virtue of the streptomycin-resistance phenotype. Excision can be further defined by the ability to restore *lacZα* function.

Thus far, P mobility has been assessed in a variety of drosophilids and nondrosophilids with the original assay (90), and in *D. melanogaster* and *Chymomyza procnemis,* a drosophilid outside the *Drosophila* genus, with the new total excision assay. We found that P could be mobilized in all the drosophilids tested as well as in a closely related ephydrid, *Paralimna decipiens* (D. A. O'Brochta & A. M. Handler, unpublished results). However, the P excision frequency decreased as a function of relatedness to *D. melanogaster,* and P mobility was not detected in tephritids, sphaerocerids, muscids, or phorids. Comparable to the original assay, assays with the new pπstrep[s] plasmid also showed a 10-fold lower total excision frequency in *C.*

procnemis relative to *D. melanogaster;* however, in both species, excisions restoring *lacZ* function accounted for only 30% of the total. The new assay also allows the unbiased recovery of excision products, and sequence analysis indicates qualitative differences between the species in terms of site preference for excision breakpoints.

While these experiments are still in progress, the results thus far indicate that the P-element is not functional in nondrosophilids, and the P gene vector system in its present state would be nonfunctional or highly inefficient in these insects. However, P mobility assays may also be used to define the basis of the dysfunction and to test modifications that may ameliorate it.

P-ELEMENT DYSFUNCTION   Several explanations for restricted P mobility in insects other than *D. melanogaster* are based on our current knowledge of P-element activity. While the transposase gene is transcribed in *A. suspensa,* Northern analysis indicates a profile of varying-sized transcripts generated from the pUChs$\pi\Delta$2–3 helper in embryos (90). This profile may indicate abnormal transcript processing, which could allow termination-codon usage, resulting in a nonfunctional truncated gene-product. At present no direct evidence indicates aberrant processing in insects. A transposase RNA splicing assay, however, using a β-galactosidase reporter gene linked in-frame to the second exon, indicates a lack of, or inefficient splicing of, intron 1 in *Anastrepha* (D. A. O'Brochta, unpublished results). In plants, aberrant processing of P has been shown directly in tobacco transformed with hs$\pi\Delta$2–3, where abundant 1.5-kb transcript results from transcript termination just beyond the second intron, and intron 1 is not processed (73).

If normal processing is not occurring in nondrosophilid insects, then transposase helpers that have introns 1 and 2 deleted, in addition to intron 3, may be functional. We began testing this notion by using as helpers pUChs$\pi\Delta$1–2–3 that had introns 2 and 3 deleted, and pUChsπcDNA that had introns 1, 2, and 3 deleted (D. A. O'Brochta & A. M. Handler, unpublished results; plasmids provided by D. Rio, Whitehead Inst.). The hs$\pi\Delta$1–2–3 helper worked well in *D. melanogaster,* but not in *Anastrepha,* though curiously, the hsπcDNA did not function in either insect. The former result could be due to a lack of intron 1 splicing in *Anastrepha* consistent with the splicing assay results, but the latter result might indicate either a mutation in the cDNA plasmid construct, or a function for intron 1 in stable transcript biogenesis. We subsequently sequenced the first and second exons of the cDNA and found a single base deletion in exon 1, which is expected to cause a frameshift resulting in a nonfunctional gene-product. Further evaluation of transposase cDNA activity in nondrosophilids would therefore require correction of the mutation or isolation of a new cDNA. If eliminating the need to process the transposase transcript relieves a major restriction on P function, then the cDNA may provide a functional helper for P-mediated transformation in nondrosophilids.

In addition to nonfunctional gene-product formation, truncated transposase products may also act to repress normal transposase function. For example, an internally deleted P-element, KP, has been associated with repression of P-mediated dysgenesis (4). It has also been proposed that the somatic product of the complete transposase gene, a truncated 66-polypeptide resulting from the unspliced third intron, has repressor activity (97). Using the excision assay, we tested this hypothesis in the *D. melanogaster* embryonic soma by co-injecting a plasmid with the full length transposase gene, phsπ, along with the indicator and helper plasmids. Repressor activity was evidenced by a dose-dependent elimination of P excision with relatively low concentrations of phsπ (D. A. O'Brochta, S. P. Gomez, & A. M. Handler, unpublished results). If processing was merely inefficient in nondrosophilids, enough repressor could be produced to inhibit any normal activity. Conceivably, a transposase cDNA would alleviate transposase-encoded repressor formation, which could be verified by an immunological analysis of transposase gene-product formed in nondrosophilid embryos.

If a cDNA is functional in *Drosophila* but fails to function in nondrosophilids, and no transposase-encoded repressor formation occurs, essential cofactors are probably missing and/or non-P-encoded repressors exist in these species. The participation of transposase-independent factors in P mobility has been inferred by our analysis of excision products, and more directly by the discovery of a protein that has binding specificity for the P-terminal sequences (98). A general approach to evaluating these possibilities is to isolate putative positive-acting factors from *Drosophila* (as DNA, RNA, or protein) and test them by co-injection in mobility assays in nondrosophilids. Conversely, putative repressors may be isolated in nondrosophilids and their negative influence tested in mobility assays in *D. melanogaster*.

New Gene Vectors

While the defect restricting P-element mobility in nondrosophilids might be corrected, leading to a functional transformation system, uncertainties still remain. For example, a negative factor may be identified but not easily eliminated, and while we may overcome restrictions in families closely related to Drosophilidae, additional restrictions may occur in families more distantly related. Thus, to achieve gene transfer in a broad range of insects, researchers will probably need to analyze other less restricted insect and noninsect transposon systems or develop methods to rapidly isolate and test species-specific transposon systems for development into gene vectors.

OTHER TRANSPOSON SYSTEMS   Transposable elements appear to be ubiquitous components of eukaryotic genomes and, as with the P system, several have been modified into gene vectors for the species in which they were discovered (24, 50). Transposable elements that are potential candidates for gene vector development should have unrestricted (or less restricted) mobility properties as indicated by an absence of tissue specificity, a broad phylogenetic distribution, or demonstrated mobility in a wide range of species. Three elements that meet these criteria are *mariner* from *D. mauritiana* (43), *Ac* from maize (29), and *Tc1* from nematodes (84).

The *mariner* element resembles P-elements structurally (51), but unlike P, *mariner* is normally active in both the germline and soma (45). Under appropriate conditions, it has a rate of mobility greater than that of P (11). Like P, *mariner* can be *trans*-mobilized by autonomous *mariner* elements (78), indicating that a binary vector-helper system, important to regulating the movement of a gene vector, may be feasible.

The *Activator/Dissociator (Ac/Ds)* transposon system from maize has thus far demonstrated the most phylogenetically unrestricted mobility properties of any known eukaryotic transposon. *Ac* elements have shown mobility, after *Agrobacterium* Ti plasmid-mediated transformation, in tobacco (1), tomatoes (120), carrots (61), and *Arabidopsis* (115). These observations indicate that either *Ac* is truly autonomous or non-*Ac*-encoded functions are highly conserved. Recent isolation of the *Ac* cDNA (44) makes possible the expression of the *Ac* transposase in animal cells. The *Tc* elements from *Caenorhabditis elegans* have also been well characterized and, like *mariner*, can be highly mobile in somatic and germline tissue (26). Interestingly, *Tc1* has a high degree of sequence similarity to the *HB* transposon family in *D. melanogaster* (46); the *Tc1* and *HB* open reading frames share about 30% amino acid homology. This homology indicates that *Tc* elements have had a long evolutionary history and/or were transmitted horizontally to insects. Although an autonomous *Tc* element has not been isolated, several candidiates have been genetically identified (85). Testing *Tc1* mobility in insects will depend upon the isolation of these elements, which should be forthcoming.

A general strategy for initially testing these elements would follow the total excision assay for P. Excision reporter plasmids might consist of transposable element terminal sequences flanking a streptomycin-sensitivity gene insertion. Helper plasmids would consist of the element-specific transposase under *hsp70*-promoter transcriptional control. Mobility would be assayed after plasmid incubation in host embryos and subsequent testing for streptomycin resistance in transformed bacteria.

NEW TRANSPOSON SYSTEMS   Transposable elements have been found in a wide range of prokaryotes (58) and eukaryotes (100) and are known to comprise a significant portion of the *D. melanogaster* genome (101, 110). They have often been identified by virtue of their mutagenic properties (39), and mutations resulting from the insertion of transposable elements often display characteristics such as a high degree of instability. Recently, insect transposon-induced mutations were recovered in baculoviruses that had infected lepidopteran cell lines (33). In these cases, the viral genome acted as a target for transposon insertion. The ability to inject, transiently maintain, and recover bacterial plasmids from insect embryos may allow the similar capture of mobile genetic elements from host insects. In a system converse to the excision assay, plasmids injected into insect embryos could harbor target sequences that would facilitate the identification of insertion events, some of which should be transposable elements. Isolating transposons by virtue of their mobility properties would expedite their subsequent analysis and development into useful gene vectors.

SITE-SPECIFIC RECOMBINATION   An alternative approach to achieving efficient gene transfer is to utilize site-specific recombination systems such as the *FLP* recombinase system of yeast (79), in which a recombinase protein facilitates specific recombination between *FRT* sequences. Insect target site strains might be created by having an *FRT* target site integrated into a host genome, possibly by random integration or inefficient gene transfer. Embryos from this strain would be co-injected with plasmids encoding the recombinase and marker genes with adjacent *FRT* sequences, which might then integrate by recombination with the chromosomal *FRT* sequences. The *FLP* recombinase system functions in *D. melanogaster* (36), and one can test its ability to promote recombination between plasmids in nondrosophilids. If successful, formidable efforts would be worthwhile to integrate *FRT* sequences into insect genomes to test this method.

Evaluation of Transformation Methodology and Selection

DNA INJECTION   The basic method for gene transformation in *Drosophila* has worked well, yet modifications will be necessary for other insects because of differences in their biology. For example, *Drosophila* eggs are injected soon after oviposition so that DNA may be introduced before cellularization and will be enveloped in the pole cells. This process occurs within 2 h of oviposition and most eggs are usually injected just before blastoderm formation. Insects with slower development have a greater preblastoderm period in which to inject, but this does not ensure that any time before cellularization is optimal. When a lepidopteran, *Plodia interpunctella,* was injected with plasmid shortly after oviposition, most of the DNA was degraded within 12 h and cellularization occurred at about 16 h (P. Shirk, personal communication). Though the cause of degradation is unknown, persistence of the injected DNA is necessary, which may need to be tested by injecting DNA at various times and assaying the transient somatic expression of a reporter gene, or determining relative rates of plasmid recovery.

Direct microinjection of preblastoderm insect eggs has been the common means of introducing DNA into germ cells. In *Drosophila*, and other species that have eggs with a relatively thin chorion that can be easily removed, this process is straightforward and survival rates can be quite high. For many other insects, this method is more challenging as a result of rigid or nonremovable chorions and sensitivity to desiccation resulting in low levels of DNA transfer or poor survival. In lepidopterans, researchers have been able to inject newly oviposited eggs that have a soft chorion (that later hardens) and seal the puncture with glue (107). Although imaginative approaches may eventually allow egg injection in most insects, the need to individually inject eggs, even for efficient transformation systems, remains a limiting factor. More universal methods allowing simultaneous DNA transfer into large numbers of eggs would certainly be preferable. Such methods would have the added benefit of allowing the testing, and possible implementation, of random DNA integration or inefficient gene vector systems.

OTHER DNA DELIVERY SYSTEMS   Several systems for introducing DNA en masse into an organism's germ cells have been successful, and some are in the preliminary stage of testing in insects. One of the first involves shooting microprojectiles coated with DNA into target cells through a shotgun type mechanism (now commercially available from DuPont). Success has been achieved in *Allium cepa* epidermis (59), yeast mitochondria (52), and *Chlamydomonas* chloroplasts (7). Initial tests in insects indicate that DNA can be delivered into eggs this way, though the mortality rate is high and transformation has yet to be demonstrated (12). Mechanical introduction of DNA into cells has also been achieved by vortexing yeast in a DNA solution with glass beads (16). Similarly, DNA has been introduced into insect eggs by vortexing them in a DNA solution with silicon carbide fibers (A. Cockburn, personal communication). A recent report of the creation of transgenic mice using sperm encoated with DNA (67) has not been repeated by others (9), though some evidence supports DNA transfer with a similar method in honey bees, which can be artificially inseminated (81).

Several procedures are also used to mediate DNA uptake into cultured cells, some of which may be modified to allow uptake into insect oocytes. For example, electroporation has been used to permeabilize tissue culture cells of various prokaryotes and eukaryotes for transfection (106). Although pulses of high voltage may result in high mortality of insect eggs, this may not be limiting considering the ability to simultaneously treat large numbers of eggs. Less apparent damage to oocytes may be achieved by the use of calcium-phosphate precipitated DNA or DNA encapsulated in liposomes. Both these DNA treatments have resulted in transfection of mammalian and insect cell lines, as well as cellular uptake in vertebrates in vivo after intraperitoneal or intravenous injection (3, 22). An analogous procedure in female insects would be the simple injection of DNA into the abdominal hemocoel, which, in *Drosophila,* has resulted in plasmid recovery from oviposited eggs (A. M. Handler & S. P. Gomez, unpublished results). Other methods may take advantage of receptor-mediated uptake where DNA is conjugated to molecules that are selectively sequestered by oocytes or cell nuclei (119). Selective uptake of DNA in liver cell nuclei has been achieved after conjugation to nonhistone chromosomal proteins and liposome fusion (53, 89). A similar approach in insects may utilize conjugation to histones, which after oocyte injection in *D. melanogaster* are selectively associated with nuclei (82), or conjugation to yolk proteins, which are selectively sequestered by oocytes (63).

Prolonging the presence of exogenous DNA in insect cells, especially through successive rounds of cell division and DNA replication, may enhance its ability to integrate either by a vector or through random integration. Extending the presence of DNA might be accomplished by linking the exogenous DNA to eukaryotic origins of replication or autonomously replicating sequences (ARS), which cause the persistance of episomal DNA. ARSs have been primarily isolated from yeast (112, 113), but similar sequences have been found in other organisms, including *Drosophila.* Interestingly, these nonyeast elements have ARS activity in yeast but lack this activity in their own cells (38, 74). Eliminating this restriction might enhance germline transformation and would almost certainly enhance transient expression analyses that depend upon persistence of exogenous DNA in somatic tissue.

SELECTION FOR TRANSFORMANTS   The identification of putative transformants depends upon the ability to accurately select for chromosomal integration of the vector. This is relatively straightforward in *Drosophila,* in which a variety of genes resulting in a selectable phenotype can be linked to the DNA to be integrated. For example, the wild-type *rosy* gene confers a wild-type red eye phenotype when integrated into a *rosy* mutant genome (102, 103). Similarly, the wild-type *alcohol dehydrogenase (adh)* gene confers the ability to metabolize ethanol when integrated into adh^- mutant hosts (35). Whereas $rosy^+$ transformants must be selected individually, the *adh* selection allows mass selection by ethanol resistance. Both of these selections, among others, are possible in *D. melanogaster* because mutant strains exist for which the wild-type gene is available as cloned DNA. In most nondrosophilid insects such mutant rescue selection is not yet possible, and so selectable marker genes must confer a dominant-acting phenotype. This may be in the form of chemical resistance or expression of a new visible phenotype. As discussed, a dominant-selection scheme has been developed for *Drosophila* using the bacterial NPT gene to confer resistance to neomycin or chemical analogs such as G418 (111). Since the NPT gene was purported to function in a wide range of eukarytic cells, this selection raised hopes that the P gene vector could be efficiently tested in various insects. NPT selection, however, has been inconsistent in its effectiveness.

Other genes confer dominant-selection phenotypes in cell lines but their usefulness in whole animals is unknown. The hygromycin-resistance gene has functioned in various eukaryotic systems including mammalian (71) and mosquito (J. Carlson & B. Beatty, personal communication) cell lines. Insect expression of dihydrofolate reductase may allow resistance to methotrexate (49), while the mammalian multiple drug resistance gene may allow resistance to various plant alkaloids (41). Although these resistance systems hold some promise, their use as selective agents may prove to be as problematic as neomycin resistance. Insecticide resistance genes may function more consistently as selective agents (87), though their practical use in transformation for field applications may be restricted.

Genes that confer a new visible selectable phenotype would have to be individually screened, but might be less prone to the inconsistencies of chemical selection. The bacterial *lacZ* gene has been transformed into *Drosophila,* resulting in high levels of β-galactosidase that yield blue coloration after a simple staining process (68). Although whole body staining would be lethal, body structures not critical to reproduction such as halteres, legs, or antennae can be dissected and stained allowing putative transformants to be mated. Similarly, the dominant expression of other detectable enzymes can be used for selection, such as chloramphenicol phosphotransferase (23) and luciferase (93). Direct screening for gene integration by DNA hybridization is feasible for efficient transformation systems (8), but this presents difficulties for insects that mate en masse, precluding the testing of siblings from established lines. This problem may be overcome by the use of polymerase chain reaction to detect integrated genes in superfluous tissues as described above. For most selections that require intermediate procedures for detection, it would probably be expedient to initially test groups of insects and subsequently focus on individuals.

Although limited, the genetic analysis of nondrosophilids has resulted in the recovery of some mutations that confer phenotypes similar to those found in *Drosophila.* Some of these may have a related genotype as well as can be complemented by cloned *Drosophila* genes. For example, the *topaz* gene in the biosynthetic pathway leading to the brown ommochrome eye pigment in *Lucilia cuprina* is homologous to the *scarlet* ommochrome gene in *D. melanogaster* (25). Prior to using these genes as germline transformation mutant rescue markers, functional complementation may be determined by transient expression experiments. For example, preliminary results indicate that transient expression of the *vermillion* gene in *Drosophila* can, at least partially, complement the *yellowish* eye mutation in *Lucilia* (P. Atkinson, personal communication).

Although selection in some insects may take advantage of gene homologies with *Drosophila,* the need for selection in a broad range of insects will depend upon more generally applicable methods. In lieu of mutant rescue, a converse selection by induction of a mutant phenotype may be eventually feasible by using ribozymes or DNA encoding antisense RNA for conserved genes having a selectable phenotype. This system would work, simply, by vector-encoded antisense RNA inhibiting the normal translation (or possibly transcription) of the complementary sense RNA from a normal resident gene, resulting in a loss of gene-product (see 40, 114a). In *Drosophila,* only mutant phenotypes visible in embryos or larvae have been induced in this way, and general inconsistencies in the activity of antisense RNA make this method prospective. Nevertheless, the potential broadbased utility of antisense RNA or ribozymes for selection should encourage their testing.

CONCLUDING REMARKS

The need to achieve efficient methods for germline transformation in economically important insects remains a high priority. Although other methods exist to analyze the expression of manipulated insect genes, none of them are optimal for critical analyses nor are they useful for stable gene integration required for biological control methods. The *Drosophila* P-element gene vector system has not facilitated gene transfer in nondrosophilids, yet limitations and inconsistencies in DNA delivery and gene-integration selection systems have impeded a fair evaluation of the P vector, at least based on germline transformation experiments. Clearly an evaluation of P, or any other gene vector or transformation methodology, will first require the development of effective and reliable ancillary methods. Nonetheless, more direct embryonic mobility assays also indicate a lack of P function in nondrosophilids, and modifications of the P-vector system will be essential. These mobility assay methods, however, also allow the testing of P modifications to possibly define and ameliorate restrictions. While the restrictions may be overcome in some insect families closely related to drosophilids, more distantly related species may have other restrictions. Similarly, the development of other gene-transfer methods may not be universal, and transformation for a broad range of insects may ultimately depend upon methods allowing the efficient isolation and testing of new gene vectors for specific groups of insects.

ACKNOWLEDGMENTS

Appreciation is extended to those who shared their unpublished results with us, to Drs. P. Greany, A. Malavasi, S. Miller, and H. Oberlander for comments on the manuscript, and to the USDA-Cooperative State Research Service for support.

Literature Cited

1. Baker, B., Schell, J., Loerz, H., Fedoroff, N. 1986. Transposition of the maize controlling element *Activator* in tobacco. *Proc. Natl. Acad. Sci. USA* 83: 4844–48

2. Bellen, H. J., O'Kane, C. J., Wilson, C., Grossniklaus, U., Pearson, R. K., Gehring, W. J. 1989. P-element-mediated enhancer detection: a versatile method to study development in *Drosophila*. *Genes Dev.* 3:1288–1300

3. Benvenisty N., Reshef, L. 1986. Direct introduction of genes into rats and expression of the genes. *Proc. Natl. Acad. Sci. USA* 83:9551–55

3a. Berg, D. E., Howe, M. M., eds. 1989. *Mobile DNA*. Washington DC: Am. Soc. Microbiol. 972 pp.

4. Black, D. M. N., Jackson, M. S., Kidwell, M. G., Dover, G. A. 1987. KP elements repress P induced hybrid dysgenesis in *D. melanogaster*. *EMBO J.* 6:4125–35

5. Blackman, R. K., Gelbart, W. M. 1989. The transposable element *hobo* of *Drosophila melanogaster*. See Ref. 3a, pp 523–29

6. Blackman, R. K., Macy, M., Koehler, D., Grimaila, R., Gelbart, W. M. 1989. Identification of a fully-functional *hobo* transposable element and its use for germ-line transformation of *Drosophila*. *EMBO J.* 8:211–17

7. Boynton, J. E., Gillham, N. W., Harris, E. H., Hosler, J. B., Johnson, A. M., et al. 1988. Chloroplast transformation in *Chlamydomonas* with high velocity microprojectiles. *Science* 240:1534–37

8. Brennan, M. D., Rowan, R. G., Dickinson, W. J. 1984. Introduction of a functional P element into the germ line of Drosophila hawaiiensis. *Cell* 38:147–51

9. Brinster, R. L., Sandgren, E. P., Behringer, R. R., Palmiter, R. D., 1989. No simple solution for making transgenic mice. *Cell* 59:239–41

10. Brookfield, J. F. Y., Montgomery, E., Langley, C. 1984. An apparent absence of transposable elements related to the P elements of *D. melanogaster* in other species of *Drosophila*. *Nature* 310:330–32

11. Bryan, G. J., Jacobson, J. W., Hartl, D. L. 1987. Heritable somatic excision of a *Drosophila* transposon. *Science* 235: 1636–38

12. Carlson, D. A., Cockburn, A. F., Tarrent, C. A. 1989. Advances in insertion of material into insect eggs via a particle shotgun technique. In *Host Regulated Developmental Mechanisms in Vector Arthropods*, ed. D. Borovsky, A. Spielman, pp. 248–52. Vero Beach, FL. Univ. Florida Press. 324 pp.

13. Casadaban, M. J., Cohen, S. N. 1979. Lactose genes fused to exogenous promoters in one step using a Mu-*lac* bacteriophage: *in vivo* probe for transcriptional control sequences. *Proc. Natl. Acad. Sci. USA* 76:4530–33

14. Caspari, E., Nawa, S. 1965. A method to demonstrate transformation in *Ephestia*. *Z. Naturforsch.* 206:281–84

15. Constantini, F., Lacy, E. 1981. Introduction of a rabbit β-globin gene into the mouse germ line. *Nature* 294:92–94

16. Costanzo, M. C., Fox, T. D. 1988. Transformation of yeast by agitation with glass beads. *Genetics* 120:667–70

17. Daniels, S. B., Peterson, K. R., Strausbaugh, L. D., Kidwell, M. G., Chovnick, A. 1990. Evidence for horizontal transmission of the P transposable element between *Drosophila* species. *Genetics* 124:339–55

18. Daniels, S. B., Strausbaugh, L. D. 1986. The distribution of P element sequences in *Drosophila:* the *willistoni* and *sultans* species groups. *J. Mol. Evol.* 23:138–48

19. Daniels, S. B., Strausbaugh, L. D., Armstrong, R. A. 1985. Molecular analysis of P element behavior in *Drosophila simulans* transformants. *Mol. Gen. Genet.* 200:258–65

20. Dean, D. 1981. A plasmid cloning vector for the direct selection of strains carrying recombinant plasmids. *Gene* 15:99–102

21. de Lozanne, A., Spudich, J. A. 1987. Disruption of the *Dictyostelium* myosin heavy chain gene by homologous recombination. *Science* 236:1086–91

22. Dubensky, T. W., Campbell, B. A., Villarreal, L. P. 1984. Direct transfection of viral and plasmid DNA into the liver or spleen of mice. *Proc. Natl. Acad. Sci. USA* 81:7529–33

23. Durbin, J. E., Falllon, A. M. 1985. Transient expression of the chloramphenicol acetyltransferase gene in cultured mosquito cells. *Gene* 36:173–78

24. Eglitis, M. A., Anderson, W. F. 1988. Retroviral vectors for introduction of genes into mammalian cells. *Biotechniques* 6:608–15

25. Elizur, A., Vacek, A. T., Howells, A. J. 1990. Cloning and characterization of the *white* and *topaz* eye color genes from the sheep blowfly *Lucilia cuprina*. *J. Mol. Evol.* 30:347–58

26. Emmons, S. W., Yesner, L., Ruan, K. S., Katzenberg, D. 1983. Evidence for a

transposon in Caenorhabditis elegans. *Cell* 32:55–65

27. Engels, W. R. 1989. P elements in *Drosophila melanogaster*. See Ref. 3a, pp. 437–84

28. Engels, W. R. 1979. Hybrid dysgenesis in *Drosophila melanogaster*; rules of inheritance of female sterility. *Genet. Res. Camb.* 33:219–36

29. Federoff, N. 1989. Maize transposable elements. See Ref. 3a, pp. 375–411

30. Foster, G. G., Maddern, R. H., Mills, A. T. 1980. Genetic instability in mass-rearing colonies of a sex-linked translocation strain of *Lucilia cuprina* (Wiedemann) (Diptera: Calliphoridae) during a field trial of genetic control. *Theor. Appl. Genet.* 58:164–75

31. Fox, A. S., Yon, S. B. 1966. Specific genetic effects of DNA in *Drosophila melanogaster*. *Genetics* 53:897–911

32. Fox, A. S., Yoon, S. B. 1970. DNA-induced transformation in *Drosophila*; locus specificity and the establishment of transformed stocks. *Proc. Natl. Acad. Sci. USA* 67:1608–15

33. Fraser, M. J. 1986. Transposon-mediated mutagenesis of baculoviruses: transposon shuttling and implications for speciation. *Ann. Entomol. Soc. Amer.* 79:773–83

34. Germeraad, S. 1976. Genetic transformation in *Drosophila* by micro-injection of DNA. *Nature* 262:229–31

35. Goldberg, D. A., Posakony, J. W., Maniatis, T. 1983. Correct developmental expression of a cloned alcohol dehydrogenase gene transduced into the drosophila germ line. *Cell* 34:59–73

36. Golic, K. G., Lindquist, S. 1989. The FLP recombinase of yeast catalyzes site-specific recombination in the Drosophila genome. *Cell* 59:499–509

37. Gordon, J. W. 1989. Transgenic animals *Int. Rev. Cytol.* 115:171–229

38. Gragerov, A. I., Danilevskaya, O. N., Didichenki, S. A., Kaverina, E. N. 1988. An ARS element from *Drosophila melanogaster* telomeres contains the yeast ARS core and bent replication enhancer. *Nucleic Acids Res.* 16:1169–80

39. Green, M. M. 1988. Mobile DNA elements and spontaneous gene mutation. See Ref. 64, pp. 41–50

40. Green, P. J., Pines, O., Inouye, M. 1986. The role of antisense RNA in gene regulation. *Annu. Rev. Biochem.* 55:569–97

41. Gros, P., Neria, Y. B., Croop, J. M., Housman, D. E. 1986. Isolation and expression of a complementary DNA that confers multiple drug resistance. *Nature* 323:728–31

42. Handler, A. M., O'Brochta, D. A. 1988. An assessment of gene transformation in insects. In *Endocrinological Frontiers in Physiological Insect Ecology*, Vol. 2, ed. F. Sehnal, A. Zabza, D. L. Denlinger, pp. 1075–84. Wroclaw, Poland: Wroclaw Tech. Univ. Press. 1087 pp.

43. Hartl, D. L., 1989. Transposable element *mariner* in *Drosophila* species. See Ref. 3a, pp. 531–36

44. Hauser, C., Fusseinkel, H., Li, J., Oellig, C., Kunze, R., et al. 1988. Overproduction of the protein encoded by the maize transposable element Ac in insect cells by a baculovirus vector. *Mol. Gen. Genet.* 214:373–78

45. Haymer, D. S., Marsh, J. L. 1986. Germ line and somatic instability of a *white* mutation in *Drosophila mauritiana* due to a transposable element. *Dev. Genet.* 6:281–91

46. Henikoff, S., Plasterik, R. H. A. 1988. Related transposons in *C. elegans* and *D. melanogaster*. *Nucleic Acid Res.* 16:6234

47. Hinnen, A., Hicks, J. B., Fing, G. R. 1978. Transformation of yeast. *Proc. Natl. Acad. Sci. USA* 75:1929–33

48. Hooper, G. H. S., Robinson, A. S., Marchand, R. P. 1987. Behaviour of a genetic sexing strain of Mediterranean fruit fly, *Ceratitis capitata*, during large scale rearing. In *Fruit Flies Proc. 2nd Int. Symp.*, ed. A. P. Economopoulos, pp. 349–62. Amsterdam: Elsevier. 590 pp.

48a. International Atomic Energy Agency. 1988. *Modern Insect Control*: Nuclear Techniques and Biotechnology. Vienna: Int. At. Energy Agency 479 pp.

49. Isola, L. M., Gordon, J. W. 1986. Systemic resistance to methotrexate in transgenic mice carrying a mutant dihydrofolate reductase gene. *Proc. Natl. Acad. Sci. USA* 83:9621–25

50. Jacobs, D., Dewerchin, M., Boeke, J. D. 1988. Retrovirus-like vectors for *Saccharomyces cerevisiae*: integration of foreign genes controlled by efficient promoters into yeast chromosomal DNA. *Gene* 67:259–69

51. Jacobson, J. W., Medhora, M. M., Hartl, D. L. 1986. Molecular structure of a somatically unstable transposable element in *Drosophila*. *Proc. Natl. Acad. Sci. USA* 83:8684–88

52. Johnston, S. A., Anziano, P. Q., Shark, K., Sanford, J. C., Butlow, R. A. 1988. Mitochondrial transformation in yeasts

by bombardment with microprojectiles. *Science* 240:1538–41

53. Kaneda, Y., Iwai, K., Uchida, T. 1989. Increased expression of DNA cointroduced with nuclear protein in adult rat liver. *Science* 243:375–78

54. Karess, R. E., Rubin, G. R. 1984. Analysis of P transposable element functions in Drosophila. *Cell* 38:135–46

55. Kidwell, M. G. 1981. Hybrid dysgenesis in *Drosophila melanogaster;* the genetis of cytotype determination in a neutral strain. *Genetics* 98:275–90

56. Kidwell, M. G. 1987. A survey of success rates using P element mutagenesis in *Drosophila melanogaster*. *Drosoph. Inf. Serv.* 66:81–86

57. Kidwell, M. G., Kidwell, J. F., Sved, J. A. 1977. Hybrid dysgenesis in *Drosophila melanogaster:* a syndrome of aberrant traits including mutation, sterility, and male recombination. *Genetics* 86:813–33

58. Kleckner, N. 1987. Transposable elements in prokaryotes. *Annu. Rev. Genet.* 15:341-404

59. Klein, T. M., Wolf, E. D., Wu, R., Sanford, J. C. 1987. High-velocity microprojectiles for delivering nucleic acids into living cells. *Nature* 327:70–73

60. Klemenz, R., Weber, U., Gehring, W. J. 1987. The *white* gene as a marker in a new P-element vector for gene transfer in *Drosophila*. *Nucleic Acids Res.* 15:3947–59

61. Knapp, S., Coupland, G., Uhrig, H., Starlinger, P., Salamini, F. 1988. Transposition of the maize transposable element *Ac* in *Solanum tuberosum*. *Mol. Gen. Genet.* 213:285–90

62. Knipling, E. F. 1955. Possibilities of insect control of eradication through the use of sexually sterile males. *J. Econ. Entomol.* 48:4559–62

63. Koller, N. C., Dhadialla, T. S., Raikhel, A. S. 1989. Selective endocytosis of vitellogenin by oocytes of the mosquito, *Aedes aegypti;* an *in vitro* study. *Insect Biochem.* 19:693-702

64. Lambert, M. E., McDonald, J. F., Weinstein, I. B., eds. 1988. *Eukaryotic Transposable Elements as Mutagenic Agents, Banbury Report 30*. Cold Spring Harbor, NY: Cold Spring Harbor Laboratory. 345 pp.

65. Lansman, R. A., Stacey, S. N., Grigliatti, T. A., Brock, H. W. 1985. Sequences homologous to the P mobile element of *Drosophila melanogaster* are widely distributed in the subgenus Sophophora. *Nature* 318:561–63

66. Laski, F. A., Rio, D. C., Rubin, G. M. 1986. Tissue specificity of Drosophila P element transposition is regulated at the level of mRNA splicing. *Cell* 44:7–19

67. Lavitrano, M., Camaioni, A., Fazio, V. M., Dolci, S., Farace, M., Spadafora, C. 1989. Sperm cells as vectors for introducing foreign DNA into eggs: genetic transformation of mice. *Cell* 57: 717–23

68. Lis, J. T., Simon, J. A., Sutton, C. A. 1983. New heat shock puffs and *β*-galactosidase activity resulting from transformation of *Drosophila* with an *hsp70-lacZ* hybrid gene. *Cell* 35:403–10

69. Liu, C. P., Kreber, R. A., Valencia, J. I. 1979. Effects on eye color mediated by DNA injection into *Drosophila* embryos. *Mol. Gen. Genet.* 172:203–10

70. Maeda, S. 1989. Expression of foreign genes in insects using baculovirus vectors. *Annu. Rev. Entomol.* 34:351–72

71. Margolskee, R. F., Kavathas, P., Berg, P. 1988. Epstein-Barr virus shuttle vector for stable episomal replication of cDNA expression libraries in human cells. *Mol. Cell. Biol.* 8:2839–47

72. Martin, P., Martin, A., Osmani, A., Sofer, W. 1986. A transient expression assay for tissue-specific gene expression of alcohol dehydrogenase in *Drosophila*. *Dev. Biol.* 117:574–80

73. Martinez-Zapater, J. M., Finkelstein, R., Somerville, C. R. 1987. *Drosophila* P-element transcripts are incorrectly processed in tobacco. *Plant Mol. Biol.* 11: 601–7

74. Marunouchi, T., Hosoya, H. 1984. Isolation of an autonomously replicating sequence (ARS) from satellite DNA of *Drosophila melanogaster*. *Mol. Gen. Genet.* 196:258–65

75. McGinnis, W., Shermoen, A. W., Beckendorf, S. K. 1983. A transposable element inserted just 5' to a Drosophila glue protein gene alters gene expression and chromatin structure. *Cell* 34:75–84

76. McGrane, V., Carlson, J. O., Miller, B. R., Beaty, B. J. 1988. Microinjection of DNA into *Aedes triseriatus* ova and detection of integration. *Am. J. Trop. Med. Hyg.* 39:502–10

77. McInnis, D. O., Tam, S. Y. T., Grace, C. R., Heilman, L. J., Courtright, J. B., Kumaran, A. K. 1988. The mediterranean fruit fly: progress in developing a genetic sexing strain using genetic engineering methodology. See Ref. 48a, pp. 251–56

78. Medhora, M. M., MacPeek, A. H., Hartl, D. L. 1988. Excision of the *Drosophila* transposable element *mariner*: identification and characterization of the Mos factor. *EMBO J.* 7:2185–89

79. Meyer-Leon, L., Senecoff, J. F., Bruck-

ner, R. C., Cox, M. M. 1984. Site-specific recombination promoted by the FLP protein of the yeast 2 micron plasmid in vitro. *Cold Spring Harbor Symp. Q. Biol.* 49:797–804

80. Miller, L. H., Sakai, R. K., Romans, P., Gwadz, R. W., Kantoff, P., Con, H. G. 1987. Stable integration and expression of a bacterial gene in the mosquito *Anopheles gambiae*. *Science* 237:779–81

81. Milne, C. P., Eishen, F. A., Collis, J. E., Jensen, T. L. 1989. Preliminary evidence for honey bee sperm-mediated DNA transfer. *Int. Symp. Mol. Insect Science, Tucson*, p. 71 (Abstr.)

82. Minden, J. S., Agard, D. A., Sedat, J. W., Aberts, B. M. 1989. Direct cell lineage analysis in *Drosophila melanogaster* by time-lapse, three dimensional optical microscopy of living embryos. *J. Cell. Biol.* 109:505–16

83. Mitsialis, S. A., Kafatos, F. C. 1986. Regulatory elements controlling chorion gene expression are conserved between flies and moths. *Nature* 317:453–56

84. Moermann, D. G., Waterston, R. H. 1989. Mobile elements in *Caenorhabditis elegans* and other nematodes. See Ref. 3a, pp. 537–56

85. Mori, I., Moermann, D. G., Waterston, R. H. 1988. Analysis of a mutator activity necessary for germline transposition and excision in *Tc1* transposable elements in *Caenorhabditis elegans*. *Genetics* 120:397–407

86. Morris, A. C., Eggelston, P., Crampton, J. M. 1989. Genetic transformation of the mosquito *Aedes aegypti* by microinjection of DNA. *Med. Vet. Entomol.* 3:1–7

87. Mouches, C. 1988. Genie genetique chez les insectes: clonage et transgenose de genes de resistance aux insecticides. See Ref. 48a, pp. 25–61

88. Nawa, S., Yamada, S. 1968. Hereditary change in *Ephestia* after treatment with DNA. *Genetics* 58:573–84

89. Nicolau, C., LePape, A., Soriano, P., Fargette, F., Juhel, M. 1983. *In vivo* expression of rat insulin after intravenous administration of the liposome-entrapped gene for rat insulin I. *Proc. Natl. Acad. Sci. USA* 80:1068–72

90. O'Brochta, D. A., Handler, A. M. 1988. Mobility of P elements in drosophilids and nondrosophilids. *Proc. Natl. Acad. Sci. USA* 85:6052–56

91. O'Hare, K., Rubin, G. R. 1983. Structure of P transposable elements and their sites of insertion and excision in the Drosophila melanogaster genome. *Cell* 34:25–35

92. O'Kane, C. J., Gehring, W. J. 1987. Detection in situ of genomic regulatory elements in *Drosophila*. *Proc. Natl. Acad. Sci. USA* 84:9123–27

93. Ow, D. W., Wood, K. V., DeLuca, M., De Wet, J. R., Helinski, D. R., Howell, S. H. 1986. Transient and stable expression of the firefly luciferase gene in plants cells and transgenic plants. *Science* 234:856–59

94. Paszkowski, J., Baur, M., Bogucki, A., Potrykus, I. 1988. Gene targeting in plants. *EMBO J.* 9:4021–26

95. Pirrotta, V. 1988. Vectors for P-mediated transformation in Drosophila. In *Vectors—A Survey of Molecular Cloning Vectors and Their Uses*, ed. R. L. Rodriguez, D. T. Denhardt, pp. 437–56. Boston; Butterworths. 578 pp.

96. Potter, S. S., Brorein, W. J., Dunsmuir, P., Rubin, G. M. 1979. Transposition of elements of *412, copia,* and *297* dispersed repeated gene families in Drosophila. *Cell* 17:415–27

97. Rio, D. C., Laski, F. A., Rubin, G. M. 1986. Identification and immunochemical analysis of biologically active Drosophila P element transposase. *Cell* 44:21–32

98. Rio, D. C., Rubin, G. M. 1988. Identification and purification of a *Drosophila* protein that binds to the terminal 31-base pair inverted repeats of the P transposable element. *Proc. Natl. Acad. Sci. USA* 85:8929–33

99. Robinson, A. S., Savakis, C., Louis, C. 1988. Status of molecular genetic studies in the medfly *Ceratitis capitata*, in relation to genetic sexing. See Ref. 48a, pp. 241–50

100. Rogers, J. H. 1985. The origin and evolution of retroposons. *Int. Rev. Cytol.* 93:187–279

101. Rubin G. M. 1983. Dispersed repetitive DNAs in *Drosophila*. In *Mobile Genetic Elements*, ed. J. A. Shapiro, pp. 329–61. London: Academic. 688 pp.

102. Rubin, G. M., Spradling, A. C. 1982. Genetic transformation of *Drosophila* with transposable element vectors. *Science* 218:348–53

103. Rubin, G. M., Spradling, A. C. 1983. Vectors for P element–mediated gene transfer in *Drosophila*. *Nucleic Acid Res.* 11:6341–51

104. Scavarda, N. J., Hartl, D. L. 1984. Interspecific DNA transformation in *Drosophila*. *Proc. Natl. Acad. Sci. USA* 81:7615–19

105. Seawright, J. A. 1988. Genetic methods for control of mosquitoes and biting flies. See Ref. 48a, pp. 179–94

106. Shigekawa, K., Dower, W. J. 1988.

Electroporation of eukaryotes and prokaryotes: a general approach to the introduction of macromolecules into cells. *Biotechniques* 6:742–51

107. Shirk, P. S., O'Brochta, D. A., Roberts, P. E., Handler, A. M. 1988. Sex-specific selection using chimeric genes: applications to sterile insect release. In *Biotechnology for Crop Protection*, ed. P.A. Hedin, J. J. Menn, R. M. Collingworth, pp. 135–46. Washington, DC: Am. Chem. Soc. 471 pp.

108. Soriano, P., Gridley, T., Jaenisch, R. 1989. Retroviral tagging in mammalian development and genetics. See Ref. 3a, pp. 927–37

109. Spradling, A. C. 1986. P element–mediated transformation. In *Drosophila: A Practical Approach,* ed. D. B. Roberts, pp. 175–97. Oxford; IRL. 295 pp.

110. Spradling, A. C., Rubin, G. M. 1981. *Drosophila* genome organization: conserved and dynamic aspects. *Annu. Rev. Genet.* 15:219–64

111. Steller, H., Pirotta, V. 1985. A transposable P vector that confers selectable G418 resistance to *Drosophila* larvae. *EMBO J* 4:167–171

112. Stinchcomb, D. H., Thomas, M., Kelly, J., Selker, E., Davis, R. W. 1980. Eukaryotic DNA segments capable of autonomous replication in yeast. *Proc. Natl. Acad. Sci. USA* 77:2651–55

113. Struhl, K., Stinchcomb, D., Scherer, S., Davis, R. W. 1979. High-frequency transformation of yeast: Autonomous replication of hybrid DNA molecules.

Proc. Natl. Acad. Sci. USA 76:1035–39

114. Thomas, K. R., Capecchi, M. R. 1987. Site-direct mutagenesis by gene targeting in mouse embryo-derived stem cells. *Cell* 51:503–12

114a. van der Krol, A. R., Mol, J. N. M., Stuitje, A. R. 1988. Modulation of eukaryotic gene expression by complementary RNA or DNA sequences. *Biotechniques* 6:958–76

115. Van Sluys, M. A., Tempe, J., Fedoroff, N. 1987. Studies on the introduction and mobility of the maize *Activator* element in *Arabidopsis thaliana* and *Daucus carota. EMBO J.* 6:3881–89

116. Walker, V. K. 1990. Gene transfer in insects. In *Advances in Cell Culture,* Vol. 7, ed. K. Maramorosch, pp. 87–124. New York: Academic

117. Weising, K., Schell, J., Kahl, G. 1988. Foreign genes in plants: transfer, structure, expression, and applications. *Annu. Rev. Genet.* 22:421–77

118. Wilson, C., Pearson, R. K., Bellen, H. J., O'Kane, C. J., Grossniklaus, U., Gehring, W. J. 1989. P-element-mediated enhancer detection: an efficient method for isolating and characterizing developmentally regulated genes in *Drosophila. Genes Dev.* 3:1301–13

119. Wu, G. Y., Wu, C. H. 1988. Receptor-mediated gene delivery and expression *in vivo. J. Biol. Chem.* 263:14621–24

120. Yoder, J. I., Palys, J., Albert, K., Lassner, M. 1988. *Ac* transposition in transgenic tomato plants. *Mol. Gen. Genet.* 213:291–96

Annu. Rev. Entomol. 1991. 36:185–203

BIOSYSTEMATICS OF THE CHEWING LICE OF POCKET GOPHERS

Ronald A. Hellenthal

Department of Biological Sciences, University of Notre Dame, Notre Dame, Indiana 46556

Roger D. Price

Department of Entomology, University of Minnesota, St. Paul, Minnesota 55108

KEY WORDS: Mallophaga, Trichodectidae, Geomyidae, Rodentia

INTRODUCTION AND OVERVIEW

One of the greatest challenges in systematic and evolutionary biology concerns the taxonomy, coevolution, and biogeography of chewing lice of the genera *Geomydoecus* and *Thomomydoecus* (Mallophaga: Trichodectidae) and their vertebrate hosts, the more than 400 recognized species and subspecies of pocket gophers (Rodentia: Geomyidae). Pocket gophers are fossorial rodents restricted geographically to North and Central America. They are relatively sedentary, exhibit low vagility, and have male and female exclusive territories, low population densities, and patchy distributions. The tendency of groups of these mammals to occur in small local populations has resulted in the development of minor character differences that led to recognition of as many as 230 subspecies within a single species (14). In addition, researchers have long known that the relationships among many of these gopher taxa are poorly understood (1).

Fortunately, pocket gophers are common enough to be well represented in major vertebrate collections and have been the subject of exhaustive morpho-

185

0066-4170/91/0101-0185$02.00

logical, chromosomal, and genic investigations. The patterns of genetic variation in these and some other fossorial rodents have received more study than those in any other mammals except humans (49). Studies have shown that pocket gophers have a high degree of both intrapopulation and interpopulation divergence in genetic systems, including extreme variability in the number of chromosomes and chromosome arms (11, 12, 38, 39, 41–46). Such chromosomal reorganization often is associated with genetic incompatibility among populations and has resulted in reproductive isolation (76–79, 81). In some cases, however, chromosomal variation apparently is independent of reproductive isolation (39). Chromosome diversity may be accompanied by low levels of intrapopulation genic variability (37).

Pocket gopher lice and their hosts present a unique opportunity for the study of host-parasite relationships and cospeciation (6). No other arthropod-vertebrate association offers such a combination of great host diversity and availability coupled with the abundant and nearly universal presence of a diverse but related parasite fauna that has little or no impact on host survival. Chewing lice of the genera *Geomydoecus* and *Thomomydoecus* are the only lice found on pocket gophers. They are restricted to pocket gophers, abundant, and easily collected from hosts or prepared skins. In the 20 years that have elapsed since the last revision of *Geomydoecus* (53), more than 30 published papers have treated the systematics and associations of these lice with their mammalian hosts. Concurrently, but generally independently, mammalogists have continued investigations of geomyid rodents. Only recently, however, mammalogists and entomologists have collaborated in examining the relationships within and between the two groups of organisms. These studies have improved understanding of both the hosts and their parasites. We offer this review as a means of encouraging and furthering cooperative associations between entomologists and biologists working on associated host animals.

TAXONOMIC HISTORY OF POCKET GOPHER LICE

Methods

Researchers long have advocated studies of the relationships between ectoparasitic invertebrates, such as lice, and their hosts (33, 88). However, to accomplish these studies, large numbers of parasites must be collected from a broad spectrum of hosts. Studies of this kind require comprehensive collections of both host and parasite taxa from the entire geographic ranges of both groups of organisms. In addition, parasites must be well understood taxonomically, and both groups of organisms must be reliably identified. To be useful in considerations of host associations, the parasites must be subjected to a thorough taxonomic treatment that includes analyses of both qualitative and quantitative characters and that employs statistical comparisons of populations and their phylogenetic relationships.

Large and intricate host taxonomic complexes, such as those among pocket gophers, present a quantity and complexity of data that can be adequately handled only with the aid of computers. In 1974, we began the development of a computerized pocket gopher–louse data management system. Identification, geographic distribution, and parasite-host association data for all pocket gopher lice that have been collected, with measurements and counts of their morphological characteristics, currently are stored as relational data bases on an IBM 370/3081 computer system at the University of Notre Dame (Table 1). A group of computer programs for host-parasite data storage, retrieval, and analysis called the BUG System (58) originally was developed for a Control Data Computer System at the University of Minnesota and subsequently was adapted for an IBM 370/3081 computer system at the University of Notre Dame. These programs are used to store and update louse, host, and locality information; define tentative taxonomic louse groups; retrieve louse data based on these groups; and analyze these data within a group or between groups. Automated techniques for examining Mallophaga populations for character heterogeneity and identifying potential taxonomic groups were developed from a variability model (Figure 1) based on estimations of variance components from a 5-level nested analysis of variance design. The investigators considered 55 quantitative morphological characters between lice from the same host, from different hosts at a single locality, from different localities within the same gopher taxon, and from different host taxa (18).

The BUG system includes a geographic mapping module (20) that produces host-louse range maps from stored collection data. When localities are added to the system, the latitude and longitude of the closest town or other landmark are entered along with the angle and distance between the landmark and the specific collection site. The system adjusts the geographic coordinates to the specific locality using calculated lengths of degrees for the appropriate parallels and meridians. In producing range maps, the system determines the geographic limits of the locality data to be used, retrieves data defining political boundaries within the limits of the data to be plotted, and then plots the localities and political boundaries as a rectangular projection of the

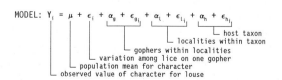

$$\text{MODEL: } Y_i = \mu + \epsilon_i + \alpha_g + \epsilon_{g_i} + \alpha_l + \epsilon_{l_{ij}} + \alpha_h + \epsilon_{h_{ij}}$$

- host taxon
- localities within taxon
- gophers within localities
- variation among lice on one gopher
- population mean for character
- observed value of character for louse

Figure 1 Model showing sources of variation for character traits in louse populations.

Table 1 Louse/locality/host computerized data base summary

Type of information	Quantity	Percentage or average
Gopher taxa	403	90% of described taxa
Louse taxa	122	100% of described taxa
Localities	3,625	8.8/29.7 per gopher/louse taxon
Gophers with lice	6,614	1.8 per locality
Host-louse associations	8,657	1.3 louse taxa per host
Adult lice identified	83,976	12.7 specimens per host
Lice measured	14,641	17.4% of lice identified
Stored character data	403,860	25–30 observations per louse

appropriate portion of North and Central America. Latitude and longitude scales are independently calculated to best display the collection sites. The precision of the system is 0.01 degree (0.75 to 1.15 km).

A separate database containing host-parasite associations for the entire order Mallophaga, which includes all known associations between pocket gophers and their hosts, is being developed (R. D. Price, K. C. Emerson, & R. A. Hellenthal, unpublished information). This database and the programs for updating and retrieving it were developed for the dBASE III Plus™ data management package on IBM PC™ microcomputers and are maintained at the University of Notre Dame.

Because a substantial proportion of the accumulated pocket gopher louse character data is mensurative and subject to both large variation and significant intercharacter correlations, we have used mostly phenetic methods for our phylogenetic evaluations of these lice. Stepwise discriminant analysis was used to identify useful characters in taxonomic separations (86), and canonical correlation has been employed to document group separation and to reduce the effect of body size in establishing relative similarities among taxonomic groups.

Systematics

In spite of the potential usefulness of studies of lice in both defining pocket gopher taxa and understanding relationships among them, intensive work on pocket gopher lice has been done only in the past 20 years. The five then-known species of pocket gopher lice had been placed in the genus *Trichodectes* until Ewing (8) described the genus *Geomydoecus*. Attempts to understand and explain louse-pocket gopher associations and relationships (50, 87) prior to a thorough taxonomic understanding of the lice have no relevance today. Before 1971, 10 species and one subspecies *(Geomydoecus chapini minor)* of lice had been recognized from all pocket gophers (89). These taxa were defined so poorly that they were essentially indistinguishable except by their association with their known hosts. Price & Emerson (53) redescribed the taxa associated with the 11 previous names, elevated *G. chapini minor* to *G. minor,* and described 34 new species and subspecies, bringing the total recognized taxa to 45. This and subsequent work through 1989 increased the number of louse taxa for specimens previously assigned to the original 11 taxa to 81; all but one *(G. chapini)* of the original species represented an assemblage of from 2 to 26 taxa each. The louse fauna of pocket gophers now includes 122 species and subspecies. These are organized as 4 species and 22 complexes containing 2 or more closely related specific or subspecific taxa (Table 2); 3 of these cross host generic or subgeneric limits. The occurrence of such a diversity of recognizable taxa of lice on pocket gophers strengthens the potential use of the lice as indicators of host relationships.

Geomydoecus was subdivided into two subgenera (*Geomydoecus* and *Thomomydoecus*) (54); members of the subgenus *Thomomydoecus* include the more slender, tapered, and smaller lice that generally occur on the same host specimens with lice of the subgenus *Geomydoecus.* The subgenera subsequently were elevated to generic status (19). Lyal (30) theorized that both *Geomydoecus* and *Thomomydoecus* are paraphyletic and that *Thomomydoecus* might be polyphyletic as well. He concluded that recognition of these genera could not be reconciled with the cladistic methods used in his study. However, his methods have been challenged (10), and the recognition of these taxa as genera has been substantiated subsequently on both morphological and genetic grounds (36).

HOST-PARASITE ASSOCIATIONS

The use of proper procedures for the collection and preparation of host material is crucial in studies of associations and relationships between ectoparasites and their hosts. Because of the amount of ectoparasite material available for study, we often could distinguish between legitimate associations of parasites with hosts and those that arise as a result of contamination or accidental transfer of parasites between hosts during collection, storage, or preparation of specimens. Measures that guard against potential transfer of lice in the collection, treatment, and preparation of hosts are particularly important in investigations of areas with host hybridization or where different host taxa occur in the same locality. For example, in examining the apparent transfer of lice among host taxa, we found that such exchanges occurred only when both host taxa were collected during the same day.

The practice of putting the carcasses of trapped hosts in a single pile prior to preparation of skins may destroy much of their value for considerations of parasitic associations. When collections of hosts and parasites are not large,

Table 2 Pocket gopher–louse associations

Host Genus (subgenus)	Gopher taxa[a] SP	SP+SSP	Louse species or complex (# taxa) Geomydoecus		Thomomydoecus		Reference
Geomys	8	38	geomydis	(8)			86
			scleritus	(2)			52, 69
			texanus	(2)			56
			truncatus	(2)			21
			dalgleishi	(1)			85
			quadridentatus	(1)			21
Orthogeomys							
(*Orthogeomys*)	2	17	alleni	(2)			65
			chiapensis	(2)			65
(*Heterogeomys*)	2	13	copei	(4)			57
(*Macrogeomys*)	6	12	panamensis	(6)			51, 68
Pappogeomys							
(*Pappogeomys*)	2	9	bulleri	(5)			66
			mcgregori	(1)			67
(*Cratogeomys*)	7	45	coronadoi	(7)			67
			mexicanus	(5)			67
			texanus	(4)			56
			mcgregori	(3)			67
			expansus	(2)			55
			telli	(1)			64
Thomomys[b]							
(*Thomomys*)	3	74	thomomyus	(8)	wardi	(3)	22, 23
(*Megascapheus*)	4[c]	231	californicus	(16)	minor	(9)	61, 63
			umbrini	(8)	neocopei	(7)	60, 62
			subcalifornicus	(2)	wardi	(1)	18, 19, 23
			oregonus	(4)			59
			tolucae	(5)			58
Zygogeomys	1	2	trichopi	(1)			67
	35	441		(102)		(20)	

[a] Counts of pocket gopher species and subspecies generally follow Hall (14), except as noted.
[b] *Thomomys* subgenera from Thaeler (80).
[c] *Thomomys (Megascapheus)* species from Patton et al (48).

this practice easily can lead to erroneous conclusions that can destroy much of the potential value of the work. We urge vertebrate field biologists to exercise great care to insure the integrity of the ectoparasite fauna on individual host specimens, even when all collected individuals appear to be a single taxon. Transfer of ectoparasites among prepared host specimens in museums appears to be a less serious problem, although common sense would suggest that grooming skins over trays containing other specimens, for example, should be avoided.

Lice have been identified from all host genera and subgenera as well as from a majority of the host species and subspecies (Table 2). The most recently published host-parasite list of the Mallophaga on mammals (5) is incomplete for pocket gopher associations with their lice and should not be used with reference to those hosts. The associations between the lice and their hosts are summarized below.

Geomys

The *Geomydoecus* taxa that occur on *Geomys* hosts show a high degree of host specificity with only a single louse taxon occurring on an individual host specimen. The eight species of the *geomydis* complex, which constitutes the largest of six complexes on *Geomys*, are restricted to the 21 taxa of *Geomys bursarius* sensu Hall (14) and are the only lice found on those gophers (86).

The only lice that occur on the eight taxa of the southeastern U.S. pocket gophers are the two species of the *scleritus* complex (52). These lice all are parthenogenetic; only several males have been collected along with nearly a thousand females.

The two species of the *truncatus* complex and both subspecies of the *Geomydoecus texanus* complex are found on *Geomys tropicalis* and on various subspecies of *Geomys personatus,* again with a high degree of specificity (21, 56). The host of *Geomydoecus truncatus* originally was thought to be *G. personatus personatus.* However, we have never collected this parasite from that host; the only host we have verified is *G. personatus fallax.* The host for *Geomydoecus neotruncatus* is *G. personatus streckeri.* One subspecies of *G. texanus* occurs on three subspecies of *G. personatus* and the other occurs on *G. tropicalis.* The remaining four members of the *texanus* complex occur on *Pappogeomys castanops.*

Geomydoecus dalgleishi is known only from *G. personatus fuscus* in southern Texas and is well isolated from all other populations of *Geomys.* Likewise, *Geomydoecus quadridentatus* occurs only on the two subspecies of *Geomys arenarius,* which are the westernmost *Geomys* and well removed from other members of this gopher genus.

Geomys subspecies with unknown louse associations are *G. bursarius ludemani, G. bursarius pratincola,* and *G. bursarius terricolus.*

Orthogeomys (Orthogeomys)

The two species in each of the *alleni* and *chiapensis* complexes, which are among the largest *Geomydoecus,* occur only on *Orthogeomys grandis* (65). The two from the *alleni* complex live on five host subspecies in Mexico. Of the *chiapensis* complex, *Geomydoecus chiapensis* is found on two subspecies in Guatemala and *Geomydoecus pygacanthi* on one subspecies in El Salvador.

Orthogeomys (Orthogeomys) taxa with unknown louse associations are *O. cuniculus, O. grandis annexus, O. grandis carbo, O. grandis engelhardi, O. grandis guerrerensis, O. grandis huixtlae, O. grandis pluto, O. grandis soconuscensis,* and *O. grandis vulcani.*

Orthogeomys (Heterogeomys)

The four species of the *copei* complex are known from 10 subspecies of *Orthogeomys hispidus* in Mexico (57). The lice fall into two distinct groups; *Geomydoecus copei* is from four subspecies in the northern third of the host range and the other three species are from six host subspecies in the southern two-thirds (57). The only discrepancy these distributions show involves the occurrence of *G. copei* in the northern ranges of the hosts *O. hispidus hispidus* and *O. hispidus torridus* and of *Geomydoecus hoffmanni* in the southern ranges of these same host subspecies, which suggests either an illogical assemblage of gopher taxa or an improper indication of host phylogeny based on these lice.

Orthogeomys (Heterogeomys) taxa with unknown louse associations are *O. hispidus hondurensis, O. hispidus latirostris,* and *O. lanius.*

Orthogeomys (Macrogeomys)

Of the six taxa of the *panamensis* complex (68), the two subspecies of *Geomydoecus panamensis* are southernmost in Costa Rica and Panama and found, respectively, on *Orthogeomys cavator* and *Orthogeomys dariensis.* *Geomydoecus costaricensis* has been taken only from the three subspecies of *Orthogeomys heterodus.* The remaining three species of this louse complex present a confused picture: *Geomydoecus setzeri* and *Geomydoecus davidhaf-neri* are both known from *Orthogeomys cherriei* and *Orthogeomys un-derwoodi,* even occurring on the same host individuals. These two species are separable only on the basis of the male's morphology. *Geomydoecus cherriei* also occurs on *O. cherriei* as well as on *Orthogeomys matagalpae.* The significance of this situation is still the subject of study.

Orthogeomys (Macrogeomys) subspecies with unknown louse associations are *O. cavator nigrescens, O. cherriei cherriei,* and *O. cherriei carlosensis.*

Pappogeomys (Pappogeomys)

The five taxa of the *bulleri* complex are associated with the eight subspecies of *Pappogeomys bulleri* (66). The most widespread of these, *Geomydoecus bulleri bulleri,* is recorded from five host subspecies, whereas the others are each associated with a single subspecies, and one of these is known only from an unidentified subspecies from Colima, Mexico. *Geomydoecus alcorni,* a member of the *mcgregori* complex whose other members occur on the other subgenus of *Pappogeomys,* is known only from *Pappogeomys alcorni.* This represents one of only three instances reported here in which members of a louse complex are found on hosts in two genera or subgenera. This observation may be explained by the fact that both collection sites for *G. alcorni* are identical to those for a *Pappogeomys tylorhinus* host of *Geomydoecus mcgregori*.

Pappogeomys (Cratogeomys)

A member of the *mexicanus* complex often occurs on the same host individual as one of the *coronadoi* complex (67). The host range for the five taxa of the *mexicanus* complex consists of five subspecies of *Pappogeomys merriami* and two of *Pappogeomys tylorhinus*, evidencing a high degree of host specificity. The range for the seven taxa of the *coronadoi* complex is somewhat comparable. It includes the same five subspecies of *P. merriami* and two of *P. tylorhinus*, but also includes *Pappogeomys zinseri*, a subspecies of *Pappogeomys gymnurus*, and two other subspecies of *P. tylorhinus*. The two subspecies of the louse *Geomydoecus polydentatus* account for this expansion of host ranges.

The remaining three taxa of the *mcgregori* complex (67), excluding *G. alcorni* discussed above, are found on *Pappogeomys fumosus*, *Pappogeomys neglectus*, *P. zinseri*, the four subspecies of *P. gymnurus*, and the six subspecies of *P. tylorhinus*. Both a member of the *mcgregori* complex and of the *coronadoi* complex often occur on the same host individual in six of these host taxa. To further complicate this picture, *Geomydoecus telli*, which is so unusual that it is placed in no other complex (64), occurs on two subspecies of *P. tylorhinus* and one of *P. gymnurus*, and in all instances was collected along with a member of the *mcgregori* complex. This occurrence of two, and occasionally even three, species of *Geomydoecus* of these four complexes living together on the same host individual aggravates the problems of identification and defies an explanation on our part.

The remaining two *Geomydoecus* complexes found on *Pappogeomys* are limited to the 25 subspecies of *Pappogeomys castanops* extending from Colorado and Kansas southward into Mexico. A study of the host associations of the *expansus* and *texanus* complexes (17) revealed a northern and southern grouping of the lice. Individuals of these two complexes often occurred together on the same host specimen. Also, the single collection of *G. tamaulipensis* appears to be from a parthenogenetic population. The *texanus* complex represents the only case in which members are shared between two gopher genera; four taxa live on *P. castanops*, and the remaining two live on *G. personatus* collected primarily along the Gulf Coast of southern Texas and northern Mexico. This host-louse association remains an enigma.

The *Pappogeomys (Cratogeomys)* species with an unknown louse association is *P. merriami peraltus*.

Thomomys (Thomomys)

The *Thomomys* subgenus of pocket gophers, composed of *Thomomys talpoides*, *Thomomys mazama*, and *Thomomys monticola*, contains a louse fauna distinct from those of all other gophers. Of the two louse complexes on this gopher group, the *Geomydoecus thomomyus* complex is by far the largest and most widely distributed both geographically and with respect to hosts (22). The eight taxa in this louse complex extend throughout much of the range of the host taxa (22). An unusual situation was encountered for eight subspecies of *T. talpoides* in southwestern Canada and adjacent areas of the US in that no lice were obtained from about 435 gopher specimens that were brushed. While this absence could be ascribed to a variety of other reasons, some populations of pocket gophers may actually have no lice. Only females have been found for *Geomydoecus betleyae* and *Geomydoecus biagiae*, which indicates that these lice are likely parthenogenetic.

The other louse complex on these gophers, the *Thomomydoecus wardi* complex, apparently is restricted to the eastern half of the host range (23). Representatives of three taxa in this complex virtually always occur on the same host individual together with specimens of the *thomomyus* complex. A fourth taxon of this complex, *Thomomydoecus byersi*, occurs on *Thomomys bottae* and is morphologically quite divergent from the other three taxa of the complex.

Thomomys (Thomomys) subspecies with unknown louse associations are *T. talpoides andersoni*, *T. talpoides cognatus*, *T. talpoides incensus*, *T. talpoides kelloggi*, *T. talpoides loringi*, *T. talpoides medius*, *T. talpoides relicinus*, *T. talpoides segregatus*, *T. talpoides shawi*, and *T. talpoides whitmani*.

Thomomys (Megascapheus)

Most gopher groups contain relatively few taxa. The subgenus *Thomomys* discussed above is the largest thus far with 74 host taxa. In contrast, the *Thomomys bottae* group of gophers that constitutes the *Megascapheus* subgenus contains 231 species and subspecies, over half of all gopher taxa. These are distributed among four gopher species, *T. bottae*, *Thomomys umbrinus*, *Thomomys bulbivorus*, and *Thomomys townsendii*. Thirty-five taxa of *Geomydoecus* lice and 17 of *Thomomydoecus* are distributed among gophers of this group. This broad diversity of both hosts and lice results in a very complicated picture of distributional associations (20), a situation that cannot be dealt with here in detail.

The *Geomydoecus californicus* complex contains 16 louse taxa and is the most widespread of all pocket gopher louse complexes, having been collected from 144 *T. bottae* and 14 *T. umbrinus* subspecies. It is generally absent from *T. bottae* subspecies of southern Oregon and California north of San Francisco Bay and is not found on *T. bulbivorus* or *T. townsendii*, except in a small hybrid zone in northeastern California. The eight taxa of the *umbrini* complex are known only from 22 subspecies of Mexican *T. umbrinus*. The two species of the *subcalifornicus* complex are distributed widely throughout the southwestern US and northern Mexico and have been collected from 43 subspecies of *T. bottae* and three of *T. umbrinus*. The lice of these three complexes form a closely knit group morphologically, and provide further support for the louse-based separation of *T. bottae* and *T. umbrinus*.

The four species of the *oregonus* complex are restricted to 21 subspecies of *T. bottae* in northern California and spottily distributed in southern California, as well as on *T. bulbivorus* and all seven subspecies of *T. townsendii*. The distribution of this complex suggests a difference between *T. bottae* taxa from these areas and those found elsewhere, with a possible affinity between the former and *T. bulbivorus* and *T. townsendii*. The five taxa of the *tolucae* complex, which have been collected from eight Mexican *T. umbrinus* subspecies and two isolated populations of *T. bottae* subspecies in Arizona, are morphologically closest to those of the *oregonus* complex.

The *Thomomydoecus* taxa nearly always occur on individual gophers that also are carrying *Geomydoecus* lice. However, *Geomydoecus* often occurs in the absence of *Thomomydoecus*. The *minor* complex is the largest of the *Thomomydoecus* complexes; its nine species were collected from 75 *T. bottae* subspecies and 18 *T. umbrinus*. The seven species of the *neocopei* complex occur on 10 subspecies of *T. umbrinus* in Mexico and southwestern New Mexico. Each louse taxon of this complex has a very restricted distribution, which may reflect the limited distribution of the hosts. The members of the *wardi* complex are principally found on *T. talpoides* group gophers, but one species occurs on two subspecies of *T. bottae* in New Mexico and southern Colorado.

Thomomys (Megascapheus) subspecies with unknown louse associations are *T. bottae aureiventris*, *T. bottae magdalenae*, *T. bottae powelli*, *T. bottae pusillus*, *T. bottae spatiosus*, *T. bottae subsimilis*, *T. bottae varus*, *T. mazama couchi*, *T. mazama melanops*, and *T. mazama premaxillaris*.

Zygogeomys

The genus *Zygogeomus* consists of only a single species and it is the host for the louse *Geomydoecus trichopi* (67). While the distribution of this pocket gopher is close to those carrying lice of the *coronadoi* and *mcgregori* complexes, *G. trichopi* is sufficiently distinct to merit standing apart from lice of the other complexes.

DISTRIBUTIONAL SIMILARITIES AND INCONGRUITIES

The procurement of large numbers of lice and the application of both univariate and multivariate morphometric analyses during the past 15 years resulted in a clearer delineation of previously known taxa of lice and the description of a large number of additional louse species and subspecies and an increased understanding of host relationships. In many cases, the lice have distributions that closely approximate those of the species and subspecies of their hosts.

This is particularly true for *Orthogeomys* and *Pappogeomys* pocket gophers inhabiting central Mexico and ranging south through Panama, where gophers occur in geographically isolated populations. While louse species in this region also have restricted ranges, often two or even three species of *Geomydoecus* occur on a single host taxon, and, for some host taxa, on the same individual. By contrast, louse distributions from *Zygogeomys* and *Geomys* typically have only one *Geomydoecus* louse taxon on a single host taxon or individual, and *Thomomydoecus* lice have never been found on these same hosts. On many *Thomomys,* a species of *Thomomydoecus* nearly always occurs with a species of *Geomydoecus* on the same gopher. However, *Geomydoecus* species often occur on hosts without *Thomomydoecus*.

In *Geomys,* northern *Pappogeomys,* and *Thomomys,* distributional associations between lice and their hosts may be much more complicated and suggest independent distributions. These host taxa, while generally allopatric, have distributions that often are contiguous, offering a greater opportunity for transfer of parasite species between host species. Thus, the interpretation of host-parasite associations in these groups rests on whether transfer of parasites among host taxa is possible or likely. While there is no intrinsic reason to believe that transfer cannot occur, an increasing body of evidence suggests that it is exceedingly rare. For example, the *T. bottae* group consists of 231 named taxa that represent over half of all of the taxa recognized in the pocket gophers. Most of the distinctions between these taxa are based on variation in adaptive pelage color and body size (14). However, the *T. bottae* group exhibits a greater degree of chromosomal variability than any other known mammalian species (39). While most of the host taxa from which lice were obtained showed a fairly uniform louse fauna, each of 70 subspecies possessed geographically separated localities with 2 or more related louse taxa. In many of these cases, the distribution of louse taxa appeared to divide the host taxa into two or more parts, indicating that some refinement of the host distributions might be appropriate. In other cases, host-louse incongruities appeared in peripheral areas of the host ranges and might indicate misidentification of the host taxa. While this observation could be cited as an example of transfer of parasites among host taxa, recent treatments of the *T. bottae* of California (47, 73) resulted in a two-thirds reduction in the number of defined taxa. These newer definitions show close conformity to the distributions of *T. bottae*–associated lice (3).

A similar but less complicated case of host-parasite distributional incongruity appears with *P. castanops*. Study of the distribution of *P. castanops* lice (17) revealed groupings of gopher taxa with major discrepancies from those proposed by Russell (72) in the revision of the genus *Pappogeomys* based on morphology. While the distribution of louse taxa conformed reasonably well to that of their hosts as given by Hall & Kelson (15), it departed substantially from the subspecies groupings defined by Russell (72) and subsequently adopted by Hall (14). Instead, the louse groupings divided hosts into northern and southern subspecies groups. This corresponded perfectly with the findings of Berry & Baker (2) that gophers north of the 25th parallel have a diploid chromosome number of 46 and those south of it have only 42. Thus, the louse associations, rather than presenting an example of host-parasite incongruity, support a classification of hosts based on genetics rather than morphology and suggest the need for additional study of this gopher group by mammalogists.

Comparison of louse distributional relationships with *G. bursarius* also showed discrepancies with the existing host classification (86). However, subsequent morphological, genetic, and biochemical work on this pocket gopher genus (16, 74) indicated a very close conformity of the louse and gopher taxonomies. As an example, the distributions of *Geomydoecus nebrathkensis* and *Geomydoecus oklahomensis* lice in Nebraska had not corresponded with the distribution of what was then recognized as *G. bursarius lutescens*. Subsequently, this host taxon was divided into two subspecies in exact agreement with the louse distributions.

One of few examples of apparent crossover and successful establishment of lice from one major host taxon to another appears among the lice on *T. talpoides levis* in Utah. Here we have seen collections from seven gophers in six localities associated with specimens of *Thomomydoecus zacatecae*, a normal inhabitant of *T. bottae*, along with the usual lice of the *thomomyus* complex. This part of Utah is an interface area between the two *Thomomys* gopher subgenera and exchanges of lice would be possible. With the exception of those occurring in hybrid zones (48), this crossover is the only one we can document even when different host species or genera occur in the same locality. For example, *Geomys* and *Thomomys* occur together in a field northwest of Las Cruces, New Mexico. C. S. Thaeler, Jr. (personal communication) has collected gophers of both genera from the same burrow system, and we have collected and examined lice from these and other hosts in this area. We can find no evidence for transfer of lice between host taxa in this or other areas where opportunities for exchange of parasites exist.

The extent to which louse movement between pocket gopher taxa may occur in areas of host hybridization has been considered (48). Genetic similarity and ectoparasite associations were compared for *T. bottae* × *T. townsendii* hybrids from Gold Run Creek in California. While both F_1 hybrid and presumptive backcross pocket gophers were apparent, no genic introgression was evident, based on five diagnostic allozyme loci present in parental populations of either taxon within a mile of the hybrid zone. Similarly, louse species unique to each parental gopher did not penetrate beyond the geographic limits of the genetic hybrid zone into the range of the opposite gopher species. Thus, a narrow zone of hydridization was defined concordantly by genetic, morphologic, and ectoparasitic parameters. On the basis of this study, these two gopher taxa were determined to be genetically, if not reproductively, isolated and should be considered separate biological species. To our knowledge, that paper (48) is the first to combine mammalian cytogenetics with parasite associations in evaluating the validity of mammalian species. Other examples of louse crossover restricted to zones of host hybridization have been documented for *T. bottae canus* × *T. townsendii relictus* hybrids (75) in the Garnier Ranch area in California and for *T. bottae modicus* × *T. umbrinus intermedius* hydrids (40) at Sycamore Canyon in Arizona.

EVOLUTION AND COSPECIATION

Studies of ectoparasites such as chewing lice, because of their conservative rates of evolution with respect to their hosts, may aid in recognition of host relationships (26, 71). While researchers generally believe that arthropod ectoparasites have a high degree of host specificity and that parasite phylogeny reflects host phylogeny (i.e. Fahrenholz's Rule or *host tracking*), Kethley & Johnston (27), drawing from their experience with quill mites but citing other examples, concluded that noncongruent host-parasite relationships, indicative of resource tracking, are much more common than previously realized. This idea is supported for the chewing lice of birds by Eveleigh & Amano (7) in their study of the lice of alcids. However, other workers continue to find evidence supporting instances of host tracking (9, 28, 29, 35, 70).

The controversy over this question extends to pocket gophers and their lice. Timm (82, 83) found evidence for host tracking among the lice of *G. bursarius*. Lyal (31, 32) claimed that these data are consistent with a resource-tracking hypothesis. Problems with underlying assumptions of Fahrenholz's Rule also are discussed by Mitter & Brooks (34) and Humphries & Seberg (24). Hafner & Nadler (13) tested Fahrenholz's Rule by constructing phylogenetic trees for both pocket gophers and their lice using protein electrophoretic data. They found a high degree of concordance in the branching patterns of trees, which suggests that a history of cospeciation exists in this host-parasite assemblage. In several cases in which the branching patterns were identical in the host and parasite phylogenies, the branch lengths also were similar. Given the assumptions of molecular clock theory, this similarity suggests that the speciation of these hosts and ectoparasites was roughly contemporaneous and causally related.

While we believe that resource tracking may have contributed to host-parasite associations in other groups of Mallophaga, evidence of genetically based associations between trichodectid lice and pocket gopher hosts is sufficient to justify a host-tracking model of cospeciation for these organisms. Pocket gopher lice already have been employed effectively to provide definitions of host taxa independent of those determined through the morphological and genetic study of host populations. They also have been successfully used as indicators of gene flow between host populations and as a means of evaluating the validity of and relationships among currently defined host taxa.

The evolutionary relationships between ectoparasitic chewing lice and their geomyid hosts appear to fit the cospeciation model described by Brooks (4) and Janzen (25). While chewing lice appear to have evolved some close, parallel behavioral traits in response to their hosts (83), pocket gophers appear to have undergone little, if any, evolutionary change in response to chewing lice. The chewing lice of pocket gophers apparently feed only on dead tissue and are not known to transmit diseases or other parasites to their hosts. Therefore, the only potential cost to the host for harboring these parasites would be increased time spent in grooming. However, this increased grooming has not been observed, and, since both grooming and feeding occur away from potential predators in enclosed tunnel systems and host populations appear to be nearly universally infested with lice, louse infestation is probably not a significant factor in host fitness or fecundity (84). Furthermore, the host as a chewing louse resource differs little within many of the large pocket gopher subspecies complexes that contain diverse allopatric louse species.

Lyal (30) investigated evolutionary relationships in the Trichodectidae, but omitted analysis of the cladistic relationships among most pocket gopher lice. These relationships have been examined phenetically for lice of *T. bottae* in California (3) and a cladistic examination of the *Geomydoecus* and *Thomomydoecus* lice is ongoing.

CONCLUSION

Although detailed studies of chromosomal and genic composition undoubtedly will yield much additional information on pocket gopher systematics and taxonomic relationships, the requirement of freshly collected specimens for preparations makes extensive studies difficult. Even though chromosomal characters have taken on an increasing importance in mammalian taxonomy, the need for taxonomic methods that are powerful and can utilize the tremendous resources contained in the preserved specimens maintained in the major museums continues. Furthermore, environmental studies often are concerned with the identification and previous distribution of taxa that may have been eliminated from areas because of human disturbance. Lice are easily collected from museum skins of hosts long dead, poorly prepared, or even from those with skulls missing. The lice may, therefore, help answer questions of distributional changes through time and assist in solving problems of host identification as well as of relationships.

ACKNOWLEDGEMENTS

For their continuing help in understanding and interpreting pocket gopher distributions and systematics, we thank R. J. Baker, H. H. Genoways, M. S. Hafner, R. S. Hoffmann, J. K. Jones, S. A. Nadler, J. L. Patton, C. S. Thaeler, Jr., and R. M. Timm. M. M. Betley, C. Biagi Chan, B. L. Clauson, K. C. Emerson, and R. M. Timm provided invaluable help in our investigations of louse systematics and host relationships. We thank B. J. Hellenthal for her assistance in the preparation of this and previous manuscripts. Partial support for our studies was supplied by grants from the National Science Foundation to the University of Minnesota (Grant No. DEB77-10179) and to the University of Notre Dame (Grant No. DEB81-17567, BSR86-14456). Our studies also have been supported partially by project No. Min-17-015 of the Minnesota Agricultural Experiment Station.

Literature Cited

1. Anderson, S. 1966. Taxonomy of gophers, especially *Thomomys* in Chihuahua, Mexico. *Syst. Zool.* 15:189–98
2. Berry, D. L., Baker, R. J. 1972. Chromosomes of pocket gophers of the genus *Pappogeomys,* subgenus *Cratogeomys. J. Mammal.* 53:303–9
3. Betley, M. M. 1985. *Chewing louse relationships and associations with California* Thomomys bottae *pocket gophers.* MS thesis. Notre Dame: Univ. Notre Dame. 94 pp.
4. Brooks, D. R. 1979. Testing the context and extent of host-parasite coevolution. *Syst. Zool.* 28:299–307
5. Emerson, K. C., Price, R. D. 1981. A host-parasite list of the Mallophaga on mammals. *Misc. Publ. Entomol. Soc. Am.* 12(1):1–72
6. Emerson, K. C., Price R. D. 1985. Evolution of Mallophaga on mammals. See Ref. 27a, pp. 233–55
7. Eveleigh, E. S., Amano, H. 1977. A numerical taxonomic study of the mallophagan genera *Cummingsiella (= Quadraceps), Saemundssonia* (Ischnocera: Philopteridae), and *Austromenopon* (Amblycera: Menoponidae) from alcids (Aves: Charadriiformes) of the northwest Atlantic with reference to host-parasite relationships. *Can. J. Zool.* 55:1788–1801

8. Ewing, H. E. 1929. *A Manual of External Parasites.* Springfield, IL: C. C. Thomas. 225 pp.
9. Fain, A. 1979. Specificity, adaptation and parallel host-parasite evolution in acarines, especially Myobiidae, with a tentative explanation for the regressive evolution caused by the immunological reactions of the host. *Proc. 5th Int. Congr. Acarol.* 2:321–28
10. Futuyma, D. J., Kim, J. 1987. Coevolution and systematics. book review. *Science* 237:441–42
11. Hafner, J. C., Hafner, D. J., Patton, J. L., Smith, M. F. 1983. Contact zones and the genetics of differentiation in the pocket gopher *Thomomys bottae* (Rodentia: Geomyidae). *Syst. Zool.* 32:1–20
12. Hafner, M. S., Hafner, J. C., Patton, J. L., Smith, M. F. 1987. Macrogeographic patterns in the pocket gopher *Thomomys umbrinus. Syst. Zool.* 36:18–34
13. Hafner, M. S., Nadler, S. A. 1988. Phylogenetic trees support the coevolution of parasites and their hosts. *Nature* 332:258–59
14. Hall, E. R. 1981. *The Mammals of North America,* Vol. 1. New York: Wiley. 690 pp. 2nd ed.
15. Hall, E. R., Kelson, K. R. 1959. *The Mammals of North America.* Vol. 1. New York: Ronald. 625 pp.

16. Heaney, L. R., Timm, R. M. 1983. Relationships of pocket gophers of the genus *Geomys* from the central and northern great plains. *Misc. Publ. Univ. Kans. Mus. Nat. Hist.* 74:1–59

17. Hellenthal, R. A., Price, R. D. 1976. Louse-host associations of *Geomydoecus* (Mallophaga: Trichodectidae) with the yellow-faced pocket gopher, *Pappogeomys castanops* (Rodentia: Geomyidae). *J. Med. Entomol.* 13:331–36

18. Hellenthal, R. A., Price, R. D. 1980. A review of the *Geomydoecus subcalifornicus* complex (Mallophaga: Trichodectidae) from *Thomomys* pocket gophers (Rodentia: Geomyidae), with a discussion of quantitative techniques and automated taxonomic procedures. *Ann. Entomol. Soc. Am.* 73:495–503

19. Hellenthal, R. A., Price, R. D. 1984a. A new species of *Thomomydoecus* (Mallophaga: Trichodectidae) from *Thomomys bottae* pocket gophers (Rodentia: Geomyidae). *J. Kans. Entomol. Soc.* 57:231–36

20. Hellenthal, R. A., Price, R. D. 1984b. Distributional associations among *Geomydoecus* and *Thomomydoecus* lice (Mallophaga: Trichodectidae) and pocket gopher hosts of the *Thomomys bottae* group (Rodentia: Geomyidae). *J. Med. Entomol.* 21:432–46

21. Hellenthal, R. A., Price, R. D. 1988. *Geomydoecus* (Mallophaga: Trichodectidae) from the Texas and desert pocket gophers (Rodentia: Geomyidae). *Proc. Entomol. Soc. Wash.* 91:1–8

22. Hellenthal, R. A., Price, R. D. 1989a. The *Geomydoecus thomomyus* complex (Mallophaga: Trichodectidae) from pocket gophers of the *Thomomys talpoides* Complex (Rodentia: Geomyidae) of the United States and Canada. *Ann. Entomol. Soc. Am.* 82:286–97

23. Hellenthal, R. A., Price, R. D. 1989b. The *Thomomydoecus wardi* complex (Mallophaga: Trichodectidae) of *Thomomys talpoides* pocket gophers (Rodentia: Geomyidae). *J. Kans. Entomol. Soc.* 62:245–53

24. Humphries, C. J., Seberg. O. 1989. Graphs and generalized tracks: some comments on method. *Syst. Zool.* 38: 69–76

25. Janzen, D. H. 1980. When is it coevolution? *Evolution* 34:611–12

26. Kellogg, V. L. 1896. New Mallophaga, I, with special reference to a collection made from maritime birds of the Bay of Monterey, California. *Proc. Calif. Acad. Sci. Ser.* 2 4:31–168

27. Kethley, J. B., Johnston, D. E. 1975. Resource tracking patterns in bird and mammal ectoparasites. *Misc. Publ. Entomol. Soc. Am.* 9:231–36

27a. Kim, K. C., ed. 1985. *Coevolution of Parasitic Arthropods and Mammals.* New York: Wiley. 752 pp.

28. Kim. K. C. 1988. Evolutionary parallelism in Anoplura and eutherian mammals. In *Biosystematics of Haematophagous Insects*, ed. M. W. Service, pp. 91–114. Oxford: Clarendon. 376 pp.

29. Kim, K. C., Repenning, C. A., Morejohn, G. V. 1975. Specific antiquity of the sucking lice and evolution of otariid seals. *Rapp. P.-V. Reun. Cons. Int. Explor. Mer.* 169:544–49

30. Lyal, C. H. C. 1985. A cladistic analysis and classification of trichodectid mammal lice (Phthiraptera: Ischnocera). *Bull. Brit. Mus. Nat. Hist. Entomol. Ser.* 51(3):187–346

31. Lyal, C. H. C. 1986. Coevolutionary relationships of lice and their hosts: a test of Fahrenholz's Rule. In *Coevolution and Systematics*, ed. A. R. Stone, D. L. Hawksworth, pp. 77–91. Oxford: Clarendon. 150 pp.

32. Lyal, C. H. C. 1987. Co-evolution of trichodectid lice (Insecta: Phthiraptera) and their mammalian hosts. *J. Nat. Hist.* 21:1–28

33. Mayr, E. 1957. Evolutionary aspects of host specificity among parasites of vertebrates. In *Premiere Symposium sur la specifité Parasitaire des Parasites de Vertebres*, pp. 7–14. Neuchatel: Univ. Neuchatel

34. Mitter, C., Brooks, D. R. 1983. Phylogenetic aspects of coevolution. In *Coevolution*, ed. D. J. Futuyma, M. Slatkin, pp. 65–98. Sunderland, MA: Sinauer Assoc. 400 pp.

35. Moss. W. W. 1979. Patterns of host-specificity and co-evolution in the Harpyrhynchidae. *Proc. 5th Int. Con. Acarol.* 2:379–84

36. Nadler, S. A., Hafner, M. S. 1989. Genetic differentiation in sympatric species of chewing lice (Mallophaga: Trichodectidae). *Ann. Entomol. Soc. Am.* 82:109–13

37. Nevo, E., Kim, Y. J., Shaw, C. R., Thaeler, C. S. Jr. 1974. Genetic variation, selection, and speciation in *Thomomys talpoides* pocket gophers. *Evolution* 28:1–23

38. Patton, J. L. 1970. Karyotypic variation following an elevational gradient in the pocket gopher, *Thomomys bottae grahamensis* Goldman. *Chromosoma* 31: 41–50

39. Patton, J. L. 1972. Patterns of geographical variation in karyotype in the pocket

gopher, *Thomomys bottae* (Eydoux and Gervais). *Evolution* 26:574–86

40. Patton. J. L. 1973. An analysis of natural hybridization between the pocket gophers, *Thomomys bottae* and *Thomomys umbrinus,* in Arizona. *J. Mammal.* 54:561–84

41. Patton, J. L. 1981. Chromosomal and genic divergence, population structure, and speciation potential in *Thomomys bottae* pocket gophers. In *Ecologia y Genetica de la Especiacion Animal,* ed. O. A. Reig. pp. 255–95. Caracas, Venezuela: Univ. Simon Bolivar

42. Patton. J. L. 1985. Population stucture and the genetics of speciation in pocket gophers, genus *Thomomys. Acta Zool. Fenn.* 170:109–14

43. Patton. J. L., Dingman, R. E. 1968. Chromosome studies of pocket gophers, genus *Thomomys.* I. The specific status of *Thomomys umbrinus* (Richardson) in Arizona. *J. Mammal.* 49:1–13

44. Patton. J. L. Dingman, R. E. 1970. Chromosome studies of pocket gophers, genus *Thomomys.* II. Variation in *T. bottae* in the American southwest. *Cytogenetics* 9:139–51

45. Patton, J. L., Feder, J. H. 1978. Genetic divergence between populations of the pocket gopher, *Thomomys umbrinus* (Richardson). *Z. Saugetierkd.* 43:17–30

46. Patton, J. L., Selander, R. K., Smith, M. H. 1972. Genic variation in hybridizing populations of gophers (genus *Thomomys*). *Syst. Zool.* 21:263–70

47. Patton, J. L. Smith, M. F. 1990. The evolutionary dynamics of the pocket gopher *Thomomys bottae,* with emphasis on California populations. *Univ. Calif. Publ. Zool.* In press

48. Patton, J. L., Smith, M. F., Price, R. D., Hellenthal, R. A. 1984. Genetics of hybridization between the pocket gophers *Thomomys bottae* and *Thomomys townsendii* in northeastern California. *Great Basin Nat.* 44:431–40

49. Patton. J. L., Yang, S. Y. 1977. Genetic variation in *Thomomys bottae* pocket gophers: macrogeographic patterns. *Evolution* 31:697–720

50. Price, R. D. 1972. Host records for *Geomydoecus* (Mallophaga: Trichodectidae) from the *Thomomys bottae-umbrinus* complex (Rodentia: Geomyidae). *J. Med. Entomol.* 9:537–44

51. Price, R. D. 1974. Two new species of *Geomydoecus* from Costa Rican pocket gophers (Mallophaga: Trichodectidae). *Proc. Entomol. Soc. Wash.* 76:41–44

52. Price, R. D. 1975. The *Geomydoecus* (Mallophaga: Trichodectidae) of the southeastern USA pocket gophers (Rodentia: Geomyidae). *Proc. Entomol. Soc. Wash.* 77:61–65

53. Price, R. D., Emerson, K. C. 1971. A revision of the genus *Geomydoecus* (Mallophaga: Trichodectidae) of the New World pocket gophers (Rodentia: Geomyidae). *J. Med. Entomol.* 8:228–57

54. Price, R. D., Emerson, K. C. 1972. A new subgenus and three new species of *Geomydoecus* (Mallophaga: Trichodectidae) from *Thomomys* (Rodentia: Geomyidae). *J. Med. Entomol.* 9:463–67

55. Price, R. D., Hellenthal, R. A. 1975a. A reconsideration of *Geomydoecus expansus* (Duges) (Mallophaga: Trichodectidae) from the yellow-faced pocket gopher (Rodentia: Geomyidae). *J. Kans. Entomol. Soc.* 48:33–42

56. Price, R. D., Hellenthal, R. A. 1975b. A review of the *Geomydoecus texanus* complex (Mallophaga: Trichodectidae) from *Geomys* and *Pappogeomys* (Rodentia: Geomyidae). *J. Med. Entomol.* 12:401–8

57. Price, R. D., Hellenthal, R. A. 1976. The *Geomydoecus* (Mallophaga: Trichodectidae) from the hispid pocket gopher (Rodentia: Geomyidae). *J. Med. Entomol.* 12:695–700

58. Price, R. D., Hellenthal, R. A. 1979. A review of the *Geomydoecus tolucae* complex (Mallophaga: Trichodectidae) from *Thomomys* (Rodentia: Geomyidae), based on qualitative and quantitative characters. *J. Med. Entomol.* 16:265–74

59. Price, R. D., Hellenthal, R. A. 1980a. The *Geomydoecus oregonus* complex (Mallophaga: Trichodectidae) of the western United States pocket gophers (Rodentia: Geomyidae). *Proc. Entomol. Soc. Wash.* 82:25–38

60. Price, R. D., Hellenthal, R. A. 1980b. The *Geomydoecus neocopei* complex (Mallophaga: Trichodectidae) of the *Thomomys umbrinus* pocket gophers (Rodentia: Geomyidae) of Mexico. *J. Kans. Entomol. Soc.* 53:567–80

61. Price, R. D., Hellenthal, R. A. 1980c. A review of the *Geomydoecus minor* complex (Mallophaga: Trichodectidae) from *Thomomys* (Rodentia: Geomyidae). *J. Med. Entomol.* 17:298–313

62. Price, R. D., Hellenthal, R. A. 1981a. The taxonomy of the *Geomydoecus umbrini* complex (Mallophaga: Trichodectidae) from *Thomomys umbrinus* (Rodentia: Geomyidae) in Mexico. *Ann. Entomol. Soc. Am.* 74:37–47

63. Price, R. D., Hellenthal, R. A. 1981b. A review of the *Geomydoecus californicus* complex (Mallophaga: Trichodecti-

dae) from *Thomomys* (Rodentia: Geomyidae). *J. Med. Entomol.* 18:1–23

64. Price, R. D., Hellenthal, R. A. 1988a. A new species of *Geomydoecus* (Mallophaga: Trichodectidae) from *Pappogeomys* (Rodentia: Geomyidae) pocket gophers in Jalisco, Mexico. *J. Entomol. Sci.* 23:212–15

65. Price, R. D., Hellenthal, R. A. 1988b. *Geomydoecus* (Mallophaga: Trichodectidae) from the central American pocket gophers of the Subgenus *Orthogeomys* (Rodentia: Geomyidae). *J. Med. Entomol.* 25:331–35

66. Price, R. D., Hellenthal, R. A. 1989a. *Geomydoecus bulleri* complex (Mallophaga: Trichodectidae) from Buller's pocket gopher, *Pappogeomys bulleri* (Rodentia: Geomyidae), in westcentral Mexico. *Ann. Entomol. Soc. Am.* 82:279–85

67. Price, R. D., Hellenthal, R. A. 1989b. *Geomydoecus* (Mallophaga: Trichodectidae) from *Pappogeomys* and *Zygogeomys* pocket gophers (Rodentia: Geomyidae) in central Mexico. *J. Med. Entomol.* 26:385–401

68. Price, R. D., Hellenthal, R. A., Hafner, M. S. 1985. The *Geomydoecus* (Mallophaga: Trichodectidae) from central American pocket gophers of the subgenus *Macrogeomys* (Rodentia: Geomyidae). *Proc. Entomol. Soc. Wash.* 87:432–43

69. Price, R. D., Timm, R. M. 1979. Description of the male of *Geomydoecus scleritus* (Mallophaga: Trichodectidae) from the southeastern pocket gopher. *J. Ga. Entomol. Soc.* 14:162–65

70. Radovsky, F. J. 1979. Specificity and parallel evolution of Mesostigmata parasitic on bats. *Proc. 5th Int. Congr. Acarol.* 2:347–54

71. Rothschild, M., Clay, T. 1952. *Fleas, Flukes and Cuckoos.* London: Collins. 305 pp.

72. Russell, R. J. 1968. Revision of the pocket gophers of the genus *Pappogeomys. Univ. Kans. Publ. Mus. Nat. Hist.* 16:581–776

73. Smith, M. F., Patton, J. L. 1988. Subspecies of pocket gophers: causal bases for geographic differentiation in *Thomomys bottae. Syst. Zool.* 37:163–78

74. Sudman, P. D., Choate, J. R., Zimmerman, E. G. 1987. Taxonomy of the chromosomal races of *Geomys bursarius lutescens* Merriam. *J. Mammal.* 68:526–43

75. Thaeler, C. S. Jr. 1968a. An analysis of three hybrid populations of pocket gophers (genus *Thomomys*). *Evolution* 22:543–55

76. Thaeler, C. S. Jr. 1968b. Karyotypes of sixteen populations of the *Thomomys talpoides* complex of pocket gophers (Rodentia: Geomyidae). *Chromosoma* 25:172–83

77. Thaeler, C. S. Jr. 1974a. Four contacts between ranges of different chromosome forms of the *Thomomys talpoides* complex (Rodentia: Geomyidae). *Syst. Zool.* 23:343–54

78. Thaeler, C. S. Jr. 1974b. Karyotypes of the *Thomomys talpoides* complex (Rodentia: Geomyidae) from New Mexico. *J. Mammal.* 55:855–59

79. Thaeler, C. S. Jr. 1976. Chromosome polymorphism in *Thomomys talpoides agrestis* Merriam (Rodentia–Gemyidae). *Southwest. Nat.* 21:105–16

80. Thaeler, C. S. Jr. 1980. Chromosome numbers and systematic relations in the genus *Thomomys* (Rodentia: Geomyidae). *J. Mammal.* 61:414–22

81. Thaeler, C. S. Jr. 1985. Chromosome variation in the *Thomomys talpoides* complex. *Acta Zool. Fenn.* 170:15–18

82. Timm, R. M. 1979. *The Geomydoecus (Mallophaga:Trichodectidae) parasitizing pocket gophers of the* Geomys *complex (Rodentia: Geomyidae).* PhD dissertation. St. Paul: Univ. Minn. 124 pp.

83. Timm, R. M. 1983. Fahrenholz's rule and resource tracking: a study of host-parasite coevolution. In *Coevolution,* ed. M. H. Nitecki, pp. 225–65. Chicago: Univ. Chicago Press. 392 pp.

84. Timm, R. M., Clauson, B. L. 1985. Mammals as evolutionary partners. See Ref. 27a, pp. 100–54

85. Timm, R. M., Price, R. D. 1979. A new species of *Geomydoecus* (Mallophaga: Trichodectidae) from the Texas pocket gopher, *Geomys personatus* (Rodentia: Geomyidae). *J. Kans. Entomol. Soc.* 52:264–68

86. Timm, R. M., Price, R. D. 1980. The taxonomy of *Geomydoecus* (Mallophaga: Trichodectidae) from the *Geomys bursarius* complex (Rodentia: Geomyidae). *J. Med. Entomol.* 17:126–45

87. Ward, R. A. 1957. Host-parasite relations of the Mallophaga (biting lice) of pocket gophers. *Bull. Entomol. Soc. Am.* 3(3):22 (Abstr.)

88. Wenzel, R. L. Tipton, V. J. 1966. Some relationships between mammal hosts and their ectoparasites. In *Ectoparasites of Panama,* pp. 677–723. Chicago: Field Mus. Nat. Hist. 861 pp.

89. Werneck, F. L. 1950. *Os Malofagos de Mamiferos. Parte II: Ischnocera (continuacao de Trichodectidae) e Rhyncophthirina.* Rio de Janeiro: Inst. Oswaldo Cruz. 207 pp.

Annu. Rev. Entomol. 1991. 36:205–28

THE FUNCTION AND EVOLUTION OF INSECT STORAGE HEXAMERS

William H. Telfer

Department of Biology, University of Pennsylvania, Philadelphia, Pennsylvania 19104-6018

Joseph G. Kunkel

Department of Zoology, University of Massachusetts, Amherst, Massachusetts 01003

KEY WORDS: arylphorin, methionine rich, larval serum protein, arthropod, hemocyanin

INTRODUCTION

The storage hexamers are a family of insect proteins with native molecular weights around 500,000 with six homologous subunits weighing generally between 70,000 and 85,000 daltons. They are synthesized and secreted by the fat body of feeding larvae and nymphs and reach extraordinary concentrations in the hemolymph just prior to metamorphosis. In holometabolous insects, they are partially recaptured by the fat body during the larval to pupal molt and stored in cytoplasmic protein granules, as well as in the hemolymph of the pupa. They disappear from these reservoirs during adult development; in many insects, both hemi- and holometabolous, they are also utilized during the nymphal or larval molts. Their amino acids are primarily incorporated into new tissues and proteins during adult development, but they may also be incorporated into cuticle as intact protein, and in one case were found to be diverted to a small degree into energy metabolism. Finally, some storage hexamers have ligand binding and transport capabilities. Functionally, developmentally, and evolutionarily, they are a complex family of proteins whose analysis promises to be exceedingly useful in the study of many basic problems in the biology of insects.

0066-4170/91/0101-0205$02.00

Scheller (102) edited a timely summary of storage hexamers in 1983. Levenbook (58) reviewed their biochemistry, and Kanost et al (37) treated their molecular biology as a component of a review on hemolymph proteins. The morphogenesis of the secretory and sequestering phases of the fat body is included in a review by Dean et al (14). Hexamers are attracting much attention as model systems for the study of developmentally and hormonally regulated gene function (18, 27, 36, 57, 85, 99, 100), and this popular topic will undoubtedly be reviewed elsewhere in the future. Emphasis is placed in this review on the storage and morphogenetic functions of hexamers and on the evolution that has been driven by these functions.

We introduce here hexamerin as a descriptive and generic term that covers all of the approximately 500-kd hexamers of arthropods. The most prominent members of the family, in addition to the insect storage hexamers, are the hemocyanins that function in oxygen transport in arthropods lacking tracheal systems (61, 126). Less widely recognized is a family of nonrespiratory hexamers in the hemolymph of Crustacea (67). Finally, several unusual hexamers with no antigenic similarity to the more widely occurring storage proteins, but with very similar developmental profiles, occur in great abundance in the hemolymph of several Lepidoptera. This proliferation suggests the recent adoption for storage purposes of cryptic proteins, presumably hexamerins, that have had a long and separate evolution. The proposed origin of hemocyanin from tyrosinase (21) is the kind of conversion that could explain the origin of novel storage hexamers.

Two other widely used terms for what we call the storage hexamers have had great utility, but as knowledge has accumulated they have become, in one case, too restrictive and, in the other, too general. In the first important summary of these proteins, Roberts & Brock (91) summarized their properties, which at that time were understood almost exclusively from studies on Diptera, and adopted the term, larval serum protein, which seemed highly appropriate for that group. But many subsequent studies on Lepidoptera focussed on pupal serum and fat body; and the term applies even less well to the Dictyoptera and Orthoptera, in which two hexamerins become prominent after dorsal closure in embryos and one remains as a major hemolymph protein of adults. A second term, storage protein (119), recognizes a major function of insect hexamerins, but includes by definition many structurally and evolutionarily unrelated proteins, such as the chromoprotein recently described in *Heliothis zea* (33). These terms also lack the advantage of pointing out the relation between the storage hexamers of insects and the hemocyanins, for which evidence is now accumulating (35, 68, 116, 131).

Hexameric structure has generally been inferred from molecular weight estimates for native and dissociated forms of the proteins in polyacrylamide gel electrophoresis (PAGE). Native gels measure, more literally, the Stoke's radius of the protein and can yield misleading results if applied to proteins that are not spherical, but this is generally not a problem in the measurement of hexamers. More serious is the problem of hexamer dissociation that occurs at the high pH often used in PAGE. Dissociative electrophoresis (SDS-PAGE) yields estimates of polypeptide length, but can be misleading if carbohydrates give the chain a branched configuration, which slows the migration rate (94), especially at high gel concentrations. Consequently, molecular weights inferred from DNA sequencing can differ markedly from those estimated in SDS-PAGE (131). Nevertheless, these methods have been remarkably useful, and any doubts about hexameric configurations have been dispelled by other methods, including analytical ultracentrifugation (76) and chemical cross-linking (49, 115), and by counting the number of native electromorphs yielded by a pool of two kinds of subunits (23).

CLASSIFICATION AND PHYLOGENETIC OCCURRENCE OF THE STORAGE HEXAMERS

Amino acid compositions have been published for 38 storage hexamers in 24 species from 6 orders of insects. These are listed in Table 1, along with the average compositions of 9 arthropod hemocyanins, three nonrespiratory hexamers of crustaceans, [and the 55 animal polypeptides analyzed by King & Jukes (44)]. Tryptophan and cystine/cysteine are omitted from the table because they are not included in many of the published reports. [Cys, when analyzed in the storage hexamers, has very low values. In these proteins, as in several other large hemolymph proteins of insects, both inter- and intrachain disulfide bridges are highly unusual (116).]

For most proteins, a simple percentage composition of amino acids is no longer an interesting topic. Most proteins conform closely to the average composition described by King & Jukes (44). Active site, sequence, secondary, tertiary, and quarternary structure have taken over as the relevant issues in protein chemistry. Composition remains a vital issue for storage hexamers, however, because it is one of the key selective features that have governed their evolution. This proposition was realized at the time of hexamerin discovery, when Munn et al (76) proposed that a high aromatic amino acid content is related to demands for these constituents during sclerotization of the cuticle. This importance is also reflected in nomenclature; *arylphorin* (115) is a designation for proteins with tyrosine and phenylalanine contents totalling more than 15%, and *methionine-rich* describes those proteins in which this amino acid exceeds 4%—in both cases the values are more than twice the average determined by King & Jukes (44) for typical polypeptides.

Two graphs based on Table 1 demonstrate the continuing importance of composition. Figure 1 *(top)* illustrates the effect of plotting the combined

Table 1 Amino acid composition of hexamerins with principal components one and two appended[a]

SUBCLASS: Order/protein group Genus species k:	asx 1	thr 2	ser 3	glx 4	pro 5	gly 6	ala 7	val 8	met 9	ile 10	leu 11	tyr 12	phe 13	lys 14	his 15	arg 16	Refs.
HEMIMETABOLA:																	
Dictyoptera/arylphorin																	
Blatta orientalis LSP	145	54	32	100	71	57	47	66	0	47	65	141	52	57	2	63	c
B. orientalis LSP	136	44	28	98	67	59	53	60	3	40	69	170	51	46	10	66	23
Blattella germanica SPI	121	34	35	118	58	44	52	79	26	38	62	90	65	81	49	47	53
Hemiptera																	
Triatoma infestans	144	44	42	100	50	63	57	78	2	33	96	72	71	71	24	55	87
Orthoptera																	
Locusta migratoria Ap	122	49	65	144	46	51	68	56	8	24	95	90	61	57	13	49	d
L. migratoria LSP-1	88	45	65	117	49	86	99	74	9	37	84	41	49	55	42	60	3
L. migratoria LHP	102	40	45	103	63	70	85	75	2	56	91	69	56	62	12	71	17
L. migratoria PSP	81	33	45	109	64	53	80	66	25	52	90	83	54	39	20	108	3
HOLOMETABOLA:																	
Diptera/calliphorin																	
Calliphora																	
erythrocephalla	116	48	43	100	31	54	33	59	45	40	58	114	111	82	33	33	16
Calliphora stygia	120	47	42	101	41	54	30	53	38	41	68	119	109	79	31	28	45
Ceratitus capitata 1	139	46	53	110	39	56	47	52	36	45	88	96	71	60	26	36	75
C. capitata 2	143	42	47	115	37	53	44	48	42	42	86	103	73	62	26	37	75
C. capitata 3	142	40	45	112	36	53	46	55	39	43	71	119	75	57	30	41	75
Drosophila melanogaster	128	52	36	103	32	70	41	47	54	37	69	93	94	69	30	42	e
Lucilia cuprina	119	43	37	106	30	51	33	63	35	38	69	118	111	82	29	34	120
Musca domestica	131	41	37	101	39	53	31	53	35	29	50	141	120	94	16	29	66
Diptera/second hexamerin																	
Calliphora stygia	140	34	62	108	71	58	46	52	20	39	87	94	62	72	50	6	45
Ceratitus capitata	134	41	42	131	50	50	54	69	19	53	66	87	72	60	36	39	74
Drosophila crucigera	155	36	77	111	50	83	57	63	20	28	48	74	61	50	56	32	e
Drosophila melanogaster	127	39	47	115	54	62	46	83	18	28	59	83	82	61	70	26	e
Drosophila mimica	140	49	58	134	36	69	59	68	21	29	63	71	65	62	38	41	e
Drosophila mulleri	150	37	74	113	47	76	52	65	12	31	62	77	71	56	44	32	e
Hymenoptera/arylphorin																	
Apis mellifera	134	42	78	93	47	58	44	32	57	51	78	96	69	54	12	53	96
Lepidoptera/arylphorin																	
Bombyx mori	121	45	40	125	61	40	54	61	27	38	69	89	98	85	13	32	123
Galleria mellonella	125	41	37	120	47	48	51	65	18	57	87	100	65	67	13	56	108

	1	2	3	4	5	6	7	8	9	10	11	12	13	14	15	16	Ref.
Heliothis zea	107	34	55	108	42	50	56	64	21	46	78	104	90	78	30	37	31
Hyalophora cecropia	111	50	49	111	73	49	44	67	15	44	69	92	87	65	36	39	115
Manduca sexta	128	29	52	107	58	49	53	76	20	32	63	103	88	77	29	35	49
Papilio polyxenes	99	64	48	113	60	48	42	67	9	45	72	113	86	73	22	37	97
Calpodes ethlius	100	49	49	72	58	73	39	77	15	32	70	127	84	98	25	32	81
Lymantria dispar	105	46	65	100	51	95	58	64	14	43	74	83	74	72	27	29	38
Lepidoptera/ high met																	
Bombyx mori	122	61	52	86	41	48*	35	86	108	46	75	55	51	70	11	55	123
Galleria mellonella	130	53	53	70	39	49	42	65	84	49	86	54	38	88	32	68	108
Hyalophora cecropia	133	59	49	95	23	54	46	70	70	45	86	51	66	85	14	56	121
H. cecropia	136	55	49	96	19	39	36	75	49	52	106	68	62	100	9	52	121
Manduca sexta	130	74	29	35	23	50	29	89	65	65	112	59	57	80	29	74	95
Papilio polyxenes	158	68	20	61	32	42	21	86	39	66	116	62	54	81	22	72	97
Lepidoptera/ other																	
Hyalophora cecropia	109	46	70	108	35	48	45	71	24	68	88	78	48	56	62	43	116
Plodia interpunctella	160	67	34	48	61	47	37	80	48	69	117	36	41	68	19	68	46
Calpodes ethlius SP1 [b]	145	67	63	87	30	47	42	59	50	57	101	53	49	63	25	61	81
C. ethlius SP2 [b]	134	61	58	30	44	44	38	79	31	54	100	93	68	78	39	50	81
NON-INSECT:																	
Arthropod/ HbCy																	
Average (9 proteins)	139	54	52	110	51	64	63	66	24	46	78	40	53	50	59	51	11,28,67
Arthropod/ non-respiratory																	
Average (3 proteins)	130	54	82	104	72	60	61	71	19	47	72	45	55	46	32	51	67
King & Jukes/ average aa																	
OAA	108	65	85	100	52	78	78	71	19	40	80	35	42	75	30	44	44
Weight factors, f_{ik} for principal component i																	
i = 2, 100 x f_{2k}:	11	13	15	-4	8	8	17	11	4	14	20	-75	-37	-16	24	20	
i = 3, 100 x f_{3k}:	-15	-18	19	50	21	20	31	-10	-45	-17	-30	-1	-2	-28	20	-18	

[a] The mole percent was recalculated on the basis of the 16 consistently analyzed amino acids, k, in standard acid hydrolysate analyses. An extra significant digit was allowed in calculations to eliminate rounding errors. Weight factors, f_{ik} (= eigenvectors) for principal components two and three are listed at the bottom of the table. Abbreviations are defined as: LSP, larval serum protein; PSP, persistant storage protein; SP, storage protein; HbCy, hemocyanin; LHP, larval hemolymph protein; OAA, observed average amino acid.

[b] The glycine content of these proteins was very high; the Gly mole percent was adjusted to that of lepidopteran LSP and the mole percents were recalculated.

[c] J. G. Kunkel & R. Duhamel, unpublished data.

[d] D. Philips & B. Loughton, unpublished data.

[e] S. Beverley & A. C. Wilson, unpublished data.

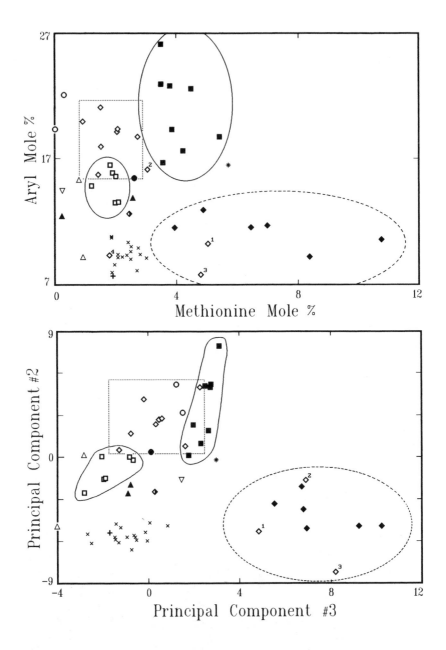

mole-percent of tyrosine and phenylalanine against that of methionine. Each order of insect is represented by a distinct symbol. The symbols cluster in well defined groups and depart markedly from the noninsect hemocyanins and from the King & Jukes polypeptide average. This focus on the aromatics and methionine is not unwarranted because a principal-components analysis of the compositions in Table 1 (Figure 1, *bottom*) revealed a heavily weighted dependence of the top principal components on the contents of these amino acids.

Phyletic and developmental characteristics tend to parallel the compositional groupings, and we have emphasized this in the organization of Table 1. These sets of properties are summarized in the following text.

The first group (Figure 1, solid squares) is the only one whose aromatic and methionine contents are both high. This dipteran protein was first isolated from extracts of *Calliphora erythrocephala* (77) and given the name calliphorin. It was shown to be synthesized by the fat body (78) and secreted into the hemolymph, beginning about half way through the larval growth period. By the end of larval feeding, it constitutes an extraordinary 60% of the soluble proteins of the insect. In wandering larvae, synthesis stops and much of the protein is endocytosed by the fat body (45) in which it is stored for the duration of metamorphosis.

A protein with very similar composition and developmental profile has been identified and variously named in many species in the suborders Cyclorrapha (8, 16, 39, 41, 45, 65, 73, 77, 132), Brachycera (77), and Nematocera (15). Homology within the Cyclorrapha and Brachycera has been confirmed by antigenic similarities to the original calliphorin. Epitope divergence has occurred for too long a period for this similarity to extend to the Nematocera, and here homology to calliphorin is suggested primarily by biochemistry.

Many Diptera contain a second hexamerin, which has much lower methionine and aromatic amino acid contents than calliphorin (Figure 1, open squares). It occurs in smaller amounts than calliphorin, but follows a very

←——

Figure 1   *(Top)* Mole percent comparisons of the methionine and aromatic amino acid (tyr plus phe) compositions of the arthropod hexamerins shown in Table 1. Each taxonomic category has its own symbol: Diptera *(square)*, Lepidoptera *(diamond)*, Dictyoptera *(circle)*, Orthoptera *(triange pointing up)*, Hemiptera *(triangle pointing down)*, Hymenoptera *(asterisk)*, noninsect arthropods (×), and King & Jukes average (+). Additional distinctions: the persistent storage hexamer of cockroach and locust (solid circle, solid triangles); calliphorin-like hexamers of Diptera *(solid squares);* high methionine hexamer of Lepidoptera (solid diamonds); footnote b, Table 1 (diamonds 1 and 2). Homologous hexamers are encircled by solid lines (Diptera) and dashed lines (Lepidoptera).   *(Bottom)* Principal component analysis of the data in Table 1. To compute the principal component i for protein j, P_{ij}, the loading factors, f_{ik}, are multiplied by their respective moles percent of amino acid k in protein j, and the product summed over all 16 amino acids, aa_{jk}:

$$P_{ij} = \text{Sum } (f_{ik} \cdot aa_{jk}) \text{ for } k = 1, 2, \ldots, 16.$$

The loading factors f_{2k} and f_{3k} for components 2 and 3 are listed at the bottom of Table 1 below their respective amino acid column. In this figure, P_{2j} is plotted against P_{3j} for each protein. These two components explain the greatest proportion of composition differences. The symbols are identical in top and bottom panels.

similar developmental profile (8, 45, 74, 77, 92). First identified by a unique electrophoretic behavior, it subsequently proved to be antigenically distinct from calliphorin. It has thus far been reported only in the suborder Cyclorrapha. Beverley & Wilson (5), using antisera against this hexamer from five species in the Drosophilidae and one species each in the Tephritidae and Calliphoridae, were able to identify a homologous protein in these seven species, as well as in ten other species of cyclorraphan flies. Only the Phoridae failed to exhibit a cross-reaction, possibly because of early divergence of this family, rather than the absence of a homologue. This protein's synthesis in *Drosophila melanogaster,* in which it is known as LSP2 (91), was recently shown to resume after the eclosion of the adult (107).

The third group (Figure 1, open diamonds) is a lepidopteran glycoprotein with high aromatic amino acid and low methionine contents (31, 38, 49, 50, 81, 97, 115, 123). It was named arylphorin (115) and thus far has been identified in every lepidopteran that has been suitably examined. Its developmental profile is very similar to that of calliphorin; synthesis begins in early or mid-larval stages and terminates in wandering larvae (86, 129), followed by sequestration by the fat body (123). As will be seen below, it seems to be the functional counterpart of calliphorin in lepidopteran cuticle deposition, and the two may well turn out to be interordinal homologues.

The fourth group (Figure 1, solid diamonds) is also a lepidopteran protein. It has high methionine and low aromatic amino acid contents (4, 95, 97, 121, 123), lacks carbohydrates, and also differs from the other three groups in developmental profile. Synthesis by the fat body and secretion into the hemolymph do not begin until close to the end of the last larval instar (86, 129). Then, the fat body avidly resequesters the protein so that its hemolymph concentration in the pupa is often lower than that of arylphorin (95). This hexamerin is the only one known to exhibit sexual dimorphism; in some species, only traces of it occur in males. It does not occur in *Lymantria dispar* (38) or *Spodoptera litura* (122) but otherwise seems to be widely distributed in the order (50). A 70-kd subunit with appropriate amino acid composition occurs in *Plodia interpunctella* (46), but the native protein has not been shown to be a hexamer.

Hexamerins also occur in the hymenopteran *Apis mellifera,* and one of these has been studied in detail (96). Antibodies to *Manduca sexta* arylphorin labelled it in immunoblots of larval hemolymph. This is one of two examples of antigenic similarity between hexamerins in different orders, the other being a monoclonal antibody against tarantula hemocyanin that cross-reacted with calliphorin (68). While the cross-reaction seems to identify the *A. mellifera* protein as an arylphorin, this protein has high methionine as well as tyrosine contents (Table 1), so that its position in Figure 1 (asterisk) is close to the calliphorin cluster. A developmental profile has yet to be published for this interesting protein.

In Dictyoptera (Figure 1, circles), a 16S protein was identified in *Blattella germanica* (53), which proved to be a high aromatic amino acid hexamer that occurs in adults as well as in larvae. A second larval protein, which disappears in the adult, was found in several species of cockroaches, and crossreacts immunologically throughout the order Dictyoptera, including the cockroaches, termites, and praying mantids (51), in confirmation of the taxonomic decision that these groups are closely related. This protein was subsequently purified and shown also to have a high aromatic amino acid content (23). Both proteins initially appear in the embryo at dorsal closure (52, 110). They subsequently cycle in rhythm with molting, reaching a peak concentration a day before each ecdysis, and a minimum shortly after ecdysis (22, 24, 51). The larval-specific hexamer (open circle) gradually disappears after adult ecdysis, while the other (solid circle) persists throughout adult life at a concentration on a par with that of lipophorin.

Locusta migratoria is the only orthopteran in which hexamerins have thus far been described, and here the pattern is much like that of the Dictyoptera (3, 17). The two larval hexamerins have intermediate levels of aromatic amino acids and low or intermediate methionine contents (Figure 1, triangles pointing up). One of these (open triangle) disappears during the adult molt, while the second (solid triangle), whose synthesis is stimulated by juvenile hormone in tandem with that of vitellogenin, persists in the adult (133).

A low methionine hexamer with intermediate aromatic amino acid content (Figure 1, triangles pointing down) was isolated from adult male hemolymph of the hemipteran *Triatoma infestans* taken five days after a blood meal (87). It was not looked for at younger stages, and therefore its role as a storage hexamer is uncertain. We indicate below that it has been implicated in lipid transport.

Several species of Lepidoptera that contain either three or four storage hexamers are exceptions to the general rule of one or two storage hexamers per species. In *Hyalophora cecropia,* fat body extracts and hemolymph yielded two antigenically and electrophoretically distinct high methionine hexamers (121), which are sufficiently similar to suggest that they may have only recently evolved away from each other. Arylphorin was later isolated from pupal hemolymph in this species (115), as was an antigenically and electrophoretically distinct storage hexamer that binds riboflavin (116). In Figure 1, this fourth hexamer (half-filled diamond) falls outside both the arylphorin and methionine-rich clusters. In a phylogenetic search, antibodies to this flavoprotein failed to detect a homologue in other Lepidoptera, including several in the family to which *H. cecropia* belongs (114). It thus appears to have only recently evolved a hemolymph storage function.

Identified storage hexamers of *Trichoplusia ni* include arylphorin and a high methionine homologue (50). The cloned cDNA of a JH-suppressible third hexamer has recently been sequenced and the inferred amino acid composition found to include only 9.3% aromatic amino acids and 1.8% methionine (35) [Figure 1 *(top)*, diamond 4].

Calpodes ethlius has four storage hexamers (62, 81). One has been identified as an arylphorin (81), and two (including diamond 1) plot in the high-methionine hexamerin cluster in Figure 1. But the fourth (diamond 2) cannot be classified as either an arylphorin or a high methionine hexamer. In *Galleria mellonella* also, arylphorin and a high methionine hexamer have been identified (4, 108), but two other hexamerin subunits (71, 84) have not been characterized. In *S. litura,* antibodies against fat body extracts were used to identify two proteins with developmental profiles in the hemolymph and fat body similar to those of the high methionine hexamer in other Lepidoptera (122). Neither protein crossreacted with antibodies to the high methionine hexamer of *Bombyx mori.* These proteins may therefore be additional examples of the recent evolution of novel hexamerins.

Finally, evidence has been presented for storage hexamers in the hemolymph and fat body of *Diatraea grandiosella* (55) and *Hyphantria cunea* (43). They are not classified here because neither amino acid analyses nor antigenic relationships are as yet available.

DEPOSITION IN THE FAT BODY OF HOLOMETABOLOUS INSECTS

Locke & Collins (63) established the fine structural basis of hemolymph protein uptake by fat body cells by tracing horse radish peroxidase (HRP) injected into last instar larvae of *C. ethlius.* HRP passed from the hemolymph into a reticulum of spaces that contained a concentrate of extracellular materials and lay between fat body cells as well as in narrow channels extending deep into the fat cells' cytoplasm. HRP was also visible in endocytotic vesicles formed at the tips of these channels and, during feeding stages, in multivesicular bodies, which were interpreted as sites where lysosomal enzymes digest it. In pharate pupae, lysosomal activity ceased, and HRP was transferred instead to the newly forming protein storage granules. Over several days of development, the granules continued to accumulate until they occupied much of the cytoplasm of the cells. Similar configurations were described in the fat body cells of pupariating *Calliphora vicina* (70).

Similarities were initially reported between major larval hemolymph proteins and those stored in pupal fat body of *Malacasoma americana* (64), *Pieris brassicae* (12), and *C. erythrocephala* (69). The hexameric nature of the counterparts was established for *H. cecropia* (121), *Bombyx mori* (123), and *C. ethlius* (62). A drop in hemolymph concentration that exactly matches a rise in fat body content implicated endocytosis for both total protein (12, 69) and specific hexamers (4, 121, 123). Investigators experimentally confirmed endocytosis by injecting labelled hemolymph proteins into larvae and showing that radioactivity is accumulated by the fat body (62, 66, 69, 72, 95, 125). In *M. sexta*, late-fifth-instar fat body sequestered in vitro a quantitative selection of proteins that had been secreted into the medium by fourth-instar fat body (10).

The possibility that storage granules arise from proteins synthesized in situ and immediately deposited without prior secretion into the hemolymph was initially discounted by a decline in the rate of amino acid incorporation by these cells during granule formation (63). In *D. melanogaster*, simultaneous injections of ^{14}C-leucine and ^{3}H-larval hemolymph proteins into late larvae resulted exclusively in ^{3}H labelling of storage granules isolated from the fat body of puparia (118). Hexamer synthesis by the fat body, and more particularly, the capacity for synthesis, implied by the presence of hexamerin gene transcripts, terminates when feeding stops (34, 56, 73, 98, 101, 105, 106, 112, 129). Other evidence, however, indicates a continued low level of synthesis of some hexamerins during the endocytotic prepupal period in some Lepidoptera (72, 86, 129).

In a search for hexamerin receptors, an 800–10,000 G centrifugal fraction of *Sarcophaga peregrina* fat body homogenates bound an iodinated storage protein with a K_D of 4×10^{-9} M (125). Binding was demonstrated in pupal but not larval fat body, unless the latter had been incubated with 20-hydroxyecdysone. The optimum pH and divalent cation dependence of binding were consistent with a receptor that is adapted to operate at the cell surface in the chemical environment of the hemolymph. Finally, evidence was presented that the binding protein is generated during ecdysone treatment from a cryptic receptor with a molecular mass of 125 kd to an active form of about 120 kd (124).

Unambiguous evidence for the selectivity of storage protein deposition in the fat body has only rarely been presented. An intuitive, and probably correct, assumption of selectivity has been based on the overriding predominance of the storage hexamers as constituents of the protein granules, but this hypothesis overlooks the hexamers' corresponding predominance as constituents of the hemolymph at the time of uptake. The process clearly has a nonselective component, as the uptake and deposition of HRP indicates. A peculiar observation from *D. melanogaster* is that fat body cultured in Schneider's medium supplemented with fetal calf serum or bovine serum albumin generates what appear to be storage granules (9, 118). Measurements of the relative contributions of nonselective and selective uptake comparable to those done for yolk deposition (113, 117) have not been reported.

In addition, a puzzling difference separates the fine structural configurations of endocytosis in fat body cells and oocytes, whose endocytosis has been more thoroughly studied. In fat body (63), most of the HRP-labelled endosomes did not exhibit the outer clathrin coat, the inner lining of adsorbed materials, and the clear central lumen characteristic of the vesicles that were found during the 1960s to abound in the cortex of vitellogenic oocytes (2, 93, 109). Only two electron micrographs of single vesicles with apparent clathrin coats have appeared in the literature from *C. ethlius* fat body studies (14, 63), and both of these are homogeneously filled and do not exhibit a clear lumen and a dense lining. Two coated vesicles with the conventional electron transluscent lumen are in fact present in an electron micrograph of *C. vicina* fat body cells during endocytosis (70), but again the frequency is orders of magnitude lower than in the vitellogenic oocyte cortex. A number of studies concerned with the remodelling of these cells during pupation make no mention of the coated vesicles (19, 130). The physiology and fine structure of the fat body's brief endocytotic phase require closer examination to clarify the questions raised by the pioneering studies described in this section.

HEXAMERIN FUNCTIONS IN MOLTING, METAMORPHOSIS, AND REPRODUCTION

Dictyopteran hexamerins and the lepidopteran arylphorins are utilized during larval molting, as well as during adult development (22, 49, 51, 115). In general, larval cycling entails a rise in arylphorin concentrations during feeding and a precipitous drop, in some cases to undetectability, during molting. Synthesis terminates in *M. sexta* at the outset of the molt (86), presumably as a response to ecdysone. Its resumption requires both the withdrawal of ecdysone and the availability of nutrients (128).

In *Blatta orientalis,* the clearing of the two hexamers during the molt results at least in part from a nonselective mechanism that can capture all hemolymph proteins (24). During the intermolt period, the storage hexamers and an equal-sized foreign protein exhibited half-lives of over 100 h, while a small foreign protein, radiolabelled bovine serum albumin, disappeared from the hemolymph with a half-life of only 4.7 h. During molting, by contrast, all three proteins exhibited short half-lives. The results imply that a size barrier separates the hemolymph from a nonselective mechanism of clearance, and that this barrier becomes less restrictive during the molt.

To investigate hexamerin functions during metamorphosis, Levenbook & Bauer (59) injected *C. vicina* larvae that had completed their growth with ^{14}C-phenylalanine–labelled calliphorin. Five days later, when the animals were half way through adult development, the injected label retained the mobility of calliphorin in SDS-PAGE, except for a small amount expired as CO_2. In four-day-old flies, in which calliphorin has disappeared, the label was widely distributed, with 46.5% in the thoracic muscles, 10.8% in the SDS-insoluble components of the cuticle, and lesser amounts in other tissues. Autoradioagrams of SDS-PAGE gels indicated that large numbers of proteins were now labelled, including especially actin and myosin. Calliphorin had thus served as a source of phenylalanine, and presumably other amino acids as well, in support of the synthesis of adult proteins.

During the first several hours after its injection into seven-day-old larvae, however, radiolabelled calliphorin was found in the cuticle (103). Antibodies to calliphorin stained the endocuticle of larvae, pupae, and adults. In a follow-up of this study (48), intact calliphorin was demonstrated in the cuticle of both third-instar larvae and pupae by immunoblotting of SDS extracts; differential extraction studies indicated that much of the protein may be covalently linked to the cuticular structure. Labelled calliphorin was incorporated into cuticle without degradation, both after injection into the animal and in isolated integuments incubated in Schneider's medium. The plausibility of a role in sclerotization was indicated by the demonstration that calliphorin and arylphorin-type hexamerins isolated from hemolymph can be cross-linked with diquinones in vitro (30).

Undefined cuticle proteins with antigenic similarities to hemolymph proteins had earlier been reported in *Periplaneta americana* (26) and *M. sexta* (47). In *L. migratoria* (83), the antigenic counterparts included the principal larval hemolymph protein, which was later shown to be a storage hexamer (17). In *M. sexta,* arylphorin in particular was shown in immunoblots of SDS-PAGE gels to be extractable from fifth-instar and pupal cuticle (86).

Are these cuticular hexamerins derived from the hemolymph, or are they secretions of the epidermis? In Lepidoptera, translatable arylphorin mRNA was detected in extracts of epidermis (86), and arylphorin itself has been detected in both apical and basal secretions of the epidermis (80). So epidermal secretion is probably the source of at least part of the cuticular form of this protein. In *C. vicina,* by contrast, the fat body appears to be the exclusive site of calliphorin mRNA (105). This observation, in combination with the labelled-protein-uptake studies (48), implicates transport from the hemolymph.

One would anticipate from these results that calliphorin is an indispensible precursor to the construction of the adult, but a genetically based test of calliphorin function devised by Roberts (89) for LSP1, the calliphorin homologue in *Drosophila,* indicated otherwise. Three genes at dispersed loci encode mRNAs for this protein, and combined deletions achieved by Roberts generated flies that completely lacked the ability to synthesize it. LSP1 normally accounts for 9% of the total protein of the larva at the puparial stage, and disappears within three days after adult emergence (92). While the mutants lacking it could complete metamorphosis, feed, mate, reproduce, and survive stress tests, the fecundity of both males and females was significantly reduced (90). This loss resulted in part from behavioral problems and in part from poor development of the gonads. The recognized metamorphic functions of the protein can apparently be taken over by other proteins, while hitherto unsuspected reproductive effects, which are not manifested until after it has disappeared, cannot.

A function in egg formation was suggested for the high methionine hexamer of Lepidoptera by its greater abundance in females than in males (4, 95, 97, 123). In *Lymantria dispar,* which lacks the high methionine hexamer, arylphorin occurs in metamorphic females in amounts seven times greater than in males (38). While proteins derived from the adult diet are the principal source of amino acids for egg formation in many insects, the lepidopteran mode is to draw on reserves carried over from the larva, and one would expect that this high protein–demanding function would result in sexual differences in the hexamerins. Ogawa & Tojo (79) interpreted the sexual difference in the high methionine hexamerin in *Bombyx mori* in this way. In particular, these authors pointed out that the chorionic proteins of *B. mori* are rich in cysteine (40), and that methionine sulfur is readily transferred metabolically to this amino acid. While this idea seems plausible, methionine is not the primary constituent of the high-methionine protein, and it must therefore serve other needs as well.

A question of paramount importance is whether hemolymph storage protein utilization during the pupal-adult molt is a consequence of general protein turnover, as Duhamel & Kunkel (24) proposed for the last larval molt of *B. orientalis* or of a selective process, such as that sequestering hexamerins in the fat body during the larval-pupal molt. General protein turnover is unquestionably important, for large amounts of foreign proteins injected into *H. cecropia* pupae disappeared from the hemolymph during adult development (82). As in *B. orientalis* nymphs (24), the protein's size was important; 45-kd ovalbumin and 58-kd human serum albumin were cleared substantially more rapidly from male hemolymph than 510-kd vitellogenin, for which the male has no selective uptake mechanism. On the other hand, selective arylphorin and hemolymph flavoprotein utilization was indicated by a more rapid clearance of both of these hexamers, particularly during the last third of the molt.

There is no published information on the site or sites of either selective or nonselective hemolymph protein clearance in Lepidoptera during adult development. However, in adult female roaches, vitellogenin that had been altered by deglycosylation or carbohydrate oxidation lost much of its affinity for the oocyte and was instead cleared primarily by the fat body and the pericardial cells (29). Clearance of foreign proteins by pericardial cells has also been demonstrated in *C. erythrocephala* (13).

A resumption of endocytosis by the fat body, followed at this stage by lysosomal digestion, rather than by storage, would simply explain selective storage hexamer clearance in the pharate adult. In *C. erythrocephala*, acid phosphatase reaches a peak in the fat body at pupation (127), which can be interpreted as formation of lysosomes for the autophagic digestion of storage granules and cytoplasm that results in the disappearance of this tissue in the adult (14). The pharate adult fat body of Lepidoptera exhibits sexual differences in this regard: in males, storage granules are digested but the cells survive, while females exibit a pronounced reduction of the tissue in association with egg formation but a retention of protein granules in the few cells surviving in the adult (7). Whether a tissue that is so deeply involved in remodelling or cell death is simultaneously the major endocytotic organ of the pharate adult is a question that requires examination.

Some hexamerin functions may not be related to amino acid storage for morphogenesis. Low levels of lipids are often present (reviewed in 58). Calliphorin binds ecdysone with a low affinity, but there is so much of the protein in the hemolymph of larvae that this activity has been proposed to have potential importance as a hormone-clearing mechanism (25). Arylphorin exhibits a relatively nonselective affinity for hydrophobic substances. This property has been proposed to be a mechanism for dealing with xenobiotics (32). And a third example of ligand binding is provided by the riboflavin content of the hemolymph flavoprotein of *H. cecropia* (116).

Finally, the occurrence of a storage hexamer in adult Dictyoptera (53), *L. migratoria* (133), and *D. melanogaster* (107), and of a putative storage hexamer in *T. infestans* (87), suggests other functions. Lipid content in the adult hexamer of the triatomid includes an unusually high proportion of free fatty acids, which suggests a hydrophobic transport function differing from that of lipophorin.

TERMINOLOGY

The nomenclature of the insect storage hexamers is needlessly chaotic. Though simple, useful terms comparable to lipophorin and vitellogenin, which facilitate communication about other hemolymph proteins throughout the Insecta, are not currently possible, the custom of giving obvious homologues different names in every species examined diffuses the impact of what should be a much more highly regarded and widely understood literature. Within each order, uniform names are clearly possible, and the people active in those domains owe it to themselves to agree on a uniform terminology. Arylphorin and high-methionine hexamer serve this function for Lepidoptera, and, for both the Dictyoptera and *L. migratoria*, larval-specific hexamer (51) and persistent storage hexamer (133) prove to be useful designations. Dipteran terminology in particular, however, needs attention.

At the same time, little has emerged from our review to suggest a more global terminology. It is not yet clear whether interordinal terms should be derived from similarities in composition and function or from evolutionary relatedness as revealed by amino acid sequencing. Ideally, the two approaches would lead to consistent suggestions, but this outcome is far from certain. Arylphorin, for instance, which is now beginning to be applied to high aromatic amino acid hexamers in orders other than the Lepidoptera, could well turn out to lump together hexamers of disparate evolutionary history. We have seen that specific hexamers can disappear, as did the high-methionine hexamers in *L. dispar* and *S. litura,* and can emerge anew from unknown corners of the genome, as did the hemolymph flavoprotein of *H. cecropia* and the *T. ni* hexamer recently sequenced by Jones et al (35). Such events, followed by convergent evolution toward common functions, could well result in compositional similarities between proteins of very different origin. As amino acid sequences become available and evolutionary histories are better understood, it may therefore become necessary to choose arbitrarily between composition and evolution as the basis for an interordinal terminology.

In making this decision, polyclonal antibodies, which have been invaluable tools in demonstrating intraordinal homologies, even between different families, will be of limited use. Cross-reactions between orders are rare, and where they do occur their relevance to the establishment of homology is not clear. Binding sites between monomers are highly conserved across a wide variety of hexamerins (131), for instance, and tarantula hemocyanin and calliphorin have recently been shown to share a common epitope (68). Structural relatedness detectable by antibodies can therefore occasionally reflect an ancestry common to the entire hexamerin family.

GENETIC STRUCTURE AND THE EVOLUTION OF THE STORAGE HEXAMERS

Roberts & Brock (91) reviewed the genetics and evolution of the two dipteran hexamers, particularly including the evidence for duplications leading to multigene families. The first indication of such a family was the complex pattern of calliphorin electromorphs in *Lucilia cuprina* (120). The genetics of the electromorphs suggested 12 closely linked genes. A similar family has been demonstrated in *C. vicina* (104), from which 11 members of a larger family (> 20) of closely linked calliphorin genes were isolated as cloned DNA segments. Fragments of three distinct *Sarcophaga bullata* genes were cloned from a genomic library and shown to be related to the mRNA expressed as LSP1 (111). Despite these indications of large multigene organization of the calliphorin gene, its homologues in *D. melanogaster* are limited to only three genes, corresponding to the alpha, beta, and gamma subunits of this protein (88, 91). And in contrast to the tight clustering of the *L. cuprina* and *C. vicina* gene families, the three *D. melanogaster* genes are located on different chromosomes, which suggests that transposable elements may figure in their duplication (91). A tandem duplication has apparently figured in the evolution of arylphorin, whose alpha and beta subunit genes in *M. sexta* are separated by 7.1 kilobases and have diverged to the extent that their amino acid sequences are now only 68% identical (131). A similar story will presumably emerge in other cases where electrophoretically different subunits are included in a single kind of hexamerin (23).

The second storage hexamer of *D. melanogaster*, LSP2, is represented by a single copy in the genome (1). Here, and in many other insect hexamerins, single subunit types are detected in SDS-PAGE, in contrast to the electrophoretic complexity seen in the multigene family products listed above. Duplications and multigene families may therefore prove not to be general features of hexamerin gene organization.

Doolittle (20) suggested that gene duplication and subsequent divergence account for the origin of most of the great multiplicity of contemporary proteins. Convincing evidence of the origin of new hexamerins in this way is seen not only in the amino acid sequence similarities discussed below, but also in exon/intron structure: the arylphorin genes of *B. mori* and *M. sexta,* and the high methionine hexamer gene of *B. mori,* all include five exons of approximately the same size (27, 100, 131). The intron structure, as might be expected, is less conservative, and shows the wide sequence and length divergence that are consistent with a more neutral role.

The fifth exon of *B. mori* high methionine hexamer is 80 bases longer than that of the arylphorin gene, and this DNA is a major source of the additional amino acids in the resultant peptide (99). The extension is particularly rich in methionine (27%), and thus of particular relevance to the unique composition of this hexamer (79).

Willott et al (131) compared the structural regions of the few insect hexamers and arthropod hemocyanin genes that have at this time been completely sequenced using alignment procedures and genetic distance matrices that can be used to construct phylogenetic trees. Such an approach provides a novel view of hexamerin and arthropod evolution. The resultant tree suggests that the two subunits of *Manduca* arylphorin diverged approximately 100 my BP, based on an assumed figure for the rate of nonsynonomous codon evolution (60). Comparison of the *B. mori* and *M. sexta* arylphorin subunits would place the divergence of the species at 140 my BP, or just before the alpha-beta divergence in the *M. sexta* line of descent. The divergence of arylphorin and high-methionine protein of *B. mori*, based on the same rate of nonsynonomous base substitution, was calculated to have occurred much deeper in evolutionary time, at 425 my BP, which would presumably preceed even the origin of the Insecta. Some caution is of course warranted in the interpretation of these interesting figures, for they are based on the average substitution rate calculated for 42 mammalian proteins. If hexamerin substitution rates have been faster than this average, the divergence times would be correspondingly more recent.

In another approach to evolutionary timing, Beverley & Wilson (5, 6) found that microcomplement fixation comparisons of the second dipteran storage hexamer in Hawaiian Drosophilidae could be used to fix the branch points in their diversification, in relation to mainland Diptera.

Percentage amino acid identities after sequence alignment of the insect hexamerins with arthropod hemocyanins range between 22 and 28% and are thus not substantially less than the 30–33% similarities between arylphorins and high-methionine protein (131). The unique storage hexamer of *T. ni* exhibited sequence identities of 28% with *M. sexta* arylphorin, 26% with *B. mori* high-methionine protein, and an almost comparable 22% with a hemocyanin amino acid sequence. Much of this sequence conservation is located in regions that have been shown in hemocyanin to form contacts between subunits and to lie near the beginnings and ends of alpha-helical segments. In all three of the sequenced storage hexamers, only one of the six copper-binding histidines is conserved.

The divergence of the hemocyanin from nonrespiratory serum proteins that have very similar size and composition may have occurred as early as 600 my BP, a currently agreed upon minimal age of the arthropods. Nonrespiratory serum proteins are known from several crustaceans and arachnids (67), and the insect hexamerins may be related more directly to these than to the hemocyanins.

The next ten years should bring much progress toward understanding the evolutionary relationships of the hexamerins. Of key importance will be deciding whether a single momentous gene duplication, deep in evolutionary time, gave rise to the paired storage hexamers of Diptera, Lepidoptera, Dictyoptera, Orthoptera, and possibly other orders whose hexamers have not yet been sufficiently examined. Or will such strict interordinal homologies fail to emerge because of evolutionary loss and appearance of new storage hexamers (processes that have clearly been active relatively recently in the Lepidoptera)?

Equally important to studies of hexamerin evolution and homologies will be a better understanding of the selective forces that have led to such great compositional diversity without disturbing the basic hexameric pattern. From the functional considerations described in this review, we may speculate that hexamer conservation has been important at least in part because a molecular size of about 500 kd is important. On the one hand, it allows the hexamerins to escape the mechanisms responsible for the rapid turnover of smaller proteins and allows the storage of large amounts of amino acids with minimal osmotic consequences. On the other hand, a size of approximately 500 kd may yield the maximum Stoke's radius, permiting diffusion through the basement membranes surrounding the fat body, epidermis, and whatever other tissues are responsible for selective hexamerin removal from the hemolymph during morphogenesis. Compositional diversity and quarternary structure conservation raise fundamental questions about insect physiology and development, and answering these will provide the biological context for understanding the evolutionary history of the hexamerins.

ACKNOWLEDGMENTS

We thank John Law, Michael Wells, and Michael Kanost for their reactions to an earlier form of this review, and Mary Telfer for bibliographic assistance. Supported by NIH grant GM-32909 to W. H. Telfer, and USDA grant 88-37251-3991 to J. G. Kunkel and D. E. Leonard.

Literature Cited

1. Akam, M., Roberts, D., Richards, G., Ashburner, M. 1978. *Drosophila:* the genetics of two major larval proteins. *Cell* 13:215–25
2. Anderson, E. 1964. Oocyte differentiation and vitellogenesis in the roach *Periplaneta americana. J. Cell Biol.* 20:131–55
3. Anksin, J. 1990. *Purification and characterization of two major hemolymph proteins from* Locusta migratoria. Ms. thesis. Kingston, Ont.: Queen's University
4. Bean, D., Silhacek, D. 1989. Changes in the titer of the female-predominant storage protein (81k) during larval and pupal development of the wax moth, *Galleria mellonella. Arch. Insect Biochem. Physiol.* 10:333–48
5. Beverley, S., Wilson, A. 1982. Molecular evolution in *Drosophila* and higher Diptera. I. Micro-complement fixation studies of a larval hemolymph protein. *J. Mol. Evol.* 18:251–64
6. Beverley, S., Wilson, A. 1985. Ancient origin for Hawaiian Drosophilinae inferred from protein comparisons. *Proc. Nat. Acad. Sci. USA* 82:4753–57
7. Bhakthan, N., Gilbert, L. 1972. Studies on the cytophysiology of the fat body of the American silkmoth. *Z. Zellforsch.* 124:433–44
8. Brock, H., Roberts, D. 1983. An immunological and electrophoretic study of the larval serum proteins of *Drosophila* species. *Insect Biochem.* 13:57–63
9. Butterworth, F., Tysell, B., Waclawaski, I. 1979. The effect of 20-hydroxyecdysone and protein on granule formation in the *in vitro* cultured fat body of *Drosophila. J. Insect Physiol.* 25:855–60
10. Caglayan, S., Gilbert, L. 1987. In vitro synthesis, release and uptake of storage proteins by the fat body of *Manduca sexta:* putative hormonal control. *Comp. Biochem. Physiol.* 87B:989–97
11. Carpenter, D., VanHolde, K. 1973. Amino acid composition, amino terminal analysis, and subunit structure of *Cancer magister* hemocyanin. *Biochemistry* 12:2231–38
12. Chippendale, G., Kilby, B. 1969. Relationship between the proteins of the haemolymph and fat body during development of *Pieris brassicae. J. Insect Physiol.* 15:905–26
13. Crossley, A. 1972. The ultrastructure and function of pericardial cells and other necrophyles in *Calliphora erythrocephala. Tiss. Cell* 4:529–60
14. Dean, R., Locke, M., Collins, J. 1985. Structure of the fat body. See Ref. 42, 3:155–210
15. deBianchi, A., Marinotti, O. 1984. A storage protein in *Rhyncosciara americana* (Diptera, Sciaridae). *Insect Biochem.* 14:453–61

16. deBianchi, A., Marinotti, O., Espinoza-Fuentes, F., Pereira, S. 1983. Purification and characterization of *Musca domestica* storage protein, and its developmental profile. *Comp. Biochem. Physiol.* 76B:861–67

17. deKort, C., Koopmanschap, A. 1987. Isolation and characterization of a larval hemolymph protein in *Locusta migratoria*. *Arch. Insect Biochem. Physiol.* 4:191–203

18. Delaney, S., Sunkel, C., Genova-Seminova, G., Davies, J., Glover, D. 1987. *Cis*-acting sequences sufficient for correct tissue and temporal specificity of larval serum protein one genes of *Drosophila*. *EMBO J.* 6:3849–54

19. dePriester, W., vanderMolen, L. 1979. Premetamorphic changes in the ultrastructure of *Calliphora* fat cells. *Cell Tiss. Res.* 198:79–83

20. Doolittle, R. 1981. Similar amino acid sequences: chance or common ancestry? *Science* 214:149–59

21. Drexel, R. 1987. Complete amino acid sequence of a functional unit from a molluscan hemocyanin *(Helix pomatia)*. *Biol. Chem. Hoppe-Seyler* 368:617–35

22. Duhamel, R., Kunkel, J. 1978. A molting rhythm for serum proteins of the cockroach, *Blatta orientalis*. *Comp. Biochem. Physiol.* 60B:333–37

23. Duhamel, R., Kunkel, J. 1983. Cockroach larval-specific protein, a tyrosine-rich serum protein. *J. Biol. Chem.* 258:14461–65

24. Duhamel, R., Kunkel, J. 1987. Moulting-cycle regulation of haemolymph protein clearance in cockroaches: possible size-dependent mechanism. *J. Insect Physiol.* 33:155–58

25. Enderle, U., Kauser, G., Reum, K., Scheller, K., Koolman, J. 1983. Ecdysteroids in the hemolymph of blowfly larvae are bound to calliphorin. See Ref. 102, pp. 40–49

26. Fox, F., Seed, J., Mills, R. 1972. Cuticle sclerotization by the American cockroach: immunological evidence for the incorporation of blood proteins into the cuticle. *J. Insect Physiol.* 18:2065–70

27. Fujii, T., Sakurai, H., Izumi, S., Tomino, S. 1989. Structure of the gene for the arylphorin-type storage protein SP 2 of *Bombyx mori*. *J. Biol. Chem.* 264:11020–25

28. Ghiretti-Magaldi, A., Nuzzolo, C., Ghiretti, F. 1966. Chemical studies on hemocyanins. I. Amino acid composition. *Biochemistry* 5:1943–51

29. Gochoco, C., Kunkel, J., Nordin, J. 1988. Experimental modifications of an insect vitellin affect its structure and its uptake by oocytes. *Arch. Insect Biochem. Physiol.* 8:179–99

30. Grun, L., Peter, M. 1983. Selective crosslinking of tyrosine-rich larval serum proteins and of soluble *Manduca sexta* cuticle proteins by nascent N-acetyldopamine quinone and N-beta-alanyldopamine quinone. See Ref. 102, pp. 102–15

31. Haunerland, N., Bowers, W. 1986. Arylphorin from the corn earworm, *Heliothis zea*. *Insect Biochem.* 16:617–25

32. Haunerland, N., Bowers, W. 1986. Binding of insecticides to lipophorin and arylphorin, two hemolymph proteins of *Heliothis zea*. *Arch. Insect Biochem. Physiol.* 3:87–96

33. Haunerland, N., Bowers, W. 1986. A larval specific lipoprotein: purification and characterization of a blue chromoprotein from *Heliothis zea*. *Biophys. Biochem. Res. Commun.* 134:580–86

34. Izumi, S., Tojo, S., Tomino, S. 1980. Translation of fat body mRNA from the silkworm, *Bombyx mori*. *Insect Biochem.* 10:429–34

35. Jones, G., Brown, N., Manczak, M., Hiremath, S., Kafatos, F. 1990. Molecular cloning, regulation and complete sequence of a hemocyanin-related, juvenile hormone-suppressible protein from insect hemolymph. *J. Biol. Chem.* In press

36. Jones, G., Hiremath, S., Hellman, G., Rhoads, R. 1988. Juvenile hormone regulation of mRNA levels for a highly abundant hemolymph protein in larval *Trichoplusia ni*. *J. Biol. Chem.* 263:1089–92

37. Kanost, M., Kawooya, J., Law, J., Ryan, R., VanHeusden, M., Ziegler, R. 1990. Insect hemolymph proteins. *Adv. Insect Physiol.* In press

38. Karpells, S., Leonard, D., Kunkel, J. 1990. Cyclic fluctuations in arylphorin, the principal serum storage protein of *Lymantria dispar*, indicate multiple roles in development. *Insect Biochem.* 20:73–82

39. Katsoris, P., Marmaras, V. 1979. Characterization of the major haemolymph proteins in *Ceratitus capitata*. *Insect Biochem.* 9:503–7

40. Kawasaki, H., Sato, H., Suzuki, M. 1971. Structural proteins in the silkworm egg-shells. *Insect Biochem.* 1:130–48

41. Kefaliakou-Bourdopoulou, M., Christodoulou, C., Marmaras, V. 1981. Storage proteins in the olive fruit fly, *Dacus oleae*: characterization and im-

munological studies. *Insect Biochem.* 11:707–11

42. Kerkut, G., Gilbert, L., eds. 1985. *Comprehensive Insect Physiology, Biochemistry, and Pharmacology.* Oxford: Pergamon

43. Kim, H., Kang, C., Mayer, R. 1989. Storage proteins of the fall webworm, *Hyphantria cunea* Drury. *Arch. Insect Biochem. Physiol.* 10:115–30

44. King, T., Jukes, J. 1969. Nondarwinian evolution. *Science* 164:788–98

45. Kinnear, J., Thomson, J. 1975. Nature, origin and fate of major haemolymph proteins in *Calliphora. Insect Biochem.* 5:531–52

46. Kling, H., Pentz, S. 1975. Isolation and chemical characterization of disc-electrophoretically pure haemolymph proteins of *Plodia interpunctella* (Indian meal moth). *Comp. Biochem. Physiol.* 50B:103–4

47. Koeppe, J., Gilbert, L. 1973. Immunochemical evidence for the transport of haemolymph protein into the cuticle of *Manduca sexta. J. Insect Physiol.* 19:615–24

48. Konig, M., Agrawal, O., Schenkel, H., Scheller, K. 1986. Incorporation of calliphorin into the cuticle of the developing blowfly, *Calliphora vicina. Wilhelm Roux's Arch. Dev. Biol.* 195:296–301

49. Kramer, S., Mundall, E., Law, J. 1980. Purification and properties of manducin, an amino acid storage protein in the haemolymph of larval and pupal *Manduca sexta. Insect Biochem.* 10:279–88

50. Kunkel, J., Grossniklaus-Buergin, C., Karpells, S., Lanzrein, B. 1990. Arylphorin of *Trichoplusia ni:* characterization and parasite-induced precocious increase in titer. *Arch. Insect Biochem. Physiol.* 13:117–25

51. Kunkel, J., Lawler, D. 1974. Larval-specific serum protein in the order Dictyoptera—immunologic characterization in larval *Blatella germanica* and cross-reaction throughout the order. *Comp. Biochem. Physiol.* 47B:697–710

52. Kunkel, J., Nordin, J. 1985. Yolk proteins. See Ref. 42, 1:83–111

53. Kunkel, J., Pan, M. 1976. Selectivity of yolk protein uptake: comparison of vitellogenins of two insects. *J. Insect Physiol.* 22:809–18

54. Law, J., ed. 1987. *Molecular Entomology.* New York: Alan R. Liss

55. Lenz, C., Venkatesh, K., Chippendale, G. 1987. Major plasma proteins of larvae of the southwestern corn borer, *Diatraea grandiosella. Arch. Insect Biochem. Physiol.* 5:271–84

56. Lepesant, J.-A., Levine, M., Garen, A., Lepesant-Kejzlarova, J., Rat, L., Somme-Martin, G. 1982. Developmentally regulated gene expression in *Drosophila* larval fat bodies. *J. Mol. Appl. Gen.* 1:371–83

57. Lepesant, J.-A., Maschat, F., Kejzlarova-Lepesant, J., Benes, H., Yanicostas, C. 1986. Developmental and ecdysteroid regulation of gene expression in the larval fat body of *Drosophila melanogaster. Arch. Insect Biochem. Physiol.* (Suppl)1:133–41

58. Levenbook, L. 1985. Insect storage proteins. See Ref. 42, 10:307–46

59. Levenbook, L., Bauer, A. 1984. The fate of the larval storage protein calliphorin during adult development of *Calliphora vicina. Insect Biochem.* 14:77–86

60. Li, W.-H., Wu, C.-I., Luo, C.-C. 1985. A new method for estimating synonomous and nonsynonomous rates of nucleotide substitution considering the relative likelihood of nucleotide and codon changes. *Mol. Biol. Evol.* 2:150–74

61. Linzen, B., Soeter, N., Riggs, A., Schneider, H.-J., Schartau, W., et al. 1985. The structure of arthropod hemocyanins. *Science* 229:519–24

62. Locke, J., McDermid, H., Brac, T., Atkinson, B. 1982. Developmental changes in the synthesis of haemolymph polypeptides and their sequestration by the prepupal fat body in *Callpodes ethlius* Stoll (Lepidoptera: Hesperiidae). *Insect Biochem.* 12:431–40

63. Locke, M., Collins, J. 1968. Protein uptake into multivesicular bodies and storage granules in the fat body of an insect. *J. Cell Biol.* 36:453–83

64. Loughton, B., West, A. 1965. The development and distribution of haemolymph proteins in Lepidoptera. *J. Insect Physiol.* 11:919–32

65. Marinotti, O., deBianchi, A. 1986. Structural properties of *Musca domestica* storage protein. *Insect Biochem.* 16:709–16

66. Marinotti, O., deBianchi, A. 1986. Uptake of storage protein by *Musca domestica* fat body. *J. Insect Physiol.* 32:819–25

67. Markl, J., Schmid, R., Czichos-Tiedt, S., Linzen, B. 1976. Hemocyanins in spiders, III. Chemical and physical properties of the proteins in *Dugesiella* and *Cupiennius* blood. *Hoppe-Seyler's Z. Physiol. Chem.* 357:1713–25

68. Markl, J., Winter, S. 1989. Subunit-specific monoclonal antibodies to tarantula hemocyanin, and a common epi-

tope shared with calliphorin. *J. Comp. Physiol.* B159:139–51

69. Martin, M., Kinnear, J., Thomson, J. 1971. Developmental changes in the late larva of *Calliphora stygia*. IV. Uptake of plasma protein by the fat body. *Aust. J. Biol. Sci.* 24:291–99

70. Marx, R. 1983. Ultrastructural aspects of protein synthesis and protein transport in larvae of *Calliphora vicina*. See Ref. 102, pp. 50–60

71. Miller, S., Silhacek, D. 1982. Identification and purification of storage proteins in tissues of the greater wax moth *Galleria mellonella* (L). *Insect Biochem.* 12:277–92

72. Miller, S., Silhacek, D. 1982. The synthesis and uptake of haemolymph storage proteins by the fat body of the greater wax moth, *Galleria mellonella* (L.). *Insect Biochem.* 12:293–300

73. Mintzas, A., Chrysanthis, G., Christodoulou, C., Marmaras, V. 1983. Translation of the mRNAs coding for the major haemolymph proteins of *Ceratitis capitata* in cell-free system. Comparison of the translatable mRNA levels to the respective biosynthetic levels of the proteins in the fat body during development. *Dev. Biol.* 95:492–96

74. Mintzas, A., Reboutsicas, D. 1984. Isolation, characterization and immunological properties of ceratitin-4, a major haemolymph protein of the Mediterranean fruit fly, *Ceratitis capitata*. *Insect Biochem.* 14:285–91

75. Mintzas, A., Rina, M. 1986. Isolation and characterization of three major larval serum proteins of the Mediterranean fruit fly *Ceratitis capitata* (Diptera). *Insect Biochem.* 16:825–34

76. Munn, E., Feinstein, A., Greville, G. 1971. The isolation and properties of the protein calliphorin. *Biochem. J.* 124:367–74

77. Munn, E., Greville, G. 1969. The soluble proteins of developing *Calliphora erythrocephala*, particularly calliphorin, and similar proteins in other insects. *J. Insect Physiol.* 15:1935–50

78. Munn, E., Price, G., Greville, G. 1969. The synthesis *in vitro* of the protein calliphorin by fat body from the larva of the blowfly, *Calliphora erythrocephala*. *J. Insect Physiol.* 15:1601–5

79. Ogawa, K., Tojo, S. 1981. Quantitative changes of storage proteins and vitellogenin during the pupal-adult development in the silkworm, *Bombyx mori* (Lepidoptera:Bombycidae). *Appl. Entomol. Zool.* 16:288–96

80. Palli, S., Locke, M. 1987. The synthesis of hemolymph proteins by the larval epidermis of an insect *Calpodes ethlius* (Lepidoptera:Hesperiidae). *Insect Biochem.* 17:711–22

81. Palli, S., Locke, M. 1987. Purification and characterization of three major hemolymph proteins of an insect, *Calpodes ethlius* (Lepidoptera:Hesperiidae). *Arch. Insect Biochem. Physiol.* 5:233–44

82. Pan, M., Telfer, W. 1990. Storage and foreign protein clearing from hemolymph of pharate adult saturniids. In *Molecular Insect Science*. ed. H. Hagedorn, J. Hildebrand, M. Kidwell, J. Law. New York: Plenum. (Abstr.) In press

83. Phillips, D., Loughton, B. 1976. Cuticle proteins in *Locusta migratoria*. *Comp. Biochem. Physiol.* 55B:129–35

84. Ray, A., Memmel, N., Kumaran, A. 1987. Developmental regulation of the larval hemolymph protein genes in *Galleria mellonella*. *Wilhelm Roux's Arch. Dev. Biol.* 196:414–20

85. Ray, A., Memmel, N., Orchekowski, R., Kumaran, A. 1987. Isolation of two cDNA clones coding for larval hemolymph proteins of *Galleria mellonella*. *Insect Biochem.* 16:603–17

86. Riddiford, L., Hice, R. 1985. Developmental profiles of the mRNAs for *Manduca* arylphorin and two other storage proteins during the final larval instar of *Manduca sexta*. *Insect Biochem.* 15:489–502

87. Rimoldi, O., Soulages, J., Gonzalez, S., Peluffo, R., Brenner, R. 1989. Purification and properties of the very high density lipoprotein from the hemolymph of adult *Triatoma infestans*. *J. Lipid Res.* 30:857–64

88. Roberts, D. 1983. The evolution of the larval serum protein genes in *Drosophila*. See Ref. 102, pp. 86–100

89. Roberts, D. 1987. The function of the major larval serum proteins of *Drosophila melanogaster*: a review. See Ref. 54, pp. 285–94

90. Roberts, D. 1987. The functions of the major serum proteins of *Drosophila* larvae. *Biol. Chem. Hoppe-Seyler* 368:572 (Abstr.)

91. Roberts, D., Brock, H. 1981. The major serum proteins of Dipteran larvae. *Experientia* 37:103–10

92. Roberts, D., Wolfe, J., Akam, M. 1977. The developmental profiles of two major haemolymph proteins from *Drosophila melanogaster*. *J. Insect Physiol.* 23:871–78

93. Roth, T. F., Porter, K. R. 1964. Yolk protein uptake in the oocyte of the mosquito *Aedes aegypti* L. *J. Cell Biol.* 20:313–32

94. Ryan, R., Anderson, D., Grimes, W., Law, J. 1985. Arylphorin from *Manduca sexta:* carbohydrate structure and immunological studies. *Arch. Biochem. Biophys.* 243:115–24

95. Ryan, R., Keim, P., Wells, M., Law, J. 1985. Purification and properties of a predominantly female-specific protein from the hemolymph of the larva of the tobacco hornworm, *Manduca sexta. J. Biol. Chem.* 260:782–87

96. Ryan, R., Schmidt, J., Law, J. 1984. Arylphorin from the haemolymph of the larval honeybee, *Apis mellifera. Insect Biochem.* 14:515–20

97. Ryan, R., Wang, X., Willott, E., Law, J. 1986. Major hemolymph proteins from larvae of the black swallowtail butterfly, *Papilio polyxenes. Arch. Insect Biochem. Physiol.* 3:539–50

98. Sakai, N., Mori, S., Izumi, S., Haino-Fukushima, K., Ogura, T., et al. 1988. Structures and expression of mRNAs coding for major plasma proteins of *Bombyx mori. Biochim. Biophys. Acta.* 949:224–32

99. Sakurai, H., Fujii, T., Izumi, S., Tomino, S. 1988. Complete nucleotide sequence of gene for sex-specific storage protein of *Bombyx mori. Nucleic Acids Res.* 16:7717–18

100. Sakurai, H., Fujii, T., Izumi, S., Tomino, S. 1988. Structure and expression of gene coding for sex-specific storage protein of *Bombyx mori. J. Biol. Chem.* 263:7876–80

101. Sato, J., Roberts, D. 1983. Synthesis of larval serum proteins 1 and 2 of *Drosophila melanogaster* by third instar fat body. *Insect Biochem.* 13:1–5

102. Scheller, K., ed. 1983. *The Larval Serum Proteins of Insects.* New York: Thieme-Stratton

103. Scheller, K., Zimmerman, H.-P., Sekeris, C. 1980. Calliphorin, a protein involved in the cuticle formation of the blowfly, *Calliphora vicina. Z. Naturforsch.* 35c:387–89

104. Schenkel, H., Kejzlarova-Lepesant, J., Berreur, P., Moreau, J., Scheller, K., et al. 1985. Identification and molecular analysis of a multigene family encoding calliphorin, the major larval serum protein of *Calliphora vicina. EMBO J.* 4:2983–90

105. Schenkel, H., Scheller, K. 1986. Stage- and tissue-specific expression of the genes encoding calliphorin, the major larval serum protein of *Calliphora vicina. Wilhelm Roux's Arch. Dev. Biol.* 195:290–95

106. Sekeris, C., Scheller, K. 1977. Calliphorin, a major protein of the blowfly: correlation between the amount of protein, its biosynthesis, and the titer of translatable calliphorin-mRNA during development. *Dev. Biol.* 59:12–23

107. Shirrar, A., Bownes, M. 1989. *cricket:* A locus regulating a number of adult functions of *Drosophila melanogaster. Proc. Natl. Acad. Sci. USA* 86:4559–63

108. Silhacek, D., Bean, D. 1988. Storage protein physiology in *Galleria mellonella.* In *Proc. Int. Conf. Endocrinol.* ed. B. Cymborowski, F. Sehnal, A. Zabza, pp. 1007–11. Wroclaw, Poland: Wroclaw Unl. Press

109. Stay, B. 1965. Protein uptake in the oocytes of the Cecropia moth. *J. Cell Biol.* 26:49–62

110. Storella, J., Wojchowski, D., Kunkel, J. 1985. Structure and embryonic degradation of two native vitellins in the cockroach, *Periplaneta americana. Insect Biochem.* 15:259–75

111. Tahara, T., Kuroiwa, A., Obinata, M., Natori, S. 1984. Multiple-gene structure of the storage protein genes of *Sarcophaga peregrina. J. Mol Biol.* 74:19–29

112. Tamura, H., Tahara, T., Kuroiwa, S., Obinata, M., Natori, S. 1983. Differential expression of two abundant messenger RNAs during development of *Sarcophaga peregrina. Dev. Biol.* 99:145–51

113. Telfer, W. 1960. The selective accumulation of blood proteins by the growing oocytes of saturniid moths. *Biol. Bull.* 118:338–51

114. Telfer, W., Canaday, D. 1987. Storage proteins of saturniid moths: interspecific similarities and contrasts. *Biol. Chem. Hoppe-Seyler* 368:571 (Abstr.)

115. Telfer, W., Keim, P., Law, J. 1983. Arylphorin, a new protein from *Hyalaphora cecropia:* Comparisons with calliphorin and manducin. *Insect Biochem.* 13:601–13

116. Telfer, W., Massey, H. 1987. A storage hexamer from *Hyalophora* that binds riboflavin and resembles the apoprotein of hemocyanin. See Ref. 54, pp. 305–14

117. Telfer, W., Pan, M. 1989. Adsorptive endocytosis of vitellogenin, lipophorin, and microvitellogenin during yolk formation in *Hyalophora. Arch. Insect Biochem. Physiol.* 9:339–55

118. Thomasson, W., Mitchell, H. 1972. Hormonal control of protein granule accumulation in fat bodies of *Drosophila melanogaster* larvae. *J. Insect Physiol.* 18:1885–99

119. Thomson, J. 1981. Speculations on the evolution of insect storage proteins. In *Evolution and Speciation.* ed. W.

Atchley, D. Woodruff, pp. 398–416. Cambridge: Cambridge Univ. Press
120. Thomson, J., Radok, K., Shaw, D., Whitten, M., Foster, G., Birt, L. 1976. Genetics of lucilin, a storage protein from the sheep blowfly, *Lucilia cuprina* (Calliphoridae). *Biochem. Genet.* 14:145–60
121. Tojo, S., Betchaku, T., Ziccardi, V., Wyatt, G. 1978. Fat body protein granules and storage proteins in the silkmoth, *Hyalophora cecropia. J. Cell Biol.* 78:823–38
122. Tojo, S., Morita, M., Agui, N., Hiruma, K. 1985. Hormonal regulation of phase polymorphism and storage-protein fluctuation in the common cutworm, *Spodoptera litura. J. Insect Physiol.* 31:283–92
123. Tojo, S., Nagata, M., Kobayashi, M. 1980. Storage proteins in the silkworm, *Bombyx mori. Insect Biochem.* 10:289–303
124. Ueno, K., Natori, S. 1984. Identification of storage protein receptor and its precursor in the fat body membrane of *Sarcophaga peregrina. J. Biol. Chem.* 259:12107–11
125. Ueno, K., Ohsawa, F., Natori, S. 1983. Identification and activation of storage protein receptor of *Sarcophaga peregrina* fat body by 20-hydroxyecdysone. *J. Biol. Chem.* 258:12210–14
126. VanHolde, K., Miller, K. 1982. Hemocyanins. *Q. Rev. Biophys.* 15:1–12

127. vanPelt-Verkuil, E. 1978. Increase in acid phosphatase activity in the fat body during larval and pharate pupal development in *Calliphora erythrocephala. J. Insect Physiol.* 24:375–82
128. Webb, B., Riddiford, L. 1988. Regulation of expression of arylphorin and female-specific protein mRNAs in the tobacco hornworm, *Manduca sexta. Dev. Biol.* 130:682–92
129. Webb, B., Riddiford, L. 1988. Synthesis of two storage proteins during larval development of the tobacco hornworm, *Manduca sexta. Dev. Biol.* 130:671–81
130. Willott, E. 1989. Manduca sexta *fat body during the larval-pupal transformation: A structural and biochemical study.* PhD thesis. Tucson: Univ. Arizona
131. Willott, E., Wang, X.-Y., Wells, M. 1989. cDNA and gene sequence of *Manduca sexta* arylphorin, an aromatic amino acid–rich larval serum protein. *J. Biol. Chem.* 264:19052–59
132. Wolfe, J., Akam, M., Roberts, D. 1977. Biochemical and immunological studies on larval serum protein 1, the major haemolymph protein of *Drosophila melanogaster* third-instar larvae. *Eur. J. Biochem.* 79:47–53
133. Wyatt, G., Kanost, M., Chin, B., Cook, K., Kawasoe, B. 1990. Juvenile hormone analog and injection effects on locust hemolymph protein synthesis. *Arch. Insect Biochem. Physiol.* In press

Annu. Rev. Entomol. 1991. 36:229–55

MANAGEMENT OF DIABROTICITE ROOTWORMS IN CORN

Eli Levine and Hassan Oloumi-Sadeghi

Office of Agricultural Entomology, University of Illinois at Urbana-Champaign and Illinois Natural History Survey, 607 East Peabody Drive, Champaign, Illinois 61820

KEY WORDS: corn rootworms, pest management, *Diabrotica virgifera virgifera, Diabrotica virgifera zeae, Diabrotica barberi*

INTRODUCTION

The western corn rootworm (WCR), *Diabrotica virgifera virgifera,* and the northern corn rootworm (NCR), *Diabrotica barberi,* are the most serious insect pests of dent corn, *Zea mays,* in the major corn-producing states of the north central United States and in Canada. Costs for soil insecticides to control larval damage to the root systems of corn and aerial sprays to reduce beetle damage to corn silks, when combined with crop losses, can approach $1 billion annually (125). Corn rootworms (CRW) also attack other types of corn including pop, flint, flour, and sweet corn.

The biology of the NCR and WCR has been extensively reviewed (30, 96). Both species are univoltine. In the corn belt, beetles are present in cornfields from July through frost, feeding on corn pollen, silks, immature kernels, and foliage as well as pollen of other plants (mainly NCR) (97). From late July through September, oviposition occurs primarily in cornfields; few eggs are laid in other crops (144). Eggs remain in the soil until the following spring. Egg hatch begins in late May and early June. The larvae can survive only on the roots of corn and on the roots of a limited number of grasses (97). Larval feeding may reduce the amount of water and nutrients supplied to developing plants (92), and extensive root damage makes plants more susceptible to lodging. Larval feeding may also facilitate infection by root and stalk rot

229

fungi, resulting in further damage (139). Yield losses also result from difficulty in harvesting lodged corn.

Pupation occurs in the soil, and adult emergence begins in early July in the Midwest. WCR and NCR adults can transmit and spread maize chlorotic mottle virus (90). Corn stalk rot fungi also have been isolated from WCR adults, which may vector the fungi to corn ears and kernels (61).

We have excluded the southern corn rootworm (SCR), *Diabrotica undecimpunctata howardi,* from our review because its life cycle and host range are very different from NCR and WCR (97) and because migrating SCR adults arrive too late in the season to cause much root damage in the corn belt (116). Our review includes the Mexican corn rootworm (MCR), *Diabrotica virgifera zeae,* because this pest is now present in Kansas, Oklahoma, and Texas as well as in Mexico and Central America (15, 96). Laboratory and field studies have shown that the WCR and MCR are sexually compatible and that their populations intergrade where their distributions overlap (100).

Management of CRW may become more challenging in the years ahead. Factors such as enhanced microbial degradation of soil insecticides, insecticide resistance, extended diapause, pesticide laws and regulations, and concern about the use of pesticides in general affect strategies for CRW management. In this review, we discuss current CRW management options, problems associated with these approaches, and new developments in CRW management.

MANAGEMENT OPTIONS AND CONSIDERATIONS

Crop Rotation

The early practice of growing corn in rotation with small grains, hay, clover, or alfalfa provided excellent control of CRW because CRW eggs are laid almost exclusively in cornfields, and larvae must feed on corn roots the following season to complete development. The practice of growing corn continuously began in the late 1940s and was largely responsible for expanding the range of NCR and WCR (13, 97). Presently, the production of continuous corn remains a popular option in many portions of the corn belt, and soil insecticides are used to protect root systems from CRW feeding.

REPORTED FAILURES OF CROP ROTATION AS A CONTROL MEASURE   Ever since CRW were elevated to pest status, there have been reports of CRW damage to first-year corn following a rotation crop. Growing corn following other crops failed to control NCR as early as 1932 (9) and 1956 (115). In the latter report, two years without corn always broke the cycle and controlled the pest (115). Severe WCR larval damage to corn following soybean grown for seed production has recently been observed (111). Greater than expected

WCR oviposition was found in these soybean fields that had very few weeds and no volunteer corn. Adults might have been attracted to the green soybean plants for oviposition; fungicides applied to the soybean foliage allowed the leaves to remain green longer than usual.

Oviposition in noncorn fields the preceding season has also been proposed as an explanation of why NCR infest first-year corn (13). Other studies fail to support this hypothesis. Chiang (27) concluded that few NCR eggs were laid outside of cornfields. In Illinois studies, oviposition in soybean and damage to corn the following season were negligible where soybean fields were essentially free of volunteer corn plants; damage was not significant even when corn followed weedy soybean fields (144). NCR infestation of first-year corn following small grains has been a frequent problem in South Dakota. But, regardless of the maturation stage of corn, NCR females laid > 80% of their eggs in corn plots rather than in small grain stubble (74).

EXTENDED DIAPAUSE   Another explanation for CRW damage in first-year corn is that eggs can remain in diapause for more than one winter. Of 676 NCR eggs buried in a Minnesota field in the fall of 1962 and left undisturbed through two winters, 2% appeared viable in the spring of 1964 (28). Two of these eggs hatched later that growing season after the completion of two winters. It was concluded that the percentage of eggs with this extended diapause characteristic was so small as to be economically unimportant.

Recently, however, approximately 40% of the eggs that hatched from a population of NCR collected in South Dakota underwent extended diapause (99). This relatively large percentage could cause significant damage to corn following a one-year rotation with another crop. Indeed, CRW problems have been reported in corn following a one-year rotation with another crop in South Dakota, Minnesota, Iowa, Illinois, and Nebraska; extended diapause has been confirmed in NCR eggs from South Dakota (98, 99), Minnesota (98), Illinois (109), and North Dakota (E. Levine & M. J. Weiss, unpublished data).

Diapause in Illinois NCR eggs has been shown to be quite variable, ranging from one to four years. In addition, significant differences in percentage of eggs showing extended diapause were found between eggs laid by different females collected at a single location, a finding that suggests a genetic component to extended diapause (112). Some populations of NCR have apparently adapted to crop rotation, i.e. a yearly rotation of corn with another crop confers a selective advantage to eggs that can remain dormant for two years (98). This observation is supported by reports of a greater incidence of extended diapause in areas of South Dakota, Minnesota, and Illinois where corn is rotated annually than in areas where corn is planted without rotation (98, 109). The early reports of CRW damage in corn following another crop are now best accounted for by extended diapause. Widespread and serious economic damage as a consequence of extended diapause does not always occur. For example, a 1986–1988 survey in Illinois showed that damage to corn following soybean was of no economic consequence in 98.8% of the 890 fields examined (153).

Until recently, extended diapause was not found in WCR (e.g. 98, 99). Investigators have suggested that its absence was due in large part to the limited diapause capabilities of WCR (98, 99). However, studies in Illinois (112) and Ontario (156) showed that < 1% of WCR eggs hatched only after passing through two simulated winters in an environmental chamber or two winters in the field. Whether prolonged diapause in WCR will come to resemble extended diapause in NCR remains to be seen.

Tillage and Soil Environment

The type of tilling equipment, seasonal timing, depth, and frequency of tillage operations may interact with such environmental factors as temperature and precipitation to affect CRW biology (64). Most studies (e.g. 69, 91) found no significant differences in either NCR or WCR oviposition among various tillage practices. CRW eggs are generally concentrated in the top 10 to 20 cm of soil. The eggs are distributed deeper under dry conditions, and WCR eggs are laid deeper than NCR eggs (69, 172). In an Iowa study, WCR egg survival increased as depth of burial increased in each of four tillage systems, with no-till having the lowest egg mortality (108, cited in 69). Another Iowa study concluded that CRW eggs may be favored by no-till and chisel-plow systems during very cold and dry winters and that the influence of tillage type on CRW populations during moderate winters was inconsequential (69). Moldboard plow and paraplow plots accumulated more negative degree days at each of four soil depths compared to chisel plow and no-till systems (108, cited in 64).

In laboratory studies, most WCR eggs held at a constant −10°C for 21 days were killed. WCR population crashes have also occurred following severe winters (72). A significant increase in mortality was also noted when WCR eggs were subjected to intermittent temperatures of freezing and thawing, such as would occur in soils without snow cover (75). Eggs of the NCR have greater cold hardiness than WCR eggs (73).

Despite similar spring egg densities, CRW larval populations and root damage were lower in no-till compared to other tillage treatments, although the overall effect of tillage on adult emergence was not significant (67, 70). Predation has been suggested as an explanation for these differences (30); ground beetles have been reported to be differentially affected by tillage (168). WCR emergence was delayed in conservation tillage compared to conventional tillage; however, by mid-August through early September, cumulative WCR emergence was similar across tillage systems (68, 167).

Emergence of NCR was affected less by tillage than was emergence of WCR (68).

In laboratory and greenhouse studies in which soil textures ranged from silty clay to loamy sand, survival of WCR larvae increased as the percentage of clay increased (165). CRW are generally not a problem in muck or sandy soils (164).

Eggs of the WCR can be submerged under water for up to 10 days in spring without incurring significant mortality (E. Levine & H. Oloumi-Sadeghi, unpublished data). When the soil remained saturated for prolonged periods during egg hatch, the establishment of larvae on corn roots was prevented (159). After larvae had become established on root systems, variations in soil moisture or brief periods of saturated soil had little effect on root damage. A single application of 5–10 cm of water when WCR were in the pupal stage reduced adult emergence by 50% compared to that in unirrigated plots (160).

Planting and Harvesting Dates

Planting date influences the relative availability of corn roots to eclosing larvae, which can survive only a few days without becoming established on a suitable host (12, 138). Delayed planting generally results in decreased root damage. If corn planting is delayed until early June, root damage is negligible and soil insecticide is probably not warranted (130). Numbers of WCR and NCR larvae and adults are significantly reduced, and the dates by which 50 and 90% of third-stage larvae, pupae, and adults occur are significantly delayed by later planting (7). In addition, later planting also increases the female to male ratio of NCR emergence (132). The effects of delayed planting on the population dynamics of CRW are probably related to the reduced availability of corn roots and the consequent increased mortality of early eclosing larvae (7).

Fields that are planted late often have a greater potential for CRW oviposition than fields that are planted early (e.g. 132). Corn grown for grain is generally harvested after CRW oviposition is finished, but corn grown for silage is cut before or during the time adults normally oviposit. When corn was harvested before September in a South Dakota study, populations of NCR and WCR larvae the next year were below economic levels (21).

TRAP CROPPING   Injury by CRW larvae, primarily WCR, in a 10-ha corn-field was reduced the year after a one-ha center strip of late-planted "trap corn" had been grown (86). However, in a similar study with a mixed population of WCR and NCR, no significant differences in oviposition were found the following year between areas with trap corn and conventional areas (174). Beetles may have returned to the conventional areas to oviposit after having fed on fresh pollen and silk in the area with trap corn. In other studies,

NCR females dispersed when food became scarce in the fields where they had emerged; however, they returned to cornfields for oviposition regardless of the quality of food present (35, 107).

Host-Plant Resistance

To date, only tolerance to larval feeding has been confirmed in some corn cultivars (31). Tolerance to feeding by CRW larvae is attributed to the massive size of root systems and to the ability of the plant to regenerate a root system that has sustained damage (16). Although susceptible corn inbreds sustained higher root damage and showed a greater tendency for lodging than tolerant inbreds, the survival of WCR larvae was much higher on tolerant than on susceptible inbreds (31). Large-scale planting of tolerant lines of corn might result in higher CRW populations that in time could overwhelm the root system. Therefore, tolerance should be considered only as a short-term management tool.

A possible means of host-plant resistance other than tolerance in three experimental corn hybrids has been reported (17). These hybrids were damaged less by WCR larvae than was a susceptible commercial hybrid and the same number of adults emerged from both types of hybrids. Root systems of the resistant hybrids were not significantly larger than those of the susceptible hybrid and root regeneration did not contribute to the resistance of the experimental hybrids. However, adults emerging from the resistant hybrids were heavier, a finding that suggests that these hybrids may have been nutritionally superior to the susceptible hybrid or that resistance was the result of an initial antixenosis followed by physiological compensation.

Zea diploperennis, a perennial discovered in Mexico in 1977, is a presumed near ancestor of corn and readily crosses with corn (136). Unfortunately, antibiosis against CRW larvae was not detected in the laboratory (14). *Tripsacum dactyloides,* another perennial relative of corn, is reported to have high levels of resistance, either antibiosis or extreme antixenosis, to WCR larvae (11). This resistance would be difficult, however, to transfer to corn (J. W. Dudley, personal communication).

Little is known about host-plant resistance to feeding by adult CRW. Corn cultivars range from those that are highly resistant to leaf feeding to those that are susceptible; this resistance is thought to be controlled by a single recessive gene and to be independent of root feeding by larvae (30). Significant differences in WCR feeding preferences among silks of 15 genetic sources of corn were found in the laboratory; however, the reasons for these differences in preference were not explored (142).

Biological Control

CRW have few natural enemies. The tachinid parasitoid *Celatoria diabroticae* has been reared from NCR adults in Kansas (30) and Illinois (52), but the efficacy of these parasitoids in controlling CRW in the field remains to be investigated. The importance of ground beetles as predators of pest species in corn has been documented (18). Radiolabeling studies suggest that some ground beetles are predators of CRW eggs and larvae (168). On the other hand, other workers (8, 95) contend that ground beetles are known as predators only by reputation and that the term "predacious ground beetles" is misapplied. Ground beetles and CRW have different habits and rarely encounter each other. Ground beetles, therefore, may play an opportunistic rather than a definite predatory role in corn agroecosystems (95).

Larvae of a soldier beetle, *Chauliognathus* sp., prey on MCR larvae. Predators and diseases seem to play a greater role in the population dynamics of MCR in Mexico than these agents play in the population dynamics of WCR in the corn belt (15). Mite species of the families Laelaptidae, Rhodacaridae, and Amerosiidae may reduce rootworm infestations by feeding on CRW eggs and larvae (29), and microarthropod predation might be a significant mortality factor for CRW eggs in the Midwest (154). The ant *Lasius neoniger* significantly reduced CRW larval populations, presumably by feeding directly on larvae (94), but did not reduce egg numbers (5).

Although NCR beetles were the most commonly consumed insect by red-winged blackbirds in New York cornfields, their usefulness as a biological control agent is questionable. The bird's reduction of NCR populations only compensated for approximately 4% of the damage done by the birds to the crop itself (10).

Steinernema feltiae (= *Neoaplectana carpocapsae*), an entomogenous nematode, may have potential for controlling soil-inhabiting insects (57). Susceptibility of WCR to *S. feltiae* varied according to the strain examined in the laboratory (89), and strain type may have been responsible for differential results in the field (129, 140). *S. feltiae* was ineffective against CRW larvae in an Illinois cornfield (152). Application techniques and edaphic factors were reported to be critical for nematode infectivity. Field trials with different strains of nematodes under various environmental conditions and at different times of application warrant further investigation.

Very little is known about entomopathogens of CRW. A fungus (presumably *Tarichium* spp.) has been reported to cause mortality in the laboratory to NCR adults collected in the field (135). CRW probably acquire infections in the larval or pupal stage. *Beauveria bassiana,* another fungus, can cause natural epizootics in CRW populations (117). Soil moisture, soil type, soil fertility, fungal strain, and presence of other soil microbes have been identified as factors greatly affecting the infectivity of *B. bassiana* in laboratory bioassays (93). In Illinois, *B. bassiana* has been studied for many years in the field, but its potential for reducing CRW feeding damage has not been proven (93).

Pathogens such as gregarine and microsporidian protozoa, fungi, bacteria, and viruslike particles as well as predaceous mites have been found in CRW laboratory colonies (88, 137); however, the role these biological agents play in the field is not known. Although the soil provides protection to biological control agents from desiccation and UV radiation, it also contains antagonistic microorganisms that may limit the effectiveness of these agents (93). Should this obstacle be overcome, microbial pathogens could play an important role in CRW management.

Controlling Larvae with Insecticides

Soil-applied insecticides have been an important tool in the management of CRW. Application of benzene hexachloride on the soil surface and the consequent reduction of root injury and plant lodging marked the first era of chemical CRW control (85). Currently, soil insecticides for CRW control are applied annually to 50–60% of the total corn acreage in the United States (125).

The persistence of soil-applied insecticides is an important factor in CRW control (1, 50). The ideal soil insecticide should persist in the soil for 6–10 weeks, the approximate length of time from insecticide application to the end of extensive larval feeding. CRW soil insecticides are typically formulated on clay, sand, or gypsum carrier granules and are generally applied at planting in an 18-cm band over the corn rows and lightly incorporated into the top three cm of soil.

Soil insecticides do not necessarily control CRW larval populations; instead, they protect root systems from injury (6, 160). For example, the numbers of beetles emerging from insecticide-treated plots were not found to differ significantly from those that emerged from untreated plots (66, 160). CRW are more prevalent in the corn belt today than ever before, even though most continuous corn acreage is treated with an insecticide at planting time (6). Because CRW insecticides protect corn roots only in a limited zone around the site of application, large numbers of larvae may survive and complete development on peripheral roots outside the treated band (47).

CRW insecticides have been generally effective in protecting corn roots, but the benefits have been much exaggerated (166). Numerous reports suggest that the field performance of soil insecticides against CRW larvae has been inconsistent. Highly variable degradation rates, enhanced biodegradation, and CRW resistance have been offered as explanations for this inconsistency (1, 33, 40).

The chlorinated hydrocarbons were efficient in controlling CRW for about two decades. The first report of inconsistent control was published in 1961 (170) followed by observations of resistance in 1962 (4). During the early 1960s, carbamates were first used for CRW control. Bufencarb was very effective when first introduced, but increasingly inconsistent performance caused its removal from the market (102). Carbofuran, which was very effective when introduced in the early 1970s, performed inconsistently by the end of the decade, especially in areas with a history of carbofuran use (50). Isofenphos, an organophosphorus compound, was removed from the market two years after its introduction in 1982, apparently in part because of enhanced microbial degradation (40).

The efficacy of soil-applied insecticides is affected by interactions among environmental characteristics (soil origin and type, soil moisture and temperature), characteristics of the insecticide (inherent toxicity, stability, adsorption, formulation, rate), biological factors (mechanisms of insecticide inactivation including biodegradation, and the behavior as well as the susceptibility of the target organism), cultural factors (planting date and tillage practices), and mechanical or operational factors (manner of insecticide application and conditions at planting including seed bed, wind speed and direction, calibration of equipment, and incorporation of formulated product into the soil) (45, 76). The importance of these factors for CRW control is discussed in the following sections.

ENVIRONMENTAL CHARACTERISTICS   Interactions among soil properties, moisture content, and soil temperature can greatly influence insecticidal bioactivity. Any process that affects insecticide concentration and distribution in the soil profile will directly affect bioactivity. Organic matter content and insecticide adsorption are positively correlated (44). Insecticides in soils with high organic matter content are generally degraded slowly; as a result, bioactivity may be prolonged. Soil bioassays, however, show that toxicity of insecticides is inversely correlated with organic matter content (45, 76). In one study, the amount of insecticide absorbed by CRW larvae was directly related to the amount desorbed from the soil (45). Other soil characteristics like pH and clay content have variable effects on insecticidal persistence. Some reports show insecticide degradation is influenced by pH (e.g. 58), but others show little effect (e.g. 23). Adsorption by clay surfaces may catalyze the breakdown of some insecticides (58).

Insecticidal activity and degradation rate may differ throughout the soil profile because soil attributes vary. During the spring, temperature and moisture in the upper three cm are warmer and lower, respectively, than in lower depths. Insecticides are degraded significantly faster as temperature increases (76). Rates of volatilization and chemical degradation are enhanced by increases in temperature and decreases in soil moisture (59). When soil is dry during egg hatch, the insecticide may remain strongly adsorbed to soil particles or the carrier granules and not be available for larval uptake. During heavy rainfall, insecticide concentrations in the root zone may be reduced by surface runoff or leaching (63).

CHARACTERISTICS OF THE INSECTICIDE   Physicochemical properties, degradability, formulation, and inherent toxicity all influence the bioactivity of soil-applied insecticides (40, 76). Water solubility, the soil adsorption coefficient (partition coefficient), and vapor pressure are among the physicochemical properties of an insecticide that affect its translocation and its absorption by larvae (44, 60). The chemicals most toxic in soil tend to be those that are least water soluble and show fumigant activity (24, 38, 79).

Insecticide degradation generally leads to the inactivation or detoxification of the insecticide. However, sometimes degradation leads to intermediate or transformation products that may be as toxic or more toxic than the original material (124). For example, terbufos and phorate degrade in soil to the sulfoxide and sulfone that, in many cases, are as toxic as the parent materials (24, 38) and are more persistent (26).

Granular formulations are generally more persistent than liquid formulations (1). Granules are more heterogeneously distributed than liquids, and the sorbed insecticides are not readily available for larval uptake.

The inherent toxicity of various insecticides to CRW larvae in soil may differ 10-fold based on the molecular structure of the insecticide (158). Based on LC_{50} response levels in laboratory experiments, 10 times more carbofuran than terbufos was required to kill equal numbers of SCR larvae in soil (38). Adult emergence from insecticide-treated plots is a poor indicator of inherent toxicity (160).

BIOLOGICAL FACTORS   Bioactivity of soil-applied insecticides may be affected by insect feeding behavior and population dynamics, susceptibility among life stages, intensity of insecticide use and development of resistance, and the degradative capabilities of soil microorganisms (45, 76).

The efficacy of soil insecticides partly depends on the time of CRW egg hatch and the movement of larvae (6). For example, poor CRW control in 1984 in Illinois was attributed in part to late egg hatching after much of the insecticide had dissipated (150). Because CRW insecticide residues are mainly confined to the top 5 cm of the soil profile in Illinois (47), larval feeding at depths > 7 cm might explain inadequate root protection (39).

Larval density may be a factor in insecticide performance (6, 150). At higher densities, larvae have more competition for roots and may feed on roots outside of the treated bands. These larvae may not contact treated soil until they reach the third instar when they are more tolerant to insecticides (159). Furthermore, laboratory studies show that WCR larvae prefer untreated soil rather than soil treated with insecticides such as terbufos (159) or fonofos (84).

Enhanced biodegradation of soil insecticides has attracted much attention in recent years and has recently been reviewed (40). Enhanced biodegradation can occur after just one application of certain insecticides to soil (80), although its effects on CRW control have generally been observed in fields only after several years of continuous use. The terms *cross-conditioning* or *cross-adaptation* have been used to describe the phenomenon in which the application of one insecticide enhances the biodegradation of another insecticide (36, 40). The most thoroughly studied insecticide in this regard has been carbofuran, and researchers have generally concluded that microbial activity is responsible for the rapid degradation of this compound in retreated soils (40). Low soil moisture, however, can inhibit microbial degradation of carbofuran and suppress the appearance of enhanced biodegradation (25, 80).

Failure of trimethacarb, isofenphos, and fonofos to control CRW has also been attributed to enhanced microbial degradation (36, 49, 80). Although all performance problems with soil insecticides are not caused by enhanced biodegradation, the phenomenon has been repeatedly documented and needs further investigation.

Production of continuous corn and the intensive use of the same or closely related CRW insecticides on the same land may provide favorable situations for the development of CRW resistance to insecticides (77). In fact, from the 1950s to the early 1970s, large-scale applications of cyclodienes for the control of WCR and NCR resulted in the development of resistance in most corn belt states (125). The mid to late 1960s were associated with presumed diazinon resistance and the early 1970s with possible resistance to bufencarb (78).

The earlier resistance studies generally relied on assays of field-collected adults. In general, the LD_{50} values of soil insecticides for adult CRW seem to have increased since they were first assayed (3, 33). However, no correlation was found between control of damage by larvae and adult response to carbofuran (48). Thus, efficacy failures in some fields where larvae were collected for experimentation could not be explained by the development of resistance (42).

CRW adults are generally more susceptible than larvae to the same insecticide. In addition, adults are relatively more susceptible to carbamates than to organophosphorus insecticides, whereas larval response to these insecticides is reversed (169). Several studies also indicate differential susceptibility of WCR and NCR to soil insecticides (33, 101, 114). Differences in adult WCR and NCR aldrin susceptibility have been attributed to diet (146). To understand the potential for resistance in CRW, studies must focus on larvae rather than adults and on mechanisms of detoxification (33, 34).

CULTURAL FACTORS   Cultural factors include planting date and tillage practices. CRW insecticides generally perform better when applied at planting to corn that is planted later (118); insecticides applied to early-planted corn may have degraded significantly by the time CRW eggs hatch (120). However, corn yields generally decline as planting date is delayed.

Soil properties like moisture content, temperature, organic matter content, pH, and microfloral composition differ among tillage systems; these properties potentially influence insecticide persistence and bioactivity (39, 154). Interactions between tillage, insecticides, and CRW have not been studied in detail. Recommended rates for currently registered CRW insecticides have generally been 1.1 kg active ingredient/ha regardless of tillage system (39). Theoretically, a tillage system such as no-till, which can increase organic matter content, may affect adsorption/desorption equilibrium and thus alter the behavior of the insecticide. Felsot et al (43) noticed no change in rate or pathway of terbufos degradation among various tillage practices. On the other hand, Stinner et al (155) reported that terbufos dissipated faster in untilled soil than in plowed soil. Contradictory results have also been reported for carbofuran in this regard (39). Tillage practices are reported to have little influence on the performance of insecticides against CRW (65, 155).

Tillage practices do not seem to affect the distribution of insecticides in the soil profile; i.e. their distribution is basically restricted to the top five to seven cm (47). Felsot el al (46) showed that total insecticide losses through surface runoff could approach 7% of applied amounts under worst-case conditions; however, reduction in tillage significantly reduced surface runoff.

MECHANICAL OR OPERATIONAL FACTORS   Application rates, placement, and incorporation of insecticides are among the mechanical factors that can influence CRW control (37, 45). Application at greater than recommended rates will not improve control. Insecticides used at reduced rates in field trials from three states often kept root damage below the economic threshold (66). Benefits of reduced or less-than-labeled application rates include less environmental contamination, fewer adverse effects on nontarget organisms, and reduced costs. However, the use of reduced rates is not recommended at present because of potential liability and many unanswered questions (6, 66).

In addition to banding, nonphytotoxic CRW soil insecticides can also be applied in the seed furrow. Some studies indicate that CRW control is slightly improved when insecticides are applied ahead of the firming wheels of the planter (151). Devices such as wind screens on insecticide banders help prevent drift, and spring tines or drag chains behind planter units incorporate granules into the soil (37). Although not as widely used, application of insecticides at cultivation time has been proposed as an alternative to applications at planting time, especially when corn is planted early in the season (121). To be effective, applications at cultivation time require favorable soil and weather conditions during a relatively short period in spring.

INSECTICIDE MANAGEMENT   The causes for inconsistent root protection by CRW soil insecticides are numerous, and the solution to the problem may be complex. Nevertheless, long-term insecticide management provides a way of coping with the problem. The subject has recently been reviewed, and several operational and technological strategies have been suggested (38, 40, 151). Operational strategies depend on techniques that are employed in insecticide application; technological strategies rely on the chemistry of the insecticide. Operational strategies include proper application, accurate calibration of application equipment, application at recommended rates, proper placement, incorporation at planting, timing of applications, and rotation of chemicals. Technological strategies include delaying biodegradation by using extenders and inhibitors, altering formulations (e.g. controlled-release formulations), and developing CRW insecticides that retain higher toxicities in the soil.

Alternation of CRW insecticide from one year to the next has been widely recommended, especially as a strategy to cope with enhanced biodegradation of insecticides and to avoid or delay the development of insect resistance (e.g. 40, 41). Enhanced biodegradation, however, can occur after a single application of an insecticide, which suggests that rotating insecticides cannot prevent the development of anti-insecticide microbial activity (80). Rotating CRW insecticides may slow the development of enhanced biodegradation but does not necessarily eliminate the phenomenon (49). Also, rotations do not have to be made strictly between classes of insecticides.

One potentially useful technological development could be the use of semiochemicals to improve the targeting in soil of insecticides to larvae. Corn seedling volatiles have been found to be important in orienting WCR larvae to roots. In a recent study, these volatiles also attracted WCR larvae to soil insecticides and significantly increased larval mortality (84).

Controlling Adults with Insecticides

A single, properly timed application of any one of several insecticides will generally reduce adult populations sufficiently for silks to regrow and pollination to take place (e.g. 149). Although it is common practice in seed production fields, such measures are needed only occasionally in commercial fields.

Reducing adult CRW populations with foliar insecticides to suppress oviposition has been evaluated in the corn belt since the 1940s. Most studies concluded that adult control could be used to minimize subsequent larval damage, but the level of control achieved was no better than that obtained with soil insecticides. Foliar applications require a higher level of management (precise timing is essential, applications must be applied when economic infestations of female beetles begin to oviposit, and treatments must remain active against immigrating gravid females and extended beetle emergence). The cost of two foliar applications would probably be higher than a soil insecticide applied at planting (6). In addition, rainfall or sprinkler irrigation can reduce the residual activity of the treatments (119). Finally, the foliar sprays often kill beneficial insects and predatory mites (173).

Since the early 1980s, a new management strategy has been under development (e.g. 106, 113, 123). This approach uses one or more semiochemicals (attractants, sex pheromones, and feeding stimulants/arrestants) to attract and encourage beetles to feed on a food source laced with small amounts of insecticide (bait). Because the insecticide is targeted, $< 5\%$ of the amount of insecticide applied to control adults by foliar applications is needed. Consequently, an insecticide in bait form should have fewer adverse effects on nontarget organisms and the environment. Dry corn cob grit, corn starch, or cereal meal baits impregnated with 0.1–0.3% cucurbitacins (nonvolatile CRW feeding stimulants/arrestants) and 0.1–0.3% insecticide (methomyl, carbaryl, carbofuran, or isofenphos) and broadcast over the corn canopy at 11–33 kg/ha significantly reduced populations of CRW adults (e.g. 113, 126).

Lance (106) evaluated responses of NCR and WCR to volatile attractants derived from plants (estragole for the WCR and eugenol for the NCR) and to the sex pheromone of both species in toxic baits containing cucurbitacin in vial traps. The addition of volatile attractants increased capture fourfold, a finding that suggests that these compounds could prove useful in improving the performance of baits or reducing their application rates. The sex pheromone was not effective at the rate studied.

More recently, semiochemical and insecticide-impregnated starch borate granules (controlled-release) have been evaluated (123). Broadcast applications of this formulation proved as effective as a conventional foliar-applied adulticide at reducing WCR populations in small corn plots, but the controlled-release formulation used 90% less insecticide and killed significantly fewer predaceous ladybird beetles. Development of a formulation with strong knockdown and good residual activity and a better understanding of beetle movement are critical to the success of this strategy.

IPM STRATEGIES

The main elements of CRW pest management include monitoring pest populations, use of economic thresholds, and integration of tactics described previously.

Monitoring and Trapping

Sampling CRW life stages has recently been reviewed (54, 143, 162). Because sampling of eggs and larvae in soil is extremely labor intensive and because satisfactory egg and larval thresholds have not been developed, most entomologists estimate adult population density to determine the potential for larval damage the following year. Visual counts are often used to provide these estimates (162); however, such counts can be imprecise because beetle activity is affected by environmental conditions, which in turn affect the number of individuals counted by the sampler. In addition, visual counts require frequent sampling of the entire cornfield during the ovipositional period, a practice that is labor intensive, especially when sex ratios must be determined. To maintain accuracy and efficiency, trained scouts should make the counts.

Sticky traps were developed to overcome some of the difficulties encountered with visual counts (82). These traps can be left in the field and retrieved after a number of days, thereby reducing variation associated with environmental conditions and beetle movement. Their use also reduces the need for highly trained scouts to perform visual plant counts. The Pherocon AM sticky trap is as effective as visual counts in predicting subsequent larval damage (although both methods accounted for only 25% of the variability in the damage) (82). Sticky traps are not without disadvantages. The sticky surfaces need to be renewed frequently, the traps are messy to handle, and determining the sex of the beetles caught on the traps is difficult. These factors have prevented sticky traps from becoming widely accepted by farmers and crop consultants.

A vial trap that contains cucurbitacin and employs no sticky material has been developed (145). The trap consists of a 60-ml perforated vial that holds an acetate strip coated with insecticide and powdered squash (*Cucurbita andreana* × *Cucurbita maxima* cross) containing cucurbitacins. Beetles enter the trap and ingest a lethal dose of the insecticide while feeding on the cucurbitacin bait (51). Traps could be left in the field for 12 days and still retain their effectiveness; furthermore, a high correlation (r = 0 .80) between trap catch and visual counts of adults on plants was found (145). Total number of beetles captured and the male-female ratio generally increased with increasing amount of bait. Consequently, the amount of bait needs to be standardized when traps are used for monitoring CRW populations (51). The trap has not been calibrated to relate catches of CRW adults to larval damage the following year.

A sex pheromone produced by WCR females has been identified (71). This pheromone is a powerful attractant for NCR and WCR males; however, visual plant counts for both species correlated poorly with numbers caught on sticky traps baited with the pheromone (122).

Recently, a large number of volatile compounds that are highly attractive to NCR and WCR adults have been identified; many have been found in the ancestral host plants of the Diabroticites—the Cucurbitaceae. These include 4-methoxycinnamaldehyde, 4-methoxycinnamonitrile, *beta*-ionone, indole, and estragole for the WCR; eugenol, isoeugenol, 2-methoxy-4-propylphenol, and cinnamyl alcohol for the NCR; and the TIC mixture (equal parts by weight of 1, 2, 4-trimethoxybenzene, indole, and *trans*-cinnamaldehyde) for both species (104, 105, 127, 128).

A simple and effective trap (110) has been developed by using an adhesive to coat 500-ml yellow plastic cups that can be inverted over a developing corn ear or a one-meter stake. The attractants are applied to dental wicks on top of the traps and are effective at dosages as low as 20–100 mg. These traps remain attractive for several weeks (110) and typically capture more females than males (2). Although the traps have collected substantial numbers of CRW when populations are virtually undetectable with the visual plant count method (110), poor correlations have generally been found between visual counts and trap catches at higher (> one per plant) population levels, especially while corn plants are in flower (2). Nevertheless, these plant volatiles may prove more useful than sex pheromones, which attract only males and may be repellant in high doses (106). Interestingly, except for *beta*-ionone, which has been found in corn silks, husks, tassels, and the whorl, and indole, which has also been found in the whorl, none of the other identified attractants have been found in corn tissues (19, 20, 55, 161). Unidentified volatile chemicals from corn silks, pollen, and corn stalks have been found to be attractive to NCR and WCR adults in laboratory bioassays (141, 157).

Economic Thresholds

ROOT PROTECTION   Root damage ratings are used a posteriori to assess larval injury by CRW (87). It has been estimated that a mean root rating > 2.5 (on a 1–6 scale) would result in economic loss (164). Other workers (56, 147) showed that root ratings were not consistent predictors of yield. Numerous factors influence the extent of damage and its impact on yield. Number and species of larvae, size of the root system, ability of the corn hybrid to regenerate roots, availability of soil moisture and nutrients, plant population, and weather are all involved in the amount of damage that may occur and the impact this damage may have on yield. If other stress levels are minimal, severe root damage may not result in significant yield losses, particularly if high winds do not cause lodging. Use of artificial infestations of WCR eggs has shown that rainfall is important in assessing the economic threshold of this pest; with ample soil moisture, corn plants tolerated higher populations of larvae without losing yield than under dry conditions (32). Regrowth of roots after injury is important in preventing yield losses and is significantly related to grain yield. In turn, regrowth is positively affected by favorable soil moisture conditions, adequate nitrogen, and moderate plant densities (147).

Control decisions are generally based on attempts to predict larval damage from estimates of the adult population present the previous summer. Because larval damage was poorly predicted by adult counts, an Iowa study concluded that applying soil insecticides prophylactically every year to continuous cornfields was more cost effective than sampling for adults (56). In large part, this finding probably results from our inadequate knowledge of the population dynamics of adult NCR and WCR, their oviposition behavior and egg survival, subsequent larval survival and feeding, and the response of plants to larval feeding (83, 133).

Entomologists have generally recommended that when the average combined density of NCR and WCR beetles at peak occurrence exceeds one per plant, a nonhost crop should be planted the following year or a soil insecticide should be used if corn is to be planted consecutively in the same field (148). As our knowledge base has increased, some recommendations have reduced this proposed threshold to 0.5 beetle per plant in fields with certain cropping sequences or plant populations (e.g. 149).

Considering the bionomics of NCR and WCR, thresholds will necessarily differ for each species. Dramatic differences in fecundity between WCR and NCR have been observed, with WCR females having a mean fecundity of > 1000 eggs and NCR females a mean fecundity around 274 eggs (131). These differences in reproductive potential may have played a critical role in the observed displacement of NCR by WCR in many parts of the country where their ranges have overlapped, and population models may need to account for species differences if they are to perform as well for mixed populations of NCR and WCR as for either species alone (131). Recent studies found that WCR survival was greater than that of the NCR and that for a given density of eggs, WCR inflicted more root damage than NCR (67, 70). In a study using artificial infestations of eggs, NCR inflicted less root damage than WCR for similar initial egg populations (53). However, reduced root damage by NCR could result from more injurious feeding by WCR because no differences were found in egg viability or for adult emergence between species (53). These observations suggest that the threshold for a population of NCR adults would probably be higher than that for a population of WCR adults.

Thresholds may have to vary depending on the prevalence of extended diapause or the cropping history of a particular field. In fields where corn is rotated with a nonhost crop and the extended diapause trait is present, four to five NCR beetles per plant may be required to produce infestations of larvae at an economically significant level when the same field is planted with corn two years later (K. R. Ostlie, personal communication; J. J. Tollefson, unpublished report). Godfrey & Turpin (62) showed that more female than male WCR beetles immigrated to first-year cornfields during the ovipositional period and that the higher proportion of female beetles found in first-year fields versus continuous cornfields resulted in an increased larval damage potential that warranted the use of an economic threshold that was 50% less than that of continuous cornfields. As fields flower progressively later in the season, populations of NCR in these fields become increasingly dominated by mature ovipositing females (132). Oviposition was completed sooner in plots that flowered earlier and, on a per-beetle basis, total egg densities were higher in plots that flowered later. Because oviposition per beetle is less in earlier-planted, earlier-flowering fields, a higher density of adult beetles could probably be tolerated by plants in those fields (134).

Use of adult sampling as a decision tool for establishing thresholds for larval control could probably be improved by considering plant density. Corn plant populations usually range from approximately 35,000 to 80,000 plants per ha, depending on the area and type of production practices. When beetle counts were adjusted for plant density, better estimates of the beetle population resulted that were significantly correlated with the number of eggs laid (171).

SILK CLIPPING   Severe silk clipping by adult NCR and WCR adults may reduce yield because of poor pollination, but such reductions seldom occur unless beetles are especially numerous at silking. Insecticide treatments to control adults are almost never applied after pollination has been completed and silks are brown. In demonstration plots, yield losses due to NCR feeding on silks occurred with as few as five beetles per plant (103), but controlled studies with WCR in irrigated Colorado corn indicated that densities of up to 20 beetles per ear did not significantly reduce yield (22). The number of beetles required to prevent pollination depends largely on the rate of silk growth. Stressful conditions (high heat and low soil moisture) can retard silk growth, and fewer beetles per plant may be required to interfere with pollination (81).

Current IPM Options

During the past dozen years, the success of soil insecticides in controlling CRW has varied. At some locations, these insecticides have provided effective control; at others, their performance has been poor. No short-term solution to this problem is in sight (149). Long-term management options include the following practices: 1. Alternate corn with another crop whenever possible, particularly in fields with a high probability of CRW damage. Longer-term rotations may be necessary where extended diapause has been a significant problem. 2. Control volunteer corn in rotational crops. When enough corn plants are present (>12,000 per ha), significant oviposition can occur in the rotational crop (149). 3. If corn must be grown after corn and beetles average 0.75 or more per plant in continuous cornfields or 0.5 or more per plant in first-year cornfields during the oviposition period, apply a soil insecticide at planting. Because plant population affects the number of beetles per ha, this factor is often considered in the decision-making process (149). Alternating CRW soil insecticides may be advantageous. 4. Consider applying a soil insecticide at the time of cultivation instead of applying it at planting time, especially if planting occurs early. Scouting for larvae may indicate that no treatment is needed. 5. Scouting for CRW beetles during the period of oviposition helps to determine the potential for CRW larval damage in a subsequent crop of corn. Fields with especially high populations should be planted with another crop; fields with lower populations can be planted with corn again if necessary. Any practice that reduces the CRW densities without increasing their annual exposure to soil insecticides should extend the useful life of these chemicals (116). 6. Consider a management program to suppress egg laying only if beetle populations and female reproductive condition can be monitored regularly and the application of foliar insecticides is properly timed. 7. Apply an insecticide to prevent silk clipping only if CRW beetles seriously compromise the pollination process.

CONCLUDING REMARKS

The history of soil insecticides in the management of CRW is marked by numerous cases of control failures. As a result, management strategies today stress an integrated approach that includes crop rotation, scouting fields to determine the need for control measures for silk clipping and root damage the following year, the use of insecticides only when necessary, crop management, and the consideration of environmental, biological, chemical, and physical factors that contribute to CRW control. NCR and WCR are distinct species. Effective management of CRW will depend on a better understanding of the significant differences between the species.

Improvement in our ability to predict CRW damage depends on our understanding of the factors influencing adult CRW population dynamics and oviposition. The importance of mortality factors and movement between fields needs to be determined, and factors that influence fecundity within field populations need to be quantified.

We know little about CRW host-finding behavior and host-mediated chemical ecology. The preference of both species for corn tissues, the change in behavior by the WCR from leaf feeding to pollen and silk feeding, and the ability of NCR to return to cornfields after feeding on the pollen of other plants suggest the existence of feeding and oviposition attractants in corn. However, except for knowledge that *beta*-ionone, indole, and certain unidentified volatile chemicals from corn tissues are attractive to NCR and WCR, it is still not known why CRW oviposit almost exclusively in cornfields. The extent to which semiochemicals influence host-plant selection and oviposition is an area for exploration that has exciting possibilities for the management of these pests. A more subtle knowledge of the relationship between feeding and oviposition could increase the accuracy of infestation predictions. Studies on the resistance of host plants could be expanded to include manipulation of volatiles produced by corn plants that in turn may influence CRW oviposition site. Studies are already underway to examine the potential of using semiochemicals to lure CRW to toxic baits or from untreated portions of cornfields to treated areas.

ACKNOWLEDGMENTS

We thank C. E. Eastman, C. R. Ellis, A. S. Felsot, J. R. Fisher, M. E. Gray, A. S. Hodgins, M. Kogan, J. L. Krysan, R. L. Lampman, D. R. Lance, J. V. Maddox, R. L. Metcalf, S. E. Naranjo, W. G. Ruesink, K. L. Steffey, G. R. Sutter, and J. J. Tollefson for their thoughtful comments on an earlier draft of the manuscript. Special appreciation is extended to A. S. Felsot for detailed discussions.

Literature Cited

1. Ahmad, N., Walgenbach, D. D., Sutter, G. R. 1979. Degradation rates of technical carbofuran and a granular formulation in four soils with known insecticide use history. *Bull. Environ. Contam. Toxicol.* 23:572–74
2. Andersen, J. F., Metcalf, R. L. 1986. Identification of a volatile attractant for *Diabrotica* and *Acalymma* spp. from blossoms of *Cucurbita maxima* Duchesne. *J. Chem. Ecol.* 12:687–99
3. Ball, H. J. 1981. Larval and adult control recommendations and insecticide resistance data for corn rootworms in Nebraska (1948–1981). *Rep. 3 (Rev.) Agric. Exp. Stn. Univ. Nebr. Lincoln.* 16 pp.
4. Ball, H. J., Weekman, G. T. 1962. Insecticide resistance in the adult western corn rootworm in Nebraska. *J. Econ. Entomol.* 55:439–41
5. Ballard, J. B., Mayo, Z B. 1979. Predatory potential of selected ant species on eggs of western corn rootworm. *Environ. Entomol.* 8:575–76
6. Bergman, M. 1987. Corn rootworm control: do we have any new solutions? *Proc. Ill. Agric. Pestic. Conf. '87*, pp. 41–49. Urbana-Champaign, IL: Univ. Ill.
7. Bergman, M. K., Turpin, F. T. 1984. Impact of corn planting date on the population dynamics of corn rootworms (Coleoptera: Chrysomelidae). *Environ. Entomol.* 13:898–901
8. Best, R. L., Beegle, C. C. 1977. Food preferences of five species of carabids commonly found in Iowa cornfields. *Environ. Entomol.* 6:9–12

9. Bigger, J. H. 1932. Short rotation fails to prevent attack of *Diabrotica longicornis* Say. *J. Econ. Entomol.* 25:196–99
10. Bollinger, E. K., Caslick, J. W. 1985. Northern corn rootworm beetle densities near a red-winged blackbird roost. *Can. J. Zool.* 63:502–5
11. Branson, T. F. 1971. Resistance in the grass tribe Maydeae to larvae of the western corn rootworm. *Ann. Entomol. Soc. Am.* 64:861–63
12. Branson, T. F. 1989. Survival of starved neonate larvae of *Diabrotica virgifera virgifera* LeConte (Coleoptera: Chrysomelidae). *J. Kans. Entomol. Soc.* 62: 521–23
13. Branson, T. F., Krysan, J. L. 1981. Feeding and oviposition behavior and life cycle strategies of *Diabrotica:* an evolutionary view with implications for pest management. *Environ. Entomol.* 10:826–31
14. Branson, T. F., Reyes, R. J. 1983. The association of *Diabrotica* spp. with *Zea diploperennis. J. Kans. Entomol. Soc.* 56:97–99
15. Branson, T. F., Reyes, R. J., Valdes M., H. 1982. Field biology of Mexican corn rootworm, *Diabrotica virgifera zeae* (Coleoptera: Chrysomelidae), in central Mexico. *Environ. Entomol.* 11: 1078–83
16. Branson, T. F., Sutter, G. R., Fisher, J. R. 1982. Comparison of a tolerant and a susceptible maize inbred under artificial infestations of *Diabrotica virgifera virgifera:* yield and adult emergence. *Environ. Entomol.* 11:371–72

17. Branson, T. F., Welch, V. A., Sutter, G. R., Fisher, J. R. 1983. Resistance to larvae of *Diabrotica virgifera virgifera* in three experimental maize hybrids. *Environ. Entomol.* 12:1509–12
18. Brust, G. E., Stinner, B. R., McCartney, D. A. 1986. Predator activity and predation in corn agroecosystems. *Environ. Entomol.* 15:1017–21
19. Buttery, R. G., Ling, L. C., Chan, B. G. 1978. Volatiles of corn kernels and husks: possible corn ear worm attractants. *J. Agric. Food Chem.* 26:866–69
20. Buttery, R. G., Ling, L. C., Teranishi, R. 1980. Volatiles of corn tassels: possible corn ear worm attractants. *J. Agric. Food Chem.* 28:771–74
21. Calkins, C. O., Kirk, V. M., Matteson, J. W., Howe, W. L. 1970. Early cutting of corn as a method of reducing populations of corn rootworms. *J. Econ. Entomol.* 63:976–78
22. Capinera, J. L., Epsky, N. D., Thompson, D. C. 1986. Effect of adult western corn rootworm (Coleoptera: Chrysomelidae) ear feeding on irrigated field corn in Colorado. *J. Econ. Entomol.* 79:1609–12
23. Chapman, R. A., Cole, C. M. 1982. Observations on the influence of water and soil pH on the persistence of insecticides. *J. Environ. Sci. Health Part B* 17:487–504
24. Chapman, R. A., Harris, C. R. 1980. Insecticidal activity and persistence of terbufos, terbufos sulfoxide and terbufos sulfone in soil. *J. Econ. Entomol.* 73:536–43
25. Chapman, R. A., Harris, C. R., Harris, C. 1986. Observations on the effect of soil type, treatment intensity, insecticide formulation, temperature and moisture on the adaptation and subsequent activity of biological agents associated with carbofuran degradation in soil. *J. Environ. Sci. Health. Part B* 21:125–41
26. Chapman, R. A., Tu, C. M., Harris, C. R., Dubois, D. 1982. Biochemical and chemical transformations of terbufos, terbufos sulfoxide, and terbufos sulfone in natural and sterile, mineral and organic soil. *J. Econ. Entomol.* 75:955–60
27. Chiang, H. C. 1965. Research on corn rootworms. *Minn. Farm Home Sci.* 23:10–13
28. Chiang, H. C. 1965. Survival of northern corn rootworm eggs through one and two winters. *J. Econ. Entomol.* 58:470–72
29. Chiang, H. C. 1970. Effects of manure applications and mite predation on corn rootworm populations in Minnesota. *J. Econ. Entomol.* 63:934–36
30. Chiang, H. C. 1973. Bionomics of the northern and western corn rootworms. *Annu. Rev. Entomol.* 18:47–72
31. Chiang, H. C., French, L. K. 1980. Host tolerance, a short-term pest management tool—maize and corn rootworm as a model. *Plant Prot. Bull. FAO* 28:137–38
32. Chiang, H. C., French, L. K., Rasmussen, D. E. 1980. Quantitative relationship between western corn rootworm population and corn yield. *J. Econ. Entomol.* 73:665–66
33. Chio, H., Chang, C. S., Metcalf, R. L., Shaw, J. 1978. Susceptibility of four species of *Diabrotica* to insecticides. *J. Econ. Entomol.* 71:389–93
34. Chio, H., Metcalf, R. L. 1979. Detoxification mechanisms for aldrin, carbofuran, fonofos, phorate, and terbufos in four species of Diabroticites. *J. Econ. Entomol.* 72:732–38
35. Cinereski, J. E., Chiang, H. C. 1968. The pattern of movements of adults of the northern corn rootworm inside and outside of corn fields. *J. Econ. Entomol.* 61:1531–36
36. Dzantor, E. K., Felsot, A. S. 1989. Effects of conditioning, cross-conditioning, and microbial growth in development of enhanced biodegradation of insecticides in soil. *J. Environ. Sci. Health Part B* 24:569–97
37. Ellis, C. R., Beattie, B. 1984. Effect of banding and incorporation on the efficacy of granular inseticides for control of corn rootworms. (Coleoptera: Chrysomelidae) in grain corn. *Proc. Entomol. Soc. Ont.* 115:31–36
38. Felsot, A. 1985. Factors affecting the bioactivity of soil insecticides. *Proc. 37th Ill. Custom Spray Oper. Train. Sch.*, pp. 134–38. Urbana-Champaign, IL: Univ. Ill.
39. Felsot, A. 1987. Fate and interactions of pesticides in conservation tillage systems. In *Arthropods in Conservation Tillage Systems*, ed. G. J. House, B. R. Stinner, pp. 35–43. College Park, MD: Misc. Publ. 65, Entomol. Soc. Am. 52 pp.
40. Felsot, A. S. 1989. Enhanced biodegradation of insecticides in soil: implications for agroecosystems. *Annu. Rev. Entomol.* 34:453–76
41. Felsot, A. 1990. An assessment of insecticide rotations for long-term management of corn rootworm feeding damage. *Proc. Ill. Agric. Pestic. Conf. '90*, pp. 71–80. Urbana-Champaign, IL: Univ. Ill.
42. Felsot, A. S., Baughman, T. A., Kuhlman, D. 1988. Monitoring corn root-

worm *(Diabrotica* spp.) larvae for resistance to soil insecticides. *Abstr. 43rd Annu. Meet. North Cent. Branch Entomol. Soc. Am. Denver Colo.* Abstr. 210

43. Felsot, A. S., Bruce, W. N., Steffey, K. L. 1987. Degradation of terbufos (Counter) soil insecticide in corn fields under conservation tillage practices. *Bull. Environ. Contam. Toxicol.* 38:369–76

44. Felsot, A., Dahm, P.A. 1979. Sorption of organophosphorus and carbamate insecticides by soil. *J. Agric. Food Chem.* 27:557–63

45. Felsot, A. S., Lew, A. 1989. Factors affecting bioactivity of soil insecticides: relationships among uptake, desorption, and toxicity of carbofuran and terbufos. *J. Econ. Entomol.* 82:389–95

46. Felsot, A., Mitchell, J. K., Kenimer, A. L. 1990. Assessment of management practices for reducing pesticide runoff from sloping cropland in Illinois. *J. Environ. Qual.* 19:539–45

47. Felsot, A. S., Steffey, K. L. 1986. Redistribution of carbofuran and terbufos in soil under different tillage conditions. *Abstr. 41st Annu. Meet. North Cent. Branch Entomol. Soc. Am. Minneapolis Minn.* Abstr. 259

48. Felsot, A. S., Steffey, K. L., Levine, E., Wilson, J. G. 1985. Carbofuran persistence in soil and adult corn rootworm (Coleoptera: Chrysomelidae) susceptibility: relationship to the control of damage by larvae. *J. Econ. Entomol.* 78:45–52

49. Felsot, A. S., Tollefson, J. J. 1990. Evaluation of some methods for coping with enhanced biodegradation of soil insecticides. In *Enhanced Biodegradation of Pesticides in the Environment,* ed. K. D. Racke, J. R. Coates, pp. 192–213. Washington, DC: Am. Chem. Soc. Symp. Ser. 426. 302 pp.

50. Felsot, A. S., Wilson, J. G., Kuhlman, D. E., Steffey, K. L. 1982. Rapid dissipation of carbofuran as a limiting factor in corn rootworm (Coleoptera: Chrysomelidae) control in fields with histories of continuous carbofuran use. *J. Econ. Entomol.* 75:1098–103

51. Fielding, D. J., Ruesink, W. G. 1985. Varying amounts of bait influences numbers of western and northern corn rootworms (Coleoptera: Chrysomelidae) caught in cucurbitacin traps. *J. Econ. Entomol.* 78:1138–44

52. Fischer, D. C. 1983. Celatoria diabroticae *Shimer and* Celatoria setosa *Coquillett: tachinid parasitoids of the diabroticite Coleoptera.* PhD thesis. Urbana-Champaign: Univ. Ill. 120 pp.

53. Fisher, J. R. 1985. Comparison of controlled infestations of *Diabrotica virgifera virgifera* and *Diabrotica barberi* (Coleoptera: Chrysomelidae) on corn. *J. Econ. Entomol.* 78:1406–408

54. Fisher, J. R., Bergman, M. K. 1986. Field sampling of larvae and pupae. See Ref. 99a, pp. 101–21

55. Flath, R. A., Forrey, R. R., John, J. O., Chan, B. G. 1978. Volatile components of corn silk *(Zea mays* L.): possible *Heliothis zea* (Boddie) attractants. *J. Agric. Food Chem.* 26: 1290–93

56. Foster, R. E., Tollefson, J. J., Nyrop, J. P., Hein, G. L. 1986. Value of adult corn rootworm (Coleoptera: Chrysomelidae) population estimates in pest management decision making. *J. Econ. Entomol.*79:303–10

57. Gaugler, R. 1981. Biological control potential of neoaplectanid nematodes. *J. Nematol.* 13:241–49

58. Getzin, L. W. 1973. Persistence and degradation of carbofuran in soil. *Environ. Entomol.* 2:461–67

59. Getzin, L. W. 1981. Dissipation of chlorpyrifos from dry soil surfaces. *J. Econ. Entomol.* 74:707–13

60. Getzin, L. W., Shanks, C. H. Jr. 1970. Persistence, degradation, and bioactivity of phorate and its oxidative analogues in soil. *J. Econ. Entomol.* 63:52–58

61. Gilbertson, R. L., Brown, W. M. Jr., Ruppel, E. G., Capinera, J. L. 1986. Association of corn stalk rot *Fusarium* spp. and western corn rootworm beetles in Colorado. *Phytopathology* 76:1309–14

62. Godfrey, L. D., Turpin, F. T. 1983. Comparison of western corn rootworm (Coleoptera: Chrysomelidae) adult populations and economic thresholds in first-year and continuous corn fields. *J. Econ. Entomol.* 76:1028–32

63. Gorder, G. W., Dahm, P. A., Tollefson, J. J. 1982. Carbofuran persistence in cornfield soils. *J. Econ. Entomol.* 75:637–42

64. Gray, M. E. 1989. Field crop pest management in sustainable systems: another look at the impacts of cover crops, tillage, and rotations. *Proc. Alt. Pest Manage. Workshop.* Peoria, Ill: Univ. Ill. at Urbana-Champaign

65. Gray, M. E., Felsot, A. S., Steffey, K. L., Levine, E. 1989. The effect of insecticides on adult emergence from different tillage systems: are we managing corn rootworm populations? *Abstr. 44th Annu. Meet. North Cent. Branch Entomol. Soc. Am. Indianapolis Ind.* Abstr. 50

66. Gray, M., Steffey, K., Kinney, K.

1990. Corn rootworm soil insecticides: are the current application rates necessary? *Proc. Ill. Agric. Pestic. Conf. '89,* pp. 35–52. Urbana-Champaign, Ill: Univ. Ill.

67. Gray, M. E., Tollefson, J. J. 1987. Influence of tillage and western and northern corn rootworm (Coleoptera: Chrysomelidae) egg populations on larval populations and root damage. *J. Econ. Entomol.* 80:911–15

68. Gray, M. E., Tollefson, J. J. 1988. Emergence of the western and northern corn rootworms (Coleoptera: Chrysomelidae) from four tillage systems. *J. Econ. Entomol.* 81:1398–1403

69. Gray, M. E., Tollefson, J. J. 1988. Influence of tillage systems on egg populations of western and northern corn rootworms (Coleoptera: Chrysomelidae). *J. Kans. Entomol. Soc.* 61:186–94

70. Gray, M. E., Tollefson, J. J. 1988. Survival of the western and northern corn rootworms (Coleoptera: Chrysomelidae) in different tillage systems throughout the growing season of corn. *J. Econ. Entomol.* 81:178–83

71. Guss, P. L., Tumlinson, J. H., Sonnet, P. E., Proveaux, A. T. 1982. Identification of a female-produced sex pheromone of the western corn rootworm. *J. Chem. Ecol.* 8:545–56

72. Gustin, R. D. 1981. Soil temperature environment of overwintering western corn rootworm eggs. *Environ. Entomol.* 10:483–87

73. Gustin, R. D. 1983. *Diabrotica longicornis barberi* (Coleoptera: Chrysomelidae): cold hardiness of eggs. *Environ. Entomol.* 12:633–34

74. Gustin, R. D. 1984. Effect of crop cover on oviposition of the northern corn rootworm, *Diabrotica longicornis barberi* Smith and Lawrence. *J. Kans. Entomol. Soc.* 57:515–16

75. Gustin, R. D. 1986. Effect of intermittent low temperatures on hatch of western corn rootworm eggs (Coleoptera: Chrysomelidae). *J. Kans. Entomol. Soc.* 59:569–70

76. Harris, C. R. 1972. Factors influencing the effectiveness of soil insecticides. *Annu. Rev. Entomol.* 17:177–98

77. Harris, C. R. 1977. Biological activity of chlorpyrifos, chlorpyrifos-methyl, phorate, and Counter in soil. *Can. Entomol.* 109:1115–20

78. Harris, C. R. 1977. Insecticide resistance in soil insects attacking crops. In *Pesticide Management and Insecticide Resistance,* ed. D. L. Watson, A. W. A. Brown, pp. 321–51. New York: Academic. 638 pp.

79. Harris, C. R., Bowman, B. T. 1981. The relationship of insecticide solubility in water to toxicity in soil. *J. Econ. Entomol.* 74:210–12

80. Harris, C. R., Chapman, R. A., Tolman, J. H., Moy, P., Henning, K., Harris, C. 1988. A comparison of the persistence in a clay loam of single and repeated annual applications of seven granular insecticides used for corn rootworm control. *J. Environ. Sci. Health Part B* 23:1–32

81. Hein, G. L., Foster, D. E. 1986. Corn rootworm management. *Pm-670 (Rev.) Coop. Ext. Serv. Iowa State Univ.* Ames.

82. Hein, G. L., Tollefson, J. J. 1985. Use of the Pherocon AM trap as a scouting tool for predicting damage by corn rootworm (Coleoptera: Chrysomelidae) larvae. *J. Econ. Entomol.* 78:200–3

83. Hein, G. L., Tollefson, J. J., Foster, R. E. 1988. Adult northern and western corn rootworm (Coleoptera: Chrysomelidae) population dynamics and oviposition. *J. Kans. Entomol. Soc.* 61:214–23

84. Hibbard, B. E., Bjostad, L. B. 1989. Corn semiochemicals and their effects on insecticide efficacy and insecticide repellency toward western corn rootworm larvae. (Coleoptera: Chrysomelidae). *J. Econ. Entomol.* 82:773–81

85. Hill, R. E., Hixson, E., Muma, M. H. 1948. Corn rootworm control tests with benzene hexachloride, DDT, nitrogen fertilizers and crop rotations. *J. Econ. Entomol.* 41:392–401

86. Hill, R. E., Mayo, Z B. 1974. Trap-crop to control rootworms. *J. Econ. Entomol.* 67:748–50

87. Hills, T. M., Peters, D. C. 1971. A method of evaluating post-plant insecticide treatments for control of western corn rootworm larvae. *J. Econ. Entomol.* 64:764–65

88. Jackson, J. J. 1986. Rearing and handling of *Diabrotica virgifera* and *Diabrotica undecimpunctata howardi.* See Ref. 99a, pp. 25–47

89. Jackson, J. J., Brooks, M. A. 1989. Susceptibility and immune response of western corn rootworm larvae (Coleoptera: Chrysomelidae) to the entomogenous nematode, *Steinernema feltiae* (Rhabditida: Steinernematidae). *J. Econ. Entomol.* 82:1073–77

90. Jensen, S. G. 1985. Laboratory transmission of maize chlorotic mottle virus by three species of corn rootworms. *Plant Dis.* 69:864–68

91. Johnson, T. B., Turpin, F. T. 1985. Northern and western corn rootworm

(Coleoptera: Chrysomelidae) oviposition in corn as influenced by foxtail populations and tillage systems. *J. Econ. Entomol.* 78:57–60

92. Kahler, A. L., Olness, A. E., Sutter, G. R., Dybing, C. D., Devine, O. J. 1985. Root damage by western corn rootworm and nutrient content in maize. *Agron. J.* 77:769–74

93. Kinney, K. K., Maddox, J. V., Dazey, D. M., McKinnis, M. W. 1989. Field evaluations of *Beauveria bassiana* for control of corn rootworm larvae: root protection and yield. In *Ill. Insectic. Eval.: Forage, Field, Veg. Crops,* pp. 28–32. Urbana-Champaign, Ill: Univ. Ill. 86 pp.

94. Kirk, V. M. 1981. Corn rootworm: population reduction associated with the ant, *Lasius neoniger. Environ. Entomol.* 10: 966–67

95. Kirk, V. M. 1982. Carabids: minimal role in pest management of corn rootworms. *Environ. Entomol.* 11:5–8

96. Krysan, J. L. 1986. Introduction: biology, distribution, and identification of pest *Diabrotica.* See Ref. 99a, pp. 1–23

97. Krysan, J. L., Branson, T. F. 1983. Biology, ecology, and distribution of *Diabrotica.* In *Proc. Int. Maize Virus Dis. Colloq. Workshop,* ed. D. T. Gordon, J. K. Knoke, L. R. Nault, R. M. Ritter, pp. 144–50. Wooster, OH: Ohio Agric. Res. Dev. Centr. 266 pp.

98. Krysan, J. L., Foster, D. E., Branson, T. F., Ostlie, K. R., Cranshaw, W. S. 1986. Two years before the hatch: rootworms adapt to crop rotation. *Bull Entomol. Soc. Am.* 32:250–53

99. Krysan, J. L., Jackson, J. J., Lew, A. C. 1984. Field termination of egg diapause in *Diabrotica* with new evidence of extended diapause in *D. barberi* (Coleoptera: Chrysomelidae). *Environ. Entomol.* 13:1237–40

99a. Krysan, J. L., Miller, T. A., eds. 1986. *Methods for the Study of Pest Diabrotica.* New York: Springer-Verlag. 260 pp.

100. Krysan, J. L., Smith, R. F., Branson, T. F., Guss, P L. 1980. A new subspecies of *Diabrotica virgifera* (Coleoptera: Chrysomelidae): description, distribution, and sexual compatibility. *Ann. Entomol. Soc. Am.* 73:123–30

101. Krysan, J. L., Sutter, G. R. 1986. Aldrin susceptibility as an indicator of geographic variability in the northern corn rootworm, *Diabrotica barberi* (Coleoptera: Chrysomelidae). *Environ. Entomol.* 15:427–30

102. Kuhlman, D. E. 1974. Results of 1973 corn rootworm control in demonstration plots. *Proc. 26th Ill. Custom Spray Oper. Train. Sch.,* pp. 56–59. Urbana-Champaign, IL: Univ. Ill.

103. Kuhlman, D. E. 1982. Silk feeding insects and economic thresholds. *Proc. 8th Annu. Ill. Crop Prot. Worshop,* pp. 83–88. Urbana-Champaign, IL: Univ. Ill.

104. Ladd, T. L. Jr. 1984. Eugenol-related attractants for the northern corn rootworm (Coleoptera: Chrysomelidae). *J. Econ. Entomol.* 77:339–41

105. Lampman, R. L., Metcalf, R. L. 1988. The comparative response of *Diabrotica* species (Coleoptera: Chrysomelidae) to volatile attractants. *Environ. Entomol.* 17:644–48

106. Lance, D. R. 1988. Potential of 8-methyl-2-decyl propanoate and plant-derived volatiles for attracting corn rootworm beetles (Coleoptera: Chrysomelidae) to toxic bait. *J. Econ. Entomol.* 81:1359–62

107. Lance, D. R., Elliott, N. C., Hein, G. L. 1989. Flight activity of *Diabrotica* spp. at the borders of cornfields and its relation to ovarian stage in *D. barberi. Entomol. Exp. Applic.* 50:61–67

108. Lawson, E. C. 1986. *Influence of tillage and depth in the soil on soil temperature and survival of overwintering western corn rootworm eggs.* MS thesis. Ames: Iowa State Univ. 63 pp.

109. Levine, E., Kuhlman, D., Steffey, K., Oloumi-Sadeghi, H. 1988. New developments regarding extended diapause in northern corn rootworms: research and survey results. *Proc. Ill. Agric. Pestic. Conf. '88,* pp. 145–53. Urbana-Champaign, IL: Univ. Ill.

110. Levine, E., Metcalf, R. L. 1988. Sticky attractant traps for monitoring corn rootworm beetles. *Ill. Nat. Hist. Surv. Rep.* 279:1–2

111. Levine, E., Oloumi-Sadeghi, H. 1988. Larval damage to corn following soybeans by the western corn rootworm, *Diabrotica virgifera virgifera,* in east central Illinois. *Abstr. 43rd Annu. Meet. North Cen. Branch Entomol. Soc. Am.* Denver Colo. Abstr. 98

112. Levine, E., Oloumi-Sadeghi, H. 1989. *Egg-hatching patterns in northern and western corn rootworms.* Presented at Natl. Meet. Entomol. Soc. Am., 36th, San Antonio, Tex., Paper 753

113. Levine, E., Oloumi-Sadeghi, H., Metcalf, R. L., Lampman, R. 1988. Dry cucurbitacin-containing baits for controlling adult western corn rootworms, *Diabrotica virgifera virgifera* (Coleoptera: Chrysomelidae), in field corn. *Cucurbit Genet. Coop.* 11:79–82

114. Lew, A. C., Sutter, G. R. 1985. Toxic-

ity of insecticides to northern corn rootworm (Coleoptera: Chrysomelidae) larvae. *J. Kans. Entomol. Soc.* 58:547–49

115. Lilly, J. H. 1956. Soil insects and their control. *Annu. Rev. Entomol.* 1:203–22

116. Luckmann, W. H. 1978. Insect control in corn—practices and prospects. In *Pest Control Strategies,* ed. E. H. Smith, D. Pimentel, pp. 137–55. New York: Academic. 334 pp.

117. Maddox, J., Kinney, K. 1989. Biological control agent of the corn rootworm. *Ill. Nat. Hist. Surv. Rep.* 287:3–4

118. Mayo, Z B. 1980. Influence of planting dates on the efficacy of soil insecticides applied to control larvae of the western and northern corn rootworm. *J. Econ. Entomol.* 73:211–12

119. Mayo, Z B. 1984. Influences of rainfall and sprinkler irrigation on the residual activity of insecticides applied to corn for control of adult western corn rootworm (Coleoptera: Chrysomelidae). *J. Econ. Entomol.* 77:190–93

120. Mayo, Z B Jr. 1986. Field evaluation of insecticides for control of larvae of corn rootworms. See Ref. 99a, pp. 183–203

121. Mayo, Z B., Peters, L. L. 1978. Planting vs. cultivation time applications of granular soil insecticides to control larvae of corn rootworms in Nebraska. *J. Econ. Entomol.* 71:801–3

122. McAuslane, H. J., Ellis, C. R., Teal, P. E. A. 1986. Chemical attractants for monitoring for adult northern and western corn rootworms. (Coleoptera: Chrysomelidae) in Ontario. *Proc. Entomol. Soc. Ont.* 117:49–57

123. Meinke, L. 1990. Potential of starch encapsulated semiochemical/insecticide formulations for corn rootworm control. *Proc. Ill. Agric. Pestic. Conf. '90,* pp. 107–11. Urbana-Champaign, IL: Univ. Ill.

124. Menzie, C. M. 1972. Fate of pesticides in the environment. *Annu. Rev. Entomol.* 17:199–222

125. Metcalf, R. L. 1986. Foreword. See Ref. 99a, pp. vii–xv

126. Metcalf, R. L., Ferguson, J. E., Lampman, R., Andersen, J. F. 1987. Dry cucurbitacin-containing baits for controlling diabroticite beetles (Coleoptera: Chrysomelidae). *J. Econ. Entomol.* 80: 870–75

127. Metcalf, R. L., Lampman, R. L. 1989. Cinnamyl alcohol and analogs as attractants for corn rootworms. (Coleoptera; Chrysomelidae). *J. Econ. Entomol.* 82: 1620–25

128. Metcalf, R. L., Lampman, R. L. 1989. Estragole analogues as attractants for corn rootworms (Coleoptera: Chrys-

omelidae). *J. Econ. Entomol.* 82:123–29

129. Munson, J. D., Helms, T. J. 1970. Field evaluation of a nematode (DD-136) for control of corn rootworm larvae. *Proc. North Cent. Branch Entomol. Soc. Am.* 25:97–99

130. Musick, G. J., Chiang, H. C., Luckmann, W. H., Mayo, Z B, Turpin, F. T. 1980. Impact of planting dates of field corn on beetle emergence and damage by the western and the northern corn rootworms in the corn belt. *Ann. Entomol. Soc. Am.* 73:207–15

131. Naranjo, S. E., Sawyer, A. J. 1987. Reproductive biology and survival of *Diabrotica barberi* (Coleoptera: Chrysomelidae): effect of temperature, food, and seasonal time of emergence. *Ann. Entomol. Soc. Am.* 80:841–48

132. Naranjo, S. E., Sawyer, A. J. 1988. Impact of host plant phenology on the population dynamics and oviposition of northern corn rootworms, *Diabrotica barberi* (Coleoptera: Chrysomelidae), in field corn. *Environ. Entomol.* 17:508–21

133. Naranjo, S. E., Sawyer, A. J. 1989. A simulation model of northern corn rootworm, *Diabrotica barberi* Smith and Lawrence (Coleoptera: Chrysomelidae), population dynamics and oviposition: significance of host plant phenology. *Can. Entomol.* 121:169–91

134. Naranjo, S. E., Sawyer, A. J. 1989. Analysis of a simulation model of northern corn rootworm, *Diabrotica barberi* Smith and Lawrence (Coleoptera: Chrysomelidae), dynamics in field corn, with implications for population management. *Can. Entomol.* 121:193–208

135. Naranjo, S. E., Steinkraus, D. C. 1988. Discovery of an entomophthoralean fungus (Zygomycetes: Entomophthorales) infecting adult northern corn rootworm, *Diabrotica barberi* (Coleoptera: Chrysomelidae). *J. Invert. Pathol.* 51:298–300

136. Nault, L. R., Findley, W. R. 1981. *Zea diploperennis:* primitive relative offers new traits for corn improvement. *Ohio Rep.* (Nov.–Dec. 1981): 90–92

137. Oloumi-Sadeghi, H., Levine, E. 1989. Controlling fungi that colonize eggs of the western corn rootworm in the laboratory. *Entomol. Exp. Applic.* 50:271–79

138. Oloumi-Sadeghi, H., Levine, E. 1989. Effect of starvation and time of egg hatch on larval survival of the western corn rootworm, *Diabrotica virgifera virgifera* (Coleoptera: Chrysomelidae), in

the laboratory. *J. Kans. Entomol. Soc.* 62:108–16

139. Palmer, L. T., Kommedahl, T. 1969. Root-infecting *Fusarium* species in relation to rootworm infestations in corn. *Phythopathology* 59:1613–17

140. Poinar, G. O. Jr., Evans, J. S., Schuster, E. 1983. Field test of the entomogenous nematode, *Neoaplectana carpocapsae,* for control of corn rootworm larvae (*Diabrotica* sp., Coleoptera). *Prot. Ecol.* 5:337–42

141. Prystupa, B., Ellis, C. R., Teal, P. E. A. 1988. Attraction of adult *Diabrotica* (Coleoptera: Chrysomelidae) to corn silks and analysis of the host-finding response. *J. Chem. Ecol.* 14:635–51

142. Reissig, W. H., Wilde, G. E. 1971. Feeding responses of western corn rootworm on silks of fifteen genetic sources of corn. *J. Kans. Entomol. Soc.* 44:479–83

143. Ruesink, W. G. 1986. Egg sampling techniques. See Ref. 99a, pp. 83–99

144. Shaw, J. T., Paullus, J. H., Luckmann, W. H. 1978. Corn rootworm oviposition in soybeans. *J. Econ. Entomol.* 71:189–91

145. Shaw, J. T., Ruesink, W. G., Briggs, S. P., Luckmann, W. H. 1984. Monitoring populations of corn rootworm beetles (Coleoptera: Chrysomelidae) with a trap baited with cucurbitacins. *J. Econ. Entomol.* 77:1495–99

146. Siegfried, B. D., Mullin, C. A. 1989. Influence of alternative host plant feeding on aldrin susceptibility and detoxification enzymes in western and northern corn rootworms. *Pestic. Biochem. Physiol.* 35:155–64

147. Spike, B. P., Tollefson, J. J. 1989. Relationship of root ratings, root size, and root regrowth to yield of corn injured by western corn rootworm. (Coleoptera: Chrysomelidae). *J. Econ. Entomol.* 82:1760–63

148. Stamm, D. E., Mayo, Z B, Campbell, J. B., Witkowski, J. F., Andersen, L. W., Kozub, R. 1985. Western corn rootworm (Coleoptera: Chrysomelidae) beetle counts as a means of making larval control recommendations in Nebraska. *J. Econ. Entomol.* 78:794–98

149. Steffey, K. L., Gray, M. E. 1989. 1990 insect pest management guide: field and forage crops. *Circular 899, Coop. Ext. Serv. Univ. Ill. Urbana-Champaign.* 38 pp.

150. Steffey, K. L., Kuhlman, D. E. 1985. Corn rootworm control: 1984 and beyond. *Proc. 37th Ill. Custom Spray Oper. Train. Sch.,* pp. 152–65. Urbana-Champaign, IL: Univ. Ill.

151. Steffey, K. L., Kuhlman, D. E., Felsot, A. S. 1986. Corn rootworm management. *Proc. 38th Ill. Custom Spray Oper. Train. Sch.* pp. 108–16. Urbana-Champaign, IL: Univ. Ill.

152. Steffey, K., Kuhlman, D., Kinney, K. 1987. Corn rootworm larval control—research and management in Illinois. *Proc. Ill. Agric. Pestic. Conf. '87,* pp. 50–62. Urbana-Champaign, IL: Univ. Ill.

153. Steffey, K., Kuhlman, D., Kinney, K., Gray, M. 1989. Management of corn rootworms: research and recommendations. *Proc. Ill. Agric. Pestic. Conf. '89,* pp. 76–92. Urbana-Champaign, IL: Univ. Ill.

154. Stinner, B. R., House, G. J. 1990. Arthropods and other invertebrates in conservation-tillage agriculture. *Annu. Rev. Entomol.* 35:299–318

155. Stinner, B. R., Krueger, H. R., McCartney, D. A. 1986. Insecticide and tillage effects on pest and non-pest arthropods in corn agroecosystems. *Agric. Ecosystems Environ.* 15:11–21

156. Stoewen, J. F. 1989. *Winter survival of corn rootworm (Coleoptera: Chrysomelidae) eggs in southern Ontario.* MS thesis. Guelph, Ont: Univ. Guelph. 110 pp.

157. Sutherlin, T. A. 1986. *Development of field and laboratory techniques to evaluate the relative attractiveness of corn plants to corn rootworm beetles.* PhD thesis. Lincoln: Univ. Nebr. 101 pp.

158. Sutter, G. R. 1982. Comparative toxicity of insecticides for corn rootworm (Coleoptera: Chrysomelidae) larvae in a soil bioassay. *J. Econ. Entomol.* 75:489–91

159. Sutter, G. R., Branson, T. F., Fisher, J. R., Elliot, N. C., Jackson, J. J. 1989. Effect of insecticide treatments on root damage ratings of maize in controlled infestations of western corn rootworms (Coleoptera: Chrysomelidae). *J. Econ. Entomol.* 82:1792–98

160. Sutter, G., Gustin, R. 1989. Environmental factors influencing corn rootworm biology and control. *Proc. Ill. Agric. Pestic. Conf. '89,* pp. 43–48. Urbana-Champaign, IL: Univ. Ill.

161. Thompson, A. C., Hedin, P. A., Gueldner, R. C., Davis, F. M. 1974. Corn bud essential oil. *Phytochemistry* 13:2029–32

162. Tollefson, J. J. 1986. Field sampling of adult populations. See Ref. 99a, pp. 123–46

163. Townsend, L. 1984. Corn rootworm beetles. *Publ. ENT-45 Coop. Ext. Serv., Univ. Ky. Lexington.* 4 pp.

164. Turpin, F. T., Dumenil, L. C., Peters, D. C. 1972. Edaphic and agronomic characters that affect potential for rootworm damage to corn in Iowa. *J. Econ. Entomol.* 65:1615–19

165. Turpin, F. T., Peters, D. C. 1971. Survival of southern and western corn rootworm larvae in relation to soil texture. *J. Econ. Entomol.* 64:1448–51

166. Turpin, F. T., York, A. C. 1981. Insect management and the pesticide syndrome. *Environ. Entomol.* 10:567–72

167. Tyler, B. M. J., Ellis, C. R. 1974. Adult emergence, oviposition and lodging damage of northern corn rootworm (Coleoptera: Chrysomelidae) under three tillage systems. *Proc. Entomol. Soc. Ont.* 105:86–89

168. Tyler, B. M. J., Ellis, C. R. 1979. Ground beetles in three tillage plots in Ontario and observations on their importance as predators of the northern corn rootworm, *Diabrotica longicornis* (Coleoptera: Chrysomelidae). *Proc. Entomol. Soc. Ont.* 110:65–73

169. Walgenbach, D. D., Sutter, G. R. 1977. Corn rootworm susceptibility to insecticides. *Proc. 29th Ill. Custom Spray Oper. Train. Sch.*, pp. 68–71. Urbana-Champaign, IL: Univ. Ill.

170. Weekman, G. T. 1961. Problems in rootworm control. *Proc. North Cent. Branch Entomol. Soc. Am.* 16:32–34

171. Weiss, M. J., Mayo, Z B. 1985. Influence of corn plant density on corn rootworm (Coleoptera: Chrysomelidae) population estimates. *Environ. Entomol.* 14:701–4

172. Weiss, M. J., Mayo, Z B, Newton, J. P. 1983. Influence of irrigation practices on the spatial distribution of corn rootworm (Coleoptera: Chrysomelidae) eggs in the soil. *Environ. Entomol.* 12:1293–95

173. Wilde, G. 1978. Corn rootworm control in Kansas. *Agric. Exp. Sta. Bull. 616.* Kans. State Univ., Manhattan. 15 pp.

174. Witkowski, J. F., Owens, J. C. 1979. Corn rootworm behavior in response to trap corn. *Iowa State J. Res.* 53:317–24

Annu. Rev. Entomol. 1991. 36:257–83

BIOLOGICAL CONTROL OF CASSAVA PESTS IN AFRICA

H. R. Herren and P. Neuenschwander

Biological Control Program, International Institute of Tropical Agriculture, B. P. 062523, Cotonou, Bénin

KEY WORDS: *Phenacoccus manihoti, Epidinocarsis lopezi, Mononychellus tanajoa,* Phytoseiidae, Coccinellidae

INTRODUCTION

Rarely have two pests almost simultaneously invaded a continent, spread at an extremely fast pace, and threatened the major staple of over 200 million people as happened in the early 1970s in Africa with the cassava mealybug *Phenacoccus manihoti* (Homoptera, Pseudococcidae) and the cassava green mite *Mononychellus tanajoa* (Acari, Tetranychidae). The cassava green mite was reported first from Uganda (108, 147). The first outbreaks of the cassava mealybug occurred in the Kinshasa (Zaire)/Brazzaville (Congo) area in 1973 (58, 114, 158). In both cases, planting material had been introduced illegally, ignoring quarantine regulations that banned introduction of vegetative plant material other than tissue cultures.

Cassava, *Manihot esculenta* (Euphorbiaceae), had been relatively free of arthropod pests in Africa (75) compared to its area of origin (South America), where approximately 200 species attack it (9). This low incidence of pests in Africa may be explained by two factors: (*a*) cassava is an exotic plant in Africa, having been introduced from South America in the 16th century, and (*b*) it possesses large quantities of cyanogenic glucosides and latex. Cassava mealybug and cassava green mite, which evidently avoid this chemical defense, soon caused direct losses of about 80 and 60%, respectively (145, 153, 181), depending on the physiological state of the plants, which is influenced by the environment and other pests and diseases. In addition, secondary losses like the reduction of healthy leaves, which are consumed in

257

0066-4170/91/0101-0257$02.00

many countries, erosion, weed invasion, and poor planting material for use in the next planting season also contributed to the problem.

The potential of the two accidental introductions to cause famine caught the attention of governments and the international donor and research communities, opening up new opportunities for biological control. Following early and unsuccessful attempts to introduce natural enemies against cassava green mite and cassava mealybug (186), the International Institute of Tropical Agriculture (IITA) started work in biological control and host plant resistance. This led to the creation of the IITA Biological Control Program (BCP), which developed a comprehensive research program and coordinated collaboration with scientists in Africa, Europe, and the Americas (72–74). The control strategy included (a) systematic exploration of the likely areas of origin of the pests, which were conducted from southern California to Paraguay; (b) rearing of the most promising natural enemies and detailed taxonomic, biological, and ecological studies of them; (c) the release of the enemies in Africa over an infested zone that is one and a half times the area of the United States of America; and (d) an analysis of the cropping system and economic impact of the pests and their indigenous and introduced natural enemies. Research based on a systems appproach became interwoven with mass-releases that were facilitated by technological innovations; both activities also served in training future biological control practitioners (112, 176).

The cassava mealybug biological control project advanced rapidly because of the early discovery of the parasitoid *Epidinocarsis lopezi* (Hymenoptera, Encyrtidae) (186), and its establishment, rapid spread (78), and impact on the cassava mealybug (61, 131, 132), which gave new confidence in biological control to policymakers and donors alike (149). A direct result of this attention is the new IITA Biological Control Centre for Africa, inaugurated in December 1988 in Cotonou, Republic of Benin (182).

The cassava green mite biological control project followed the same pattern of research as that for the cassava mealybug. The general lack of experience of mite biological control in field populations and the large number of candidate predators called for preliminary studies to assess their potential. Also, special quarantine procedures had to be developed. Because results from field testing of mite predators are still preliminary, a detailed review of the cassava green mite project is left to the future. Nor is any attempt made to discuss the contribution and the potential of host plant resistance against cassava green mite and cassava mealybug.

SPREAD OF CASSAVA MEALYBUG AND CASSAVA GREEN MITE

Following its presumed accidental introduction in the Congo (114, 158) and Zaire (58), the cassava mealybug spread rapidly through Zaire where widespread outbreaks occurred in the Bas-Zaire, Bandundu, Shaba, and Kivu regions over the next nine years (71). The pest gained new footholds in Senegal and Gambia in 1976, in Nigeria and Benin in 1979 (3), and in Sierra Leone in 1985. In West Africa, it spread up to 300 km per year (78). In East Africa, the Rift Valley proved a formidable barrier and the mealybugs' spread in the highlands was usually less spectacular and rather irregular. As described in numerous unpublished reports, cassava mealybug infestations, always accompanied by dispersed or newly released E. *lopezi*, have progressed in Zambia since about 1983, in Rwanda since 1984, and in Burundi since 1987. New foci, many hundreds of km away from the closest infestation, were reported in Malawi (1985), Mozambique (1986), Tanzania (1987), and Kenya (1989). By early 1990, the entire cassava belt, with the main exceptions of Uganda and Madagascar, was infested by the cassava mealybug (Figure 1).

Within an infested cassava field, the cassava mealybug shows a highly aggregated distribution, which differs between seasons. This led to the development of different sampling plans (155) for fields, which have to be chosen by unbiased procedures (61, 130).

From its site of accidental introduction in Uganda in 1971, the cassava

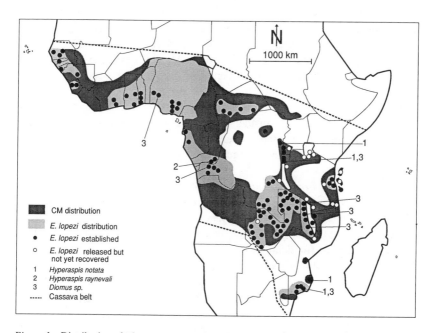

Figure 1   Distribution of *Phenacoccus manihoti* (CM) and its introduced parasitoid *Epidinocarsis lopezi* in Africa, together with recovery sites of some exotic coccinellid predators (*Hyperaspis* spp. and *Diomus* sp.), in 1990.

green mite spread, slowly at first, then swept through the cassava belt at the rate of several hundred km per year (108, 178, 181). By 1990, it was homogeneously distributed throughout the cassava belt, with the major exceptions of Senegal, Gambia, Guinea-Bissau, and Madagascar.

Sampling plans based on mite counts have been developed for Africa (177) and South America (20). The widely used scoring scales do not properly represent mite numbers, unless they are calibrated for different cassava varieties and seasons (185).

FOREIGN EXPLORATION, TAXONOMIC STUDIES, AND QUARANTINE

Following the scientific description of the cassava mealybug, which was thought to be of Neotropical origin (113), the CAB International Institute of Biological Control (IIBC) started exploration in the Carribean and the northern part of South America. The parasitoids that were found, on what was at the time considered to be *P. manihoti,* failed to reproduce on the cassava mealybug in the Congo and Zaire (11, 106). In 1980, BCP started its own exploration with a systematic search. Because 30 years of research on cassava in the Americas had not led to the identification of the cassava mealybug, close relatives of cassava in the family Euphorbiaceae were searched as well. Exploration was done in areas of rich species diversity of the cassava genus *Manihot* (150), namely in central/western Mexico, Yucatan, northeastern Brazil, Peru, and Paraguay, and covered *Phenacoccus* spp. and other mealybugs (175). While the exploration was in progress, the cassava mealybug found in the Guyanas and northern Brazil was described as *Phenacoccus herreni* (24). The previous misidentification as *P. manihoti* explains the lack of success in the earlier establishment attempts in the Congo and Zaire, and underscores the need for accurate identification in successful biological control programs.

P. manihoti was finally discovered in 1981 in Paraguay by A. C. Bellotti of the Centro Internacional de Agricultura Tropical (CIAT). Subsequently, BCP made arrangements with IIBC for foreign exploration in Paraguay and Bolivia, which yielded, among other parasitoids, *E. lopezi* (186). In 1982, BCP sent its own team for detailed exploration of all cassava growing areas of Paraguay, Brazil, and Bolivia and for biological and ecological studies of the host and its natural enemies (104, 106). Cassava mealybug proved to have a very limited distribution in the Paraguay River basin, with generally very low and erratically occurring populations. Up to 1987, 18 species of natural enemies were found in the area of origin of the cassava mealybug. Eight species were sent for quarantine (see section on mass rearing); another two species with interesting traits, the encyrtid *Parapyrus manihoti* (141) and the syrphids *Ocyptamus* spp., could not be reared successfully. Two additional predators were introduced through CIAT from Colombia, where they fed on *P. herreni*.

From South America, these beneficials went to the IIBC quarantine laboratory in the United Kingdom. Processing met the requirements established by the Organization of African Unity/Inter-African Phytosanitary Council (OAU/IAPSC), which regulates the introduction of plants and animals into Africa and which accredited the IIBC laboratory. The exotic natural enemies were then sent to BCP, all introductions being covered by import permits issued by the national quarantine authorities of Nigeria and, later, Benin. Under the same OAU umbrella, beneficials reared at BCP were and still are being shipped to various African countries upon their request and under cover of their own quarantine permits.

In South America, the cassava green mite is only one of numerous tetranychid mites attacking cassava (21, 38). Because it was already known as a minor pest in parts of the continent, the exploration for cassava green mite natural enemies followed a slightly different course than that for the cassava mealybug (181). In order to carry out an efficient and systematic search for natural enemies of the cassava green mite in the Neotropics, ecologically homologous areas were identified on both continents using the conditions of potential release areas in Africa as criteria for choosing exploration routes in South America (179). Natural enemies have an important effect on mite populations in the Neotropics, especially in traditional farming systems where locally adapted cultivars are grown (9, 19, 152, 187). They include phytoseiid, blattisocid, and tydeid mites, staphylinid and coccinellid beetles, as well as cecidomyiid midges, syrphid flies, green lace wings, thrips, and the pathogens *Hirsutella thompsonii* (7, 21) and *Triplosporium* sp. (1).

Phytoseiid mites are considered the most important natural enemies. First priority in the search for biological control agents has therefore been given to this group (116, 120, 185). Among the about 50 species attacking phytophagous mites on cassava in South America (7, 8, 21, 123, 124), eight species (see section on mass rearing), which occur frequently and are regularly associated with the cassava green mite, have been introduced according to the quarantine regulations described above. The actual quarantine processing, originally done by IIBC directly, is executed by the Subfaculty of Experimental and Applied Biology of the University of Amsterdam under the umbrella of IIBC.

The taxonomic status of the cassava green mite in Africa remained unclear until recently, but it is now accepted that all cassava green mites from Africa, despite their polymorphism, are *M. tanajoa* (139, 151, 181; for a different opinion, see 55). The difficulties in identifying the introduced species indicate, as in the case of cassava mealybug, the importance of taxonomy for selecting specific natural enemies.

BIOLOGICAL AND ECOLOGICAL STUDIES

The Cassava Host Plant

Cultivation and plant characteristics of cassava have been reviewed (23, 57, 159) and aspects of cassava yield formation have been studied in several continents (citations in 53, 153). For Africa, suitable quantitative analysis of the growth of the cassava plant, which is needed for a better understanding of biological control of cassava pests, was lacking; yields at time of harvest were often the only data available. Dynamics of yield formation were therefore analyzed as the first trophic level of a multitrophic system (153). The allocation of photosynthates and biomass, and the effects of nitrogen under different water regimes, were investigated and incorporated in a cassava system simulation model that runs on the basis of real weather data (53). Done in collaboration with the University of California, Berkeley, and the Swiss Federal Institute of Technology, Zürich, it is based on the distributed delay dynamics model (109, 172), and the functional response model of Frazer-Gilbert (39), which are extended across trophic levels. The model reflects the wide range of responses by cassava to different environmental factors in the various ecological zones and can be extended to include different varieties. The model suggests, and recent field data (F. Schulthess, personal communication) demonstrate, that a highly branching IITA variety has a relatively sink-limited growth of tubers, while the idiotypical cassava plant from CIAT shows more a source limitation (101).

The Cassava Mealybug

The cassava mealybug is parthenogenetic with three nymphal instars and was first described from African material (113). Since its appearance in Africa, the cassava mealybug has often been studied in the field (29, 30, 32, 35, 100, 146, 153). The cassava mealybug attacks *Manihot* spp. and is found on other plants only accidentally. By stunting the cassava tips, it creates for itself a protected environment and usually develops the largest populations during the dry season. Cassava mealybug populations collapse during the rainy season, but sometimes—when the plant substrate becomes unacceptable—even before the rains (31). During the rainy season, cassava mealybugs suffer mortality from the washing effect of the rains (29, 96, 153). Because of the renewed growth of the tips during this period, cassava mealybugs might also risk a higher exposure to the adverse effects of rains and natural enemies.

  Dispersal of the cassava mealybugs over long distances is by crawlers that are picked up by air currents and by infested planting material transported by man. Within a plant, cassava mealybugs move to the stems and up to the tips (165) so that, at low population levels, almost all cassava mealybugs on a cassava plant are found in the tips (130, 154).

  Life table statistics were assessed in the laboratory (81, 92, 95, 142, 146,

154; using a new method for calculating the intrinsic rate of increase, see 98). They have been used to develop a simulation model (51) and may be summed up as follows (154): cassava mealybugs have a low thermal threshold of 14.7°C, an optimal temperature of about 28°C, no development above 35°C, and a net reproductive rate of about 500 eggs. All life table parameters are marginally better on stressed plants. They are not affected by the age of the leaf, but differ between varieties.

Cassava mealybug infestation seems to influence the distribution of cyanide between leaves and tubers (4) and first results with *P. herreni* (173) suggest biochemical, physiological, and morphological changes at the cellular level caused by the mealybug. The lack of knowledge concerning such insect-plant interactions is considered to be a major limiting factor for further evaluation of biological control of the cassava mealybug.

Natural Enemies of the Cassava Mealybug

The cassava mealybug is attacked by the usual guild of polyphagous or oligophagous predators and parasitoids of mealybugs (6, 119), which switched over to this new food source when the cassava mealybug invaded Africa. Entomologists in the Congo (36, 113), Zaire (143), Nigeria (3, 82), Gabon (15), and other countries compiled lists of the new insects on cassava and noted, in some cases, their ecological roles. Following the arrival of *E. lopezi,* the cassava mealybug food web was investigated again over the whole continent, and it now comprises about 130 species (13, 135, 143). Many are opportunistic visitors that do not reproduce on the cassava mealybug; some are attracted more to the bunchy tops with the rich organic material from live and dead cassava mealybugs. Only about 20 species are common and seem to have some impact. Because of the reduction in cassava mealybug populations due to *E. lopezi,* many species became less abundant after the establishment of the exotic parasitoid (130).

Coccinellids are the most important predators that regularly occur at high cassava mealybug densities (15, 33, 130, 135), when these beetles sometimes suffer from a high degree of parasitism (163). Most of the species found on the cassava mealybug in Africa have been described or revised by Fürsch (41–44; see 34 for a different interpretation for species of the genus *Exochomus*). Some life table parameters of indigenous species have been measured (33, 161, 163, 164) together with those of the exotic *Hyperaspis raynevali* (which is probably synonymous to *Hyperaspis jucunda* as used in IITA texts) (86, 87, 142, 144). This exotic species did not have a higher intrinsic rate of increase than a local *Hyperaspis senegalensis* (37). As is often the case with large predators with high food-intake requirements (56), the indigenous coccinellids do not respond to low host populations (132). In olfactometer studies, however, the introduced and locally established *Diomus* sp. reacted more strongly to cassava mealybug, honeydew and exuviae than the indigenous *Exochomus* sp. (168). Life-table studies on this exotic species are lacking.

Other predators, with an erratic occurrence, include *Dicrodiplosis manihoti*, a host-specific cecidomyiid (65), and the lycaenid *Spalgis lemolea*, which intrigues and frightens farmers because its pupa resembles the face of a monkey.

Hymenopterous parasitoids of different life styles, reared from cassava mealybug–infested cassava tips in Africa, can be identified with a simple key (17). A few indigenous parasitoids of the encyrtid genus *Anagyrus* that attack African mealybugs like *Phenacoccus madeirensis* have adapted to the cassava mealybug. Though parasitization rates over two months were rather similar between *Anagyrus* spp. and *E. lopezi* (37), the most common of them, *Anagyrus niombae* (16), suffered from competitive displacement by *E. lopezi* and is now rarely obtained from the cassava mealybug (13, 18). It is not clear whether the difference in specificity alone is responsible for the superior competitiveness of *E. lopezi,* or whether different preferences for host instars or differences in the ability to locate the host, tip the competitive balance in favor of *E. lopezi.*

Wherever *E. lopezi* was established, it became the most important parasitoid and was more abundant than coccinellids (61, 130). Although at the onset of the biological control project in Africa it was only known from the original description of the adults (26), *E. lopezi* is now one of the best known encyrtids. The morphology of the eggs and larvae (103, 128) and of the female genitalia (84) have been described in detail. *E. lopezi* is a strictly solitary internal parasitoid, and is generally considered absolutely specific to the cassava mealybug.

Immature development of *E. lopezi* in their preferred third instar hosts takes only two weeks, with males developing faster than females (12, 90, 105, 148). However, it is slowed down considerably in second instar hosts and, probably, where encapsulated larvae can extricate themselves from the melanized sheet (102, 105, 163a).

Encapsulation and melanization of young solitary larvae, often including injured host tissue, was observed in about 13% of all cases at 28°C (128, 157). It is lower at other temperatures (128). In cassava mealybugs with multiple stings by *E. lopezi,* encapsulated parasitoid larvae are always much more frequent than in those with one sting only (128, 157). These larvae probably died from mutual competition and were only then encapsulated. Thus, the proposition that *E. lopezi* is inefficient because of encapsulation (127) seems untenable.

The adult *E. lopezi* survives only a few days at ambient temperatures—especially when crowded—even when fed adequately with honey or sugar grains (134). It has a highly variable reproductive capacity of between about

40 and 90 offspring, or about 10 per day. Optimum temperature, as judged from the intrinsic rate of increase, is at 27°C; but total oviposition is higher and life-span longer at lower temperatures of about 23°C (12, 105, 148).

Behavioral studies, in collaboration with the University of Leiden, indicated that females mate immediately upon emergence and are attracted by unidentified synomones to leaves from infested plants (126). They respond to the wax of the cassava mealybugs by increasing search time and decreasing walking speed (91, 165). For oviposition, they prefer third instar hosts and lay mostly male eggs into younger hosts (90, 105). Field experiments with caged *E. lopezi* demonstrate that mating is random and that local mate competition (22, 59, 174) does not play a role in sex ratio determination (170). Mostly, the age composition of the host colonies determines the sex ratio of the offspring (169a). Because female wasps from large hosts live longer and lay more eggs (90), and small males perform as well as large males, *E. lopezi* maximizes its fitness by using the larger available hosts for laying female eggs.

Though *E. lopezi* females can discriminate between unparasitized and parasitized hosts, they often superparasitize depending on the circumstances, e.g. when host numbers in the laboratory are limited (12, 84, 90). In addition, older hosts frighten off a wasp by wiggling their bodies; this is common if they previously experienced an oviposition sting (84, 90, 102). However, not all stings by the female *E. lopezi* lead to oviposition and the ultimate death of the host. Particularly at higher temperatures, host feeding is common; it is mainly on the second instars on which it is a more important mortality factor than oviposition. Host feeding increases longevity and oviposition, at least under some conditions. In addition, many stung hosts without eggs die from the wounds, or they develop more slowly and lay only a small fraction of the normal egg complement (102, 105, 137, 163a).

In biological control theory, positive density-dependent reaction of a beneficial parasitoid or predator to the population size of its host has been considered a key element (115; see differing opinion in 125). In the laboratory, parasitism rates of *E. lopezi* tend to decrease with increasing host densities, which has been interpreted as a sign of inefficiency (148). However, if very low host densities are offered in a seminatural setting, parasitism rates increase at low densities (102). In the field, these rates fluctuate widely and remain mostly below 30% (12, 61, 62). Generally, parasitism rates are reduced on larger cassava mealybug colonies (143). However, reevaluation of four years of field data from Nigeria on a per tip basis demonstrated that, while 89% of all sampled tips did not have any cassava mealybugs, parasitism rates increased marginally but significantly on the 7% of tips containing 1–10 third and fourth instar cassava mealybug. On the 4% of tips containing colonies larger than 10 cassava mealybugs, parasitism rates declined rapidly with increasing colony size. This result indicates that *E. lopezi* responds in a density-dependent manner to its host population on 64% of all colonies (61). Similarly, in a field experiment, adult females were observed to prefer infested tips over uninfested ones and to aggregate significantly in a density-dependent manner on larger host colonies (64, 165).

In conclusion, *E. lopezi* has the following characteristics that might explain why it is a successful biological control agent: (*a*) a short generation time and consequently a power of increase larger than that of its host; (*b*) the capacity for host-feeding, mutilation, and—where the host is not killed directly—inhibiting the successful reproduction of the cassava mealybug; (*c*) high specificity and searching capacity for its host, allowing it to survive on very low host populations; and (*d*) a density-dependent aggregation and reproduction on most host-population densities encountered in the field. These traits should be contrasted with those of other, unsuccessful species like the exotic *Epidinocarsis diversicornis* or the indigenous *Anagyrus* spp. Though *E. lopezi* does not seem to be capable of substantially lowering high-density cassava mealybug populations, it can very well keep low-density host populations, as found at the end of the rainy season, at a low level. By having a high searching and dispersal capacity, its life strategy seems to ensure an ecological niche without competition from less specialized competitors, which have a higher r_m, but lower searching efficiency.

E. lopezi is attacked by unspecific hyperparasitoids, transferring mainly from *Anagyrus* spp., even in the first generation after its release. About a dozen species have been reared, the most common ones are solitary *Prochiloneurus* spp. (Encyrtidae) and a gregarious *Chartocerus* sp. (Signiphoridae) (13, 15, 135). High hyperparasitism rates in situations with high cassava mealybug populations led to the fear that hyperparasitoids would compromise the efficiency of *E. lopezi* (83). In general, the impact of hyperparasitoids is not clearly understood (10, 156). Computer simulation models predict slightly higher average densities of pest populations but also more stability in systems with hyperparasitoids than those without (66, 107). Many biological control projects succeeded with a relatively high degree of hyperparasitism. Our field data from large areas indicate a strongly density-dependent response by the hyperparasitoids, i.e., high percentage hyperparasitism where *E. lopezi* is abundant on high-density cassava mealybug populations and low percentage hyperparasitism where cassava mealybug and *E. lopezi* populations are low (130–132). In addition, at least *Prochiloneurus insolitus* has a negative influence directly on the phytophagous host when it stings and mutilates a cassava mealybug that lacks an *E. lopezi* larva (G. Goergen & P. Neuenschwander, unpublished results).

In addition to insect predators and parasitoids, the entomophthoral fungus *Neozygites fumosa* also attacks the cassava mealybug. It is mainly active on adult cassava mealybugs in the rainy season, that is, when the relative humidity is above 90%, the canopy preferably wet, and temperature above 20°C (94, 99).

The Cassava Green Mite and Its Phytoseiid Predators

The cassava green mite has four active stages: a six-legged larva, two nymphal instars, and the adult, and it is biologically similar to other agronomically important tetranychids (69). In Africa, life-table studies showed that immature developmental times (12.3 days at 27°C) were similar on cassava leaves of different ages and from different seasons. Length of adult life, fecundity, and consequently the intrinsic rate of increase were enhanced on young leaves of young plants (180). In the field, cassava green mite populations increase on new leaf growth during the early dry season, decline on the old leaves during the later parts of the dry season, then increase again on the new flush of leaves after the first rains (184). Following this, rainfall decreases densities of mite populations (2, 21, 147, 184). The combined effects of rainfall, drought stress via the host plant, food availability (production and persistence of new foliage), and leaf quality (nitrogen concentration) on cassava green mite populations and tuber yields were assessed in a simulation model. The simulation data suggest that the cassava green mite would be a more severe problem on irrigated and fertilized cassava (54).

In Africa, insect predators like the staphylinid *Holobus* sp., the coccinellid *Stethorus* spp., the thrips *Scolothrips* spp., and the cecidomyiid *Arthrocnodax* sp., are associated mainly with very high cassava green mite populations (185), and a considerable number of phytoseiid mite species has been recovered in the cassava ecosystem (120–122, 160, 185). In addition, the fungal pathogens *Entomophthora* sp. and *Hirsutella* sp. were recovered from cassava green mites (5). Regular cassava green mite outbreaks indicate, however, that indigenous natural enemies in Africa are less well adapted for controlling this exotic species than those in South America (110, 111, 183).

Some of the imported phytoseiid mite predators have been studied at different temperatures. At 25°C, development to the adult stage takes less than five days, i.e. less than half of that of their cassava green mite host (7). Some feeding preferences have been determined (7, 167). *Neoseiulus (Amblyseius) idaeus* has proved to be particularly well adapted to dry conditions (27, 28). The intrinsic rates of increase of many more species are being studied and generally found to be higher than those of the prey. At the same time, the behavior of phytoseiid mites, their relation to the cassava plant, and their prey preferences and coincidence in the habitat are being investigated. These characteristics might give more important insights into the efficiency of a given species than do classical life-table studies.

MASS REARING AND RELEASES

The following insects from South America are reared at BCP for use as biological control agents against the cassava mealybug: the encyrtids *E. lopezi*, *E. diversicornis*, the platygasterid *Allotropa* sp., the coccinellids *Hyperaspis* ?*jucunda* (up to 1987 only), *Hyperaspis notata* (strains from Brazil, Colombia, and their cross), two different undescribed species of *Hyperaspis*, *Diomus* sp., and the hemerobiid *Sympherobius maculipennis*.

E. *lopezi* is, or will soon be, reared also in insectaries in Burundi, Gabon, Mozambique, Tanzania, and Zaire. *H. raynevali*, which is probably a synonym of *H.* ?*jucunda*, is in culture in Congo.

The following phytoseiid mite species are reared at BCP (185): *Amblyseius aerialis*, *Euseius concordis*, *Neoseiulus anonymus*, *Neoseiulus californicus*, *Neoseiulus idaeus*, *Typhlodromalus limonicus*, *Galendromus annectens*, and *Galendromus helveolus*. Colombian and Brazilian strains of *A. aerialis*, *N. idaeus*, and *T. limonicus* are kept separately. With the exception of *T. limonicus*, which seems to be specific to the cassava green mite, phytoseiid mite predators of the cassava green mite are reared on the tetranychid *Tetranychus urticae* on leguminous plants. Host mites are brushed or washed off and served to the phytoseiids on disks surrounded by water (7, 40, 46, 117), and mass production of phytoseiids can be monitored with the aid of a simulation model (88).

Most insects are reared in cages on cassava mealybugs on potted cassava plants, and are used for studies as well as experimental releases. To cater to the many requests from countries for predators and parasites to be delivered urgently and often in large quantities, new rearing technologies have been developed and tested for *E. lopezi*, and later adapted to phytoseiid mites (67, 68, 134). They are based on hydroponic cultures of cassava. Rearing is done in mechanized rearing chambers or in units, called "cassava trees," made to a large extent with local material. A comparison of the production costs for *E. lopezi*, obtained with different rearing techniques, showed that this wasp is produced more economically in "cassava trees" than in cages or chambers, and that production costs are as low as those for insects of similar life styles from commercial insectaries (133).

The high quality of *E. lopezi*, as witnessed by its successful establishment all over Africa, is attributed to (*a*) the high quality of the original insects brought from South America, (*b*) the rearing on the original host and original host plant, (*c*) careful technical manipulations and constant supervision in the insectary, (*d*) the good size and the high number of rearing units, and (*e*) infestation schedules that consider all known facets of the biologies of the plant, the cassava mealybug, and *E. lopezi* (133).

Because of the size of and often difficult access to the area to be covered with natural enemies, the BCP aircraft transported the reared beneficials. Releases were mostly made from the ground; however, for speedy delivery to remote areas, an aerial release technique was developed that allows the ejection of viable natural enemies while flying over cassava growing areas (14, 76). This system has proven effective for *E. lopezi* and coccinellids, and is now being tested for phytoseiid mites.

Between 1981 and 1990, *E. lopezi* was released across the African cassava belt in over 100 areas indicated in Figure 1, in which each dot may represent releases in several villages and on several occasions (60, 78, 136). Frequently, other natural enemies of the cassava mealybug were also released, usually in separate fields. Since 1984, experimental releases of several species and strains of phytoseiid mites have been carried out in different ecological zones in 10 countries (118; J. S. Yaninek, personal communication).

MONITORING FOR ESTABLISHMENT AND SPREAD

Only a few of the species released against the cassava mealybug have been established permanently (Figure 1). Among the cassava mealybug predators, only *Diomus* sp. in Kinshasa (70) and Malawi (R. Borowka, H. Hammans, E. H. Kapeya, unpublished results) and *H. notata* in Burundi (M. A. Autrique, S. Muyango, E. Nkubaye, unpublished results) and the Kivu province of Zaire (N. H. D. Nsiama She, unpublished results) show long-term establishment and are recovered in substantial numbers. Some recoveries of these and other species have been made in other countries, including *H. raynevali* in the Congo (37). Of the three exotic parasitoids released by the BCP, two were recovered only occasionally and in very low numbers, while *E. lopezi* was established across all ecological zones where the cassava mealybug occurs.

First releases of *E. lopezi* were made in southwestern Nigeria in 1981–1982 (Figure 1) (77, 93). After five months, i.e. about 10 generations, *E. lopezi* was recovered from almost all sampled fields within a 100 km distance from the respective release site, and as far as 170 km north of it, in the Guinea savannah. Spread in the rain forest zone to the south and southeast of the release site was slower. Three years after these releases, *E. lopezi* was found in 70% of all fields on more than 200,000 km^2 (78). This dispersal occurred mostly in traditional farming environments on local cassava varieties. The contribution of the different modes of transport (active flight, passive transport by wind, or by man transporting infested cassava cuttings and leaves) is not known.

Outside Nigeria, establishment and spread of *E. lopezi* have been documented in Ghana (89, 132), Gabon (18), Zaire (71), and the Congo (12, 45, 63). In addition to the releases and establishments listed earlier (60, 78, 136), detailed distribution and dispersal records from Burundi, Malawi,

Mozambique, Tanzania, and Zambia, obtained by entomologists in the national programs and BCP, exist but are yet to be published. Figure 1 indicates the approximate distribution of *E. lopezi* according to the latest surveys. At the edge of the distribution, occurrence of *E. lopezi* is sometimes erratic both in time and space. By 1990, *E. lopezi* was established in the following 25 African countries (year of first release in parentheses, countries listed from northwest to southeast) (Figure 1): Senegal (1984, present distribution uncertain), Gambia (1984), Guinea-Bissau (1984), Guinea-Conakry (release 1988, but *E. lopezi* was already present through dispersal from neighboring countries), Sierra Leone (1985), Côte d'Ivoire (1986), Ghana (1984), Togo (release 1984, but *E. lopezi* was already present through dispersal from Benin), Benin (dispersal from Nigeria), Nigeria (1981), Cameroon (dispersal from Nigeria), Niger (dispersal from Nigeria), Central African Republic (1988), Gabon (1986), Equatorial Guinea (dispersal to Annobón Island), Congo (1982), Zaire (1982), Angola (dispersal from Zaire, present distribution uncertain), Rwanda (1985), Burundi (1988), Tanzania (1988), Malawi (1985), Zambia (1984), Mozambique (1989), and the Republic of South Africa [1990, dispersal from Mozambique (B. Beck, personal communication)]. By 1990, *E. lopezi* had spread over an estimated area of 2.7 million km^2, of which about 1% is under cassava cultivation. The distribution thus exceeds that of any other agent introduced into African for biological control of insect pests, and the adaptability of *E. lopezi* to different ecological conditions is without precedent in Africa (48).

In high cassava mealybug populations of West and Central Africa, maximum dispersal rates by *E. lopezi* of more than 100 km per dry season were observed in several countries. In East Africa, the cassava mealybug and, consequently, *E. lopezi* spread more slowly, which is partly attributable to the impact of *E. lopezi* at the spreading front of the cassava mealybug.

IMPACT ASSESSMENT

Impact assessment beyond recording establishment of exotic species is an important, but often neglected, part of biological control projects. Methods for evaluating the effectiveness of biological control agents have been reviewed (79, 85, 129, 171).

Physical and chemical exclusion experiments demonstrated the efficiency of *E. lopezi* under the conditions of southwestern Nigeria (138), without giving insights into the mechanisms involved. Similarly, under rain forest conditions in Ghana, when cassava mealybugs were protected from *E. lopezi* either totally by sleeves or partially through interference with ants, their populations were much higher than with *E. lopezi* (A. R. Cudjoe, M. Copland, P. Neuenschwander, unpublished results).

The clearest demonstration of a parasitoid's impact on its host is usually obtained from data on population dynamics. Seven years of continuous monitoring in numerous fields in two areas of southwestern Nigeria revealed that mean cassava mealybug population peaks never reached the height (means of up to 90 cassava mealybugs/tip) and the duration (7 months with over 10 cassava mealybugs/tip) observed during the first season of release (61, 62). Though occasional sharp peaks of up to 30 cassava mealybugs per tip were registered, E. lopezi maintains a high level of biological control of the cassava mealybug in this area. The same data also demonstrate the presence of a significant positive density-dependent reaction by the parasitoid in the field, thus offering one mechanism that explains E. lopezi's efficiency.

Similar, but much shorter term and unpublished data on population dynamics come from Burundi (M. A. Autrique, S. Muyango, E. Nkubaye) Malawi (G. Phiri, G. K. C. Nyirenda, D. C. Munthali, R. F. N. Sauti) Mozambique (L. Santos,), and Zambia (C. Klinnert). In the Congo, short-term studies on population dynamics after the establishment of E. lopezi were compared with the situation 10 years ago, i.e. before the releases. Though parasitism levels increased and an indigenous parasitoid was displaced competitively, the investigators concluded that E. lopezi had no strong impact on the cassava mealybug population levels (83, 97). Inexplicably, the indicated cassava mealybug population peaks seemed several times higher than those measured before the releases (30). A recent survey, however, covering the Congo and the neighboring countries of Gabon and Zaire, in which fields were chosen at regular intervals and shoot tips were collected at random, produced rather different conclusions. Infestation levels were relatively low, similar to those obtained in West Africa, and did not differ substantially among the three countries (63). Biological control was judged to be effective, and it was shown (63) that the unbiased sampling procedures employed were not comparable to those used previously in the Congo (83, 97).

In contrast to these studies on population dynamics with frequent sampling occasions, surveys can cover large areas, but give only a snapshot of fluctuating and ever-changing conditions. The key to a successful extrapolation of the data consists of a rigorous sampling procedure with an unbiased choice of fields and tips to be sampled. This was used to show that, within two years of E. lopezi's establishment, average percentage of tip stunting fell from 88 to 23%, leaving mean population densities at the end of the dry season, i.e. the time of maximum cassava mealybug infestations, at about 10 mealybugs per tip (130). A follow-up survey four years later, covering all of Nigeria and Benin, confirmed overall low cassava mealybug infestation levels of below 10 mealybugs per tip, with only 3.2% of all tips being stunted. Infested fields were concentrated on leached out soils without mulch, that is on 4.8% of all fields. In neighboring fields with mulch, however, infestations were low.

Thus, agronomic factors, mediated through the condition of the plant, influenced biological control (131).

Reductions of cassava mealybug populations have been documented in Zambia (J. Purakal, unpublished results) and Malawi (P. Neuenschwander, R. Borowka, H. Hammans, E. H. Kapeya, unpublished results). Field data, based on a strictly unbiased sampling procedure, are not yet available from all ecological zones; but those that exist indicate that *E. lopezi* lowered cassava mealybug population levels overall to satisfactory levels except under the worst agronomic conditions, where the reduction achieved is not sufficient to prevent damage.

On the basis of laboratory data assessing percentage parasitism, several authors concluded that *E. lopezi* cannot efficiently reduce cassava mealybug populations (37, 148). Percentage parasitism as a yardstick for measuring efficiency is, however, often misleading, and the behavior of the female wasp and her interaction with the host and the host plant have to be considered (127, 166, 169). When most of the knowledge on *E. lopezi* presented previously is introduced into the simulation model on cassava growth (53) using time-varying life tables (49, 50, 80, 109, 172), the overriding influence of *E. lopezi* becomes apparent (51). *E. lopezi* populations in this model are shown to react in a density-dependent manner to their host populations. The simulations indicate the strong influence of rains and a relatively low importance for local coccinellids, which are active during a period when damage to the plant is already completed. Though indigenous coccinellids have been observed to reduce local cassava mealybug populations spectacularly at the end of an outbreak (162), long-term monitoring (61) confirms the conclusions from the simulation model (51) about the limited long-term impact on the level of biological control achieved by indigenous coccinellids.

Yield loss experiments indicated up to 84% loss due to the cassava mealybug, with early infestations causing higher losses than later ones (145). Such high losses occur if harvest is at the beginning of the rainy season when the plant mobilizes reserves from the roots for regrowth. Later, losses are partially compensated for (153). In a large-scale survey across different ecological zones in Ghana and Côte d'Ivoire, econometric multiple regression analysis showed that the loss due to the cassava mealybug was reduced significantly by an average of 2500kg/ha in the savannah region in areas where *E. lopezi* had been present for most of the planting season, compared with areas where *E. lopezi* was not yet established (132). This figure gives an idea of the high economic yield of this biological control project. It can be used to recalculate a previous estimate of the cost:benefit ratio of 1:149 (140). This estimate is based on subjective assessments and the assumption of diminishing returns. Because farmers cannot easily switch to another staple and in view of the life strategy of *E. lopezi,* we believe that returns should be calculated as being constant over time.

Despite the considerable positive impact of the cassava mealybug biological control campaign as demonstrated by entomological data, professional and lay perceptions of this impact have varied greatly from one country to another. The following problems were encountered in countries that we judged to have areas under good biological control: First, the cassava mealybug is still spreading in several countries. Consequently, nation-wide damage remains high, even if *E. lopezi* has brought the cassava mealybug under control in the first release sites. Second, within a large area under the umbrella of biological control by *E. lopezi,* individual fields or corners of fields always have comparatively high infestations. Most of these infestations have been shown to be a consequence of bad farming practices. Third, biological control activities are free to the farmer and, most often, to the government as well. They sometimes lead to the funding of a project or are coupled with food aid to the farmers. Therefore, socioeconomic interests in declaring cassava mealybug infestations a continued disaster exist. Fourth, ignorance about mechanisms of pest impact and biological control has sometimes led to false expectations. While the cassava mealybug produces quite noticeable symptoms at the end of the dry season at levels too low to measurably affect yield, memory of the really devastating cassava mealybug infestations encountered before the release of *E. lopezi* is dwindling.

CONCLUSIONS AND PERSPECTIVES

Biological control of the cassava mealybug involved the classical approach, from foreign exploration, biological studies, mass rearing, and release to monitoring and impact studies. The special attention given to quantify observations on all trophic levels in a holistic approach transformed the discovery of an efficient biological control agent into a success for biological control, which has implications for the future of sustainable agriculture world-wide.

In an undertaking of the size of the cassava pest biological control program, no single institution has the capacity to handle all aspects of the required research, training, and implementation alone. BCP has, therefore, taken the lead in organizing a network of collaborators in Africa, Europe, North, Central, and South America to complement its own activities (72, 74). These collaborators are with universities (at present 23), research institutions, and national biological control programs (NBCPs, at present 24). In addition, as a member of the Consultative Group on International Agricultural Research (CGIAR), IITA is itself part of a world-wide system of 13 research institutes, from which it draws logistical and scientific support. The collaboration covers all aspects of research from exploration to economic impact analysis.

The success of the cassava mealybug project provides the momentum necessary to consider control of the cassava green mite and other pest problems such as the maize stem and cob borers, insects on cowpeas, and other crop pests and weeds (47). As a dividend of the infrastructure and training invested in the NBCPs, six countries could be assisted by the BCP in a biological control project against the mango mealybug, initiated by the Togo Crop Protection Service.

Because several exotic pests have recently invaded Africa and others are likely to be introduced in the future, classical biological control remains high on the priority list of the BCP. With or without the use of exotic beneficials, research and training efforts will be invested toward a thorough understanding of the ecosystem. This approach leads to the identification of possible imbalances in the ecosystem that create or enhance pest and weed problems, and the development of appropriate corrective measures to avoid or reduce pest incidence (25, 74).

ACKNOWLEDGMENTS

This work was financed by aid agencies of Austria, Belgium, Canada, Denmark, Federal Republic of Germany, Italy, Norway, Sweden, Switzerland, The Netherlands, and the European Economic Community, as well as the African Development Bank, the Food and Agriculture Organization of the United Nations, the International Fund for Agricultural Development, the United Nations Development Program, and—indirectly—the CGIAR donors. We also thank our colleagues from IITA, particularly R. H. Markham and J. S. Yaninek, and J. J. M. van Alphen from the University of Leiden, for reviewing the manuscript.

Literature Cited

1. Agudelo-Silva, P. 1986. A species of *Triplosporium* (Zygomycetes: Entomophthorales) infecting *Mononychellus progressivus* (Acari:Tetranychidae) in Venezuela. *Fla. Entomol.* 69:444–46
2. Akinlosotu, T. A. 1981. Seasonal trend of green spider mite *Mononychellus tanajoa* population on cassava, *Manihot esculenta* and its relationship with weather factors at Moor Plantation. *Insect Sci. Appl.* 3:251–54
3. Akinlosotu, T. A., Leuschner, K. 1981. Outbreak of two new cassava pests *(Mononychellus tanajoa* and *Phenacoccus manihoti)* in southwestern Nigeria. *Trop. Pest Manage.* 27:247–50
4. Ayanru, D. K. G., Sharma, V. C. 1984/85. Changes in total cyanide content of tissues from cassava plants infested by mites *(Mononychellus tanajoa)* and mealybugs *(Phenacoccus manihoti).* *Agric. Ecosyst. Environ.* 12:35–46
5. Bartkowski, J., Odindo, M. O., Otieno, W. A. 1988. Some fungal pathogens of the cassava green spider mites *Monony-chellus* spp. (Tetranychidae) in Kenya. *Insect Sci. Appl.* 9:457–59
6. Bartlett, B. R. 1978. Pseudococcidae. In *Introduced Parasites and Predators of Arthropod Pests and Weeds: A World Review,* ed. C. P. Clausen, *USDA Agric. Handbk.* 480:137–70, Washington: USDA
7. Bellotti, A. C., Mesa, N., Serrano, M., Guerrero, J. M., Herrera, C. J. 1987. Taxonomic inventory and survey activity for natural enemies of cassava green mites in the Americas. *Insect Sci. Appl.* 8:845–49
8. Bellotti, A. C., Reyes, J. A., Guerrero, J. M., Varela, A. M. 1985. The mealybug and cassava green spider mite complex in the Americas: problems of and potential for biological control. In *Cassava: Research. Production and Utilization,* ed. J. H. Cock, J. A. Reyes, pp. 393–416. Cali: CIAT, UNDP
9. Bellotti, A. T., van Schoonhoven, A.

1978. Mite and insect pests of cassava. *Annu. Rev. Entomol.* 23:39–67

10. Bennett, F. D. 1981. Hyperparasitism in the practice of biological control. See Ref. 151a, pp. 43–49

11. Bennett, F. D., Yaseen, M. 1980. Investigations on the natural enemies of cassava mealybugs (*Phenacoccus* spp.) in the Neotropics. *CIBC Rep.* 19 pp.

12. Biassangama, A., Fabres, G., Nénon, J. P. 1988. Parasitisme au laboratoire et au champ d'*Epidinocarsis (Apoanagyrus) lopezi* (Hym.: Encyrtidae) auxiliaire exotique introduit au Congo pour la régulation de l'abondance de *Phenacoccus manihoti* (Hom.: Pseudococcidae). *Entomophaga* 33:453–65

13. Biassangama, A., le Rü, B., Iziquel, Y., Kiyindou, A., Bimangou, A. S. 1989. L'entomocénose inféodée à la cochenille du manioc, *Phenacoccus manihoti* (Homoptera: Pseudococcidae), au Congo, cinq ans après l'introduction d'*Epidinocarsis lopezi* (Hymenoptera: Encyrtidae). *Ann. Soc. Entomol. Fr.* 25:315–20

14. Bird, T. J. 1987. Fighting African cassava pests from the air. *Aerogram* 4:6–7

15. Boussienguet, J. 1986. Le complexe entomophage de la cochenille du manioc, *Phenacoccus manihoti* (Hom. Coccoidea Pseudococcidae) au Gabon. I. Inventaire faunistique et relations trophiques. *Ann. Soc. Entomol. Fr.* 22:35–44

16. Boussienguet, J. 1988. Morphologie et biologie d'*Anagyrus nyombae,* n. sp., parasite de *Phaenacoccus manihoti,* au Gabon (Hymenoptera, Encyrtidae; Homoptera, Pseudococcidae). *Rev. Fr. Entomol.* 10:277–83

17. Boussienguet, J., Neuenschwander, P. 1989. Le complexe entomophage de la cochenille du manioc en Afrique. 3. -Clé annotée pour la détermination des hyménoptères parasitoides associés à ce ravageur. *Rev. Zool. Afr.* 103:395–403

18. Boussienguet, J., Neuenschwander, P., Herren, H. R. 1990. Le complexe entomophage de la cochenille du manioc au Gabon. 4. Etablissement du parasitoide *Epidinocarsis lopezi.* *Entomophaga.* In press

19. Braun, A. R., Bellotti, A. C., Guerrero, J. M., Wilson, L. T. 1989. Effect of predator exclusion on cassava infested with tetranychid mites (Acari: Tetranychidae). *Environ. Entomol.* 18:711–14

20. Braun, A. R., Guerrero, J. M., Bellotti, A. C., Wilson, L. T. 1989. Within-plant distribution of *Mononychellus tanajoa* (Bondar) (Acari: Tetranychidae) on cassava: effect of clone and predation on aggregation. *Bull. Entomol. Res.* 79: 235–49

21. Byrne, D. H., Bellotti, A. C., Guerrero, J. M. 1983. The cassava mites. *Trop. Pest Manage.* 29:378–94

22. Charnov, E. L. 1982. *The Theory of Sex Allocation.* Princeton: Princeton Univ. Press. 355 pp.

23. Cock, J. H. 1985. *Cassava, New Potential for a Neglected Crop.* Boulder/ London: Westview Press. 191 pp.

24. Cox, J., Williams, D. J. 1981. An account of cassava mealybug (Hemiptera: Pseudococcidae) with a description of a new species. *Bull. Entomol. Res.* 71:247–58

25. Delucchi, V. 1989. Integrated pest management vs systems management. See Ref. 182, pp. 51–67

26. De Santis, L. 1963. Encírtidos de la Republica Argentina (Hymenoptera: Chalcidoidea). *An. Com. Invest. Cient. Prov. Buenos-Aires Gobern La Plata* 4:9–422

27. Dinh, N. V., Janssen, A., Sabelis, M. W. 1988. Reproductive success of *Amblyseius idaeus* and *A. anonymus* on a diet of two-spotted spider mite. *Exp. Appl. Acarol.* 4:41–51

28. Dinh, N. V., Sabelis, M. W., Janssen, A. 1988. Influence of humidity and water availability on the survival of *Amblyseius idaeus* and *A. anonymus* (Acarina: Phytoseiidae). *Exp. Appl. Acarol.* 4:27–40

29. Fabres, G. 1981. Première quantification du phénomène de gradation des populations de *Phenacoccus manihoti* Matile-Ferrero (Hom. Pseudococcidae) en République Populaire du Congo. *Agronomie* 1:483–86

30. Fabres, G. 1982. Bioécologie de la cochenille du manico (*Phenacoccus manihoti* Hom. Pseudococcidae) en République Populaire du Congo. II. Variations d'abondance et facteurs de régulation. *Agron. Trop.* 36:369–77

31. Fabres, G. 1989. Influence de la "capacité limite" dans la régulation de l'abondance d'un phytophage: Le cas de la cochenille du manioc au Congo. *Bull. Soc. Zool. Fr.* 114:35–42

32. Fabres, G., Boussienguet, J. 1981. Bioécologie de la cochenille du manioc *(Phenacoccus manihoti* Hom. Pseudococcidae) en République Populaire du Congo. *Agron. Trop.* 36:82–89

33. Fabres, G., Kiyindou, A. 1985. Comparaison du potentiel biotique de deux coccinelles (*Exochomus flaviventris* et *Hyperaspis senegalensis hottentotta,* Col. Coccinellidae) prédatrices de *Phe-*

nacoccus manihoti (Hom. Pseudococcidae) au congo. *Acta Oecol. Oecol. Appl.* 6:339–48

34. Fabres, G., Kiyindou, A., Epouna-Mouinga, S. 1981. Les entomophages inféodés à la chenille du manioc *Phenacoccus manihoti* (Hom. Pseudococcidae) en République Populaire du Congo. II. Etude morphologique comparative de trois espèces dominantes de Coccinellidae (Col.). *Cah. ORSTOM Sèr. Biol.* 44:3–8

35. Fabres, G., Le Rü, B. 1986. Etude des relations plante-insecte pour la mise au point de méthodes de régulation des populations de la cochenille du manioc. In *La Cochenille du Manioc et sa Biocoenose au Congo 1979–1984*. Travaux de l'équipe Franco-Congolaise ORSTOM-DGRS, pp. 57–71. Paris: ORSTOM

36. Fabres, G., Matile-Ferrero, D. 1980. Les entomophages inféodés à la cochenille du manioc, *Phenacoccus manihoti* (Hom. Coccoidea Pseudococcidae) en République Populaire du Congo. I. Les composantes de l'entomocoenose et leurs inter-relations. *Ann. Soc. Entomol. Fr.* 16:509–15

37. Fabres, G., Nénon, J. P., Kiyindou, A., Biassangama, A. 1989. Réflexions sur l'acclimatation d'entomophages exotiques pour la régulation des populations de la cochenille du manioc au Congo. *Bull. Soc. Zool. Fr.* 114:43–48

38. Flechtmann, C. H. W. 1986. Taxonomy of the cassava green spider mite complex, *Mononychellus* spp. (Tetranychidae). See Ref. 76a, pp. 70–80

39. Frazer, B. D., Gilbert, N. 1976. Coccinellids and aphids: a quantitative study of the impact of adult ladybirds (Coleoptera: Coccinellidae) preying on field populations of pea aphids (Homoptera: Aphididae). *J. Entomol. Soc. B. C.* 73:33–56

40. Friese, D. D., Mégevand, B., Yaninek, J. S. 1987. Culture maintenance and mass production of exotic phytoseiids. *Insect. Sci. Appl.* 8:875–78

41. Fürsch, H. 1961. Revision der afrikanischen Arten um *Exochomus flavipes* Thunb. Col. Cocc. *Entomol. Arb. Mus. G. Frey Tutzing München* 12:68–91

42. Fürsch, H. 1966. Die *Scymnus*-Arten Westafrikas (Col. Cocc.). *Entomol. Arb. Mus. G. Frey Tutzing München* 17:135–92

43. Fürsch, H. 1972. Die *Hyperaspis* Arten Afrikas mit Ausnahme des Mittelmeergebietes (Coleoptera Coccinellidae). *Mus. R. Afr. Cent. Tervuren Belg. Ann. Sér. Octavo. Sci. Zool.* 201:1–45

44. Fürsch, H. 1987. Neue afrikanische Scymnini-Arten (Coleoptera Coccinellidae) als Fressfeinde von Manihot-Schädlingen. *Rev. Zool. Afr.* 100:387–94

45. Ganga, T. 1984. Possibilités de régulations de la cochenille du manioc *Phenacoccus manihoti* Mat.-Ferr. (Hom. Pseudococcidae) par un entomophage exotique *Epidinocarsis lopezi* De Santis (Hym. Encyrtidae) en République Populaire du Congo. ORSTOM. June 1984 25 pp.

46. Gilstrap, F. E. 1977. Table top production of tetranychid mites (Acarina) and their phytoseiid natural enemies. *J. Kans. Entomol. Soc.* 50:229–33

47. Greathead, D. J., Waage, J. K. 1983. Opportunities for biological control of agricultural pests in developing countries. *World Bank Tech. Pap.* 11:1–44

48. Greathead, D. J., Lionnet, J. F. G., Lodos, N., Whellan J. A. 1971. A review of biological control in the Ethiopian Region. *Commonw. Agric. Bur. Tech. Commun.* 5. 162 pp.

49. Gutierrez, A. P., Baumgärtner, J. W. 1984. Multitrophic level models of predator-prey energetics: 1. Age-specific energetic models—pea aphid *Acyrtosiphon pisum* (Harris) (Homoptera: Aphididae) as an example. *Can. Entomol.* 116:924–32

50. Gutierrez, A. P., Baumgärtner, J. W. 1984. Multitrophic level models of predator-prey energetics: II. A realistic model of plant-herbivore-parasitoid-predator interactions. *Can. Entomol.* 116:933–49

51. Gutierrez, A. P., Neuenschwander, P., Schulthess, F., Herren, H. R., Baumgärtner, J. W., et al. 1988. Analysis of biological control of cassava pests in Africa: II. Cassava mealybug *Phenacoccus manihoti. J. Appl. Ecol.* 25:921–40

52. Gutierrez, A. P., Schulthess, F., Wilson, L. T., Villacorta, A. M., Ellis, C. K., Baumgärtner, J. W. 1987. Energy acquisition and allocation in plants and insects: a hypothesis for the feeding patterns in insects. *Can. Entomol.* 199:109–29

53. Gutierrez, A. P., Wermelinger, B., Schulthess, F., Baumgärtner, J. U., Herren, H. R., et al. 1988. Analysis of biological control of cassava pests in Africa: I. Simulation of carbon, nitrogen and water dynamics in cassava. *J. Appl. Ecol.* 25:901–20

54. Gutierrez, A. P., Yaninek, J. S., Wermelinger, B., Herren, H. R., Ellis,

C. K. 1988. Analysis of biological control of cassava pests in Africa: III. Cassava green mite *Mononychellus tanajoa*. *J. Appl. Ecol.* 25:941–50
55. Gutierrez, J. 1987. The cassava green mite in Africa: One or two species? (Acari: Tetranychidae). *Exp. Appl. Acarol.* 3:163–68
56. Hagen, K. S. 1976. Role of nutrition in insect management. *Proc. Tall Timbers Conf. Ecolo. Animal Control by Habitat Manage. Gainesville, 1974*, pp. 221–61. Tallahassee: Tall Timbers Res. Stn.
57. Hahn, S. K., Terry, E. R., Leuschner, K., Akobundu, I. O., Okali, C., Lal, R. 1979. Cassava improvement in Africa. *Field Crops Res.* 2:193–226
58. Hahn, S. K., Williams, R. J. 1973. Enquête sur le manioc en République du Zaire, March 1973: *Rep. Minist. Agric. Republic of Zaire*, Ibadan: IITA. 12 pp.
59. Hamilton, W. D. 1967. Extraordinary sex ratios. *Science* 156:477–88
60. Hammond, W. N. O., 1988. *Ecological assessment of Natural Enemies of the Cassava Mealybug Phenacoccus manihoti Mat.-Ferr. (Hom. Pseudococcidae) in Africa.* PhD thesis. Leiden: Univ. Leiden. 109 pp.
61. Hammond, W. N. O., Neuenschwander, P. 1990. Sustained biological control of the cassava mealybug *Phenacoccus manihoti* (Hom: Pseudococcidae) by *Epidinocarsis lopezi* (Hym.: Encyrtidae) in Nigeria. *Entomophaga* 35: In press
62. Hammond, W. N. O., Neuenschwander, P., Herren, H. R. 1987. Impact of the exotic parasitoid *Epidinocarsis lopezi* on cassava mealybug *(Phenacoccus manihoti)* populations. *Insect. Sci. Appl.* 8:887–91
63. Hammond, W. N. O., Nsiama She, H. D., Boussienguet, J., Ganga, T., Reyd, G. 1989. Status of biological control of the cassava mealybug, *Phenacoccus manihoti* Mat.-Ferr. (Hom., Pseudococcidae) under various ecological conditions in Central Africa. *Proc. 4th Triennial Symp. Int. Soc. Trop. Root Crops, Africa Branch, Kinshasa, 1989.* Ohowa: IDRC In press
64. Hammond, W. N. O., van Alpen, J. J. M., Neuenschwander, P., van Dijken, M. J. 1990. Aggregative foraging by field populations of *Epidinocarsis lopezi* (De Santis) (Hym.: Encyrtidae), a parasitoid of the cassava mealybug *Phenacoccus manihoti* Mat.-Ferr. Hom.: Pseudococcidae). *Oecologia* In press
65. Harris, K. M. 1981. *Dicrodiplosis manihoti*, sp.n. (Diptera: Cecidomyiidae), a predator on cassava mealybug. *Phenacoccus manihoti* Matile-Ferrero (Hom-

optera: Coccoidea: Pseudococcidae) in Africa. *Ann. Soc. Entomol. Fr.* 17:337–44
66. Hassell, M. P., Waage, J. K. 1984. Host-parasitoid population interactions. *Annu. Rev. Entomol.* 29:89–114
67. Haug, T., Herren, H. R., Nadel, D. J., Akinwumi, J. B. 1987. Technologies for the mass-rearing of cassava mealybugs, cassava green mites and their natural enemies. *Insect Sci. Appl.* 8:879–81
68. Haug, T., Mégevand, B. 1989. Development of technologies in support of contemporary biological control. See Ref. 182, pp. 141–46
69. Helle, W., Sabelis, M. W., eds. 1985. *Spider Mites. Their Biology, Natural Enemies and Control*, Vols. 1, 2. Amsterdam: Elsevier. 405 pp., 458 pp.
70. Hennessey, R. D., Muaka, T. 1987. Field biology of the cassava mealybug, *Phenacoccus manihoti*, and its natural enemies in Zaire. *Insect Sci. Appl.* 8:899–903
71. Hennessey, R. D., Neuenschwander, P., Muaka, T. 1990. Spread and current distribution of the cassava mealybug, *Phenacoccus manihoti* (Homoptera: Pseudococcidae), in Zaire. *Trop. Pest Manage.* In press
72. Herren, H. R. 1981. Cassava mealybug; an example of international collaboration. *Biocontrol News Info. CAB* 3:1
73. Herren, H. R. 1987. A review of objectives and achievements. *Insect Sci. Appl.* 8:837–40
74. Herren, H. R. 1989. The biological control program of IITA: From concept to reality. See Ref. 182. pp. 18–30
75. Herren, H. R., Bennett, F. D. 1984. Cassava pests, their spread and control. In *Advancing Agricultural Production in Africa, Proc. CAB 1st Sci. Conf. Arusha, Tanzania, Feb.*, ed. D. L. Hawksworth, pp. 110–14. Slough: CAB
76. Herren, H. R., Bird, T. J., Nadel, D. J. 1987. Technology for automated aerial release of natural enemies of the cassava mealybug and cassava green mite. *Insect Sci. Appl.* 8:883–85
76a. Herren, H. R., Hennessey, R. D., Bitterli, R., eds. 1986. *Biological Control and Host Plant Resistance to Control the Cassava Mealybug and Green Mite in Africa. Proc. Int. Workshop IFAD OAU/ STRC, Dec. 1982.* Ibadan: IITA
77. Herren, H. R., Lema, K. M. 1982. CMB-first successful releases. *Biocontrol News Inf. CAB* 3:185
78. Herren, H. R., Neuenschwander, P., Hennessey, R. D., Hammond, W. N. O. 1987. Introduction and dispersal of *Epidinocarsis lopezi* (Hym., Encyrtidae),

an exotic parasitoid of the cassava mealybug *Phenacoccus manihoti* (Hom., Pseudococcidae), in Africa. *Agric. Ecosyst. Environ.* 19:131–44
79. Hodek, I. K., Hagen, K. S., van Emden, H. F. 1972. Methods for studying effectiveness of natural enemies. In *Aphid Technology,* ed. H. F. van Emden, pp. 147–88. London: Academic
80. Hughes, R. D. 1963. The population dynamics of the cabbage aphid, *Brevicoryne brassicae* (L.). *J. Anim. Ecol.* 32:393–426
81. Iheagwam, E. U. 1981. The influence of temperature on increase rates of the cassava mealybug *Phenacoccus manihoti* Mat.-Ferr. (Homoptera, Pseudococcidae). *Rev. Zool. Afr.* 95:959–67
82. Iheagwam, E. U. 1981. Natural enemies and alternative host plant of the cassava mealybug, *Phenacoccus manihoti* (Hom., Pseudococcidae) in southeastern Nigeria. *Rev. Zool. Afr.* 95:433–38
83. Iziquel, Y., Le Rü, B., 1989. Influence de l'hyperparasitisme sur les populations d'un hyménoptère encyrtidae, *Epidinocarsis lopezi,* parasitoide de la cochenille du manioc *Phenacoccus manihoti* introduit au Congo. *Entomol. Exp. Appl.* 52:239–47
84. Iziquel, Y., le Ralec, A., Nénon, J. P. 1988. *Epidinocarsis lopezi* (Hymenoptera; Encyrtidae): ovipositeur, types de piqûres et nature du parasitisme sur *Phenacoccus manihoti* (Homoptera; Pseudococcidae). *Nat. Can. (Rev. Ecol. Syst.)* 115:355–66
85. Kiritani, K., Dempster, J. P. 1973. Different approaches to the quantitative evaluation of natural enemies. *J. Appl. Ecol.* 10:323–30
86. Kiyindou, A. 1989. Seuil thermique de développement de trois coccinelles prédatrices de la cochenille du manioc au Congo. *Entomophaga* 34:409–15
87. Kiyindou, A., Fabres, G. 1987. Etude de la capacité d'accroissement chez *Hyperaspis raynevali* (Col.: Coccinellidae) prédateur introduit au Congo pour la régulation des populations de *Phenacoccus manihoti* (Hom.: Pseudococcidae). *Entomophaga* 32:181–89
88. Kläy, A. 1990. The cassava green spider mite biological control programme: The use of population models to monitor mass-production of natural enemies. *Workshop on Status and Prospect of Integrated Pest Management for Root and Tuber Crops in the Tropics.* Ibadan: IITA In press
89. Korang-Amoakoh, S., Cujoe, R. A., Adjakloe, R. K. 1987. Biological control of cassava pests in Ghana. *Insect Sci. Appl.* 8:905–7
90. Kraaijeveld, A. R., van Alpen, J. J. M. 1986. Host-stage selection and sex allocation by *Epidinocarsis lopezi* (Hymenoptera; Encyrtidae), a parasitoid of the cassava mealybug, *Phenacoccus manihoti* (Homoptera; Pseudococcidae). *Meded. Fac. Landbouwwet. Rijksuniv. Gent.* 51:1067–78
91. Langenbach, G. E. J., van Alpen, J. J. M. 1986. Searching behavior of *Epidinocarsis lopezi* (Hymenoptera; Encyrtidae) on cassava: effect of leaf topography and a kairomone produced by its host, the cassava mealybug *(Phenacoccus manihoti). Meded Fac. Landbouwwet. Rijksuniv. Gent* 51:1057–65
92. Lema, K. M., Herren, H. R. 1985. The influence of constant temperature on population growth rates of the cassava mealybug, *Phenacoccus manihoti. Entomol. Exp. Appl.* 38:165–69
93. Lema, K. M., Herren, H. R. 1985. Release and establishment in Nigeria of *Epidinocarsis lopezi* a parasitoid of the cassava mealybug, *Phenacoccus manihoti. Entomol. Exp. Appl.* 38:171–75
94. Le Rü, B. 1986. Etude de l'évolution d'une mycose à *Neozygites fumosa* (Zygomycetes, Entomophthorales) dans une population de la cochenille du manioc *Phenacoccus manihoti* (Hom.: Pseudococcidae). *Entomophaga* 31:79–89
95. Le Rü, B., Fabres, G. 1987. Influence de la température et de l'hygrométrie relative sur la capacité d'accroissement et le profil d'abondance des populations de la cochenille du manioc, *Phenacoccus manihoti* (Hom., Pseudococcidae) au Congo. *Acta Oecol. Oecol. Appl.* 8:165–74
96. Le Rü, B., Iziquel, Y. 1988. Evaluation de l'incidence des pluies, à l'aide d'un simulateur de pluies, sur la dynamique des populations de la cochenille du manioc, *Phenacoccus manihoti.* In *La Cochenille du Manioc et sa Biocenose au Congo 1985–1987.* Travaux de l' équipe Franco-Congolaise ORSTOM-DGRS, pp. 19–42. Paris: ORSTOM
97. Le Rü, B., Iziquel, Y., Biassangama, A., Kiyindou, A. 1990. Comparaison des effectifs de la cochenille du manioc *Phenacoccus manihoti* avant et après introduction d'*Epidinocarsis lopezi* Encyrtidae Americain, au Congo en 1982. *Entomol. Exp. Appl.* In press
98. Le Rü, B., Papierok, B. 1987. Taux intrisèque d'accroissement naturel de la cochenille du manioc, *Phenacoccus manihoti* Matile-Ferrero (Homoptères,

Pseudococcidae). Intérêt d'une méthode simplifiée d'estimation de r_m. *Acta Oecol. Oecol. Appl.* 8:3–14

99. Le Rü, B., Silvie, P., Papierok, B. 1985. L'entomophthorale *Neozygites fumosa* pathogène de la cochenille du manioc, *Phenacoccus manihoti* (Hom.: Pseudococcidae), en République Populaire du Congo. *Entomophaga* 30:23–29

100. Leuschner, K. 1978. Preliminary observations on the mealybug (Hemiptera Pseudococcidae) in Zaire and a projected outline for subsequent work. See Ref. 145a, pp. 15–19.

101. Lian, T. S., Cock, J. H. 1979. Branching habit as a yield determinant in cassava. *Field Crop Res.* 2:281–89

102. Löhr, B., Neuenschwander, P., Varela, A. M., Santos, B. 1988. Interactions between the female parasitoid *Epidinocarsis lopezi* De Santis (Hym., Encyrtidae) and its host the cassava mealybug, *Phenacoccus manihoti* Matile-Ferrero (Hom., Pseudococcidae). *J. Appl. Entomol.* 105:403–12

103. Löhr, B., Santos, B., Varela, A. M. 1989. Larval development and morphometry of *Epidinocarsis lopezi* (De Santis) (Hym., Encyrtidae), parasitoid of the cassava mealybug, *Phenacoccus manihoti* Mat.-Ferr. (Hom., Pseudococcidae). *J. Appl. Entomol.* 107:334–43

104. Löhr, B., Varela, A. M. 1986. The cassava mealybug, *Phenacoccus manihoti* Mat.-Ferr., in Paraguay: Further information on occurrence and population dynamics of the pest and its natural enemies. See Ref 76a, pp. 57–69

105. Löhr, B., Varela, A. M., Santos, B. 1989. Life-table studies of *Epidinocarsis lopezi* (De Santis) (Hym. Encyrtidae), a parasitoid of the cassava mealybug. *Phenacoccus manihoti* Mat.-Ferr. (Hom., Pseudococcidae). *J. Appl. Entomol.* 107:425–34

106. Löhr, B., Varela, A. M., Santos, B. 1990. Exploration for natural enemies of the cassava mealybug, *Phenacoccus manihoti* (Hemiptera: Pseudococcidae), in South America for the biological control of this introduced pest in Africa. *Bull. Entomol. Res.* In press

107. Luck, R. F., Messenger, P. S., Barbieri, J. F. 1981. The influence of hyperparasitism on the performance of biological control agents See Ref. 151a, pp. 34–42

108. Lyon, W. F. 1973. A plant feeding mite *Mononychellus tanajoa* (Bondar) (Acarina: Tetranychidae) new to the African continent threatens cassava *(Manihot-*

esculenta Crantz) in Uganda, East Africa. *Pest Artic. News Summ.* 19:36–37

109. Manetsch, T. J. 1976. Time varying distributed delays and their use in aggregative models of large systems. *IEEE Trans. Syst. Man Cybern.* 6:547–53

110. Markham, R. H., Robertson, I. A. D., eds. 1987. *Cassava Green Mite in Eastern Africa: Yield losses and integrated Control. Proceedings of a Regional Workshop, Nairobi, May, 1986.* Nairobi: CABI. 186 pp.

111. Markham, R. H., Robertson, I. A. D., Kirby, R. A. 1987. Cassava green mite in East Africa: A regional approach to research and control. *Insect Sci. Appl.* 8:909–14

112. Markham, R. H., Sicely, E. M. 1989. Appropriate support for national programs: Training, research, administration, and funding. See Ref. 182, pp. 156–65

113. Matile-Ferrero, D. 1977. Une cochenille nouvelle nuisible au manioc en Afrique équatoriale, *Phenacoccus manihoti* n. sp. (Hom. Coccoidea Pseudococcidae). *Ann. Soc. Entomol. Fr.* 13:145–52

114. Matile-Ferrero, D. 1978. Cassava mealybug in the People's Republic of Congo. See Ref. 145a, pp. 29–46

115. May, R. M., Hassell, M. P. 1988. Population dynamics and biological control. *Philos. Trans. R. Soc. London Ser. B* 318:129–69

116. McMurtry, J. A. 1982. The use of phytoseiids for biological control: progress and future prospects. In *Recent Advances in Knowledge of the Phytoseiidae. Proc. Formal Conf. Acarology Soc. Am. held at the Entomol. Soc. Am. Meet. San Diego, 1981,* ed. M. A. Hoy, pp. 23–48. Berkeley: Univ. Calif. Publ. 3284

117. McMurtry, J. A., Scriven, G. T. 1985. Insectary production of phytoseiid mites. *J. Econ. Entomol.* 58:282–84

118. Mégevand, B., Yaninek, J. S., Friese, D. D. 1987. Classical biological control of the cassava green mite. *Insect Sci. Appl.* 8:871–74

119. Moore, D. 1988. Agents used for biological control of mealybugs (Pseudococcidae). *Biocontrol News Info.* 9:209–25

120. Moraes, G. J., McMurtry, J. A., Denmark, H. A. 1986. *A Catalog of the Mite Family, Phytoseiidae: References to Taxonomy, Synonymy, Distribution and Habitat.* Brasília EMBRAPA Dep. Difusão Technol. 353 pp.

121. Moraes, G. J., McMurtry, J. A., van den Berg, H., Yaninek, J. S. 1989. Phy-

toseiid mites (Acari: Phytoseiidae) of Kenya, with descriptions of five new species and complementary descriptions of eight species. *Int. J. Acarol.* 15:79–93

122. Moraes, G. J., McMurtry, J. A., Yaninek, J. S. 1989. Some phytoseiid mites (Acari: Phytoseiidae) from tropical Africa with a description of a new species. *Int. J. Acarol.* 15:95–102

123. Moraes, G. J., Mesa, N. C. 1988. Mites of the family Phytoseiidae (Acari) in Colombia, with descriptions of three new species. *Int. J. Acarol.* 14:71–88

124. Moraes, G. J., Mesa, N. C., Reyes, J. A. 1988. Some phytoseiid mites (Acari: Phytoseiidae) from Paraguay, with description of a new species. *Int. J. Acarol.* 14:221–23

125. Murdoch, W. W., Chesson, J., Chesson, P. L. 1985. Biological control in theory and practice. *Am. Nat.* 125:344–66

126. Nadel, H., van Alpen, J. J. M. 1987. The role of host- and host-plant odours in the attraction of a parasitoid, *Epidinocarsis lopezi*, to the habitat of its host, the cassava mealybug, *Phenacoccus manihoti*. *Entomol. Exp. Appl.* 45:181–86

127. Nénon, J. P. 1989. Interactions réciproques entre Hyménoptères entomoparasitoides et hôtes; nature et succès du parasitisme. *Bull. Soc. Zool. Fr.* 114:1–11

128. Nénon, J. P., Guyomard, O., Hémon, G. 1988. Encapsulement des oeufs et des larves de l'Hyménoptère Encyrtidae *Epidinocarsis* (=*Apoanagyrus*) *lopezi* par son hôte Pseudococcidae *Phenacoccus manihoti:* effet de la température et du superparasitisme. *CR Acad. Sci.* 306:325–41

129. Neuenschwander, P., Gutierrez, A. P. 1989. Evaluating the impact of biological control. See Ref. 182, pp. 147–55

130. Neuenschwander, P., Hammond, W. N. O. 1988. Natural enemy activity following the introduction of *Epidinocarsis lopezi* (Hymenoptera, Encyrtidae) against the cassava mealybug, *Phenacoccus manihoti* (Homoptera, Pseudococcidae), in southwestern Nigeria. *Environ. Entomol.* 17:894–902

131. Neuenschwander, P., Hammond, W. N. O., Ajuonu, O., Gado, A., Echendu, N., et al. 1990. Biological control of the cassava mealybug, *Phenacoccus manihoti* (Hom. Pseudococcidae) by *Epidinocarsis lopezi* (Hym., Encyrtidae) in West Africa, as influenced by climate and soil. *Agric. Ecosyst. Environ.* 22:In press

132. Neuenschwander, P., Hammond, W. N. O., Gutierrez, A. P., Cudjoe, A. R., Baumgärtner, J. U. et al. 1989. Impact assessment of the biological control of the cassava mealybug, *Phenacoccus manihoti* Matile-Ferrero (Hemiptera: Pseudococcidae) by the introduced parasitoid *Epidinocarsis lopezi* (De Santis) (Hymenoptera: Encyrtidae). *Bull Entomol. Res.* 79:579–94

133. Neuenschwander, P., Haug, T. 1990. New technologies for rearing *Epidinocarsis lopezi*, a biological control agent against the cassava mealybug. In *Advances and Applications in Insect Rearing*, ed. T. A. Anderson, N. Leppla. Boulder: Westview. In press

134. Neuenschwander, P., Haug, T., Ajuonu, O., Davis, H., Akinwumi, B., et al. 1989. Quality requirements in natural enemies used for inoculative release. Practical experience from a successful biological control programme. *J. Appl. Entomol.* 108:409–20

135. Neuenschwander, P., Hennessey, R. D., Herren, H. R. 1987. Food web of insects associated with the cassava mealybug, *Phenacoccus manihoti* Matile-Ferrero (Hemiptera: Pseudococcidae), and its introduced parasitoid *Epidinocarsis lopezi* (De Santis) (Hymenoptera: Encyrtidae), in Africa. *Bull. Entomol. Res.* 77:177–89

136. Neuenschwander, P., Herren, H. R. 1988. Biological control of the cassava mealybug, *Phenacoccus manihoti*, by the exotic parasitoid *Epidinocarsis lopezi* in Africa. *Philos. Trans. R. Soc. London Ser. B* 318:319–33

137. Neuenschwander, P., Madojemu, E. 1986. Mortality of the cassava mealybug *Phenacoccus manihoti* Mat.-Ferr. (Hom., Pseudococcidae) associated with an attack by *Epidinocarsis lopezi* (Hym., Encyrtidae). *Mitt. Schweiz. Entomol. Ges.* 59:57–62

138. Neuenschwander, P., Schulthess, F., Madojemu, E. 1986. Experimental evaluation of the efficiency of *Epidinocarsis lopezi*, a parasitoid introduced into Africa against the cassava mealybug *Phenacoccus manihoti*. *Entomol. Exp. Appl.* 42:133–38

139. Nokoe, S., Rogo, L. M. 1988. A discriminant function of the short-and long-setaed forms of the *Mononychellus* (Acari: Tetranychidae) species complex. *Insect Sci. Appl.* 9:429–32

140. Norgaard, R. B. 1988. The biological control of cassava mealybug in Africa. *Am. J. Agric. Econ.* 70:366–71

141. Noyes, J. S. 1984. A new genus and

species of encyrtid (Hymenoptera: Chalcidoidea) parasitic on the cassava mealybug, *Phenacoccus manihoti* Matile-Ferrero (Hemiptera: Pseudococcidae). *Bull. Entomol. Res.* 74:529–33

142. Nsiama She, H. D. 1985. *The Bioecology of the Predator* Hyperaspis jucunda *Muls. (Coleoptera: Coccinellidae) and the Temperature Responses of its Prey, the Cassava Mealybug* Phenacoccus manihoti *Mat.-Ferr.* PhD thesis. Ibadan: Univ. Ibadan. 300 pp.

143. Nsiama She, H. D. 1987. Progrès enrégistré en matière de lutte biologique contre la cochenille farineuse du manioc au Zaïre, *Séminaire sur les maladies et les ravageurs des principales cultures vivrières d'Afrique centrale, Bujumbura,* pp. 256–65. Wegeningen: CTA; Bruxelles: AGCD

144. Nsiama She, H. D., Odebiyi, J. A., Herren, H. R. 1984. The biology of *Hyperaspis jucunda* (Col.: Coccinellidae) an exotic predator of the cassava mealybug *Phenacoccus manihoti* (Hom.: Pseudococcidae) in southern Nigeria. *Entomophaga* 29:87–93

145. Nwanze, K. F. 1982. Relationships between cassava root yields and infestations by the mealybug, *Phenacoccus manihoti. Trop. Pest Manage.* 28:27–32

145a. Nwanze, K. F., Leuschner, K., eds. 1978. *Proc. Int. Workshop on the Cassava Mealybug* Phenacoccus manihoti *Mat.-Ferr. (Pseudococcidae). M'vuazi, Zaire, June 1977.* Ibadan: IITA

146. Nwanze, K. F., Leuschner, K., Ezumah, H. C. 1979. The cassava mealybug, *Phenacoccus* sp. in the Republic of Zaire. *Pest Artic. News Summ.* 25:125–30

147. Nyiira, Z. M. 1972. *Report of Investigation of Cassava Mite,* Mononychellus tanajoa *Bondar.* Kawanda Res. Stn. Kampala. 14 pp.

148. Odebiyi, J. A., Bokonon-Ganta, A. H. 1986. Biology of *Epidinocarsis (=Apoanagyrus) lopezi* (Hymenoptera: Encyrtidae) an exotic parasite of cassava mealybug, *Phenacoccus manihoti* (Homoptera: Pseudococcidae) in Nigeria. *Entomophaga* 31:251–60

149. Ouayogodé, B. V. 1989. Assistance needed by national institutions in developing sustainable pest management capacity. See Ref. 182, pp. 107–11

150. Renvoize, B. S. 1973. The area of origin of *Manihot esculenta* as a crop plant—a review of the evidence. *Econ. Bot.* 26:352–60

151. Rogo, L. M., Oloo, W., Nokoe, S.,

Magalit, H. 1988. A study of the *Mononychellus* (Acari: Tetranychidae) species complex from selected cassava growing areas of Africa using principal component analysis. *Insect Sci. Appl.* 9:593–99

151a. Rosen, D., ed. 1981. *The Role of Hyperparasitism in the Practice of Biological Control: A Symposium.* Berkeley: Univ. Calif. Div. Agric. Sci.

152. Samways, M. J. 1979. Immigration, population growth and mortality of insects and mites on cassava in Brazil. *Bull. Entomol. Res.* 69:491–505

153. Schulthess, F. 1987. *The Interactions between Cassava Mealybug* (Phenacoccus manihoti *Mat.-Ferr.) Populations and Cassava* (Manihot esculenta *Crantz) as Influenced by Weather.* PhD thesis. Zurich: Swiss Fed. Inst. Technol. 136 pp.

154. Schulthess, F., Baumgärtner, J. U., Herren, H. R. 1987. Factors influencing the life table statistics of the cassava mealybug *Phenacoccus manihoti. Insect Sci. Appl.* 8:851–56

155. Schulthess, F., Baumgärtner, J. U., Herren, H. R. 1989. Sampling *Phenacoccus manihoti* in cassava fields in Nigeria. *Trop. Pest Manage.* 35:193–200

156. Sulllivan, D. J. 1988. Hyperparasites. In *Aphids, Their Biology, Natural Enemies and Control,* ed. A. K. Minks, P. Harrewijn, B:189–203. Amsterdam: Elsevier

157. Sullivan, D. J., Neuenschwander, P. 1988. Melanization of eggs and larvae of the parasitoid, *Epidinocarsis lopezi* (Hymenoptera: Encyrtidae), by the cassava mealybug, *Phenacoccus manihoti* (Homoptera: Pseudococcidae). *Can. Entomol.* 120:63–71

158. Sylvestre, P. 1973. *Aspects Agronomiques de la Production du Manioc à la ferme d'État de Mantsumba (Rép. Pop. Congo).* Paris: Inst. Rech. Agron. Trop. 35 pp.

159. Sylvestre, P., Arraudeau, M. 1983. Le manioc. In *Techniques Agricoles et Production Tropicales,* Vol. 32, ed. R. Coste, Paris: Maisonneuve & Larose. 262 pp.

160. Ueckermann, E. A., Loots, G. C. 1988. The African species of the subgenera *Anthoseius* De Leon and *Amblyseius* Berlese. *Entomol. Mem. South Africa* 73:1–168

161. Umeh, E. D. N. N. 1982. Biological studies on *Hyperaspis marmottani* Fairm. (Col., Coccinellidae), a predator of the cassava mealybug *Phenacoccus manihoti* Mat.-Ferr. (Hom., Pseudo-

coccidae). *J. Appl. Entomol.* 94:530–32

162. Umeh, E. D. N. N. 1983. Toward a biological control of cassava mealybug, *Phenacoccus manihoti* Mat.-Ferr. (Homoptera: Pseudococcidae). *Rev. Zool. Afr.* 97:60–64

163. Umeh, E. D. N. N. 1984. Biological control of the cassava mealybug *Phenacoccus manihoti* Mat.-Ferr. (Homoptera: Pseudococcidae) by *Hyperaspis marmottani* Fairm. (Coleoptera: Coccinellidae). Presented at *17th Annu. Conf. Entomol. Soc. Nigeria, Ibadan.* Pap. 27

163a. Umeh, E. D. N. 1988. Development, oviposition, host feeding and sex determination in *Epidinocarsis lopezi* (De Santis) (Hymenoptera : Encyrtidae). *Bull. Entomol. Res.* 78:605–11

164. Umeh, E. D. N. 1990. *Exochomus troberti* Mulsant (Coleoptera: Coccinellidae): a predator of cassava mealybug, *Phenacoccus manihoti* Mat.-Ferr. (Homoptera: Pseudococcidae) in southeastern Nigeria. *Insect Sci. Appl.* 11:In press

165. van Alpen, J. J. M., Neuenschwander, P., van Dijken, M. J., Hammond, W. N. O., Herren, H. R. 1989. Insect invasions: the case of the cassava mealybug and its natural enemies evaluated. *Entomologist* 108:38–55

166. van Alpen, J. J. M., Vet, L. E. M. 1986. An evolutionary approach to host finding and selection. In *Insect Parasitoids*, ed. J. Waage, D. Greathead, pp. 23–61. London: Academic

167. van den Berg, H., Markham, R. H. 1987. Prey preference and reproductive success of the predatory mite *Neoseiulus idaeus* on the prey species *Mononychellus tanajoa* and *Tetranychus lombardinii. Insect Sci. Appl.* 8:867–69

168. van den Meiracker, R. A. F., Hammond, W. N. O., van Alpen, J. J. M. 1988. The role of kairomones in prey finding in the coccinellids *Diomus* sp. and *Exochomus* sp., predators of the cassava mealybug, *Phenacoccus manihoti. Meded. Fac. Landbouwwet. Rijksuniv. Gent.* 53:1063–77

169. van Driesche, R. G. 1983. Meaning of "percent parasitism" in studies of insect parasitoids. *Environ. Entomol.* 12:1611–22

169a. van Dijken, M. J., Neuenschwander, P., van Alphen J. J. M., Hammond, W. N. O. 1990. Sex ratios in field populations of *Epidinocarsis lopezi*, an exotic parasitoid of the cassava mealybug, *Phenacoccus manihoti*, in Africa. *Ecol. Entomol.* In press

170. van Dijken, M. J., van Alphen, J. J. M., van Stratum, P. 1989. Sex allocation in *Epidinocarsis lopezi:* local mate competition. *Entomol. Exp. Appl.* 52:249–55

171. van Lenteren, J. C. 1980. Evaluation of control capabilities of natural enemies: Does art have to become science? *Neth. J. Zool.* 30:369–81

172. Vansickle, J. 1977. Attrition in distributed delay models. *IEEE Trans. Syst. Man Cybern.* 7:635–38

173. Vargas, O. H., Bellotti, A. C., El-Sharkawy, M., Hernandez, A., del Pilar 1989. Calcium extraction by *Phenacoccus herreni:* Symptoms and effects on cassava photosynthesis. *Cassava Newsl. CIAT* 13:8–10

174. Waage, J. K., Godfray, H. C. J. 1985. Reproductive strategies and population ecology of insect parasitoids. In *Behavioural Ecology. Ecological Consequences of Adaptive Behaviour*, ed. R. Sibly, R. H. Smith, pp. 449–70. Oxford: Blackwell Scientific

175. Williams, D. J. 1986. Mealybugs (Homoptera: Pseudococcidae) on cassava with special reference to those associated with wild and cultivated cassava in the Americas. See Ref. 76a, pp. 49–56

176. Wodageneh, A., Herren H. R. 1987. International co-operation: Training and initiation of national biological control programmes. *Insect Sci. Appl.* 8:915–18

177. Yaninek, J. S. 1985. *An Assessment of the Phenology, Dynamics and Impact of Cassava Green Mites on Cassava Yields in Nigeria: A Component of Biological Control.* PhD thesis. Berkeley: Univ. Calif. 166 pp.

178. Yaninek, J. S. 1988. Continental dispersal of the cassava green mite, an exotic pest in Africa, and implications for biological control. *Exp. Appl. Acarol.* 4:211–24

179. Yaninek, J. S., Bellotti, A. C. 1987. Exploration for natural enemies of cassava green mites based on agrometeorological criteria. In *Proceedings of the Seminar on Agrometeorology and Crop Protection in the Lowland Humid and Sub-humid Tropics*, ed. D. Rijks, G. Mathys, pp. 69–75. Geneva: World Metorological Organization

180. Yaninek, J. S., Gutierrez, A. P., Herren, H. R. 1989. Dynamics of *Mononychellus tanajoa* (Acari: Tetranychidae) in Africa: Experimental evidence of the effects of temperature and host plant on population growth rates. *Environ. Entomol.* 18:633–40

181. Yaninek, J. S., Herren, H. R. 1988.

Introduction and spread of the cassava green mite, *Mononychellus tanajoa* (Bondar) (Acari: Tetranychidae), an exotic pest in Africa and the search for appropriate control methods: a review. *Bull Entomol. Res.* 78:1–13

182. Yaninek, J. S., Herren, H. R., eds. 1989. *Biological Control: A Sustainable Solution to Crop Pest Problems in Africa.* Ibadan: IITA. 210 pp.

183. Yaninek, J. S., Herren, H. R., Gutierrez, A. P. 1987. The biological basis for the seasonal outbreak of cassava green mites in Africa. *Insect Sci. Appl.* 8:861–65

184. Yaninek, J. S., Herren, H. R., Gutierrez, A. P. 1989. Dynamics of *Mononychellus tanajoa* (Acari: Tetranychidae) in Africa: Seasonal factors affecting phenology and abundance. *Environ. Entomol.* 18:625–32

185. Yaninek, J. S., Moraes, G. J., Markham, R. H. 1989. *Handbook on the Cassava Green Mite* Mononcychellus tanajoa *in Africa: A Guide to Their Biology and Procedures for Implementing Classical Biological Control.* Ibadan: IITA. 140 pp.

186. Yaseen, M. 1986. Exploration for natural enemies of *Phenacoccus manihoti* and *Mononychellus tanajoa:* The challenge, the achievements. See Ref. 76a, pp. 81–102

187. Yaseen, M., Bennett, F. D. 1977. Distribution, biology, and population dynamics of the green cassava mite in the neotropics. In *Proc. 4th Symp. Int. Soc. Trop. Root Crops, CIAT, Cali, Aug. 1976,* ed. J. Cock, R. MacIntyre, M. Graham, pp. 197–202. Ottawa: IDRC

Annu. Rev. Entomol. 1991. 36:285–304

SAMPLING AND ANALYSIS OF INSECT POPULATIONS

Eizi Kuno

Entomological Laboratory, College of Agriculture, Kyoto University, Kyoto 606, Japan

KEY WORDS: dynamics, spatial patterns, variance-mean relationship, sequential census, key factor analysis

INTRODUCTION

In the past decades, methodology for sampling and analyzing insect populations has developed markedly and has played a significant role in recent advances in the study of insect population dynamics. As a result, we now have a wide variety of techniques, which range from mechanical devices for observing, collecting, or marking living insects to mathematical procedures for analyzing dynamic population processes based on life-table data, that can be applied to diverse aspects of insect population studies.

At present, several comprehensive textbooks (73,97, 109, 118) and review articles (33, 81, 117, 121, 143) have been published about this subject. After reading these, one can grasp an outline of population techniques together with the history of their development. I therefore restrict my present review to the following three specified topics that have either advanced markedly or been actively debated in recent years: (*a*) analysis of spatial distribution patterns based on quadrat counts; (*b*) design of labor-saving sampling procedures for population estimation and decision making; and (*c*) detection of key factors and density regulation from life table data.

285

0066-4170/91/0101-0285$02.00

ANALYSIS OF SPATIAL PATTERNS

Patterns of Frequency Distribution

The analysis of spatial distribution patterns is now recognized as an indispensable procedure for studies of insect populations and provides basic information for interpreting the spatial structures and for designing efficient sampling programs for population estimation and pest management (43, 127). The usual first step is to divide the habitat area for study into many units or *quadrats* of equal size, and then to describe the pattern as a frequency distribution of the number of individuals observed in each quadrat using appropriate mathematical distribution models.

The simplest of these models is the well-known Poisson distribution with a single parameter, mean m, i.e. $P(x) = m^x e^{-m}/x!$ [$P(x)$ is the probability of finding x individuals in a given quadrat]. This model assumes purely random or independent disposition of individuals over the homogeneous habitat so that it serves as the criterion for classifying the observed data set into either of three categories, *uniform, random,* and *patchy* or *(contagious)* distributions.

If so classified, however, most of the observed insect distributions may fall into the last of the three, with the variance larger than the mean. For such patchy distributions, several two-parameter mathematical models have been proposed, such as Neyman's (88), Thomas' (137), Polya's (114), Poisson-binomial (74), and the negative binomial (5, 13). Most of these model randomly distributed clumps or colonies that have different size distributions (see 127 for a review). Among these models, ecologists now use only one, the negative binomial distribution,

$$P(x) = \binom{-k}{x} \left(\frac{-m}{k + m}\right)^x \left(\frac{k}{k + m}\right)^k$$

where x, m, and k are the number of individuals at each quadrat, mean density, and non-negative characteristic exponent, respectively. This equation describes populations so well that one might rarely encounter field data that show significant departure from this model because of the inherent plasticity characterizing it, i.e. that it can be derived from several quite different assumptions or processes such as habitat heterogeneity, random distribution of colonies, true contagion, and so on (5).

The model's mathematical properties as well as the methodology for fitting it to field data and calculating its specific parameters have been well investigated (5, 13, 14). Although the original negative binomial is defined only for $k > 0$, the same model can be readily extended to include $k < 0$ (more strictly, k is a negative integer) (60), which corresponds to the model known so far as the *positive binomial* (17). Namely, because of this generalization, the above formula can now describe distributions of all three categories, random ($1/k = 0$), patchy ($1/k > 0$), and uniform ($1/k < 0$), by varying its parameter k.

To conclude, adoption of this extended negative binomial distribution may eliminate the need in most studies of insect populations to consider any other models of spatial distribution, at least for the purpose of simple description.

Indices of Crowding and Patchiness

Once the general pattern of individual sample distributions is pictured by using some mathematical frequency model such as the negative binomial, the next step is to describe the quantitative properties in terms of some ecologically meaningful parameters, which are classified conceptually into two types: the indexes of crowding and of patchiness (43, 94).

For measuring the degree of crowding, an appropriate index is widely used, i.e. Lloyd's (64) *mean crowding,* which represents the mean number per individual of other individuals coexisting in the same quadrat, a most straightforward expression of the crowding intensity that individuals actually face. It is a simple function of mean m and variance σ^2 written as $m^* = m + (\sigma^2/m) - 1$, so that one can readily estimate m* by substituting sample mean and variance ($\bar{x}$ and s^2) for m and σ^2 in this equation. For specific cases in which the quadrat size is not constant but fluctuates remarkably from one quadrat to another, a modified formula to calculate the adjusted mean crowding is also available (148). Furthermore, if the distribution can be assumed, as is usual, to follow the negative binomial, the variance of this sample estimate $\hat{m}^*$ for a sample of size n can also be estimated from the following relation (127) by replacing m and k with the sample estimates $\bar{x}$ and $\hat{k}$ $[= (\bar{x}^2 - s^2/n)/(s^2 - \bar{x})]$:

$$V(\hat{m}^*) \simeq \frac{1 + 1/k}{n} \left\{ \frac{m^2}{k} \left(1 + \frac{3}{k} \right) + m \left(1 + \frac{5}{k} \right) + 2 \right\}.$$

Patchiness, on the other hand, is a statistical concept that corresponds to the relative magnitude of spatial, quadrat-to-quadrat variations of population density (94), and now we have three indices appropriate to measure it: Morisita's (76–78) I_δ based on the probability that two randomly selected individuals will be in the same quadrat; Kuno's (53) c_A based on the concept of relative variance; and Lloyd's (64) m^*/m based on the crowding index just described.

These three indices are relative ones in the sense that they can measure patchiness independently of the mean density, taking the value of either 1 (I_δ and m^*/m) or 0 (c_A) in the random (Poisson) distribution as the null case (94). Use of these indices is sometimes criticized because in field data their values often show significant correlations with mean densities (127, 131). But this correlation does not invalidate the use of these indices; rather, it indicates that patchiness itself changes according to the density. The density independence of these indices per se may be evident from the fact that random removal of individuals does not change the index value (53).

Interestingly, these three indices have mathematical forms that are essentially the same despite the differences in the process of derivation. All are simple functions of mean m and variance σ^2, i.e. $c_A = (\sigma^2 - m)/m^2$ and $m^*/m = 1 + (\sigma^2/m^2) - (1/m) \simeq I_\delta$ so that $c_A = (m^*/m) - 1 \simeq I_\delta - 1$ (94). These relations also show that the degree of patchiness thus measured is closely related to the degree of crowding, described as $m^* = (c_A + 1)m = I_\delta m$. The sample estimators of these patchiness indices can all be obtained by replacing m, σ^2, and m^2 in these expressions with their unbiased sample estimators, $\bar{x}$, s^2, and $\bar{x}^2 - (s^2/n)$. If the distribution is known to follow the negative binomial, the common asymptotic variance of these three mathematically equivalent statistics is calculable from the following relation (5) by replacing m and k with $\bar{x}$ and $\hat{k}$ $[= (\bar{x}^2 - s^2/n)/(s^2 - \bar{x}) = 1/\hat{c}_A]$:

$$V(\hat{c}_A) = V(\hat{I}_\delta) = V(\hat{m}^*/m) \simeq \frac{2}{n}\left\{\left(1 + \frac{1}{k}\right)\left(\frac{1}{m} + \frac{1}{k}\right)^2\right\}.$$

Also, if several estimates of the same patchiness index (e.g. c_A) are obtained from the same population, one can determine the unbiased, weighted average based on the variances, invoking the convenient procedure devised for calculating the reciprocal of a common k of the negative binomial (14).

Variance-Mean Relationship

We have seen above that the indices of patchiness and crowding that characterize individual frequency distributions are all simple functions of the first two moments, m and σ^2. It may follow that, if we have an adequate series of samples from populations of a given species, the analysis of the variance-mean relationship based on these samples may provide us with comprehensive information on the species' spatial pattern, together with the statistical basis of sampling design for the estimation and management of the population.

In discrete distributions, variance inherently has a close correlation with the mean. This correlation of course applies to all the theoretical models so far discussed, such as the Poisson, negative binomial, positive binomial, Neyman's, etc (see Figure 1A). In general, the relationship of organisms in actual populations is similarly close, as one can see in Figure 1B, in which the σ^2/m relations observed for 13 species of organisms from humans to an orchid (and including four insect species) are redrawn on the same graph after Taylor et al's extensive study (Figure 6 in 131). One may find that, despite extremely wide diversity of these 13 species as to taxonomy, morphology, ecology, quadrat size, etc, the overall relationship of variance to mean is surprisingly definite and consistent, showing monotonic increase with fairly good fit to the curve of the tentatively drawn negative binomial ($k = 1$).

Of course it may be nonsensical to imagine from this superficial correlation, which might also have been assisted by stabilizing effects of log-transformation on both the axes, that the distributions of all these different organisms have been generated by any common underlying mechanisms. Presumably this high correlation may be interpreted simply as manifesting the inherent, statistical tendency underlying any discrete distribution, which is strong enough to conceal definite differences in the distribution-generating processes among such widely divergent species.

So far two types of equations, empirical and deductive, have been used as general models to describe the variance-mean relationship of biological populations. One is Taylor's (125) empirical power equation (TP), $\sigma^2 = am^b$, whereas the other is Iwao & Kuno's (48) deductive quadratic equation (IQ), $\sigma^2 = Am + Bm^2$, which is derived from Iwao's (42) linear regression model of $m*$ on m and includes the classical negative binomial equation $\sigma^2 = m + (m^2/k)$ as a special case. Researchers have debated which of the two is superior as an ecological model (41, 42, 49, 126, 127, 131). Here we review this dispute by thoroughly comparing both merits and drawbacks of the two models in three aspects: descriptive ability, simplicity, and ecological rationality.

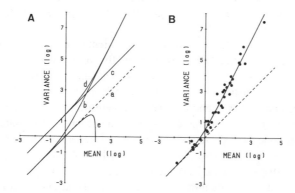

Figure 1   Relationship of variance (σ^2) to mean (m) in some theoretical distribution models (*A*) and in actual populations of organisms (*B*). Lines *a, b, c, d,* and *e* in *A* indicate the relations for the Poisson, negative binomial, Neyman (random distribution of colonies), contagious distribution of colonies, and positive binomial distribution, respectively. The data points in *B* are for 13 species from humans to an orchid, plotted together after selecting three points with highest, lowest, and intermediate *m* values for each species from Figure 6 of Reference 131.

Using as many as 156 sets of extensive data covering a wide variety of animals and plants (including many insects), Taylor et al (131) compared the descriptive ability of both the models for biological populations and concluded that TP is superior to IQ because the greater part of the data sets (106/156) fit TP better than IQ. Taylor et al's method of comparison based on either linear (TP) or nonlinear (IQ) regression on the same log-log graph is no doubt fair and the data seem indeed to be ample enough, so that we may safely conclude that, generally speaking, their empirical model has somehow higher descriptive ability than the deductive IQ model. To make an unbiased judgement from their analysis, however, one should also know that this difference between the two models is just relative and by no means wide. As seen in their table (Appendix B of 131), in most of the data sets, r^2-values are so high for both the models that either model may suffice for the practical purpose of description.

Fundamentally, TP and IQ do not differ in mathematical simplicity; both include just two parameters. The formula for variance-stabilizing transformation for ANOVA is also simple in either case: $y = x^{1-b/2}$ (125) for TP and $y = \sinh^{-1} \sqrt{Ax/B}$ (48) for IQ. The operational convenience for parameter estimation of TP may be somewhat superior to IQ because both a and b in TP can be readily estimated by simple linear regression on a log-log graph. Parameters A and B of IQ, meanwhile, are usually obtained by linear m^*/m regression, $m^* = \alpha + \beta m$, as $A = \alpha + 1$, $B = \beta - 1$ (42). If the range of m is very large, this procedure may produce some bias in the parameter estimates because the variance of m^* for a given sample size n generally tends to increase with m (131). For such a case, therefore, one should calculate A and B based on the nonlinear regression of log σ^2 on log m, though the procedure may become somewhat more troublesome.

Concerning theoretical rationality or consistency, an obvious difference separates the two models. The IQ was originally derived with close reference to theoretical distribution models (42, 49) and actually all five curves shown in Figure 1A, for instance, correspond exactly to this model for different value sets of parameters A and B. One can, therefore, easily interpret ecological implications of the parameters as far as the observed distribution can be related to any one of such models in the actual generating processes of the distribution (49). Furthermore, this approach will lead to more comprehensive systematic analysis of spatial population structure by changing the quadrat size (44), as well as to quantitative analysis of spatial association between different species (47).

In contrast, as a purely empirical model, TP has no such definite theoretical background and does not hold in those theoretical models except for when $b = 1$, i.e. for either the Poisson or the generalized Poisson distribution as Neyman's. Thus, examination of the model over a wide range of mean m may reveal ecologically unreasonable situations, such that any population that shows a clumped spatial pattern at high m (i.e. $b > 1$) should inevitably change its distribution to a uniform one with negative c_A or even negative m^* as m becomes very low, since, in this case, $c_A = am^{b-2} - m^{-1}$ and $m^* = m + am^{b-1} - 1$. In fact, this situation might rarely, if ever, be observed; the slope in actual log σ^2/log m regression generally tends to approach unity as $m \rightarrow 0$, as seen in Figure 1B. Taylor (127) tried to explain this discrepancy by assuming that any population has its own threshold density level below which the population would follow the Poisson irrespective of its original a and b values. But this argument is not persuasive enough because this assumption seems to have no rational basis.

The empirical nature of TP makes attaching any definite ecological meaning to its characteristic parameters a and b difficult. Taylor and collaborators have repeatedly claimed that b in this law is a definite species-specific characteristic that reflects the mode of density-dependent dispersal of that species, being entirely independent of other factors such as sampling scale, quadrat size, and so on, whereas a, in contrast, reflects exclusively artificial factors that are nonspecific to the species (95, 125–132). But this argument has been criticized by many authors and from various angles (3, 23, 25, 31, 32, 41, 49, 108, 116, 138). Indeed, little ecological evidence supports the argument of the significance of b as a species-specific characteristic. Hence, to accept them for ecological interpretation of this model is difficult.

Therefore, the TP model may be used best when it is adopted as a purely descriptive model over a wide range of m, taking care to avoid misleading effects of extrapolation. IQ, in contrast, can be useful as an analytical tool in intensive population studies with limited, specified ranges of density and with other specified conditions to assess the ecological processes characterizing the spatial pattern under study. But even here the description of variance-mean relationship by such a simple model may better be regarded as just an initial step for detailed analyses because insect distributions in nature are likely to be determined by complicated processes involving surely more than two, and perhaps many other, factors.

The m^*/m regression analysis of spatial pattern on which this IQ is based is sometimes criticized rightly because it does not consider density-dependent changes in spatial pattern that might induce some nonlinearity to the regression (127, 131). A natural way for answering this criticism and thereby advancing to the next step of distribution analysis may be to improve the original IQ by adding a new parameter that incorporates density dependence instead of looking for other two-parameter models like TP. No doubt the two factors already involved in the original IQ represent the basic two aspects of spatial pattern of biological populations that cannot be neglected, i.e. the innate tendency for aggregation in terms of $\alpha (= A - 1)$ as an index of basic contagion and the habitat heterogeneity usually involved in $\beta (= B + 1)$ as a density-contagiousness coefficient (42, 49).

SAMPLING TECHNIQUES

Sequential Population Estimation

In sequential estimation, the census is started without determining sample size, carried on through continuous feedback of the interim result, and terminated as soon as the precision of estimation reaches the prescribed level (4, 6). This technique may be of practical value since it provides us with a rational way to cope with the inherent difficulty of population estimation that the sample size necessary for securing a given precision level invariably changes in close relation to the mean m or to the very parameter to be estimated by that census.

The first application of sequential estimation procedure to population censuses (54, 56) adopted the aforementioned IQ model for variance-mean relationship and relative sampling error, $D = SE/m$ ($SE = \sigma^2/n$), as the criterion of the precision level to be prescribed. Later, Green (29) presented a modified version of this graphical method using the TP instead of IQ model to describe variance. The original plan based on IQ has been further extended to include multi-stage sampling (58). Statistical problems concerning practical application of these plans have also been discussed in some detail (36, 56, 106, 140).

In recent years, these fixed-precision-level plans of sequential estimation have been applied to the efficient management of a variety of insect pests, some using the original plan (1, 19, 26, 61, 72, 89, 100, 110, 113, 123, 134, 135, 145, 150, 155), and others using Green's modified version (2, 30, 38, 72, 75, 107, 140, 156). For estimation by simple random sampling, use of either the original or Green's modified plan may be similarly justified, as far as the underlying model, IQ or TP, is realistic enough for the probable range of mean density.

In some situations (for example, when the count of the insects sampled needs to be made in the laboratory), it might be difficult or even impossible to follow this type of sampling procedure faithfully. For such cases, the plan called *double sampling,* in which sampling is replicated only twice, the first time just for determining the final sample size (18, 56), is the recommended alternative. In principle, this procedure is regarded as a degenerate sequential plan, and hence it readily applies to any such cases, using the same graph as that for the original sequential estimation (56).

Sequential Test of Hypothesis

Population surveys for insect pest management often require statistical tests of some hypotheses in terms of density. Two types of sequential plans for such a test have so far been used; one is Wald's (146) sequential probability ratio test

(SPRT) for decision making, and the other is Iwao's (45) sequential difference test (SDT) for density level classification.

SPRT can be applied specifically for making decisions about control performance. Here two critical density levels, m_1 and m_2, are settled, and the boundary lines for making sequential decisions on the graph are defined in such a way that the probabilities of making wrong decisions (i.e. control despite $m < m_1$ and no control despite $m > m_2$) are kept automatically below their predetermined critical levels (α_1, and α_2, respectively) if the plan is strictly performed. To apply this method, the form of frequency distribution as well as its parameters should be rigidly defined beforehand, and, in insect surveys, the plan has been applied widely when based on the negative binomial distribution, in which the boundary lines dividing the whole plane into the three regions, *control, continue,* and *no control,* are two parallel straight lines (1, 26, 66, 67, 72, 79, 112, 124, 135, 139, 149).

The value of this elegant method lies in its very high efficiency because of the test being restricted to a yes/no decision. Thus, when such a decision is really needed, as it is in pest control programs, the plan is a preferable method. If, however, the real purpose of the census is just to classify the density level itself with respect to some criterion, application of this SPRT can be misleading and hence should be avoided. Namely, termination of the census in the *control* region, for example, does not necessarily mean that $m > m_2$ because, even when $m_1 < m < m_2$, the census should come sooner or later to the conclusion of either *control* or *no control.* SPRT has been misused in this respect in past studies (e.g. 79), in which proposals were made to use two sets of boundary lines of SPRT drawn in the same test chart to classify the density or infestation into three categories such as *heavy, moderate,* and *light.*

The other method, SDT, in contrast, is a more general sequential test well suited to simple classification of density level. Statistically, it is a sequential version of the conventional test of difference from a critical level m_0 using fiducial limits for a fixed probability, so that SDT may naturally be less efficient and less strict (91) than SPRT. The original plan of this test (45) assumes the IQ model for the σ^2/m relationship, but a modified version that uses TP instead of IQ has also been proposed (6), and recently both the original (71, 111, 113, 123, 134, 136, 141, 147, 150) and the modified (24, 136) plans have been widely used in pest management practices. In either case, SDT may be used, unlike SPRT, for classifying the density into three or more categories by combined use of two or more sets of boundary lines with different m_0s.

Presence-Absence Sampling

Presence-absence sampling is a labor-saving technique to estimate the density simply from the proportion p of quadrats occupied by the insect (frequency index) instead of the real number occupying the quadrat. Its first application to insect population censuses was made by Kono & Sugino (52) using an empirical equation (KS model), $p = 1 - \exp(-am^b)$, which was later reproposed independently by Gerard & Chiang (28) and also by Nachman (86). The method using another equation (NB model) based on the negative binomial, i.e. $p = 1 - [1 + (m/k)]^{-k}$ was then proposed by Pielou (96). Besides these two equations, we now have two others for use in this technique. Both are based on the negative binomial distribution but have different assumptions as to the σ^2/m relationship, i.e one based on the TP (153) and the other based on the IQ model (46).

In recent years, this convenient census method has attracted entomologists' attention for use in integrated pest management systems, and many attempts have been made using either KS (11, 52, 86, 87, 101), NB (39, 61, 96, 151), or TP (30, 51, 90, 152, 153).

When comparing these four models, one usually does not find definite differences in their descriptive abilities for the actual p/m relationship. The curve in either case increases monotonically from 0 toward 1 with a gradually decreasing rate. Also, the latter two, i.e. the negative binomial models combined with TP and IQ, are mathematically more complex and much harder to treat. Accordingly, for practical use in most studies, consideration of the simpler two models, NB and KS, may be sufficient, the former with a single parameter at first, and then the latter if the fitness to NB is unsatisfactory (62).

For those two models, one can now calculate asymptotic variances of the density estimates based on a standard procedure that Seber (109) calls the "delta method," then determine the necessary sample size for attaining a given, desired level of precision, and furthermore design either sequential or double sampling plans for density estimation with a fixed precision level (62). As for the KS model, a more specified method of variance calculation is also available in which the error of fitting the model to the observed p/m relation is incorporated in addition to the sampling error (87), which is more tedius but may give more precise results if applicable (92).

Application of this presence-absence sampling to sequential hypothesis tests is also quite easy because it was originally a technique based on the simple binomial distribution. The SPRT plan has already been applied to decision making (61, 124, 152) as has the SDT plan to density classification (11), in which the frequency index p is taken instead of m as the criterion for statistical judgement.

In some specific surveys in pest control programs, such as those for quarantine inspection or those for confirming the success of a pest eradication project, one may need to judge if the density is so low that it should be regarded as virtually zero. For use in such a survey, a different type of simple sequential test is available (59). In this test, sampling inspection is continued sequentially unless disrupted by the presence of an individual of the species under study in the sample. Thus, if the succession of "absence" or zeros reaches the predetermined criterion, $n_0 = \log \alpha/\log(1 - p_0)$, then one may stop the census and conclude at the confidence probability, $1 - \alpha$, that the density is virtually zero (or more precisely, p is less than the very low predetermined critical level p_0).

KEY FACTOR ANALYSIS

Detection of Key Factors

Key factor analysis is a simple correlation-regression technique for intensive studies of field population dynamics, which was developed originally by Morris (80, 82, 83) and Varley & Gradwell (142). This technique is used to assess the factors that cause major fluctuations in population size (the key factors) and those that regulate it (the regulating factors) on the basis of a series of life-tables for successive generations or different places.

The procedure to detect key factors of population changes has now been well established. Varley's & Gradwell's (142) graphical analysis (VG) uses correlations or relative contributions of individual mortality rates, k_i (= $-\log s_i$; s_i is the survival rate at stage i), to total generation mortality, K (= Σ k_i). This approach has long been accepted and used as a standard method in insect ecology (118, 143). Quantitative confirmation of the result, if necessary, can also be made readily on the basis of slope b_i in the k_i-on-K regression (98).

This standard method requires an estimation of the species' potential fecundity for calculating K, which may in some cases be difficult. It may therefore be worthwhile to propose here a modified version, which may be regarded as a hybrid of VG (142) with Morris' (83) method. 1. Divide the overall log population-change rate into several basic components (in log), as, say, $I = R + s_E + s_L + s_P$, where R and s values are realized fecundity per emerging adult and survival rates in the egg, larval, and pupal stages, respectively. Examine the correlation to I of each component first visually as in the VG method and then numerically (if the data are sufficient) by calculating the slope as the b value in its regression on I (note that sum total of all the b values is always 1 here) to evaluate relative importance of each of these major components in determining population changes. 2. If necessary and possible divide each component further into a number of subcomponents (in log), e.g. $R = s_A + s_R + f$ (s_A is adult survival rate until reproduction; s_R is proportion of females; f is realized fecundity per surviving female), and examine using a procedure similar to that for the population-change rate the relative contributions of individual subcomponents within each component. Branching-type analysis like this may meaningfully avoid mis-interpretation, especially if the whole process has many components.

The procedure to detect key factors involves the implicit assumption of mutual independence (zero covariance) of individual components, so that interpretation of the result of such an analysis must be made with care and with consideration of possible covariances if these components appear to have any significant correlations (84). This is usually the case when the population is regulated by some density-dependent processes (see below), or when the population estimates for use in the analysis have considerable sampling errors (see 55, 61 for evaluation of the effects of sampling errors and relevant suggestions for sampling design).

Detection of Density Dependence and Regulation

The detection of density dependence in field populations has long been of interest to ecologists because of the long-lasting controversies on mechanisms of animal population regulation. More recently, there have been active discussions on the significance of density dependence in reference to key factor analysis (12, 15, 20, 21, 34, 35, 40, 57, 63, 85, 103–105, 117, 120, 122, 154).

A straightforward approach to detecting the overall density dependence in generation-to-generation population changes is to examine the slobe b in the regression of either I or K described above on the initial density, N_t, under the null hypothesis ($\beta = 0$) of density independence. However, now well known is that, in this conventional regression analysis, spurious density dependence [i.e. negative (positive) b value in the regression of $I(K)$ on N_t] may often arise from purely statistical causes, even if the population change is really density independent. This can occur when one uses time-series data taken at the same place for successive generations (57, 65, 68, 119) and/or population estimates that are subjected to considerable sampling errors (40, 55).

Because of this defect in some density-dependence tests, alternative, stricter test methods specified to time-series analysis have been proposed (16, 22, 27, 99, 102, 115, 144). For usual analyses, however, I again suggest that one adopt the conventional approach after making some simple modifications, rather than using these stricter tests. This suggestion is for two reasons: (*a*) these strict methods usually require that the survey cover a fairly large number of generations—a condition not easily attained in our usual population studies; and (*b*) our main concern in most studies is to elucidate how consistently, at what stages, and in what manner the density-dependent processes are really working, rather than to get strict, qualitative proof as to whether or not the population is ultimately governed by density-dependent processes.

Namely, the procedure proposed here is first to examine the overall density dependence by calculating the adjusted slope for I-on-N_i regression resulting from Bartlett's (9, 10) correction for time-series data as a first-order Markov process, as

$$b_\mathrm{c} = b + \frac{1 + 2\,(1 + b)}{n} \pm \frac{\mu_\gamma\sqrt{1 - (1 + b)^2}}{\sqrt{n}},$$

where b is the slope calculated by the ordinary least squares method and n is the number of generations or data points. The last term of this equation represents asymptotic fiducial limits of b_c for a prescribed probability where μ_γ is the normal variate (e.g. $\mu_\gamma = 1.96$ for $\gamma = 0.95$). The b_c value thus obtained may indicate the consistency of overall density dependence in the population dynamics. The departure from the null hypothesis of density independence can then be tested by checking whether the upper limit value for the prescribed probability (e.g. 0.95) is below 0.

   If the overall density dependence proves to be significant, the relative contributions of individual stages or factors in such overall density dependence may then be analyzed by comparing the b values for their regression on the densities at the corresponding stages (69). At this stage of analysis, comparison of variance in the density among different stages is also to be made, since it may tell us the stage(s) at which density-dependent stabilization or regulation is actually occurring, as well as those at which density-independent destabilization or disturbance comes into ation (57, 117).

   The effects of sampling error may appear in various forms in density-dependence tests, too. But they are not serious in usual situations, and are readily appraisable for correction as long as the errors for the individual estimates are known in terms of the aforementioned D value (= SE/mean) (55).

EFFECTS OF DISPERSAL

Finally, I point out another factor that may sometimes quite seriously disturb the result of key factor analysis but yet has received little attention since it was first suggested; This is the migration of individuals from and into the study area (55). The disturbance due to migration may arise when (a) the whole habitat under study consists of several patches with widely varying conditions for reproduction and survival; (b) the interpatch migration of individuals actively occurs in each generation as random dispersal and results in rather even re-allocation of the individuals; and (c) the census is always made at a fixed, restricted part within the whole habitat (typically, only at one fixed patch).

   Migration has the same statistical effect as sampling error (55), but may become much more serious in some situations, distorting not only the result of the density dependence test but also the analysis for key factor identification. This distortion may be immediately realized if one makes some simulation experiments based on simple population models in which the above assumptions are incorporated. A high degree of migration is likely to occur rather commonly in our population studies. Unfortunately, however, we have no practical way to detect or adjust the effect of this disturbing factor once involved. At the beginning of any population study, special attention should therefore be paid to selecting an area large enough to cover the usual dispersal range for the population under study.

CONCLUDING REMARKS

The three topics reviewed here do not represent all the major problems fundamental to the methodology of sampling and analyzing insect populations. Some were not discussed here because the techniques have already been well developed and given thorough descriptions and reviews. Quotable examples of such topics are the mark-and-recapture analysis for population estimation, which has so far made invaluable contributions to the study of populations composed of actively moving individuals (109, 118), and the analysis of stage-frequency data to compose stage-specific life tables, which makes possible applications of key factor analysis from otherwise intractable data of successive population censuses (70).

Besides these established problems, several others are important but await detailed studies for further development. The intensive analysis of dispersal on individual basis, for example, is no doubt a procedure that is indispensable to interpreting the spatial dynamics of a population as a real entity moving in nature (93, 133). Mark-and-capture procedures also should provide necessary information for the analysis of individual movements. But we have as yet rather few reliable techniques for use in such an intensive study, except for Inoue's (37) regression method for basic analysis.

Also important may be the extension of key factor analysis for use in detailed studies of within-population variations of reproductive or surviving performance among different types or groups of individuals. The significance and interest of studying qualitative heterogeneity or individual variation in insect population ecology are now increasing more and more, particularly in response to recent progress in evolutionary aspects of this topic. Nevertheless, few methodological techniques have contributed to this line since Iwao & Wellington's (50) pioneer work on detecting key factors responsible for differential mortalities imposed on different types of individuals.

The multivariate regression technique recently proposed by evolutionary ecologists for measuring the components of reproductive success to compare the opportunity for natural selection (7, 8) suggests a promising approach to such extension of key-factor analysis. With some modifications, this approach will be applicable in various phases of our ecological study on within-population variations, which may bring new advances to insect population dynamics, especially if this regression technique is combined with intensive analysis of movements of individuals.

Literature Cited

1. Allen, J., Gonzalez, D., Gokhale, D. V. 1972. Sequential sampling plans for the bollworm, *Heliothis zea*. *Environ. Entomol.* 1:771–80
2. Allisopp, P. G., Bull, R. M. 1989. Spatial patterns and sequential sampling plans for melolonthine larvae (Coleoptera: Scarabaeidae) in southern Queensland sugarcane. *Bull. Entomol. Res.* 79:251–58
3. Anderson, R. M., Gordon, D. M., Crawley, M. J., Hassell, M. P. 1982. Variability in the abundance of animal and plant species. *Nature* 296:245–58
4. Anscombe, F. J. 1949. Large sample theory of sequential estimation. *Biometrika* 36:455–58
5. Anscombe, F. J. 1950. Sampling theory of the negative binomial and logarithmic series distributions. *Biometrika* 37:358–82
6. Anscombe, F. J. 1953. Sequential estimation. *J. R. Stat. Soc. Ser. B* 15:1–21
7. Arnold, S. J., Wade, M. J. 1984. On the measurement of natural and sexual selection: theory. *Evolution* 38:709–19
8. Arnold, S. J., Wade, M. J. 1984. On the measurement of natural and sexual selection: applications. *Evolution* 38:720–34
9. Bartlett, M. S. 1953. Approximate confidence intervals II. More than one unknown parameters. *Biometrika* 40:306–17
10. Bartlett, M. S. 1955. Approximate confidence intervals III. A bias correction. *Biometrika* 42:201–4
11. Bechinski, E. J., Stoltz, R. L. 1985. Presence-absence sequential decision plans for *Tetranychus urticae* (Acari: Tetranychidae) in garden-seed beans, *Phaseolus vulgaris*. *J. Econ. Entomol.* 78:1475–80
12. Benson, J. F. 1973. Some problems of testing for density-dependence in animal populations. *Oecologia* 13:183–90
13. Bliss, C. I., Fisher, R. A. 1953. Fitting the negative binomial distribution to biological data and a note on the efficient fitting of the negative binomial. *Biometrics* 9:176–200

14. Bliss, C. I., Owen, A. R. G. 1958. Negative binomial distributions with a common *k*. *Biometrika* 45:39–58
15. Brockelman, W. Y., Fagen, R. M. 1972. On modeling density-independent population change. *Ecology* 53:944–48
16. Bulmer, M. G. 1975. The statistical analysis of density dependence. *Biometrics* 31:901–11
17. Cassie, R. M. 1962. Frequency distribution models in the ecology of plankton and other organisms. *J. Anim. Ecol.* 31:65–92
18. Cox, D. R. 1952. Estimation by double sampling. *Biometrika* 39:217–27
19. Davis, P. M., Pedigo, L. P. 1989. Analysis of spatial patterns and sequential count plans for stalk borer (Lepidoptera: Noctuidae). *Environ. Entomol.* 18:504–9
20. Dempster, J. P. 1983. The natural control of populations of butterflies and moths. *Biol. Rev.* 58:461–81
21. Den Boer, P. J. 1986. Density dependence and the stabilization of animal numbers. *Oecologia* 69:507–12
22. Den Boer, P. J., Reddingius, J. 1989. On the stabilization of animal numbers. Problems of testing 2. Confrontation with data from the field. *Oecologia* 79:143–49
23. Downing, J. A. 1986. Spatial heterogeneity: evolved behaviour or mathematical artifact? *Nature* 323:255–57
24. Ekbom, B. D. 1985. Spatial dispersion of *Rhopalosiphum padi* in spring cereals in Sweden and its importance for sampling. *Environ. Entomol.* 14:312–16
25. Fitt, G. P., Zalucki, M. P., Twine, P. 1989. Temporal and spatial patterns in pheromone trap catches of *Heliothis* spp. (Lepidoptera: Noctuidae) in cotton-growing areas of Australia. *Bull. Entomol. Res.* 79:145–61
26. Foster, R. E., Tollefson, J. J., Steffy, K. L. 1982. Sequential sampling plans for adult corn rootworms (Coleoptera: Chrisomeridae). *J. Econ. Entomol.* 75:791–93
27. Gaston, K. J., Lawton, J. H. 1987. A

test of statistical techniques for detecting density dependence in sequential censuses of animal populations. *Oecologia* 74:404–10

28. Gerard, D. J., Chiang, H. C. 1970. Density estimation of corn rootworm egg populations based upon frequency of occurrence. *Ecology* 51:237–45

29. Green, R. H. 1970. On fixed precision level sequential sampling. *Res. Popul. Ecol.* 12:249–51

30. Grout, T. G. 1985. Binomial and sequential sampling of *Euseius tularensis* (Acrai: Phytoseiidae), a predator of citrus red mite (Acari: Tetranychidae) and citrus thrips (Thysanotera: Thripidae). *J. Econ. Entomol.* 78:567–70

31. Hanski, I. 1982. On patterns of temporal and spatial variation in animal populations. *Ann. Zool. Fenn.* 19:21–37

32. Hanski, I. 1987. Cross-correlation in population dynamics and the slope of spatial variance-mean regression. *Oikos* 50:148–51

33. Harcourt, D. G. 1969. The development and use of life tables in the study of natural insect populations. *Annu. Rev. Entomol.* 14:175–96

34. Hassell, M. P. 1985. Insect natural enemies as regulating factors. *J. Anim. Ecol.* 54:323–34

35. Hassell, M. P. 1987. Detecting regulation in patchily distributed animal populations. *J. Anim. Ecol.* 56:705–13

36. Hutchson, W. D., Hogg, D. B., Poswal, M. A., Berberet, R. C., Cuperus, G. W. 1988. Implications of the stochastic nature of Kuno's and Green's fixed precision stop lines: sampling plans for the pea aphid (Homoptera: Aphididae) in alfalfa as an example. *J. Econ. Entomol.* 8:749–58

37. Inoue, T. 1978. A new regression method for analyzing animal movement patterns. *Res. Popul. Ecol.* 20:141–63

38. Iperti, G., Lapchin, L., Ferran, A., Rabasse, J. M., Lyon, J. P. 1988. Sequential sampling of adult *Coccinella septempunctata* in wheat fields. *Can. Entomol.* 120:773–8

39. Itô, Y. 1962. Distribution of the overwintering arrowhead scale, *Prontaspis yanonensis*, on the Satsuma orange leaves. *Jpn. J. Appl. Entomol. Zool.* 6:183–89. (In Japanese, English summary)

40. Itô, Y. 1972. On the methods for determining density dependence by means of regression. *Oecologia* 10:347–72

41. Itô, Y., Kitching, R. L. 1986. The importance of non-linearity: a comment on the views of Taylor. *Res. Popul. Ecol.* 28:39–42

42. Iwao, S. 1968. A new regression method for analyzing the aggregation pattern of animal populations. *Res. Popul. Ecol.* 10:1–20

43. Iwao, S. 1970. Problems of spatial distribution in animal population ecology. In *Random Counts in Biomedical and Social Sciences.* ed. G. P. Patil. pp. 117–49. Univ. Park/London: Penn. State Univ. Press

44. Iwao, S. 1972. Application of the m*-m method to the analysis of spatial patterns by changing the quadrat size. *Res. Popul. Ecol.* 14:97–128

45. Iwao, S. 1975. A new method of sequential sampling to classify populations relative to a critical density. *Res. Popul. Ecol.* 16:281–88

46. Iwao, S. 1976. Relation of frequency index to population density and distribution pattern. *Physiol. Ecol. Japan* 17:457–63

47. Iwao, S. 1977. Analysis of spatial association between two species based on the interspecies mean crowding. *Res. Popul. Ecol.* 18:243–60

48. Iwao, S., Kuno, E. 1968. Use of the regression of mean crowding on mean density for estimating sample size and the transformation of data for the analysis of variance. *Res. Popul. Ecol.* 10:210–14

49. Iwao, S., Kuno, E. 1971. An approach to the aggregation pattern in biological populations. See Ref. 93a, pp. 461–513

50. Iwao, S., Wellington, W. G. 1970. The western tent caterpiller: qualitative differences and the action of natural enemies. *Res. Popul. Ecol.* 12:81–99

51. Jones, V. P., Parrella, M. P. 1986. Development of sampling strategies for larvae of *Liriomyza trifolii* (Diptera: Agromyzidae) in chrysanthemums. *Environ. Entomol.* 15:268–73

51a. Kogan, M., Herzog, D. C., eds. 1980. *Sampling Methods in Soybean Entomology.* New York/Berlin: Springer-Verlag

52. Kono, T., Sugino, T. 1958. On the estimation of the density of rice stems infested by the rice stem borer. *Jpn. J. Appl. Ent. Zool.* 2:184–88 (In Japanese, English summary)

53. Kuno, E. 1968. Studies on the population dynamics of rice leafhoppers in a paddy field. *Bull. Kyushu Agric. Exp. Stat.* 14:131–246. In Japanese, English summary

54. Kuno, E. 1969. A new method of sequential sampling to obtain the population estimates with a fixed level of precision. *Res. Popul. Ecol.* 11:127–36

55. Kuno, E. 1971. Sampling error as a misleading artifact in key factor analysis. *Res. Popul. Ecol.* 13:28–45

56. Kuno, E. 1972. Some notes on population estimation by sequential sampling. *Res. Popul. Ecol.* 14:58–73
57. Kuno, E. 1973. Statistical characteristics of the density-independent population fluctuation and the evaluation of density-dependence and regulation in animal populations. *Res. Popul. Ecol.* 15:99–120
58. Kuno, E. 1976. Multi-stage sampling for population estimation. *Res. Popul. Ecol.* 18:39–56
59. Kuno, E. 1976. On the assessment of low rate of pest infestation based on successive zero samples. *Jpn. J. Appl. Entomol. Zool.* 22:45–46 (In Japanese)
60. Kuno, E. 1983. Factors governing dynamical behaviour of insect populations: a theoretical inquiry. *Res. Popul. Ecol.* 3:27–45 (Suppl.)
61. Kuno, E. 1984. The design of sampling for pest population forecasting and control. In *Pest and Pathogen Control— Strategic, Tactical, and Policy Models.* ed. G. R. Conway. pp. 254–71. Chichester/New York: Wiley
62. Kuno, E. 1986. Evaulation of statistical precision and design of efficient sampling for the population estimation based on frequency of occurrence. *Res. Popul. Ecol.* 28:305–19
63. Kuno, E. 1987. Principles of predator-prey interaction in theoretical, experimental, and natural population systems. *Adv. Ecol. Res.* 16:249–337
64. Lloyd, M. 1967. "Mean crowding". *J. Anim. Ecol.* 36:1–30
65. Luck, R. F. 1971. An appraisal of two methods of analyzing insect life tables. *Can. Entomol.* 103:1261–71
66. Luna, J. M., Fleischer, S. J., Allen, W. A. 1983. Development and validation of sequential sampling plans for potato leafhopper (Homoptera: Cicadellidae). *Environ. Entomol.* 12:1690–94
67. Lye, B. H., Story, R. N. 1989. Spatial dispersion and sequential sampling plan of the southern green stink bug (Hemiptera: Pentatomidae) on fresh market tomatoes. *Environ. Entomol.* 18:139–44
68. Maelzer, D. A. 1970. The regression of $\log N_{n+1}$ on $\log N_n$ as a test of density dependence: an exercise with computer-constructed density-independent populations. *Ecology* 51:810–22
69. Manly, B. F. J. 1977. The detection of key factors from life table data. *Oecologia* 31:111–17
70. Manly, B. F. J. 1989. A review of methods for the analysis of stage-frequency data. In Ref. 73, pp. 3–69
71. Martel, P., Belcourt, J., Choquette, D., Boivin, G. 1986. Spatial dispersion and sequential sampling plan for the Colorado potato beetle (Coleoptera: Chrysomeridae). *J. Econ. Entomol.* 79:414–17
72. McAuslane, H. J., Ellis, C. R., Allen, O. B. 1987. Sequential sampling of adult northern and western corn rootworms (Coleoptera: Chrysomeridae) in southern Ontario. *Can. Entomol.* 119:577–85
73. McDonald, L. L., Manly, B. F. J., Lockwood, J. A., Logan, J. A., eds. 1989. *Estimation and Analysis of Insect Populations.* Berlin: Springer-Verlag
74. McGuire, J. V., Brindley, T. A., Bancroft, T. A. 1957. The distribution of European corn borer larvae *Pyrausta nubilalis* (Hbn.) in field corn. *Biometrics* 13:65–78
75. Mollet, J. A., Trumble, J. T., Walker, G. P., Sevacherian, V. 1984. Sampling schemes for determining population intensity of *Tetranychus cinnabarices* (Boisduval) (Acarina: Tetranychidae) in cotton. *Environ. Entomol.* 13:1015–17
76. Morisita, M. 1959. Measuring the dispersion of individuals and analysis of the distributional patterns. *Mem. Fac. Sci. Kyushu Univ. Ser. Biol.* 2:215–35
77. Morisita, M. 1962. I_δ-index, a measure of dispersion of individuals. *Res. Popul. Ecol.* 4:1–7
78. Morisita, M. 1971. Composition of the I_δ-index. *Res. Popul. Ecol.* 13:1–27
79. Morris, R. F. 1954. A sequential technique for spruce budworm egg surveys. *Can. J. Zool.* 32:302–13
80. Morris, R. F. 1959. Single factor analysis in population dynamics. *Ecology* 40:580–88
81. Morris, R. F. 1960. Sampling insect populations. *Annu. Rev. Entomol.* 5:243–64
82. Morris, R. F. 1963. Predictive population equations based on key factors. *Mem. Entomol. Soc. Can.* 32:16–21
83. Morris, R. F. 1963. The dynamics of epidemic spruce budworm populations. *Mem. Entomol. Soc. Can.* 31:1–332
84. Mott, D. G. 1966. The analysis of determination in population systems. In *Systems Analysis in Ecology.* ed. K. E. F. Watt. pp. 179–94. London/New York: Academic
85. Mountford, M. D. 1988. Population regulation, density dependence, and heterogeneity. *J. Anim. Ecol.* 57:845–58
86. Nachman, G. 1981. A mathematical model of the functional relationship between density and spatial distribution of a population. *J. Anim. Ecol.* 50:453–60
87. Nachman, G. 1984. Estimates of mean population density and spatial distribu-

tion of *Tetranychus ulticae* (Acarina: Tetranychidae) and *Phytoseiulus persimilis* (Acarina: Phytoseiidae) based upon the proportion of empty sampling units. *J. Appl. Ecol.* 21:903–13

88. Neyman, J. 1939. On a new class of "contagious" distribution, applicable in entomology and bacteriology. *Ann. Math. Stat.* 10:35–57

89. Nishino, M. 1974. Studies on biology and forecasting of occurrence of the arrowhead scale, *Unaspis yanonensis* Kuwana. *Spec. Bull. Shizuoka Pref. Citrus Exp. Stat.* 2:1–101 (In Japanese, English summary)

90. Nowierski, R. M., Gutierrez, A. P. 1986. Numerical and binomial sampling plans for the walnut aphid, *Chromaphis juglandicola* (Homoptera: Aphididae). *J. Econ. Entomol.* 79:868–72

91. Nyrup, J. P., Simmons, G. A. 1984. Errors involved when using Iwao's sequential decision plans in insect sampling. *Environ. Entomol.* 13:1459–65

92. Nyrup, J. P., Angello, A. M., Kovach, J., Reissig, W. H. 1989. Binomial sequential classification sampling plans for European red mite (Acari: Tetranychidae) with special reference to performance criteria. *J. Econ. Entomol.* 82:482–90

93. Okubo, A. 1980. *Diffusion and Ecological Problems: Mathematical Models*. Berlin: Springer-Verlag

93a. Patil, G. P., Pielou, E. C., Waters, W. E., eds. 1971. *Statistical Ecology,* Vol. 1. Univ. Park/London: Penn. State Univ. Press

94. Patil, G. P., Stiteler, W. M. 1974. Concept of aggregation and their quantification. *Res. Popul. Ecol.* 15:121–37

95. Perry, J. N. 1988. Some models for spatial variability of animal species. *Oikos* 51:124–30

96. Pielou, D. P. 1960. Contagious distribution in the European red mite, *Panonychus ulmi* (Koch) and a method of grading population densities from a count of mite-free leaves. *Can. J. Zool.* 38:645–53

97. Pielou, E. C. 1974. *Population and Community Ecology—Principles and Methods*. New York: Gordon and Breach

98. Podoler, H., Rogers, D. 1975. A new method for the identification of key factors from life-table data. *J. Anim. Ecol.* 44:85–114

99. Pollard, E., Lakhani, K. H. 1987. The detection of density dependence from a series of animal censuses. *Ecology* 68:2046–55

100. Poston, F. L., Whitworth, R. J., Welch, S. M., Loera, J. 1983. Sampling western corn borer populations in postharvest corn. *Environ. Entomol.* 12:33–36

101. Raworth, D. A. 1986. Sample statistics and a sampling scheme for the twospotted spider mite, *Tetranychus urticae* (Acari: Tetranychidae), on strawberries. *Can. Entomol.* 118:807–14

102. Reddingius, J., Den Boer, P. J. 1989. On the stabilization of animal numbers: problems of testing. *Oecologia* 78:1–8

103. Royama, T. 1977. Population persistence and density dependence. *Ecological Monographs* 47:1–35

104. Royama, T. 1981. Fundamental concepts and methodology for the analysis of animal population dynamics, with particular reference to univoltine species. *Ecol. Monogr.* 51:473–93

105. Royama, T. 1981. Evaluation of mortality factors in insect life table analysis. *Ecol. Monogr.* 51:495–505

106. Rudd, W. G. 1980. Sequential estimation of soybean arthropod population densities. See Ref. 51a, pp. 94–103

107. Salifu, A. B., Hodgson, C. J. 1987. Dispersion patterns and sequential sampling plans for *Megalurothrips sjostedti* (Trybom) (Thysanoptera: Thripidae) in cowpeas. *Bull. Entomol. Res.* 77:441–49

108. Sawyer, A. J. 1989. Inconsistency of Taylor's b: simulated sampling with different quadrat sizes and spatial distributions. *Res. Popul. Ecol.* 31:11–24

109. Seber, G. A. F. 1982. *The Estimation of Animal Abundance and Related Parameters*. London: Griffin. 2nd ed.

110. Sekita, N., Yamada, M. 1972. Applicability of a new sequential sampling method in the field population surveys. *Appl. Entomol. Zool.* 8:8–17

111. Shaw, P. B., Kidd, H., Flaherty, D. L., Barnett, W. W., Andris, H. L. 1983. Spatial distribution of infestation of *Platynota stultana* (Lepidoptera: Tortricidae) in California vineyards and a plan for sequential sampling. *Environ. Entomol.* 12:60–65

112. Shepard, M. 1980. Sequential sampling plans for soybean arthropods. See Ref. 51a, pp. 94–103

113. Shepherd, R. F., Otvos, I. S. 1984. Pest management of Douglas-fir tussock moth (Lepidoptera: Lymantridae): A sequential sampling method to determine egg mass density. *Can. Entomol.* 116:1041–49

114. Skellam, J. G. 1952. Studies in statistical ecology I. Spatial pattern. *Biometrika* 39:346–62

115. Slade, N. A. 1977. Statistical detection of density dependence from a series of

sequential censuses. *Ecology* 58:1094–1102

116. Soberon, J. M., Loevinsohn, M. 1987. Patterns of variation and the biological foundations of Taylor's law of the mean. *Oikos* 48:249–52

117. Solomon, M. E. 1964. Analysis of processes involved in the natural control of insects. *Adv. Ecol. Res.* 2:1–58

118. Southwood, T. R. E. 1978. *Ecological Methods.* London: Chapman and Hall. 2nd ed.

119. St. Amant, J. L. 1970. The detection of regulation in animal populations. *Ecology* 51:823–28

120. Stiling, P. 1988. Density-dependent processes and key factors in insect populations. *J. Anim. Ecol.* 57:581–93

121. Strickland, A. H. 1961. Sampling crop pests and their hosts. *Annu. Rev. Entomol.* 6:201–20

122. Stubbs, M. 1977. Density dependence in the life cycles of animals and its importance in K- and r-strategies. *J. Anim. Ecol.* 46:677–88

123. Sweeney, J. D., Miller, G. F. 1989. Distribution of *Barbara colfaxiana* (Lepidoptera: Tortricidae) eggs within and among Douglas-fir crowns and methods for estimating egg densities. *Can. Entomol.* 121:569–78

124. Sylvester, E. S., Cox, E. L. 1961. Sequential plans for sampling aphids on sugar beets in Kern County, California. *J. Econ. Entomol.* 54:1080–85

125. Taylor, L. R. 1961. Aggregation, variance and the mean. *Nature* 189:732–35

126. Taylor, L. R. 1971. Aggregation as a species characteristic. See Ref. 93a, pp. 353–77

127. Taylor, L. R. 1984. Assessing and interpreting the spatial distribution of insect populations. *Annu. Rev. Entomol.* 29:321–57

128. Taylor, L. R., Perry, J. N., Woiwoid, I. P., Taylor, R. A. J. 1988. Specificity of the spatial power-law exponent in ecology and agriculture. *Nature* 332:721–22

129. Taylor, L. R., Taylor, R. A. J. 1979. Aggregation, migration and population mechanics. *Nature* 265:415–21

130. Taylor, L. R., Taylor, R. A. J., Woiwoid, I. P., Perry, J. N. 1983. Behavioural dynamics. *Nature* 30:801–4

131. Taylor, L. R., Woiwoid, I. P., Perry, J. N. 1978. The density dependence of spatial behaviour and the rarity of randomness. *J. Anim. Ecol.* 47:383–406

132. Taylor, L., R., Woiwoid, I. P., Perry, J. N. 1979. The negative binomial as a dynamic ecological model for aggregation, and the density dependence of k. *J. Anim. Ecol.* 48:289–304

133. Taylor, R. A. J., Taylor, L. R. 1979. A behavioural model for the evolution of spatial dynamics. In *Population Dynamics,* ed. R. M. Anderson, B. D. Turner, L. R. Taylor, pp. 1–27. Oxford/London: Blackwell

134. Terry, L. I., DeGrandi-Hoffman, G. 1988. Monitoring western flower thrips (Thysanoptera: Thripidae) in "Granny Smith" apple blossom clusters. *Can. Entomol.* 120:1003–16

135. Terry, I., Bradley, J. R. Jr., Van Duyn, J. W. 1989. *Heliothis zea* (Lepidoptera: Noctuidae) eggs in soybeans: within-field distribution and precision level sequential count plans. *Environ. Entomol.* 18:908–16

136. Thistlewood, H. M. A. 1989. Spatial dispersion and sampling of *Campylomna verbasci* (Heteroptera: Miridae) on apple. *Environ. Entomol.* 18:398–402

137. Thomas, M. 1949. A generalization of Poisson's binomial limit for use in ecology. *Biometrika* 36:18–25

138. Thorarinsson, K. 1986. Population density and movement: a critique of Δ-models. *Oikos* 46:70–81

139. Todd, J. W., Herzog, D. C. 1980. Sampling phytophagous Pentatomidae on soybean. See Ref. 51a, pp. 438–78

140. Trumble, J. T., Brewer, M. J., Shelton, A. M., Nyrup, J. P. 1989. Transportability of fixed precision level sampling plans. *Res. Popul. Ecol.* 31:325–42

141. Turgeon, J. J., Regniere, J. 1987. Development of sampling techniques for the spruce budmoth, *Zeiraphera canadensis* Mut. and Free. (Lepidoptera: Tortricidae). *Can. Entomol.* 119:239–49

142. Varley, G. C., Gradwell, G. R. 1960. Key factors in population studies. *J. Anim. Ecol.* 29:399–401

143. Varley, G. C., Gradwell, G. R. 1970. Recent advances in insect population dynamics. *Annu. Rev. Entomol.* 15:1–24

144. Vickery, W. L., Nudds, T. D. 1984. Detection of density-dependent effects in annual duck censuses. *Ecology* 65:96–104

145. Wada, T., Kobayashi, M. 1985. Distribution pattern and sampling techniques of the rice leafroller, *Cnapharocrosis medinalis,* in a paddy field. *Jpn. J. Appl. Entomol. Zool.* 29:230–35 (In Japanese, English summary)

146. Wald, A. 1945. Sequential tests of statistical hypothesis. *Ann. Math. Stat.* 16:117–86

147. Walgenbach, J. F., Wyman, J. A., Hogg, D. B. 1985. Evaluation of sampling methods and development of sequential sampling plans for potato

leafhopper (Homoptera: Cicadellidae) on potatoes. *Environ. Entomol.* 14:231–36

148. Watanabe, N. 1988. A new proposal for measurement of the adjusted mean crowding through consideration of size variability in habitat units. *Res. Popul. Ecol.* 30:215–25

149. Waters, W. E. 1955. Sequential sampling in forest insect surveys. *For. Sci.* 1:68–79

150. Weinzierl, R. A., Berry, R. E., Fisher, G. C. 1987. Sweep-net sampling for western spotted cucumber beetle (Coleoptera: Chrisomeridae) in snap beans: spatial distribution, economic injury level, and sequential sampling plans. *J. Econ. Entomol.* 80:1278–83

151. Wilson, L. F., Gerard, D. J. 1971. A new procedure for rapidly estimating European pine sawfly (Hymenoptera: Diprionidae) population levels in young pine plantations. *Can. Entomol.* 103:1315–22

152. Wilson, L. T., Gonzalez, D., Leigh, T. F., Maggi, V., Foristiere, C., Goodell, P. 1983. Within-plant distribution of spider mites (Acari: Tetranychidae) on cotton: a developing implementable monitoring program. *Environ. Entomol.* 12:128–34

153. Wilson, L. T., Room, P. M. 1983. Clumping patterns of fruit and arthropods in cotton, with implications for binomial sampling. *Environ. Entomol.* 12:50–54

154. Wolda, H. 1989. The equilibrium concept and density dependence tests. What does it all mean? *Oecologia* 81:430–32

155. Yano, E. 1983. Spatial distribution of greenhouse whitefly (*Trialeurodes vaporariorum* Westwood) and a suggested sampling plan for estimating its density in greenhouses. *Res. Popul. Ecol.* 25:309–20

156. Zehnder, G. W., Trumble, J. T. 1985. Sequential sampling plans with fixed levels of precision for *Liriomyza* species (Diptera: Agromyzidae) in fresh market tomatoes. *J. Econ. Entomol.* 78:138–42

Annu. Rev. Entomol. 1991. 36:305–30

ARTHROPOD BEHAVIOR AND THE EFFICACY OF PLANT PROTECTANTS

Fred Gould

Department of Entomology, North Carolina State University, Raleigh, North Carolina
27695–7634

KEY WORDS: arthropod, insect, behavior, deterrents, pesticide

INTRODUCTION

If the members of a phytophagous insect population were not mobile and were distributed uniformly over the plant surfaces in their environment, coverage of 90% of the plant surface area with a contact insecticide would theoretically kill 90% of the insects. In this situation, percent mortality would be independent of the type of plant surfaces that remained untreated (e.g. buds, fruits, stems, upper or lower leaves). However, most insects are not randomly distributed over plant surfaces. Thus, 90% insecticide coverage could kill 0% of the insects, or 1% coverage could kill 100% of the insects. Furthermore, most insects are mobile, so they could potentially avoid or seek out areas already treated with insecticide. Thus, insect behaviors have the potential to significantly alter insecticide efficacy as well as the efficacy of other plant protectants such as plant-produced toxins, attractants, and deterrents.

Agricultural entomologists have long recognized that an understanding of arthropod behavior could help increase the efficacy of insecticides. The use of feeding stimulants in insecticide formulations and the selective treatment of trap crops with insecticides are not new approaches. Recently, however, we

305

0066-4170/91/0101-0305$02.00

have seen the development of new, potent arthropod attractants, the advent of new technologies that can efficiently target the placement of pesticides and other plant protectants, and increased public pressure to limit the amount of pesticides used in agriculture. One challenge in agricultural entomology is to use our knowledge of arthropod behavior in developing efficacious, environmentally benign, sustainable control tactics.

Literature reviews on the behavioral responses of arthropods to specific types of plant protectants are available (23, 42, 43, 62, 83, 102, 126, 138), so an exhaustive review here is unnecessary. Instead, this review attempts to treat all types of plant protectants within the same conceptual framework, drawing on empirical information available on arthropod response to one kind of protectant to assist in understanding arthropod response to another kind of protectant (also see 52, 102). For example, arthropods have been exposed to plant-produced toxins for millions of years, while insecticides are comparatively novel. The evolutionary patterns found in arthropod responses to natural plant toxins may therefore assist us in predicting the evolution of arthropod behavioral response to synthetic pesticides.

As we develop new strategies for the use of plant protectants such as toxin-producing, genetically engineered plants and powerful but highly specific arthropod attractants, we will need a conceptual framework to help predict how genetic and ontogenetic variability in arthropod behavior will affect the efficacy and sustainability of these tactics. In this review, I examine how arthropod behavior has affected the efficacy of existing tactics, and I use this empirical evidence to develop hypotheses related to how arthropod behavior may affect future approaches.

For the purposes of this review, behavior is defined broadly to encompass actions on the part of the arthropods that are motivated by sensory inputs, natural physiological feedback mechanisms, and environmental factors that disrupt the sensory or physiological systems and result in altered behavior. A chemical may therefore affect behavior even if it does not directly stimulate sensory receptors.

BEHAVIORAL RESPONSE TO ENDOGENOUS PLANT PROTECTANTS

Within-Plant Distribution of Protectants and Behaviors

The parts of a single plant can differ enormously in their potential to support the growth and reproduction of an arthropod. This disparity in part results from divergent physical properties of various structures such as leaves and stems and the unequal production, transport, and storage of nutrients (122, 124). However, these differences in suitability can also result from variation in the concentration of toxins and other compounds that impede feeding (86).

The phytochemical literature provides many cases in which plant compounds present in one structure are entirely absent from other structures (e.g. 86, 89). Even more studies report that the concentration of a compound differs among plant structures (58, 86, 107, 132). A study by Schultz (120) demonstrated that even leaves on a single branch can differ significantly in tanning coefficient and presumably in suitability to lepidopteran larvae. Such variation in chemistry could result from internal constraints (38, 86), could be a specific component of an evolved defensive system, or could be a combination of the two. Of specific concern within the context of this review is the question of how arthropods respond behaviorally to this variation and how this response affects plant fitness.

On a dry weight basis, young leaves have often been found to contribute more to plant fitness than old leaves, based on net contribution of photosynthate over the remainder of their respective life spans (6, 21, 57, 92). Thus, consumption of one g of young tissue is thought to decrease plant fitness more than consumption of one g of old leaf material. Young leaves of most plants have higher water, nitrogen, and soluble protein contents than do older leaves (21, 38, 122), and they are not as tough as older leaves (38, 121). This generally makes them nutritionally superior for herbivorous arthropods (122, 124).

If arthropods could choose plant parts with optimal levels of water and nutrients, they would be expected to prefer young leaves over old leaves. A survey of polyphagous insects by Cates (20) reveals just the opposite tendency. Since polyphagous insects probably do not avoid young leaves to minimize their negative impact on their host's fitness, other factors such as differential concentration of toxins and deterrents have been hypothesized to cause generalists to avoid young leaves. The same survey (20) found that specialist herbivores preferred young leaves, which was hypothesized to result from the specialists' tolerance of the toxins produced by their specific hosts.

While broad surveys are useful in revealing patterns, they are far from conclusive. A number of detailed case studies back up the general survey findings and also provide useful exceptions. American holly (*Ilex opaca*), retains some of its leaves through the winter, so in April and early May, herbivores have a choice between newly flushed leaves and tough leaves that are one or more years old. A study by Potter & Kimmerer (103) demonstrated that very young leaves had much higher nitrogen, soluble protein, and water contents than did year-old leaves; however, the saponin content of the young leaves was also higher. When the relatively polyphagous southern red mite, *Oligonychus ilicis*, was fed very young leaves, survival was 14%, while on old leaves survival was 84%. Three-week egg production was 49 times higher on the old leaves. Given a choice of young, expanding leaves or tough, year-old leaves, 100% of the mites quickly chose the old leaves.

In contrast to the behavior of this mite, adult leafminers of the species *Phytomyza ilicicola* that specializes on holly, feed and oviposit almost exclusively on the partially expanded leaves (104). The two arthropods in this system therefore follow the pattern predicted by Cates's (20) general survey.

Not all specialized herbivorous insects prefer young leaves. Two monophagous sawflies (*Neodiprion rugifrons* and *Neodiprion swainei*), that feed on jack pine (*Pinus banksiana*) prefer needles that are one or more years old over young needles. This preference is related to higher concentrations of two compounds in the juvenile foliage (69). The larch sawfly (*Prisiphora erichsonii*) presents a curious case in which eggs are deposited on the single needles of young shoots but larvae migrate to feed on the tufts of 15–45 needles found on old shoots. Eclosing larvae initially take a few bites out of the single needles before moving on (96, 135). Other cases may be found in the literature in which specialist or generalist herbivores avoid plant parts that are rich in nutrients but contain toxins or deterrents (92).

Plant Protectants Can Lead to Increased Plant Damage

In the examples above, plant protectants caused some insects to avoid plant parts that generally contribute the most to plant fitness (i.e. young leaves). Efficacy of the protectants, therefore, can be increased by an insect's behavioral response (i.e. avoidance). When insects fail to respond to a plant protectant, the presence of the compound is neutral in terms of plant damage. At the opposite extreme are situations in which presence of a "plant protectant" leads to more plant damage than would have occurred in its absence. In a number of documented cases, specialized insects utilized the odor of an internally produced plant protectant as an aid in host finding. Some of this literature has been reviewed recently (25, 39, 109).

Behavioral response to the taste of a plant protectant can also lead to negative effects on plants. As an example, cucurbitacins are triterpenoids commonly found in cucurbits and appear to be toxic to some arthropods (18, 27, 50). These high-molecular-weight compounds with low volatility are unlikely to serve as olfactory cues. However, one group of insects, beetles in the genus *Diabrotica,* are physiologically unaffected by these compounds and respond to them as feeding stimulants (89). The beetles sequester the cucurbitacins (41) and these sequestered compounds may protect them from some predators (40, 56), presumably leading to larger populations and more plant damage.

Effects on Arthropod Fitness

The discussion above has primarily been concerned with how behavioral response of arthropods to plant protectants affects the plant's fitness. These behaviors may also affect the arthropod's fitness. When an arthropod moves from a highly nutritious plant part to one with lower nutritive value, growth rate and/or fecundity may decrease, or arthropods may compensate for lower nutritional quality by increased consumption rates (123).

However, the plant parts containing the highest levels of toxic compounds are not always the most nutritious. In *Diplacus aurantiacus,* old leaves that have lower resin content than younger leaves are preferentially fed upon by larvae of the checkerspot butterfly, *Euphydras chalcedona* (92). From April through June, nitrogen content of these old leaves is about equal to the nitrogen content of younger leaves. The plant may profit by larval avoidance of young leaves because the young leaves produce more photosynthate than old leaves (57), but the larvae are probably not abandoning a superior resource.

Also, the quantity of a plant protectant is not necessarily related to the nutritional quality of the plant parts when the level of the protectant increases in response to insect feeding or other types of damage to the plant. The cucurbitacins described above are considered a prime example of a quantitatively inducible defense (18, 132). Cucurbitacin levels in the leaves of a squash variety were shown to increase from undetectable quantities to $9.0\mu g/g$ wet weight after mechanical damage (131). This induction of cucurbitacins is partially systemic in that leaves adjacent to the damaged leaves also have increased cucurbitacin levels. Similar responses in cucurbitacin levels have been associated with damage caused by the feeding of two beetle species (*Diabrotica undecimpunctata howardi* and *Epilachna borealis*) (132). *Epilachna* beetles are negatively affected by cucurbitacins and avoid them by trenching an area within which they subsequently feed; tissue within the trenched area does not exhibit the inducible response. When exposed to already-induced leaf material, these beetles are deterred but seem capable of moving to other areas or to other plants (18). While long-term effects of this induction on *Epilachna* population dynamics have not been studied, the short-term effects seem to be circumvented by behavior. In contrast, the phytophagous mite, *Tetranychus urticae,* does not avoid cucurbitacin-containing leaves even when they are presented in a choice situation (50). Lacking an avoidance behavior, these mites will feed on the cucurbitacin-containing leaves until they die (50).

Some arthropods that feed on other plants with inducible plant protectants have also been shown to avoid the induced tissue, usually by crawling away from the area to undamaged leaves (10, 43, 61, 72). The question remains of whether these mobile, responsive arthropods have totally circumvented the effects of the plant protectants or if these plant protectants have an indirect beneficial effect on the plant. It is feasible that a little damage to many leaves is less damaging than a lot of damage to a few leaves. A second possibility is that the utility of these inducible protectants is specifically in causing a behavioral response in the arthropod. One hypothesis is that the increased movement by the arthropod makes it more obvious to predators and parasites and also increases the chance of the arthropod encountering pathogens that reside on the leaf surface (43). To date, field tests of the enhanced-enemy-encounter hypothesis have produced negative results, but the systems used in the experiments were not optimal for testing the hypothesis (9, 10, 72).

What Came First, the Protectant or the Behavior?

There is substantial documentation of arthropod behavioral response to plant protectants in natural plant communities. In most of the systems examined, the ecological relationship between the chemical classes of plant protectants and the arthropod lineages may be millions of years old. In such systems, it is difficult to determine what came first, the protectant compound or the ability to behaviorally respond to the protectant. Did most plant protectants that affect insect behavior evolve because they stimulated arthropod sensory organs, or did the arthropods evolve sensory receptors that enabled them to identify and avoid toxic compounds? Berenbaum (8) has argued that nontoxic compounds that deter insects based solely on their interaction with an insect's sensory system would be rapidly circumvented by genetic changes in the insect's sensory system. Bernays & Chapman (12) argue that deterrents without toxic properties may be abundant. While conceding that in some cases avoidance of plant secondary compounds may have specifically evolved as a means of avoiding toxins, the authors stress the fact that deterrent responses may simply be a consequence of an initially broad range of sensitivity of the insect sensilla, and hence insects avoid unusual situations. They suggest that the deterrent properties of some secondary compounds are a by-product of evolution rather than specifically selected characteristics.

An empirical question arising from the hypotheses of Berenbaum (8) and Bernays & Chapman (12) is "how strong a correlation is there between the toxicity and deterrency of secondary plant compounds?" Addressing this question experimentally is not simple because of the difficulty in separating toxicity from deterrency (79). Does an animal not eat a compound because it is sick or because of sensory inputs? Cottee et al (26) addressed this problem by administering secondary compounds to the fifth stadium of the locusts *Schistocerca gregaria* and *Locusta migratoria* in three ways: (*a*) normal feeding, (*b*) cannulating the compounds into the crop of the insect with a thin tube, and (*c*) direct injection into the hemocoel. A general positive correlation was found between toxicity by injection and deterrency for the two grasshoppers tested. Results using the cannulation technique were difficult to interpret because there was not enough mortality in any cases. While one could argue that the lack of mortality indicates that the compounds were not toxic when ingested, the treatments involved large individuals that are generally more tolerant of toxins and the duration of the treatment was very short. The only surprising result was that one compound, azadirachtin, was more deterrent to the grasshopper that physiologically tolerated it better.

This study involved a small number of compounds and only two insect species. Other studies in progress (E. A. Bernays, personal communication) indicate that the overall correlation between deterrency and toxicity may be very weak. However, even if the correlation is zero, insect avoidance of nontoxic compounds may still be a primary evolutionary response. Insects may have evolved responses to these specific compounds in order to avoid toxic compounds, found in the same plants, that could not be perceived (8, 12).

From a different perspective, we must question what the evolutionary difference is between a compound that increases plant fitness by killing insects or by deterring insects. In environments where alternative hosts are available, insects could be less heavily selected to adapt to the deterrent compound than a toxic compound (52). Thus, compounds that were only deterrent might be evolutionarily more stable in some environments than compounds that were just toxic.

BEHAVIORAL RESPONSE TO EXOGENOUS PLANT PROTECTANTS

Pesticides that Affect Arthropod Behavior

While some synthetic plant protectants are chemically related to natural plant protectants (e.g. pyrethroids, carbamates, modified avermectins), others could be considered ecological novelties in terms of insect physiological and behavioral responses. Some arthropods may have evolved mechanisms for sensing and avoiding the plant toxin pyrethrum and therefore avoid related pyrethroids, but we have no adaptive reason to expect arthropods to have sensory receptors for malathion. However, examination of the literature on behavioral response of arthropods to pesticides indicates that they are often deterred by malathion and other organophosphate pesticides. In a survey of the literature on behavioral effects of pesticides, Lockwood et al (83) found that a total of 35 pesticides had some record of behavioral effects on arthropods.

PRECEDENTS IN MEDICAL AND VETERINARY ENTOMOLOGY   While the existence of behavioral effects caused by pesticides was accepted by the entomological community, the importance of these insect responses as determinants of insecticide efficacy has been debated for the past 50 years. Most of the early debate involved insects of medical importance (especially mosquitoes) and is discussed here as background information. The initial assertion was that although mosquitoes left households that had been treated with DDT, they had already received a lethal dose of pesticide (e.g. 15). Kennedy (73), Muirhead-Thompson (94), and Busvine (14) discredited this idea by presenting empirical evidence that pesticide deterrency allowed mosquitoes to enter a treated household, take a blood meal, and leave without receiving a toxic dose.

Hutzel (67) found that male German cockroaches that received a topical dose of natural pyrethrins exhibited short periods of rapid locomotion. Work by Ebeling et al (35, 36) demonstrated that bioassays of blatticides that did not take roach behavior into account could overestimate the efficacy of a compound. They developed a method for testing roaches in more realistic arenas where avoidance of toxins was possible. They found that compounds like boric acid were more effective than acute toxins such as diazinon and chlordane because roaches left the treated areas in response to these acute toxins, but not in response to the boric acid. More recent studies of newer compounds such as pyrethroids have found similar behavioral approaches important in assessing efficacy (76, 117). Ebeling et al (35, 36) postulated that learning was involved in pesticide avoidance, but follow-up studies have not been conducted on this specific issue. A recent study by Miall & le Patourel (90) showed that pyrethroids and propoxur can change the roach's orientation to light and that this may in some cases explain avoidance of toxins that are placed in dark hiding places.

Another insect group in which avoidance has generally been considered detrimental to pesticide efficacy is the Isoptera. In evaluating the efficacy of compounds for control of the Formosan subterranean termite, *Coptotermes formosanus*, Su et al (128) categorized insecticides into three types. Type 1 insecticides appear to be direct repellents. Termite tunnels leading to these insecticides are sealed off within one day and there is minimal contact. Type 2 insecticides do not cause initial avoidance, but once some termites begin to die, tunnels leading to the insecticides and dead termites are sealed off or avoided by other termites. Type 3 insecticides do not cause avoidance, presumably because the termites do not die near the insecticide. In a later study, Su & Scheffrahn (127) found that, when presented in a choice situation, a novel insecticide (a dihaloalkyl arylsulfone) did not produce a linear dose-response curve. Low and high doses resulted in lower mortality than intermediate doses. This response was considered to be the result of avoidance and was postulated to involve associative learning.

The termite example demonstrates that the effect of insect behavior on pesticidal efficacy is goal dependent. While avoidance decreases the utility of an insecticide in destroying a termite colony, such avoidance could increase effectiveness of the same insecticide in protecting simple wooden objects like fence posts.

BEHAVIORAL EFFECTS OF PESTICIDES IN AGRICULTURAL SYSTEMS   In agricultural systems, the importance of behavior in determining the efficacy of insecticides has only received intermittent attention until recently, possibly because routes of escape from a pesticide-treated 40-acre field are hard to envision, or simply because of the reasonably good agreement between the results of standard laboratory toxicity assays and field efficacy trials. In the 1970s, attention was drawn to the possible deterrent effects of the new pyrethroid insecticides (116). Initial anecdotal reports of deterrency were followed by more carefully documented studies on a still growing list of arthropod/pesticide combinations (83, 126). Phytophagous and predatory mites have become a focus for studies of deterrent effects of pyrethroids and other synthetic pesticides.

Early tests with pyrethroids in orchards indicated that good suppression of many phytophagous insects could be attained, but that outbreaks of phytophagous mites appeared to be enhanced (65, 110). Researchers initially recognized that these outbreaks at least in part resulted from high mortality of predacious mites and other natural enemies (65, 114) and may also have involved changes in plant physiology (59, 65). Subsequent studies demonstrated that effects of the pyrethroids on the behavior of both phytophagous and predacious mites may have contributed to the outbreaks.

A number of initial laboratory experiments (2, 59, 99, 101) demonstrated that the twospotted spider mite, *T. urticae,* was deterred by leaves treated with pyrethroids. The researchers hypothesized that this deterrence would lead mites to seek out areas in the orchard that received low or no doses of the pyrethroid (59) and that these refuges could serve as nurseries for mites that would recolonize the entire orchard as residues decreased. Further testing (68, 99) demonstrated that *T. urticae* was extremely responsive to pyrethroid residues and would walk from leaf to leaf, settling in refuges as small as one-half of a single leaf. Because pesticide coverage in orchards and in row crops is far from uniform (1, 125, 133), this behavior could certainly decrease pesticide efficacy as long as the mites did not pick up lethal doses of a pyrethroid while moving along the plant surfaces.

T. urticae's behavioral repertoire for leaving pyrethroid-treated plants is not limited to walk-off. A number of studies have found that mites will spindown on silks (2, 99) or posture themselves to be blown off plants by wind (84). They will also lay their eggs in silks or on trichomes when leaf surfaces are treated (32). Which of these tactics they use in response to a treated area may depend on their physiological state, their spatial position, the specific compound, and abiotic factors.

Penman et al (101) found that the predacious mite *Amblyseius fallacis* avoided fenvalerate residues and was deterred from feeding on eggs that had been treated with fenvalerate. This contrasted with the lack of avoidance of eggs and surfaces treated with azinphosmethyl. In comparisons of *T. urticae* and the predacious mite *Typhlodromus occidentalis*, Riedl & Hoying (110) found that pyrethroids were not only more toxic (five times) to predacious mites than phytophagous mites, but that the predacious mites were behaviorally more sensitive to fenvalerate than *T. urticae*. In tests with leaf discs from pesticide-treated trees, they found that even 24 weeks after spraying, when residues did not cause any significant predator mortality from toxicity, over 50% of the predaceous mites died from trying to run off the leaf discs. The leaf discs deterred *T. urticae* for only one week post spraying.

The high sensitivity of these predators to pyrethroids is not unique to this class of pesticide. Preliminary work by Jackson & Ford (70) indicated that a fungicide (captan) and malathion also deter the predatory mites. Further work by Hislop et al (63) demonstrated that *A. fallacis* rejected prey eggs that had been treated with field rates of nonpyrethroid insecticides and fungicides.

Interactions of pyrethroids with the behavior and physiology of prey and predator mites is probably the best studied example of arthropod behavioral response to pesticides in an agricultural system. However, even with the number of studies on the toxicological and behavioral effects of pesticides on spider mites and predacious mites, we are still far from solving the puzzle of which factor or combination of factors is responsible for outbreaks of the phytophagous mites in treated orchards. A similar quandry was reached in the 1950s and 60s in trying to determine why DDT caused mite outbreaks (31, 66). This compound also has deterrent effects on phytophagous mites (28), although at that time the behavioral response was not examined in detail.

Although behavior may contribute to mite outbreaks in orchard systems, behavioral responses of mites to pyrethroids are not always detrimental to pesticide efficacy. Aerts (2), working with greenhouse populations of *T. urticae*, concluded that the spin-down of mites from treated tomatoes in response to pyrethroids was helpful in establishing control.

For several other plant-feeding arthropods whose behavior is affected by pesticides, laboratory data is suggestive of potential effects on field efficacy. Robb & Parrella (111) compared the feeding and oviposition behavior of the leafminer *Liriomyza trifoli* in arenas containing chrysanthemums that had been treated with a variety of insecticides as well as untreated controls. Both no-choice and choice tests were conducted. In choice tests using foliar insecticide applications of permethrin and methyl parathion, each compound deterred feeding and oviposition. When the two compounds were combined, adults fed approximately four times as much on control plants as on treated plants after one day, and about twice as much after seven days. Systemic insecticides had no effect on adult feeding, but three of the five systemic compounds significantly deterred oviposition seven days after pesticide treatment. In general, the most toxic compounds caused the highest deterrence, with the exception of methamidophos, which was highly toxic but not de-terrent. A follow-up study (98) found even stronger deterrent effects of the methyl parathion-permethrin combination in choice tests; L. *trifoli* adults fed about five times as much on controls after seven days. Two of three additional pesticides deterred feeding and oviposition.

In a uniformly treated field, a segment of the adult leafminer population might leave the field in response to the deterrency of a pesticide. However, if the foliar application or the systemic action was uneven, the females would probably lay their eggs in areas of the field or in parts of the plants with the lowest pesicide concentrations. Only careful, manipulative field experiments could assess the economic impact of these insect responses.

Reissig et al (108) examined the effect of azinphosmethyl on apple maggot behavior and survival. While their test was set up in a no-choice design, they did find that the females spent less time on treated apples than controls, even when the treated apples only had sublethal levels of the pesticide. This decrease in residence time on treated apples could translate into more rapid emigration from the orchard to wild hosts in the area. Given the apple maggots' demonstrated ability to learn host charcteristics (97, 105), it would be interesting to know if aversion learning would occur when females re-ceived sublethal pesticide doses in attempts to oviposit on large commercial apples, but not on wild hosts.

Several lines of evidence indicate the effects of the microbial pesticide *Bacillus thuringiensis* on the behavior of lepidopteran larvae. Initial evidence came from rapid cessation of feeding when high doses of *B. thuringiensis* were administered (34) and from the lack of linearity in dose-response studies (64, 91). Recent experiments in which *Heliothis virescens* larvae were given a choice of treated and untreated diets showed significant larval avoidance of *B. thuringiensis* and its purified delta-endotoxins at both lethal and sublethal concentrations (55).

EFFECT OF BEHAVIOR ON PICK-UP AND DISPOSAL OF PESTICIDES   As indicated above, irritation by pesticides can result in pests settling in refuges, but it is not clear whether they pick up lethal doses of pesticides before arriving at a refuge. While it seems obvious that movement over a plant surface results in pick-up of pesticides, little work has quantified this effect (42). Salt & Ford (118) provide one detailed study of the interaction between leaf surface, pesticide residues, and lepidopteran larvae. They masterfully combined a stochastic mathematical model with empirical experiments. Re-sults of their work indicate that larvae could pick up significantly higher doses of pesticide from crawling compared to feeding. However, they found that the amount of pesticide picked up varied considerably depending on the size of pesticide droplets and the age of residues. Permethrin, the specific pesticide they tested, became significantly less available for pick-up by crawling larvae even two hours after spraying. The total leaf residue had not decreased significantly during this short period. They attributed the decreased pick-up to absorption of the pesticide by the leaf surface waxes, especially in the case of small droplets.

Salt's & Ford's (118) computer simulations used parameters appropriate for lepidopteran larvae with five sets of large prolegs. The relative amount of pesticide picked up during movement of such an insect may be much greater than that picked up by a leaf-feeding beetle. Therefore, decreased feeding and increased movement could lead to higher mortality of lepidopteran larvae but decreased mortality of some other insects. The work of Salt & Ford (118) indicates a need for more empirical information to address this issue.

Insect behavior can influence the acquisition of pesticide doses, but it can also influence the disposal of these acquired doses. Golenda & Forgash (49) found that pesticide-induced grooming behavior could result in the removal of more than 10% of a topically applied dose of pesticide, but whether this removal would increase survival was not clear. Moore et al (93) found that adult diamondback moths had a rather unconventional method for removing pesticides from their bodies. When they picked up fenvalerate by walking on a treated surface, a significant proportion of the moths disarticulated, then dropped one or both metathoracic legs. This behavior resulted in a 21% decrease in pesticide residue and increased survival by 13% compared to moths that retained their legs. The only insect that rivals the diamondback moth's pesticide-related behavior for uniqueness is a euglosine bee in Brazil that is highly attracted to DDT and actively collects DDT from treated houses, presumably for use as a pheromone (112).

Behavior and the Efficacy of Deterrents and Attractants

DETERRENTS   While some plant protectants are classified as toxins, the above discussion makes clear that they may also affect behavior. Conversely, compounds regarded as deterrents often have direct, adverse physiological effects (4, 60). Furthermore, a compound that primarily has behavioral effects on one arthropod may have primarily physiological effects on another arthropod, as has been demonstrated with azadirachtin (119).

While compounds that are primarily insecticidal are ubiquitous in modern agriculture, those with primarily behavioral effects have rarely been used on a commercial scale. Bernays (11) stated that "the use of antifeedants in pest-management programmes has enormous intuitive appeal. It satisfies the need to protect specific crops while avoiding damage to nontarget organisms, so that the potential value is great, but so too are the pitfalls." These pitfalls have been discussed in detail (5, 11). Aside from the practical problems of finding and producing a chemically stable product are the problems of arthropod behavior in the field. Unless a formulation of a deterrent has systemic action, plant growth will produce new, vulnerable tissue. For some arthropods, the concentration gradient of a deterrent that develops as leaves expand and new ones are formed could serve as a pest's road map to the most suitable plant parts. With some highly mobile insects, uneven concentrations may not pose a problem if, after a quick assessment of the plant surfaces, they leave the field or orchard. Unfortunately, no one has done the rigorous experiments that are needed to determine if and when either of these responses occurs in the field (22).

Another problem in developing efficacious deterrents is insect habituation. In the absence of a correlated toxic effect, a number of insects appear capable of habituating to single deterrent compounds (8, 71, 129). Landis & Gould (80) found that while the fungicide thiram caused southern corn rootworms to avoid treated corn seedlings in short-term laboratory bioassays, it was not effective in the field or in long-term laboratory studies. However, when this deterrent was combined with a sublethal dose of a toxin (dieldrin), its long-term deterrency was enhanced. A number of studies have shown aversive learning behavior in invertebrates, including insects (13, 29, 30, 46, 81), when a toxicant is combined with a marginally suitable food, but not when the food is highly suitable. Thus the combination of a deterrent with a low level toxicant or use of a deterrent that is itself partially toxic may lead to increased deterrency over time.

ATTRACTANTS   The use of attractants in combination with toxicants has recently received increased attention. Work by Metcalf and co-workers (3, 78, 87) identified volatiles and combinations of volatiles that were attractive to corn rootworm adults. When these compounds were combined with the feeding stimulant cucurbitacin and an insecticide, they proved effective in the field for control of the northern and western corn rootworms (G. Sutter, personal communication). Some competition between the toxic baits and naturally attractive components of corn tassels could compromise the bait's effectiveness, but the short-term outlook is good. If successful on a commercial scale, the use of attractants could potentially reduce the amount of insecticide used in the Midwest by an order of magnitude.

Other types of attractants such as pheromones have been tested as plant protectants. While some successes are notable (e.g. 47), reviews of the literature indicate limitations in their unilateral use as protectants (17, 113). Space does not allow a detailed discussion of this area in this review.

As we learn more about the details of insect interactions with behaviorally active compounds, it may be possible to develop more potent attractant and repellant formulations. A number of recent studies have found synergistic interactions between compounds in stimulating insects to oviposit or feed (33, 39, 78, 95). Mixtures of such compounds are likely to be species-specific and may require development on a case-by-case basis.

GENETIC CHANGES IN BEHAVIORAL RESPONSES TO PLANT PROTECTANTS

Behavioral Adaptation to Endogenous Plant Toxins

Some of the behaviors of phytophagous species discussed in the first section of this review indicate that at a general level insect species have developed behaviors for circumventing the effects of naturally occurring toxins. Examples in which females oviposit on highly toxic hosts (24) and in which feeding stages do not prefer suitable plant tissue over toxic tissue (50) indicate that herbivore genetic systems are not always optimally tuned with regard to avoidance behavior. Chew (24) argues that one reasonable explanation of the lack of avoidance of *Pieris napi macdunnoughii* of the toxic plant *Thalaspi arvense* is that the insect has not had enough time to evolve an avoidance mechanism since *T. arvense* is a recently introduced plant.

How long does it take evolutionary forces to adjust behavior to host suitability? The claim is often made that behavior is more labile than physiology (e.g. 85). Berenbaum's (8) statement that a deterrent chemical that was not also toxic would not provide lasting protection implicitly assumes that evolution of a useful and heritable change in an insect's sensory system is not difficult. It also assumes that insects evolve to avoid specific toxins because physiological toxicity provides a more formidable evolutionary barrier. Are these assumptions well founded? The only formal test of these assumptions offers support. In a study of the bean weevil, *Callosobruchus maculatus,* Wasserman & Futuyma (137) found that, when the insects were reared for 11 generations in an arena where females had a choice of ovipositing on two types of seeds differing in suitability, a genetic shift occurred in oviposition preference toward the more suitable seeds. In this choice regime and in selection regimes in which beetles had to develop on a single host, neither survivorship nor reproduction increased. The tentative conclusion drawn from this study was that one could genetically alter behavior more easily than physiology.

More tests of this type would be useful, especially because in the above case it was not known whether the growth retardation resulted from a toxin or a nutrient deficiency. Many laboratory studies in which insects were selected for physiological tolerance of natural or synthetic toxins have resulted in significant, genetically based adaptation (48, 51, 53, 115). Thus, results with the bean weevil may not be representative. Of additional interest is Tabashnik's (130) finding that a *Colias* species that invaded a new habitat showed physiological adaptation to a new host, but preference did not change.

In assessing the probability that behavior and/or physiology of a herbivore will become altered genetically, we must consider two separate and significant factors. First, one must determine whether more genetic variance is present in the species for the behavioral change or for the physiological change. Second, and at least equally important, one must decide whether, in the circumscribed ecological situation, a behavioral change or a physiological change would increase the arthropod's fitness the most. For the *Colias* butterfly population, Tabashnik (130) concluded that, in the new environment, a change in the degree of behavioral preference of females would not have been selected for as strongly as a change in larval growth ability because the females could already accept the new host.

For the bean weevils, the hosts were intermingled and therefore oviposition choice affected fitness. Certainly an increase in survival on the less suitable host would have increased fitness as well. Indeed, had the weevils quickly adapted to utilize the less suitable beans, there may have been no selective pressure to alter behavior. Mathematical models have been developed that demonstrate that in cases where either a genetic change in behavior or physiology could increase fitness, the initial level of genetic variance for the two types of traits could determine the evolutionary trajectory (19, 52, 106).

Should we expect more genetic variance for physiological or behavioral adaptation in natural systems? Once we divorce this question from the question of selective pressure, it becomes obvious that we have little information upon which to base an answer. Additionally, researchers have assumed that physiological adaptation to one toxin must come at some cost to other physiological processes and thus be evolutionarily constrained. Empirical support for this assumption has, however, been hard to find (45, 53, 134).

The assumption of evolutionary constraints or trade-offs is not generally made when we consider behavioral adaptation. Does a change in sensory perception have no ecological costs? Single receptors carry out multiple functions, so increasing a receptor's ability to sense one compound could decrease or increase its ability to sense other ecologically important compounds. If changes in receptors are not very specific in their effects, selection for avoiding one toxic plant species could lead to avoidance of other, more suitable plant species (44, 82). Few empirical studies address this question. Waldvogel & Gould (136) found that higher preference for one species within a taxonomic family of plants by strains of the moth *H. virescens* was not linked to higher preference for other species in that plant family. This preliminary information indicates that fine-tuning of host-related behavior at the plant-species level may be possible, but more empirical studies are needed in this area. We comfortably assume that arthropods may evolve behaviors to avoid toxins because these arthropods are not capable of physiological adaptation to the toxin. We must also examine the possibility that insects have adapted to toxins because they did not have the proper genetic variance needed for evolving avoidance of these toxins.

Behavioral Adaptation to Synthetic Pesticides

After early debates regarding the innate ability of insect disease vectors to behaviorally avoid pesticides in treated households, further debate developed concerning the possibility that genetic changes in the innate behavior of a species could significantly decrease or increase the efficacy of a pesticide (94). While the importance of innate species behaviors that allowed escape from pesticides was accepted, the issue of whether or not populations could be expected to evolve enhanced pesticide-avoidance abilities is still not resolved. While most work on short-term evolution of arthropod resistance to pesticides has assumed the primacy of physiological adaptation, there has always been some interest in behavioral adaptation to pesticides. This topic has been reviewed recently (83, 102, 126).

Often when heightened avoidance of a pesticide is found in one strain or population of arthropods compared to a second strain or population, a concomitant decrease in physiological tolerance is found. The general explanation for this result has been that avoidance is directly related to "poisoning." The less tolerant a strain was to a pesticide, the more likely its members were to become "irritated" and move away from the source of irritation (48).

This direct relationship, though common, has many important exceptions. Most of these exceptions involve laboratory strains of insects, but a few natural populations also provide evidence (83, 102). For mosquitoes, investigators found that, after eight years of heavy treatment of households with DDT, *Aedes gambiae* (94) became much less prone to enter households whether or not they had been treated with DDT. This constituted an important evolutionary shift in behavior that could protect the mosquito from the toxin while protecting citizens from malaria. Muirhead-Thompson (94) discussed several other cases in which genetically based behavioral changes were at least implicated in mosquito resistance to control measures.

The most striking genetically based behavioral shift documented in a natural population involved houseflies in Georgia (USA) that were subjected to control with malathion-laced baits (37, 74). Over time, this population evolved both physiological resistance and avoidance of the baits. Unlike many cases of pesticide avoidance, these flies could sense the toxin before establishing physical contact with it. In a field test of flies from two dairies, a total of 101 flies at dairy 1 were attracted to the malathion baits, while zero flies were attracted at dairy 2, in spite of a higher density of flies at dairy 2. In contrast, when a DDVP carbamate bait was used at both dairies, more flies from dairy 2 were attracted to the bait, reflecting the higher fly density. Laboratory tests confirmed the difference between the two strains.

More recently, a case of behavioral avoidance of pyrethroid-impregnated ear tags was described (16, 83). A field survey of horn flies on cattle revealed that resistant flies were more likely to be found on parts of the cattle that received the least pesticide exposure (e.g. the belly area) (16). As with the houseflies, the resistant horn flies had both behavioral and physiological resistance to the pesticide.

Examples of field populations of crop pests that have evolved enhanced avoidance of pesticides independent of physiological sensitivity are unavailable. Sparks et al (126) discuss a case in which a laboratory-selected, pyrethroid-resistant strain of tobacco budworm picks up less permethrin than a susceptible strain when third instar larvae are placed in glass vials coated with permethrin. Accompanying experiments indicate that the permethrin-resistant strain may actually decrease movement when it contacts permethrin residues (also see 7). Sparks et al (126) argue that this is a true behavioral adaptation that is not associated with the physiological mechanism of resistance. More detailed work on this system using a field-selected strain would be useful.

Another laboratory example that suggests that field populations of herbivores could evolve behavioral pesticide resistance that is not associated with physiological sensitivity involves spider mites. Penman et al (100) succeeded in selecting a T. urticae population for both increased and decreased behavioral response to the pyrethroid flucythrinate. After 11 cycles of selection, the difference in flucythrinate doses needed to cause equivalent dispersal of the two strains was more than 20-fold. There was no accompanying change in the physiological resistance of either strain. This observation suggests that if, under field conditions, behavioral avoidance of pyrethroids is advantageous for mites, the field populations should evolve a change in behavior.

To date no evidence indicates that field strains of T. urticae vary in behavioral response to pyrethroids in a manner independent of their variance in physiological tolerance (K. Suiter, unpublished data; 77). Although a thorough examination of many field populations has not been conducted, these preliminary negative results are interesting. In the field, fitness-associated consequences of a mite's decision to spin down, walk off, or blow off a leaf are difficult to estimate. Spinning down to the soil below a soybean canopy could have consequences different from those encountered by a mite that crawled to another leaf, and both of these behaviors could be more detrimental than not moving at all. The information available to date does not allow us to reject the hypothesis that mite species innately display optimal behavior in pesticide-treated fields.

Mites are not the only herbivorous arthropods that could be expected to genetically alter their behavior in the presence of recurrent pesticide residues. It is useful to ask why many pests that currently oviposit on wild hosts and on heavily sprayed crops have not evolved the ability to avoid ovipositing in response to pesticide residues or to the secondary chemical profiles of consistently sprayed crops. Perhaps more nontoxic refuges exist in sprayed orchards and fields than we think and therefore oviposition in a treated crop is not a complete dead end. Another possibility is that on wild hosts, a combination of lower nutritive value coupled with high levels of predation and parasitism lead to higher mortality rates than those caused by intensive pesticide use, which eliminates many natural enemies.

FUTURE DIRECTIONS

Determining Factors Influencing Evolution of Behavioral Resistance

While we have many examples of behavioral responses of arthropods to diverse types of plant protectants, we have few examples in which the effect of these behaviors on plant damage or arthropod fitness have been quantified in the field. Although there are some cases like the avoidance of terpene-containing young leaves that are likely to benefit both the plant and arthropod involved, in many situations such as in the response to induced defenses and the avoidance of pesticide residues, the answers are far from clear. Detailed behavioral observations, manipulative experiments, and development of life tables for field populations of arthropod strains that differ only in behavioral response to plant protectants could be helpful in clarifying this picture. Comparisons of the in-field mortality caused by compounds with similar physiological but different behavioral effects on an arthropod (75) would also be useful.

As pointed out repeatedly (83, 102), documenting physiological resistance to pesticides is much easier than documenting behavioral resistance (83, 102). We have laboratory evidence suggestive of physiological resistance in hundreds of species, while well-documented instances of behavioral resistance are few. Either we are missing many cases in which behavioral resistance has evolved or it is indeed a rare occurrence. If it is a rare occurrence, we should find out whether it is rare because most populations lack the necessary genetic variation for altering these behaviors or because genotypes with altered response to pesticides usually do not have a significant selective advantage.

Most cases in which an arthropod species' normal behavioral response to a chemical causes an obvious change in that chemical's efficacy involve chemicals whose dispersion is spatially heterogeneous. Chemicals that fall in that category are phytochemicals that only occur in specific plant parts and synthetic chemicals that are set out as baits or are in other ways localized. Most documented cases of genetic alteration of behavioral response to chemicals by arthropod populations also involve situations in which the chemicals are heterogeneously distributed in the environment (e.g. ear tags).

Future Pest Management Tactics and Arthropod Behavior

In our efforts to minimize the use of pesticides, we are devising some systems for pest control that will rely heavily on toxins with localized placement. Examples of these include the use of toxin-laced feeding stimulants and long-to-short distance attractants. The beauty of these approaches is that they may permit us to limit pest populations with very low input of pesticides that, because of their placement, do not kill natural enemies. One haunting question that arises here asks whether what we know about arthropod behavioral interactions with other plant protectants indicates that arthropods are likely to evolutionarily circumvent these tactics.

As an example, if genetic variance for response to synergistic combinations of attractants is present in the *Diabrotica* beetles, aren't they expected to decrease their response to compounds that are set out in combination with very toxic pesticides? We currently see some competition between toxic baits and corn tassles in attracting these beetles. Don't we expect natural selection to favor individuals that are more attracted to corn tassles than toxic baits?

In the case of synergistically acting mixtures, the mechanism by which beetles evolutionarily decrease their response could be in discriminating against the specific ratio of compounds used in the commercial attractant. If this is the case, countering their adaptation may simply involve adjustments in the relative concentration of the mixture's components. On the other hand, if beetles alter their behavior by decreasing their absolute response to components of the mixture, counter strategies would be more complex.

Evolution of behavioral resistance to these attractants requires that (*a*) genetic variation needed for the decrease in response to the attractants is available, and (*b*) that such a genetically based decrease in response would significantly increase the pest's fitness. In terms of assumption *a*, response to the attractants in the bait may be genetically labile as indicated by some of the studies referred to in this review. However, at least in terms of the feeding stimulant used in baits (cucurbitacin), genetic variation may be lacking. The fact that *Diabrotica* beetle species that feed solely on monocots are still attracted to cucurbitacins only produced in their ancestral host may indicate a lack of evolutionary flexibility for response to this compound (88, but see 41, 56). Laboratory selection experiments might be useful in developing an empirical estimate of the genetic variance for such behavioral traits.

In terms of assumption *b*, avoidance of toxin-laced baits would obviously increase one component of a beetle's fitness. How a decrease in response to the attractants or feeding stimulants in the bait would affect other components of the beetle's fitness is not so obvious. If response to these attractants helped the beetles locate good food sources or oviposition sites, a decrease in response could compromise fitness. Given the potential of this tactic for pest management, researchers should gather information relevant to assumptions *a* and *b* before initiating a large-scale program using this tactic.

Another approach being developed that has the potential for limiting the use of pesticides involves genetically engineered plants that produce their own toxins. The simplest strategy for engineering such plants is to have toxins produced in all plant parts all the time by linking a constitutive promotor to the genetic sequence that codes for the toxin. However, this approach has two potential problems: (*a*) it creates intense selection pressure for physiological adaptation (54); and (*b*) the presence of certain toxins in edible plant parts may not be acceptable to regulatory agencies.

One way to circumvent these problems is to use promotor sequences that only activate gene expression in specific plant parts or only trigger activation at certain times in the growing season. These strategies could mimic the types of heterogeneous toxin production found in natural plant systems and would have both the advantages and disadvantages of these systems. Larvae feeding on a plant that only produced a protectant compound in vulnerable young leaves and reproductive structures might move to and feed on older leaves that were not as relevant to yield. This could offer crop protection while not selecting heavily for pest adaptation. Unfortunately, some toxins (e.g. *B. thuringiensis* delta-endotoxin) being considered are mostly potent against young larvae. If behavior enabled the young larvae to survive on unprotected tissues until they grew larger and more tolerant of the protectant compound, these later more damaging larval stages could return to feed on the more valuable plant parts.

Before research resources are put into the development of engineered plants with heterogeneous gene expression, molecular geneticists will want empirical information to give them assurance that the efficacy of a specific type of toxin-gene expression in a plant will not be circumvented by extant insect behavior and physiology. Gathering such information will require careful laboratory and field experiments that look to natural systems and past experience with conventional pesticides for guidance.

ACKNOWLEDGMENTS

D. Landis and M. Villani influenced the gestation of ideas presented in this paper. I appreciate helpful comments on this manuscript from E. Bernays, R. Chapman, P. Follett, G. Kennedy, J. Lockwood, D. Potter, A. Sheck, K. Suiter, and D. Tallamy. C. Satterwhite's patience throughout the typing and referencing of this review lightened my load considerably. I acknowledge omitting references to many relevant studies for the sake of brevity.

Literature Cited

1. Adams, A. J., Hall, F. 1989. Influence of bifenthrin spray deposit quality on mortality of *Trichoplusia ni* (Lepidoptera: Noctuidae) on cabbage. *Crop Prot.* 8:206–11
2. Aerts, J. 1978. Repellent-werking van decamethrine tegen kasspint bij tomaten. *Meded. Fac. Landbouwwet. Rijksuniv. Gent* 43:655–60
3. Anderson, J. F., Metcalf, R. L. 1986. Identification of a volatile attractant for *Diabrotica* and *Acalymma* spp. from blossoms of *Cucurbita maxima* Duchesne. *J. Chem. Ecol.* 12:687–99
4. Ascher, K. R. S. 1979. Fifteen years (1963–1978) of organotin antifeedants: a chronological bibliography. *Phytoparasitica* 7:117–37
5. Ascher, K. R. S. 1987. Plant-derived insect antifeedants: problems and prospects. *Int. Pest Control* 29:131–33
6. Aslam, M., Lowe, S. B., Hunt, L. A. 1977. Effect of leaf age on photosynthesis and transpiration of cassava (*Manihot esculenta*). *Can. J. Bot.* 55:2288–95
7. Benedict, J. H., Treacy, M. F., Camp, B. J. 1989. Behavior of pyrethroid-susceptible and resistant *Heliothis virescens* larvae on cotton treated with insecticides. *Proc. Beltwide Cotton Conference*
8. Berenbaum, M. 1986. Postingestive effects of phytochemicals on insects: on paracelsus and plant products. In *Insect-Plant Interactions*, ed. J. R. Miller, T. A. Miller, pp. 121–53. New York: Springer-Verlag
9. Bergelson, J., Fowler, S., Hartley, S. 1986. The effects of foliage damage on casebearing moth larvae, *Coleophora serratella*, feeding on birth. *Ecol. Entomol.* 11:241–50
10. Bergelson, J. M., Lawton, J. H. 1988. Does foliage damage influence predation on the insect herbivores of birch? *Ecology* 69:434–45
11. Bernays, E. A. 1983. Antifeedants in crop pest management. In *Natural Products for Innovative Pest Management*, ed. D. L. Whitehead, W. S. Bowers, pp. 259–71. New York: Pergamon
12. Bernays, E. A., Chapman, R. 1987. The evolution of deterrent responses in plant-feeding insects. See Ref. 23, pp. 159–73
13. Bernays, E. A., Lee, J. C. 1988. Food aversion learning in the polyphagous grasshopper, *Schistocerca americana*. *Physiol. Entomol.* 13:131–37
14. Busvine, J. R. 1964. The significance of DDT-irritability tests on mosquitoes. *Bull. WHO* 31:645–56
15. Buxton, P. A. 1945. The use of DDT in relation to problems of tropical medicine. *Trans. R. Soc. Trop. Med. Hyg.* 38:267–393
16. Byford, R. L., Lockwood, J. A., Smith, S. M., Franke, D. E. 1987. Redistribution of behaviorally resistant horn flies (Diptera: Muscidae) on cattle treated with pyrethroid-impregnated ear tags. *Environ. Entomol.* 16:467–70
17. Campion, D. G. 1984. Survey of pheromone uses in pest control. In *Techniques in Pheromone Research*, ed. H. E. Hummel, T. A. Miller, pp. 405–49. New York: Springer-Verlag
18. Carroll, D. R., Hoffman, C. A. 1980. Chemical feeding deterrent mobilized in response to insect herbivory and counter adaptation by *Epilachnus tredecimnotata*. *Science* 209:414–16
19. Castillo-Chavez, C., Levin, S. A., Gould, F. 1988. Physiological and behavioral adaptation to varying environments: a mathematical model. *Evolution* 42:986–94
20. Cates, R. G. 1980. Feeding patterns of monophagous, oligophagous, and polyphagous insect herbivores: the effect of resource abundance and plant chemistry. *Oecologia* 46:22–31
21. Chabot, B. F., Hicks, D. J. 1982. The ecology of leaf life spans. *Annu. Rev. Ecol. Syst.* 13:229–59
22. Chapman, R. F., Bernays, E. A. 1989. Insect behavior at the leaf surface and learning as aspects of host plant selection. *Experientia* 45:215–22
23. Chapman, R. F., Bernays, E. A., Stoffolano, J. G. Jr. eds. 1987. *Perspectives in Chemoreception and Behavior*. New York: Springer-Verlag
24. Chew, F. S. 1977. Coevolution of pierid butterflies and their cruciferous foodplants. II. The distribution of eggs on potential foodplants. *Evolution* 31:568–79
25. Chew, F. S. 1988. Searching for defensive chemistry in the Cruciferae, or, do glucosinolates always control interactions of Cruciferae with their potential herbivores and symbionts? See Ref. 126a, pp. 81–112
26. Cottee, K., Bernays, E. A., Mordue, A. J. 1988. Comparisons of deterrency and toxicity of selected secondary plant compounds to an oligophagous and a polyphagous acridid. *Entomol. Exp. Appl.* 46:241–47

27. Da Costa, C. P., Jones, C. M. 1971. Cucumber beetle resistance and mite susceptibility controlled by the bitter gene in *Cucumis sativus* L. *Science* 172:1145–46

28. Davis, D. W. 1952. Some effects of DDT on spider mites. *J. Econ. Entomol.* 45:1011–19

29. Dethier, V. G. 1980. Food-aversion learning in two polyphagous caterpillars, *Diacrisia virginica* and *Estigmene congrua*. *Physiol. Entomol.* 5:321–25

30. Dethier, V. G. 1988. Induction and aversion-learning in polyphagous arctiid larvae (Lepidoptera:) in an ecological setting. *Can. Entomol.* 120:125–131

31. Dittrich, V., Streibert, P., Bathe, P. A. 1974. An old case reopened: mite stimulation by insecticide residues. *Environ. Entomol.* 3:534–40

32. Donahue, D. J., McPherson, R. M. 1990. Oviposition response of *Tetranychus urticae* (Acari: Tetranychidae) to direct treatment and residue of pyrethroids on soybean. *J. Entomol. Sci.* 25:158–69

33. Doss, R. P. 1983. Activity of obsure root weevil, *Sciopithes obscurus* (Coleoptera: Curculionidae), phagostimulants individually and in combination. *Environ. Entomol.* 12:848–51

34. Dulmage, H. T., Graham, H. M., Martinez, E. 1978. Interactions between the tobacco budworm, *Heliothis virescens*, and the delta-endotoxin produced by the HD-1 isolate of *Bacillus thuringiensis* var. *kurstaki*: relationship between length of exposure to the toxin and survival. *J. Invert. Pathol.* 32:40–50

35. Ebeling, W., Reierson, D. A., Wagner, R. E. 1968. The influence of repellency on the efficacy of blatticides. III. Field experiments with German cockroaches with notes on three other species. *J. Econ. Entomol.* 61:751–61

36. Ebeling, W., Wagner, R. E., Reierson, D. A. 1966. Influence of repellency on the efficacy of blatticides. I. Learned modification of behavior of the German cockroach. *J. Econ. Entomol.* 59:1374–88

37. Fay, R. W., Kilpatrick, J. W., Morris, G. C. III. 1958. Malathion resistance studies on the house fly. *J. Econ. Entomol.* 51:452–53

38. Feeny, P. P. 1970. Seasonal changes in oak leaf tannins and nutrients as a cause of spring feeding by winter moth caterpillars. *Ecology* 51:565–81

39. Feeny, P. P. 1987. The roles of plant chemistry in associations between swallowtail butterflies and their host plants. In *Insects—Plants, Proc. 6th Int. Symposium*, ed. V. Labeyrie, G. Fabres, D. Lachaise, pp. 353–59. Dordrecht, Netherlands: Junk

40. Ferguson, J. E., Metcalf, R. L. 1985. Cucurbitacins. Plant-derived defense compounds for diabroticites (Coleoptera: Chrysomelidae). *J. Chem. Ecol.* 11:311–18

41. Ferguson, J. E., Metcalf, R. L., Fisher, D. C. 1985. Disposition and fate of cucurbitacin-B in five species of Diabroticites. *J. Chem. Ecol.* 11:1307–21

42. Ford, M. G., Salt, D. W. 1987. Behaviour of insecticide deposits and their transfer from plant to insect surfaces. In *Pesticides on Plant Surfaces*, ed. H. J. Cottrell, pp. 26–81. New York: Wiley

43. Fowler, S. V., Lawton, J. H. 1985. Rapidly induced defenses and talking trees: the devil's advocate position. *Am. Nat.* 126:181–95

44. Futuyma, D. J. 1983. Selective factors in the evolution of host choice by phytophagous insects. In *Herbivorous Insects: Host-Seeking Behavior and Mechanics*, ed. S. Ahmad, pp. 227–44. New York: Academic

45. Futuyma, D. J., Philippi, T. E. 1987. Genetic variation and covariation in responses to host plants by *Alsophila pometaria* (Lepidoptera: Geometridae). *Evolution* 41:269–79

46. Gelperin, A. 1975. Rapid food-aversion learning by a terrestrial mollusk. *Science* 189:567–70

47. Gentry, C. R. 1987. Peachtree borer (Lepidoptera: Sesiidae): control by mass trapping with synthetic sex pheromone. *Misc. Publ. Entomol. Soc. Am. Fruit Tree and Nut Management*

48. Georghiou, G. P. 1972. The evolution of resistance to pesticides. *Annu. Rev. Ecol. Syst.* 3:133–68

49. Golenda, C. F., Forgash, A. J. 1986. Grooming behavior in response to fenvalerate treatment in pyrethroid-resistant house flies. *Entomol. Exp. Appl.* 40:169–75

50. Gould, F. 1978. Resistance of cucumber varieties to *Tetranychus urticae*: genetic and environmental determinants. *J. Econ. Entomol.* 71:680–83

51. Gould, F. 1979. Rapid host range evolution in a population of the phytophagous mite, *Tetranychus urticae* Koch. *Evolution* 33:791–802

52. Gould, F. 1984. Role of behavior in the evolution of insect adaptation to insecticides and resistant host plants. *Bull. Entomol. Soc. Am.* 30:34–41

53. Gould, F. 1988a. Genetics of pairwise and multispecies plant-herbivore coevolution. See Ref. 126a, pp. 13–55

54. Gould, F. 1988b. Evolutionary biology and genetically engineered crops. *BioScience* 38:26–33

55. Gould, F., Anderson, A., Landis, D., Van Mellaert, H. 1990. Feeding behavior and growth of *Heliothis virescens* larvae (Lepidoptera: Noctuidae) on diets containing *Bacillus thuringiensis* formulations or endotoxins. *Entomol. Exp. Appl.* In press

56. Gould, F., Massey, A. 1984. Cucurbitacins and the biological control of *Diabrotica undecimpunctata howardi*. *Entomol. Exp. Appl.* 36:273–78

57. Gulmon, S. L., Chu, C. C. 1981. The effects of light and nitrogen on photosynthesis, leaf characteristics, and dry-matter allocation in the chaparral shrub, *Diplacus aurantiacus*. *Oecologia* 49: 207–12

58. Gustaffson, A., Gadd, I. 1965. Mutations and crop improvement. II. The genus *Lupinus* (Leguminosae). *Hereditas* 53:15–37

59. Hall, F. R. 1979. Effects of synthetic pyrethroids on major insect and mite pests of apple. *J. Econ. Entomol.* 72:441–46

60. Hare, J. D. 1984. Suppression of the Colorado potato beetle, *Leptinotarsa decimplineata* (Say) (Coleoptera: Chrysomelidae), on Solanaceous crops with a copper-based fungicide. *Environ. Entomol.* 13:1010–14

61. Haukioja, E., Niemela, P. 1977. Retarded growth of a geometrid larva after mechanical damage to leaves of its host tree. *Ann. Zool. Fennici* 14:48–52

62. Haynes, K. F. 1988. Sublethal effects of neurotoycin insecticides on insect behavior. *Annu. Rev. Entomol.* 33:149–68

63. Hislop, R. G., Auditore, P. J., Weeks, B. L., Prokopy, R. J. 1981. Repellency of pesticides to the mite predator *Amblyseius fallacis*. *Prot. Ecol.* 3:253–57

64. Hornby, J. A., Gardner, W. A. 1987. Dosage/mortality response of *Spodoptera frugiperda* (Lepidoptera: Noctuidae) and other noctuid larvae to beta-endotoxin of *Bacillus thuringiensis*. *J. Econ. Entomol.* 80:925–29

65. Hoyt, S. C., Westigard, P. H., Burts, E. C. 1978. Effects of two synthetic pyrethroids on the codling moth, pear psylla, and various mite species in Northwest apple and pear orchards. *J. Econ. Entomol.* 71:431–35

66. Huffaker, C. B., Spitzer, C. H. Jr. 1950. Some factors affecting red mite populations on pears in California. *J. Econ. Entomol.* 43:819–31

67. Hutzel, J. M. 1942. The activating effect of pyrethrum upon the German cockroach. *J. Econ. Entomol.* 35:929–33

68. Iftner, D. C., Hall, F. R. 1983. Effects of fenvalerate and permethrin on *Tetranychus urticae* Koch (Acari: Tetranychidae) dispersal behavior. *Environ. Entomol.* 12:1782–86

69. Ikeda, T., Matsumura, F., Benjamin, D. M. 1977. Chemical basis for feeding adaptation of pine sawflies. *Science* 197:497–99

70. Jackson, G. J., Ford, J. B. 1973. The feeding behaviour of *Phytoseiulus persimilis* (Acarina: Phytoseiidae), particularly as affected by certain pesticides. *Ann. Appl. Biol.* 75:165–71

71. Jermy, T. 1987. The role of experience in the host selection of phytophagous insects. See Ref. 23, pp. 143–57

72. Karban, R., Myers, J. H. 1989. Induced plant responses to herbivory. *Annu. Rev. Ecol. Syst.* 20:331–48

73. Kennedy, J. S. 1947. The excitant and repellent effects on mosquitos of sublethal contacts with DDT. *Bull. Entomol. Res.* 37:593–607

74. Kilpatrick, J. W., Schoof, H. F. 1958. A field strain of malathion-resistant house flies. *J. Econ. Entomol.* 51:18–19

75. Knowles, C. O. 1987. Effects of formamidines on Acarine dispersal and reproduction. In *Sites of Action for Neurotoxic Pesticides*, ed. R. M. Hollingworth, M. B. Green, pp. 174–79. Washington, DC: ACS Symp. Ser. 356

76. Koehler, P. G., Patterson, R. S. 1988. Suppression of German cockroach (Orthoptera: Blattellidae) populations with cypermethrin and two chlorpyrifos formulations. *J. Econ. Entomol.* 81: 845–49

77. Kolmes, S. A., Dennehy, T. J., Frisicano, L. 1990. Behavioral aspects of dicofol resistance in the twospotted spider mite *Tetranychus urticae* (Acari: Tetranychidae). *Entomol. Exp. Appl.* In press

78. Lampman, R. L., Metcalf, R. L. 1987. Multicomponent kairomonal lures for southern and western corn rootworms (Coleoptera: Chrysomelidae: *Diabrotica* spp.). *J. Econ. Entomol.* 80:1137–42

79. Landis, D. A. 1987. *Assessing the utility of the feeding deterrent approach to crop protection*. PhD dissertation. Raleigh: N.C. State Univ.

80. Landis, D. A., Gould, F. 1989. Investigating the effectiveness of feeding deterrents against the southern corn rootworm, using behavioral bioassays and

toxicity testing. *Entomol. Exp. Appl.* 51:163–74

81. Lee, J. C., Bernays, E. A. 1988. Declining acceptibility of a food plant for the polyphagous grasshopper, *Schistocerca americana:* the role of food aversion learning. *Physiol. Entomol.* 13:291–301

82. Levins, R., MacArthur, R. 1969. An hypothesis to explain the incident of monophagy. *Ecology* 59:910–11

83. Lockwood, J. A., Sparks, T. C., Story, R. N. 1984. Evolution of insect resistance to insecticides: a reevaluation of the roles of physiology and behavior. *Bull. Entomol. Soc. Am.* 30:41–51

84. Margolies, D. C., Kennedy, G. G. 1988. Fenvalerate-induced aerial dispersal by the twospotted spider mite. *Entomol. Exp. Appl.* 46:233–40

85. Mayr, E. 1970. *Populations, Species and Evolution,* pp. 363–64. Cambridge: Harvard Univ. Press

86. McKey, D. 1979. The distribution of secondary compounds within plants. In *Herbivores. Their Interaction with Secondary Plant Metabolites,* ed. G. A. Rosenthal, D. H. Janzen, pp. 55–133. New York: Academic

87. Metcalf, R. L., Lampman, R. L. 1989. Cinnamyl alcohol and analogs as attractants for corn rootworms (Coleoptera: Chrysomelidae). *J. Econ. Entomol.* 82:1620–25

88. Metcalf, R. L., Metcalf, R. A., Rhodes, A. M. 1980. Cucurbitacins as kairomones for diabroticite beetles. *Proc. Natl. Acad. Sci. USA* 17:3769–72

89. Metcalf, R. L., Rhodes, A. M., Metcalf, R. A., Ferguson, J., Metcalf, E. R., Lu, P.-Y. 1982. Cucurbitacin contents and diabroticite (Coleoptera: Chrysomelidae) feeding upon *Cucurbita* spp. *Environ. Entomol.* 11:931–37

90. Miall, S. M., le Patourel, G. N. J. 1989. Response of the German cockroach *Blattella germanica* (L.) to a light source following exposure to surface deposits of insecticides. *Pestic. Sci.* 25:43–51

91. Mohd-Salleh, D. K., Lewis, L. C. 1982. Feeding deterrent response of corn insects to delta-endotoxin of *Bacillus thuringiensis. J. Invert. Pathol.* 39:323–28

92. Mooney, H. A., Williams, K. S., Lincoln, D. E., Ehrlich, P. R. 1981. Temporal and spatial variability in the interaction between the checkerspot butterfly, *Euphydryas chalcedona* and its principle food source, the California shrub, *Diplacus aurantiacus. Oecologia* 50:195–98

93. Moore, A., Tabashnik, B. E., Stark, J.

D. 1989. Leg autotomy: a novel mechanism of protection against insecticide poisoning in diamondback moth (Lepidoptera: Plutellidae). *J. Econ. Entomol.* 82:1295–98

94. Muirhead-Thompson, R. C. 1960. The significance of irritability, behaviouristic avoidance and allied phenomena in malaria eradication. *Bull. Org. Mond. Sante: Bull. WHO* 22:721–34

95. Nishida, R., Fukami, H. 1989. Oviposition stimulants of an aristolochiacae-feeding swallowtail butterfly, *Atrophaneura alcinous. J. Chem. Ecol.* 15:2565–75

96. Ohigoshi, H., Wagner, M. R., Matsumura, F., Benjamin, D. M. 1981. Chemical basis of differential feeding behavior of the larch sawfly, *Pristiphora erichsonii* (Hartig). *J. Chem. Ecol.* 7:599–614

97. Papaj, D. R., Prokopy, R. J. 1988. The effect of prior adult experience on components of habitat preference in the apple maggot fly *(Rhagoletis pomonella). Oecologia* 76:538–43

98. Parrella, M. P., Robb, K. L., Virzi, J. K. 1988. Analysis of the impact of abamectin on *Liriomyza trifolii* (Burgess) (Diptera: Agromyzidae). *Can. Entomol.* 120:831–37

99. Penman, D. R., Chapman, R. B. 1982. Fenvalerate-induced distributional imbalances on two-spotted spider mite on bean plants. *Entomol. Exp. Appl.* 31:71–78

100. Penman, D. R., Chapman, R. B., Bowie, M. H. 1988. Selection for behavioral resistance in twospotted spider mite (Acari: Tetranychidae) to flucythrinate. *J. Econ. Entomol.* 81:40–44

101. Penman, D. R., Chapman, R. B., Jesson, K. E. 1981. Effects of fenvalerate and azinphosmethyl on two-spotted spider mite and phytoseiid mites. *Entomol. Exp. Appl.* 30:91–97

102. Plethero, F. G., Singh, R. S. 1984. Insect behavioural responses to toxins: practical and evolutionary considerations. *Can. Entomol.* 116:57–68

103. Potter, D. A., Kimmerer, T. W. 1989. Inhibition of herbivory on young holly leaves: evidence for the defensive role of saponins. *Oecologia* 78:322–29

104. Potter, D. A., Redmond, C. T. 1989. Early spring defoliation, secondary leaf flush, and leafminer outbreaks on American holly. *Oecologia* 81:192–97

105. Prokopy, R. J., Averill, A. L., Cooley, S. S., Roitberg, C. A. 1982. Associative learning in egglaying site selection by apple maggot flies. *Science* 218:76–77

106. Rauscher, M. 1990. The evolution of

habitat preference. III. The evolution of avoidance and adaptation. In *Evolution of Insect Pests: Patterns of Variations,* ed. K. C. Kim. New York: Wiley. In press

107. Rees, C. J. C. 1969. Chemoreceptor specificity associated with choice of feeding site by the beetle, *Chrysolina brunsvicensis* on its foodplant, *Hypericum hirsutum. Entomol. Exp. Appl.* 12:565–83

108. Reissig, W. H., Stanley, B. H., Valla, M. E., Seem, R. C., Bourke, J. B. 1983. Effects of surface residues of azinphosmethyl on apple maggot behavior, oviposition, and mortality. *Environ. Entomol.* 12:815–22

109. Renwick, J. A. A. 1988. Comparative mechanisms of host selection by insects attacking pine trees and crucifers. See Ref. 126a, pp. 303–16

110. Riedl, H., Hoying, S. A. 1983. Toxicity and residual activity of fenvalerate to *Typhlodromus occidentalis* (Acari: Phytoseiidae) and its prey *Tetranychus urticae* (Acari: Tetranychidae) on pear. *Can. Entomol.* 115:807–13

111. Robb, K. L., Parrella, M. P. 1985. Antifeeding and oviposition-deterring effects of insecticides on adult *Liriomyza trifolii* (Diptera: Agromyzidae). *J. Econ. Entomol.* 78:709–13

112. Roberts, D. R., Alecrim, W. D., Heller, J. M., Ehrhardt, S. R., Lima, J. B. 1982. Male *Eufriesia purpurata,* a DDT-collecting euglossine bee in Brazil. *Nature* 297:62–63

113. Rothschild, G. H. L. 1981. Mating disruption of lepidopterous pests: current status and future prospects. In *Management of Insect Pests with Semiochemicals,* ed. E. R. Mitchell, pp. 207–28. New York: Plenum

114. Roush, R. T., Hoy, M. A. 1978. Relative toxicity of permethrin to a predator, *Metoseiulus occidentalis,* and its prey, *Tetranychus urticae. Environ. Entomol.* 7:287–88

115. Roush, R. T., McKenzie, J. A. 1987. Ecological genetics of insecticide and acaricide resistance. *Annu. Rev. Entomol.* 32:361–80

116. Ruscoe, C. N. E. 1977. The new NRDC pyrethroids as agricultural insecticides. *Pestic. Sci.* 8:236–42

117. Rust, M. K., Reierson, D. A. 1978. Comparison of the laboratory and field efficacy of insecticides used for German cockroach control. *J. Econ. Entomol.* 71:704–8

118. Salt, D. W., Ford, M. G. 1984. The kinetics of insecticide action. III. The use of stochastic modeling to investigate the pick-up of insecticides from ULV-treated surfaces by larvae of *Spodoptera littoralis* Boisd. *Pestic. Sci.* 15:382–410

119. Schmutterer, H. 1990. Properties and potential of natural pesticides from the neem tree, *Azadirachta indica. Annu. Rev. Entomol.* 35:271–97

120. Schultz, J. C. 1983. Habitat selection and foraging tactics of caterpillars in heterogeneous trees. In *Variable Plants and Herbivores in Natural and Managed Systems,* pp. 61–86. New York: Academic

121. Schultz, J. C., Nothnagle, P. J., Baldwin, I. T. 1982. Seasonal and individual variation in leaf quality of two northern hardwood tree species. *Am. J. Bot.* 69:753–59

122. Scriber, M., Slansky, F. 1981. The nutritional ecology of immature insects. *Annu. Rev. Entomol.* 26:183–11

123. Slansky, F., Feeny, P. P. 1977. Stabilization of the rate of nitrogen accumulation by larvae of the cabbage butterfly on wild and cultivated food plants. *Ecol. Monogr.* 47:209–28

124. Slansky, F., Rodriguez, T. G. 1987. *Nutritional Ecology of Insects, Mites, Spiders, and Related Invertebrates.* New York: Wiley Interscience

125. Southwick, L. M., Clower, J. P., Clower, D. F., Graves, J. B., Willis, G. H. 1983. Effects of ultra-low-volume and emulsifiable concentrate formulations on permethrin coverage and persistence on cotton leaves. *J. Econ. Entomol.* 76: 1442–47

126. Sparks, T. C., Lockwood, J. A., Byford, R. L., Graves, J. B., Leonard, B. R. 1989. The role of behavior in insecticide resistance. *Pestic. Sci.* 26: 383–99

126a. Spencer, K., ed. 1988 *Chemical Mediation of Coevolution.* New York: Academic

127. Su, N.-Y., Scheffrahn, R. H. 1988. Toxicity and feeding deterrency of a dihaloalkyl arylsulfone biocide, A-9248, against the Formosan subterranean termite (Isoptera: Rhinotermitidae). *J. Econ. Entomol.* 81:850–54

128. Su, N.-Y., Tamashiro, M., Yates, J. R., Haverty, M. I. 1982. Effect of behavior on the evaluation of insecticides for prevention of or remedial control of the Formosan subterranean termite. *J. Econ. Entomol.* 75:188–93

129. Szentesi, A., Bernays, E. A. 1984. A study of behavioural habituation to a feeding deterrent in nymphs of *Schistocerca gregaria. Physiol. Entomol.* 9:329–40

130. Tabashnik, B. E. 1983. The shift from

native legume hosts to alfalfa by the butterfly, *Colias philodice eriphyle*. *Evolution* 37: 150–62

131. Tallamy, D. W. 1985. Squash beetle feeding behavior: an adaptation against induced cucurbit defenses. *Ecology* 66: 1574–79

132. Tallamy, D. W., Krischik, V. A. 1989. Variation and function of cucurbitacins in *Cucurbita:* an examination of current hypotheses. *Am. Nat.* 133:766–86

133. Uk, S., Courshee, R. J. 1982. Distribution and likely effectiveness of spray deposits within a cotton canopy from fine ultralow-volume spray applied by aircraft. *Pestic. Sci.* 13:529–36

134. Via, S. 1990. Ecological genetics and host adaptation in herbivorous insects: the experimental study of evolution in natural and agricultural systems. *Annu. Rev. Entomol.* 35:421–46

135. Wagner, M. R., Ikeda, T., Benjamin, D. M., Matsumura, F. 1979. Host derived chemicals: the basis for preferential feeding behavior of the larch sawfly, *Pristiphora erichsonii* (Hymenoptera: Tenthredinidae), on tamarack, *Larix laricina. Can. Entomol.* 111:165–69

136. Waldvogel, M. G., Gould, F. 1990. Variation in oviposition preference in strains of *Heliothis virescens* (Lepidoptera: Noctuidae). *Evolution.* In press

137. Wasserman, S. S., Futuyma, D. J. 1981. Evolution of host plant utilization in laboratory populations of the southern cowpea weevil, *Callosobruchus maculatus* Fabricius (Coleoptera: Bruchidae). *Evolution* 35:605–17

138. Whitehead, D. L., Bowers, W. S. 1983. *Natural Products for Innovative Pest Management.* New York: Pergamon

Annu. Rev. Entomol. 1991. 36:331–54

SENSILLA OF IMMATURE INSECTS

Russell Y. Zacharuk and Vonnie D. Shields

Department of Biology, University of Regina, Regina, Saskatchewan, Canada S4S 0A2

KEY WORDS: distribution, morphology, morphogenesis, electrophysiology, behavior

INTRODUCTION

Most of the insect pest species of significance to human economies in agriculture, forestry, homes, and gardens are most destructive in the immature (larval or nymphal) stage. Insects that are pests primarily in the adult stage, such as those that affect human health and welfare either directly or as transmitters of diseases, spend a considerable part of their life cycle in immature stages. During development to adults, the larvae are generally restricted in habitat and mobility, and are generally more favored targets for the management and control of pest-insect populations. To sustain their growth and development, immature insects are voracious feeders. Knowledge of the sensory armature involved in their food-finding, feeding behavior, and ingestion is paramount in the development and application of those pest control agents, in particular, that require ingestion (i.e. microbial agents) or that will prevent feeding (antifeedants).

Exterosensilla in adult insects monitor the environment for cues to find conspecific mates, suitable oviposition sites, food, and suitable temperature and humidity levels. They also control feeding behavior and provide visual and tactile cues for orientation and protection. Because of the greater mobility of adult insects, short- and long-range olfaction is a primary need. The sensory requirements of immature stages are more limited. The eggs are usually deposited on or near suitable food sources for the early developmental stage. Thus, olfaction is usually restricted to short-range orientation to optimally nutritious food and, in some insects, to aggregation or trail pher-

331

omones that are also of a limited range. Taste and tactile cues for food selection and gustation are of primary importance to the immature stage, along with sensilla that provide temperature, humidity, tactile, and some visual monitoring for orientation and protection. Immature stages develop through several molts, at which times cuticular parts of existing sensilla are replaced and new sensilla may develop. Immatures also require external and internal proprioceptors, as do adults, for monitoring position and movement of body parts.

This review primarily emphasizes the distribution and functional morphology of various types of chemo-, hygro-, thermo-, and mechanosensilla in larval or nymphal insects, and their morphogenesis at a molt. These are related to the food selection and feeding behavior of selected insect groups. Space does not permit consideration of their visual or subepidermal Type II neuronal sensory systems. The former is included in recent reviews on structure and function of insect eyes (32, 71, 94, 133), and the latter is in reviews on other sensory systems (102, 145). Several recent reviews that detail the structure and function of insect sensilla include those of immature forms. The information on the sensilla of immatures is summarized and updated here, as is that from some recent reviews on food selection and feeding behavior.

TYPES OF SENSILLA

Ten types of insect sensilla have been categorized on the basis of the morphology of their cuticular parts (118). Nine of these are represented in immature insects. Three functional types have been identified: aporous (AP) that are mostly mechanosensilla (MS) but include many thermo-hygrosensilla (T-HS); uniporous (UP) that may be gustatory chemosensilla (GCS) or chemo-mechanosensilla (GC-MS); and multiporous (MP) olfactory chemosensilla (OCS) (5, 143). The following list incorporates the above morphological and functional typologies, both of which are in common use in current literature: (*a*) sensilla chaetica are heavy, thick-walled bristles or spines that are AP (tactile) or UP (usually GC-MS); (*b*) sensilla trichodea are hairs that may be AP (MS, T-HS, or rarely, OCS), UP (commonly GC-MS), or MP (OCS); (*c*) sensilla basiconica are peg-like and, in addition to the sensory characteristics of the trichodea, may be UP osmosensors; (*d*) sensilla coeloconica are pegs set in shallow pits and are usually AP (T-HS) or MP (OCS); (*e*) sensilla ampullacea are pegs in deep pits with sensory characteristics similar to the coeloconica; (*f*) sensilla campaniformia are usually AP domes within the cuticle that are MS; (*g*) sensilla placodea are UP or MP plates that are generally level with the cuticular surface and are GCS or OCS; (*h*) sensilla styloconica are mostly UP (GCS) but are sometimes AP (MS) pegs set on an elongated style; (*i*) sensilla scolopalia and the variant scolopophora are largely subcuticular with either a dendritic insertion into the cuticle or an attachment to the cuticle through an accessory cell, usually have no surface cuticular manifestation, and are MS. The scale-like sensilla squamiformia have not been identified in immature insects.

Sensilla with more uniquely shaped sensory cuticles have been described as large conical sensory appendices (124), domes (127), knobs (49), partitioned plates (128), knobbed rods (26), flower-shaped domes (93), ear-like sensilla auricularia (26), clubbed hairs (24), and coniform complexes (91). These are either modifications of one of the above ten basic types, or represent composite sensilla that incorporate the cellular elements of a number of adjacent sensilla of one of these basic types under one sensory cuticle.

GENERALIZED STRUCTURE AND FUNCTION

All but certain scolopalial sensilla in immature, as in adult, insects consist of a specialized sensory cuticle innervated by the dendrites of one or more sensory neurons and usually three or four accessory cells that ensheath the neurons and associated sinuses (102, 145). The form and position of the sensory cuticle varies with sensillar type as noted previously. A sheath of cuticular origin encases the terminal one- to two-thirds of the outer segments of the dendrites. It is continuous with or has a connection to the surface cuticulin layer of the cuticle (145) and is shed with the exuvium at each molt (141).

Sensory Cuticle and Dendritic Terminations

MECHANOSENSILLA   In AP cuticular MS, the sensory cuticle transmits the stimulus physically through the dendritic sheath to a specialized, tightly encased termination of the dendrite by movement, lamellar stress, swelling, contraction, or stretch (77). The dendritic termination contains closely spaced microtubules in a dense matrix; deformation of the surrounding membrane is believed to initiate neuronal activity (77, 102).

Scolopalial sensilla are positioned subcuticularly to a greater or lesser extent. Those with dendritic insertions in the cuticle, as in the lacinia of a catopid larva (56) and the scolopophorous sensilla in larval mandibular teeth of many insect groups (2, 43, 55, 96, 114, 146), have terminally closed sheaths like those in the MS. These tightly encase the dendritic terminations and fuse with the cuticle terminally. A surface depression (25) or demarcated cuticle (146) may denote externally the insertion points of the sheaths. The mandibular organs respond to pressure stresses in the cuticular lamellae of the teeth (146). The dendrites of most scolopalial sensilla terminate below the cuticle in a scolopale cap, which is believed to be the homologue of the dendritic sheath of exterosensilla (145). While the cap is no longer inserted in the cuticle, it remains connected to it by a microtubule-filled attachment cell

(56, 102). In some scolopalia, the capped dendrite has lost all connections with the cuticle; instead, it inserts through an attachment cell into an extracellular connective tissue matrix around sensillar cell bundles in head appendages (13, 34) or other internal body organs (102).

The scolopalia typically respond to stretch during movements of body parts or to vibrations transmitted to intersegmental joints (77, 102, 104), and may also be proprioceptive in orientation to gravity (87). Some external body hairs and campaniformia also function in proprioception (102) and in response to gravity (87) and vibrations (101, 132). The digitiform organs inserted in the sidewall of maxillary and labial palps (2, 18, 70, 83, 147) have been shown to also respond to vibrations, as well as to movement of the sensory cuticle (147).

THERMO-HYGROSENSILLA   The sensory cuticle of most T-HS is an AP peg. The tips of the dendrites of the humidity-sensitive neurons are presumed to be the ones that are tightly encased within the peg (8). They have densely packed microtubules somewhat resembling those in a tubular body (11, 102, 130) and may respond to deformative mechanical changes in the cuticle caused by changes in humidity (8). These pegs are usually also thermosensitive. The neuron with the lamellated subcuticular dendritic termination is believed to be the thermosensor (8, 36, 58, 97). Some MP OCS also respond to changes in temperature or humidity, but the dendritic terminations and mechanisms involved are not known (8).

CHEMOSENSILLA   Most chemosensilla (CS) are innervated by one or more neurons in a single ensheathed bundle. However, large composite MP sensilla occur on the antennae of many larval endopterygotes (17, 48, 52, 121, 124, 150). The sensory cuticle of these is innervated by several discrete neuronal bundles, which suggests that it consists of the fused cuticles of several sensilla (145). An intermediate stage in the formation of such sensilla may be the antennal coniform complex of plecopteran larvae (91), in which a number of pegs are closely appressed but not fused. Only one reference was seen to a composite UP sensillum, two of which occur in the pharynx of *Drosophila melanogaster* larvae (127).

The stimulating molecules typically diffuse to the dendritic terminations through a single terminal pore (UP) or the many pores (MP) of a chemical conduction system. The form and function of the sidewall pore canal system is maintained in AP and UP sensilla, but is largely modified into a pore tubule system in many but not all MP sensilla (145). In Diptera, for example, it is absent in the large composite antennal sensillum of a mosquito (89, 150) but is present in the homologous sensillum of the house fly (48) and an MP antennal sensillum of a simuliid (60) larva. It is distinct in an elaterid (142) but not in a catopid (52) larva.

Many UP peg and hair sensilla have one neuron whose dendrite ends in a tubular body at the base of the sensory cuticle. These are considered to be dually chemo-mechanosensory (145). No reference was seen to a mechanosensory neuron being present in MP OCS. While a distinction is commonly made between olfactory and taste sensilla in insects, the latter have been shown to respond to odors in a lepidopteran larva (113, 131). This distinction is often made on the basis of structural characteristics but is more uncertain at the functional level, particularly in aquatic insects.

Gustatory sensilla   The terminal pore in UP sensilla varies in shape, size, and nature of the surrounding cuticle, which may regulate its size or degree of closure (145). These characteristics may affect its permeability to different chemicals. The contents of the underlying pore canal may be of greater importance to its selective permeability. In some aquatic larvae (148) and in cibarial sensilla (137), the pore contents may be simply a mucoid exudate from the dendritic channel. The pores of most exterosensilla contain, in addition to dendritic channel liquor, pore tubules (22) or plugs of fenestrated fibrils of cuticular or other origin (78, 99; V. D. Shields & R. Y. Zacharuk, in preparation). Such plugs could confer a greater stimulant selectivity based on size of interfibrillar spaces, nature of binding sites, and charge characteristics of ionized side groups in the matrix of the plug. Differences in permeability to Co, Hg, and Pb ions occur between the lateral and medial styloconic pegs on the galea of *Manduca sexta* larvae (R. Y. Zacharuk & J. L. Frazier, in preparation), and in both of these pegs between different species of lepidopteran larvae (V. D. Shields & R. Y. Zacharuk, in preparation). Such differences may be of significance particularly in the interpretation of UP sensillar responses to chemicals applied to the terminal pore. Response may not be affected, either because that chemical has no reactive site or because it is simply not getting through the pore matrix to the active site. The pore matrix could also affect impedance in tip recording with micropipette electrodes.

The sidewall pore canal system around the terminal pore and the distal portion of the dendritic sheath are also readily permeable to the divalent cations in UP pegs of some lepidopteran larvae but not others (V. D. Shields & R. Y. Zacharuk, in preparation). The functional significance of the permeability of the sidewall pore canal system into the sensillar sinus is not known. Slifer et al (129) first demonstrated dye permeability into this outer sinus in UP pegs of a grasshopper; they suggested that the dendritic sheath was permeable to the dyes that entered through the terminal pore. This permeability was similarly suggested for the labellar taste hairs of adult *Lucilia caesar* when Ag^+ was applied to the terminal pore (108). A permeable dendritic sheath would enable the sensillar sinus to act as a reservoir of ions and biochemicals required by the dendrites in their sensing function, as has been suggested for taste hairs of adult flies (40). Figure 1 shows, based on the above, the cation permeant sites into and within the terminal portion of a UP galeal styloconic peg of lepidopteran larvae. At this time, however, the structural or physiological basis for the intra- and interspecific differences in permeabilities into homologous UP sensilla is not clear.

Olfactory sensilla   The thin aporous cuticle that covers the dendritic terminations in the composite antennal cone in mosquito larvae is presumed to be permeable to chemicals and to be olfactory (89, 150). The pore tubules of the porous homologous sensillum in an elaterid larva are lipoidal and probably hydrophobic, but a thin central core may be hydrophilic (142). These features could be characteristic of pore tubules in similar MP sensilla of other insects. In the antennal cone of a cave beetle larva (52) and in the unique flower-shaped sensilla (93) and the coniform complex (91) of stonefly nymphs, the

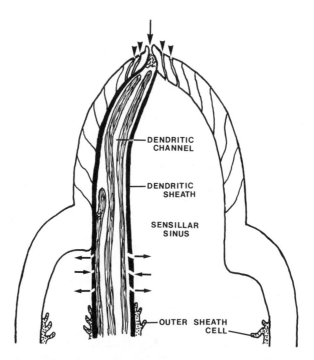

Figure 1   Permeable sites into the peg through the terminal pore *(large arrow)* and surrounding sidewall pore canals *(arrowheads)* and through the dendritic sheath between the dendritic channel (showing the mechanosensory and two of the four chemosensory dendrites) and sensillar sinus *(small arrows)* in a galeal styloconic peg of a lepidopteran larva (redrawn from R. Y. Zacharuk & J. L. Frazier, in preparation).

medium that conducts the stimulant through the pores of the sensory cuticle is most likely the largely hydrophilic mucoid liquor from the dendritic channel. MP sensilla with pore tubules on the antennae and maxillary palps of some lepidopteran larvae are not permeant to Co^{++} in the living or glutaraldehyde-fixed states (R. Y. Zacharuk, unpublished data). Similar sensilla with an undetermined conduction system stain with aqueous crystal violet and reduced $AgNO_3$ in a coleopteran larva (16). These differences in the stimulant conduction systems and permeabilities of MP sensory cuticles may confer an initial discrimination of odors at this peripheral site.

Some atypical antennal AP trichoid hairs in mosquito larvae have been described as chemo-mechanosensory (89, 148). They have one unbranched dendrite terminating near the tip within the hair shaft and another that ends in a tubular body at the base of the sidewall. In mealworms, the unbranched dendrites of both neurons extend into the hair shaft and are putatively chemosensory (35). Altner et al (10) noted smooth-walled hairs with fine, sidewall pore canal filaments on the antennae of *Periplaneta americana* that are innervated within the hair shaft by branched dendrites from two neurons. The authors confirmed an olfactory function for these hairs.

OSMOSENSILLA   Sensilla described as UP and putatively identified as osmosensory occur on the antennae of mosquito (149) and black fly (60) larvae. These are innervated by single dendrites with concentric lamellations and terminal branches in the former, and three dendrites with similar lamellations in the latter. The flower-shaped MP sensilla of stonefly nymphs, innervated by individual branched dendrites, may be osmoregulatory (93) and perhaps also osmosensory. One would expect osmosensilla to have some structural similarities to CS, but no known physiological studies have confirmed their identity in insects to date.

Sensory Neurons

All insect integumental sensilla are innervated by typical Type I bipolar neurons; the dendrite directly innervates the sensory cuticle and the axon usually makes direct synaptic connections in the CNS. The dendrite consists of a thin outer ciliary segment that terminates in or beneath the cuticle and inserts basally into the tip of a thicker proximal segment, usually through two centriolelike basal bodies and associated ciliary rootlets. The ciliary segment usually contains only longitudinally oriented microtubules with a $9 \times 2 + 0$ configuration near its basal insertion and numerous evenly spaced, individual microtubules distally (102, 145). In scolopalial sensilla (102) and in one neuron of some T-HS (8), the basal microtubule configuration is maintained along the entire length of the cilium. The proximal segment is generally thicker than the ciliary, contains organelles similar to those in the cell body cytoplasm (102, 145), and may be more electron-dense in chemosensory than in mechanosensory dendrites (148).

Individual vesicles and multivesicular bodies were described earlier among the microtubules of the dendritic ciliary segments in a composite MP sensillum fixed conventionally (125). Synapticlike vesicles were shown recently in multivesicular bodies near the base of the ciliary segments in cryo-fixed galeal UP styloconic pegs of *M. sexta* larvae (R. Y. Zacharuk, in preparation). The significance of these vesicles is not known. They seem to be formed in the more proximal neuronal cytoplasm for transport to the dendritic terminations (125).

APOROUS SENSILLA   Sensilla that are solely mechanosensory are innervated by a single neuron with a terminal tubular body (102), but a MS with two neurons, one without a tubular body, was noted on the maxillary palp of a coleopteran larva (57). Larval scolopalial sensilla have one or two neurons, and mandibular scolopophorous sensilla have two neurons in an individual sensing unit; neither have a tubular body (102). T-HS usually have three or four neurons, two with endings in the sensory cuticle, one with a subcuticular lamellate termination and, when present, one with a scolopalial-like cilium ending at the base of the dendritic sheath (8, 10, 36). One neuron with a branched or lamellated dendritic termination innervates digitiform sensilla (57, 70, 147). Except for the lamellated dendrites in the above sensilla, mechanosensory dendrites are unbranched.

UNIPOROUS SENSILLA   Taste sensilla are innervated by two to eight neurons (6, 48); usually there are four or five, one of which may insert with a tubular body at the base of the sensory cuticle (145). In the sensillum with eight neurons, one contains dense-cored vesicles and may be neurosecretory (6). The dendritic terminations in UP sensilla have been described as unbranched (145) except in osmosensilla (60, 149). However, some terminal branching was confirmed recently in the galeal styloconic pegs of *M. sexta* larvae by cryo-fixation-substitution (R. Y. Zacharuk, in preparation) and identified in another lepidopteran species in conventionally fixed preparations (V. D. Shields & R. Y. Zacharuk, in preparation) using transmission electron microscopy (TEM).

MULTIPOROUS SENSILLA   MP OCS may have 1 (92) to 3 (53, 57, 60, 70) to as many as 9 to 17 (21) neurons in an individual unit. The smallest composite sensillum noted to date is a cone on the antennal flagellum of the spruce budworm, which has 3 units with a total of 8 neurons (R. Y. Zacharuk & P. J. Albert, in preparation). The largest is a placoid sensillum on the antennal pedicel of *Tribolium* spp., which has 10–11 units with a total of about 130 neurons (21). The dendrites in all MP sensilla of immature insects for which descriptions were seen have multibranched terminations (145).

ELECTROPHYSIOLOGY   Each sensory neuron in a cuticular sensillum produces a train of nerve impulses usually with a characteristic and reproducible waveform, amplitude, and temporal frequency in response to one or more specific stimuli (108). Interpretation is simplified in sensilla, such as most MS, that are innervated by a single neuron. In T-HS, responses to cold, warm, wet, and dry have been monitored but not related to any one specific neuron of the three or four present other than by inference from their ultrastructure (8), as is also true for multineuronal CS.

The most studied CS of immature insects are the galeal UP styloconic pegs of Lepidoptera. In *Mamestra brassicae,* the five neurons in these sensilla respond to glycosides, sugars, cations, anions, and mechanical stimuli, respectively (138). However, Ma (99) concluded that, in *Pieris brassicae,* "absolute receptor specificity could not be assumed under all conditions of stimulation." In other lepidopteran species, individual neurons have a limited degree of specificity to individual or a group of chemicals and a differential sensitivity between two sensilla in one individual and between homologous sensilla intra- and interspecifically; synergism occurred nearly as often as inhibition (68). Similar differential sensitivities in neuronal responses to food plant saps (63) or their constituents (135) were noted in a variety of larval species.

More limited studies of a similar nature on taste sensilla of other insects gave similar results for neuronal specificity to phagostimulants and inhibitors. Sugar-sensitive cells (4, 106) and sugar-binding sites (20) are now being characterized in larval or nymphal sensilla. Considerable interest currently focuses on plant constituents, or chemicals based on these, that stimulate a deterrent cell in taste sensilla (31, 66, 73, 105). Such compounds could become viable and environmentally friendly crop protection agents with a degree of specificity for pest insect species. Inhibitions seem to affect general receptor function and are largely not related to receptor specificity (122). Other literature on feeding inhibitors was recently reviewed by Dethier (66) and Frazier (75).

Characterization of neurons in MP OCS has been very limited. Different cells in these sensilla can be characterized by spike amplitude and frequency (59, 65, 69, 121). The cells respond differently to different odors. Discrimination of plant odors may be coded in the temporal patterns of spike activity (69) or may be based on the absolute number of cells responding and/or the ratios of their frequencies (65). Recent studies examine the effect of inputs from antennal olfactory sensilla on the activity of interneurons in the brain of larval *M. sexta* and the integration of mechanosensory inputs with the olfactory in these interneurons (88).

Associated Cells

Cuticular sensilla typically have three sheath cells and a basal glial cell (145). A tubular inner sheath (thecogen) cell encases the neuronal dendrites and cell bodies from the base of the dendritic sheath, to which it is tightly appressed, to near the base of the cell bodies. The thecogen is expanded around the base of the ciliary segments where it encloses a liquor-filled sinus into which microvilli are extended. The glial cell usually ensheathes the bases of the cell bodies and axons before the inner sheath cell terminates proximally. The proximal but not the distal dendritic segments, the cell bodies, and the axons are usually individually wrapped and separated from one another by the thecogen and the glial cell (102, 145). In mosquito larvae, however, parts of the cell bodies and axons are unsheathed and exposed to the hemocoel (150), a characteristic of some other aquatic insects (R. Y. Zacharuk, unpublished).

The intermediate (trichogen) and outer (tormogen) sheath cells successively wrap the inner and a basal portion of the dendritic sheath. They subtend a large, liquor-filled outer sinus that surrounds the terminal portion of the dendritic sheath and extends into the sensory cuticle (Figure 1). The outer sheath cell is tightly appressed to the integumental cuticle around the sinus (145). Microvilli or lamellae extend from both cells into the sinus and often have coated vesicles associated with their basal membranes (35), which is indicative of high secretory activity. The sidewall pore canals in AP and UP and the pore tubule system in MP sensilla extend from this sinus to the surface of the sensory cuticle (145).

The sensillar sinus is exceptionally large and elongated in the galeal styloconic pegs of two species of lepidopteran larvae (V. D. Shields & R. Y. Zacharuk, in preparation). It contains a much higher level of K^+ than is present in the hemolymph (R. Y. Zacharuk, unpublished data). In the labial palp of an elaterid larva, the outer sinuses of all but one MP peg in the apical cluster are confluent. Gobletlike hemolymph sinus cells secrete into this common sinus (23). In a unique UP peg of a damselfly larva, the intermediate and outer sheath cells each enclose a separate sinus, resulting in three separate sinuses (19).

A number of sensilla have only two sheath cells (21, 70, 148), an inner that encloses the ciliary sinus and an outer that subtends the sensillar sinus. Some sensilla have four sheath cells; the additional cell is identified as a second intermediate cell (35, 36). In the antennal composite MP sensillum of an elaterid larva, each of the 12 bundles of neurons is ensheathed by an individual inner sheath cell; 12 intermediate and 12 outer sheath cells subtend and enclose a common outer sinus around the neuronal bundles (124). In the homologous sensillum of the mosquito larva, each of the six neuronal bundles is wrapped by an individual inner sheath cell and, basally, by an outer sheath cell (150). The latter are confluent distally around the common sensillar sinus, with no other cells present.

MORPHOGENESIS DURING A MOLT

In existing sensilla, the cellular elements are maintained through a molt, but the sensory cuticle and attached dendritic sheath are shed with the exuvium and replaced in the new cuticle (12, 30, 79–81, 141). At the onset of apolysis, the inner sheath cell secretes a thin proximal extension to the dendritic sheath that is initially folded many times (123). The dendritic sheath is usually much longer than the original when shed (141). It is pulled out as the ecdysial space forms by retraction of the hypodermis; this is presumably accompanied by an elongation of the ciliary dendritic segment to maintain a functional attachment to the old sensory cuticle (79, 80). The vibration-sensitive thoracic filiform hairs of a caterpillar maintain maximum sensitivity except for 30–60 min during actual ecdysis when they do not respond (82). The intermediate sheath cell forms an elongated process prior to secretion of new cuticle; this process forms the sensory cuticle of AP tactile sensilla (79–81) or MP OCS (12). MP OCS cuticle formation includes the pore tubule system that is shed with the exuvium. The outer sheath cell concurrently forms the basal cuticle (socket) around the sensory cuticle (79, 80).

Most sensilla maintain their external form through a molt, but the flower-shaped sensillum of a stonefly nymph becomes more elaborate at each molt (93). The dendrites of two of the eight neurons in a collembolan UP peg are subcuticular and do not participate in the molting or reformation events as do the other six (6). A molting pore or scar usually remains in the new cuticle after the old exuvial sheath ecdyses (145).

Additional sensilla are often formed during a molt in exopterygotes (42, 47, 95, 115, 116). In antennae, this formation is usually coupled with an increase in the number and length of the flagellar subsegments (41, 47, 115, 116). Cells of these sensilla originate by division of one epidermal cell to form four cells in a MS and eight in a CS (95). A *numb* gene that controls differentiation of neurons and associated cells in *D. melanogaster* embryos has now been isolated (134). Increased levels of juvenile hormone are associated with the molting process through the immature instars (80), but these levels inhibit formation of antennal OCS and the expression of sexual dimorphism (increased sensillar number in males) in the final nymphal molt (117). The added sensilla would not be expected to have a molting scar or pore.

The larvae of cyclorrhaphan Diptera have no distinct cephalic appendages, but have sensillar organs that are related to the antennae and mouthparts of other endopterygote larvae (48–50). The dorsal, terminal, and ventral organs are purported to develop from the embryonic antennal, maxillary, and labial placodes, respectively (1). An earlier study suggested that the terminal organ is of mixed segmental origin in *D. melanogaster* (76). In *Cuterebra horripilum* (14) and *Haematobia irritans* (15), the labium has two lateral lobes, each of which has a terminal sensillum in *H. irritans,* in addition to the pair of ventral organs dorsal to the mouth. A possible interpretation based on the above is that the ventral organ is part of a maxillary complex and not of labial origin. In the tsetse larva, which is free-living for less than one hour, the sensory organs are reduced to a single pair of antenno-maxillary processes (74).

The unsheathed dendritic terminations of Type II bipolar or multipolar neurons have associations with various cephalic sensilla (51, 109, 110, 144), as well as with epidermal cells (111). These terminations contain synapticlike vesicles at points of apposition to sensillar sheath (144) or epidermal (110) cells. These neurons may have a neurosecretory role in controlling the secretory activities of these cells during and after a molt (144, 145), as well as a possible role in proprioceptive mechanosensing (111).

DISTRIBUTION OF SENSILLA AND BEHAVIOR

The number and location of the various types of sensilla on one or more of the cephalic appendages are described using light microscopy and scanning electron microscopy (SEM) for many immature insects. Much more limited are detailed studies using TEM, which provide a more accurate account of innervating neurons, associated cells, and, with correlated electrophysiology, identity of function. Most studies are on pest species of Lepidoptera, Coleoptera, Diptera, and Orthoptera. In the endopterygotes, the number of sensilla remains fairly constant through the immature stages; major change to the adult patterns occurs in the pupal stage. Sensillar numbers usually change at each molt in exopterygotes; the greatest increase occurs in the terminal molt, at which time sexual dimorphism and changes in behavior pattern are expressed maximally.

Distribution of Sensilla

ANTENNAE   The antennae of endopterygote larvae generally have a distinct scape and pedicel, but the flagellum is one-segmented and much reduced (17, 35, 60, 121, 140, 150). The sensilla are primarily on the tips of the pedicel and flagellum. Of these, 1–9 (3–45 neurons) are UP and 1–3 (9–130 neurons) are MP sensilla. Some of the former are also mechanosensitive and at least one of the latter is a composite sensillum containing most of the antennal olfactory neurons. The antenna is reduced to a dome with one composite MP

Table 1 Number of sensilla/total neurons on one head appendage of larval endoptery-gotes, excluding AP MS

Family	AP/T-HS	UP/CS/MS	MP/comp[a]	MP/peg[b]	Reference
Antenna					
Elateridae	2/9	4/17/3	1/36	1/2	123, 124
Tenebrionidae	2/5	8/34/1	1/130		21
	2/8	9/45/7			35, 36
Catopidae	3/9	7/28/7	1/12	2/4	52, 59
Sphingidae			3/16		121
Tortricidae	2/6	2/5/2	3/41		In prep.[c]
Cecidomyiidae	1/3	1/3/1	1/63		130
Muscidae		6/8/4	1/21		48
Drosophilidae		6/8/4	1/21		127
Simuliidae		3/11/		3/9	60
Culicidae		1/3/1	1/13		148, 150
		1/3/1	1/11		89
Maxilla: palp					
Catopidae		10/45/8		1/3	57
Noctuidae		8/25/8		3/7	70
Muscidae		9/28/6		2/2	49
Drosophilidae		9/28/6		2/2	127
Culicidae		5/14/			103
Maxilla: galea					
Catopidae		2/5/2			57
Noctuidae	1/2?	2/8/2			70
Sphingidae	1/3?	2/8/2			121
Muscidae		1/2/			50
Drosophilidae		1/4/			127
Culicidae		2/6/			103
Labium: palp					
Elateridae		20/39/19		3/9	22
Catopidae	1/1	5/17/4		1/3	53

[a] Composite olfactory sensilla.
[b] Individual olfactory sensilla.
[c] R. Y. Zacharuk & P. J. Albert, in preparation.

(21 neurons) circled by six UP sensilla (8 neurons) in most of the cyclor-rhaphan Diptera (Table 1).

In exopterygotes, the flagellum is elongated and usually has many sub-segments. The total number of sensilla on these increases most rapidly in the later instars (Figures 2, 3) but may decrease in the earlier instars (Figure 3) (25, 41, 47, 112, 115, 128). This fluctuation results primarily from changes in the number of OCS. The central to distal segments develop more MP sensilla than do the proximal; the trichoid UP sensilla are more uniformly distributed from the scape to the flagellar tip (Figure 3). Most of the coelocon-

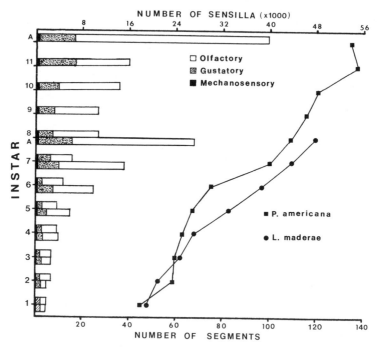

Figure 2 Average number of antennal flagellar subsegments *(lines)* and totals of three types of sensilla on them *(bars)* in the nymphal instars and adult males of *Leucophaea maderae* (115) and *Periplaneta americana* (116).

ic pegs of acridids are olfactory (44), but a small number are thermo-hygrosensitive (9, 112).

MAXILLAE   Each maxilla in larval endopterygotes has a palp with an apical cluster of pegs, a galea, and a lacinia. The latter two lobes are fused to a greater (2, 17, 83) or lesser (140) extent in lepidopteran and coleopteran larvae. The entire maxilla is reduced to a cluster of sensilla on and near a small protuberance in cyclorrhaphan Diptera (49, 127). The palp generally has 5–10 UP pegs (14–45 GCS and 6–8 MS neurons) and 1–3 MP pegs (2–7 neurons). The galea-lacinia has 1–2 UP pegs, perhaps 1 T-HS, and some tactile hairs (Table 1) (2, 84, 140).

In gryllids, the palp, galea, and lacinia have numerous sensilla, many of which appear to be tactile (114). The palp in acridids has numerous trichoid and basiconic sensilla along its length, with a cluster of the latter at the tip. The pegs increase and the hairs decrease in number at each molt (25). Most of the apical pegs are UP (28), but a few are MP (29). The galea has trichoid

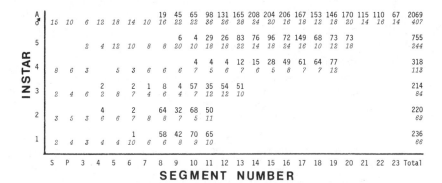

Figure 3 Distribution of olfactory *(roman type)* and trichoid putative contact chemo-mechanosensilla *(italic type)* on the antennal segments, and the number of flagellar subsegments, in nymphal instars and adult males of *Hypochlora alba* (25).

sensilla covering the surface and, on the inner face, papillae similar to the epipharyngeal taste sensilla (98). No reference was seen to the nature of sensilla on the lacinia; most are probably tactile.

LABIUM   In gryllids, the labium bears a pair of palps, paraglossae, and glossae (114). The latter two pairs of lobes are fused to varying extents in other insects, fusing completely into a ligula in coleopteran larvae (140). The ligula contains the spinneret in Lepidoptera (2). In Lepidoptera, the tip of each palp has only two MS (2, 70). In coleopterans, 12–23 pegs cluster at the tip of each labial palp; 3–4 are MP (3–9 neurons) and the remainder are mostly UP (17–39 neurons) (Table 1). In acridid nymphs, the distribution of types and changes in number of sensilla on the labial palps at each molt are similar to those on the maxillary palps, but labial palps have fewer sensilla (25). The paraglossae in *Locusta migratoria* together have 74 sensilla (296 neurons) in the first, and 184 (736 neurons) in the fifth instar (44); these are multineuronal and probably chemosensitive. Similar sensilla occur on the glossae and paraglossae of blattids and gryllids (44).

CLYPEO-LABRUM   The anterior surface of the clypeo-labrum of larval endopterygotes seems to have only MS (2, 54, 140). Exopterygotes may have a few CS, possibly OCS, on this surface as well (42). The posterior surface (epipharynx) usually has two or more types of MS and at least one type of GCS. Lepidopteran larvae have one pair of the latter and five of the former (2, 99). Elaterid larvae have five pairs (140) and catopid larvae ten pairs (54) of multineuronal UP sensilla, along with several campaniform and trichoid MS on the epipharynx. *D. melanogaster* larvae have nine UP and AP sensilla in two groups on the dorsal and one UP sensillum on the ventral wall of the pharynx (127). The latter and one of the former are unique composite UP sensilla. Numerous UP and AP sensilla occur on the epipharynx of gryllids (42), acridids (25, 51), and blattids (109). These are usually arranged in groups in paired fields (42).

A unique thrust MS on the clypeo-labrum of a predatory beetle larva is coupled to the mandibular motor system. When deflected, it initiates closure of the mandibles over the prey (7).

MANDIBLES   Only MS occur on the mandibles of immature insects (2, 55, 114, 146). Most have two (146) to many (114) trichoid hairs; usually many campaniform sensilla in the cuticle, particularly of the outer wall (114, 146); and there are several (2) to many (114) scolopophorous sensilla inserted into the cuticle of the teeth. The latter two types of sensilla monitor stresses on the mandibular walls and teeth during biting or chewing (146).

Sensilla and Behavior

Earlier studies related individual or groups of sensilla to the feeding responses of immature insects after ablation. Currently, research increasingly emphasizes correlation of feeding behavior with electrophysiological responses of selected sensilla, particularly those involved in gustation. Similar studies on olfaction are more limited. Most studies are on Lepidoptera and, to a lesser extent, Coleoptera and Orthoptera. Dethier (64) has delightfully summarized the past and future studies on sensilla in insect–host plant relations.

OLFACTION   MP sensilla, while not nearly as numerous as in adults, play an important role in food finding by immature insects (90, 136). In Lepidoptera, those on the antennae are the most important in olfactory discrimination between different plants (65, 119); those on the maxillary palps contribute also, but to a lesser extent (62, 85, 119). In elaterid larvae, the olfactory pegs on the maxillary and, to a lesser extent, labial palps are involved in orientation to CO_2; MP sensilla on the antennae are not (72). House fly larvae respond to odors only if their dorsal organs, which have a composite MP sensillum, are intact (38). Onion fly larvae require either the dorsal or the anterior organs, or one of each pair, to orient to their host plant; in addition to the dorsal organ, the contact chemosensilla on the anterior organ are presumed to be involved in this orientation (139). The antennal OCS of nymphal blattids perceive a spectrum of food odors and aggregation and possibly also aggression-stimulating pheromones; these sensilla in acridid nymphs discriminate between different foods, and may also perceive social pheromones (44). However, perception of volatile trail pheromones, which is generally by olfaction, is by contact chemoreception by GCS on the maxillary palps of larval *Malacosoma americana* and *M. neustria* and *Yponomeuta cagnagellus* (113).

GUSTATION   UP sensilla are present on the antennae of most immature insects, but those on the maxillae, labium, and epipharynx are primarily involved in the biting and ingestion responses. In larval Lepidoptera, some pegs at the tip of the maxillary palp have a gustatory function in perceiving specific phagostimulants and feeding deterrents (119), but the two styloconic pegs on the galea play the major role in discriminating between plant constituents or their derivatives in the biting response (31, 33, 85, 99, 100, 119, 135). The pair of pharyngeal taste sensilla control swallowing responses (99) and are the final peripheral sensilla that discriminate between foods in the gustatory process (33, 119). Whether food is swallowed or not seems to be based on the presence of deterrents rather than phagostimulants (61). Feeding intensity is strongly correlated with receptor neuron activity (33). Impulses from deterrent cells counteract those from phagostimulated cells in the CNS; one impulse from the former neutralizes 2.5 impulses from the latter (120). Prior experience, such as deprivation (39), nutritional state (100, 126), or food previously ingested (3), affects subsequent feeding behavior and may result in induction or aversion learning (3, 67, 99).

The primary UP sensilla involved in food discrimination and gustation in acridids and blattids are pegs and hairs at the tips of the maxillary and labial palps, and papillae on the epipharynx (86), inner face of the galeae, and outer face of the paraglossae (44). Differences in distribution and numbers of sensillar types in acridids are related to the nature and spectrum of their normal host food plants (27, 37, 45, 46).

No reference was seen to electrophysiological studies on UP sensilla of dipteran larvae. A behavioral assay indicated *D. melanogaster* larvae have a sugar cell with two different binding sites, a salt cell with anionic and cationic sites, and an amino acid cell in their sensilla (107). The larvae are attracted to sugars and low levels of salts but avoid the amino acids.

CONCLUDING REMARKS

Many behavioral, electrophysiological, and morphological studies using light microscopy and SEM are in the literature on antennal and mouthpart sensilla of immature insects, but surprisingly few sufficiently detailed or complete accounts of these are based on TEM. Of particular interest are the nature of immature insects' stimulus conduction mechanisms, number of innervating neurons, and types of dendritic terminations and the relationship of associated cells and cuticular structures to these components. This information is scanty even for sensilla that are much studied behaviorally or physiologically, such as the styloconic pegs on the galea and the diverse sensilla on the antennae and maxillary palps of lepidopteran larvae. Where detailed mophology exists, modality is often inferred by comparison to structure and putative function of similar sensilla in the literature rather than confirmed by direct testing. More collaborative or corroborative studies by morphologists and physiologists would alleviate this deficiency.

Virtually nothing in the literature deals with the nature of stimulant receptor sites, ion constituents, channels and pumps, or probable enzymes and second messengers in sensilla of immature insects. The effects of pharmacological agents or physiological manipulations on sensillar response are, therefore, difficult to interpret, and correlative studies on the effects of these on the ultrastructure of the organelles involved in the transduction process, which could contribute to such interpretation, are lacking. Knowledge of the above is basic to further research on and development of agents and protocols for immature insect pest management and crop protection. For example, it could have immediate application in the design of synthetic feeding inhibitors based on natural plant products for crop protection. Progress in such research would be greatly enhanced by a collaborative interdisciplinary team approach.

ACKNOWLEDGMENTS

Support was provided by E. I. DuPont de Nemours and Company through its Agricultural Products Department and the Natural Sciences and Engineering Research Council of Canada. We dedicate this review to P. W. Riegert, Professor Emeritus, University of Regina, on the occasion of his recent retirement; he continues to write, teach, discuss, and practice Entomology as much as before, and we thank him for his constructive comments on this review.

Literature Cited

1. Ajidagba, P. A., Bay, D. E., Pitts, C. W. 1985. Morphogenesis of the external features of the first-stage larva of the stable fly (Diptera: Muscidae). *J. Kansas Entomol. Soc.* 58:569–77
2. Albert, P. J. 1980. Morphology and innervation of mouthpart sensilla in larvae of the spruce budworm, *Choristoneura fumiferana* (Clem.) (Lepidoptera: Tortricidae). *Can. J. Zool.* 58:842–51
3. Albert, P. J., Parisella, S. 1985. Tests for induction of feeding preferences in larvae of eastern spruce budworm using extracts from three host plants. *J. Chem. Ecol.* 11:809–17
4. Albert, P. J., Parisella, S. 1988. Physiology of a sucrose-sensitive cell on the galea of the eastern spruce budworm larva, *Choristoneura fumiferana*. *Physiol. Entomol.* 13:243–47
5. Altner, H. 1977. Insect sensillum specificity and structure: an approach to a new typology. In *Olfaction and Taste VI*, ed. J. LeMagnen, P. MacLeod, 6:295–303. Washington: Information Retrieval
6. Altner, H., Altner, I. 1985. Multicellular antennal sensilla containing a sensory cell with a short dendrite and dense-core granules in the insect, *Hypogastrura socialis* (Collembola): intermolt and molting stages. *Cell Tissue Res.* 241:119–28
7. Altner, H., Bauer, T. 1982. Ultrastructure of a specialized, thrust-sensitive, insect mechanoreceptor: Stimulus-transmitting structures and sensory apparatus in the rostral horns of *Notiophilus biguttatus*. *Cell Tissue Res.* 226:337–54
8. Altner, H., Loftus, R. 1985. Ultrastructure and function of insect thermo- and hygroreceptors. *Annu. Rev. Entomol.* 30:273–95
9. Altner, H., Routil, Ch., Loftus, R. 1981. The structure of bimodal chemo-,

thermo-, and hygroreceptive sensilla on the antenna of *Locusta migratoria*. *Cell Tissue Res.* 215:289–308

10. Altner, H., Sass, H., Altner, I. 1977. Relationship between structure and function of antennal chemo-, hygro-, and thermoreceptive sensilla in *Periplaneta americana*. *Cell Tissue Res.* 176:389–405

11. Altner, H., Schaller-Selzer, L., Stetter, H., Wohlrab, I. 1983. Poreless sensilla with inflexible sockets. A comparative study of a fundamental type of insect sensilla probably comprising thermo- and hygroreceptors. *Cell Tissue Res.* 234:279–307

12. Altner, H., Thies, G. 1972. Reizleitende Strukturen und Ablauf der Häutung an Sensillen einer euedaphischen Collembolenart. *Z. Zellforsch. Mikrosk. Anat.* 129:196–216

13. Altner, H., Thies, G. 1984. Internal proprioceptive organs of the distal antennal segments in *Allacma fusca* (L.) (Collembola: Sminthuridae): proprioceptors phylogenetically derived from sensillum-bound exteroceptors. *Int. J. Insect Morphol. Embryol.* 13:315–30

14. Baker, G. T. 1986. Morphological aspects of the egg and cephalic region of the first instar larva of *Cuterebra horripilum* (Diptera: Cuterebridae). *Wasmann J. Biol.* 44:66–72

15. Baker, G. T. 1987. Morphological aspects of the third instar larva of *Haematobia irritans*. *Med. Vet. Entomol.* 1:279–83

16. Baker, G. T. 1987. Apical sensilla on the adult and larval labial and maxillary palpi of *Odontotaenius disjunctus* (Illiger) (Coleoptera: Passalidae). *Proc. Entomol. Soc. Wash.* 89:682–86

17. Baker, G. T., Chan, W. P. 1987. Sensilla on the antennae and mouthparts of the larval and adult stages of *Olethreutes cespitana* (Lepidoptera: Tortricidae). *Ann. Soc. Entomol. Fr.* (N. S.) 23:387–97

18. Baker, G. T., Ellsbury, M. M. 1989. Morphology of the mouth parts and antenna of the larva of the clover stem borer, *Languria mozardi* Latreille (Coleoptera: Languriidae). *Proc. Entomol. Soc. Wash.* 91:15–21

19. Bassemir, U., Hansen, K. 1980. Single-pore sensilla of damselfly-larvae: representatives of phylogenetically old contact chemoreceptors? *Cell Tissue Res.* 207:307–20

20. Becker, A., Peters, W. 1989. Localization of sugar-binding sites in contact chemosensilla of *Periplaneta americana*. *J. Insect Physiol.* 35:239–50

21. Behan, M., Ryan, M. F. 1978. Ultrastructure of antennal sensory receptors of *Tribolium* larvae (Coleoptera: Tenebrionidae). *Int. J. Insect Morphol. Embryol.* 7:221–36

22. Bellamy, F. W. 1973. *Ultrastructure of the labial palp and its associated sensilla of the prairie grain wireworm* Ctenicera destructor *(Brown) (Elateridae: Coleoptera)*. PhD thesis. Regina, Saskatchewan: University of Saskatchewan, Regina Campus. 361 pp.

23. Bellamy, F. W., Zacharuk, R. Y. 1976. Structure of the labial palp of a larval elaterid (Coleoptera) and of sinus cells associated with its sensilla. *Can. J. Zool.* 54:2118–28

24. Bernays, E. A., Cook, A. G., Padgham, D. E. 1976. A club-shaped hair found on the first-instar nymphs of *Schistocerca gregaria*. *Physiol. Entomol.* 1:3–13

25. Bland, R. G. 1982. Morphology and distribution of sensilla on the antennae and mouthparts of *Hypochlora alba* (Orthoptera: Acrididae). *Ann. Entomol. Soc. Am.* 75:272–83

26. Bland, R. G. 1983. Sensilla on the antennae, mouthparts, and body of the larva of the alfalfa weevil, *Hypera postica* (Gyllenhal) (Coleoptera: Curculionidae). *Int. J. Insect Morphol. Embryol.* 12:261–72

27. Bland, R. G. 1989. Antennal sensilla of Acrididae (Orthoptera) in relation to subfamily and food preference. *Ann. Entomol. Soc. Am.* 82:368–84

28. Blaney, W. M. 1975. Behavioural and electrophysiological studies of taste discrimination by the maxillary palps of larvae of *Locusta migratoria* (L.). *J. Exp. Biol.* 62:555–69

29. Blaney, W. M. 1977. The ultrastructure of an olfactory sensillum on the maxillary palps of *Locusta migratoria* (L.). *Cell Tissue Res.* 184:397–409

30. Blaney, W. M., Chapman, R. F., Cook, A. G. 1971. The structure of the terminal sensilla on the maxillary palps of *Locusta migratoria* (L.), and changes associated with moulting. *Z. Zellforsch. Mikrosk. Anat.* 121:48–68

31. Blaney, W. M., Simmonds, M. S. J. 1988. Food selection in adults and larvae of three species of Lepidoptera: a behavioural and electrophysiological study. *Entomol. Exp. Appl.* 49:111–21

32. Blest, A. D. 1988. The turnover of phototransductive membrane in compound eyes and ocelli. *Adv. Insect Physiol.* 20:1–53

33. Blom, F. 1978. Sensory activity and food intake: a study of input-output rela-

tionships in two phytophagous insects. *Neth. J. Zool.* 28:277–340

34. Bloom, J. W., Zacharuk, R. Y., Holodniuk, A. E. 1981. Ultrastructure of a terminal chordotonal sensillum in larval antennae of the yellow mealworm, *Tenebrio molitor* L. *Can. J. Zool.* 59:515–24

35. Bloom, J. W., Zacharuk, R. Y., Holodniuk, A. E. 1982. Ultrastructure of the larval antenna of *Tenebrio molitor* L. (Coleoptera: Tenebrionidae): structure of the trichoid and uniporous peg sensilla. *Can. J. Zool.* 60:1528–44

36. Bloom, J. W., Zacharuk, R. Y., Holodniuk, A. E. 1982. Ultrastructure of the larval antenna of *Tenebrio molitor* L. (Coleoptera: Tenebrionidae): structure of the blunt-tipped peg and papillate sensilla. *Can. J. Zool.* 60:1545–56

37. Blust, M. H., Hopkins, T. L. 1987. Gustatory responses of a specialist and a generalist grasshopper to terpenoids of *Artemisia ludoviciana. Entomol. Exp. Appl.* 45:37–46

38. Bolwig, N. 1946. Senses and sense organs of the anterior end of the house fly larvae. *Vidensk. Medd. Dan. Naturhist. Foren. Khobenhavn* 109:81–217

39. Bowdan, E. 1988. The effect of deprivation on the microstructure of feeding by the tobacco hornworm caterpillar. *J. Insect Behav.* 1:31–50

40. Broyles, J. L., Hanson, F. E., Shapiro, A. M. 1976. Ion dependence of the tarsal sugar receptor of the blowfly *Phormia regina. J. Insect Physiol.* 22:1587–1600

41. Campbell, F. L., Priestley, J. D. 1970. Flagellar annuli of *Blattella germanica* (Dictyoptera: Blattellidae).—Changes in their numbers and dimensions during postembryonic development. *Ann. Entomol. Soc. Am.* 63:81–88

42. Carline, T., Kubra, K., Brown, V. K., Beck, R. 1984. Comparison of the distribution and nervous innervation of the sensilla on the labrum of *Gryllus bimaculatus* (De Geer) and *Acheta domesticus* (L.) (Orthoptera: Gryllidae), and an account of their development in *A. domesticus. Int. J. Insect Morphol. Embryol.* 13:81–103

43. Chan, W. P., Baker, G. T., Ellsbury, M. M. 1988. Sensilla on the larvae of four *Hypera* species (Coleoptera: Curculionidae). *Proc. Entomol. Soc. Wash.* 90:269–87

44. Chapman, R. F. 1982. Chemoreception: the significance of receptor numbers. *Adv. Insect. Physiol.* 16:247–356

45. Chapman, R. F. 1988. Sensory aspects of host-plant recognition by Acridoidea:

questions associated with the multiplicity of receptors and variability of response. *J. Insect Physiol.* 34:167–74

46. Chapman, R. F., Fraser, J. 1989. The chemosensory system of the monophagous grasshopper, *Bootettix argentatus* Bruner (Orthoptera: Acrididae). *Int. J. Insect Morphol. Embryol.* 18:111–18

47. Chapman R. F., Greenwood, M. 1986. Changes in distribution and abundance of antennal sensilla during growth of *Locusta migratoria* L. (Orthoptera: Acrididae). *Int. J. Insect Morphol. Embryol.* 15:83–96

48. Chu, I.-W., Axtell, R. C. 1971. Fine structure of the dorsal organ of the house fly larva, *Musca domestica* L. *Z. Zellforsch. Mikrosk. Anat.* 117:17–34

49. Chu-Wang, I.-W., Axtell, R. C. 1972. Fine structure of the terminal organ of the house fly larva, *Musca domestica* L. *Z. Zellforsch. Mikrosk. Anat.* 127:287–305

50. Chu-Wang, I.-W., Axtell, R. C. 1972. Fine structure of the ventral organ of the house fly larva, *Musca domestica* L. *Z. Zellforsch. Mikrosk. Anat.* 130:489–95

51. Cook, A. G. 1972. The ultrastructure of the A1 sensilla on the posterior surface of the clypeo-labrum of *Locusta migratoria migratorioides* (R and F). *Z. Zellforsch. Mikrosk. Anat.* 134:539–54

52. Corbière, G. 1969. Ultrastructure et électrophysiologie du lobe membraneux de l'antenne chez la larve du *Speophyes lucidulus* (Coléoptère). *J. Insect Physiol.* 15:1759–65

53. Corbière-Tichané, G. 1969. Ultrastructure du labium de la larve du *Speophyes lucidulus* Delarouzeei (Coleoptera, Catopidae). *Z. Morphol. Tiere* 66:73–86

54. Corbière-Tichané, G. 1970. Ultrastructure du labre des larves du *Speophyes lucidulus* (Delarouzeei), coléoptère cavernicole de la sous-famille des Bathysciinae (Coleoptera, Catopidae). *Z. Morphol. Tiere* 67:86–96

55. Corbière-Tichané, G. 1971. Ultrastructure de l'équipement sensoriel de la mandibule chez la larve du *Speophyes lucidulus* Delar. (Coléoptère cavernicole de la sous-famille des Bathysciinae). *Z. Zellforsch. Mikrosk. Anat.* 112:129–38

56. Corbière-Tichané, G. 1971. Ultrastructure des organes chordotonaux des pièces céphaliques chez la larve du *Speophyes lucidulus* Delar. (Coléoptère cavernicole de la sous-famille des Bathysciinae). *Z. Zellforsch. Mikrosk. Anat.* 117:275–302

57. Corbière-Tichané, G. 1971. Ultra-

structure du système sensoriel de la maxille chez la larve du coléoptère cavernicole *Speophyes lucidulus* Delar. (Bathysciinae). *J. Ultrastruct. Res.* 36:318–41

58. Corbière-Tichané, G. 1973. Sur les structures sensorielles et leurs fonctions chez la larve de *Speophyes lucidulus*. *Ann. Spéléol.* 28:247–65

59. Corbière-Tichané, G., Bermond, N. 1971. Ultrastructure et électrophysiologie des styles antennaires de la larve de *Speophyes lucidulus* Delar. (Coléoptère Bathysciinae). *Ann. Sci. Nat. Zool. Paris.* 13:505–41

60. Craig, D. A., Batz, H. 1982. Innervation and fine structure of antennal sensilla of Simuliidae larvae (Diptera: Culicomorpha). *Can. J. Zool.* 60:696–711

61. De Boer, G., Dethier, V. G., Schoonhoven, L. M. 1977. Chemoreceptors in the preoral cavity of the tobacco hornworm, *Manduca sexta*, and their possible function in feeding behaviour. *Entomol. Exp. Appl.* 21:287–98

62. DeBoer, G., Hanson, F. E. 1987. Differentiation of roles of chemosensory organs in food discrimination among host and non-host plants by larvae of the tobacco hornworm, *Manduca sexta*. *Physiol. Entomol.* 12:387–98

63. Dethier, V. G. 1973. Electrophysiological studies of gustation in lepidopterous larvae. II. Taste spectra in relation to food-plant discrimination. *J. Comp. Physiol.* 82:103–34

64. Dethier, V. G. 1978. Studies on insect/host plant relations—past and future. *Entomol. Exp. Appl.* 24:759–66

65. Dethier, V. G. 1980. Responses of some olfactory receptors of the eastern tent caterpillar *(Malacosoma americanum)* to leaves. *J. Chem. Ecol.* 6:213–20

66. Dethier, V. G. 1987. Discriminative taste inhibitors affecting insects. *Chem. Senses* 12:251–63

67. Dethier, V. G. 1988. Induction and aversion-learning in polyphagous arctiid larvae (Lepidoptera) in an ecological setting. *Can. Entomol.* 120:125–31

68. Dethier, V. G., Kuch, J. H. 1971. Electrophysiological studies of gustation in lepidopterous larvae. I. Comparative sensitivity to sugars, amino acids, and glycosides. *Z. Vgl. Physiol.* 72:343–63

69. Dethier, V. G., Schoonhoven, L. M. 1969. Olfactory coding by lepidopterous larvae. *Entomol. Exp. Appl.* 12:535–43

70. Devitt, B. D., Smith, J. J. B. 1982. Morphology and fine structure of mouthpart sensilla in the dark-sided cutworm *Euxoa messoria* (Harris) (Lepidoptera:

Noctuidae). *Int. J. Insect Morphol. Embryol.* 11:255–70

71. Devoe, R. D. 1985. The eye: electrical activity. See Ref. 93a, pp. 277–354

72. Doane, J. F., Klingeer, J. 1978. Location of CO_2-receptive sensilla on larvae of the wireworms *Agriotes lineatusobscurus* and *Limonius californicus*. *Ann. Entomol. Soc. Am.* 71:357–63

73. Fellows, L. E., Kite, G. C., Nash, R. J., Simmonds, M. S. J., Scofield, A. M. 1989. Castanospermine, swainsonine and related polyhydroxy alkaloids: structure, distribution and biological activity. In *Plant Nitrogen Metabolism*, ed. J. E. Poulton, J. T. Romeo, E. E. Conn, pp. 395–427. New York: Plenum

74. Finlayson, L. H. 1972. Chemoreceptors, cuticular mechanoreceptors, and peripheral multiterminal neurones in the larva of the tsetse fly *(Glossina)*. *J. Insect Physiol.* 18:2265–76

75. Frazier, J. L. 1986. The perception of plant allelochemicals that inhibit feeding. In *Molecular Aspects of Insect-Plant Associations*, ed. L. B. Brattsten, S. Ahmad, pp. 1–42. New York: Plenum

76. Frederick, R. D., Denell, R. E. 1982. Embryological origin of the antennomaxillary complex of the larva of *Drosophila melanogaster* Meigen (Diptera: Drosophilidae). *Int. J. Insect Morphol. Embryol.* 11:227–33

77. French, A. S. 1988. Transduction mechanisms of mechanosensilla. *Annu. Rev. Entomol.* 33:39–58

78. Gaffal, K. P. 1979. An ultrastructural study of the tips of four classical bimodal sensilla with one mechanosensitive and several chemosensitive receptor cells. *Zoomorphologie* 92:273–91

79. Gnatzy, W. 1978. Development of the filiform hairs on the cerci of *Gryllus bimaculatus* Deg. (Saltatoria, Gryllidae). *Cell Tissue Res.* 187:1–24

80. Gnatzy, W. 1980. Morphogenesis of mechanoreceptor and epidermal cells of crickets during the last instar, and its relation to molting-hormone level. *Cell Tissue Res.* 213:369–91

81. Gnatzy, W., Schmidt, K. 1972. Die Feinstruktur der Sinneshaare auf den Cerci von *Gryllus bimaculatus* Deg. (Saltatoria, Gryllidae). IV. Die Häutung der kurzen Borstenhaare. *Z. Zellforsch. Mikrosk. Anat.* 126:223–39

82. Gnatzy, W., Tautz, J. 1977. Sensitivity of an insect mechanoreceptor during moulting. *Physiol. Entomol.* 2:279–88

83. Grimes, L. R., Neunzig, H. H. 1986. Morphological survey of the maxillae in last stage larvae of the suborder Ditrysia

(Lepidoptera): palpi. *Ann. Entomol. Soc. Am.* 79:491–509

84. Grimes, L. R., Neunzig, H. H. 1986. Morphological survey of the maxillae in last-stage larvae of the suborder Ditrysia (Lepidoptera): mesal lobes (laciniogaleae). *Ann. Entomol. Soc. Am.* 79: 510–26

85. Hanson, F. E., Dethier, V. G. 1973. Role of gustation and olfaction in food plant discrimination in the tobacco hornworm, *Manduca sexta. J. Insect Physiol.* 19:1019–34

86. Haskell, P. T., Schoonhoven, L. M. 1969. The function of certain mouth part receptors in relation to feeding in *Schistocerca gregaria* and *Locusta migratoria migratorioides. Entomol. Exp. Appl.* 12: 423–40

87. Horn, E. 1985. Gravity. See Ref. 93a, pp. 557–76

88. Itagaki, H., Hildebrand, J. G. 1990. Olfactory interneurons in the brain of the larval sphinx moth *Manduca sexta. J. Comp. Physiol. A.* In press

89. Jez, D. H., McIver, S. B. 1980. Fine structure of antennal sensilla of larval *Toxorhynchites brevipalpis* Theobald (Diptera: Culicidae). *Int. J. Insect Morphol. Embryol.* 9:147–59

90. Jones, O. T., Coaker, T. H. 1978. A basis for host plant finding in phytophagous larvae. *Entomol. Exp. Appl.* 24:472–84

91. Kapoor, N. N. 1987. Fine structure of the coniform sensillar complex on the antennal flagellum of the stonefly nymph *Paragnetina media* (Plecoptera: Perlidae). *Can. J. Zool.* 65:1827–32

92. Kapoor, N. N. 1989. Distribution and innervation of sensilla on the mouthparts of the carnivorous stonefly nymph, *Paragnetina media* (Walker) (Plecoptera: Perlidae). *Can. J. Zool.* 67:831–38

93. Kapoor, N. N., Zachariah, K. 1984. Scanning and transmission electron microscopy of the developmental stages of the flower-shape sensillum of the stonefly nymph, *Thaumatoperla alpina* Burns and Neboiss (Plecoptera: Eustheniidae). *Int. J. Insect Morphol. Embryol.* 13: 177–89

93a. Kerkut, G. A., Gilbert, L. I., eds. 1985. *Comprehensive Insect Physiology, Biochemistry and Pharmacology,* Vol. 6, *Nervous System: Sensory.* New York: Pergamon

94. Land, M. F. 1985. The eye: optics. See Ref. 93a, pp. 225–75

95. Lawrence, P. A. 1966. Development and determination of hairs and bristles in the milkweed bug, *Oncopeltus fasciatus*

(Lygaeidae, Hemiptera). *J. Cell Sci.* 1:475–98

96. Le Berre, J. R., Louveaux, A. 1969. Equipement sensoriel des mandibules de la larve du premier stade de *Locusta migratoria* L. *C. R. Acad. Sci. Paris* 268:2907–10

97. Loftus, R., Corbière-Tichané, G. 1981. Antennal warm and cold receptors of the cave beetle, *Speophyes lucidulus* Delar., in sensilla with a lamellated dendrite. I. Response to sudden temperature change. *J. Comp. Physiol. A* 143:443–52

98. Louveaux, A. 1973. Etude d'une plage d'organes sensoriels sur les galea de *Locusta migratoria* L. *C. R. Acad. Sci. Paris* 277:1353–56

99. Ma, W.-C. 1972. *Dynamics of feeding responses in* Pieris brassicae Linn. *as a function of chemosensory input: A behavioural, ultrastructural and electrophysiological study.* PhD thesis. Wageningen: Meded. Landbouwhogesch. 72–11. 162 pp.

100. Ma, W.-C. 1976. Experimental observations of food-aversive responses in larvae of *Spodoptera exempta* (Wlk.) (Lepidoptera, Noctuidae). *Bull. Entomol. Res.* 66:87–96

101. Markl, H., Tautz, J. 1975. The sensitivity of hair receptors in caterpillars of *Barathra brassicae* L. (Lepidoptera, Noctuidae) to particle movement in a sound field. *J. Comp. Physiol.* 99:79–87

102. McIver, S. B. 1985. Mechanoreception. See Ref. 93a, pp. 71–132

103. McIver, S., Siemicki, R. 1982. Fine structure of maxillary sensilla of larval *Toxorhynchites brevipalpis* (Diptera: Culicidae) with comments on the role of sensilla in behavior. *J. Morphol.* 171: 293–303

104. Michelsen, A., Larsen, O. N. 1985. Hearing and sound. See Ref. 93a, pp. 495–556

105. Mitchell, B. K. 1978. Some aspects of gustation in the larval red turnip beetle, *Entomoscelis americana*, related to feeding and host plant selection. *Entomol. Exp. Appl.* 24:540–49

106. Mitchell, B. K., Gregory, P. 1979. Physiology of the maxillary sugar sensitive cell in the red turnip beetle, *Entomoscelis americana. J. Comp. Physiol. A* 132:167–78

107. Miyakawa, Y. 1982. Behavioural evidence for the existence of sugar, salt and amino acid taste receptor cells and some of their properties in *Drosophila* larvae. *J. Insect Physiol.* 28:405–10

108. Morita, H., Shiraishi, A. 1985. Chemoreception physiology. See Ref. 93a, pp. 133–70

109. Moulins, M. 1971. Ultrastructure et physiologie des organes épipharyngiens et hypopharyngiens (Chimiorécepteurs cibariaux) de *Blabera craniifer* Burm. (Insecte, Dictyoptère). *Z. Vgl. Physiol.* 73:139–66

110. Moulins, M., Noirot, Ch. 1972. Morphological features bearing on transduction and peripheral integration in insect gustatory organs. In *Olfaction and Taste IV*, ed. D. Schneider, pp. 49–55. Stuttgart: Wissenschaftliche Verlagsgesellschaft

111. Osborne, M. P. 1964. Sensory nerve terminations in the epidermis of the blowfly larva. *Nature* 201:526

112. Riegert, P. W. 1960. The humidity reactions of *Melanoplus bivittatus* (Say) (Orthoptera, Acrididae): antennal sensilla and hygro-reception. *Can. Entomol.* 92:561–70

113. Roessingh, P., Peterson, S. C., Fitzgerald, T. D. 1988. The sensory basis of trail following in some lepidopterous larvae: contact chemoreception. *Physiol. Entomol.* 13:219–24

114. Rościszewska, M., Fudalewicz-Niemczyk, W. 1974. The peripheral nervous system of the larva of *Gryllus domesticus* L. (Orthoptera). Part II. Mouth parts. *Acta Biol. Cracov. Ser. Zool.* 17:19–39

115. Schafer, R. 1973. Postembryonic development in the antenna of the cockroach, *Leucophaea maderae*: growth, regeneration, and the development of the adult pattern of sense organs. *J. Exp. Zool.* 183:353–63

116. Schafer, R., Sanchez, T. V. 1973. Antennal sensory system of the cockroach, *Periplaneta americana*: Postembryonic development and morphology of the sense organs. *J. Comp. Neurol.* 149:335–53

117. Schafer, R., Sanchez, T. V. 1976. The nature and development of sex attractant specificity in cockroaches of the genus *Periplaneta*. II. Juvenile hormone regulates sexual dimorphism in the distribution of antennal olfactory receptors. *J. Exp. Zool.* 198:323–36

118. Schneider, D. 1964. Insect antennae. *Annu. Rev. Entomol.* 9:103–22

119. Schoonhoven, L. M. 1972. Plant recognition by lepidopterous larvae. In *Insect/Plant Relationships,* ed. H. F. van Emden, 6:87–99. London: Symp. R. Entomol. Soc. London

120. Schoonhoven, L. M., Blom, F. 1988. Chemoreception and feeding behaviour in a caterpillar: towards a model of brain functioning in insects. *Entomol. Exp. Appl.* 49:123–29

121. Schoonhoven, L. M., Dethier, V. G. 1966. Sensory aspects of host-plant discrimination by lepidopterous larvae. *Arch. Néerl. Zool.* 16:497–530

122. Schoonhoven, L. M., Fu-Shun, Y. 1989. Interference with normal chemoreceptor activity by some sesquiterpenoid antifeedants in an herbivorous insect *Pieris brassicae*. *J. Insect Physiol.* 35:725–28

123. Scott, D. A. 1969. *Fine structure of sensilla on the antenna of* Ctenicera destructor *(Brown) (Elateridae:Coleoptera) with special reference to chemoreceptors.* PhD thesis. Regina, Saskatchewan: University of Saskatchewan, Regina Campus. 184 pp.

124. Scott, D. A., Zacharuk, R. Y. 1971. Fine structure of the antennal sensory appendix in the larva of *Ctenicera destructor* (Brown) (Elateridae: Coleoptera). *Can. J. Zool.* 49:199–210

125. Scott, D. A., Zacharuk, R. Y. 1971. Fine structure of the dendritic junction body region of the antennal sensory cone in a larval elaterid (Coleoptera). *Can J. Zool.* 49:817–21

126. Simpson, S. J., Simonds, M. S. J., Blaney, W. M., Jones, J. P. 1990. Compensatory dietary selection in larval *Locusta migratoria* but not *Spodoptera littoralis* after a single deficient meal during *ad libitum* feeding. *Physiol. Entomol.* 15:235–42

127. Singh, R. N., Singh, K. 1984. Fine structure of the sensory organs of *Drosophila melanogaster* Meigen larva (Diptera: Drosophilidae). *Int. J. Insect Morphol. Embryol.* 13:255–73

128. Singleton-Smith, J., Chang, J. F., Philogéne, B. J. R. 1978. Morphological differences between nymphal instars and descriptions of the antennal sensory structures of the nymphs and adults of *Psylla pyricola* Foerster (Homoptera: Psyllidae). *Can. J. Zool.* 56:1576–84

129. Slifer, E. H., Prestage, J. J., Beams, H. W. 1957. The fine structure of the long basiconic sensory pegs of the grasshopper (Orthoptera, Acrididae) with special reference to those on the antenna. *J. Morphol.* 101:359–97

130. Solinas, M., Nuzzaci, G., Isidoro, N. 1989. Antennal sensory structures and their ecological-behavioural meaning in Cecidomyiidae (Diptera) larvae. *Entomologica* 22:165–84

131. Städler, E., Hanson, F. E. 1975. Olfactory capabilities of the "gustatory" chemoreceptors of the tobacco hornworm larvae. *J. Comp. Physiol.* 104:97–102

132. Tautz, J. 1978. Reception of medium vibration by thoracal hairs of caterpillars

of *Barathra brassicae* L. (Lepidoptera, Noctuidae). II. Response characteristics of the sensory cell. *J. Comp. Physiol. A* 125:67–77

133. Trujillo-Cenóz, O. 1985. The eye: development, structure and neural connections. See Ref. 93a, pp. 171–223

134. Uemura, T., Shepherd, S., Ackerman, L., Jan, L. Y., Jan, Y. N. 1989. *numb*, a gene required in determination of cell fate during sensory organ formation in *Drosophila* embryos. *Cell* 58:349–60

135. van Drongelen, W. 1979. Contact chemoreception of host plant specific chemicals in larvae of various *Yponomeuta* species (Lepidoptera). *J. Comp. Physiol. A* 134:265–79

136. Visser, J. H. 1986. Host odor perception in phytophagous insects. *Annu. Rev. Entomol.* 31:121–44

137. Wensler, R. J., Filshie, B. K. 1969. Gustatory sense organs in the food canal of aphids. *J. Morphol.* 129:473–91

138. Wieczorek, H. 1976. The glycoside receptor of the larvae of *Mamestra brassicae* L. (Lepidoptera, Noctuidae). *J. Comp. Physiol. A* 106:153–76

139. Yamada, Y., Ishikawa, Y., Ikeshoji, T., Matsumoto, Y. 1981. Cephalic sensory organs of the onion fly larva, *Hylemya antiqua* Meigen (Diptera: Anthomyiidae) responsible for host-plant finding. *Appl. Entomol. Zool.* 16:121–28

140. Zacharuk, R. Y. 1962. Sense organs of the head of larvae of some Elateridae (Coleoptera): their distribution, structure and innervation. *J. Morphol.* 111:1–33

141. Zacharuk, R. Y. 1962. Exuvial sheaths of sensory neurones in the larva of *Cte-nicera destructor* (Brown) (Coleoptera, Elateridae). *J. Morphol.* 111:35–47

142. Zacharuk, R. Y. 1971. Fine structure of peripheral terminations in the porous sensillar cone of larvae of *Ctenicera destructor* (Brown) (Coleoptera, Elateridae), and probable fixation artifacts. *Can. J. Zool.* 49:789–99

143. Zacharuk, R. Y. 1980. Ultrastructure and function of insect chemosensilla. *Annu. Rev. Entomol.* 25:27–47

144. Zacharuk, R. Y. 1980. Innervation of sheath cells of an insect sensillum by a bipolar type II neuron. *Can. J. Zool.* 58:1264–76

145. Zacharuk, R. Y. 1985. Antennae and sensilla. See Ref. 93a, pp. 1–69

146. Zacharuk, R. Y., Albert, P. J. 1978. Ultrastructure and function of scolopophorous sensilla in the mandible of an elaterid larva (Coleoptera). *Can. J. Zool.* 56:246–59

147. Zacharuk, R. Y., Albert, P. J., Bellamy, F. W. 1977. Ultrastructure and function of digitiform sensilla on the labial palp of a larval elaterid (Coleoptera). *Can. J. Zool.* 55:569–78

148. Zacharuk, R. Y., Blue, S. G. 1971. Ultrastructure of the peg and hair sensilla on the antenna of larval *Aedes aegypti* (L.). *J. Morphol.* 135:433–55

149. Zacharuk, R. Y., Blue, S. G. 1971. Ultrastructure of a chordotonal and a sinusoidal peg organ in the antenna of larval *Aedes aegypti* (L.). *Can. J. Zool.* 49:1223–29

150. Zacharuk, R. Y., Yin, L. R.-S., Blue, S. G. 1971. Fine structure of the antenna and its sensory cone in larvae of *Aedes aegypti* (L.). *J. Morphol.* 135:273–97

Annu. Rev. Entomol. 1991. 36:355–81

TRANSMISSION OF RETROVIRUSES BY ARTHROPODS

L. D. Foil and C. J. Issel

Department of Entomology and Departments of Veterinary Science and Veterinary Microbiology and Parasitology, Louisiana State University Agricultural Center, Baton Rouge, Louisiana 70803

KEY WORDS: tabanid, mechanical transmission, insect, lentivirus, equine infectious anemia

INTRODUCTION

Currently, no evidence supports the possibility of propagation of vertebrate-associated retroviruses in arthropods that would lead to amplification or biological transmission. This review presents a primer on retrovirus replication as a perspective on the possible interactions of retroviruses and insect cells. Because retroviruses mutate at relatively high rates, mutants could replicate in insect cells. Therefore, longitudinal studies must be conducted to determine the mechanisms by which arthropods interact with and prevent the replication of retroviruses. Knowledge of the replication strategies of the different retroviruses is also important in judging whether mechanical transmission (MT) of the agent by arthropods can occur. Mechanical transmission of ungulate retroviruses frequently occurs.

The first diseases now known to be induced by retroviruses were described in pioneering studies on equine infectious anemia (96) and Rous sarcoma (81). Research on animal retroviruses blossomed when the knowledge of the deleterious effects of retroviruses in outbred populations (cats, cattle) coincided with the war on cancer in the United States in the 1970s. Although the immediate search for oncogenic human retroviruses was not productive, the infusion of research support into retroviruses yielded invaluable information. Highlights are: (*a*) retroviruses contain RNA genomes but have DNA-

355

0066-4170/91/0101-0355$02.00

dependent steps in their replication; (*b*) purified retrovirus preparations contain reverse transcriptase activity; (*c*) retrovirus nucleic acid (provirus) can integrate with host cell DNA; (*d*) retroviruses that can transform host cells at high rates contain gene sequences similar to those found in normal host cells (viral oncogenes and cellular proto-oncogenes); (*e*) retroviral gene sequences can expand, be modified, and move within the host cell chromosome; (*f*) human retroviruses can cause leukemia and immune deficiencies; and (*g*) retroviruses and their genetic material can be used to carry and insert foreign genetic material into host-cell chromosomes.

Although the reverse transcriptase enzyme was first discovered in retroviruses, it is known to play an important role in a variety of systems. Today, the retroviruses are often referred to in the larger context of *retroid* viruses, a group of viruses that utilize reverse transcriptase during their replication—the hepadnaviruses, cauliflower mosaic virus, and the retroviruses. Retroviruses are also considered one of a class of *retrotransposons,* a group of transposable genetic elements that utilize reverse transcriptase (6, 91). These elements have been found in plant and animal cell systems.

RETROVIRUS REPLICATION

The following scheme describing retrovirus replication strategy is designed to be illustrative relative to the subject of this text. Obviously, other sources are more comprehensive, and the reader is referred to these for detailed references (14, 16, 45, 67, 92, 101, 102, 109). The pathology associated with the retroviruses covered in this text is discussed only as it relates to transmission. When retroviruses encounter host cells that contain the necessary receptor molecules for viral attachment, their fate is not sealed. They may attach to receptor molecules of more than one cell and cause fusion of the cells without multiplying. They more frequently attach to receptors on one cell and then penetrate it through receptor-mediated endocytosis or through fusion of the viral envelope with the host-cell membrane. The viral nucleic acid is then delivered in the cytoplasm of the host cell and the multifunctional viral reverse transcriptase: (*a*) catalyzes the formation of a full-length copy of DNA from viral RNA genome, (reverse transcriptase funtion); (*b*) catalyzes the destruction of the input viral RNA strand (ribonuclease H function); (*c*) aids in the formation of a double stranded (ds) viral DNA; and (*d*) aids in the integration of the dsDNA provirus into the host cell genome (integrase function), although integration may not be a required step in replication. This complex scheme effectively delivers the virion RNA, which is very unstable in the environment, into a form, namely dsDNA, that can persist indefinitely within the genome of the host cell.

In some cases, the proviral DNA may remain unexpressed for long periods. One class of retroviruses known as the endogenous retroviruses is carried in the normal genetic material of many vertebrate species. These endogenous sequences are transmitted to daughter cells when the proviral DNA replicates with that of the host cell. The endogenous retrovirus genetic material is vertically transmitted in the germline of individuals of the species. Several types of endogenous retroviruses are known from several species, e.g. chickens, cats, and mice. Although these endogenous sequences are generally transcriptionally silent, occasionally they are expressed and virus is released; other individuals who had not carried these specific endogenous sequences can be infected and pass these new endogenous sequences on to their progeny. Thus, vertical and horizontal transmission of the endogenous retroviruses occurs. Humans have multiple gene sequences analogous to known retroviruses. The origins and functions of these endogenous sequences are not known. These human endogenous retroviral sequences are all defective in one or more coding regions and do not appear to result in mature retroviruses if and when expressed. In contrast, the cells of all cats contain sequences that code for feline endogenous viruses. When the endogenous viral genes are transcribed, they cause the synthesis of proteins that form structures that enclose the full-length endogenous viral RNA to make fully infectious retrovirions, the host range of which may be very wide. As a sidelight, the first putative oncogenic human retrovirus proved to be a contaminant endogenous feline retrovirus in cultured cells from a human rhabdomyosarcoma.

Once the proviral DNA is within the nucleus of the host cell, its transcription is usually under the control of that cell. The site of integration in the host-cell chromosome may play a role in the rate of expression, but random integration is the rule. The viral LTR (long terminal repeat) plays a most important role in determining the rate of transcription. This sequence contains enhancerlike elements that can increase the rate of transcription of viral and/or contiguous cellular sequences. In some retrovirus systems, proviral gene sequences code only for virion structural proteins and viral enzymes that cleave the structural proteins as they are synthesized, folded, and assembled into virions. In more complex systems, e.g. the human immunodeficiency virus, proviral genes are found in a variety of contiguous and noncontiguous open reading frames and code for nonstructural proteins that can have specialized regulatory functions: upregulate or downregulate virus expression, transactivate other genes, or change the normal host-cell genome expression. These proteins may function in the nucleus, cytoplasm, or cell membrane.

In relatively straightforward systems, the viral mRNA species are transcribed, spliced, capped, and translated. They code for internal viral proteins (*gag* gene), reverse transcriptase (*pol* gene), proteases (*pro* gene), and envelope proteins (*env* gene products). Internal or core proteins accumulate in intracellular locations where viral envelope proteins have been inserted in cell membranes. In many cases, viral particles are only observed in infected cells by transmission electron microscopy as they are budding from infected cell membranes. Assembly and maturation occurs simultaneously in these systems.

In most cases, retrovirus replication proceeds in the host cell without causing its destruction. Cell transformation may occur rapidly following infection of cells with retroviruses whose genetic material contains a complete viral oncogene, but is less frequent if the oncogene is missing or defective. Although most retrovirus–host cell interactions are relatively innocuous and result in persistent cellular infections, several retroviruses destroy their host cells.

Both cell-associated and cell-free forms of retroviruses exist within infected hosts. The environmental persistence of retroviruses within infected cells is much greater than those free in the extracellular space, even if cell viability is lost, because of the protective effect of coagulated proteins on viral infectivity. In some retrovirus infections, e.g. bovine leukemia virus (BLV), cell-free virus is not demonstrable in vivo although cell-free virus can be produced in some cell-culture systems or after blast transformation of infected lymphocytes in vitro. The infection is transmitted through the transfer of infected cells that contain BLV provirus without necessarily containing infectious virions. In fact, how the infected cow develops immune responses to the structural and regulatory proteins of BLV is not known because complete replication of BLV has not been documented in vivo.

In other retrovirus systems, cell-free virus can be demonstrated in high levels in the plasma or in secretions of infected hosts, e.g. during acute infections with equine infectious anemia virus (EIAV) and feline leukemia virus (FeLV), before immune responses can be detected. In the case of FeLV, virus in secretions and excretions is efficiently transmitted by direct contact with cats (74). Risk of infection increases as a result of cats biting and scratching in multiple-cat households. In a percentage of FeLV infections, the suppression of the immune system by FeLV proteins permits virus replication to proceed in a relatively unchecked fashion. These cats continue to shed FeLV in secretions until their death.

The lentivirus caprine arthritis-encephalitis virus (CAEV) is efficiently transmitted through the transfer of infected cells in colostrum and milk. The majority of milk goats in the United States are infected with this lentivirus (67).

Contact transmission of EIAV is thought to occur rarely, even though EIAV can be found in most secretions and excretions of horses with acute signs of disease (51). During the acute bouts, virus is found in relatively high concentrations in the plasma (up to 10^6 ID_{50}/ml) but may be present only within infected cells once immune responses that control the spread of the input virus have been produced. During persistent infections with EIAV, antigenic variants arise through mutation of the *env* gene (65). These variants can be released from infected cells in the presence of immune factors that limit the spread of the original virus. The continued clinical bouts observed in the chronic form of EIA are associated with the continued production and release of novel antigenic variants of the virus. Transfer of blood from the EIAV-infected horse to the recipient horse by man or by hematophagous insects effectively maintains the infection in nature.

In summary, retroviruses have evolved several strategies to survive with their hosts. Generally, viral replication does not kill the host cell and virus released from cells can spread in a rapid fashion. In some cases, the virus does not multiply in the host cell and the infection spreads only through the proviral DNA that replicates when cellular DNA replicates during cell division. In others, the proviral DNA causes the synthesis of oncogene proteins that induce the host cell to divide at a more rapid rate than normal, and the result is cancer, sometimes even without virus production.

Insects can interfere with retrovirus replication at several steps in the replication cycle. Cell-free virus would be most susceptible to the action of salivary gland secretions, gut secretions, factors in the hemocoel or hemolymph, or shearing or mechanical forces created by structures such as cibarial teeth (18). These factors could strip critical surface determinants and render the virion unable to attach to or penetrate host cells or may cause the destruction of cells containing provirus. The restriction of retrovirus multiplication in insect cells may be greatly influenced by the inability of viral-surface determinants to attach to specific cell receptors. Other inhibitors could stop virus replication at one of the several functions of viral reverse transcriptase, could interfere with transcription, translation, or any other stage of assembly or maturation.

MECHANICAL TRANSMISSION

The magnitude of mechanical transmission of agents by hematophagous arthropods in nature is difficult to assess, with the possible exception of myxomatosis (21). In this review, we discuss transmission trials with several retroviruses. Bear in mind that successful transmission in these trials only represents a possibility of the event occurring; the probability of the event occurring in nature is determined by many factors. Tabanids (horse flies, deer flies, clegs) are large telmophagous insects that inflict painful bites and are strong fliers. They have been associated with the transmission of over 35 pathogenic agents (28, 57). Epizootics of livestock diseases have been frequently associated with explosive populations of tabanids. Thus, tabanids are considered as prototypic vectors of mechanically transmitted diseases. Studies conducted on the epizootiology of agents in which MT by tabanids is considered important have been summarized recently (28, 31).

The variables that contribute to the probability of MT are categorized into factors that contribute the most toward: (a) an interrupted feeding, (b) a mixed feeding, (c) the quantity of the agent transferred between hosts (inoculum), and (d) the ultimate successful infection in a new host. These categories are discussed in the following four paragraphs and are also presented in Figure 1.

Factors that contribute to the occurrence of an interrupted feeding have been reviewed recently (20) and included the pain of the bite and the feeding behavior of the arthropod. Foil (27) measured the relative feeding persistence of tabanids using mark-recapture techniques and found that the smaller tabanids (e.g. *Tabanus lineola*) fed on more than one host (four horses each 9 m apart) only 2% of the time, while larger species fed on more than one host from 7.1% to 12.3% of the time. Host-defensive behavior also contributes to interrupted feeding; Warnes & Finlay (103) found that livestock that responded with tail, foot, and head maneuvers to the feeding of stable flies had

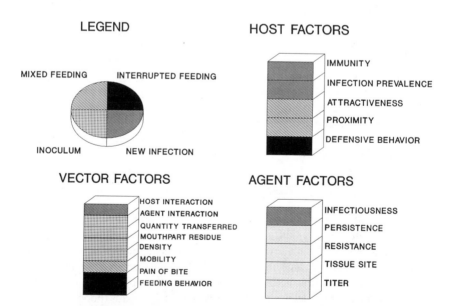

Figure 1   The variables that contribute to the probability of mechanical transmission of an agent to a new host by arthropods can be subcategorized (*legend*) into those that: contribute to the event of an interrupted feeding, contribute to the event of a mixed feeding (two hosts), determine the quantity of the inoculum, and contribute to establishing an infection in a new host. For example, factors that contribute most to the interrupted feed (*black*) are the defensive behavior of the host and the feeding behavior and pain of the bite of the vector.

fewer flies than more placid contemporaries. Extrapolations of these observations to disease situations suggest that, although an acutely ill host may be the best potential source of an infectious agent, it may not expend energy for defensive movements. As a result, interrupted feeding and MT would not occur.

Factors that contribute to the percentage of mixed feedings at a particular site include vector mobility, host proximity, and host attractiveness. There is a positive linear relationship between the percentage of tabanids that complete blood-feeding on the initial host and the distance between the initial host and an available second host (27). Up to 19% of the human blood meals of anopheline mosquitoes have been estimated to represent interrupted feedings (mixed meals). Differential host attractiveness and differential defensive behavior are considered to contribute to this rate (9). Foil et al (35) reported that foals had as low as 1.4% of the tabanid burden of mares. This correlated with a low incidence of EIAV infection in foals of test-positive mares (31, 49).

Factors that contribute to the quantity of the infectious agent deposited in a second host include the titer of the agent in the initial host, the time between feedings, and the stability of the agent (21). The virus titer in the blood of donors in MT trials for retroviruses can be used to predict the types and numbers of insects required to transfer an infection (Figure 2). These trials are discussed later in the text. Friend leukemia virus (FLV) reaching titers of 10^6 to 10^9 infectious doses (ID)/ml of blood can be transmitted by a single mosquito (25); EIAV at 10^6 ID/ml can be transmitted by a single horse fly (43); and BLV with approximately 10^4 ID/ml of blood can be transmitted by at least one horse fly in ten. The titer of different retroviruses and their distribution in host blood varies in nature. The BLV causes a persistent lymphocytosis (PL) in some animals, and this increase in circulating cells that contain BLV provirus creates a stable infectious reservoir for transmission. Animals that are infected with BLV but do not develop PL are less important in MT by insects (32, 34, 97, 99). The EIAV causes destruction of infected macrophages and up to 10^6 ID/ml of blood may be found in the plasma in the first febrile episode; during this transient high viremia, MT is favored (33, 43). If equids survive the first febrile episode, a persistent infection of cells is established and the quantity of these infected cells in the circulation will determine whether the horse is an important source for MT during these periods (50, 56). If a constant rate of interrupted feeding were present, then an infectious source with 10^6 ID/ml for one day would be the equivalent of a stable source with 10^4 ID/ml for 100 days in the same environment.

The factors associated with vectors that contribute to the amount of infectious agent transferred between hosts include vector density and vector interaction with the agent, e.g. degradation by enzymes or other lytic compounds in saliva or digestive fluids (77, 86). The quantity of blood on the

INFECTIVITY OF DONOR BLOOD

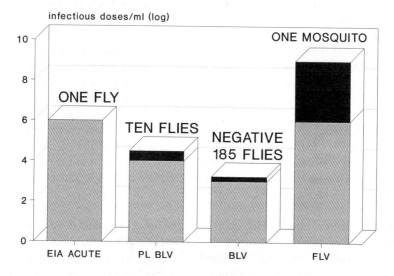

Figure 2 The viremia of the donor most often determines the numbers and types of vectors required to transfer infection in retrovirus mechanical transmission trials. These trials are discussed in the text. Friend Leukemia Virus (FLV) has been transmitted by a single mosquito in trials where the donor titers ranged from 10^6 to 10^9 infectious doses (ID)/ml (26). Equine infectious anemia virus (EIAV) reaches 10^6 ID/ml in the febrile donor and a single fly (*Tabanus fuscicostatus*) has been shown to transfer the infection (43). Bovine leukemia virus (BLV) can exceed 10^4 ID/ml and transmission by tabanids becomes possible (32). Transmission of BLV by tabanids using donors in the range of 10^3 ID/ml has not been demonstrated in limited trials (34).

mouthparts of insects following an interrupted feeding and the amount that is deposited in a second host are obviously important, and both of these factors may largely depend upon whether the insect is a vessel feeder (solenophagous) or a pool feeder (telmophagous) (30, 68).

Factors that influence the ultimate successful infection in a new host include the site of deposition, the invasiveness of the agent, and the immune status (innate or active) of the second host. If interrupted feeding occurred at a standard rate, then increasing the number of infectious sources (prevalence) would also increase the probability of an agent reaching a new host. Also, studies have indicated that host responses to salivary gland components of insects can enhance the infectivity of certain agents (94).

EQUINE INFECTIOUS ANEMIA VIRUS

The equine infectious anemia virus (EIAV) is found worldwide in the family Equidae. Because of the early description and the importance of the disease,

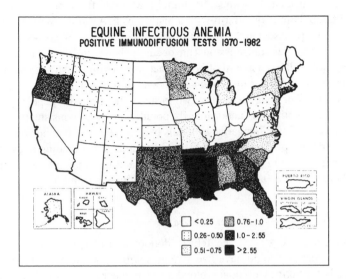

Figure 3   Percentage of positive EIA immunodiffusion test, 1970 to 1982. The results are biased toward the negative as a result of repeated testing of horses where negative tests were required at congregation points at about 12 month intervals and biased toward positive as a result of confirmatory retests. Reprinted from the Journal of the VMA (73) with permission of the editor.

the virus has been researched for approximately a century. We confine our observations on epidemiology of EIA to the US for the sake of brevity, although equivalent activity exists throughout the world (53). In the acutely infected febrile donor, EIAV can reach approximately 10^6 horse infectious doses (HID)/ml of blood. Early transmission studies quickly identified the parenteral transfer of blood (e.g. transmission via hypodermic needle stick) as a most effective route of transmission (84). Transplacental transfer can occur and is usually associated with leaky placentation (47, 49). The environmental (the common name for EIA is *swamp fever*) and temporal aspects of EIA transmission are indicative of insect transmission. In the US the percentage of reported positive antibody tests is highest in the Gulf Coast states (Figure 3). This distribution is presumably associated with high vector populations over long periods of time (48, 73). The environmental stresses of the hot humid summers could also cause recrudescence of disease, coinciding with high vector populations.

By the early 1900s, biting insects were considered important in the transmission of EIAV (64). In the absence of iatrogenic transmission, epizootics have occurred consistently during periods of insect activity (38, 90, 95). Observations on tabanid activity frequently accompany reports on epizootics, but other Diptera (e.g. stable flies) are obviously important alternate vectors.

In south Louisiana, Hawkins (40) found low stable fly populations and high tabanid populations on a farm with a history of EIAV transmission. Epizootics usually are associated temporally with the presence of horses with acute signs of EIA, but transmission can occur in herds in which no disease is evident. We followed a herd of horses that were seropositive but inapparently infected with EIAV pastured with seronegative sentinel horses over an eight-year period (four years in each of two farms). The mosquito populations were often high at both areas and stable fly populations were consistently low. On the first farm, 8 of 27 negative sentinels became seropositive; epizootics were associated with peaks of tabanid activity documented at landing rates of up to 1,000 per hour. On the second farm, which had less than 2% of the tabanid populations of the first farm, none of 36 sentinels became infected (29, 31, 49).

Biological Transmission Studies

Unless otherwise stated, the febrile donors in EIAV transmission studies discussed below should have had viremias with approximately 10^6 ID/ml of blood, but titrations of donor blood were not performed in all cases. The results of early transmission studies were based upon the clinical signs of recipients or the clinical signs of horses subinoculated from recipients. As summarized by Stein et al (88) negative reports regarded the potential for EIAV propagation in black flies, bot flies, mosquitoes, ticks, or the progeny of ticks and mosquitoes collected from EIAV-infected donors. One study did indicate evidence of virus in the eggs of biting lice collected on infected horses, and one reported transmission following injection of homogenates of blood-engorged mosquitoes collected from stables.

The introduction of a specific diagnostic test (the agar gel immunodiffusion test) for antibody to EIAV in 1970 allowed a second generation of studies on insect transmission of EIAV (17). Contemporary studies on the potential for biological transmission of EIAV have been limited because of the lack of any field evidence that biological transmission occurs, i.e. new cases do not occur once the test-positive horses are removed. In studies designed to detect vertical transmission of EIAV in insects, the progeny of approximately 650 Culex pipiens quinquefasciatus reared from adults that blood-fed on an acutely ill pony were fed in groups of 100–250 on three recipient ponies, and no transmission occurred (7). Kemen et al (56) inoculated a pony intravenously with a homogenate prepared from approximately 250 one-day-old Tabanus quinquevittatus larvae obtained from three female flies that had been fed on an acutely affected pony; no transmission occurred.

In studies designed to detect EIAV propagation in insects fed on infected donors, Kemen et al (56) fed tabanids (T. quinquevittatus; 25 bites per pony) 24 hours postoviposition (approximately 6 days post-feeding on febrile or afebrile donors) on 2 recipient ponies with negative results. Hawkins et al (41) fed *Tabanus fuscicostatus* on febrile ponies; 6–8 days later, 2 ponies received 25 and 19 bites and no transmission occurred. Hawkins (40) also fed stable flies on febrile donors and fed groups of 14–25 on 4 ponies 3–10 days later with negative results. Williams et al (107) found no transmission when feeding 42 *Aedes sollicitans* or 25 *Psorophora columbiae* held for 14 and 7 days, respectively, after feeding on febrile donors.

Breaud et al (8) reported that a tissue culture of *C. pipiens quinquefasciatus* ovarian cells supported the growth of EIAV. The serum from a donor with approximately 10^6 HID was diluted with the tissue culture media to a 10^{-1} dilution, and subsequent passages each week resulted in a tenfold dilution. Ponies inoculated with the second, seventh, ninth, and eleventh weekly passage were infected. Williams et al (107) repeated these studies with similar results; extensive microscopic studies failed to reveal viruslike particles or morphological changes in *Culex* ovarian cells. Williams et al (107) did detect EIAV antigen by radioimmunoassay in *P. columbiae* fed on acutely affected donors at 1, 7, and 14 days after feeding but not at 2, 3, and 5 days. In a study using a significant number of pony recipients (48 total), Shen et al (85) failed to show any propagation of EIAV in mosquitoes or *Culicoides variipennis*. Mosquitoes (*Culex tarsalis, Aedes aegypti, Culiseta inornata*, and *Anopheles freeborni*) were inoculated intrathoracically with the Wyoming strain of EIAV in bovine serum ($10^{6.5}$ TCID$_{50}$/ml). Homogenates of 10–11 mosquitoes from day 0 and from mosquitoes held from 3–18 days were inoculated into recipient ponies; only day-0 homogenates were infective. The gnats were fed a blood meal containing $10^{5.5}$ TCID$_{50}$ of EIAV; homogenates from day 0 were infective, but homogenates from flies held for 6–26 days were not. Therefore, although *Culex* ovarian cells in tissue culture may support the growth of EIAV, the progeny of *Culex* mosquitoes did not transmit EIAV while feeding (7). Furthermore, intrathoracic injections of mosquitoes failed to indicate propagation (85). If dipteran ovarian tissues could support EIAV propagation in vivo, all studies to date indicate that barriers prevent infective virus from reaching that site.

Mechanical Transmission Studies

Because reviews on mechanical transmission of EIAV by insects are available (28, 31, 52), the discussion is limited to the studies most instructive on vector potential. We use the term *interrupted feeding* to describe trials in which the insect(s) is allowed to initiate feeding on the donor, and after taking a partial bloodmeal, is transferred to the recipient to complete engorgement.

MOSQUITOES   Early transmission studies were interpreted by the symptomatology of recipients or through round-robin subinoculation studies; results of these studies are often difficult to interpret. For example, in the only study that reported transmission of EIAV by mosquitoes, Stein et al (89) fed 186 *P. columbiae* on a febrile donor and then on a susceptible horse. The recipient was negative 86 days later and a second horse was inoculated with 100 ml of the recipient's blood. The second recipient was negative after 123 days. A third recipient was inoculated with blood from the second recipient. The last horse developed clinical signs three days post exposure (the incubation period for acute signs of EIA with the high doses of the strain of EIAV used in that study is rarely less than nine days). Studies performed after a specific serologic test was developed (17) have been conducted with interrupted feedings of two groups of 51 and 103 *P. columbiae* and three groups of 189, 201, and 307 *Aedes taeniorhynchus* (19), and two groups of 42 and 100 *P. columbiae* and two groups of 94 and 104 *A. sollicitans* (107). All have generated negative results.

Williams et al (107) found that EIAV can survive on needles for up to 96 hours but on mosquito mouthparts for only about one hour. The authors considered that grooming by the live mosquitoes could be responsible for their unsuccessful attempts to find virus on the mouthparts, but they suggested that an inhibitor of EIAV was associated with the mouthparts. Inhibitors of the infectivity of other retroviruses in the heads or mouthparts of mosquitoes did not appear to reduce the ability of mosquitoes to transmit FLV or to retain BLV-infected cells (13, 26).

STABLE FLIES   Scott (83) placed three infected horses and three healthy horses in a large cage that contained a breeding population of stable flies, and two of the three healthy horses developed EIA. Stein et al (88) reported transmission of EIAV by interrupted feeding with 233 stable flies. Hawkins et al (42) failed to transmit EIAV with interrupted feeding of three groups of stable flies (200–203) fed on febrile donors. However, transmission was later demonstrated with groups of 224–400 stable flies (19) and groups of 52–100 stable flies (33). Considering the number of negative trials (five with over 100 stable flies) (19, 33, 42) using donors with up to 10^6 HID/ml, it is not surprising that similar trials on BLV with groups of 50–75 flies and with donors with less than 10^5 bovine infectious doses (BID)/ml have been negative (12, 105). Negative results of these transmission trials also indicate either that regurgitation, which has been reported to occur in stable flies feeding on in vitro systems (11), does not occur with high frequency under natural conditions or that viral inhibitors are present in the gut of stable flies.

TABANIDS   Many studies have demonstrated EIAV transmission using groups of tabanids in interrupted feeding trials (52). Hawkins et al (43) transmited EIAV with a single horse fly *(T. fuscicostatus)* in one of seven trials; Kemen et al (56) could not transmit EIAV with single *T. quinquevittatus* in three trials. Foil et al (33) reported transmission with six *Chrysops flavidus* (the lowest number used). Hawkins et al (43) also obtained transmission of EIAV with groups of 25 *T. fuscicostatus* held for 3–30 minutes between the initial interrupted meal and the second feeding but not with groups held for 4 and 24 hours. Foil et al (30) reported a bloodmeal residue of approximately 10 nl on the mouthparts of *T. fuscicostatus* dissected immediately after feeding on a pony. Mouthparts dissected immediately after feeding but not processed until 0.5–4.0 hours postfeeding had residues of approximately 7.3 nl, while mouthparts dissected from flies kept alive for 0.5–4.0 hours postfeeding had residues of approximately 4.2 nl. Therefore, grooming would not necessarily account for the absence of transmission when tabanids were held 4 hours (43).

The only reports of successful transmission trials using afebrile donors have used horse flies. Kemen et al (56) reported transmission via interrupted feeding with three *Hybomitra lasiophthalma* but not with groups of 75 *T. quinquevittatus* or 17 *Tabanus sulcifrons;* the donor was a pony that had been inoculated previously with Wyoming strain EIAV but had survived and was afebrile during the trial. Issel et al (50) used groups of 25 *T. fuscicostatus* in interrupted feeding on 10 afebrile test-positive donors with no known history of clinical EIA; no transmission occurred. However, transmission was reported from a horse that had recovered from a clinical attack of EIA nine months previously. Viremia in the afebrile EIAV-infected horse is known to fluctuate up to 1,000-fold without any associated clinical signs in the animal (50). Unfortunately, the titer of EIAV in the blood of the two afebrile donors used in the above tirals was not determined.

HUMAN IMMUNODEFICIENCY VIRUSES

The transfer of blood or other body fluids containing the human immunodeficiency viruses (HIV) between humans is the major route of HIV transmission. Although EIAV, a closely related lentivirus, is transferred between hosts primarily by the transfer of blood by insects, nothing has indicated that insect transmission could be a major factor in the spread of HIV (37). In the mid 1980s, several groups publicly insisted that insects were involved in AIDS transmission. A workshop organized in 1987 and coordinated by L. Miike of the Office of Technology Assessment for the US Congress was attended by 17 scientists (including these authors) to address this issue (62). Highlights of the reasoning that led to the meeting and data that were shared are relevant to this article.

One disease transmitted among humans by the transfer of blood and other body fluids is Hepatitis B, caused by Hepatitis B virus (HBV). Blumberg et al (5), after describing the relationship between Australia antigen (Au) and HBV, found Au in mosquitoes collected in Africa where Au frequencies were high. Bedbugs were also found to have a high incidence of Au (108). Further studies indicated that, although HBV did not replicate in the bedbug, crushing of bugs, contaminated feces, and interrupted feeding apparently could result in mechanical transmission of HBV (55, 69). None of the reports on the potential for HBV transmission by arthropods proposed this as a major pathway of transmission. Although the infectious titer of HIV is considered to be generally 10 to 100-fold less than HBV based on needle-stick injury, the two viruses have mechanisms of transmission in common (62).

Biggar et al (4) initially reported a correlation between the prevalence of antibody against human retroviruses and the level of antibodies against malaria *(Plasmodium falciparum)*. The primary author of this early survey later reported a high number of false positives for human retrovirus antibodies associated with "sticky sera" from patients in malarious areas (3). Subsequent surveys by this group do not support a role for arthropods in HIV transmission (61). Lockwood & Weber (58) speculated that increased frequency of *Leishmania* infections could be expected in AIDS victims as a result of opportunistic infections of the immunocompromised host. Therefore, correlation of antibody prevalence between retroviruses and known vector-borne diseases does not necessarily indicate insect transmission of retroviruses.

In 1985, oral reports of several AIDS cases from Belle Glade, Florida, related the infection to insect transmission. The high incidence of AIDS was investigated subsequently and seroepidemiologic studies convincingly demonstrated that the incidence of the disease was associated with sexual activity and intravenous drug use (15). Although no firm data implicate insects in AIDS transmission, the fear of rapid spread of HIV in human populations has warranted a careful examination of the potential for insect transmission.

Biological Transmission Studies

Studies in this area have been limited because of the absence of epidemiologic evidence for biological transmission of HIV. Most of the questions regarding HIV and insects have been associated with mechanical transmission. The studies that created initial interest in biological transmission were conducted by a French research group; HIV-like nucleic acid sequences were found in insects collected from two areas with endemic HIV in central Africa but not from insects from Paris (2). Cockroaches, tsetse, and antlions from Zaire and mosquitoes, ticks, and bedbugs from the Central African Republic were all positive in assays for HIV proviral DNA using fluorescent probes for HIV. Groups of insects were negative, but the percentage of positive insects was high, and some insect species that do not bloodfeed were positive. These results led to questions regarding the potential for flaws in the experimental techniques. Several laboratory studies have been conducted subsequently to address this issue.

Srinivasan et al (87) introduced molecularly cloned HIV proviral DNA into tick, moth, and fruitfly cell lines. The cells were also cultivated with phytohemagglutinin-stimulated peripheral blood lymphocytes or a human T4-

lymphocyte cell line. Reverse transcriptase activity was not detected in any of these systems. A chimeric plasmid including the HIV long-terminal repeat sequences was constructed and the cell lines were transfected; HIV replication was not indicated.

Laboratory studies with HIV and insects have had conflicting results. Jupp & Lyons (54) fed bedbugs (*Cimex lectularius* and *Cimex hemipterus*) and mosquitoes (*A. aegypti formosus*) on a blood-virus mixture (4.4 × 10^5 TCID/ml) through membranes. Reverse transcriptase activity was only recorded for up to 4 hours in pools of *C. lectularius*, for 1–2 hours in *C. hemipterus*, and not recorded from mosquito pools collected from 1 hour to 1 day postfeeding. Webb et al (104) injected the mosquito *Toxorhynchites amboinensis*, a sensitive host for a variety of arboviruses (80), intrathoracically with a solution containing concentrated HIV; reverse transcriptase was recorded from pools of triturated mosquitoes cultured with the MT-4 cell line at 1 and 2 hours but not 24 hours–28 days postinjection. Bedbugs *(C. hemipterus)* were injected intra-abdominally and assayed by the techniques used for the mosquitoes; reverse transcriptase activity was detected from insect homogenates immediately after injection but not 7–28 days postinjection. Bedbugs were also fed a meal consisting of five parts fresh mouse blood to one part Alsever's solution with HIV added. In one trial, reverse transcriptase activity was recorded in a pool of five insects 8 days postfeeding. The virus was not detected in bedbug feces collected throughout the experiment. Two HIV-specific gene probes were used to confirm the persistence of HIV sequences in blood meals for up to 3 days. The authors concluded that the pool of bedbugs on day 8 retained at least 1 TCID of virus that replicated when added to MT-4 cells, i.e. that HIV survived but did not replicate in the bug (104).

Considering the genetic similarities between EIAV and HIV and the relative paucity of information available on the interactions of insects and retroviruses, a combined summary follows. Neither EIAV nor HIV has been shown to survive when injected into the hemocoel of insects (85, 104). Infective EIAV appears to survive up to 10 passages in *Culex* ovarian cells (8); HIV did not replicate in tick, moth, or fruitfly cell lines (87). Antigen for EIAV was detected in mosquitoes 7–14 days after feeding on an acutely infected pony, but these mosquitoes did not transmit EIAV to recipients (107). Infective HIV can survive for up to 8 days in bedbugs but replication is not indicated (104). Therefore, laboratory studies support epidemiologic evidence that EIAV and HIV are not biologically transmitted by insects.

Mechanical Transmission Studies

No evidence indicates that insects are biological vectors of HIV or that insects are important in the epidemiology of this disease. Questions remain concerning the potential for mechanical transmission of HIV, which may account for a small percentage of cases in people who do not appear to have (or who do not report) high-risk behavior. Since the transfer of blood between hosts can transfer the infection, an obvious approach would compare the volume of blood required to transfer the infection to the volume of blood that different insects transfer between hosts during an interrupted feeding. These are, in fact, the variables that we have used successfully in the study of horse fly transmission of BLV and EIAV (30, 32, 34). However, other factors must be considered. For example, mosquitoes may be able to rapidly inactivate EIAV and HIV (54, 104, 107).

Jupp & Lyons (54) failed to demonstrate MT of HIV by initially feeding four groups of 100 bedbugs (*C. lectularius*) on a blood-virus mixture (4.4 × 10^5 TCID/ml), transferring them to an uninfected membrane feeder to complete the meal, and assaying the feeder fluid for virus. Webb et al (104) partially fed 50 bedbugs (*C. hemipterus*) on a blood meal containing 2.0 × 10^6 cpm RT activity/ml and let these bugs feed to repletion on a 1.5-ml second bloodmeal; no virus was detected in the recipient blood or membrane. These two studies used concentrated HIV. Under natural conditions, with much lower concentrations of HIV, the bedbug is probably not an effective mechanical vector of HIV. Although preliminary studies indicated that the bedbug mouthpart blood-meal residue was approximately 7×10^{-5} ml (62), Ogston (68) reported the amount of blood actually transferred by bedbugs during an interrupted in vitro feeding was $1.5–1.8 \times 10^{-10}$ ml.

Recent estimates declared that the amount of HIV in 1 μl blood from an HIV-infected but asymptomatic person would average 0.06 TCID compared to 7.00 TCID from a symptomatic patient (46). The quantity of blood transferred between hosts has been estimated to be less than a picoliter (10^{-12}) for individual vessel feeding insects, e.g. bedbugs or mosquitoes (62, 68). Successful tabanid transmission of BLV has been described from donors with infectious titers an order of magnitude lower than those described for HIV. However, the types and number of insect pests of livestock and humans and their interactions are very different, e.g. a landing rate of 1,000 tabanids per hour on horses reported by Foil et al (31); humans have developed multiple strategies to avoid extreme attack by hematophagous insects.

BOVINE RETROVIRUSES

The bovine immunodeficiency virus (BIV) has been recently described, and, although epidemiological studies are incomplete, iatrogenic transmission is considered the primary mechanism of spread (67). BIV and bovine leukemia virus (BLV) can be found in mixed and separate infections in cattle herds (D. G. Luther, personal communication).

Although true vertical transmission of BLV does not seem to occur, transmission via milk and in utero transmission have been described. The prevalence of these modes of transmission has been difficult to estimate because of the presence of colostrum-derived antibody (79, 93, 100). Ferrer (25) reviewed studies conducted on this subject and suggested that less than 20% of calves of BLV test-positive dams were infected before birth and that horizontal transmission among adult animals was more important than transmission via milk under natural conditions. Recent studies support these views (24).

Several modes of iatrogenic transmission have been described in association with management procedures that involve contact with body fluids, especially blood, e.g. tattooing, dehorning, rectal palpation, and vaccinating (22, 44, 59). Contact among adult animals is also important in horizontal transmission. Although epizootics have been described during a winter housing period (106), epidemiologic evidence for a role of insects in BLV transmission has been described (1). Recently, Manet et al (60) reported the results of a 2-year prospective serological study on 3,328 cattle in 3 areas of France. The authors found a significant correlation between the density of tabanids and the incidence of BLV. Furthermore, the highest rates of seroconversion were correlated with the seasonal activity of tabanids.

BLV is one of the animal models for the human retrovirus HTLV-I (14), even though BLV is more associated with B lymphocytes than is HTLV-I. HTLV-I has been associated with adult T-cell leukemia, tropical spastic paraperesis, and HTLV-I-associated myelopathy (76). Miller et al (63) conducted a seroepidemiologic survey in an urban community of Trinidad. The authors reported significant relationships between seropositivity for HTLV-I antibody and poor-quality housing and proximal water courses; they concluded that insects may be important in HTLV-I transmission under certain circumstances. Murphy et al (66) responded to the above study by reporting results from a survey conducted in Jamaica. Serum samples from 200 (90 HTLV-I positive and 110 HTLV-I negative) volunteers were screened for 5 arboviruses reported to occur in Jamaica. No correlation was found between prevalence of HTLV-I and arboviruses. In Jamaica, the major routes of transmission for HTLV-I appear to be similar to those of HIV, although breast-feeding appears to play a role in HTLV-I but not HIV infection (76).

Transmission Studies

In the only published work testing for biological transmission of BLV, Okada (70) found no viruslike particles in an electron micrographic study of salivary ducts, mesentery, malpighian tubules, and ovaries of *Tabanus nipponicus* at 24, 48, and 120 hours after engorgement on BLV test–positive cattle. Van Der Maaten (97) reviewed routes of transmission for BLV and indicated that the most sensitive portal of entry was an intradermal or intracutaneous inoculation (compared to intraperitoneal, intranasal, intratracheal, aerosol, oral, and intrauterine). The most infectious bovine tissue, secretion, or excretion was blood. With persistent lymphocytosis (PL), which is the result of proliferation of infected lymphocyte clones, the volume of blood required to transfer infection can be as low as 50–100 nl (23, 32, 82).

Bech-Nielsen et al (1) recovered viable BLV-infected lymphocytes from the midgut of horse flies (*Tabanus nigrovittatus*) that had been fed on a BLV-infected cow. Romero et al (78) recovered viable BLV-infected cells from the guts of female *Boophilus microplus* ticks immediately after removal from BLV-infected heifers and infected two of four sheep with inoculations of these cells (4×10^6 leucocytes). Okada (70) found up to 2,000 lymphocytes on the mouthparts of *T. nipponicus* that had fed on leukemic cattle; 2,500 lymphocytes is considered to be an appropriate intradermal infectious dose for cattle (98).

Oshima et al (72) were the first to demonstrate that tabanids could mechanically transmit BLV; the virus was transferred by interrupted feeding of 131–140 tabanids (more than 90% *T. nipponicus*) from an infected cow with PL to 2 sheep. In a second study, horizontal transmission of BLV by tabanids was related to the development of lymphosarcoma in 3 of 12 recipient sheep (71). A total of 9 of 12 sheep seroconverted for BLV antibody after exposure to groups of 10–140 tabanids (*T. nipponicus* or *T. trigeminus*) that had fed on cows with PL.

Foil et al (34) demonstrated BLV transmission to goats and sheep by groups of 50–100 *T. fuscicostatus* fed on a BLV-seropositive cow with PL. However, transmission did not occur when 185 flies (*T. fuscicostatus*) were first fed on an aleukemic BLV-seropositive cow with 9,500 white blood cells/mm^3 and then on 4 goats. In subsequent studies, BLV was transmitted by horse flies (*T. fuscicostatus*) initially fed on a cow with PL; as few as 10 and 20 flies transmitted BLV to goats and dairy calves, respectively (32). Mouthpart-residue estimates (10 nl) for *T. fuscicostatus* predicted that groups of 10–20 flies would transfer from 50–200 nl of blood to recipients (30); 100 nl of donor blood was shown to be infectious for calves and goats (32). BLV was also transmitted by groups of 150 and 100 *T. fuscicostatus* first fed on a cow with a lymphocyte count of 14,600/mm^3 and then fed on two beef calves; groups of 250, 200, 75, 50 and 25 flies failed to transfer BLV from the donor to beef calves. These studies confirmed that tabanids could be involved in horizontal transmission of BLV among cattle and supported the concept of eliminating cattle with PL to reduce the horizontal transmission of BLV (32). Procedures for reducing the potential for mechanical transmission of pathogenic agents by tabanids are described elsewhere (28, 31), but traditional fly control with insecticides would probably have little impact on the epizootiology of BLV (99).

Weber et al (105) conducted studies on stable flies as vectors of BLV.

Calves were housed with BLV-infected cows or calves inoculated with BLV in the presence of introduced stable flies. Contact among animals was prevented and trials lasted up to 3 months. Calves received between an estimated 24,500 and 200,000 bites during these trials with no BLV transmission recorded. The authors did not estimate the number of feeding flies that would have been interrupted, but they did observe that the flies that were disturbed while feeding would normally cease feeding or return to the original host. A similar experiment with fewer flies reported the same results (12). Weber et al (105) fed groups of 50 stable flies on each of three calves following an interrupted feeding on a BLV-seropositive cow. One of the three calves became seropositive for BLV antibody at 81 days following exposure, but this was 24 days following seroconversion of a stallmate. The authors concluded that the transmission did not result from the stable-fly feeding and also concluded that the stable fly should not be considered a primary vector of BLV (105). Buxton et al (12) failed to transfer BLV infection to 4 calves with groups of 75 stable flies initially fed on a cow with 8,300 lymphocytes/mm^3 of blood. Buxton et al (12, 13) also reported results of studies in which they inoculated animals with extracts prepared from mosquito, horn fly, stable fly, and tabanid mouthparts and/or heads after the flies fed on infective blood or culture medium, but the significance of studies using homogenates of dissected insects is difficult to interpret.

The relative importance of insects in the epizootiology of BLV most probably is related to geographic factors. Burny et al (10) suggested that arthropods do not substantially contribute to BLV transmission in temperate climates. However, experimental transmission and epidemiological studies indicate that tabanids are involved in the natural horizontal transmission of BLV (32, 60, 71).

FELINE RETROVIRUSES

Feline leukemia virus (FeLV) is the most common of five feline retroviruses and is transmitted horizontally among cats (75). Hardy et al (39) recovered FeLV-infected leukocytes from a mosquito that had fed on an infected cat that had lymphosarcoma. Francis et al (36) studied the infectious titer of FeLV in the saliva and plasma of nine nonleukemic FeLV-infected cats and found ranges of 5.4×10^3 to 2.0×10^6 focus forming units (ffu)/ml in saliva and 9.4×10^2 to 8.4×10^5 ffu/ml in plasma. Urine, feces, and 10 pools of 10 unidentified fleas vacuumed off the cats were all negative for virus. The male free-roaming cat is the highest-risk category for the feline lentivirus, feline immunodeficiency virus (FIV) (67, 75). Therefore, it appears that FIV and FeLV are both transmitted primarily by cat bites.

FRIEND (MURINE) LEUKEMIA VIRUS

Fisher et al (26) conducted studies on the arthropod transmission of Friend leukemia virus (FLV). The strain of FLV that the authors used achieved $10^{4.8}$–$10^{8.9}$ ID_{50}/ml in the blood of donor mice. Transmission studies were conducted with stable flies and four mosquito species. Insects were fed on donors and then transferred in groups of 1–20 to recipient suckling mice. The minimum number of insects necessary to transfer infection and the maximum titer reported for donor groups were: (a) 1 Aedes triseriatus at $10^{6.2}$, (b) 2 Anopheles quadrimaculatus at $10^{5.1}$, (c) 1 C. pipiens quinquefasciatus at $10^{8.9}$, (d) 20 C. pipiens pipiens (no transmission) at $10^{8.9}$, (e) 1 Stomoxys calcitcans at $10^{7.5}$. The number of positive transmissions by groups ranged from 0% to 5.26% (mosquitoes) and 11.84% (stable flies). Crushing engorged mosquitoes on scarified skin transferred infection in over 90% of the trials. Assuming the massive number of recipients (2,335) did not lead the researchers to confuse the individuals, the transmission of FLV by 1 A. triseriatus at a maximum $10^{6.2}$ ID_{50}/ml donor titer does not fit the current attitude toward the vector potential (mechanical) for mosquitoes (62).

SUMMARY AND FUTURE STUDIES

Traditional study of insects as biological vectors for viruses involves isolating the virus from unengorged, wild-caught specimens in areas where the disease is occurring, demonstrating that the insect in question is feeding on the host species in question, and verifying in the laboratory that the insect can be infected and can transmit the agent after a period of extrinsic incubation. No data document retrovirus replication in insect cell lines or within intact insects. Although reports have documented the long-term survival of infectivity in several retrovirus-insect or retrovirus–insect cell systems, field programs designed to control or eradicate EIA or BLV that exclude vector control have been successful. Together, these data suggest that if biological arthropod vectors for retroviruses exist, their current recognized role is minimal. Evidence also indicates that EIAV and HIV are both quickly inactivated following intrathoracic injections of insects and that retrovirus inhibitors may be associated with the mouthparts of mosquitoes (85, 104, 107). The possibility of novel antiviral compounds occurring in insects should be explored. Because compartmentalized factors could affect viral infectivity in addition to enzyme inhibitors, the use of whole-body homogenates to describe retrovirus persistence in insects in future studies should be discouraged.

Retrovirus–arthropod cell interactions could be dramatically altered as a result of viral mutations that permitted efficient viral attachment to arthropod cell receptors. Because certain surface determinants of lentiviruses (EIAV,

HIV) mutate at high rates, the likelihood of such changes is high. The next logical question involves the ability of the virus to replicate in the host cells after attachment occurs. Preliminary studies using transfection with HIV plasmids in a limited number of insect cell lines suggest that replication is restricted at one or more levels of expression. Whether mutations that would permit more efficient attachment could also lift the restrictions could be addressed. Viral pseudotypes that contained retroviral nucleic acid inside the envelope of a known arthropod-borne virus that multiples in insects or insect cells could be created. If the viral nucleic acid were delivered into the cytosol of the insect cell by using pseudotypes or through effective transfection with provirus, one could study the ability of the viral nucleic acid to survive and replicate in the insect cell.

More exhaustive systematic surveys and studies of retrovirus-arthropod cell interactions need to be conducted. The finding of gene sequences analogous to known pathogenic retroviruses in a variety of hosts, including invertebrates, is not unexpected. If gene sequences found in invertebrates were determined to be identical to those of HIV and were found to possess the entire HIV genome, one would still be faced with the question of whether the proviral gene was silent or occasionally expressed. The powerful techniques of molecular biology, which can amplify a segment of DNA a millionfold in hours, must be carefully applied to retrovirus-insect considerations. The immediate response of the lay community to the finding of HIV-like gene sequences in insects in Africa (2) was that insects were widely infected with the causative agent of AIDS. Although a satisfactory explanation for these results has not yet been promulgated in the scientific community, we must remember that the sensitivity of the tools currently available to us may greatly exceed our ability to interpret their specificity.

The importance of mechanical transmission of agents by arthropods has to be considered at a more focal level than biological transmission. Even in the epizootiology of EIAV, we tend to rule out all other transmission routes before defaulting to arthropod transmission. In geographic areas where explosive tabanid populations are consistently associated with EIAV epizootics, new infections established by alternate routes still account for a certain percentage of background transmission. In geographic areas where potential vector density is low, MT by arthropods can be a certain percentage of background for EIAV transmission. For BLV (and probably BIV), horizontal transmission is more important than vertical transmission, but horizontal transmission by insects is less effective than for EIAV. However, seroepizootilogic and entomologic studies are sensitive enough to establish a role for MT by arthropods in geographic areas with appropriate vector populations (60). In a global rather than focal view of bovine retrovirus epizootiology, MT by arthropods is probably mostly background. No epizootiologic tool is available to determine the level of background MT for retrovirus diseases by arthropods in which vertical or contact transmission is effective, e.g. maedi-visna and murine leukemia (67). Furthermore, no epidemiological tool is available to detect the arthropod MT of infectious agents of humans that can be transmitted horizontally by the transfer of blood but for which much more effective routes of transmission occur. Therefore, advances in the future understanding of the MT of retroviruses in nature will most likely come from models of MT.

This review summarized most of the variables that contribute to the probability of mechanical transmission of agents by arthropods (Figure 1). The variables were all equally weighted for graphic purposes, but analytical models could be established that would consider an ultimate probability of the event of transmission (20). The importance of such models for the study of livestock diseases is obvious. However, other benefits are certain. Mechanical transmission trials are often difficult to relate to what actually occurs in nature even when live insects are given an interrupted feeding on an appropriate donor and recipient. When insects are dissected, homogenized, or fed on artificially spiked donors (animals or membranes), the application of the results to real situations is reduced. For example, the two studies reporting a lack of MT of HIV by arthropods used concentrated virus solutions (54, 104). Although negative results are reassuring, positive results would not have indicated a role for arthropods in HIV transmission and the authors would have probably conducted studies with more dilute solutions. However, more concentrated solutions or different types of insects could have been used to transfer HIV from one membrane to another. Currently, the interpretation of such results would be left up to intuitive logic, which is not uniformly distributed. Models to determine the probability rather than the possibility of MT could be used to interpret the relative importance of MT trials and could also be used to assess the potential for background transmission of blood-borne agents by arthropods.

ACKNOWLEDGMENTS

Approved for publication by the Director of the Louisiana State Agricultural Experiment Station as manuscript no. 90-17-4132. We thank the individuals who responded to requests for copies of recent publications and the many cooperators at LSU, in particular W. V. Adams, Jr., who have assisted in the cited studies. We also thank Kim Kubricht, Eric Chris, Wendy Lavendar, Daniel Leprince, Larry Hribar, Beth Elkins, Lacey Wieser-Schimpf and Karen Chalona for their technical help. Recent studies on mechanical transmission of diseases by insects have been partially supported by the Grayson-Jockey Club Research Foundation, Lexington, KY.

Literature Cited

1. Bech-Nielsen, S., Piper, C. E., Ferrer, J. F. 1978. Natural mode of transmission of the bovine leukemia virus: role of bloodsucking insects. *Am. J. Vet. Res.* 39:1089–92
2. Becker, J. L., Hazan, U., Nugeyre, M. T., Rey, F., Spire, B., et al. 1986. Infection de cellules d'insectes en culture par le virus HIV, agent du SIDA. et mise en evidence d'insectes d'orgine africaine contamines par ce virus. *C. R. Acad. Sci.* 303:303–6
3. Biggar, R. J. 1986. The AIDS problem in Africa. *Lancet* 1:79–82
4. Biggar, R. J., Gigase, P. L., Melbye, M., Kestens, L., Sarin, P. S., et al. 1985. ELISA HTLV retrovirus antibody reactivity associated with malaria and immune complexes in healthy Africans. *Lancet* 2:520–23
5. Blumberg, B. S., Wills, W., Millman, I., London, W. T. 1973. Australia antigen in mosquitoes. Feeding experiments and field studies. *Res. Comm. Chem. Pathol. Pharmacol.* 6:719–32
6. Boeke, J. D. 1988. Retrotransposers. In *RNA Genetics*, ed. E. Domingo, J. J. Holland, P. Ahlquist, 2:59–103. Boca Raton, Florida: CRC
7. Breaud, T. P. 1973. *An examination of the virus-vector relationship between the southern house mosquito,* Culex pipiens quinquefasciatus *Say, and equine infectious anemia.* MS thesis. Baton Rouge: La. State Univ. 65 pp.
8. Breaud, T. P., Steelman, C. D., Roth, E. E., Adams, W. V. Jr. 1976. Apparent propagation of the equine infectious anemia virus in a mosquito (*Culex pipiens quinqufasciatus* Say) ovarian cell line. *Am. J. Vet. Res.* 48:1069–70
9. Burkot, T. R. 1988. Non-random host selection by anopheline mosquitoes. *Parasit. Today* 4:156–62
10. Burny, A., Bruck, C., Cleuter, Y., Couez, D., Gregoire, D., et al. 1986. Bovine leukemia virus as an inducer of bovine leukemia. See Ref. 81a, 1:107–19
11. Butler, J. F., Kroft, W. J., DuBose, L. A., Kroft, E. S. 1977. Re-contamination of food after feeding a ³²P food source to biting Muscidae. *J. Med. Entomol.* 13:567–71
12. Buxton, B. A., Hinkle, N. C., Schultz, R. D. 1985. Role of insects in the transmission of bovine leukosis virus: potential for transmission by stable flies, horn flies, and tabanids. *Am. J. Vet. Res.* 46:123–26
13. Buxton, B. A., Schultz, R. D., Collins, W. E. 1982. Role of insects in the transmission of bovine leukosis virus: potential for transmission by mosquitoes. *Am. J. Vet. Res.* 43:1458–59
14. Cann, A. J., Chen, I. S. Y. 1985. Human T-cell leukemia virus types I and II. See Ref. 25a, 2:150–27
15. Castro, K. G., Lieb, S., Jaffe, H. W., Narkunas, J. P., Calisher, C. H., et al. 1988. Transmission of HIV in Belle Glade, Florida: lessons for other communities in the United States. *Science* 239:193–97
16. Coffin, J. M. 1985. Retroviridae and their replication. See Ref. 25a, 2:1437–1500
17. Coggins, L., Norcross, N. L. 1970. Immunodiffusion reaction in equine infectious anemia. *Cornell Vet.* 60:330–35
18. Coluzzi, M., Concetti, A., Ascoli, F. 1982. Effect of cibaral armature of mosquitoes (Diptera, Culicidae) on bloodmeal haemolysis. *J. Insect. Physiol.* 28:885–88
19. Cupp, E. W., Kemen, M. J. 1980. The role of stable flies and mosquitoes in the transmission of equine infectious anemia virus. *Proc. Annu. Meet. US Animal Health Assoc., 84th, Louisville,* pp. 362–67. Richmond, VA: Spencer
20. Davies, C. R. 1990. Interrupted feeding: causes and effects. *Parasit. Today* 6:19–22
21. Day, M. R., Fenner, F., Woodroofe, G. M. 1956. Further studies on the mechanism of mosquito transmission of myxomatosis in the European rabbit. *J. Hyg.* 54:258–83
22. DiGiacomo, R. F., Darlington, R. L., Evermann, J. F. 1985. Natural transmission of bovine leukemia virus in dairy calves by dehorning. *Can. J. Comp. Med.* 49:340–42
23. Evermann, J. F., DiGiacomo, R. F., Ferrer, J. F., Parish, S. M. 1986. Transmission of bovine leukosis virus by blood inoculation. *Am. J. Vet. Res.* 9:1885–87
24. Evermann, J. F., DiGiacomo, R. F., Hopkins, S. G. 1987. Bovine leukosis virus: understanding viral transmission and the methods of control. *Vet. Med.* 82:1051–58
25. Ferrer, J. F. 1979. Bovine leukosis: natural transmission and principles of control. *J. Am. Vet. Med. Assoc.* 175:1281–86
25a. Fields, B. N., Knipe, D. M., eds.

Fields Virology, Vol. 2. New York: Raven. 2336 pp.

26. Fischer, R. G., Luecke, D. H., Rehacek, J. 1973. Friend leukemia virus (FLV) activity in certain arthropods. *Neoplasma* 20:255–60

27. Foil, L. D. 1983. A mark-recapture method for measuring effects of spatial separation of horses on tabanid (Diptera) movement between hosts. *J. Med. Entomol.* 20:301–5

28. Foil, L. D. 1989. Tabanids as vectors of disease agents. *Parasit. Today* 5:88–96

29. Foil, L. D., Adams, W. V. Jr., Issel, C. J., Pierce, R. 1984. Tabanid (Diptera) populations associated with an equine infectious anemia outbreak in an inapparently infected herd of horses. *J. Med. Entomol.* 21:28–30

30. Foil, L. D., Adams, W. V. Jr., McManus, J. M., Issel, C. J. 1987. Bloodmeal residues on mouthparts of *Tabanus fuscicostatus* (Diptera: Tabanidae) and the potential for mechanical transmission of pathogens. *J. Med. Entomol.* 24:613–16

31. Foil, L. D., Adams, W. V. Jr., McManus, J. M., Issel, C. J. 1988. Quantifying the role of horse flies as vectors of equine infectious anemia. *Equine Infectious Diseases. Proc. 5th Int. Conf.* ed. D. G. Powell. pp. 185–95. Lexington: Univ. Press Ky.

32. Foil, L. D., French, D. D., Hoyt, P. G., Issel, C. J., Leprince, D. J., et al. 1989. Transmission of bovine leukemia virus by *Tabanus fuscicostatus*. *Am. J. Vet. Res.* 50:1771–73

33. Foil, L. D., Meek, C. L., Adams, W. V. Jr., Issel, C. J. 1983. Mechanical transmission of equine infectious anemia virus by deer flies (*Chrysops flavidus*) and stable flies (*Stomoxys calcitrans.*) *Am. J. Vet. Res.* 44:155–56

34. Foil, L. D., Seger, C. L., French, D. D., Issel, C. J., McManus, J. M., et al. 1988. The mechanical transmission of bovine leukemia virus by horse flies (Diptera: Tabanidae). *J. Med. Entomol.* 25:374–76

35. Foil, L. D., Stage, D., Adams, W. V. Jr., Issel, C. J. 1985. Observations of tabanid feeding on mares and foals. *Am. J. Vet. Res.* 46:1111–13

36. Francis, D. P., Essex, M. 1977. Excretion of feline leukaemia virus by naturally infected pet cats. *Nature* 269:252–54

37. Friedland, G. H., Klein, R. S. 1987. Transmission of the human immunodeficiency virus. *New Engl. J. Med.* 317:1125–34

38. Hall, R. F., Pursell, A. R., Cole, J. R., Youmans, B. C. 1988. A propagating epizootic of equine infectious anemia on a horse farm. *J. Am. Vet. Med. Assoc.* 193:1082–84

39. Hardy, W. D., Old, L. J., Hess, P. W., Essex, M., Cotter, S. 1973. Horizontal transmission of feline leukaemia virus. *Nature* 244:266–69

40. Hawkins, J. A. 1973. *The role of the horse fly* Tabanus fuscicostatus *Hine and the stable fly* Stomoxys calcitrans *L. in the transmission of equine infectious anemia in Louisiana*. MS thesis. Baton Rouge: La. State Univ. 48 pp.

41. Hawkins, J. A., Adams, W. V. Jr., Cook, L., Wilson, B. H., Roth, E. E. 1972. Transmission of equine infectious anemia with the horse fly *Tabanus fuscicostatus* Hine. *Proc. Annu. Meet. US Animal Health Assoc., 76th,* pp. 227–30. Richmond, VA: Spencer

42. Hawkins, J. A., Adams, W. V. Jr., Cook, L., Wilson, B. H., Roth, E. E. 1973. Role of horse fly (*Tabanus fuscicostatus* Hine) and stable fly (*Stomoxys calcitrans* L.) in transmission of equine infectious anemia to ponies in Louisiana. *Am. J. Vet. Res.* 34:1583–86

43. Hawkins, J. A., Adams, W. V. Jr., Wilson, B. H., Issel, C. J., Roth, E. E. 1976. Transmission of equine infectious anemia by *Tabanus fuscicostatus*. *J. Am. Vet. Med. Assoc.* 168:63–64

44. Henry, E. T., Levine, J. F., Coggins, L. 1987. Rectal transmission of bovine leukemia virus in cattle and sheep. *Am. J. Vet. Res.* 48:634–36

45. Hirsch, M. S., Curran, J. 1985. Human immunodeficiency viruses: biology and medical aspects. See Ref. 25a, 2:1545–70

46. Ho, D. D., Moudgil, T., Alam, M. 1989. Quantitation of human immunodeficiency virus type 1 in the blood of infected persons. *New Engl. J. Med.* 321:1621–25

47. Ishii, S. 1938. On the existence of virus in semen and the possibility of infection by the copulation in the horse affected with infectious anemia. *Jpn. J. Appl. Vet. Med.* 11:647–52

48. Issel, C. J., Adams, W. V. Jr. 1982. Detection of equine infectious anemia virus in a horse with an equivocal agar gel immunodiffusion test reaction. *J. Am. Vet. Med. Assoc.* 180:276–78

49. Issel, C. J., Adams, W. V. Jr., Foil, L. D. 1985. Prospective study of the progeny of inapparent equine carriers of equine infectious anemia virus. *Am. J. Vet. Res.* 46:1114–16

50. Issel, C. J., Adams, W. V. Jr., Meek, L., Ochoa, R. 1982. Transmission of equine infectious anemia virus from

horses without clinical signs of disease. *J. Am. Vet. Med. Assoc.* 180:272–75
51. Issel, C. J., Coggins, L. 1979. Equine infectious anemia: current knowledge. *J. Am. Vet. Med. Assoc.* 174:727–33
52. Issel, C. J., Foil, L. D. 1984. Studies on equine infectious anemia virus transmission by insects. *J. Am. Vet. Med. Assoc.* 184:293–97
53. Johnson, A. W. 1976. Equine infectious anemia: the literature 1966–1975. *Vet. Bull.* 46:559–74
54. Jupp, P. G., Lyons, S. F. 1987. Experimental assessment of bedbugs (*Cimex lectularius* and *Cimex hemipterus*) and mosquitoes (*Aedes aegypti formosus*) as vectors of human immunodeficiency virus. *AIDS* 1:171–74
55. Jupp, P. G., McElligott, S. E., Lecatsas, G. 1983. The mechanical transmission of hepatitis B virus by the common bedbug (*Cimex lectularius* L.) in South Africa. *S. Afr. Med. J.* 63:77–81
56. Kemen, M. J., McClain, D. S., Matthysse, J. G. 1978. Role of horse flies in transmission of equine infectious anemia from carrier ponies. *J. Am. Vet. Med. Assoc.* 172:360–62
57. Krinsky, W. L. 1976. Animal disease agents transmitted by horse flies and deer flies (Diptera: Tabanidae). *J. Med. Entomol.* 13:225–75
58. Lockwood, D. N. J., Weber, J. N. 1989. Parasite infections in AIDS. *Parasit. Today* 5:310–16
59. Lucas, M. H., Roberts, D. H., Wibberley, G. 1985. Ear tattooing as a method of spread of bovine leukosis virus infection. *Br. Vet. J.* 141:647–49
60. Manet, G., Guilbert, X., Roux, A., Vuillaume, A., Parodi, A. L. 1989. Natural mode of horizontal transmission of bovine leukemia virus (BLV): the potential role of tabanids (*Tabanus* spp.) *Vet. Immunol. Immunopathol.* 22:255–63
61. Melbye, M., Njelasani, E. K., Bayley, A., Mukelabai, K., Manuwele, J. K., et al. 1986. Evidence for heterosexual transmission and clinical manifestations of human immunodeficiency virus infection and related conditions in Lusaka, Zambia. *Lancet* 2:1113–15
62. Miike, L. 1987. *Do insects transmit AIDS?* Staff paper of Office of Technology Assessment. Washington, DC. 43 pp.
63. Miller, G. J., Pegram, S. M., Kirkwook, B. R., Beckles, G. L. A., Byam, N. T. A., et al. 1986. Ethnic composition, age and sex, together with location and standard of housing as determinants of the HLTV-I infection in an urban Tri-

nidadian community. *Int. J. Cancer* 38:801–8
64. Mohler, J. R. 1908. Swamp fever of horses. *Annu. Rep. Bur. Animal Ind. USDA* p. 225
65. Montelaro, R. C., Ball, J. M., Rwambo, P. M., Issel, C. J. 1989. Antigenic variation during persistent lentivirus infections and its implications for vaccine development. Immunobiology of proteins and peptide V. *Adv. Exp. Med. Biol.* 251:251–72
66. Murphy, E. L., Calisher, C. H., Figueroa, J. P., Gibbs, W. N., Blatner, W. A. 1989. HTLV-I infection and arthropod vectors. *New Engl. J. Med.* 321:1146
67. Narayan, O., Clements, J. E. 1985. Lentiviruses. See Ref. 25a, 2:1571–89
68. Ogston, C. W. 1981. Transfer of radioactive tracer by the bedbug *Cimex hemipterus* (Hemiptera: Cimicidae): a model for mechanical transmission of Hepatitis B virus. *J. Med. Entomol.* 18:107–11
69. Ogston, C. W., London, W. T. 1980. Excretion of hepatitis B surface antigen by the bedbug *Cimex hemipterus* Fabr. *Trans. R. Soc. Trop. Med. Hyg.* 74: 823–25
70. Okado, K., Oshima, K., Numakunai, S., Shidara, S., Ikeda, T., Mitsui, T. 1981. Ultrastructural studies on bloodsucking tabanid flies transmitting bovine leukemia virus. *Jpn. J. Vet. Med. Assoc.* 34:116–20
71. Oshima, K., Iigo, Y., Okado, K., Numakunai, S. 1984. Bovine lymphosarcoma experimentally infected with bovine leukemia virus by tabanid flies. *Proc. Congr. FAVA, 4th, Tampei, Taiwan, ROC.* pp. 63–69. Tampei: Chinese Soc. Vet. Sci.
72. Oshima, K., Okada, K., Numakunai, S., Yoneyama, Y., Sato, S., Takahashi, K. 1981. Evidence on horizontal transmission of bovine leukemia virus due to blood-sucking tabanid flies. *Jpn. J. Vet. Sci.* 43:79–81
73. Pearson, J. E., Knowles, R. C. 1984. Standardization of the equine infectious anemia immunodiffusion test and its application to the control of the disease in the United States. *J. Am. Vet. Med. Assoc.* 184:298–301
74. Pederson, N. C. 1987. Feline leukemia virus. In *Virus Infections of Carnivores,* ed. M. J. Appel, 1:299–320. New York: Elsevier Science Publishers.
75. Pedersen, N. C., Ho, E. W., Brown, M. L., Yamamoto, J. K. 1987. Isolation of a T-lymphotropic virus from domestic

cats with an immunodeficiency-like syndrome. *Science* 235:790–93
76. Quinn, T. C., Zacarias, F. R. K., St. John, R. K. 1989. HIV and HTLV-I infections in the Americas: A region perspective. *Medicine* 68:189–209
77. Ribeiro, J. M. C. 1987. Role of saliva in blood-feeding by arthropods. *Annu. Rev. Entomol.* 32:463–78
78. Romero, C. H., Abaracon, D., Rowe, C. A., Silva, A. G. 1984. Bovine leukosis virus infectivity in *Boophilus microplus* ticks. *Vet Rec.* 115:440–41
79. Romero, C. H., Cruz, G. B., Rowe, C. A. 1983. Transmission of bovine leukaemia virus in milk. *Trop. Animal Health Prod.* 15:215–18
80. Rosen, L. 1981. The use of *Toxorhynchites* mosquitoes to detect and propagate dengue and other arboviruses. *Am. J. Trop. Med. Hyg.* 30:177–83
81. Rous, P. 1911. Transmission of a malignant new growth by means of a cell-free filtrate. *J. Am. Med. Assoc.* 56:198
81a. Salzman, L. A., ed. 1986. *Animal Models of Retrovirus Infection and Their Relationship to AIDS*, Vol. 1. New York: Academic. 470 pp.
82. Schultz, R. D., Manning, T. O., Rhyan, J. C., Buxton, B. A., Panangala, B. S., et al. 1986. Immunologic and virologic studies on bovine leukosis. See Ref. 81a, 1:301–23
83. Scott, J. W. 1915. Twenty-fifth report of the parasitologist. *Annu. Rep. Wyo. Agric. Exp. Sta.* p. 104
84. Scott, J. W. 1924. The experimental transmission of swamp fever or infectious anemia of horses. *Univ. Wyo. Agric. Exp. Sta. Bull.* 138:34–39
85. Shen, D. T., Gorham, J. R., Jones, R. H., Crawford, T. B. 1978. Failure to propagate equine infectious anemia virus in mosquitoes and *Culicoides variipennis*. *Am. J. Vet. Res.* 39:875–76
86. Spates, G. E., Mayer, R. T. 1984. Midgut hemolytic activity of the horn fly, *Haematobia irritans* (Diptera: Muscidae). *J. Med. Entomol.* 21:58–62
87. Srinivasan, A., York, D., Bohan, C. 1987. Lack of HIV replication in arthropod cells. *Lancet* 1:1094–95
88. Stein, C. D., Lotze, J. C., Mott, L. O. 1942. Transmission of equine infectious anemia by the stablefly *Stomoxys calcitrans*, the horsefly, *Tabanus sulcifrons* (Macquart), and by injection of minute amounts of virus. *Am. J. Vet. Res.* 3:183–93
89. Stein, C. D., Lotze, J. C., Mott, L. O. 1943. Evidence of transmission of inapparent (subclinical) form of equine infectious anemia by mosquitoes (*Psor-*

ophora columbiae), and by injection of the virus in extremely high dilution. *J. Am. Vet. Med. Assoc.* 102:163–69
90. Stein, C. D., Mott, L. O. 1947. Equine infectious anemia in the United States with special reference to the recent outbreak in New England. *Proc. US Livest. Sanit. Assoc.* 51:37–52
91. Temin, H. M. 1985. Reverse transcription in the eukaryotic genome: retroviruses, pararetroviruses, retrotransposers, and retrotranscripts. *Mol. Biol. Evol.* 6:455–68
92. Temin, H. M. 1989. Retrovirus vectors: promise and reality. *Science* 246:9
93. Thurmond, M. C., Burridge, M. J. 1982. Application of research to control of bovine leukemia virus infection and to exportation of bovine leukemia virus-free cattle and semen. *J. Am. Vet. Med. Assoc.* 181:1531–34
94. Titus, R. G., Ribeiro, J. M. C. 1988. Salivary gland lysates from the sand fly *Lutzomyia longipalpis* enhance *Leishmania* infectivity. *Science* 239:1306–8
95. Umphenour, N. W., Kemen, M. J., Coggins, L. 1974. Equine infectious anemia: a retrospective study of an epizootic. *J. Am. Vet. Med. Assoc.* 164:66–69
96. Vallee, H., Carre, H. 1904. Sur la nature infectieuse de l'anemie du cheval. *C. R. Acad. Sci.* 139:331
97. Van Der Maaten, M. J. 1986. Pathogenesis of bovine retrovirus infection. See Ref. 81a, 1:213–22
98. Van Der Maaten, M. J., Miller, J. M. 1977. Susceptibility of cattle to bovine leukemia virus infection by various routes of exposure. In *Advances in Comparative Leukemia Research*, ed. P. Bentuelzen, pp. 29–32. Amsterdam: Elsevier/North-Holland Biomedical
99. Van Der Maaten, M. J., Miller, J. M. 1979. Appraisal of control measures for bovine leukosis. *J. Am. Vet. Med. Assoc.* 175:1287–90
100. Van Der Maaten, M. J., Miller, J. M., Schmerr, J. F. 1981. In utero transmission of bovine leukemia virus. *Am. J. Vet. Res.* 42:1052–54
101. Varmus, H. 1988. Retroviruses. *Science* 240:1427–34
102. Varmus, H., Brown, P. 1989. Retroviruses. In *Mobile DNA*, ed. M. Howe, D. Berg, 1:53–108. Washington DC: Am. Soc. Microbiol.
103. Warnes, M. L., Finlayson, L. H. 1987. Effect of host behavior on host preference in *Stomoxys calcitrans*. *Med. Vet. Entomol.* 1:53–57
104. Webb, P. A., Happ, C. M., Maupin, G. O., Johnson, B. J. B., Ou, C., Monath,

T. P. 1989. Potential for insect transmission of HIV: experimental exposure of *Cimex hemipterus* and *Toxorhyncites amboinensis* to human immunodeficiency virus. *J. Infect. Dis.* 160:970–77

105. Weber, A. F., Moon, R. D., Sorensen, D. K., Bates, D. W., Meiske, J. C., et al. 1988. Evaluation of the stable fly (*Stomoxys calcitrans*) as a vector of enzootic bovine leukosis. *Am. J. Vet. Res.* 49:1543–49

106. Wilesmith, J. W., Straub, O. C., Lorenz, R. J. 1980. Some observations on the epidemiology of bovine leucosis virus infection in a large dairy herd. *Res. Vet. Sci.* 28:10–16

107. Williams, D. L., Issel, C. J., Steelman, C. D., Adams, W. V. Jr., Benton, C. V. 1981. Studies with equine infectious anemia virus: transmission attempts by mosquitoes and survival of virus on vector mouthparts and hypodermic needles, and in mosquito tissue culture. *Am. J. Vet. Res.* 42:1469–73

108. Wills, W., Larouze, B., London, W. T., Millman, I., Werner, B. G., et al. 1977. Hepatitis B virus in bedbugs (*Cimex hemipterus*) from Senegal. *Lancet* 2:217–20

109. Wong-Staal, F. 1985. Human immunodeficiency viruses and their replication. See Ref. 25a, 2:1529–43

Annu. Rev. Entomol. 1991. 36:383–406

ECOLOGY AND MANAGEMENT OF TURFGRASS INSECTS

Daniel A. Potter

Department of Entomology, University of Kentucky, Lexington, Kentucky 40546-0091

S. Kristine Braman

Department of Entomology, University of Georgia, Georgia Experiment Station, Griffin, Georgia 30223-1797

KEY WORDS:　turf insects, urban integrated pest management, lawn insects, insecticide, low input sustainable agriculture

PERSPECTIVES AND OVERVIEW

"The grasses are the least noticed of the flowering plants. They seem to be taken for granted like air and sunlight, and the general run of people never give them a thought" (73). Sixty years later, this view of grasses has changed. Turfgrasses have become the most widely used and intensively managed urban plantings, and the management of turfgrass insect pests is an increasingly important concern. In this first review of turfgrass entomology printed by the *Annual Review of Entomology,* we critically appraise the status of biological and integrated pest management (IPM) research on turfgrass insects, discuss some of the problems of implementing IPM in this unique system, and identify areas of promise for future research.

Turfgrass may be defined as a uniform stand of grass or a mixture of grasses maintained at a relatively low height and used for recreational or functional purposes or to enhance and beautify human surroundings (177). The culture of mowed lawns dates back at least to medieval times when turf was used for outdoor sports and recreation and was considered a symbol of the

0066-4170/91/0101-0383$02.00

rich and powerful (78). Use of turf in the United States skyrocketed during the 1950s and 1960s as large tracts of land were developed to accomodate the growing urban population. Turfgrasses now cover an estimated 10.1 to 12.1 million hectares in the US. About 81% of this area is subdivided among more than 50 million lawns; the remainder is in parks, golf courses, athletic fields, cemeteries, sod farms, roadsides, and other sites (156). Turfgrass culture, in its many forms, is at least a $25 billion per year industry in the US. About 500,000 people make their living directly from establishment and mainte-nance of turf (156, 177). Commercial lawn-care receipts increased at an average annual rate of 22% from 1977–1984 (110). In 1987, more than 20 million US golfers played about 445 million rounds (156).

Turfgrass has functional, aesthetic, and monetary value. Attractive land-scaping conveys a favorable impression in a business setting and increases the value of residential property by as much as 15% (82). Living turf supplies oxygen, reduces erosion and surface runoff, glare, and noise pollution, moderates surface temperatures, and filters dust from the air (14, 82). Turf-grasses enhance the safety and enjoyment of sports and leisure activities, and they help to make urban areas a more pleasant place to live and work. These benefits are increasingly important to the physical health and mental well-being of the urban population (167).

Demand for high quality turfgrass has been accompanied by growing public concern about the negative aspects of pesticides, especially ground water contamination (129) and potential risks to human health (104). Urban pesti-cide usage has become a volatile political issue (58, 215). Beset with in-creased litigation and rising insurance costs, loss of registrations, and local ordinances restricting pesticide use, the turfgrass industry will be forced to make greater use of alternative pest control methods (58).

Commitment of personnel, extramural funding, and other resources to turfgrass entomology is woefully inadequate. In 1980, the total allocation of professional assignments to university research and extension in the US was the equivalent of only 7.2 and 6.7 full-time positions, respectively (121). Much of that effort was divided among persons who had 10% or less of their time assigned to turfgrass. Twenty-five states had no research assignment at all in this area. Only 41 of the 8,418 research papers (0.49%) presented at national conferences of the Entomological Society of America during 1984–1989 dealt specifically with turfgrass insects (D. A. Potter & S. K. Braman, unpublished survey).

THE TURFGRASS SYSTEM

Turfgrass consists of the roots, stems, and leaves of individual grass plants, together with a tightly intermingled layer of living and dead roots, stolons, and organic debris commonly called thatch (14). Turfgrass species grown in the US are classified on the basis of their climatic adaptations (14). Cool-season grasses, including bluegrasses, fescues, ryegrasses, and bentgrasses, are the main species grown in the northern two-thirds of the US. Warm-season grasses, including bermudagrasses, zoysiagrass, St. Augustinegrass, bahiagrass, and centipedegrass, predominate in the southeast and in the warm, semiarid zones of the south and southwest. A transitional climatic zone extends south from Delaware to central Georgia and west from southern Kentucky to Oklahoma (14).

According to Beard (14), the six basic components of turfgrass quality are uniformity, density, texture, growth habit, smoothness, and color. The relative importance of these factors varies with individual perceptions and with the purpose for which the turf is used. Color and uniformity are desirable for golf fairways and home lawns, whereas smoothness is essential for golf putting greens. Aesthetic standards are often so high that even limited insect damage is considered intolerable. Consequently, insecticide treatments are often viewed as insurance, especially by the lawn-care and golf-course industries.

ARTHROPOD PESTS OF TURFGRASS

Arthropod pests of turfgrass include soil-inhabiting species that feed upon roots or damage turf through their burrowing activity, those that consume leaves and stems, and those that suck plant juices. Here, we briefly discuss the biology and damage caused by the main species. Tashiro (177) provides more complete accounts and citations.

Soil-Inhabiting Turfgrass Pests

MOLE CRICKETS   The tawny mole cricket, *Scapteriscus vicinus,* and the southern mole cricket, *Scapteriscus acletus,* are major pests of warm season turfgrasses, especially in the southeastern US. Neither species is native; both were apparently first introduced between 1899 and 1904 (119, 207, 209).

Male mole crickets stridulate from within specially constructed burrows in the soil that amplify their species-specific songs (51, 118, 193, 194). Individual variation in intensity of calls may lead to greater mating success by larger males (51). Louder male calls may also be indicative of superior oviposition sites where higher moisture favors both efficient burrow construction for calling and for incubation of eggs (51, 53, 69).

Calling songs of mole crickets have been used in species separation (52), together with such characters as foreleg morphology, pronotal markings, maxillae, and cuticular lipids (177). Acoustical traps have been used for population studies of the pests (206) and their parasitoids (54). While crickets may be attracted by the thousands to recorded or electronically synthesized male calls (206), population control via sound trapping has not proven effective. In fact, such callers may increase populations in the vicinity of the trap by attracting individuals that escape capture (113).

The southern mole cricket is primarily carnivorous, damaging turf mainly through its burrowing activity that mechanically dislodges roots and leaves piles of soil on the surface. The tawny mole cricket causes similar damage, but, because it also feeds on leaves, stems, and roots, it is considered the more significant pest (193). Both species are particularly damaging to bahia-grass and bermudagrass, especially on light sandy or loamy soils (177). Use of radioisotopes to trace underground movements of mole crickets has been explored (75).

Dispersal flights of females, which occur during spring and fall, have contributed to the rapid range extension of these two species (208, 210). Both species are univoltine throughout most of their range; in south Florida the southern mole cricket may be bivoltine (53, 210).

WHITE GRUBS   Scarabaeid grubs feed on the roots of all species of com-monly used turfgrasses, and they are the most important insect pests of cool-season grasses in the US (154, 155, 177). Heavily infested turf develops irregular dead patches that can be lifted or rolled back like a carpet. Avian or mammalian predators (e.g. moles, skunks) often cause further damage by tearing up the turf in their search for grubs. The subterranean habits of white grubs make them especially difficult to control because insecticides applied to the surface must move through the thatch and into the soil (120, 177). At least 10 species are pests of turfgrasses; many also do damage as larvae or adults in other agricultural settings (49, 154).

Several of the turf-infesting white grubs are introduced species. The most significant is the Japanese beetle, *Popillia japonica* (49, 50), followed (in order of importance) by: the European chafer, *Rhizotrogus majalis* (63, 179); oriental beetle, *Anomala orientalis* (1); and asiatic garden beetle, *Maladera castanea* (56, 64). The latter three species are of regional importance mainly in the Northeast. Since the discovery of the Japanese beetle in New Jersey in 1916, it has spread throughout the eastern US as far south as Georgia and Alabama and west to Iowa and Wisconsin. Efforts to eradicate localized infestations in California and Oregon are ongoing (177). Adult Japanese beetles feed upon foliage of more than 300 plant species (49).

Several widely distributed native species, including the northern and south-ern masked chafers, *Cyclocephala borealis* and *Cyclocephala lurida* [former-ly *C. immaculata* (42)] (134, 154); the black turfgrass ataenius, *Ataenius spretulus* (213); and many of the 152 species of *Phyllophaga* (68, 106, 155), can also be very destructive turf pests. Green June beetle larvae, *Cotinus nitida*, feed mainly on detritus (26) but damage turf by their burrowing activities.

Temperature, soil type, and especially soil moisture influence scarabaeid oviposition, egg and larval survival, adult emergence and flight activity, and expression of damage (35, 59, 63, 96, 134, 135, 143, 161). Eggs absorb water from the soil and cannot survive below critical moisture thresholds (139, 145); vertical movement of grubs is also governed by soil moisture and temperature (198). Noninvasive radiography techniques have recently been used to study white grub behavior below the soil surface (198).

Most turfgrass-infesting scarabaeids have one-year life cycles, hatching from eggs laid in June and July and inflicting their greatest damage in the late summer and early fall before moving deeper into the soil for overwintering (177). *A. spretulus* completes two generations per year in some areas (213), and portions of the population of other species (e.g. European chafers, oriental beetle) or species at the northern limits of their range (e.g. Japanese beetle) may require two years to complete development. Life cycles of *Phyllophaga* spp. vary from one to four years depending upon species and location; three-year life cycles are most common in cool-season turf (68, 155). *Phyllophaga crenata* and *Phyllophaga latifrons* are annual grubs that cause extensive damage to turf in Texas and Florida, respectively.

BILLBUGS   The two most widespread and important curculionid beetles that damage turfgrasses in the US are the bluegrass billbug, *Sphenophorus parvulus*, a pest of cool-season grasses (65, 89, 184), and the hunting billbug, *Sphenophorus venatus*, considered a pest mainly of warm-season grasses in the southern US and Hawaii (86). However, recent work on this poorly known group revealed a complex of four abundant species, including both the bluegrasss billbug and the hunting billbug, that damage cool-season grasses in New Jersey (81). This work also yielded a key to the eight known billbug turf pests in the US.

Both the bluegrass and the hunting billbug overwinter as adults in protected areas, emerging in early spring. Adults feed on and deposit eggs in stems, leaf sheaths, and crowns; young larvae tunnel in the stem and crown, causing the damaged plants to break off easily. Older larvae enter the soil and feed on the roots before pupating and becoming adults by fall. Numbers of adults migrating over adjacent paved surfaces in spring may be an indicator of subsequent larval damage (184). Another curculionid, *Listronotus maculicollis* (formerly *Hyperodes* sp. near *anthracina*), damages annual bluegrass, *Poa annua*, and can be a pest on golf courses in the northeastern US (203).

Insects that Consume Leaves and Stems

SOD WEBWORMS   About 100 species of sod webworms (Lepidoptera: Pyralidae) are recognized in North America (16), dozens of which infest turfgrasses in the US (3, 112, 189). Ainslee (3–7) described the biology of important native species including: the striped sod webworm, *Fissicrambus mutabilis;* the silver striped webworm, *Crambus praefectellus;* the larger sod webworm, *Pediasia trisecta;* and the bluegrass webworm, *Parapediasia teterrella.* The western lawn moth, *Tehama bonifetella,* and *Crambus sperryellus* are the main webworm pests of California lawns (16), while *Herpetogramma phaeopteralis,* the tropical sod webworm, is important in Florida (87). The grass webworm, *Herpetogramma licarsisalis,* is the principal turf pest in Hawaii (174). The cranberry girdler, *Chrysoteuchia topiaria,* damages grasses grown for commercial seed production in the Pacific Northwest (83) and lawns in northern Illinois and Michigan.

Most temperate-region sod webworms have one to four generations per year depending upon species and location (e.g. 16, 189). Larvae overwinter in the thatch or soil. First instars feed only on the surface tissues of leaves and stems; older larvae construct silk-lined burrows from which they emerge at night to feed upon grass blades or shoots. Adult emergence, mating, and oviposition are nocturnal events for some species (e.g. 174), while in others these behaviors are crepuscular (13) or diurnal (32). Temperature and relative humidity affect development and survival of eggs (70, 115). Sculpturing of the chorion is species-specific and was used to construct a key to the eggs of 15 species in Tennessee (111).

CUTWORMS, ARMYWORMS, AND SKIPPERS   Tashiro (177) recognized five noctuid species as pests of turfgrasses in the continental US: the black cutworm *Agrotis ipsilon,* the variegated cutworm *Peridroma saucia,* the bronzed cutworm *Nephelodes minians,* the armyworm *Pseudaletia unipuncta,* and the fall armyworm *Spodoptera frugiperda.* Cutworms and armyworms feed mainly on graminaceous hosts; many are pests of field crops (105, 205).

Cutworms are named for the larval habit of severing food plants at their base and pulling them into a subterranean burrow before feeding (205). The black cutworm, a widespread and destructive species (153), is especially damaging to golf greens. Larvae stay in aeration holes and emerge at night to feed. It overwinters in the pupal stage at Ohio and Illinois latitudes, with an additional influx of migrating adults from the south resulting in several annual population peaks arising from asynchronous ancestries (190).

The fall armyworm annually invades much of the continental US and southern Canada, yet is unable to survive the winter in the temperate zone (105, 165). Sporadic outbreaks of fall armyworms cause significant damage to field crops and turf, especially in the Southeast and Gulf states. The insects deposit egg masses indiscriminately on buildings, golf carts, and other objects as well as on host plants. Mating behavior, developmental biology, and feeding preferences of the fall armyworm have been extensively studied (25, 105, 165). The fiery skipper, *Hylephila phyleus,* is another species that sometimes damages bermudagrass and other grasses. Although widely distributed, it is considered a turf pest mainly in Hawaii and California (16, 182).

FRIT FLY   Larvae of *Oscinella frit* (Diptera: Chloropidae) cause occasional damage to golf-course greens. The maggots damage cool-season grasses by killing their growing points (8), and the adults are a nuisance because they are attracted to and land on white golf balls (177). Its biology as a turf pest has recently been studied (187, 188).

Arthropods that Suck Plant Juices

CHINCH BUG COMPLEX   Chinch bugs are well known pests of Graminocea (100, 101). Two species, the hairy chinch bug, *Blissus leucopterus hirtus,* and the southern chinch bug, *Blissus insularis,* are important turfgrass pests (177). The chinch bug, *B. leucopterus leucopterus,* is a migratory species that feeds mainly on small grains and corn (74) but occasionally damages turf.

The hairy chinch bug attacks cool-season turfgrasses and zoysiagrass in the northeastern US and Canadian border provinces west to Minnesota and south to Virginia. Aggregations of nymphs and adults suck sap from stems and crowns, causing localized injury that may coalesce into large patches of dead and dying turf. Damage, which may be compounded by moisture stress, is reportedly greatest on sandy soils and in full sunlight (114, 177). The hairy chinch bug overwinters as an adult, completing one generation per year in southern Ontario (103) and two in the southern parts of its range (108, 114, 120). Developmental biology of the hairy chinch bug has been studied in the laboratory (11).

The southern chinch bug is the most injurious pest of St. Augustinegrass in Florida and the Gulf Coast region (88, 150), where repeated spraying to control as many as 7–10 generations per year has led to widespread insecticidal resistance (151). Densities of southern chinch bug frequently exceed 500–1000/0.1 m^2; open, sunny areas, heavily fertilized lawns, and those with thick, spongy thatch are reportedly preferred (150). Brachypterous adults predominate; dispersal is mainly short-range by walking (88, 100).

ACARINE PESTS   Bermudagrass mite, *Eriophyes cynodoniensis,* was first reported from the US in 1959 (192), but its occurrence throughout much of the southern US in the 1960s (37) suggests an earlier introduction. The mites overwinter beneath leaf sheaths of dormant plants. Oviposition begins in the spring. Multiple generations are completed in a growing season. Heavy infestations cause shortening of the internodes, rosetting, browning, and dieback that encourages subsequent encroachment by weeds. The mites reportedly prefer well-fertilized lawns, and moisture stress aggravates their damage (192).

Winter grain mite, *Penthaleus major*, is a widely distributed pest of small grains, vegetables, and several species of cool-season turfgrasses (23). It feeds by rasping grass blades, causing damage that resembles winter desiccation. Eggs hatch in October and develop into adult females by November; all life stages may be present during the winter months. Mites are most active at night or on dark, cloudy days. Two generations have been reported; second-generation females deposit eggs that aestivate and hatch the following fall (172, 173). The clover mite, *Bryobia praetiosa*, and the Banks grass mite, *Oligonychus pratensis*, also occasionally damage turf (177).

SPITTLEBUGS   The twolined spittlebug, *Prosapia bicincta*, can be a serious pest of pastures, especially in the southeastern US (43, 130). Spittlebugs also damage southern lawns, especially bermuda and centipedegrass, where dense uniform turf, combined with regular irrigation, may favor egg hatch and nymphal survival. Eggs overwinter on soil, in vegetation, or on plant debris; two generations occur annually in the Southeast. Adults and nymphs feed upon plant sap, causing withering and phytotoxemia (169).

APHIDS, SCALES, AND MEALYBUGS   The greenbug, *Schizaphis graminum*, is mainly a pest of small grains (204), but damage to bluegrass lawns was reported as long ago as 1907 (212). Since 1970, the aphid has caused sporadic, severe damage to cool-season grasses in the midwestern states, and host resistance studies suggest the probability of new biotypes (127). In northern states, where most of its damage to turf occurs, greenbugs overwinter mainly as eggs adhering to grass blades, fallen tree leaves, and debris (127). The first spring generation consists of wingless females that give rise to as many as 15 additional generations per year. Damage results from withdrawal of phloem sap and from phytotoxemia in response to salivary secretions.

Several other sucking insects can be pests of warm season turfgrasses, especially bermudagrass, or tall fescue in the southern US. Feeding by rhodesgrass mealybug, *Antonina graminis* (24), and bermudagrass scale, *Odonaspis ruthae* (186), causes loss of vitality and browning that may be compounded by drought. Ground pearls, *Margarodes* spp., are subterranean scale insects that infest the roots of warm season grasses (94). Eggs are deposited in a clump enclosed in a waxy sac; the newly hatched nymphs secrete a pearl-like shell or cyst. Ground pearls may occur as deep as 25 cm in the soil, precluding practical means of control (120).

Other Pests

Numerous other arthropods, e.g. ants (especially imported red fire ant, *Solenopsis invicta*), crane fly larvae, wasps, and fleas, may sometimes become pests in turfgrass by feeding upon the plants themselves, by producing burrows or mounds, or by being a nuisance to people. Tashiro (177) discusses their biology in this context.

CURRENT STATUS OF INTEGRATED PEST MANAGEMENT

Insecticides

Conventional insecticides are the mainstay of turfgrass insect control, and their efficacy is being constantly evaluated (e.g. 10, 122, 199, 201). At present, use of an insecticide is often the only practical way to prevent significant damage from unexpected or heavy pest infestations.

Highly persistent cyclodiene insecticides, including chlordane, dieldrin, aldrin, and heptachlor, were used very effectively in the 1950s and 1960s. A single application often provided residual control for many years (175). By the early 1970s, the effectiveness of cyclodienes had become limited by increased insect resistance (147, 176), and environmental concerns resulted in cancellation of cyclodiene registrations for turf. Subsequently, organophosphates (OPs) such as chlorpyrifos, diazinon, trichlorfon, ethoprop, and isazophos and carbamates, including bendiocarb and carbaryl, came into general use on turf. Their versatility is limited by relatively short residual toxicity (95, 158, 175) and by their sometimes inconsistent performance under differing edaphic conditions (67, 181). Isofenphos, an OP with relatively long residual toxicity in soil (175, 201), became widely used in the 1980s, and the synthetic pyrethroids fluvalinate and cyfluthrin were more recently labelled for use on turf. Resistance to OPs or carbamates has been documented for chinch bugs and greenbugs (147) and, in at least one instance, for white grubs (2). Few new insecticides are targeted for turfgrass because the market is relatively small and offers limited opportunity for recovery of the massive development and registration costs. Effective insect growth regulators, chitin inhibitors, and other so-called third generation insecticides have not yet been labelled for use on turf.

INFLUENCE OF THATCH   Immediate post-treatment irrigation is usually recommended when OPs or carbamates are applied for control of root-feeding insects, ostensibly to leach the insecticide through the thatch and into the soil. Even with irrigation, most of the residues may become bound in the highly adsorptive thatch (122, 125, 126, 158) where they are rapidly degraded by chemical hydrolysis or microbial decomposition (17). Contact with and ingestion of residues at the thatch-soil interface or in the thatch itself may be the primary source of the lethal dose (122, 126). Further study of the mobility of insecticide residues in thatch and soil, and of movement of white grubs and other insects in response to soil moisture (e.g. 198) is needed to better understand the factors that limit control of these pests. Binding of insecticides in thatch probably reduces their potential for leaching from turfgrass into ground water or for run-off into benthic systems.

ENHANCED BIODEGRADATION   Enhanced biodegradation, in which pesticides are degraded at an accelerated rate by microorganisms in soils that have been conditioned by repeated exposure to a pesticide, has been reported for isofenphos, diazinon, ethoprop, and other insecticides used on turf (45). Enhanced biodegradation has been implicated in reduced residual effectiveness of isofenphos on golf courses previously treated with that chemical (123).

ENVIRONMENTAL SIDE-EFFECTS   Excessive thatch is a common problem on highly maintained turfgrass, contributing to reduced water infiltration, shallow rooting, and increased vulnerability to stress and pest problems (14). Use of certain pesticides or fertilizers may encourage thatch accumulation by adversely affecting earthworms and other soil organisms that are important to decomposition processes (137, 138, 140, 142). A single spring application of bendiocarb, carbaryl, ethoprop, or the fungicide benomyl to Kentucky bluegrass resulted in 60–90% reductions in earthworm abundance; effects lasted for at least 20 weeks (21, 137a). Insecticides may also suppress populations of predators and parasitoids (9, 29, 30, 196) and apparent resurgences or secondary outbreaks of certain pests (e.g. chinch bugs, winter grain mite) have been reported (146, 170, 172).

Certain turfgrass pesticides (e.g. diazinon) are very toxic to birds and fish (131). Application of chlorpyrifos to freshwater ponds resulted in poisoning of waterfowl, apparently from ingestion of moribund, chlorpyrifos-contaminated insects (79). Birds may gorge upon dead or moribund mole crickets (18) or Cotinus grubs (D. A. Potter, unpublished observation) that have come to the turf surface following an insecticide treatment. Mortality of waterfowl following foraging on insecticide-contaminated turfgrass or invertebrates contributed to recent cancellation of diazinon usage on golf courses and sod farms.

Sampling, Monitoring, and Risk Assessment

Turfgrass insects often go unnoticed until feeding damage becomes obvious or the more active and visible adult stages emerge. Simple, reliable, and cost-effective survey techniques other than visual inspections are lacking for most species. This absence of practical methods of risk assessment, together with low damage thresholds, are factors that contribute to insecticide use.

Direct population survey techniques for turfgrass insects include soil sampling, often with a golf-cup cutter or motorized sod-cutter (117), flotation (107), irritant drenches (183), pit-fall traps (81), sweep net or suction sampling, sound traps (206), and heat extraction from thatch and soil (120). These techniques are useful for research purposes, but in practical usage they are often prohibitively destructive and time consuming, especially for the lawn-care industry. Moreover, their value in decision-making is limited by the general lack of established damage thresholds (136). Spatial distributions have been studied and sequential sampling plans developed for several turfgrass pests (77, 103, 109, 116, 117, 188), but relationships between population density of the damaging life stage and injury to turf have been studied for only a few species (28, 135).

Pheromones or sex attractants have been identified for a number of important turfgrass insects including the Japanese beetle (191), fall armyworm, armyworm, cranberry girdler, western lawn moth (for review, see 84), black cutworm (72), and bluegrass webworm (27) and demonstrated for several others including the larger sod webworm (12), masked chafers (133), and the green June beetle (38). Other adult-trapping and survey methods include blacklight traps for night-flying species (180, 189), food-type baits (98), and sticky plastic sheets (46).

Potential uses for semiochemicals and other trapping methods include monitoring for the purpose of treatment timing, detection and evaluation of pest populations, and direct suppression by mass trapping (91) or disruption of sexual communication (84). Semiochemicals could be used to target high-risk lawns or golf fairways for selective treatments, but little work relates trap captures of adults to subsequent larval populations or damage. This approach would work better for species in which adults tend to remain in the area where they fed as larvae (e.g. sod webworms, masked chafers) than for species capable of long-distance flight to traps (e.g. Japanese beetles, armyworms).

Models based on accumulated degree-days that allow prediction of phenology or adult emergence have been developed for several turfgrass pests, including chinch bugs (103), Japanese beetles (144), masked chafers (134), sod webworms (189), and frit flies (187). Timing of emergence, flight, or egg-laying of some species, e.g. European chafer and *A. spretulus,* has been related to dates of flowering of common plants (178, 212).

Biological Control

The fact that insect outbreaks are relatively uncommon in low-maintenance turfgrass suggests that many pests are normally held in check by indigenous natural enemies. Several surveys document the diverse community of entomophagous invertebrates inhabiting turfgrass in the US (31, 76, 146, 171) and Europe (33, 85). Practically every turfgrass pest has one or more natural enemies associated with it (177), but manipulation of these beneficial arthropods as components of IPM has progressed little (92, 136).

The ecological, economic, and sociological difficulties of implementing biological control in urban settings have been discussed elsewhere (47, 55, 128, 136). The urban landscape, with its diversity of pests and plantings, its frequent modifications and disruptions, and its kaleidescope of public attitudes and expectations, poses formidable challenges to biocontrol. Conservation of natural enemies and other beneficials should be a factor in pesticide selection, but comparative data (e.g. 29) upon which to base such decisions are sparse.

Between 1920–1933, about 49 species of parasitoids and predators were imported to the US from the Orient and Australia and released into Japanese beetle–infested areas (48). Only a few of these became established. The most important are two species of tiphiid wasps that parasitize the grubs. Their present impact is limited (48). An introduced parasitoid, the tachinid *Hyperecteina aldrichi,* was recovered from about 20% of the adult Japanese beetles sampled in central Connecticut in 1979 (177).

A sphecid wasp, *Larra bicolor,* was imported into Puerto Rico from Brazil in the late 1930s for suppression of mole crickets (216), and efforts to establish this and other imported parasitoids in Florida were recently reviewed (76). Importation of an encyrtid parasitoid from India to Texas provided almost complete biological control of the rhodesgrass mealybug, with reported savings of $17 million/year (36).

Propagation and dissemination of milky disease bacteria, mainly *Bacillus popilliae,* by the USDA and cooperating state agencies during the 1930s and 1940s was credited with reducing Japanese beetle populations over much of the eastern US (48, 92, 97, but see 40, 66). Other species or strains of milky disease bacteria attack other white grubs, including *Cyclocephala* spp. (211, 214), *A. spretulus* (166), and *R. majalis* (185), but commercial formulations are presently marketed only for control of Japanese beetle larvae.

Aspects of the pathology, host specificity, and application of milky disease bacteria for area-wide suppression of Japanese beetles have been extensively studied (for reviews, see 41, 90, 157), and a 1976 bibliography listed 239 references on the subject (93). Nevertheless, presently no published data document the performance of milky disease bacteria in replicated field trials or in the urban landscape. The turfgrass industry has not made greater use of milky disease bacteria for reasons that include its relatively high cost, difficulty of application, limited availability, inconsistent performance, lack of field-efficacy data, and the long establishment period required for control. Commercial milky disease bacteria formulations have been relatively expensive because their production required hand collection and inoculation of individual grubs. Early in vitro propagation efforts were disappointing (19, 157), but recent advances in large-scale fermentation production of *B. popilliae* (152) may increase the availability and reduce the cost of milky disease bacteria formulations. In vitro production methods may also make producing milky disease bacteria formulations that are infectious to other grub species commercially feasible.

Bacillus thuringiensis formulations are labelled for lepidopteran turf pests, but they are not widely used and are frequently omitted from control recommendations. Current *B. thuringiensis kurstaki* strains have limited activity on soil-dwelling caterpillars (159). Naturally occurring epizootics of the fungus *Beauveria bassiana* may suppress populations of southern and hairy chinch bugs (108, 146), but the value of the fungus in IPM is limited by its requirement for specific environmental conditions. Other microbial pathogens infect particular turf pests (66, 92, 177), but none has yet been successfully manipulated for IPM.

Commercially available steinernematid and heterorhabditid nematodes have provided satisfactory control of white grubs in turf (159, 160, 197), but efficacy was variable and required irrigation following application and moderate to high soil moisture for nematode establishment. Bait and spray formulations containing nematodes controlled black cutworms and tawny mole crickets in laboratory tests (60, 159). Nematodes offer a promising alternative to conventional insecticides, but problems of availability, storage, cost, handling, and reliability must be resolved if they are to become more widely used by the turfgrass industry.

Cultural Control

Routine cultural practices such as fertilization, irrigation, and mowing affect pest populations and their damage (14, 61, 96, 135), but published accounts are mostly anecdotal and data are sparse. Leafhoppers and flea beetles may favor fertilized pasture and turfgrass (9, 141). Frequent mowing did not affect leafhopper populations, but adult frit flies displayed short-term attraction to recently mowed plots (44). Use of a heavy roller on pastures in New Zealand gave more than 60% control of scarabaeid grubs, with negligible mortality of earthworms (168). Rainfall and irrigation patterns affect the distribution and abundance of grubs on golf courses and home lawns (59, 139, 143).

Work conducted in Ohio in the 1950s suggested that Japanese beetle oviposition preference and grub survival were adversely affected by high soil pH (132), but, in more-recent studies, manipulation of soil pH had little effect on Japanese beetles or European chafer grubs (200, 202). Because most turfgrasses grow best at pH 6–7 (14), manipulation of pH outside of this range would probably be impractical even if it did affect insect populations.

Host Plant Resistance

Laboratory and field screening has identified turfgrass genotypes that are relatively resistant, tolerant, or less preferred by particular insects or mites (for review, see 148). Notably, little work has focused on host resistance to mole crickets and almost none on scarabaeid grubs. Variation in cultivar damage ratings between trials and locations (e.g. 80) suggests that expression of resistance may be modified by different management regimes or by environmental conditions.

Turfgrass breeders have rarely attempted to combine genetic insect resistance with other desirable traits. Genotypes that have been released as new cultivars include Floratam and Floralawn St. Augustinegrass, which are resistant to the southern chinch bug (39, 149), and Bell rhodesgrass, resistant to rhodesgrass mealybug (148). Recently, some southern chinch bug populations in Florida have caused extensive damage to previously resistant Floratam and Floralawn (20), underscoring the potential for genetically variable pest populations to overcome host resistance.

The endophytic fungi, *Acremonium lolii* and *Acremonium coenophialum*, form mutualistic symbioses with perennial ryegrass and tall fescue, respectively (for reviews, see 34, 163, 164). Both endophytes are carried as intracellular hyphae in infected plants and are transmitted by seed via the maternal parent. Endophytes are not presently known from Kentucky bluegrass. The endophytes are associated with production of neurotoxins that cause toxicoses in livestock that graze on infected pastures (164), and efforts to remove the fungi to improve forage quality first revealed their importance in host defense.

Endophyte-enhanced resistance has been demonstrated for perennial ryegrass, tall fescue, and hard and Chewings fescues; cultivars are being marketed with high endophyte levels that enhance resistance to webworms, billbugs, chinch bugs, and other pests (57, 163). Endophyte-associated alkaloids are concentrated in stems, leaves, and seeds (163), but low levels in the roots may also deter some feeding by white grubs (D. A. Potter, unpublished data). Viability of *A. lolii* and *A. coenophialum* mycelium in seed declines during storage at ambient temperature, so seed of infected grasses will require refrigerated storage and certification to guarantee endophyte viability (164). Techniques now exist for transferring endophytic fungi between host grasses and to grasses not known to be hosts via inoculation (99) or by plant breeding (57). Augmentation of endophyte levels in grass species or cultivars with other desirable characteristics may provide a broad-based mechanism for developing new turfgrasses with multiple resistance to insects.

Prospects for Integrated Pest Management

Implementation of IPM for turfgrass faces many of the same hurdles that have been identified for other urban settings (15). Only a small number of specific, alternative control methods are available for turfgrass insects, and these often are slower acting, less reliable, or more difficult to use than conventional insecticides. From the industry perspective, aesthetic standards for lawns and golf courses leave little margin for error. Homeowners with little agronomic experience may be unable or unwilling to translate IPM information into appropriate action. Progress in IPM implementation will depend as much upon modification of public attitudes and expectations as on research advances and new technology. Indeed, it would be naive to assume that the public is ready to accept lower standards for lawns and sports turf in exchange for reduced pesticide usage and IPM programs developed by entomologists (22).

A small number of pilot IPM programs for turfgrass has been developed and implemented under special circumstances and on a limited scale (62, 71, 162). Acceptance of comprehensive IPM will probably be fastest among golf superintendents and other skilled landscape managers, many of whom already monitor their landscapes and use cultural tactics to minimize pest problems. The highly competitive lawn-care business poses special challenges because the time required to monitor and sample individual lawns may be prohibitive and because clients are perceived as more likely to cancel service if told that treatments are unnecessary. Traditionally, it has been simpler, more reliable, and more profitable for companies to apply routine, scheduled treatments than to hire or train qualified IPM consultants to engage in sampling, monitoring, and decision-making.

Nevertheless, growing public concern about potential environmental and human health risks of pesticides, and associated political and legal issues surrounding urban pesticide usage will almost certainly mandate much more limited and selective pesticide use on turf. Such concerns are creating a new market of pesticide-conscious consumers, and unprecedented incentives for the turfgrass industry to explore pest-control alternatives (58, 215).

Existing information on the ecology of turfgrass insects is meager. To target insecticides more efficiently, we must better understand how environmental factors affect pest populations. We presently know little about why particular sites are attractive to insects, or even how such routine practices as watering or fertilization affect insect abundance and damage. More work is needed on genetic and endophyte-enhanced host resistance, especially for key pests such as white grubs, chinch bugs, billbugs, and mole crickets. Simpler monitoring and sampling methods and better decision-making guidelines are needed to support the transition toward IPM. Relationships between the more easily sampled adult stages of holometabolous insects and subsequent larval densities would be especially useful.

Extramural funding for turfgrass entomology research has been and continues to be largely in the form of grants-in-aid from the agrichemical industry. Not surprisingly, insecticide-related topics have dominated the literature. Still more work is needed on the factors that limit insecticide performance, on new application methods (e.g. high pressure soil injection), on pesticide movement into soil and ground water, and on compatibility of pesticides with the beneficial fauna. Nematodes, microbial agents, and insect growth regulators will play a bigger role in IPM, but the factors that presently limit their performance must be better understood. Conventional insecticides will no doubt remain essential components of turfgrass IPM for the foreseeable future. However, there must be much greater research emphasis on the biology and ecology of turfgrass insects to reduce reliance on insecticides, and to increase the efficacy of those applications that are necessary.

CONCLUSIONS

By the year 2025, more than 85% of the North American population will reside in urban areas (195). Public perceptions of entomology will be governed more and more by our capability to provide safe, effective solutions to everyday pest problems. Demand for quality turfgrass with less usage of pesticides is providing strong motivation for the turfgrass industry to explore IPM and putting greater demands on the entomological profession to support this transition. Present meager levels of personnel and funding for turfgrass entomology must be expanded if these challenges are to be met.

ACKNOWLEDGMENTS

We are grateful to P. P. Cobb, K. F. Haynes, W. G. Hudson, D. J. Shetlar, and K. V. Yeargan for their comments on drafts of this review, and to our colleagues who provided reprints, unpublished manuscripts, and suggestions. We especially thank Gwyn Ison for her help in preparing the manuscript. This contribution (no. 90-7-10) is in connection with a project of the Kentucky Agricultural Experiment Station and is published with the approval of the Director.

Literature Cited

1. Adams, J. A. 1949. The Oriental beetle as a turf pest associated with the Japanese beetle in New York. *J. Econ. Entomol.* 42:366–71
2. Ahmad, S., Ng, Y. S. 1981. Further evidence for chlorpyrifos tolerance and partial resistance by the Japanese beetle (Coleoptera: Scarabaeidae). *J. N. Y. Entomol. Soc.* 89:34–39
3. Ainslee, G. G. 1923. The crambinae of Florida. *Fla. Entomol.* 6:49–55
4. Ainslee, G. G. 1923. Striped sod webworm, *Crambus mutabilis* Clemens. *J. Agric. Res.* 24:399–414
5. Ainslee, G. G. 1923. Silver-striped webworm, *Crambus praefectellus* Zinchen. *J. Agric. Res.* 24:415–26
6. Ainslee, G. G. 1927. The larger sod webworm. *USDA Tech. Bull. No.* 31. 17 pp.
7. Ainslee, G. G. 1930. The bluegrass webworm. *USDA Tech. Bull. No.* 173. 25 pp.
8. Aldrich, J. M. 1920. European frit fly in North America. *J. Agric. Res.* 18:451–73
9. Arnold, T. B., Potter, D. A. 1987. Impact of a high-maintenance lawn-care program on nontarget invertebrates in Kentucky bluegrass turf. *Environ. Entomol.* 16:100–5
10. Baker, P. B. 1986. Responses by Japanese and oriental beetle grubs (Coleoptera: Scarabaeidae) to bendiocarb, chlorpyrifos, and isophenphos. *J. Econ. Entomol.* 79:452–54
11. Baker, P. B., Ratcliffe, R. H., Steinhauer, A. L. 1981. Laboratory rearing of the hairy chinch bug. *Environ. Entomol.* 10:226–29
12. Banerjee, A. C. 1969. Sex attractants in sod webworms. *J. Econ. Entomol.* 62:705–8
13. Banerjee, A. C., Decker, G. C. 1966. Studies on sod webworms. I. Emergence rhythm, mating, and oviposition behavior under natural conditions. *J. Econ. Entomol.* 59:1237–44
14. Beard, J. B. 1973. *Turfgrass: Science and Culture.* Englewood Cliffs, NJ: Prentice-Hall. 658 pp.
15. Bennett, G. W., Owens, J. M., ed. 1986. *Advances in Urban Pest Management.* New York: Van Nostrand Reinhold. 399 pp.
16. Bohart, R. M. 1947. Sod webworms and other lawn pests in California. *Hilgardia* 17:267–307
17. Branham, B. E., Wehner, D. J. 1985. The fate of diazinon applied to thatched turf. *Agron. J.* 77:101–4
18. Brewer, L. W., Driver, C. J., Kendall, R. J., Lacher, T. E. Jr., Galindo, J. C., et al. 1988. Avian response to a turf application of Triumph 4E. *Environ. Toxicol. Chem.* 7:391–401
19. Bulla, L. A., Costilow, R. N., Sharpe, E. S. 1978. Biology of *Bacillus popilliae. Adv. Appl. Microbiol.* 23:1–18
20. Busey, P., Center, B. J. 1987. Southern chinch bug (Hemiptera: Heteroptera: Lygaeidae) overcomes resistance in St. Augustinegrass. *J. Econ. Entomol.* 80:608–11
21. Buxton, M. C., Potter, D. A. 1989. Pesticide effects on earthworm populations and thatch breakdown in Kentucky bluegrass turf. *Ky. Turfgrass Res. Prog. Rep.* 319:37–39
22. Byrne, D. N., Carpenter, E. H. 1986. Attitudes and actions of urbanites in managing household arthropods. See Ref. 15, pp. 13–24
23. Chada, H. L. 1956. Biology of the winter grain mite and its control in small grains. *J. Econ. Entomol.* 49:515–20
24. Chada, H. L., Wood, E. A. 1960. Biology and control of the rhodesgrass scale. *USDA Tech. Bull. No.* 1221. 21 pp.
25. Chang, N. T., Wiseman, B. R., Lynch, R. E., Habeck, D. H. 1986. Growth and development of fall armyworm (Lepidoptera: Noctuidae) on selected grasses. *Environ. Entomol.* 15:182–89
26. Chittenden, F. H., Fink, D. E. 1922. The green June beetle. *USDA Agric. Bull. No.* 891. 52 pp.
27. Clark, J. D., Haynes, K. F. 1990. Sex attractant for the bluegrass webworm (Lepidoptera: Pyralidae). *J. Econ. Entomol.* 83:856–59
28. Cobb, P. P., Mack, T. P. 1989. A rating system for evaluating tawny mole cricket, *Scapteriscus vicinus* Scudder, damage (Orthoptera: Gryllotalpidae). *J. Entomol. Sci.* 24:142–44
29. Cockfield, S. D., Potter, D. A. 1983. Short-term effects of insecticidal applications on predaceous arthropods and oribatid mites in Kentucky bluegrass turf. *Environ. Entomol.* 12:1260–64
30. Cockfield, S. D., Potter, D. A. 1984. Predation on sod webworm (Lepidoptera: Pyralidae) eggs as affected by chlorpyrifos application to Kentucky bluegrass turf. *J. Econ. Entomol.* 77:1542–44
31. Cockfield, S. D., Potter, D. A. 1985. Predatory arthropods in high- and low-maintenance turfgrass. *Can. Entomol.* 117:423–29

32. Crawford, C. S. 1967. Oviposition rhythm studies in *Crambus topiarius* (Lepidoptera: Pyralidae: Crambinae). *Ann. Entomol. Soc. Am.* 60:1014–18

33. Czechowski, W. 1980. Sampling of Carabidae (Coleoptera) by Barber's traps and biocenometric method in urban environment. *Ser. Sci. Biol. Cl. II.* 27: 461–65

34. Dahlman, D. L., Eichenseer, H., Siegel, M. R. 1990. Chemical perspectives on endophyte-grass interactions and their implications to insect herbivory. In *Multi-trophic Level Interactions Among Microorganisms, Plants, and Insects,* ed. P. Barbarosa, V. Krischik, C. G. Jones. New York: Wiley. In press

35. Davidson, R. L., Wiseman, J. R., Wolfe, V. J. 1972. Environmental stress in the pasture scarab *Sericesthis nigrolineata* Biosd. II. Effects of soil moisture and temperature on survival of first instar larvae. *J. Appl. Ecol.* 9:799–806

36. Dean, H. A., Schuster, M. F., Boling, J. C., Riherd, P. T. 1979. Complete biological control of *Antonina graminis* in Texas with *Neodusmetia sangwani* (a classic example). *Bull. Entomol. Soc. Am.* 25:262–67

37. Denmark, H. A. 1964. The bermudagrass stunt mite. *Fla. Dept. Agric. Div. Plant Ind. Entomol. Circ.* No. 11. 1 p.

38. Domek, J. M., Johnson, D. T. 1988. Demonstration of semiochemically induced aggregation in the green June beetle, *Cotinus nitida* (L.) (Coleoptera: Scarabaeidae). *Environ. Entomol.* 17: 147–49

39. Dudeck, A. E., Reinert, J. A., Busey, P. 1986. Floralawn St. Agustinegrass. *Fla. Agric. Exp. Sta. Circ.* S-327

40. Dunbar, D. M., Beard, R. L. 1975. Present status of milky disease of Japanese beetle and Oriental beetles in Connecticut. *J. Econ. Entomol.* 68:453–57

41. Dutky, S. R. 1963. The milky diseases. In *Insect Pathology, an Advanced Treatise,* ed. E. A. Steinhaus, pp. 75–115. London: Academic. 861 pp.

42. Endrodi, S. 1985. *The Dynastinae of the World*. Akademiai Kiado, Budapest: Junk. 800 pp.

43. Fagan, E. B., Kuitert, L. C. 1969. Biology of the two-lined spittlebug, *Prosapia bicincta,* on Florida pastures (Homoptera: Cercopidae). *Fla. Entomol.* 52:199–206

44. Falk, J. H. 1982. Response of two turf insects, *Endria inimica* and *Oscinella frit* to mowing. *Environ. Entomol.* 11: 29–31

45. Felsot, A. S. 1989. Enhanced biodegradation of insecticides in soil. *Annu. Rev. Entomol.* 34:453–76

46. Fiori, B. J. 1983. Sticky plastic sheet traps as survey tools for the European chafer (Coleoptera: Scarabaeidae). *Can. Entomol.* 115:1429–31

47. Flanders, R. V. 1986. Potential for biological control in urban environments. See Ref. 15, pp. 95–127

48. Fleming, W. E. 1968. Biological control of the Japanese beetle. *USDA Tech. Bull.* No. 1383. 78 pp.

49. Fleming, W. E. 1972. Biology of the Japanese beetle. *USDA Tech. Bull.* No. 1449. 129 pp.

50. Fleming, W. E. 1976. Integrating control of the Japanese beetle—A historical review. *USDA Tech. Bull.* No. 1545. 65 pp.

51. Forrest, T. G. 1981. Acoustic communication and baffling behaviors of crickets. *Fla. Entomol.* 65:33–44

52. Forrest, T. G. 1983. Phonotaxis and calling in Puerto Rican mole crickets (Orthoptera: Gryllotalpidae). *Ann. Entomol. Soc. Am.* 76:797–99

53. Forrest, T. G. 1986. Oviposition and maternal investment in mole crickets (Orthoptera: Gryllotalpidae): effects of season, size, and senescence. *Ann. Entomol. Soc. Am.* 79:918–24

54. Fowler, H. G. 1987. Field behavior of *Euphasiopteryx depleta* (Diptera: Tachinidae) phonotactically orienting parasitoids of mole crickets (Orthoptera: Gryllotalpidae: Scapteriscus). *J. NY Entomol. Soc.* 95:474–80

55. Frankie, G. W., Ehler, L. E. 1978. Ecology of insects in urban environments. *Annu. Rev. Entomol.* 23:367–87

56. Friend, R. B. 1929. The Asiatic beetle in Connecticut. *Conn. Agr. Expt. Sta. Bull.* 304:585–664

57. Funk, C. R., Clarke, B. B., Johnson-Cicalese, J. M. 1989. Role of endophytes in enhancing the performance of grasses used for conservation and turf. See Ref. 102, pp. 203–10

58. Funk, R. C. 1989. Lawn service industry: transition in services. See Ref. 102, pp. 97–105

59. Gaylor, M. J., Frankie, G. W. 1979. The relationship of rainfall to adult flight activity; and of soil moisture to oviposition behavior and egg and first instar survival in *Phyllophaga crinita*. *Environ. Entomol.* 8:591–94

60. Georgis, R., Poinar, G. O. 1989. Field effectiveness of entomophilic nematodes Neoaplectana and Heterorhabditis. See Ref. 102, pp. 213–24

61. Graber, L. F., Fluke, C. L., Dexter, S.

T. 1931. Insect injury of bluegrass in relation to the environment. *Ecology* 12:547–66

62. Grant, M. D. 1989. Integrated pest management in the golf course industry: A case study and some general considerations. See Ref. 102, pp. 85–92

63. Gyrisco, G. G., Whitcomb, W. H., Burrage, R. H., Logothetis, C., Schwardt, H. H. 1954. Biology of European chafer, *Amphimallon majalis* Razoumowsky (Scarabaeidae). *Cornell Univ. Agr. Exp. Sta. Mem.* No. 328. 35 pp.

64. Hallock, H. C., Hawley, I. M. 1936. Life history and control of the Asiatic garden beetle. *USDA Circ.* No. 246. 20 pp.

65. Hansen, J. D. 1987. Seasonal history of bluegrass billbug, *Sphenophorus parvulus* (Coleoptera: Curculionidae), in a range grass nursery. *Environ. Entomol.* 16:752–56

66. Hanula, J. L., Andrealis, T. G. 1988. Parasitic microorganisms of Japanese beetle (Coleoptera: Scarabaeidae) and associated scarab larvae in Connecticut soils. *Environ. Entomol.* 17:709–14

67. Harris, C. R. 1982. Factors influencing the toxicity of insecticides in soil. See Ref. 124, pp. 47–52

68. Hayes, W. P. 1925. A comparative study of the history of certain phytophagous scarabaeid beetles. *Kans. Agr. Exp. Sta. Tech. Bull.* 16. 145 pp.

69. Hayslip, N. C. 1943. Notes on biological studies of mole crickets at Plant City Florida. *Fla. Entomol.* 26:33–46

70. Heinricks, E. A., Matheny, E. L. 1969. Hatching of sod webworm eggs in relation to low temperatures. *J. Econ. Entomol.* 62:1344–47

71. Hellman, J. L., Davidson, J. A., Holmes, J. 1982. Urban ornamental and integrated pest management in Maryland. See Ref. 124, pp. 31–38

72. Hill, A. S., Rings, R. W., Swier, S. R., Roelofs, W. L. 1979. Sex pheromone of the black cutworm moth, *Agrotis ipsilon*. *J. Chem. Ecol.* 5:439–57

73. Hitchcock, A. S. 1931. *Old and New Plant Lore*, Vol. II. Washington, DC: Smithsonian Institute. 249 pp.

74. Horton, J. R., Satterthwait, A. F. 1922. The chinch bug and its control. *USDA Farmers Bull.* No. 1223

75. Hudson, W. G., Cromroy, H. L. 1985. Radioisotope labelling of mole crickets. *Fla. Entomol.* 68:349–50

76. Hudson, W. G., Frank, J. H., Castner, J. L. 1988. Biological control of *Scapteriscus* spp. mole crickets (Orthoptera: Gryllotalpidae) in Florida. *Bull. Entomol. Soc. Am.* 34:192–98

77. Hudson, W. G., Saw, J. G. 1987. Spatial distribution of the tawny mole cricket, *Scapteriscus vicinus*. *Entomol. Exp. Appl.* 45:99–102

78. Huffine, W. W., Grau, F. V. 1969. History of turfgrass usage. In *Turfgrass Science*, ed. A. A. Hanson, F. V. Juska, pp. 1–8. Madison, WI: Am. Soc. Agron. 715 pp.

79. Hurlbert, S. H., Mulla, M. S., Keith, J. O., Westlake, W. E., Dusch, M. E. 1970. Biological effects and persistence of Dursban in freshwater ponds. *J. Econ. Entomol.* 63:43–52

80. Johnson-Cicalese, J. M., Hurley, R. H., Wolfe, G. W., Funk, C. R. 1989. Developing turfgrasses with improved resistance to billbugs. *Proc. 6th Internat. Turfgrass Res. Conf.*, pp. 107–11. Tokyo, Japan: Int. Turfgrass Soc. & Jpn. Soc. Turfgrass Sci.

81. Johnson-Cicalese, J. M., Wolfe, G. W., Funk, C. R. 1990. Biology, distribution, and taxonomy of billbug turf pests (Coleoptera: Curculionidae). *Environ. Entomol.* 19:1037–46

82. Kageyama, M. E. 1982. Industry's contribution to turfgrass entomology. See Ref. 124, pp. 133–38

83. Kamm, J. A. 1973. Biotic factors that affect sod webworms in grass fields in Oregon. *Environ. Entomol.* 2:94–96

84. Kamm, J. A. 1982. Use of insect sex pheromones in turfgrass management. See Ref. 124, pp. 39–41

85. Kausnitzer, B., Richter, K., Koeberlein, C., Koeberlein, K. 1980. Faunistic studies of soil arthropods of two Leipzig, East Germany, city parks with particular reference to Carabidae and Staphylinidae. *Wiss. Z. Karl-Marx-Univ. Leipzig Math. Naturwiss. Reihe.* 29:583–97

86. Kelsheimer, E. G. 1956. The hunting billbug, a serious pest of zoysia. *Proc. Fla. Hort. Soc.* 69:415–18

87. Kerr, S. H. 1955. Life history of the tropical sod webworm *Pachyzancla phaeopteralis* Guenee. *Fla. Entomol.* 38:3–11

88. Kerr, S. H. 1966. Biology of the lawn chinch bug *Blissus insularis*. *Fla. Entomol.* 49:9–18

89. Kindler, S. D., Spomer, S. M. 1986. Observations on the biology of the bluegrass billbug, *Sphenophorus parvulus* Gyllenhal (Coleoptera: Curculionidae) in an eastern Nebraska sod field. *J. Kans. Entomol. Soc.* 59:26–31

90. Klein, M. G. 1981. Advances in the use of *Bacillus popilliae* for pest control. In *Microbial Control of Pests and Plant Diseases 1970–1980*, ed. H. D. Burges. pp. 183–92. London: Academic. 949 pp.

91. Klein, M. G. 1981. Mass trapping for suppression of Japanese beetles. In *Management of Insect Pests With Semiochemicals*, ed. E. R. Mitchell, pp. 183–90. New York: Plenum

92. Klein, M. G. 1982. Biological suppression of turf insects. See Ref. 124, pp. 91–97

93. Klein, M. G., Johnson, C. H., Ladd, T. L. Jr. 1976. A bibliography of the milky disease bacteria (*Bacillus* spp.) associated with the Japanese beetle, *Popillia japonica* and closely related Scarabaeidae. *Bull. Entomol. Soc. Am.* 22:305–10

94. Kouskolekas, C. A., Self, R. L. 1974. Biology and control of the ground pearl in relation to turfgrass infestation. In *Proc. 2nd Int. Turfgrass Res. Conf. Amer. Soc. Agron.*, ed. E. C. Roberts, pp. 421–23. Madison, WI: Am Soc. Agron.

95. Kuhr, R. J., Tashiro, H. 1978. Distribution and persistence of chlorpyrifos and diazinon applied to turf. *Bull. Environ. Contam. Toxicol.* 20:652–56

96. Ladd, T. L., Burriff, C. R. 1979. Japanese beetle: influence of larval feeding on bluegrass yields at two levels of soil moisture. *J. Econ. Entomol.* 72:311–14

97. Ladd, T. L., McCabe, P. J. 1967. Persistence of spore of *Bacillus popilliae*, the causal agent of type A milky disease of Japanese beetle larvae. *J. Econ. Entomol.* 60:493–95

98. Ladd, T. L., McGovern, T. P. 1980. Japanese beetle: a superior attractant, phenethyl propionate + eugenol + geraniol, 3:7:3. *J. Econ. Entomol.* 73:689–91

99. Latch, G. C. M., Christensen, M. J. 1985. Artificial infection of grasses with endophytes. *Ann. Appl. Biol.* 107:17–24

100. Leonard, D. E. 1966. Biosystematics of the "*Leucopterus* complex" of the genus *Blissus* (Heteroptera: Lygaeidae). *Conn. Agric. Exp. Sta. Bull.* No. 677

101. Leonard, D. E. 1968. A revision of the genus *Blissus* (Heteroptera: Lygaeidae) in eastern North America. *Ann. Entomol. Soc. Am.* 61:239–50

102. Leslie, A. R., Metcalf, R. L., eds. 1989. *Integrated Pest Management for Turfgrass and Ornamentals*. Washington, DC: US EPA. 337 pp.

103. Liu, H. J., McEwen, F. L. 1979. The use of temperature accumulations and sequential sampling in predicting damaging populations of *Blissus leucopterus hirtus*. *Environ. Entomol.* 8:512–15

104. Lowengart, R. A., Peters, J. M.,

Cicioni, C., Buckley, J., Berstein, L., et al. 1987. Childhood leukemia and parent's occupational and home exposures. *J. Natl. Cancer Inst.* 79:39–46

105. Luginbill, P. 1928. The fall armyworm. *USDA Tech. Bull.* No. 34

106. Luginbill, P., Painter, H. R. 1953. May beetles of the United States and Canada. *USDA Tech. Bull.* No. 1060. 102 pp.

107. Mailloux, G., Streu, H. T. 1979. A sampling technique for estimating hairy chinch bug (*Blissus leucopterus hirtus* Montandon, Hemiptera: Lygaeidae) populations and other arthropods from turfgrass. *Ann. Entomol. Soc. Quebec* 24:139–43

108. Mailloux, G., Streu, H. T. 1981. Population biology of the hairy chinch bug (*Blissus leucopterus hirtus* Montandon, Hemiptera: Lygaeidae). *Ann. Entomol. Soc. Quebec* 26:51–90

109. Mailloux, G., Streu, H. T. 1982. Spatial distribution of hairy chinch bug (*Blissus leucopterus hirtus* Montandon: Hemiptera: Lygaeidae) populations in turfgrass. *Ann. Entomol. Soc. Quebec* 27:111–31

110. Maras, E. 1984. Lawn care receipts vault to more than $2 billion. *Lawn Care Ind.* 8(6):1,8

111. Matheny, E. L., Heinrichs, E. A. 1972. Chorion characteristics of sod webworm eggs. *Ann. Entomol. Soc. Am.* 65:238–47

112. Matheny, E. L., Heinrichs, E. A. 1975. Seasonal abundance and geographical distribution of sod webworms (Lepidoptera: Pyralidae: Crambinae) in Tennessee. *J. Tenn. Acad. Sci.* 50:33–35

113. Matheny, E. L. Jr., Kepner, R. L., Portier, K. M. 1983. Landing distribution and density of two sound-attracted mole crickets (Orthoptera: Gryllotalpidae: *Scapteriscus*). *Ann. Entomol. Soc. Am.* 76:278–81

114. Maxwell, K. E., MacLeod, G. F. 1936. Experimental studies of the hairy chinch bug. *J. Econ. Entomol.* 29:339–43

115. Morrison, W. P., Pass, B. C., Crawford, C. S. 1972. Effect of humidity on eggs of two populations of the bluegrass webworm. *Environ. Entomol.* 1:218–21

116. Ng, Y. S., Trout, J. R., Ahmad, S. 1983. Sequential sampling plans for larval populations of the Japanese beetle (Coleoptera: Scarabaeidae) in turfgrass. *J. Econ. Entomol.* 76:251–53

117. Ng, Y. S., Trout, J. R., Ahmad, S. 1983. Spatial distribution of the larval populations of the Japanese beetle (Coleoptera: Scarabaeidae) in turfgrass. *J. Econ. Entomol.* 76:479–83

118. Nickerson, J. C., Snyder, D. E., Oliver,

C. C. 1979. Acoustical burrows constructed by mole crickets. *Ann. Entomol. Soc. Am.* 72:438–40

119. Nickle, D. A., Castner, J. L. 1984. Introduced species of mole crickets in the United States, Puerto Rico, and the Virgin Islands (Orthoptera: Gryllotalpidae). *Ann. Entomol. Soc. Am.* 77:450–65

120. Niemczyk, H. D. 1981. *Destructive Turf Insects*. Fostoria, OH: Gray Printing. 48 pp.

121. Niemczyk, H. D. 1982. The status of USDA-SEA-AR and U.S. university input of professional personnel to turfgrass entomology—1980. See Ref. 124, pp. 127–132

122. Niemczyk, H. D. 1987. The influence of application timing and posttreatment irrigation on the fate and effectiveness of isofenphos for control of Japanese beetle (Coleoptera: Scarabaeidae) larvae in turfgrass. *J. Econ. Entomol.* 80:465–70

123. Niemczyk, H. D., Chapman, R. A. 1987. Evidence of enhanced degradation of isofenphos in turfgrass thatch and soil. *J. Econ. Entomol.* 80:880–82

124. Niemczyk, H. D., Joyner, B. G., ed. 1982. *Advances in Turfgrass Entomology*. Piqua, OH: Hammer Graphics. 150 pp.

125. Niemczyk, H. D., Krueger, H. R. 1982. Binding of insecticides on turfgrass thatch. See Ref. 124, pp. 61–63

126. Niemczyk, H. D., Krueger, H. R. 1987. Persistence and mobility of isazofos in turfgrass thatch and soil. *J. Econ. Entomol.* 80:950–52

127. Niemczyk, H. D., Moser, J. R. 1982. Greenbug occurrence and control on turfgrasses in Ohio. See Ref. 124, pp. 105–11

128. Olkowski, W., Olkowski, H., Kaplan, A. I., Van den Bosch, R. 1978. The potential for biological control in urban areas: shade tree pests. In *Perspectives in Urban Entomology*, ed. G. W. Frankie, C. S. Koehler. pp. 311–47. New York: Academic

129. Parsons, D. W., Witt, J. M. 1988. *Pesticides in Groundwater in the United States of America*. Corvallis, OR: Oreg. State Univ. Ext. Serv.

130. Pass, B. C., Reed, J. K. 1965. Biology and control of the spittlebug *Prosapia bicincta* in coastal bermudagrass. *J. Econ. Entomol.* 58:275–78

131. Pimentel, D. 1971. *Ecological Effects of Pesticides on Nontarget Species*. Washington, DC: Office of Science and Technology

132. Polivka, J. B. 1960. Grub population in turf varies with pH levels in Ohio soils. *J. Econ. Entomol.* 53:860–63

133. Potter, D. A. 1980. Flight activity and sex attraction of northern and southern masked chafers in Kentucky turfgrass. *Ann. Entomol. Soc. Am.* 73:414–17

134. Potter, D. A. 1981. Seasonal emergence and flight of northern and southern masked chafers in relation to air and soil temperature and rainfall patterns. *Environ. Entomol.* 10:793–97

135. Potter, D. A. 1982. Influence of feeding by grubs of the southern masked chafer on quality and yield of Kentucky bluegrass. *J. Econ. Entomol.* 75:21–24

136. Potter, D. A. 1986. Urban landscape pest management. See Ref. 15, pp. 219–52

137. Potter, D. A., Bridges, B. L., Gordon, F. C. 1985. Effect of N fertilization on earthworm and microarthropod populations in Kentucky bluegrass turf. *Agron. J.* 77:367–72

137a. Potter, D. A., Buxton, M. C., Redmond, C. T., Patterson, C. G., Powell, A. J. 1990. Toxicity of pesticides to earthworms (Oligochaeta: Lumbricidae) and effect on thatch degradation in Kentucky bluegrass turf. *J. Econ. Entomol.* 83: In press

138. Potter, D. A., Cockfield, S. D., Morris, T. A. 1989. Ecological side effects of pesticide and fertilizer use on turfgrass. See Ref. 102, pp. 33–44

139. Potter, D. A., Gordon, F. C. 1984. Susceptibility of *Cyclocephala immaculata* (Coleoptera: Scarabaeidae) eggs and immatures to heat and drought in turf grass. *Environ. Entomol.* 13:794–99

140. Potter, D. A., Powell, A. J., Smith, M. S. 1989. Decomposition of turfgrass thatch by earthworms and other soil invertebrates. *J. Econ. Entomol.* 83:205–11

141. Prestidge, R. A. 1982. The influence of nitrogen fertilizer on the grassland Auchenorrhyncha (Homoptera). *J. Appl. Ecol.* 19:735–49

142. Randell, R. J., Butler, J. D., Hughes, T. D. 1972. The effect of pesticides on thatch accumulation and earthworm populations in Kentucky bluegrass turf. *HortScience* 7:64–65

143. Regniere, J. R., Rabb, R. L., Stinner, R. E. 1981. *Popillia japonica* (Coleoptera: Scarabaeidae): distribution and movement of adults in heterogeneous environments. *Can. Entomol.* 115:287–94

144. Regniere, J., Rabb, R. L., Stinner, R. E. 1981. *Popillia japonica*: Stimulation of temperature-dependent development of the immatures, and prediction of adult emergence. *Environ. Entomol.* 10:290–96

145. Regniere, J., Rabb, R. L., Stinner, R. E. 1981. *Popillia japonica:* Effect of soil moisture and texture on survival and development of eggs and first instar grubs. *Environ. Entomol.* 10:654–60
146. Reinert, J. A. 1978. Natural enemy complex of the southern chinch bug in Florida. *Ann. Entomol. Soc. Am.* 71: 728–31
147. Reinert, J. A. 1982. Insecticide resistance in epigeal insect pests of turfgrass: 1. A review. See Ref. 124, pp. 71–76
148. Reinert, J. A. 1982. A review of host resistance in turfgrasses to insects and acarines with emphasis on the southern chinch bug. See Ref. 124, pp. 3–12
149. Reinert, J. A., Dudeck, A. E. 1974. Southern chinch bug resistance in St. Augustinegrass. *J. Econ. Entomol.* 67: 275–77
150. Reinert, J. A., Kerr, S. H. 1973. Bionomics and control of lawn chinch bugs. *Bull. Entomol. Soc. Am.* 19:91–92
151. Reinert, J. A., Portier, K. M. 1983. Distribution and characterization of organophosphate-resistant southern chinch bugs (Heteroptera: Lygaeidae) in Florida. *J. Econ. Entomol.* 76:1187–90
152. Reuter Laboratories, Inc. 1989. In vitro method for producing infective bacterial spores and spore-containing insecticidal compositions. *US Patent No. 4,824,671*
153. Rings, R. W., Arnold, F. J., Keaster, A. J., Musick, G. J. 1974. A worldwide, annotated bibliography of the black cutworm, *Agrotis ipsilon. Ohio Agric. Res. Dev. Cent. Res. Circ.* No. 198
154. Ritcher, P. O. 1940. Kentucky white grubs. *Ky. Agr. Exp. Sta. Bull.* 401:71–157
155. Ritcher, P. O. 1966. *White Grubs and Their Allies: a Study of North American Scarabaeid Larvae.* Corvallis: Oregon State Univ. Press. 219 pp.
156. Roberts, E. C., Roberts, B. C. 1987. *Lawn and Sports Turf Benefits.* Pleasant Hill, TN: The Lawn Institute. 31 pp.
157. St. Julian, G., Bulla, L. A. 1973. Milky disease. In *Current Topics in Comparative Pathobiology,* Vol. 2, ed. T. C. Cheng, pp. 57–85. New York: Academic. 334 pp.
158. Sears, M. K., Chapman, R. A. 1979. Persistence and movement of four insecticides applied to turfgrass. *J. Econ. Entomol.* 72:272–74
159. Shetlar, D. J. 1989. Entomogenous nematodes for control of turfgrass insects with notes on other biological control agents. See Ref. 102, pp. 223–53
160. Shetlar, D. J., Suleman, P. E., Georgis, R. 1988. Irrigation and use of entomogenous nematodes, *Neoplectana* spp. and *Heterorhabditis helipothidis* (Rhabditida: Steinernematidae and Heterorhabditidae), for control of Japanese beetle (Coleoptera: Scarabaeidae) grubs in turfgrass. *J. Econ. Entomol.* 81:1318–22
161. Shorey, H. H., Burrage, R. H., Gyrisco, G. G. 1960. The relationship between environmental factors and the density of the European chafer *(Amphimallon majalis)* larvae in permanent pasture sod. *Ecology* 41:253–58
162. Short, D. E., Reinert, J. A., Atilano, R. A. 1982. Integrated pest management for urban turfgrass culture—Florida. See Ref. 124, pp. 25–30
163. Siegel, M. R., Dahlman, D. L., Bush, L. P. 1989. The role of endophytic fungi in grasses: new approaches to biological control of pests. See Ref. 102, pp. 169–79
164. Siegel, M. R., Latch, G. C. M., Johnson, M. C. 1987. Fungal endophytes of grasses. *Annu. Rev. Phytopathol.* 25: 293–315
165. Sparks, A. N. 1979. A review of the biology of the fall armyworm. *Fla. Entomol.* 62:82–86
166. Splittstoesser, C. M., Tashiro, H. 1977. Three milky disease bacilli from a scarabaeid, *Ataenius spretulus. J. Invert. Pathol.* 30:436–38
167. Starkey, D. G. 1979. Trees and their relationship to mental health. *J. Arboric.* 5:153–54
168. Stewart, K. M., Van Toor, R. 1983. Control of grass grub [*Costelytra zealandica* (White)] by heavy rolling. *N. Z. J. Exper. Agric.* 11:265–70
169. Stinmann, M. W., Taliaferro, C. M. 1969. Resistance of selected accessions of bermudagrass to phytotoxemia caused by adult twolined spittlebugs. *J. Econ. Entomol.* 62:1189–90
170. Streu, H. T. 1969. Some cumulative effects of pesticides in the turfgrass ecosystem. In *Proc. Scotts Turfgrass Res. Conf. I. Entomology,* pp. 53–59. Marysville, OH: Scott & Sons
171. Streu, H. T. 1973. The turfgrass ecosystem: impact of pesticides. *Bull. Entomol. Soc. Am.* 19:89–91
172. Streu, H. T., Gingrich, J. B. 1972. Seasonal activity of the winter grain mite in turfgrass in New Jersey. *J. Econ. Entomol.* 65:427–30
173. Streu, H. T., Niemczyk, H. D. 1982. Pest status and control of winter grain mite. See Ref. 124, pp. 101–4
174. Tashiro, H. 1976. Biology of the grass webworm, *Herpetogramma licarsisalis*

(Lepidoptera: Pyraustidae) in Hawaii. *Ann. Entomol. Soc. Am.* 69:797–803

175. Tashiro, H. 1981. Limitations of organophosphate soil insecticides on turfgrass scarabaeid grubs and resolution with isofenphos. *Proc. 4th Int. Turfgrass Res. Conf.* pp. 425–32. Guelph, Ontario: Ontario Agric. Coll. & Int. Turfgrass Soc.

176. Tashiro, H. 1982. The incidence of insecticide resistance in soil-inhabiting turfgrass insects. See Ref. 124, pp. 81–84

177. Tashiro, H. 1987. *Turfgrass Insects of the United States and Canada.* Ithaca, NY: Cornell Univ. Press. 391 pp.

178. Tashiro, H., Gambrell, F. L. 1963. Correlation of European chafer development with the flowering period of common plants. *Ann. Entomol. Soc. Am.* 56:239–43

179. Tashiro, H., Gyrisco, G. G., Gambrell, F. L., Fiori, B. J., Breitfield, H. 1969. Biology of the European chafer, *Amphimallon majalis* (Coleoptera: Scarabaeidae) in the northeastern United States. *NY State Agr. Exp. Sta. Bull.* No. 828. 71 pp.

180. Tashiro, H., Hartsock, J. G., Rohwer, G. G. 1967. Development of blacklight traps for European chafer surveys. *USDA Tech. Bull.* No. 1366. 53 pp.

181. Tashiro, H., Kuhr, R. J. 1978. Some factors influencing the toxicity of soil applications of chlorpyrifos and diazinon to European chafer grubs. *J. Econ. Entomol.* 71:904–7

182. Tashiro, H., Mitchell, W. C. 1985. Biology of the fiery skipper, *Hylephila phyleus* (Lepidoptera: Hesperiidae), a turfgrass pest in Hawaii. *Proc. Hawaiian Entomol. Soc.* 25:131–38

183. Tashiro, H., Murdoch, C. L., Mitchell, W. C. 1983. Development of a survey technique for larvae of the grass webworm and other lepidopterous species in turfgrass. *Environ. Entomol.* 12:1428–32

184. Tashiro, H., Personius, K. E. 1970. Current status of the bluegrass billbug and its control in western New York home lawns. *J. Econ. Entomol.* 63:23–29

185. Tashiro, H., White, R. T. 1954. Milky diseases of the European chafer. *J. Econ. Entomol.* 47:1087–92

186. Tippins, H. H., Martin, P. B. 1982. Seasonal occurrence of bermudagrass scale. *J. Ga. Entomol. Soc.* 17:319–21

187. Tolley, M. P., Niemczyk, H. D. 1988. Seasonal abundance, oviposition activity, and degree-day prediction of adult frit fly (Diptera: Chloropidae) occur-

rence on turfgrass in Ohio. *Environ. Entomol.* 17:855–62

188. Tolley, M. P., Niemczyk, H. D. 1988. Spatial distribution of frit fly, *Oscinella frit* (Diptera: Chloropidae) occurrence on turfgrass in Ohio. *Entomol. News* 99:267–71

189. Tolley, M. P., Robinson, W. H. 1986. Seasonal abundance and degree-day prediction of sod webworm (Lepidoptera: Pyralidae) adult emergence in Virginia. *J. Econ. Entomol.* 79:400–4

190. Troester, S. J., Ruesink, W. G., Rings, R. W. 1982. A model of the black cutworm *(Agriotis ipsilon)* Development: Description, uses, and implications. *Ill. Agric. Exp. Stn. Bull.* No. 774

191. Tumlinson, J. H., Klein, M. G., Doolittle, R. E., Ladd, T. L., Proveaux, A. T. 1977. Identification of the female Japanese beetle sex pheromone: inhibition of male response by an enantiomer. *Science* 197:789–92

192. Tuttle, D. M., Butler, G. D. Jr. 1961. A new eriophyid mite infesting bermudagrass. *J. Econ. Entomol.* 54:836–38

193. Ulagaraj, S. M. 1975. Mole crickets: ecology, behavior, and dispersal flight (Orthoptera: Gryllotalpidae: *Scapteriscus). Environ. Entomol.* 4:265–73

194. Ulagaraj, S. M., Walker, T. J. 1973. Phonotaxis of crickets in flight: attraction of male and female crickets to male calling songs. *Science* 182:1278–79

195. United Nations. 1985. *World Population Trends, Population and Development Interrelations and Population Policies,* Vol. 1. *Population Trends.* New York: United Nations

196. Vavrek, R. C., Niemczyk, H. D. 1990. The impact of isofenphos on non-target invertebrates in turfgrass. *Environ. Entomol.* 19:In press

197. Villani, M. G., Wright, R. J. 1988. Entomogenous nematodes as biological control agents of European chafer and Japanese beetle (Coleoptera: Scarabaeidae) larvae infesting turfgrass. *J. Econ. Entomol.* 81:484–87

198. Villani, M. G., Wright, R. J. 1988. Use of radiography in behavioral studies of turfgrass-infesting scarab grub species (Coleoptera: Scarabaeidae). *Bull. Entomol. Soc. Am.* 34:132–44

199. Villani, M. G., Wright, R. J., Baker, P. B. 1988. Differential susceptibility of Japanese Beetle, oriental beetle, and European chafer (Coleoptera: Scarabaeidae) larvae to five soil insecticides. *J. Econ. Entomol.* 81:785–88

200. Vittum, P. J. 1984. Effect of lime applications on Japanese beetle (Coleop-

tera: Scarabaeidae) grub populations in Massachusetts soils. *J. Econ. Entomol.* 77:687–90

201. Vittum, P. J. 1985. Effect of timing of application on effectiveness of isofenphos, and diazinon on Japanese beetle (Coleoptera: Scarabaeidae) grubs in turf. *J. Econ. Entomol.* 78:172–80

202. Vittum, P. J., Tashiro, H. 1980. Effect of soil pH on survival of Japanese beetle and European chafer larvae. *J. Econ. Entomol.* 73:577–79

203. Vittum, P. J., Tashiro, H. 1987. Seasonal activity of *Listronotus maculicollis* (Coleoptera: Curculionidae) on annual bluegrass. *J. Econ. Entomol.* 80:773–78

204. Wadley, F. M. 1931. Ecology of *Toxoptera graminum,* especially as to factors affecting importance in the northern United States. *Ann. Entomol. Soc. Am.* 24:325–95

205. Walkden, H. H. 1950. Cutworms, armyworms, and related species attacking cereal and forage crops in the central great plains. *USDA Agric. Circ.* No. 849

206. Walker, T. J. 1982. Sound traps for sampling mole cricket flights (Orthoptera: Gryllotalpidae: *Scapteriscus*). *Fla. Entomol.* 65:105–10

207. Walker, T. J., ed. 1984. Mole crickets in Florida. *Univ. Fla. Agr. Exp. Sta. Bull.* No. 846

208. Walker, T. J., Fritz, G. N. 1983. Migratory and local flights in mole crickets,

Scapteriscus spp. (Gryllotalpidae). *Environ. Entomol.* 12:953–58

209. Walker, T. J., Nickle, D. A. 1981. Introduction and spread of pest mole crickets: *Scapteriscus vicinus* and *S. acletus* reexamined. *Ann. Entomol. Soc. Am.* 74:158–63

210. Walker, T. J., Reinert, J. A., Schuster, D. J. 1983. Geographical variation in flights of the mole cricket, *Scapteriscus* spp. (Orthoptera: Gryllotalpidae). *Ann. Entomol. Soc.* 76:507–17

211. Warren, G. W., Potter, D. A. 1983. Pathogenicity of *Bacillus popilliae* (*Cyclocephala* Strain) and other milky disease bacteria in grubs of the southern masked chafer (Coleoptera: Scarabaeidae). *J. Econ. Entomol.* 76:69–73

212. Webster, F. M., Phillips, W. J. 1912. The spring grain aphid, or "greenbug". *US Bur. Entomol. Bull.* No. 110. 153 pp.

213. Wegner, G. S., Niemczyk, H. D. 1981. Bionomics and phenology of *Ataenius spretulus.* *Ann. Entomol. Soc. Am.* 74:374–84

214. White, R. T. 1947. Milky disease infecting *Cyclocephala* larvae in the field. *J. Econ. Entomol.* 40:912–14

215. Wilkinson, J. F. 1989. Current and future regulatory concerns for lawn care operators. See Ref. 102, pp. 45–49

216. Wolcott, G. N. 1941. The establishment in Puerto Rico of *Larra americana* Saussure. *J. Econ. Entomol.* 34:53–56

Annu. Rev. Entomol. 1991. 36:407–30

BEHAVIORAL ECOLOGY OF PHEROMONE-MEDIATED COMMUNICATION IN MOTHS AND ITS IMPORTANCE IN THE USE OF PHEROMONE TRAPS

Jeremy N. McNeil

Départment de biologie, Université Laval, Ste Foy, Québec, Canada G1K 7P4

KEY WORDS: mating behavior, Lepidoptera, integrated pest management

INTRODUCTION

The realization that chemical insecticides, as the quick and easy solution to insect problems, are not without considerable cost on several fronts (19, 45, 88, 151) has led to a serious reevaluation of our pest management philosophy. The acceptance that we must use more ecologically rational approaches to insect control, while respecting economic constraints, has resulted in the dawning of what Metcalf (88) called the Era of Integrated Pest Management. This realization has understandably led to a flurry of research efforts in areas of promising alternative control strategies that could be employed in intelligent management programs.

Among the wide array of potential tools available, semiochemicals, and in particular sex pheromones, have figured quite prominently. Research in the field has flourished, as evidenced by a plethora of papers dealing wth semiochemicals in many insect orders that appear regularly in scientific journals. Our understanding of many basic aspects of pheromone biology, including behavior (100), physiology and biochemistry (31, 100–102, 106, 107, 136,

407

0066-4170/91/0101-0407$02.00

139), and genetics (26, 113) advances at an exciting pace. Similarly, from a practical perspective, pheromones have proved useful in a variety of ways including monitoring emergence patterns (9), evaluating population numbers to determine if and when insecticide sprays should be applied (2, 91), assessing levels of insecticide resistance in pest populations (51, 52), supression of populations through extensive trapping of males (81), or mating disruption (23, 24).

However, despite the numerous successes, semiochemical use in integrated pest management has not always lived up to early expectations (76), possibly because we set ambitious goals with inadequate data bases. This observation does not question the obvious potential of semiochemicals, but rather underlines the need to increase our knowledge of the factors that influence their efficacy. Tumlison (144) has indicated four areas where additional research will improve our use of semiochemicals: 1. Elucidation of the complete pheromone blend in many species. The identification and synthesis of complex pheromone blends continues, greatly aided by the rapidly improving quality of analytical equipment and techniques (62), and reports on laboratory and field trials of new or improved pheromone blends abound. 2. A better understanding of the mechanisms and pathways of pheromone synthesis (which may well aid in the identification of the complete blend) (78). 3. Increase in knowledge of the mechanisms of semiochemical reception. 4. Increase in research efforts on other semiochemicals of potential use in pest management, which include epideictic pheromones (103), kairomones that influence the behavior of natural enemies (152), and host plant volatiles that influence oviposition (35, 109).

A basic premise of effective integrated pest management is an understanding of the life system of the pest species (104), and Roitberg & Angerilli (115) have argued that a behavioral-ecological approach would be fruitful for many aspects of pest management, including the use of sex pheromones. The current dirth of information on behavioral ecology with respect to semiochemically mediated communication is somewhat ironic, given the importance of semiochemicals in both intraspecific and interspecific (within and between trophic levels) processes that influence an individual's ultimate fitness. In this paper, I first examine the different biotic and abiotic factors that may influence the emission and reception of the sex pheromone in moths, and then discuss how a better understanding of these behavioral and ecological components may also help in the interpretation of trap-catch data. Limiting the scope of this review to a discussion of pheromone traps for moths reflects both personal research interests and constraints of time and space. However, I believe that the rationale could be applied to the use of any semiochemicals in different management programs.

FACTORS AFFECTING EMISSION AND
RECEPTION OF PHEROMONES

Age

Several aspects of female pheromone biology may change significantly with age. In some species, most females initiate calling within 24 h of emergence (159, 162) while in others individuals vary considerably in the age at which calling starts (58, 60, 65, 135, 142, 145). Furthermore, the time females spend calling may increase (20, 60, 145, 159, 162) or decrease (12) on successive nights. Increased calling may result from females initiating calling behavior earlier in the scotophase (60, 145) or by extending the calling period later into the scotophase (162) on successive nights of calling. Several authors have suggested that an advance in the mean onset time of calling by older females is an adaptive behavior that will increase the probabilities of older females attracting a mate through reduced competition with young ones early in the scotophase. For example, older females of the omnivorous leaf roller, *Platynota stultana,* which have considerably less pheromone in their glands (159) and attract considerably fewer males than younger individuals under field conditions (3), initiate calling significantly earlier than younger ones (159). However, no well-controlled experiments have been carried out on this or any other species to determine whether such behavioral changes actually do afford any reproductive advantage to older virgin females.

In several species, pheromone titer in the gland changes both with age and the time during the scotophase (33, 108, 121, 125, 159). However, we really do not know if, or how, such changes influence the rate of pheromone emission and the relative attractiveness of different-aged females during the calling period. While older females of the cabbage looper, *Trichoplusia ni,* call for a considerably shorter time and release significantly less total pheromone per night than younger ones, they have a much higher rate of pheromone release (12). However, it is not known if this rate renders them more attractive during their short period of calling than a younger female. Bjostad et al (12) also reported that the rate of pheromone release in four-day-old *T. ni* females decreases over time from the onset of calling, and a similar situation has been observed in the arctiid *Holomelina lamae* (121). Are females therefore more attractive at the beginning of their calling period, and are the series of short bouts observed at this time in certain species (134, 138, 145, 162) an important aspect of mating?

Receptivity of males to a pheromone source may vary with age; individuals generally exhibit increasing levels of response over the first few days following emergence (8, 123, 124, 135, 141, 142, 148, 161). Under field conditions, a significantly higher proportion of marked two-day-old gypsy moth males were recaptured than one-day-old individuals (39). While periodicity in male responsiveness to a pheromone source may be age dependent (8, 148), no studies have looked to see if the width of the receptivity window, which is generally wider than the female calling period (58, 71, 127), changes with age.

Presence of Conspecific Pheromone

The presence of the female sex pheromone in the atmosphere may modify conspecific calling behavior. When exposed to a pheromone source, spruce budworm virgin females initiated calling several hours earlier than the controls, and at any time during the calling window the proportion of females calling was higher in permeated than clean air (99). The inverse situation has been reported for the small tea tortrix and the oriental tea tortrix in which, under both laboratory and field conditions, the presence of conspecific pheromone resulted in a delay of up to 40 min in the onset of calling in virgins (95).

Male pheromones may also influence the behavior of conspecifics. Hirai et al (57) concluded that the contents of *Pseudaletia unipuncta* male hair pencil serve as a male-to-male inhibitor to reduce competition for receptive males. If true, this could modify trap catch, especially at high population densities. However, subsequent experiments found no evidence of true armyworm male upwind flight to a source of female sex pheromone being inhibited by hairpencil semiochemicals (43).

Mating Status

A refractory period in both the expression of calling behavior (60, 120) and pheromone synthesis (108) has been observed following mating, although, in species that mate more than once, these activities resume several days after mating (60; J. N. McNeil, J. Delisle, & S. M. Fitzpatrick, unpublished data). As observed for virgins, the mean onset time of calling by mated Bertha armyworm, *Mamestra configurata,* females is earlier on successive nights (60). However, mated females oviposit throughout much of the night and only initiate calling late in the scotophase, several hours after similar-aged virgin females started. Male quality has been reported to influence calling and pheromone synthesis in *Plodia interpunctella* (80), while in *T. ni* only calling was affected (48).

Male Lepidoptera have the capacity to mate repeatedly (37, 131), and, in a mark-recapture study, gypsy moth males that had mated approximately 16 h prior to release were recaptured at the same frequency as unmated individuals (40). The presence of a dark colored fluid only found in the primary simplex of virgin males of certain species provides a potential means of determining mating status of individuals captured in pheromone traps (10, 11, 56).

Researchers using this technique estimated that 68% of *Heliothis zea* males captured were mated (56), while, in spruce budworm, the percentage of mated males passed from 36 to 50% over the flight period (11). However, to my knowledge, no experiments have been carried out under controlled conditions in a wind tunnel to determine the effect of mating status, as well as the time since last mating, on male responsiveness to different blends and concentrations of sex pheromones.

Host Plants

Available evidence suggests that female Lepidoptera do not use specific secondary plant compounds obtained during larval feeding as pheromone precursors (90), as had once been hypothesized (55), but rather synthesize them de novo (114). However, host plant quality may still influence pheromone production. Replacing rice bran by tripalmitin in synthetic diet of the rice stem borer resulted in a noticeable increase in pheromone titers (150), while the elimination of either the soy bean or tea leaf powder from the smaller tea tortrix moth's synthetic diet resulted in significant decreases in pheromone production (130). In contrast, when the true armyworm, *P. unipuncta,* a polyphagous species, was reared on a variety of host plants, no differences were observed in the pheromone titer in the glands of virgin females on their second night of calling (86). Even if larval or adult food sources do not influence pheromone titer, however, they can affect other aspects of female pheromone biology, such the age at which females call for the first time following emergence and the periodicity of calling behavior (86).

While reports that specific host plant volatiles affected the prereproductive behavior of the polyphemus moth (110, 111) were not supported in a subsequent study (17), evidence now indicates that host plant volatiles significantly influence female pheromone biology in several species of Lepidoptera. Females of the sunflower moth, *Homoeosoma electellum,* preferentially oviposit in newly opened sunflowers (36, 140), a choice modulated by the presence of pollen (35), which is an essential food source for neonate larvae (116). Virgin females in the presence of pollen and sugar water initiate calling at a significantly younger age following emergence and had a significantly longer daily calling period than those with sugar water only (87). A similar situation has been reported for several species of ermine moth when held with or without leaves of their host plants (54). In two *Heliothis* species, females never call in the absence of suitable host plant volatiles (105).

Male sex pheromones are often compounds sequestered directly from host plants (86 and references therein), but, as these semiochemicals are not used in management programs, this aspect is not considered here. Information concerning the effects of host plants on male receptivity is even more sparse than that pertaining to female calling behavior. Traps placed in cotton fields accounted for a higher proportion of the total number of pink bollworm males caught in a pheromone trapping network when cotton had fruiting bodies than when it did not (64). Similarly, the number of male sunflower moths caught in pheromone traps placed in sunflower plots, relative to those placed in grassland pastures, increased significantly when sunflower anthesis occurred (149). In both cases, the increased number of males caught in the traps could have resulted from an increased density of calling females, themselves attracted to the plant volatiles emitted from suitable oviposition sites. However, the volatiles from suitable host plants may modify male receptivity to pheromones. Experiments carried out on the gelechiid *Sitotroga cerealella* clearly demonstrated that the number of males captured in traps baited with corn extract (which alone made the traps no more attractive than empty traps) and pheromone were more attractive than those containing only pheromone (129). Furthermore, the greatest difference between the two traps occurred when the standing corn crop was reaching maturity. Male responsiveness to host plant volatiles may increase with age (74), as it does with pheromones, and it is possible that the response to the plant volatiles and pheromone combined may be age dependent.

Pathogens

Despite the knowledge that sublethal pathogenic infections may have a marked effect on various adult life history parameters, their potential influence on semiochemically mediated communication has been largely neglected. While it has been demonstrated that infected females can attract males (120a), I have found no published works comparing patterns of calling behavior, or pheromone synthesis in infected and healthy females, and only two papers examining the effect of a pathogen on male receptivity. Sweeny & McLean (132) reported that the presence of sublethal infections of *Nosema fumiferanae* significantly decreased the number of males completing different steps in the behavioral sequence that are normally exhibited when *Choristoneura occidentalis* males were exposed to a synthetic pheromone source in a wind tunnel. Furthermore, the number of males exhibiting the different steps in the sequence decreased as *N. fumiferanae* spore load increased. They found that the decrease in responsiveness was not associated with the peripheral sensory system because electroantennogram (EAG) responses were similar for infected and control males. Saunders & Wilson (120a) found no significant difference in the flight duration of infected and uninfected spruce budworm males when the insects were exposed to a pheromone source in a wind tunnel. However, flight speed was not examined and this parameter may be of greater importance in male-male competition for receptive females than flight duration. Given the importance of pathogens in the population dynamics of many species, considerably more attention should be given to the possible effects that sublethal pathogenic infections have on synthesis and emission of female pheromones, as well as on male receptivity.

Temperature

Temperature conditions during pupal development (146) or adult life (34, 46, 146) may affect the age at which females initiate calling for the first time after emergence: generally, the time between emergence and the onset of calling increases with a decrease in temperature. However, in the few cases that have been studied, the length of the delay differs markedly, and is generally longer in suspected migrant species. A prolonged delay in the onset of mating may be associated with a seasonal migration (34), and one may be able to separate resident and migrant species using the length of temperature-induced delays in calling (84).

When females are held under different constant temperature regimes from emergence (34, 46, 146) or are subjected to temperature fluctuations at some time following emergence (8, 16, 20, 33, 47, 50, 138, 157, 159), significant changes in calling periodicity have been observed. In nocturnal species, females initiate calling earlier in the evening or night at lower temperatures (1, 14, 16, 20, 33, 34, 46, 68, 72, 120, 128, 138, 146, 157, 159, 160), while, under cool conditions, females of diurnal species delay calling until later in the day when the ambient temperature has increased or body temperatures may be raised through basking (47). Temperature conditions experienced during the previous calling period (8, 33) or previous photophase (50) may also determine the periodicity of calling. Changes in temperature may also modify the actual calling behavior. Arctiids, many of which advance the time of calling at lower temperatures, exhibit a characteristic extrusion and retraction of the ovipositor during calling, and the rate of ovipositor pumping varies with temperature (29).

While no major or consistent changes have been observed in the time spent calling by females held under different constant temperature regimes (34, 36, 146), this parameter may change significantly when females are subjected to temperature fluctuations (20, 33, 34, 46, 47, 50). As clearly shown in studies on the omnivorous leaf roller *P. stultana* (157, 159), the duration of the calling period within a given cycle may be extended or shortened by either temperature increases or decreases, depending on when the fluctuations occurred. Also, species differences will determine to what extent fluctuations influence the duration of calling. Females of the Bertha armyworm (46) and the true armyworm (34), species that generally call in the latter half of the scotophase, spend significantly more time calling than controls when subjected to temperature decrease, while a temperature increase during the same period results in no change or a shorter calling period. In comparison, the calling period of the artichoke plume moth, which generally calls earlier in the scotophase than the above mentioned armyworm species, is usually extended with either an increase or decrease in temperature (50).

To date very few experiments have tested the effect of temperature on the age at which males respond to a pheromone source. Bollinger et al (13) demonstrated that *T. ni* males reared at 15°C required at least nine days to exhibit maximum responsiveness, measured by the increased level of activity in the 30 s following the introduction of a pheromone source, while those reared at 25°C required only three to five days. Similar results were reported for the true armyworm when the entire up-wind behavioral sequence was studied in a wind tunnel (148). Furthermore, Dumont (38) showed that, even after 20 days at 10°C, < 25% of true armyworm males undertook upwind flight towards a pheromone source. This delay in the onset of male receptivity by low temperature conditions could explain why males are captured in light traps but not in pheromone traps in late summer and fall (85). A similar situation has been reported for males of the black cutworm, another suspected migrant species (69).

Temperature conditions during the flight period may influence male selectivity with respect to pheromone blends. Males of the oriental fruit moth and the pink bollworm respond to a fairly narrow range of blend-dose combinations of their respective pheromones when tested at 20°C in a wind tunnel, while at 26°C male specificity in both species markedly decreases (77). Temperature conditions may also modify other aspects of male receptivity, such as the latency period (15), as the proportion of marked gypsy moth males that were recaptured increased with an increase in temperature (39, 40).

Landolt & Curtis (72) reported seasonal changes in the time during the scotophase at which navel orange worm males were captured in traps. However, as the traps were baited with virgin females, whose calling behavior advances under lower temperatures (72), we cannot determine to what extent, if any, temperature-induced changes in male receptivity contributed to the observed seasonal shifts. However, data obtained using traps baited with synthetic pheromone sources indicate that male behavior is affected by temperature because males of several crepuscular and nocturnal species are generally captured earlier under cool temperatures than under warmer ones (13, 16, 27, 147, 156). Song & Riedl (126) clearly demonstrated a diel periodicity in general activity of codling moth males and that activity occurs earlier at colder temperatures. Thus, earlier captures under cool conditions may entirely result from an advance in overall flight activity and subsequent inhibition of flight as temperatures continue to decline. However, the maximum response of cabbage looper (13) and rice stem borer (68) males reared at lower temperatures occurred several hours earlier than for males reared at higher ones, suggesting that the actual periodicity of male receptivity to the sex pheromone may also change. On the other hand, while a drop in temperature caused oriental fruit moth and codling moth females to call earlier, no shifts in male responsiveness were observed (8, 20). Considerably more work is required to improve our understanding of temperature effects on periodicity of male receptivity, especially as the majority of the above-mentioned bioassays were carried out in arenas where free flight was not possible and examined only one or two specific male behaviors, such as wing fanning (7), rather than the entire sequence of behaviors.

Daylength

The influence of daylength on the age at which females initiate calling has been studied in three noctuid species. The precalling period was significantly extended when females of the true armyworm (32) and the oriental armyworm, *Mythimna separata* (49), were exposed to short-day conditions, but remained unchanged in the Bertha armyworm (46). The first two species occur annually in northern temperate zones where they cannot overwinter. It is hypothesized that they undertake a southward migration in the fall. In such a situation, daylength would be a reliable cue for seasonal habitat deterioration (137) and a delay in the onset of reproduction would facilitate migration. In contrast, the Bertha armyworm is considered a nonmigrant, overwintering as a diapausing pupa, and hence delaying reproductive behavior has no evident advantage.

Similarly, the effect of daylength on the periodicity and duration of calling behavior varies considerably between species. The spruce budworm (120), the cabbage looper (128), and the arctiid *Holomelina immaculata* (16) called at the same time following reception of the photoperiodic cue responsible for entraining the periodicity of calling regardless of daylength, while the mean onset time of calling following the lights-off signal varies with photoperiodic conditions in the artichoke plume moth (50), the rice borer (67), the Bertha armyworm (46), and the true armyworm (32). Haynes & Birch (50) suggested that daylength would be more important in the periodicity of multivoltine species, in which the photoperiodic conditions are markedly different during successive flight periods. Similarly, Delisle & McNeil (32) suggested that daylength would also be an important cue for species that experience rapid changes in daylength following long-distance migration. Photoperiodic conditions not only affected the mean onset time of calling in Bertha armyworm females but also the length of the calling period; the mean time spent calling increased as daylength decreased (46). A similar situation was observed with the rice stem borer (67), while the duration of the calling period in the true armyworm did not change with daylength, except when the scotophase was extremely short (32).

At 25°C, short-day conditions induce a 24 h delay in the sexual maturation of *P. unipuncta* males, when compared with long day controls (38), although by the fifth day no significant differences appeared between treatments. When tested at 10°C, the photoperiodic effect on males was minor compared with that of temperature (38), a situation similar to that observed with female calling (34). A similar photoperiodic response may exist in other species, especially in suspected migrants, because a decline in male responsiveness late in the season, as reported for the true armyworm (85), has been observed in the black cutworm (93 and references therein).

Few studies have examined the periodicity of male responsiveness under different photoperiodic conditions, and the findings are quite different. Kanno (67) found that rice stem borer males held under short-day conditions responded later in the scotophase than under long days, while peak activity of the true armyworm occurred 7 h following the lights-off signal under both 12L:12D and 16L:8D photoperiodic regimes (38).

Light Intensity

The incidence of calling is inversely proportional to light intensity in several nocturnal species (66, 127, 141). However, the upper light intensity threshold changes with prevailing temperature conditions; females held at low temperatures exhibit calling behavior at light intensities that were inhibitory at higher temperatures. For example, at 30°C light intensities above 40 lux inhibit calling in rice stem borer females, while at 15°C some females call at intensities up to 1200 lux (66). Changes in temperature may also modify light intensity thresholds in the rice stem borer (66) and *H. immaculata* (16). In both species, females held at a constant temperature (24–25°C) and under continuous light do not call. However, a 9–10°C drop in temperature resulted in the expression of calling even though the light intensity remained unchanged. This adaptation would be of particular importance for insects that are active around sundown, for at lower temperatures the mean onset time of calling advances and will occur at the end of the photophase. Similarly, for insects that call at the end of the scotophase, increasing light intensity may be an important factor in determining the duration of the calling period (160), undoubtedly in conjunction with prevailing temperature conditions.

Whether diurnal species have temperature-dependent light-intensity thresholds for calling is not known, but, if present, they would be expected to vary depending on the time of the calling period. One would predict a lower intensity threshold that must be exceeded before the onset of calling in species that call early in the photophase and that this threshold would increase with decreasing temperature. The only information available for species calling in the morning pertains to the Mediterranean flour moth (143), which begins calling at lights-on and finishes by midphotophase under temperature conditions ranging from 20–27°C. This calling pattern did not change over a 1–200 lux range of photophase intensities. Species that call late in the day may have a lower light-intensity threshold, which again increases with decreasing temperature and below which calling will not occur. Females of the oriental fruit moth, which only call in the latter part of the photophase under a 16L:8D photoperiodic cycle at 25°C, called during the first hours of the scotophase when transferred from 25 to 31°C (8), suggesting a possible influence of temperature on the effect of light intensity. However, given the extremely limited data available, considerably more research is required to obtain any understanding of the relative importance of light intensity on diurnal species.

In several species, calling continues throughout much of the 24 h cycle (22, 96), suggesting that light intensity is not of major importance in the expression of calling behavior. The diel periodicty in the rate of pheromone emission by female gypsy moth does not appear to be influenced by changes in light intensity during the photoperiodic cycle (22). In contrast, *P. interpunctella* females released considerably less pheromone during the photophase, a period when they exhibited weak calling behavior, than during the scotophase when they expressed strong calling behavior (96, 97). Whether these changes are directly or indirectly related to light-intensity levels is presently unknown.

The level of male activity elicited by a given concentration of pheromone (at temperatures 23–25°C) has been shown to decrease with an increase in light intensity in the cabbage looper (123), the codling moth (20), the rice stem borer (66), and *Mamestra brassicae* (141). However, in complete darkness, *M. brassicae* males were inactive, leading the authors (141) to suggest that some light is essential for responsiveness. The light intensity threshold for male responsiveness in the rice stem borer is modified by temperature; males are totally inhibited above 20 lux at 30°C, while at 15°C some males are still receptive at 700 lux (66). In this species, the females called over a wider range of light intensities than those in which they mated, suggesting that the influence of light intensity on male receptivity may be of considerable importance. The responsiveness of codling moth males, while decreasing with increasing light intensity ranging from 1.5 to 1,500 lux at 24°C, showed no evident trend at 16°C and generally remained constant (20). Increased responsiveness at higher light intensities when temperatures are low would be expected as these species are active in the late afternoon or at dusk under cool conditions. For diurnal species, working hypotheses similar to those proposed for female calling would be a reasonable point of departure, but to my knowledge no data are presently available.

Relative Humidity

Very little research has examined the influence of relative humidity on any aspect of pheromone-mediated communication in Lepidoptera.

The age at which European corn borer females initiated calling for the first time following emergence did not differ significantly when adults were held from emergence at three different constant relative humidity conditions, although at the lowest condition (40%), some females delayed calling for more than three days (L. Royer & J. N. McNeil, unpublished data). Webster & Cardé (158) reported a decrease in the proportion of females calling from both pheromone strains of the European corn borer when the insects were transferred to low humidity conditions on the second or third day following emergence, and that calling behavior was initiated later in the scotophase under dry conditions. In contrast, no effect of relative humidity on calling behavior of the oriental fruit moth (8) or the codling moth (20) was evident.

Comeau et al (27) reported that the average time of redbanded leafroller male activity, estimated from captures in traps baited with pheromone lures placed in apple orchards, was not significantly affected by relative humidity. Miller & McDougall (89) found a negative correlation between trap catch and relative humidity in a 12-year trapping study on the spruce budworm. However, traps were baited with virgin females; as the authors pointed out, one cannot separate out male and female effects. Under controlled conditions, the responsiveness of gypsy moth males to a given concentration of pheromone in a wind tunnel is reportedly greater at high (70%) than at low (30%) relative humidity (C. Linn, personal communication). The responsiveness of European corn borer also varies with relative humidity conditions: the proportion of males flying upwind and touching the source declined at both very high and very low humidity conditions (L. Royer & J. N. McNeil, unpublished data). Furthermore, preliminary data indicate that the optimal humidity conditions may change with pheromone concentration, which, if true, could markedly effect trap efficiency.

Wind Speed

Kaae & Shorey (63) found that the proportion of *T. ni* females calling was lower at 0 and 3 m/s than at 0.3 and 1 m/s, and that calling was inhibited at velocities > 4 m/s. They also noted that the proportion of females fanning their wings while calling decreased as wind velocity increased. They argued that the higher incidence of wing fanning at the low wind speeds would help pheromone dispersal by increasing the velocity of air passing over the exposed gland. Conner & Best (28), using the arctiid *Spilosoma congrua,* showed that the characteristic calling postures assumed by females of many species may not only increase the rate of air flow over the pheromone gland but also reduces turbulence near the gland surface. Thus, for species such as *T. ni,* appropriate air movement over the gland may be attained at higher wind speeds without wing fanning, which would reduce the energetic costs of pheromone release and possibly reduce the turbulence of the pheromone plume.

The only other paper examining calling behavior with respect to wind speed noted that the frequency of ovipositor pumping in *Utetheisa ornatrix* increased with an increase in wind speed, over a range of velocities up to 1.2 m/s, but remained unaltered in five other arctiids (29). The significance of this change as it relates to mating success, and why such changes do not occur in all species, still has to be determined.

Kaae & Shorey (63) reported that *T. ni* males appeared to have difficulty in finding the pheromone source without detectable wind and that trap catches declined on nights when wind speeds approached 2–3 m/s. These observations suggest the obvious, yet no detailed studies examine how male responsiveness is affected by wind speed. The obvious importance of air velocity in both the dispersion of, and the ability of males to locate, sex pheromones in field studies (42 and references therein), makes it difficult to understand why so little research has been carried out under controlled conditions.

Atmospheric Conditions

The results obtained from the analysis of data collected during a 12-year study suggested that maximum and minimum captures of the spruce budworm occurred during periods of maximum decreases and maximum increases in pressure, respectively (89), but as traps were baited with virgin females, we cannot factor out the relative importance of changes in female calling behavior and in male responsiveness. However, anyone working for any length of time on moth flight in wind tunnels has observed day to day variability in male responses to a given concentration of pheromone even though all other conditions (such as temperature, daylength, light intensity, wind speed, male age, etc) were identical. One possible explanation is that certain atmospheric conditions modify the level of male responsiveness. Appropriate analysis of carefully collected weather data and trap catches could help determine to what extent, if any, meteorological conditions influence male responsiveness. I believe that such an approach is necessary, with an emphasis on the ecological implications of any observable behavioral changes, as suggested by Lanier & Burns (73).

IMPLICATIONS FOR THE INTERPRETATION OF TRAP-CATCH DATA

We know that trap design affects plume structure (75), which, in turn, influences male flight toward the pheromone source, so that an understanding of these mechanisms (6 and references therein) could provide information leading to more efficient traps. It has also been well documented that the relative efficiency of traps may vary with respect to their placement (locations and/or height) within the habitat (e.g. 61, 83, 117, 147). The density, spacing, and size of different objects in the habitat determines the degree to which these objects physically deflect the plume and influence the degree of turbulence observed at different sites. Combined, these factors determine the rate of directional change of a pheromone plume as well as the probability that, and the distance over which, males will be able to locate the source (42, and references therein). The physical nature of the habitat may also affect the degree to which pheromone is taken up and reemitted by vegetation, a factor that may modify male behavior (154, 155). Visual cues, including color, may play an important role in close-range mate location as males approach a pheromone source (18, 21, 53, 92, 118, 133). Whether the relative importance of visual cues for a given species is related to the complexity of its habitat, and therefore the integrity of the pheromone plume, merits attention.

The daily changes in local climatic conditions resulting from habitat structure will not only influence the structure and trajectory of plumes, but may modify adult behavior to an extent that could be of considerable importance to trap efficiency. Webb & Berisford (156) noted that nantucket pine tip moth males concentrated their searching activity higher in the tree canopy as the temperature close to the ground declined towards the 10°C flight threshold. Therefore, the relative efficacy of traps placed at different heights, especially in forest ecosystems, could vary considerably from day to day if insects partition their flight activity within the habitat in response to changes in vertical temperature gradients. Thus, an understandig of how both male and female distributions within a given habitat change in response to temperature and the other factors influencing mating, particularly during the female calling period, would help in the interpretation of trap captures. This could be of particular importance in multivoltine species, especially if trap lines are maintained at constant heights throughout the season.

The relative efficacy of a given pheromone trap is influenced by the number and position of other competing pheromone sources present in the habitat, be they lures (25, 41, 83, 153) or caged virgin females (30, 40, 59). Thus, comprehending that factors that determine the degree of competition between the traps and calling females is of fundamental importance when interpreting trap-catch data. Protandry (male emergence preceding that of females) and temporal changes in the density of virgin females have been identified as two of these factors (70, 94). However, competition is not simply a case of relative densities of males to females or of mated to unmated females. A major consideration is the actual width of the competition window, which is determined by the time over which females within the population call during any given trapping period. As seen above, the calling period of an individual female may change with factors such as age, the presence or absence of suitable host plants, as well as prevailing climatic conditions. Thus, the width of the calling window could differ markedly depending on the age structure of virgin females in the populations and the phenology of the available host plants, as well as fluctuations in weather. The presence of both calling conspecifics and synthetic pheromone sources could also change the width of the competition window. In the tea tortrix moths (95), the competition period may be reduced as females delay calling; in contrast, increased calling by spruce budworm females (99) could result in an extended feral female–trap competition. However, higher pheromone concentrations may also increase female spruce budworm flight activity (98, 119), and, if they disperse out of the habitat, the level of competition could decrease.

The degree to which the male response and the female calling windows overlap will also influence the level of competition. A comparison of traps baited with pheromone lures and virgin females showed that the response window of polyphemus (*Antheraea polyphemus*) males was considerably wider than that of calling females, and that maximum captures with the lures occurred early in the female calling window (71). If for any reason the male response window was shorter and more synchronous with that of female calling, then the degree of competition would undoubtedly increase. As seen above, many of the factors affecting calling do affect male receptivity (e.g. temperature, photoperiod, light intensity, host plant volatiles), but, for a given species, do they influence the width of the male response window in the same way as they do for the female calling window? While this idea may appear attractive, it may not always be the case: in two moth species, a decrease in temperature resulted in an advance in the timing of female calling behavior but did not change the timing of male responsiveness (8, 20). To date we really know very little about the factors influencing the width of the male response window, and we urgently need to increase our knowledge in this area.

Even during the female calling period, several factors may influence the intensity of competition. If the pheromone identification has been incomplete, then the degree of competition between feral females and traps will probably be lower than with a lure containing the complete pheromone. The threshold hypothesis (112) predicts that a higher release of an off-blend will be at least as effective as a lower release of the natural blend, and temperature-related changes in release rate of pheromone from lures was proposed to explain seasonal changes in the most effective blend (5, 44). However, the recent findings of Linn et al (77) clearly show that temperature conditions also affect male blend specificity directly. Thus, the relative efficacy of a given pheromone trap may not only vary seasonally but also during a given flight period as a result of day to day temperature fluctuations influencing both release rates (82) and male selectivity.

When interpreting trap catch data, we assume that the males caught are representative of the breeding population, although the validity of such an assumption has not really been examined. The low numbers of mated spruce budworm males that Bergh & Seabrook (11) caught early in the season would be expected, given that the species is protandrous, but the low proportion captured at the end of adult flight is somewhat surprising. As suggested by the authors, some of the males may have incorrectly been classed as virgins because in the spruce budworm the pigment regenerates several days after mating. However, if the observed patterns reflect the actual field situation, then many males still remain unmated at the end of the season. If this is the case, an explanation must be sought because such trapping data would not give a very precise measure of the male breeding population and would not permit accurate estimates of subsequent larval densities. Eastern spruce bud-worm populations may have high levels of *N. fumiferanae* infection (163), and a poor competitive performance of infected males might explain the high incidence of virgins captured in pheromone traps throughout the season (11). Recent work on the oblique banded leafroller showed that adults with a sublethal dose of *N. fumiferanae* had heavily infected flight muscles and that male flight capacity of such individuals was significantly reduced (M. Gardi-ner, personal communication). Therefore, if infected spruce budworm males have a lowered ability to locate and compete for receptive females, they could be over-represented in traps, as the pheromone source is not "removed" by the earlier arrival of a competing, healthy male.

Although pheromone traps are often used in management programs for species that have very large geographic distributions, little consideration is given to the important ecological differences that exist over the species range. Recently, several studies reported geographic pheromone dialects within a species (4, 79), and it would not be unreasonable to expect significant variation in the response to geographic differences in the abiotic and biotic cues modulating semiochemically mediated communication. For example, populations at high latitudes would experience markedly different daylengths, as well daily changes in light intensity at dawn and dusk, than populations of the same species at lower latitudes. Given these differences, are the windows of female calling and male responsiveness the same at both sites? Is the degree of competition between traps and feral females the same? Such information is essential if we are to make rational management recommendations.

CONCLUDING REMARKS

I believe that pheromones, and indeed all semiochemicals, will play an even more important role in future integrated pest management programs than they do already. However, we must obtain a sound ecological understanding to accompany the advances in the fields of chemistry, biochemistry, physiology, and genetics. The available data clearly indicate that many factors influence both female calling behavior and male receptivity. Furthermore, the relative importance of the different factors varies between species and different populations of the same species, undoubtedly reflecting the different ecological conditions to which they are normally subjected. A species active in open habitats during the day experiences very different conditions than a nocturnal one that limits its activity to dense vegetation. Therefore, researchers should be able to establish generalized classes or species groupings with respect to ecological requirements, but this classification will require a much larger data base than is presently available. Such an approach would help us determine which of the different potentially important abiotic and biotic factors most influence feral female-trap competition for a given species. This information could then serve to develop refined, species-group models that integrate the effects that daily and seasonal changes in these parameters have on trap catch when interpreting field data. While I do not believe that a better understanding of the behavioral ecology of pheromone-mediated mating will permit us to "make the insects jump through a hoop" (122), such information would permit the more efficacious use of pheromones and further reduce our dependence on "environmentally unfriendly" means of control.

ACKNOWLEDGMENTS

I thank Jacques Brodeur, Ring Cardé, Johanne Delisle, Charlie Linn, Larry Marshall, Jacques Regnière, Lucie Royer, and Reggie Webster for helpful comments and suggestions on earlier versions of this text.

Literature Cited

1. Alford, A. R., Hammond, A. M. Jr. 1982. Temperature modification of female sex pheromone release in *Trichoplusia ni* Hübner) and *Pseudoplusia includens* (Walker) (Lepidoptera: Noctuidae). *Environ. Entomol.* 11:889–92
2. Alford, D. V., Carden, P. W., Dennis, E. B., Gould, H. J., Vernon, J. D. R. 1979. Monitoring codling and tortix moths in United Kingdom apple orchards using pheromone traps. *Ann. Appl. Biol.* 91:165–78
3. AliNiazee, M. T., Stafford, E. M. 1971. Evidence of a sex pheromone in the omnivorous leaf roller, *Platynota stultana* (Lepidoptera: Tortricidae): Laboratory and field testing of male attraction to virgin females. *Ann. Entomol. Soc. Am.* 64:1330–35
4. Arn, H., Esbjerg, P., Bues, R., Tóth, M., Szöcs, G., et al. 1983. Field attraction of *Agrotis segetum* males in four European countries to mixtures containing three homologous acetates. *J. Chem. Ecol.* 9:267–76
5. Baker, J. L., Hill, A. S., Roelofs, W. L. 1978. Seasonal variation in pheromone trap catches of male omnivorous leafroller moths, *Platynota stultana. Environ. Entomol.* 7:399–401
6. Baker, T. C. 1989. Sex pheromone communication in the Lepidoptera: New research progress. *Experientia* 45:248–62
7. Baker, T. C., Cardé, R. T. 1979. Analysis of pheromone-mediated behaviors in male *Grapholitha molesta*, the Oriental fruit moth (Lepidoptera: Tortricidae). *Environ. Entomol.* 8:956–68
8. Baker, T. C., Cardé, R. T. 1979. Endogenous and exogenous factors affecting periodicities of female calling

424 McNEIL

and male sex pheromone response in *Grapholitha molesta* (Busck). *J. Insect Physiol.* 25:943–50

9. Baker, T. C., Cardé, R. T., Croft, B. A. 1980. Relationship between pheromone trap capture and emergence of adult Oriental fruit moths, *Grapholitha molesta* (Lepidoptera: Tortricidae). *Can. Entomol.* 112:11–15

10. Bergh, J. C., Seabrook, W. D. 1986. A simple technique for indexing the mating status of male spruce budworm, *Choristoneura fumiferana* (Lepidoptera: Tortricidae). *Can. Entomol.* 118:37–41

11. Bergh, J. C., Seabrook, W. D. 1986. The mating status of spruce budworm males, *Choristoneura fumiferana* (Clem.) (Lepidoptera: Tortricidae), caught in pheromone-baited traps. *J. Entomol. Sci.* 21:254–62

12. Bjostad, L. B., Gaston, L. K., Shorey, H. H. 1980. Temporal pattern of sex pheromone release by female *Trichoplusia ni*. *J. Insect Physiol.* 26:493–98

13. Bollinger, J. F., Shorey, H. H., Gaston, L. K. 1977. Effect of several temperature regimes on the development and timing of responsiveness of males of *Trichoplusia ni* to the female sex pheromone. *Environ. Entomol.* 6:311–14

14. Cardé, R. T., Comeau, A., Baker, T. C., Roelofs, W. L. 1975. Moth mating periodicity: Temperature regulates the circadian gate. *Experientia* 31:46–48

15. Cardé, R. T., Hagaman, T. E. 1983. Influence of ambient and thoracic temperatures upon the sexual behaviour of the gypsy moth, *Lymantria dispar*. *Physiol. Entomol.* 8:7–14

16. Cardé, R. T., Roelofs, W. L. 1973. Temperature modification of male sex pheromone response and factors affecting female calling in *Holomelina immaculata* (Lepidoptera: Arctiidae). *Can. Entomol.* 105:1505–12

17. Cardé, R. T., Taschenberg, E. F. 1984. A reinvestigation of the role of (E)-2-hexenal in female calling behaviour of the polyphemus moth *(Antheraea polyphemus)*. *J. Insect Physiol.* 30:109–12

18. Carpenter, J. E. Sparks, A. N. 1982. Effects of vision on mating behavior of the male corn earworm. *J. Econ. Entomol.* 75:248–50

19. Carson, R. 1962. *Silent Spring*. Boston: Houghton Mifflin Co. 368 pp.

20. Castrovillo, P. J., Cardé, R. T. 1979. Environmental regulation of female calling and male pheromone response periodicities in the codling moth *(Laspeyresia pomonella)*. *J. Insect Physiol.* 25:659–67

21. Castrovillo, P. J., Cardé, R. T. 1980. Male codling moth *(Laspeyresia pomonella)* orientation to visual cues in the presence of pheromone and sequences of courtship behaviors. *Ann. Entomol. Soc. Am.* 73:100–5

22. Charlton, R. E., Cardé, R. T. 1982. Rate and diel periodicity of pheromone emission from female gypsy moths, *(Lymantria dispar)* determined with a glass-adsorption collection system. *J. Insect Physiol.* 28:423–30

23. Charmillot, P.-J., Bloesch, B. 1987. La technique de confusion sexuelle: un moyen spécifique de lutte contre le carpocapse *Cydia pomonella* L. *Rev. Suisse Vitic. Arboric. Hortic.* 19:129–38

24. Charmillot, P.-J., Bloesch, B., Schmid, A., Neumann, U. 1987. Lutte contre cochylis de la vigne *Eupoecilia ambiguella* Hb., par la technique de confusion sexuelle. *Rev. Suisse Vitic. Arboric. Hortic.* 19:155–64

25. Charmillot, P.-J., Schmid, A. 1981. Influence de la densité des pièges sexuels sur les captures de capua, la tordeuse de la pelure *(Adoxophyes orana* F. v. R.). *Rev. Suisse Vitic. Arboric. Hortic.* 13:93–97

26. Collins, R. D., Cardé, R. T. 1985. Variation in and heritability of aspects of pheromone production in the pink bollworm moth, *Pectinophora gossypiella* (Lepidoptera: Gelechiidae). *Ann. Entomol. Soc. Am.* 78:229–34

27. Comeau, A., Cardé, R. T., Roelofs, W. L. 1976. Relationship of ambient temperatures to diel periodicities of sex attraction in six species of Lepidoptera. *Can. Entomol.* 108:415–18

28. Conner, W. E., Best, B. A. 1988. Biomechanics of the release of sex pheromone in moths: Effects of body posture on local airflow. *Physiol. Entomol.* 13:15–20

29. Conner, W. E., Webster, R. P., Itagaki, H. 1985. Calling behaviour in arctiid moths: The effects of temperature and wind speed on the rhythmic exposure of the sex attractant gland. *J. Insect Physiol.* 31:815–20

30. Croft, B. A., Knight, A. L., Flexner, J. L., Miller, R. W. 1986. Competition between caged virgin female *Argyrotaenia citrana* (Lepidoptera: Tortricidae) and pheromone traps for capture of released males in a semi-enclosed courtyard. *Environ. Entomol.* 15:232–39

31. Cusson, M., McNeil, J. N. 1989. Involvement of juvenile hormone in the regulation of pheromone release activities in a moth. *Science* 243:210–12

32. Delisle, J., McNeil, J. N. 1986. The effect of photoperiod on the calling behaviour of virgin females of the true armyworm, *Pseudaletia unipuncta* (Haw.) (Lepidoptera: Noctuidae). *J. Insect Physiol.* 32:199–206

33. Delisle, J., McNeil, J. N. 1987. Calling behaviour and pheromone titre of the true armyworm *Pseudaletia unipuncta* (Haw.) (Lepidoptera: Noctuidae) under different temperature and photoperiodic conditions. *J. Insect Physiol.* 33:315–24

34. Delisle, J., McNeil, J. N. 1987. The combined effect of photoperiod and temperature on the calling behaviour of the true armyworm, *Pseudaletia unipuncta*. *Physiol. Entomol.* 12:157–64

35. Delisle, J., McNeil, J. N., Underhill, E. W., Barton, D. 1989. *Helianthus annuus* pollen, an oviposition stimulant for the sunflower moth, *Homoeosoma electellum. Entomol. Exp. Appl.* 50:53–60

36. DePew, L. J. 1983. Sunflower moth (Lepidoptera: Pyralidae): Oviposition and chemical control of larvae on sunflowers. *J. Econ. Entomol.* 76:1164–66

37. Drummond, B. A. 1984. Multiple mating and sperm competition in the Lepidoptera. In *Sperm Competition and the Evolution of Animal Mating Systems,* ed. R. L. Smith, pp. 291–370. New York: Academic

38. Dumont, S. 1989. *L'influence de l'âge et des conditions abiotiques sur la réponse des mâles à la phéromone sexuelle femelle chez Pseudaletia unipuncta (Lepidoptera: Noctuidae).* MS thesis. Laval: Université Laval. 64 pp.

39. Elkinton, J. S., Cardé, R. T. 1980. Distribution, dispersal and apparent survival of male gypsy moths as determined by capture in pheromone-baited traps. *Environ. Entomol.* 9:729–37

40. Elkinton, J. S., Cardé, R. T. 1984. Effect of wild and laboratory-reared female gypsy moths, *Lymantria dispar* L. (Lepidoptera: Lymantriidae), on the capture of males in pheromone-baited traps. *Environ. Entomol.* 13:1377–85

41. Elkinton, J. S., Cardé, R. T. 1988. Effect of intertrap distance and wind direction on the interaction of gypsy moth (Lepidoptera: Lymantriidae) pheromone-baited traps. *Environ. Entomol.* 17:764–69

42. Elkinton, J. S., Schal, C., Ono, T., Cardé, R. T. 1987. Pheromone puff trajectory and upwind flight of male gypsy moths in a forest. *Physiol. Entomol.* 12:399–406

43. Fitzpatrick, S. M., McNeil, J. N., Dumont, S. 1988. Does male phero-mone effectively inhibit competition among courting true armyworm males (Lepidoptera: Noctuidae)? *Anim. Beh.* 36:1831–35

44. Flint, H. M., Smith, R. L., Forey, D. E., Horn, B. R. 1977. Pink bollworm: Response of males to (Z, Z)- and Z, E)-isomers of gossyplure. *Environ. Entomol.* 6:274–75

45. Flint, M. L., Van den Bosch, R. 1981. *Introduction to Integrated Pest Management.* New York: Plenum. 240 pp.

46. Gerber, G. H., Howlader, M. A. 1987. The effects of photoperiod and temperature on calling behaviour and egg development on the bertha armyworm, *Mamestra configurata* (Lepidoptera: Noctuidae). *J. Insect Physiol.* 33:429–36

47. Gorsuch, C. S., Karandinos, M. G., Koval, C. F. 1975. Daily rhythm of *Synanthedon pictipes* (Lepidoptera: Aegeriidae) female calling behavior in Wisconsin: Temperature effects. *Entomol. Exp. Appl.* 18:367–76

48. Hagan, D. V., Brady, U. E. 1981. Effects of male photoperiod on calling, pheromone levels, and oviposition of mated female *Trichoplusia ni. Ann. Entomol. Soc. Am.* 74:286–88

49. Han, E.-N. 1988. *Laboratory studies on the regulation of migration in the Oriental armyworm,* Mythimna separata *(Walker) (Lepidopera: Noctuidae).* PhD dissertation. Bangor: Univ. Wales

50. Haynes, K. F., Birch, M. C. 1984. The periodicity of pheromone release and male responsiveness in the artichoke plume moth, *Platyptilia carduidactyla. Physiol. Entomol.* 9:287–95

51. Haynes, K. F., Miller, T. A., Staten, R. T., Li, W.-G., Baker, T. C. 1986. Monitoring insecticide resistance with insect pheromones. *Experientia* 42:1293–95

52. Haynes, K. F., Miller, T. A., Staten, R. T., Li, W.-G., Baker, T. C. 1987. Pheromone trap for monitoring insecticide resistance in the pink bollworm moth (Lepidoptera: Gelechiidae): New tool for resistance management. *Environ. Entomol.* 16:84–89

53. Hendricks, D. E., Hollingsworth, J. P., Hartstack, A. W. Jr. 1972. Catch of tobacco budworm moths influenced by color of sex-lure traps. *Environ. Entomol.* 1:48–51

54. Hendrikse, A., Vos-Bunnemeyer, E. 1987. Role of host-plant stimuli in sexual behaviour of small ermine moths *(Yponomeuta). Ecol. Entomol.* 12:363–71

55. Hendry, L. B., Wichmann, J. K.,

Hindenlang, D. M., Mumma, R. O., Anderson, M. E. 1975. Evidence for origin of insect sex pheromones: Presence in food plants. *Science* 188:59–62
56. Henneberry, T. J., Clayton, T. E. 1984. Time of emergence, mating, sperm movement, and transfer of ejaculatory duct secretory fluid by *Heliothis virescens* (F.) (Lepidoptera: Noctuidae) under reversed light-dark cycle laboratory conditions. *Ann. Entomol. Soc. Am.* 77:301–5
57. Hirai, K., Shorey, H. H., Gaston, L. K. 1978. Competition among courting male moths: male-to-male inhibitory pheromone. *Science* 202:644–5
58. Hirano, C., Muramoto, H. 1976. Effect of age on mating activity of the sweet potato leaf folder, *Brachmia macroscopa* (Lepidoptera: Gelechiidae). *Appl. Entomol. Zool.* 11:154–59
59. Howell, J. F. 1974. The competitive effect of field populations of codling moth on sex attractant trap efficiency. *Environ. Entomol.* 3:803–7
60. Howlader, M. A., Gerber, G. H. 1986. Effects of age, egg development, and mating on calling behavior of the bertha armyworm, *Mamestra configurata* Walker (Lepidoptera: Noctuidae). *Can. Entomol.* 118:1221–30
61. Hoyt, S. C., Westigard, P. H., Rice, R. E. 1983. Development of pheromone trapping techniques for male San Jose scale (Homoptera: Diaspididae). *Environ. Entomol.* 12:371–75
62. Hummel, H. E., Miller, T. A., eds. 1984. *Techniques in Pheromone Research.* New York: Springer-Verlag. 464 pp.
63. Kaae, R. S., Shorey, H. H. 1972. Sex pheromones of noctuid moths. XXVII. Influence of wind velocity on sex pheromone releasing behavior of *Trichoplusia ni* females. *Ann. Entomol. Soc. Am.* 65:437–40
64. Kaae, R. S., Shorey, H. H., Gaston, L. K., Sellers, D. 1977. Sex pheromones of Lepidoptera: Seasonal distribution of male *Pectinophora gossypiella* in a cotton-growing area. *Environ. Entomol.* 6:284–86
65. Kanno, H. 1979. Effects of age on calling behaviour of the rice stem borer, *Chilo suppressalis* (Walker) (Lepidoptera: Pyralidae). *Bull. Entomol. Res.* 69:331–35
66. Kanno, H. 1981. Mating behaviour of the rice stem borer moth, *Chilo suppressalis* Walker (Lepidoptera: Pyralidae). V. Critical illumination intensity for female calling and male sexual response

under various temperatures. *Appl. Entomol. Zool.* 16:179–85
67. Kanno, H. 1981. Mating behaviour of the rice stem borer moth, *Chilo suppressalis* Walker (Lepidoptera: Pyralidae). VI. Effects of photoperiod on the diel rhythms of mating behaviours. *Appl. Entomol. Zool.* 16:406–11
68. Kanno, H., Sato, A. 1979. Mating behaviour of the rice stem borer moth, *chilo suppressalis* Walker (Lepidoptera: Pyralidae). II. Effects of temperature and relative humidity on mating activity. *Appl. Entomol. Zool.* 14:419–27
69. Kaster, L. V., Showers, W. B. 1982. Evidence of spring immigration and autumn reproductive diapause of the adult black cutworm in Iowa. *Environ. Entomol.* 11:306–12
70. Knight, A. L., Croft, B. A. 1987. Temporal patterns of competition between a pheromone trap and caged female moths for males of *Argyrotaenia citrana* (Lepidoptera: Tortricidae) in a semienclosed courtyard. *Environ. Entomol.* 16:1185–92
71. Kochansky, J. P., Cardé, R. T., Taschenberg, E. F., Roelofs, W. L. 1977. Rhythms of male *Antheraea polyphemus*. Attraction and female attractiveness, and improved pheromone synthesis. *J. Chem. Ecol.* 3:419–27
72. Landolt, P. J., Curtis, C. E. 1982. Effects of temperature on the circadian rhythm of navel orangeworm sexual activity. *Environ. Entomol.* 11:107–10
73. Lanier, G. N., Burns, B. W. 1978. Barometric flux. Effects on the responsiveness of bark beetles to aggregation attractants. *J. Chem. Ecol.* 4:139–47
74. Lecomte, C., Thibout, E. 1981. Attraction d'*Acrolepiopsis assectella*, en olfactométre, par des substances allélochimiques volatiles d'*Allium porrum*. *Entomol. Exp. Appl.* 10:293–300
75. Lewis, T., Macaulay, E. D. M. 1976. Design and elevation of sex-attractant traps for pea moth, *Cydia nigricana* (Steph.) and the effect of plume shape on catches. *Ecol. Entomol.* 1:175–87
76. Lewis, W. J. 1981. Semiochemicals: Their role with changing approaches to pest control. See Ref. 97a, pp. 3–12
77. Linn, C. E., Campbell, M. G., Roelofs, W. L. 1988. Temperature modulation of behavioural thresholds controlling male moth sex pheromone response specificity. *Physiol. Entomol.* 13:59–67
78. Linn, C. E. Jr., Roelofs, W. L. 1989. Response specificity of male moths to

multicomponent pheromones. *Chem. Senses* 14:421–37
79. Löfstedt, C., Löfqvist, J., Lanne, B. S., Van Der Pers, J. N. C., Hansson, B. S. 1986. Pheromone dialects in European turnip moths *Agrotis segetum*. *Oikos* 46:250–57
80. Lum, P. T. M., Brady, U. E. 1973. Levels of pheromone in female *Plodia interpunctella* mating with males reared in different light regimens. *Ann. Entomol. Soc. Am.* 66:822–23
81. MacLellan, C. R. 1976. Suppression of codling moth (Lepidoptera: Olethreutidae) by sex phromone trapping of males. *Can. Entomol.* 108:1037–40
82. McDonough, L. M., Brown, D. F., Aller, W. C. 1989. Insect sex pheromones. Effects of temperature on evaporation rates of acetates from rubber septa. *J. Econ. Entomol.* 15:779–90
83. McNally, P. S., Barnes, M. M. 1981. Effects of codling moth pheromone trap placement, orientation and density on trap catches. *Environ. Entomol.* 10:22–26
84. McNeil, J. N. 1986. Calling behavior: Can it be used to identify migratory species of moths? *Fla. Entomol.* 69:78–84
85. McNeil, J. N. 1987. The true armyworm, *Pseudaletia unipuncta*: A victim of the pied piper or a seasonal migrant? *Insect Sci. Applic.* 8:591–97
86. McNeil, J. N., Delisle, J. 1989. Are host plants important in pheromone-mediated mating systems of Lepidoptera? *Experientia* 45:236–40
87. McNeil, J. N., Delisle, J. 1989. Host plant pollen influences calling behavior and ovarian development of the sunflower moth, *Homoeosoma electellum*. *Oecologia* 80:201–5
88. Metcalf, R. L. 1980. Changing role of insecticides in crop protection. *Annu. Rev. Entomol.* 25:219–56
89. Miller, C. A., McDougall, G. A. 1973. Spruce budworm moth trapping using virgin females. *Can. J. Zool.* 51:853–58
90. Miller, J. R., Baker, T. C., Cardé, R. T., Roelofs, W. L. 1976. Reinvestigation of oak leaf roller sex pheromone components and the hypothesis that they vary with diet. *Science* 192:140–43
91. Minks, A. K., de Jong, D. J. 1975. Determination of spraying dates for *Adoxophyes orana* by sex pheromone traps and temperature recordings. *J. Econ. Entomol.* 68:729–32
92. Mitchell, E. R., Agee, H. R., Heath, R. R. 1989. Influence of pheromone trap color and design on capture of male velvetbean caterpillar and fall armyworm moths (Lepidoptera: Noctuidae). *J. Chem. Ecol.* 15:1775–84
93. Mulder, P. G., Showers, W. B., von Kaster, L., Vanschaik, J. 1989. Seasonal activity and response of the male black cutworm (Lepidoptera: Noctuidae) to virgin females reared for one or multigenerations in the laboratory. *Environ. Entomol.* 18:19–23
94. Nakamura, K. 1982. Competition between females and pheromone traps: Time lag between female mating activity and male trap captures. *Appl. Entomol. Zool.* 17:292–300
95. Noguchi, H., Tamaki, Y. 1985. Conspecific female sex pheromone delays calling behavior of *Adoxopyes* sp. and *Homona magnanima* (Lepidoptera: Tortricidae). *Jpn. J. Appl. Entomol. Zool.* 29:113–18
96. Nordlund, D. A., Brady, U. E. 1974. The calling behavior of female *Plodia interpunctella* (Hübner) (Lepidoptera: Pyralidae) under two light regimes. *Environ. Entomol.* 3:793–96
97. Nordlung, D. A., Brady, U. E. 1974. Factors affecting release rate and production of sex pheromone by female *Plodia interpunctella* (Hübner) (Lepidoptera: Pyralidae). *Environ. Entomol.* 3:797–802
97a. Nordlund, D. A., Jones, R. L., Lewis, W. J., eds. 1981. *Semiochemicals. Their Role in Pest Control*. New York: Wiley. 306 pp.
98. Palanisawamy, P., Seabrook, W. D. 1978. Behavioral responses of the female eastern spruce budworm, *Choristoneura fumiferana* (Lepidoptera: Tortricidae) to the sex pheromone of her own species. *J. Chem. Ecol.* 4:649–55
99. Palanisawamy, P., Seabrook, W. D. 1985. The alteration of calling behaviour by female *Choristoneura fumiferana* when exposed to the synthetic sex pheromone. *Entomol. Exp. Appl.* 37:13–16
100. Payne, T. L., Birch, M. C., Kennedy, C. E. J., eds. 1986. *Mechanisms in Insect Olfaction*. Oxford: Clarendon. 364 pp.
101. Prestwich, G. D., Blomquist, G. J., eds. 1987. *Pheromone Biochemistry*. New York: Academic. 565 pp.
102. Prestwich, G. D., Graham, S. McG., Handley, M., Latli, B., Streinz, L., Tasayco J. M. L. 1989. Enzymatic processing of pheromones and pheromone analogs. *Experientia* 45:263–70
103. Prokopy, R. J. 1981. Epideictic pheromones that influence spacing patterns

of phytophagous insects. See Ref. 97a, pp. 181–213

104. Rabb, R. L. 1970. Introduction to the conference. In *Concepts of Pest Management*, ed. R. L. Rabb, F. E. Guthrie, pp. 1–5. Raleigh: N.C. State Univ. 242 pp.

105. Raina, A. K. 1988. Host plant, hormone interaction and sex pheromone production and release in *Heliothis* species. In *Endocrinological Frontiers in Physiological Insect Ecology*, ed. F. Sehnal, A. Zabza, D. L. Denlinger, pp. 33–36. Wroclaw: Wroclaw Tech. Univ. Press

106. Raina, A. K., Jaffe, H., Kempe, T. G., Keim, P., Blacher, R. W., et al. 1989. Identification of a neuropeptide hormone that regulates sex pheromone production in female moths. *Science* 244:796–98

107. Raina, A. K., Klun, J. A. 1984. Brain factor control of sex pheromone production in the female corn earworm moth. *Science* 225:531–33

108. Raina, A. K., Klun, J. A., Stadelbacher, E. A. 1986. Diel periodicity and effect of age and mating on female sex pheromone titer in *Heliothis zea* (Lepidoptera: Noctuidae). *Ann. Entomol. Soc. Am.* 79:128–31

109. Renwick, J. A. A. 1989. Chemical ecology of oviposition in phytophagous insects. *Experientia* 45:223–28

110. Riddiford, L. M. 1967. Trans-2-hexenal: Mating stimulant for polyphemus moths. *Science* 158:139–41

111. Riddiford, L. M., Williams, C. M. 1967. Volatile principle from oak leaves: Role in sex life of the polyphemus moths. *Science* 155:589–90

112. Roelofs, W. L. 1978. Threshold hypothesis for pheromone perception. *J. Chem. Ecol.* 4:685–99

113. Roelofs, W. L., Glover, T., Tang, X.-H., Sreng, I., Robbins, P., et al. 1987. Sex pheromone production and perception in European corn borer moths is determined by both autosomal and sex-linked genes. *Proc. Natl. Acad. Sci. USA* 84:7585–89

114. Roelofs, W. L., Wolf, W. A. 1988. Pheromone biosynthesis in Lepidoptera. *J. Chem. Ecol.* 14:2019–31

115. Roitberg, B. D., Angerilli, N. P. D. 1986. Management of temperate-zone deciduous fruit pests: Applied behavioural ecology. In *Agricultural Zoology Reviews*, ed. G. E. Russell, pp. 137–65. Newcastle upon Tyne: Intercept

116. Rossiter, M., Gershenzon, J., Mabry, T. J. 1986. Behavioral and growth responses of specialist herbivore, *Homoeosoma electellum*, to major ter-

penoid of its host, *Helianthus* spp. *J. Chem. Ecol.* 12:1505–21

117. Saario, C. A., Shorey, H. H., Gaston, L. K. 1970. Sex pheromones of noctuid moths. XIX. Effect of environmental and seasonal factors on captures of males of *Trichoplusia ni* in pheromone-baited traps. *Ann. Entomol. Soc. Am.* 63:667–71

118. Sanders, C. J. 1978. Evaluation of sex attractant traps for monitoring spruce budworm populations (Lepidoptera: Tortricidae). *Can. Entomol.* 110:43–50

119. Sanders, C. J. 1987. Flight and copulation of female spruce budworm in pheromone-permeated air. *J. Chem. Ecol.* 13:1749–58

120. Sanders, C. J., Lucuik, G. S. 1972. Factors affecting calling by female eastern spruce budworm, *Choristoneura fumiferana* (Lepidoptera: Tortricidae). *Can. Entomol.* 104:1751–62

120a. Sanders, C. J., Wilson, G. G. 1990. Flight duration of male spruce budworm (*Choristoneura fumiferana* [Clem.]) and attractiveness of female spruce budworm are unaffected by microsporidian infection or moth size. *Can. Entomol.* 122:419–22

121. Schal, C., Charlton, R. E., Cardé, R. T. 1987. Temporal patterns of sex pheromone titers and release rates in *Holomelina lamae* (Lepidoptera: Arctiidae). *J. Chem. Ecol.* 13:1115–29

122. Shorey, H. H. 1977. Interaction of insects with their chemical environment. In *Chemical Control of Insect Behavior*, ed. H. H. Shorey, J. J. McKelvey, Jr., pp. 1–5. New York: Wiley

123. Shorey, H. H., Gaston, L. K. 1964. Sex pheromone of noctuid moths. III. Inhibition of males responses to the sex pheromone of *Trichoplusia ni* (Lepidoptera: Noctuidae). *Ann. Entomol. Soc. Am.* 57:775–79

124. Shorey, H. H., Morin, K. L., Gaston, L. K. 1968. Sex pheromones of noctuid moths. XV. Timing of development of pheromone-responsiveness and other indicators of reproductive age in males of eight species. *Ann. Entomol. Soc. Am.* 61:857–61

125. Snir, R., Dunkelblum, E., Gothilf, S., Harpaz, I. 1986. Sexual behavior and pheromone titre in the tomato looper, *Plusia chalcites* (Esp.) (Lepidoptera: Noctuidae). *J. Insect Physiol.* 32:735–39

126. Song, Y. H., Riedl, H. 1985. Effects of temperature and photoperiod on male activity in *Laspeyresia pomonella* (L.) in

New York. *Korean J. Plant Prot.* 24:71–77

127. Sower, L. L., Shorey, H. H., Gaston, L. K. 1970. Sex pheromones of noctuid moths. XXI. Light:dark cycle regulation and light inhibition of sex pheromone release by females of *Trichoplusia ni. Ann. Entomol.Soc. Am.* 63:1090–92

128. Sower, L. L., Shorey, H. H., Gaston, L. K. 1971. Sex pheromones of noctuid moths. XXV. Effects of temperature and photoperiod on circadian rhythms of sex pheromone release by females of *Trichoplusia ni. Ann. Entomol. Soc. Am.* 64:488–92

129. Stockel, J. P., Boidron, J. N. 1981. Influence d'extraits aromatiques de grains de Maïs sur l'activité reproductrice de l'alucite des céréales *Sitotroga cerealella* (Lépidoptère: Gelechiidae) en conditions naturelles. *C. R. Acad. Sci. Paris* 292:343–46

130. Sugie, H., Yamazaki, S., Tamaki, Y. 1976. On the origin of the sex pheromone components of the smaller tea tortrix moth, *Adoxophyes fasciata* Walsingham (Lepidoptera: Tortricidae). *Appl. Entomol. Zool.* 11:371–73

131. Svärd, L., Wiklund, C. 1989. Mass and production rate of ejaculates in relation to monandry/polyandry in butterflies. *Behav. Ecol. Sociobiol.* 24:395–402

132. Sweeney, J. D., McLean, J. A. 1987. Effect of sublethal infection levels of *Nosema* sp. on the pheromone-mediated behavior of the western spruce budworm, *Choristoneura occidentalis* Freeman (Lepidoptera: Tortricidae). *Can. Entomol.* 119:587–94

133. Swier, S. R., Rings, R. W., Musick, G. J. 1976. Reproductive behavior of the black cutworm, *Agrotis ipsilon. Ann. Entomol. Soc. Am.* 69:546–50

134. Swier, S. R., Rings, R. W., Musick, G. J. 1977. Age-related calling behavior of the black cutworm, *Agrotis ipsilon. Ann. Entomol. Soc. Am.* 70:919–24

135. Szöcs, G., Tóth, M. 1979. Daily rhythm and age dependence of female calling behaviour and male responsiveness to sex pheromone in the gamma moth, *Autographa gamma* (L.) (Lepidoptera: Noctuidae). *Acta Phytopathol. Acad. Sci. Hung.* 14:453–59

136. Tang, J. D., Charlton, R. E., Jurenka, R. A., Wolf, W. A., Phelan, P. L., et al. 1989. Regulation of pheromone biosynthesis by a brain hormone in two moth species. *Proc. Natl. Acad. Sci. USA* 86:1806–10

137. Tauber, M. J., Tauber, C. A., Masaki, S. 1986. *Seasonal Adaptations of In-sects.* Oxford: Oxford Univ. Press. 411 pp.

138. Teal, P. E. A., Byers, J. R. 1980. Biosystematics of the genus *Euxoa* (Lepidoptera: Noctuidae). XIV. Effect of temperature on female calling behavior and temporal partitioning in three sibling species of *Declarata* group. *Can. Entomol.* 112:113–17

139. Teal, P. E. A., Tumlinson, J. H., Oberlander, H. 1989. Neural regulation of sex pheromone biosynthesis in *Heliothis* moths. *Proc. Natl. Acad. Sci. USA* 86:2488–92

140. Teetes, G. L., Randolf, N. M. 1969. Chemical and cultural control of the sunflower moth in Texas. *J. Econ. Entomol.* 62:1444–47

141. Tomescu, N., Stan, G., Chis, V., Jeleriu, S., Pastinaru, C. 1981. Influence of light and age on the response of males of *Mamestra brassicae* L. (Lepidoptera: Noctuidae) to sexual pheromone. *Stud. Univ. Babes* 26:43–47

142. Tóth, M. 1979. Pheromone-related behaviour of *Mamestra suasa* (Schiff.): Daily rhythm and age dependence. *Acta Phytopathol. Acad. Sci. Hung.* 14:189–94

143. Traynier, R. M. M. 1970. Sexual behaviour of the mediterranean flour moth, *Anagasta kühniella:* Some influences of age, photoperiod, and light intensity. *Can. Entomol.* 102:534–40

144. Tumlinson, J. H. 1988. Contemporary frontiers in insect semiochemical research. *J. Chem. Ecol.* 14:2109–30

145. Turgeon, J., McNeil, J. 1982. Calling behaviour of the armyworm, *Pseudaletia unipuncta. Entomol. Exp. Appl.* 31:402–8

146. Turgeon, J. J., McNeil, J. N. 1983. Modifications in the calling behaviour of *Pseudaletia unipuncta* (Lepidoptera: Noctuidae), induced by temperature conditions during pupal and adult development. *Can. Entomol.* 115:1015–22

147. Turgeon, J. J., McNeil, J. N., Roelofs, W. L. 1983. Field testing of various parameters for the development of a pheromone-based monitoring system for the armyworm, *Pseudaletia unipuncta* (Haworth) (Lepidoptera: Noctuidae). *Environ. Entomol.* 12:891–94

148. Turgeon, J. J., McNeil, J. N., Roelofs, W. L. 1983. Responsiveness of *Pseudaletia unipuncta* males to the female sex pheromone. *Physiol. Entomol.* 8:339–44

149. Underhill, E. W., Rogers, C. E., Chisholm, M. D., Steck, W. F. 1982. Monitoring field populations of the sun-

flower moths, *Homoeosoma electellum* (Lepidoptera: Pyralidae), with its sex pheromone. *Environ. Entomol.* 11:681–84

150. Usui, K., Uchiumi, K., Kurihara, M., Fukami, J.-I., Tatsuki, S. 1988. Sex pheromone content in female *Chilo suppressalis* Walker (Lepidoptera: Pyralidae) reared on artificial diets. *Appl. Entomol. Zool.* 23:97–99

151. Van Den Bosch, R. 1978. *The Pesticide Conspiracy.* New York: Doubleday. 226 pp.

152. Vinson, S. B. 1977. Behavioral chemicals in the augmentation of natural enemies. In *Biological Control by Augmentation of Natural Enemies,* ed. R. L. Ridgway, S. B. Vinson, pp. 237–79. New York: Plenum

153. Wall, C., Perry, J. N. 1980. Effects of spacing and trap number on interactions between pea moth pheromone traps. *Entomol. Exp. Appl.* 28:313–21

154. Wall, C., Perry, J. N. 1983. Further observations on the responses of male pea moth, *Cydia nigricana,* to vegetation previously exposed to sex attractant. *Entomol. Exp. Appl.* 33:112–16

155. Wall, C., Sturgeon, D. M., Greenway, A. R., Perry, J. N. 1981. Contamination of vegetation with synthetic sex-attractant released from traps for the pea moth, *Cydia nigricana. Entomol. Exp. Appl.* 30:111–15

156. Webb, J. W., Berisford, C. W. 1978. Temperature modification of flight and response to pheromones in *Rhyacionia frustrana. Environ. Entomol.* 7:278–80

157. Webster, R. P. 1988. Modulation of the expression of calling by temperature in the omnivorous leafroller moth, *Platynota stultana* (Lepidoptera: Tortricidae) and other moths: A hypothesis. *Ann. Entomol. Soc. Am.* 81:138–51

158. Webster, R. P., Cardé, R. T. 1982. Influence of relative humidity on calling behaviour of the female European corn borer *(Ostrinia nubilalis). Entomol. Exp. Appl.* 32:181–85

159. Webster, R. P., Cardé, R. T. 1982. Relationships among pheromone titre, calling and age in the omnivorous leafroller moth *(Platynota stultana). J. Insect Physiol.* 28:925–33

160. Webster, R. P., Conner, W. E. 1986. Effects of temperature, photoperiod, and light intensity on the calling rhythm in arctiid moths. *Entomol. Exp. Appl.* 40:239–45

161. Werner, R. A. 1977. Behavioral responses of the spear-marked black moth, *Rheumaptera hastata,* to a female-produced sex pheromone. *Ann. Entomol. Soc. Am.* 70:84–86

162. West, R. J., Teal, P. E. A., Laing, J. E., Grant, G. M. 1984. Calling behavior of the potato stem borer, *Hydraecia micacea* Esper (Lepidoptera: Noctuidae), in the laboratory and the field. *Environ. Entomol.* 13:1399–1404

163. Wilson, G. G. 1987. Observations on the level of infection and intensity of *Nosema fumiferanae* (Microsporida) in two different field populations of the spruce budworm, *Choristoneura fumiferana. Inf. Report FPM-X-79. Ontario, Canadian Forestry Service.* 15 pp.

Annu. Rev. Entomol. 1991. 36:431–57
Copyright © 1991 by Annual Reviews Inc. All rights reserved

WHITEFLY BIOLOGY

David N. Byrne

Department of Entomology, University of Arizona, Tucson, Arizona 85721

Thomas S. Bellows, Jr.

Department of Entomology, University of California, Riverside, California 92521

KEY WORDS: Aleyrodidae, Homoptera, life history, population dynamics, migration

INTRODUCTION

The importance of whiteflies as economic pests seems to expand continually (31, 78, 120). These homopteran insects damage crops by extracting large quantities of phloem sap, which can result in greater than 50% yield reductions (119). The honeydew excreted by these insects serves as a medium for sooty mold fungi (e.g. *Capnodium* spp.) that discolor parts of the plants used for food and fiber (141, 142). Finally, a few species serve as vectors of several economically important viral plant pathogens (133).

This article reviews characteristics of whiteflies that set them apart from other members of Sternorrhyncha and presents a review of whitefly literature to obtain a better understanding of their biology and assess the status of whitefly research.

The family Aleyrodidae is considered to have two subfamilies. The Aleurodicinae, endemic primarily to Central and South America, may be considered the more primitive taxa because of more complex wing venation (81, 131). However, the increased venation may be necessary because Aleurodicinae members are generally larger than whiteflies of the subfamily Aleyrodinae. This increased size [> 2.0 mm long (81)] likely requires greater wing support. The Aleyrodinae is larger in terms of number of species and is also more widespread.

431

Enderlein (72) suggested a third subfamily, Udamoselinae on the basis of one specimen of a South American species, a male with a body length of 7 mm. Today the existence of the subfamily is thought to be dubious (131).

HISTORICAL PERSPECTIVE

Whiteflies are considered the tropical equivalent of aphids owing to their ordinal characteristics and their scarcity in temperate climates. Réaumur first described whiteflies in 1736, although he mistakenly placed *Aleyrodes proletella* in Lepidoptera (64). In 1795, Latreille correctly placed them in Homoptera (64). The majority of the literature from the 19th century was taxonomic in nature and consisted primarily of descriptions of pupal characteristics (e.g. 63, 121). Pupal cases are still generally of more value than other life stages when making taxonomic decisions. Maskell (121) reports that, from the time of the insects' first description in 1736 until 1895, little more than 50 noteworthy articles were published on them. Many of these papers were taxonomic treatments with little information about whitefly biology.

Quaintaince (149) in 1900 solved many of the taxonomic problems concerning the whiteflies of North America. Work by Bemis (20), who helped bring the total number of described North American species to 62, also did much to end taxonomic confusion. Trehan (162) and Mound & Halsey (131) played a large role in correcting global whitefly taxonomy. To date, however, little has been accomplished concerning whitefly systematics. The most comprehensive work available is *Whitefly of the World* (131); this very useful publication presents a list of genera and species but says little about systematic relationships. Data on whitefly cladistics appear to be unavailable.

Information on basic biology has been even slower in developing, and early research was in part restricted to pest species. Back (14) was concerned about the presence of *Aleurothrixus floccosus* on citrus. Lloyd (119) published a report on the biology of *Trialeurodes vaporariorum*, but his primary intent was to describe greenhouse management strategies. *Bemisia tabaci* was first noticed on cotton in India in 1905 (97). Bionomic investigations did not begin in earnest, however, until several years later, when this whitefly was shown to seriously damage crops.

After the turn of the century, more attention was focused on basic biology. Morrill & Back (127) examined the biology of *Dialeurodes citrifolii* and *Dialeurodes citri*. Hargreaves' (89) examination of *T. vaporariorum* in 1915 was another early report on whitefly bionomics and Garman & Jewitt (76) studied whitefly biology in greenhouses.

Because of concentration on pest species, particularly *B. tabaci* and *T. vaporariorum*, much of our data relate to a limited number of the > 1,200 known species. The most closely examined species are those that feed on a wide variety of herbaceous hosts. Polyphagy, however, is not usually recorded in Aleyrodidae. As a result, drawing conclusions about the family based on what is known about a few species might not be appropriate. Fortunately, we have learned a great deal about the biology of whiteflies from work that concentrates on nonpest species. Information is available concerning > 25 others, most of which are monophagous or oligophagous animals associated with woody perennial hosts. Bibliographies of the family include those by Trehan & Butani (163), Butani (26), Thompson & Reinert (161), and Cock (45).

ORIGIN AND DISTRIBUTION

The geographic origin of many whitefly species is largely speculation. The process of making these determinations, while always difficult, becomes particularly hard with a family that has not had the attention of entomologists for a long period of time. For example, *Dialeurodes chittendeni* is thought to have a Himalayan origin because of its connection with rhododendrons, which also originated there (115). Although such botanical associations provide some information concerning origin, such evidence should not be considered conclusive unless it is connected to additional evidence such as the presence of sibling species. Identifying whitefly distribution is troublesome, and these determinations are becoming even more difficult as humans facilitate whitefly movement. Nevertheless, the following description offered by Mound & Halsey (131) is the best available and was undoubtedly accurate when published:

> Three of the largest whitefly genera, *Aleuroplatus, Aleurotrachelus* and *Tetraleurodes,* are more or less artificial assemblages of species with black pupal cases reported from many parts of the world. Many of the other genera, however, have a more restricted distribution. The genera of the subfamily Aleurodicinae are almost entirely confined to the Neotropics, and this is also true of a few genera of the Aleyrodinae such as *Aleurocerus, Aleurothrixus, Bellitudo* and *Crenidorsum.* The genus *Trialeurodes* has most of its species in the New World as does *Aleuroparadoxus.* In contrast *Africaleurodes, Aleurolonga, Aleuropteridis, Corbettia* and *Dialeurolonga* are recorded only from Africa and Madagascar. *Acaudaleyrodes, Aleurocanthuys, Aleurolobus, Aleurotuberculatus, Dialeuropora* and the largest genus, *Dialeurodes,* are widely distributed in the Ethiopian and Oriental Regions. *Pealius, Odontaleyrodes* and *Rhachisphora* are also particularly common in the Oriental and Austro-Oriental regions, whereas *Orchamoplatus* is apparently most common in the Pacific.

PHYSICAL CHARACTERISTICS

Whiteflies share many characteristics with other homopterans. All are plant feeders with piercing, sucking mouthparts. They are opisthognathus. Additionally, adults of both sexes have four membranous wings. Finally, members of the family undergo incomplete metamorphosis (with certain complications).

Vasiform Orifice

One set of features that sets the Aleyrodidae apart from related familes is the presence of a vasiform orifice with its operculum and lingula (81). The vasiform orifice is generally located on the dorsum of the ninth abdominal segment of males. In females, it extends to the eighth abdominal segment (87). It is not the anus, but rather the depression into which the anus empties the contents of the digestive tract (honeydew). A dorsal anus is also found in psyllid adults (rarely in the nymphs) and it also functions to rid these insects of honeydew. In psyllids the anus seems to have been displaced by enlarged genitalia (130). The dorsal placement of the anus in Aleyrodidae permits the effective handling of the honeydew that these phloem-feeding insects produce in copious amounts. Honeydew can present problems, particularly for sessile nymphs, because it is viscous and associated with sooty mold fungi. Whitefly nymphs cannot walk away from their honeydew droplets. They flip their excreta away: the honeydew fills the vasiform orifice; the lingula is cocked down into the liquid; and, when the lingula is released, the honeydew is catapulted away.

Waxes

Whiteflies are distinctive in that all life stages, except the egg, can produce extracuticular waxes that cover the body. Some of the waxes are similar in form to those found on coccids (122); others are not. In whitefly nymphs, wax can appear as a gelatinous mass, as plumes, as columns, or as setae-like projections (81), and comes in two colors (81). One is clear or colorless, although it may reflect the color of the surrounding leaf surface. The other is brilliant white. The transparent form usually appears as a very thin layer over the entire dorsal surface. Transparent wax may also appear as a marginal fringe or as dorsal spikelike wax rays such as those found in *Trialeurodes* spp. The number of these rays in some nymphs differs when the whiteflies are reared on different hosts (129). This variance results from changes in the number of papillae (56, 151). Opaque white wax may be produced as a marginal fringe (e.g. in *Tetraleurodes* spp.), dorsally as tufts *(Aleurotrachelus jelinekii* and *Aleuroplatus coronata),* or as flocculent or woolly mats *(A. floccosus* and *Aleurodicus dispersus).* Other colors rarely, if ever, occur in this family although the marginal fringes in some species of *Trialeurodes* tend to have light yellow or orange tints (81).

Wax in the adults takes on a different appearance. In *B. tabaci* and *T. vaporariorum,* the wax forms tight curls of threads approximately 1 μm in diameter (35). The material is extruded from wax plates that consist of rows of microtrichia found on the ventro-lateral abdominal surface. Each microtrichium is associated with a wax canal. Females have two pairs of these wax plates; males have four. The material is extruded as a continuous ribbon but is broken off as curly particles when the animal's hind tibiae pass over the plates (35). Hind and forelegs distribute the particles over the wings and the rest of the body (except the eyes). The wax of the two species consists primarily of triacylglycerols (65–75%) with a trace of wax esters, free fatty acids, alcohols, and hydrocarbons (35).

Fourth Nymphal Instar

In the literature, the fourth nymphal instar is commonly referred to as a pupa (20, 81). The term implies that whiteflies exhibit a degree of holometabolism. Hinton (91, 92) was certain whiteflies had a pupal stage in the sense that this stage serves as a mold for some of the imaginal muscles.

Although we do not consider Hinton's interpretation to be correct, we believe that the word pupa has been so inculcated into whitefly literature that its replacement would only cause confusion. What takes place during the whitefly's fourth nymphal instar is, in fact, distinct from what occurs in holometabolous families. Pupation has also never been reported to occur in other homopterous families. Nechols & Tauber (134), discussing *T. vaporariorum,* stated that the fourth nymphal instar has three morphologically distinct forms. The early fourth instar is flattened and transluscent. This form according to Gill (81) feeds and so clearly is not a pupa. The next form, the transitional substage, is expanded and opaque-white with dorsal and lateral waxy, spinelike processes. At this point, apolysis takes place. The last stage, when the pharate adult form is present, has the red eyes and the yellow body pigment of the adult. At this stage, apolysis is complete and the adult cuticle is laid down. J. R. Nechols (personal communication) states, however, that even after careful histological examination, he was unable to observe a distinct stadium that intervened between the last nymphal stage and the adult. For functional purposes, we would reserve the term pupa for the last, nonfeeding portion of the last nymphal stadium found after apolysis has occurred and refer to the earlier portion of the last stadium as the fourth nymphal instar.

LIFE HISTORY

Size

Gill (81) reports that the physically largest genus is *Aleurodicus* spp., which has an adult body length of > 2 mm and a wing expanse of > 3.5–4.0 mm. The dimensions of most whitefly adults, however, more closely approximate a range with *B. tabaci* at the lower end, *T. vaporariorum* in midrange, and *D.*

citri at the upper end (Table 1). Sexual dimorphism in adult body dimension is usual for members of Aleyrodidae; males are smaller (Table 1). This observation applies to late nymphal instars of *D. citri* (T. S. Bellows, unpublished) as well as *D. chittendeni* (177) and *Tetraleurodes acaciae* (66).

Oogenesis

The internal reproductive system and oogenesis in whiteflies is similar to that of other homopterans. Gameel (75) depicts 15 ovarioles in *B. tabaci;* each ovariole contains one or more follicles and a germarium.

Eggs

Whitefly eggs generally are pyriform or ovoid and possess a pedicel that is a peglike extension of the chorion. In most species, eggs assume an erect stance, but the eggs of *Aleurocybotus occiduus* lie on their side (144). The pedicel is either inserted into a slit made by the ovipositor in the leaf surface or into a stomatal opening. The literature describes 13 species that insert their pedicels into stomata (138, 144) and four that utilize slits in the leaf epidermis (58, 86, 138, 144). Poinar (144) speculated that egg pedicles of *A. occiduus* were inserted into the stomata because the epidermis of grasses and sedges contains large amounts of silica and lignin and would be difficult for females to penetrate.

Quaintance & Baker (150) believed the pedicel, in addition to providing a means of attachment, served as a guide for spermatozoa during fertilization, during which time the pedicel is filled with protoplasm. After fertilization, the protoplasm withdraws and the pedicel becomes a hollow tube.

Weber (174) observed that *T. vaporariorum* secretes a gluelike substance around the pedicel. When the pedicel is inserted into parenchyma cells, very little of this material is present. Pedicels inserted into interstitial spaces are associated with a much larger quantity of the glue. Weber postulated that

Table 1 Body measurements (mm) of three species of Aleyrodidae (n = 5)

Species	Body length	Wing expanse
Bemisia tabaci		
male	0.85 ± 0.05	1.81 ± 0.06
female	0.91 ± 0.04	2.13 ± 0.06
Trialeurodes vaporariorum		
male	0.99 ± 0.03	2.41 ± 0.06
female	1.06 ± 0.04	2.65 ± 0.12
Dialeurodes citri		
male	1.20 ± 0.03	2.49 ± 0.07
female	1.20 ± 0.05	3.01 ± 0.05

water passes osmotically across this colloidal mass and enters the egg through the pedicel. Gameel (75) reports a similar substance surrounding the egg pedicle of *B. tabaci,* and associates the production of this material with a "cement gland."

Several other authors have also suggested that aleyrodids use the pedicel as a means of absorbing water into the egg. Citing Weber (174) and Wiggelsworth (175), Hinton (93) made the claim that the pedicel was involved in the transfer of water into the egg but offered no empirical evidence. Poinar (144) used the fact that eggs of *A. occiduus* dried up when removed from the leaf to support the argument that the pedicel must absorb water from the plant. Recently Byrne et al (33) assayed eggs raised on plants irrigated with tritiated water and demonstrated that water extracted from plant tissue accounted for approximately 50% of the mass of a mature whitefly egg.

Azab et al (13) examined oviposition rates of *B. tabaci* reared on cotton in Egypt in an open-air insectary. The maximum number of eggs per female varied from 48 to 394. During July and August when the daily maximum temperature (DMT) was 28.5°C, the mean rate of oviposition was 252 eggs per female. During October and November, the mean was 204 the (DMT 22.7°C). In December and January (DMT 14.3°C), the mean rate was 61 eggs. *B. tabaci* from Arizona cotton laid no eggs at 14.9°C, but produced 81 eggs at 26.7°C and 72 at 32.2°C (29). These results occurred under conditions of constant temperature and light. We interpret Gameel's (75) results on *B. tabaci* in Sudanese cotton fields to reveal a fecundity of 160.4 eggs during early fall. For the same species, Husain & Trehan (99) report a fecundity of 43 eggs on cotton in India, and Avidov (10) gives a fecundity value of approximately 50 eggs on eggplant. Von Arx et al (168) calculated the fecundity of a Sudanese strain to be 127.5 eggs. However, Dittrich et al (61) found that Sudanese whiteflies have a fecundity on cotton of 344 eggs. Sudanese strains of *B. tabaci* appear to be much more prolific than strains from other parts of the world (80).

Oviposition rates also vary greatly in other species and are affected by environmental conditions and host plant. Burnett (25) reports that the fecundity of *T. vaporariorum* at 18°C was 319.5, and that it fell to 5.5 at 33°C and to zero at 9°C. Fecundity increased fivefold when the insects were placed on a new leaf (25). The fecundity of *Aleurocanthus woglumi* varied with the nymphal host plant, ranging from 8.3 to 39.6 eggs per female (69).

Ovipositional Habits

Oviposition habits differ somewhat between species. Whiteflies such as *B. tabaci* deposit a few eggs on the leaf upon which they emerge as adults and then move to newer growth. The eggs are laid indiscriminately. *Parabemisia myricae* limits oviposition on citrus to very young leaves and oviposits on the underside of the leaves as well as on the leaf margins. *Tetraleurodes stanfordi* lays its eggs on both sides of the leaf (D. N. Byrne, personal observation). *A. woglumi* lays its eggs in a spiral arrangement (86). Some whiteflies oviposit their eggs in an arc or circle while their mouthparts remain inserted. *T. vaporariorum* lays its eggs in a circular fashion on glabrous leaves, although it abandons that pattern on pubescent leaves (47). *A. floccossus* and *Paraleyrodes* sp. also oviposit eggs in a circular fashion.

Crawlers

Following completion of development, the egg cracks at the apical end along a longitudinal line of dehiscence. As the first-instar nymph of *B. tabaci* begins to emerge, it bends in half until its front legs can clasp the leaf, after which it walks away from the spent chorion. Emergence behavior of *A. occiduus* crawlers is slightly different because these eggs are laid on their side (144).

Crawlers (first-instar nymphs) of Aleyrodinae have functional walking legs (with three apparent segments) and antennae (with two apparent segments). Legs and antennae of the second- and third-instar nymphs appear to have only one segment (81). Fourth-instar nymphs have distinct legs and antennae, although the divisions are indistinct. Domenichini (62) suggests an additional segmentation or suture in legs of second- through fourth-instar nymphs. Nymphs of all nymphal stadia of the Aleurodicinae have three-segmented legs. Some species possibly use the nymphal legs to assist in casting exuviae (81).

In warmer summer conditions, crawlers walk rather quickly over the leaf surface in search of an available minor vein, usually on the same leaf upon which the egg was laid. Some crawlers, such as those of *A. woglumi,* were found to move between plants (67). The portion of the population moving between host plants, however, was small (approximately 0.3%) and the distances traveled were short (< 30 mm). While the nymphs usually settle in a few hours, the process can take several days in cooler weather (10, 155). After settling, they insert their mouthparts into the phloem tissue and begin extracting sap.

Crawler mortality has been attributed to several plant characteristics, including cuticular thickness and nutritional factors. Walker (169) reported a high degree of crawler mortality for *P. myricae* on mature lemon leaves. He attributed this mortality to the fact that the thick cuticle of mature leaves prevents penetration. Walker also suggested that the possible presence of a probing deterrent might interfere with feeding. Byrne & Draeger (34), comparing crawler survival of *B. tabaci* on young (5-leaf stage) vs mature (> 25-leaf stage) lettuce, found 100% mortality on the mature lettuce and 58.1% on the young lettuce. They ascribed this difference to changes in the nutritional quality of the plant because crawlers were equally successful in reaching the phloem tissue of both plant stages.

If crawlers successfully reach the phloem of an appropriate host, they remain sessile until they reach the adult stage, except for brief periods during molts.

Whitefly nymphs have shallow breathing folds in the ventral body wall, two thoracic and one caudal (R. J. Gill, personal communication). These form a passage to the spiracles and may assist in conduction of air. The folds probably developed in response to the fact that nymphs have a flattened body and lie closely appressed to the leaf surface.

Middle Instars

Most second- and third-instar whitefly nymphs have an oval or elongate-oval body. Some nymphs may be circular or nearly heart-shaped.

While some members of the *Trialeurodes* spp. possess dorsal papillae—*B. tabaci* possesses dorsal setae and *Siphoninus* spp. have dorsal siphunculi— most species have a relatively simple dorsum. Gill (81) provides a recent review of the morphology of nymphs, pupae, and adults for the family.

Adult Eclosion

Under a constant temperature of $29.5 \pm 0.6°C$ and a photoperiod of $14:10$ LD, 90% of *B. tabaci* adults emerged from their pupal cases between 0600 and 0930 hours (lights on occurred at 0600 hours) (94). Few emerged during hours of darkness, and the peak time of emergence was delayed when temperatures were fluctuated. Under a series of constant temperatures, a significant inverse correlation was found between the time of median emergence (i.e. eclosion of 50% of the total number of adults) and temperature. No emergence was observed at temperatures below $17 \pm 0.3°C$. Emergence patterns persisted under conditions of continuous light or darkness, suggesting the presence of a circadian system. Similar patterns of emergence were found for *B. tabaci* and *T. vaporariorum* (38). *A. floccosus* was reported to eclose primarily between 0600 and 0900 hours in Hawaii (139).

The teneral period (between final ecdysis and first fight) for *B. tabaci* and *T. vaporariorum* is approximately 4 h 10 min at 27°C (38). For *A. proletella*, the teneral period was inversely related to temperature (57.8 h, at 10°C and 16.4 h at 25°C) and was longer when the animals were reared on young leaves (16.4 h) than on older ones (5.7 h) (103). This difference suggests that feeding takes place during the teneral period.

NUTRITION AND EXCRETION

Nutrition

As far as is known, whiteflies are phloem feeders. A great deal of information is available about the nutritional requirements of other homopterans, particularly aphids (7–9, 52–54, 125, 126, 156), but until recently almost nothing was known about the needs of whiteflies. Byrne & Miller (37) analyzed the carbohydrate and amino acid contents of pumpkin and poinsettia phloem sap, and of the honeydew produced by *B. tabaci* feeding on these host plants. Of the 14 or 15 amino acids found in the phloem sap, approximately half were found at significantly lower levels in the honeydew produced by *B. tabaci* feeding on the two hosts. This result indicates that these amino acids were metabolized either by the whiteflies or by the symbionts housed in their mycetocytes (96). Additionally, six amino acids not found in the phloem sap were found in the honeydew of whiteflies feeding on both hosts. The predominant amino acid in honeydew was glutamine ($> 50\%$ of the total amino acid content).

Carbohydrates found in the phloem sap were common transport sugars and their constituents (e.g. sucrose, glucose, and fructose in poinsettia and these plus stachyose and raffinose in pumpkin) (37). The honeydew of whiteflies on both hosts also contained melezitose, a trisaccharide common in the honeydew of aphids (137). The most noteworthy discovery was that of a disaccharide, trahalulose (1-0-α-D-glucopyranosyl-D-fructofuranose) (37). Although trahalulose has been synthesized in the laboratory and appears to be created by certain microorganisms (12, 42, 132), its production has never before been associated with members of the Insecta.

Excretion

No uric acid or hypoxanthine was found in the honeydew of *B. tabaci* feeding on poinsettia, although xanthene occurred at a low level ($0.16 \pm 0.04\%$) (37). This lack suggests that certain amino acids may be used to discharge nitrogenous compounds. Although several may be involved, glutamine is a likely candidate; it was present in high concentrations in the honeydew, is relatively inexpensive energetically to produce (particularly in the presence of so much carbon), and has a high nitrogen to carbon ratio.

ADULT CHARACTERISTICS

Mating Behavior

During summer months, copulation takes place within 1 to 8 h following eclosion by *B. tabaci* from the pupal case. During the fall and spring, copulation takes place during the three days following eclosion (10). Butler (27) stated that *Aleurodes brassicae* copulate even before their wings have dried and pigmentation is complete.

These pairings follow a complex mating behavior (117). *B. tabaci* females are attracted to, but avoid, males within the first 10 hours following pupal eclosion. After that time, males are allowed to initiate courtship. During phase I, the male encircles the female several times before placing a foretarsus or antenna on the edge of her wing. If she doesn't move, and another male doesn't interfere, he then moves parallel to the female. The female may flap her wings or push the male away to discourage him at this point. In phase II, the male is situated parallel to the female, and antennae of both sexes are held at a 24° angle to the horizontal axis of the head. The male drums the flagellum of the female antennae with his antennae while moving his abdomen up and down in synchrony with antennal drumming. Phase III involves pushing the female with the side of the male's body. This occurs in about 15% of the observed pairings. During phase IV, antennuation ceases and the male raises the pair of wings closest to the female. The male then places his abdomen beneath the female at a 25° angle. The male claspers are open and the aedeagus protruded as he tries to clasp the female terminalia. During this phase, the female often rejects the male by flapping her wings or pushing the male away with her mesothoracic legs. When females accept males, the terminal flap that covers the female gonophore is pulled open. The abdomen of the male is bent upward at nearly a 90° angle and the aedeagus also is bent at a 90° angle, bringing the aedeagus parallel to the longitudinal axis of the female body when it is inserted into the gonopore. Copulation lasts from 125 to 265 s. The female terminates copulation by prying the male free with her meso- and metathoracic legs. Polygyny and polyandry occur in *T. vaporariorum* (6) and *B. tabaci* (117).

A comparison of the sexual behavior of *B. tabaci* with that of *T. vaporariorum* (114, 116) reveals a great many similarities and some differences. Li & Maschwitz (116) state that if pheromones are involved in the courtship behavior of *B. tabaci*, they are active over only a few millimeters. These authors (116) also state that pheromones are much more important during *T. vaporariorum* courtship and that males can detect females from some distance (at least 5 cm).

Reproduction

Most whiteflies reproduce by arrhenotoky. However, one species, *P. myricae*, apparently reproduces by thelytoky and its populations consist entirely of females. Unlike aphids, whiteflies do not experience host alternation or seasonal sexual phases. Unmated females, except in *P. myricae*, produce male offspring (XO). Mated females may produce both males and females (XO and XX).

Hargreaves (89) and Williams (176), working with *T. vaporariorum* in Merton, England, reported that unfertilized eggs laid by virgin females gave rise exclusively to females. Other investigations indicated that *T. vaporariorum* had two races, an American race showing arrhenotokous parthenogenesis (which seems to be the more usual mode of reproduction) and an

English race with thelytokous parthenogenesis (153). In no current cultures of
T. vaporariorum do virgin females give rise to females (6).

The American race was shown to have a haploid chromosome number of 11
and a diploid number of 22 (153). All eggs show two maturation divisions and
undergo reduction. Unfertilized eggs give rise to males, which are haploid.
Fertilized eggs develop with the diploid number into females. In spermatogenesis the haploid chromosome number is retained without further
reduction.

The English race did not exhibit normal reduction and fertilization but
rather a form of apomixis. In thelytokous *T. vaporariorum* females, oogenesis
occurs, up to a certain point, in the same way as in arrhenotokous strains
(153). The diploid chromosome number was the same with a normal meiosis
through prophase. At this stage, pseudoreduction results in only 11 chromosomes through the maturation divisions. After the second division, the 11
chromosomes divide, and 22 chromosomes are found in all segmentation
nuclei and later on in the nymphal mitoses.

Sex Ratios

The ratio of male to female whitefly adults constantly changes throughout the
course of the year. Adult field populations reported in two cases (83, 127) had
sex ratios of 2 females:1 male. This may nevertheless follow a 1:1 primary
sex ratio because females live longer in the population as adults (130, 164).

HOST PLANT INTERACTIONS

Few quantitative studies have been conducted on the degree of polyphagy
among members of Aleyrodidae (e.g. 68); rather, many of our assumptions
concerning host plant associations are derived from published host lists. This
information provides a potentially biased view of host plant associations
because more information usually is available for pest species. Such censor-
ing of data may be responsible for the fact that the majority of the records are
for pest species that appear to be polyphagous; however, polyphagous species
are more likely to become pests. Many whiteflies are known from a single
host and so would be identified as monophagous, but these records are often
for species for which very few collections have been made (131). Within the
limitations imposed by such records, polyphagy is pronounced in only a few
whiteflies that feed on herbaceous plants, e.g. *T. vaporariorum* and *B. tabaci*.
Most whiteflies are known primarily from woody angiosperms, and
oligophagy may be the most common host plant association (131). A lack of
information on host plants also applies to oligophagous whiteflies, however,
so the assumption of oligophagy should also be viewed with caution. The only

record of a whitefly from a gymnosperm is *T. vaporariorum* from *Dioon* sp. (131).

Polyphagous species have a range of fitness on different hosts. Thus, in *T. vaporariorum*, reproduction rates were higher on tomato than on *Brassica* spp. (104), and, in *A. woglumi*, survival and reproductive rates vary widely among several host plants (68, 69). Different populations of the same species may show different host ranges, as in *B. tabaci* (22, 49, 130). A population of *Aleurodicus cocois*, a species widely reported as a pest of coconut in the Neotropics, reportedly attacks and severely damages cashew in Brazil (57a, 57b, 83), but this particular population is not found on coconut grown in the area (57b, 59).

Many populations that achieve high population densities may utilize plants as hosts that would not be utilized under lower population densities. *A. woglumi* reportedly is widely polyphagous in the Neotropics where it was introduced and where it had attained extraordinary population densities (60). It is known only from *Citrus* spp., however, in the Orient (44). In such situations, the action of natural enemies or other mortality factors in the species' native range may sufficiently reduce the fitness of populations on marginal hosts such that the whiteflies rarely use these hosts.

HOST PLANT SELECTION

Once whiteflies enter an area containing suitable plant hosts, most species respond to color as a cue to select landing sites for feeding and oviposition (48, 65, 99, 128). Coombe (48) discovered that *T. vaporariorum* while migrating responds initially to light with a wavelength of approximately 400 nm (that of the blue sky), but eventually responds to wavelengths of approximately 550 nm (the green/yellow of plants). Woets and van Lenteren (165, 178) demonstrated that leaf shape, structure, and odor do not play a role in initial host finding for *T. vaporariorum*, although the insects did respond to color. The same is true for *B. tabaci* (164). While color brings whiteflies into contact with the plant, it does not correlate well with offspring survival (164). The only report of a cue other than color is the finding that *A. proletella* responds to the odor of crushed cabbage leaves (27).

After a host is selected, entry into the leaf by the stylet bundle is a complicated process similar to that used by aphids. The mouth parts are typical for Homoptera. In the dorsal (anterior) wall of the pharynx is the cribriform organ, which functions in tasting the phloem sap as it enters the mouth (86, 89). Walker & Gordh (172), examining the apex of the labium of adults for six species of whiteflies, found that each had seven pairs of sensilla, one of each pair on either side of the labial groove. They suggest that three of these have a chemosensory or mechano-chemosensory function. Walker (170) described how the labium is rubbed or tapped on the plant surface prior to insertion of the stylets. Apparently, some decisions concerning host-plant quality are made before the leaf is fully penetrated.

Most whitefly mouth parts enter the plant by piercing the epidermal cells (145, 169). After penetration, the stylets of *P. myricae* for the most part follow an intercellular path, although pierced cells can sometimes be observed (169).

MIGRATION

The literature offers no absolute evidence of long-range whitefly migration similar to that described for other homopterans (24, 160). Although it may occur, we assume that such journeys are rigorous for these small, soft-bodied animals with their high surface-to-volume ratio. Movement of more than a few hundred meters is likely assisted by humans. Nevertheless, short-range migration takes place regularly. Apparently, much of the short-range movement by *B. tabaci* occurs near ground level (below 10 cm) (39, 79). Flight of *B. tabaci* occurs during the morning and midday hours and has one peak (19, 38). Flight of *P. myricae* is concentrated in the early morning and evening hours (124).

Byrne et al (32) found that whiteflies have relatively low wing loading values (.00174 to .00532 g/cm^2) and relatively high wingbeat frequencies (165.6 to 224.2 Hz) when compared to aphids. This finding was unexpected because many other animals (5, 85) have high wing-beat frequencies to compensate for the smallness of their wings in relation to their body mass. When comparing data on whiteflies and aphids to data for other insects, Byrne et al (32) found that more massive insects have significantly and positively correlated wing loading and wingbeat frequencies, indicating they do compensate for high wing loading by increasing wingbeat frequency. This was not true for insects weighing < 0.03 g, a group that includes whiteflies and aphids. Flying strategies for these smaller insects are believed to be different, e.g. whiteflies employ a "clap and fling" strategy (179). It is probably appropriate to characterize them as poor fliers.

Two morphs have been found to exist within populations of *B. tabaci*—a migratory and a trivial-flying morph (36). While even the migrators may be poor fliers, single individuals have reportedly traveled distances of up to 7 km (46). The question remains as to whether or not such whiteflies survive such a long journey. We do know that *B. tabaci* are routinely seen flying over fallow ground in extremely high numbers at least 150 m away from any vegetation (39, 79). *A. woglumi* reportedly disperses over distances of up to 150 m (65, 123).

Short-range migration is apparently all that whiteflies need once they are established in an area that has crop and weed hosts available all year. In the Near East (77) and India (100), *B. tabaci* populations overwinter on a variety of cultivated and wild vegetation such as vegetables and cheeseweed and move to such spring hosts as potato and cultivated sunflower. Wherever these whiteflies are a serious problem, these wild and cultivated hosts grow in close proximity to one another.

Gerling & Horowitz (79) and Byrne & Houck (36) surmised that subsets of whitefly populations leave their original habitat in response to deteriorating conditions in search of better feeding or oviposition sites. Having left the original habitat, they have little control over what happens. Adult whiteflies were routinely observed in the upper part of a cotton crop canopy, while the majority of whiteflies migrating between habitats moved quickly to the ground after leaving the field (39). At this level, because they are poor fliers, they tumble along on the ground boundary level. The direction of their flight is primarily dictated by the wind as they drift about in the manner of aerial plankton. They land on particular plants mostly by chance, electing to stay on suitable hosts and moving away from those that are not (164).

DEMOGRAPHY AND POPULATION DYNAMICS

The number of species for which life tables have been constructed is relatively limited (Table 2). *T. vaporariorum* and *B. tabaci* have received the most attention in the literature regarding laboratory life tables (for reviews, see 80, 104, 164). Many researchers report adult longevity, fecundity, and pre-imaginal developmental and survival rates. Some generalizations can be made from such studies, particularly regarding developmental threshold tempera-tures, which appear to occur near 10°C for these two species. More es-pecially, however, the vital rates reported for various whiteflies appear to vary widely. These variations are attributable in part to the use of different populations of a particular species; different populations can have markedly different vital rates (80, 164). Additional important differences in develop-ment, survival, and fecundity are caused by rearing on different host plants (51, 68, 69, 80, 164). Finally, preimaginal survival of *B. tabaci* varies inversely with relative humidity; it may be 2–80% in the range of 31–90% relative humidity (80).

Voltinism and Overwintering

Whiteflies generally appear to be multivoltine, with two to six yearly genera-tions. Most species are recorded from tropical or subtropical regions, and such species may develop and breed continually so long as temperature conditions permit [e.g. *Siphoninus phillyreae* (18), *B. tabaci* (51, 77)]. *P. myricae* overwinters as both adults and nymphs on avocado in Israel (158).

Table 2 Life history parameters for field populations of whiteflies

Species	Location	Number gen./yr	Developmental time (days)	Adult Longevity (days)	Fecundity (eggs/female)	Egg-adult survival (%)	Density	Reference
Aleurocanthus husaini	India	2						98
Aleurocanthus spinosus	Japan	3						108
Aleurocanthus spiniferus[a]	Japan	4			17–22			111
	Guam	5–6						143
Aleurocanthus woglumi	Cuba	5			40			101
	South Africa	3						16
	India							98
	Venezuela	4	54–103			17		23
Aleurocybotus occidus	California		36–50	23–32				144
Aleurodicus cocois anacardi[b]	Brazil		36	16		10		83
Aleurodicus pimentae	Jamaica		50					84
Aleurolobus barodensis	India	9[c]	28–35					88
	India						50–929/lf	154
	Pakistan						330–425/lf[d]	105
	Pakistan						70/lf[e]	105
Aleurothrixus flocossus	Spain	6						140
	Oman	6	30					173
	Hawaii[f]		27.4	36.4	53.2			139
Aleurotrachelus jelinekii[g]	USSR	1		140				113
							< 5/lf	90

Aleyrodes fragariae[h]	Norway	3						73
Aleyrodes proletella[i]	France							15
Asterobemisia carpini[j]	Italy	2						102
Bemisia tabaci	India		17–65		20–35			99
	Israel				300			10
	Arizona[k]				71–82			29
Dialeurodes citri	Florida	2–5	41–333		149		2206/lf[l]	127
	Florida	3						180
	USSR			10			48/cm²[m]	11
	USSR						10/cm²[n]	11
	France	3					22/cm²[o]	159
	India	2						98
	Britain	1						177
Dialeurodes chittendeni	India	2				58		98
Dialeurolonga elongata	Pakistan							109
Neomaskellia andropogonis[p]	India		14–20	2–3	120–150			146
Neomaskellia bergii[q]	India		25–30					110
Parabemisia myricae	Japan	6–7						158
	Israel					45[r]		171
	California					0–49[s]		147
Siphoninus phillyreae[t]	Egypt	2–3						66
Tetraleurodes acaciae[u]	Florida	8						
Tetraleurodes semilunaris							6/cm²	2

Table 2 (Continued)

Species	Location	Number gen./yr	Developmental time (days)	Adult Longevity (days)	Fecundity (eggs/female)	Egg-adult survival (%)	Density	Reference
Trialeurodes abutiloneus	Arizona[v]		25–32					28
Trialeurodes lauri[h]	USSR	1			220			113
Trialeurodes lauri	Sudan		21	13	100			71
Trialeurodes vaporariorum[w]			21–30	9–50	5–319	69–93		25

[a] 64–96% egg hatch; overwinter in last nymphal stage.
[b] Sex ratio 2 females: 1 male; preoviposition period 3.4 days.
[c] Nine gen./yr in laboratory study.
[d] Sprayed—low natural enemy-caused mortality.
[e] Unsprayed—22–71% parasitism.
[f] Emergence primarily between 0600–0900 h.
[g] Overwinter as nymphs.
[h] Adults overwinter.
[i] Oviposition in February, overwinter as adults.
[j] Nymphs ("pupae") overwinter among fallen leaves.
[k] Laboratory study.
[l] Hosted on Umbrella tree.
[m] Absence of natural enemies.
[n] Following introduction of beetle and fungi.
[o] Nymphs of second and succeeding generations diapause.
[p] 60–100% parasitism (*Encarsia* sp. and *Eretmocerus* sp.).
[q] Cage house study. 45% parasitism by *Eretmocerus* sp. and *Encarsia* sp.
[r] Winter nymphal survival on avocado leaves; adults and nymphs overwinter on trees.
[s] Laboratory study—35–49% survival on young and middle-aged leaves, 0% on old leaves.
[t] Authors suggest overwinter as adults.
[u] Year-round development.
[v] Greenhouse study.
[w] Laboratory study—values depend on temperature.

Other multivoltine species (e.g. *A. spiniferus* and *D. citri*) overwinter in temperate latitudes as nymphs on evergreen hosts (Table 2).

Developmental times for multivoltine whiteflies vary usually with the season, but most species reported develop from egg to adult in from 25 to 50 days under field conditions (Table 2). Developmental time for *D. citri* can vary widely for individuals from a single cohort of eggs from 48 to 333 days (127).

Some species are reported as univoltine; these are primarily known from temperate latitudes and evergreen hosts (Table 2). Some of these species overwinter as nymphs on foliage (e.g. *D. chittendeni* and *A. jelinekii*), and *Asterobemisia carpini* reportedly overwinters in the last nymphal stage on fallen leaves (102). Three species of *Aleyrodes* overwinter as adults (3, 15, 73, 82).

Overwintering *A. proletella* adults experience a phase resembling ovarian diapause (3). No true refractory stage occurs, however, and ovarian development simply takes place at a greatly reduced rate (4).

Population Densities

Many records of population densities are probably not characteristic of whitefly populations in natural settings and generally refer to whitefly species introduced into new areas lacking natural enemies or to populations under pesticide treatment. In such settings, populations often appear to increase unchecked except by the limitation of suitable foliage (31). High nymphal densities of $20\text{--}40/\text{cm}^2$ or up to several hundred per leaf are reported in such circumstances (Table 2). Nymphal densities in more natural settings are generally much lower, ranging from 10 to less than $1/\text{cm}^2$ (T. S. Bellows, unpublished data). These differences may result from the action of natural enemies; densities of *A. woglumi* in Texas, for example, were reduced from 5–10 nymphs per leaf to 1 nymph per 1000 leaves following the introduction of parasitic wasps (157).

Population Dynamics

Studies of the population dynamics of whiteflies in natural settings are almost entirely lacking. Most reports concern species whose populations increase dramatically on being introduced into regions. In nearly every case, an introduced whitefly population reproduces so rapidly that its populations reach enormous densities, causing leaf chlorosis, leaf withering, premature dehiscence, defoliation, and plant death (1, 18, 21, 50, 57, 74, 101, 106, 107, 112, 118, 147, 148, 152). This general outcome indicates that whitefly populations have the potential for rapid, perhaps exponential increase under favorable conditions of climate and host-plant availability.

Whitefly introductions have often been followed by introductions of natural enemies, usually parasitic wasps in the families Aphelinidae and Platygasteridae and coccinellid predators. Fungi have also been introduced in fewer cases. [Other natural enemies recorded for the Aleyrodidae include predacious mites, Neuroptera, and Hemipterous predators (45, 131).] In nearly every case, the natural enemy introductions have resulted in substantial reductions in whitefly population size (e.g. 55, 70, 136, 157). Whitefly natural enemies have been considered in other reviews (e.g. 55, 45, 78, 166, 167).

Climatic factors (temperature, wind, rain, relative humidity) also may play a role in certain populations [e.g. *B. tabaci* (10), *A. floccosus* (135)]. In populations that attain high densities, intraspecific competition among nymphal stages can be significant (135), and competition among adults for oviposition sites may also occur [e.g. reports of 2000 eggs/cm^2 (135)].

Long-term studies of the population dynamics of *A. jelinekii* were conducted in England (90, 155). This atypical species is univoltine and was apparently introduced into England. The studies covered 17 generations, and the maximum average density reached by the populations was approximately 5 nymphs per leaf. The principal mortalities affecting the populations occurred between adult emergence and egg deposition and during the fourth nymphal instar. The authors identified seven factors acting on the population between egg deposition and adult emergence: egg mortality, crawler mortality, predation, fungal disease of the nymphs, parasitism, fallen fourth instar nymphs, and unidentified deaths of each nymphal stage. Analysis of these factors demonstrated density dependence (90), most frequently in crawler mortality. In this population, internal regulatory processes, rather than the action of natural enemies, may be important.

Several studies have examined *B. tabaci* populations (17, 95, 181). Populations in cotton appear to increase nearly exponentially during the middle part of the growing season. This may in part result from the adaptation of the population to cotton as a host (164) and in part from the suppression of natural-enemy activity by insecticide application (17).

CONCLUSION

Much of what we know about aleyrodid biology comes from reports concerning pest species, primarily *B. tabaci* and *T. vaporariorum*. These two species are polyphagous and therefore may not be typical of the family. Some information is available on the life history of approximately 25 other species. This represents but a small fraction of the total number of described species, and undoubtedly many more species are as yet undiscovered. As a result, in this chapter we make statements about what is known currently about whiteflies, realizing that as more research is conducted on this amazing group of insects our perception of their life history may change.

ACKNOWLEDGMENTS

The authors are indebted to H. S. Costa, R. J. Gill, and H. E. DeVries for early review of this manuscript. We also thank E. A. Draeger and J. M. Collins for technical assistance in its preparation. This work was supported in part (T. S. B.) by grant BSR-8604546 from the US National Science Foundation. This is journal article number 7266 of the University of Arizona Agricultural Experiment Station.

Literature Cited

1. Abbas, H. M., Khan, M. S., Haque, H. 1955. Black fly of *Citrus* (*Aleurocanthus woglumi*, Ashby) in Sind and its control. *Agric. Pak.* 6:5–23
2. Abraham, C. C., Joy, P. J. 1978. New record of *Tetraleurodes similunaria* Corbett (Aleurodidae: Hemiptera) as a pest of lemon grass *Cymbopogon flexuosus* (Steud.). *Entomologist* 3:313–14
3. Adams, A. J. 1985. The critical field photoperiod inducing ovarian diapause in the cabbage whitefly, *Aleyrodes proletella* (Homoptera: Aleyrodidae). *Physiol. Entomol.* 10:243–49
4. Adams, A. J. 1986. The control of ovarian development during adult diapause in the cabbage whitefly, *Aleyrodes proletella*. *Physiol. Entomol.* 11:117–24
5. Ahmad, A. 1984. A comparative study on flight surface and aerodynamic parameters of insects, birds and bats. *Indian J. Exp. Biol.* 22:270–78
6. Ahman, I., Ekbom, B. S. 1981. Sexual behaviour of the greenhouse whitefly (*Trialeurodes vaporariorum*): orientation and courtship. *Entomol. Exp. Appl.* 29:330–38
7. Auclair, J. L. 1959. Feeding and excretion by the pea aphid, *Acyrthosiphon pisum* (Harr.) (Homoptera: Aphididae), reared on different varieties of peas. *Entomol. Exp. Appl.* 2:279–86
8. Auclair, J. L. 1963. Aphid feeding and nutrition. *Annu. Rev. Entom.* 8:439–90
9. Auclair, J. L., Maltais, J. B., Cartier, J. J. 1957. Factors in resistance of peas to the pea aphid, *Acyrthosiphon pisum* (Harr.) (Homoptera: Aphididae). II. Amino acids. *Can. Entomol.* 89:457–67
10. Avidov, Z. 1956. Bionomics of the tobacco whitefly (*Bemisia tabaci* Gennad.) in Israel. *Ktavim* 7:25–41
11. Avidzba, N. S. 1983. Bioecology of the citrus whitefly and its integrated management. In *10th Int. Congr. Plant Prot. 1983*, Vol. 3, *Proceeding of a Conference Held at Brighton, England, 20–25 November, 1983. Plant Protection for Human Welfare*. Croydon, UK: British Crop Protection Council. 1021 pp.

12. Avigad, G. 1959. Synthesis of glucosyl-fructoses by the action of a yeast glucosidase. *Biochemistry* 73:573–87
13. Azab, A. K., Megahed, M. M., El-Mirsawi, D. H. 1971. On the biology of *Bemisia tabaci* (Genn.). *Bull. Soc. Entomol. Egypte* 55:305–15
14. Back, E. A. 1909. A new enemy of the Florida orange. *J. Econ. Entomol.* 2:448–49
15. Bardie, A. 1923. Remarques sur l'*Aleurodes chelidonii*, Latr. *P. V. Soc. Linn. Bordeaux* 74:152–53
16. Bedford, E. C. G., Thomas, E. D. 1965. Biological control of citrus black-fly, *Aleurocanthus woglumi* (Ashby) (Homoptera: Aleyrodidae) in South Africa. *J. Entomol. Soc. South. Afr.* 28:117–32
17. Bellows, T. S. Jr., Arakawa, K. 1988. Dynamics of preimaginal populations of *Bemisia tabaci* (Homoptera: Aleyrodidae) and *Eretmocerus* sp. (Hymenoptera: Aphelinidae) in southern California cotton. *Environ. Entomol.* 17:483–87
18. Bellows, T. S., Paine, T. D., Arakawa, K. Y., Meisenbacher, C., Leddy, P., Kabashima, J. 1990. Biological control sought for ash whitefly. *Calif. Agric.* 44:4–6
19. Bellows, T. S. Jr., Perring, T. M., Arakawa, K., Farrar, C. A. 1988. Patterns in diel flight activity of *Bemisia tabaci* (Homoptera: Aleyrodidae) in cropping systems in southern California. *Environ. Entomol.* 17:225–28
20. Bemis, F. E. 1904. The aleyrodids, or mealy-winged flies, of California, with references to other American species. *Proc. US Nat. Mus.* 27:471–53
21. Benfatto, D. 1982. Lo mosca bianca lanosa (*Aleurothrixus floccosus* [Mask.]) si stadiffondendo negli agrumeti siciliani. *Inf. Agric.* 38:23187–90
22. Bird, J. 1981. Relationships between whiteflies and whitefly-borne diseases. *Abstracts of the 1981 Int. Workshop on Pathogens Transmitted by Whiteflies*, Oxford, England: Assoc. Appl. Biol. Wellesbourne

23. Boscan de Martinez, N. 1983. Biologia de la mosca prieta de los citricos *Aleurocanthus woglumi* Ashby (Homoptera: Aleyrodidae) en el campo. *Agron. Trop.* 31:211–18

24. Broadbent, L. 1948. Aphis migration and the efficiency of the trapping method. *Appl. Biol.* 35:379–94

25. Burnett, T. 1949. The effect of temperature on an insect host-parasite population. *Ecology* 30:113–34

26. Butani, D. K. 1970. Bibliography of Aleyrodidae II. *Beitr. Entomol.* 20:317–35

27. Butler, C. G. 1938. On the ecology of *Aleurodes brassicae* Walk (Hemiptera). *Trans. R. Entomol. Soc. London* 87:291–311

28. Butler, G. D. Jr. 1967. Development of the banded-wing whitefly at different temperatures. *J. Econ. Entomol.* 60:877–78

29. Butler, G. D. Jr., Henneberry, T. J., Clayton, T. E. 1983. *Bemisia tabaci* (Homoptera: Aleurodidae): development, oviposition and longevity in relation to temperature. *Ann. Entomol. Soc. Am.* 76:310–13

30. Butler, G. D., Henneberry, T. J., Wilson, F. D. 1986. *Bemisia tabaci* (Homoptera: Aleyrodidae) on cotton: adult activity and cultivar oviposition preference. *J. Econ. Entomol.* 79:350–54

31. Byrne, D. N., Bellows, T. S., Parrella, M. P. 1990. Whiteflies in agricultural systems. See Ref. 78, pp. 227–61

32. Byrne, D. N., Buchmann, S. L., Spangler, H. G. 1988. Relationship between wing loading, wingbeat frequency and body mass in homopterous insects. *J. Exp. Biol.* 135:9–23

33. Byrne, D. N., Cohen, A. C., Draeger, E. A. 1990. Water uptake from plant tissue by the egg pedicel of the greenhouse whitefly, *Trialeurodes vaporariorum* (Westwood) (Homoptera: Aleyrodidae). *Can. J. Zool.* 68:1193–95

34. Byrne, D. N., Draeger, E. A. 1989. Effect of plant maturity on oviposition and nymphal mortality of *Bemisia tabaci* (Homoptera: Aleyrodidae). *Environ. Entomol.* 18:429–32

35. Byrne, D. N., Hadley, N. F. 1988. Particulate surface waxes of whiteflies: morphology, composition and waxing behaviour. *Physiol. Entomol.* 13:267–76

36. Byrne, D. N., Houck, M. A. 1990. Morphometric identification of wing polymorphism in *Bemisia tabaci* (Gennadius) (Homoptera: Aleyrodidae). *Ann. Entomol. Soc. Am.* 83:487–93

37. Byrne, D. N., Miller, W. B. 1990. Carbohydrate and amino acid composition of phloem sap and honeydew produced by *Bemisia tabaci*. *J. Insect Physiol.* 36:433–39

38. Byrne, D. N., von Bretzel, P. K. 1987. Similarity in flight activity rhythms in coexisting species of Aleyrodidae, *Bemisia tabaci* and *Trialeurodes abutilonea*. *Entomol. Exp. Appl.* 43:215–19

39. Byrne, D. N., von Bretzel, P. K., Hoffman, C. J. 1986. Impact of trap design and placement when monitoring for the bandedwinged whitefly and the sweetpotato whitefly. *Environ. Entomol.* 15:300–4

40. Deleted in proof

41. Deleted in proof

42. Cheetham, P. S. J. 1984. The extraction and mechanism of a novel isomaltulose-synthesizing enzyme from *Erwinia rhapontici*. *Biochem. J.* 220:213–20

43. Clausen, C. P., ed. 1978. Introduced parasites and predators of arthropod pests and weeds: A world review. *USDA Handb.* 480. 545 pp.

44. Clausen, C. P., Berry, P. A. 1932. The citrus blackfly in Asia, and the importation of its natural enemies into Tropical Ameria. *USDA Tech. Bull.* 320. 58 pp

45. Cock, M. J. W., ed. 1986. *Bemisia tabaci*—a literature survey on the cotton whitefly with an anotated bibliography. *CAB Int. Inst. Biol. Control.* Ascot, Berks., UK:CAB 121 pp.

46. Cohen, S., Ben-Joseph, R. 1986. Preliminary studies of the distribution of whiteflies (*Bemisia tabaci*), using fluorescent dust to mark the insects. *Phytoparasitica* 14:152–53

47. Collman, G. L., All, J. N. 1980. Quantification of the greenhouse whitefly life cycle in a controlled environment. *J. Ga. Entomol. Soc.* 15(4):432–38

48. Coombe, P. E. 1982. Visual behaviour of the greenhouse whitefly, *Trialeurodes vaporariorum*. *Physiol. Entomol.* 7:243–51

49. Costa, A. S., Russell, L. M. 1975. Failure of *Bemisia tabaci* to breed on cassava plants in Brazil (Homoptera-Aleyrodidae). *Cienc. Cult.* 27:388–90

50. Costacos, T. A. 1963. On a severe attack by *Siphoninus phillyreae* Halliday supsb. *inaequalis* Gautier on fruit trees and its control. *Geoponika* 105:3–7 (in Greek)

51. Coudriet, D. L., Prabhaker, N., Kishaba, A. N., Meyerdirk, D. E. 1985. Variation in developmental rate on different hosts and overwintering of the sweetpo-

tato whitefly, *Bemisia tabaci* (Homoptera: Aleyrodidae). *Environ. Entomol.* 14:516–19
52. Dadd, R. H. 1973. Insect nutrition: current developments and metabolic implications. *Annu. Rev. Entomol.* 18:381–420
53. Dadd, R. H. 1985. Nutrition: organisms. In *Comprehensive Insect Physiology Biochemistry and Pharmacology*, ed. G. A. Kerkut, L. I. Gilbert. 4:313–90. New York: Pergamon
54. Dadd, R. H., Krieger, D. L. 1968. Dietary amino acid requirements of the aphid, *Myzus persicae*. *J. Insect Physiol.* 14:741–64
55. Dalgarno, W. T. 1935. Notes on the biological control of insect pests in the Bahamas. *Trop. Agric.* 12:78
56. David, B. V., Ananthakrishnan, T. N. 1976. Host correlated variation in *Trialeurodes rara* Singh and *Bemisia tabaci* (Gennadius) (Aleyrodidae: Homoptera: Insecta). *Curr. Sci.* 45:223–25
57. Debach, P. H., Rose, M. 1976. Biological control of woolly whitefly. *Calif. Agric.* 30(5):4–7
57a. de Carvalho, M. B., Arruda, E. C., Pereirade Arruda, G. 1976. Uma possivel raca hospedeira do *Aleurodicus cocois* (Curtis, 1846) (Homoptera, Aleyrodidae). *An. Soc. Entomol. Bras.* 5:243–45
57b. de Carvalho, M. B., de O. Frietas, A., Pereira de Arruda, G. 1971. Algumas consideraçoes sôbre o *Aleurodicus cocois* (Curtis, 1846) (Homoptera: Aleyrodidae), 'mosca branca' do cajueiro, no Estado de Pernambuco. *Bol. Tec. Inst. Pasqui. Agron.*, Recife 18. 20 pp.
58. Deshpande, V. G. 1936. Miscellaneous observations on the biology of Aleurodidae (*Aleyrodes brassicae*). *J. Bombay Nat. Hist. Soc.* 39:190–93
59. de Souza Leão Viega, A. F., Pereira de Arruda, G., Porta de Carvalho, E. 1970. Os inimigos naturais da mosca branca no Estado Pernambuco. SBE. Resumos da II Reunião Anual, Recife, Pernambuco, Brasil, 1 a 6 dezembro, 1969. *Recife, Soc. Brasileira Entomol.*, p. 68
60. Dietz, H. F., Zetek, J. 1920. The blackfly of citrus and other subtropical plants. *USDA Bull.* 855. pp. 1–55
61. Dittrich, V., Hassan, S. O., Ernst, G. H. 1985. Development of a new primary pest of cotton in the Sudan: *Bemisia tabaci* (Gennadius) in cotton fields in Israel. *Ecol. Appl.* 5:221–33
62. Domenichini, G. 1982. Strutture di

Trialurodes vaporariorum (Westw.) e loro funzioni (Homoptera Aleyrodidae). *Mem. Soc. Entomol. Ital.* 60:169–76
63. Douglas, J. W. 1879. The genus *Aleurodes*. *Entomol. Mon. Mag.* 16:43
64. Douglas, J. W., Rye, E. C., McLachlan, R., Stainton, H. T. 1877–78. Notes on the genus Aleurodes. *Entomol. Mon. Mag.* 16:230–33
65. Dowell, R. V. 1979. Host selection by the citrus blackfly *Aleurocanthus woglumi* (Homoptera: Aleyrodidae). *Entomol. Exp. Appl.* 25:289–96
66. Dowell, R. V. 1983. Biology of *Tetraleurodes acaciae* (Quaintance) (Homoptera: Aleyrodidae). *Pan-Pac. Entomol.* 58:312–18
67. Dowell, R. V., Fitzpatrick, G. E., Howard, F. W. 1978. Activity and dispersal of first instar larvae of the citrus blackfly. *J. N. Y. Entomol. Soc.* 86:121–22
68. Dowell, R. V., Steinberg, B. 1979. Development and survival of immature citrus blackfly (Homoptera: Aleyrodidae) on twenty-three plant species. *Ann. Entomol. Soc. Am.* 72:721–24
69. Dowell, R. V., Steinberg, B. 1990. Influence of host plant on fecundity of citrus blackfly *Aleurocanthus woglumi* Ashby (Homoptera: Aleyrodidae). *Panpac. Entomol.* In press
70. Edwards, W. H. 1936. Pests attacking *Citrus* in Jamaica. *Bull. Entomol. Res.* 27:335–37
71. El-Khidir, E., Khalifa, A. 1962. A new aleyrodid from the Sudan. *Proc. R. Entomol. Soc. London Ser. B* 31:47–51
72. Enderlein, G. 1909. *Udamoselis*, eine neue Aleurodiden-Gattung. *Zool. Anz.* 34:230–33
73. Fjelddalen, J. 1955. Jordbaermjollus *Aleyrodes fragariae* Walk. *Frukt Baer* 8:17–21
74. Food and Agriculture Organization. 1954. Outbreaks and new records. *FAO Plant Prot. Bull.* 2:187–89
75. Gameel, O. I. 1974. Some aspects of the mating and oviposition behaviour of the cotton whitefly *Bemisia tabaci* (Genn.). *Rev. Zool. Afr.* 88:784–88
76. Garman, H., Jewett, H. H. 1922. The whiteflies of hothouses. *Ky. Agric. Exp. Stn. Bull. No. 241* pp. 77–111
77. Gerling, D. 1984. The overwintering mode of *Bemisia tabaci* and its parasitoids in Israel. *Phytoparacitica* 12:109–18
78. Gerling, D., ed. 1990. *Whiteflies: Their Bionomics, Pest Status and Management*. Wimborne, UK: Intercept. 348 pp.

79. Gerling, D., Horowitz, A. R. 1984. Yellow traps for evaluating the population levels and dispersal patterns of *Bemisia tabaci* (Gennadius) (Homoptera: Aleyrodidae). *Ann. Entomol. Soc. Am.* 77:753–59

80. Gerling, D., Horowitz, A. R., Baumgaertner, J. 1986. Autecology of *Bemisia tabaci. Agric. Ecosyst. Environ.* 17:5–19

81. Gill, R. J. 1990. The morphology of whiteflies. See Ref. 78, pp. 13–46

82. Gillespie, D. R. 1985. Endemic Aleyrodidae (Homoptera) and their parasites (Hymenoptera) on southern Vancouver Island, British Columbia. *J. Entomol. Soc. B. C.* 82:12–13

83. Gondim, M. T. P., Sales, F. J. M. 1983. Ciclo Biologico da mosca branca do cajueiro. Nota Previa. *Fitossanidade* 5:38

84. Gowdy, C. C. 1928. Report of the government entomologist. *Annu. Rep. Dep. Agric. Jamaica, 1927.* pp. 20–21

85. Greenewalt, C. H. 1962. Dimensional relationships for flying animals. *Smithson. Misc. Collect.* 144:1–46

86. Guillot, F. S., Droste, T., Hart, W. 1979. Egg attachment of citrus blackfly *Aleurocanthus woglumi* Ashby to citrus leaves. *Southwest. Entomol.* 4:167–69

87. Gupta, P. C. 1972. External morphology of *Bemisia gossypiperda* (M.& L.) a vector of plant virus diseases (Homoptera: Aleurodidae). *Zool. Beitr.* 18:1–20

88. Harbans Singh, K. A. N., Kabra, A. N., Sandhu, J. S. 1956. Sugarcane whitefly (*Aleurolobus barodensis* Mask.) and its control. *Indian Sugar* 5:689–93

89. Hargreaves, E. 1915. The life-history and habits of the greenhouse white fly (*Aleyrodes vaporariorum* Westd.). *Ann. Appl. Biol.* 1:303–34

90. Hassell, M. P., Southwood, T. R. E., Reader, P. M. 1987. The dynamics of the Viburnum whitefly (*Aleurotrachelus jelinekii)*: a case study of population regulation. *J. Anim. Ecol.* 56:283–300

91. Hinton, H. E. 1948. On the origin and function of the pupal stage. *Trans. R. Entomol. Soc. London* 99:395–409

92. Hinton, H. E. 1976. Notes on neglected phases in metamorphosis, and a reply to J. M. Whitten. *Ann. Entomol. Soc. Am.* 69:560–66

93. Hinton, H. E., ed. 1981. *Biology of Insect Eggs*, Vols. 1, 2. New York: Pergamon. 778 pp.

94. Hoffman, C. J., Byrne, D. N. 1986. Effects of temperature and photoperiod upon adult eclosion of sweetpotato whiteflies. *Entomol. Exp. Appl.* 42:139–43

95. Horowitz, A. R., Podoler, H., Gerling, D. 1984. Life table analysis for the tobacco whitefly *Bemisia tabaci* (Gennadius) in cotton fields in Israel. *Ecol. Appl.* 5:221–33

96. Houk, E. J., Griffiths, G. W. 1980. Ultracellular symbiotes of the Homoptera. *Annu. Rev. Entomol.* 25:161–87

97. Husain, M. A. 1931. A preliminary note on the white-fly of cottons in the Punjab. *Agric. J. India* 25:508–26

98. Husain, M. A., Khan, A. W. 1945. The citrus Aleyrodidae (Homoptera) in the Punjab and their control. *Mem. Entomol. Soc. India.* 1. 41 pp.

99. Husain, M. A., Trehan, K. N. 1933. Observations on the life-history, bionomics and control of the white-fly of cotton (*Bemisia gossypiperda* M & L). *Indian J. Agric. Sci.* 3:701–53

100. Husain, M. A., Trehan, K. N., Verma, P. M. 1936. Studies on *Bemisia gossypiperda*, M. & L. *Indian J. Agric. Sci.* 6:893–903

101. Hutson, J. C. 1918. Some insect pests in Cuba. *Agric. News Barbados* 17(no. 421, 15 June 118):186–87

102. Iaccarina, F. M., Viggiani, G. 1983. Osservazioni morfo-biologiche sull'*Asterobemisia avellanae* Signoret (Homoptera, Aleyrodidae) e i suoi entomoparassiti. In *Atti XIII Congr. Naz. Ital. Entomol.*, pp. 81–88. Turin, Italy: Instituto di Entomologia Agraria e Apicoltura, Universita di Torino

103. Iheagwam, E. U. 1979. Teneral stage and take-off with age in the cabbage whitefly, *Aleyrodes brassicae* (Homoptera: Aleyrodidae). *Entomol. Exp. Appl.* 25:349–53

104. Iheagwam, E. U. 1980. Comparative studies on the increase rates of the greenhouse whitefly, *Trialeurodes vaporariorum* Westwood, and the cabbage whitefly, *Aleyrodes brassicae* Wlk. (Homoptera: Aleyrodidae). *Appl. Entomol. Zool.* 15:106–8

105. Inayatullah, C. 1984. Sugar-cane aleurodids, *Aleurolobus barodensis* (Maskell) and *Neomaskellia andropogonis* Corbett (Hom.: Alerodidae), and their natural enemies in Pakistan. *Insect. Sci. Applic.* 5:279–82

106. Issac, P. V., Misra, C. S. 1933. The chief insect pests of sugarcane and methods for their control. *Agric. Live-Stock India* 3:315–24

107. Jenkins, C. F. H., Shedley, D. G. 1953. Insect pests and their control. The citrus whitefly (*Aleuroplatus citri* Tak.). *J. Dept. Agric. W. Aust.* (3)2:49–51

108. Kato, T. 1970. Life cycles and over-

wintering stages of the spiny blackfly, *Aleurocanthus spiniferus* Quaintance. *Jpn. J. Appl. Entomol. Zool.* 14:12–18 (in Japanese)

109. Khan, A. G., Mohyuddin, A. I. 1982. Two new aleyrodids on sugarcane in Pakistan. *Entomol. Newsl. Int. Soc. Sugar Cane Tech.* 12:12

110. Kobayashi, G. 1929. On *Bemisia myricae* Kuwana, a pest of mulberry. *Insect World* 33:9–13

111. Kodama, G. 1931. *Studies on* Aleurocanthus spiniferus *Quaint.* Kyushu, Japan: Kagoshima-Ken. 38 pp. (in Japanese)

112. Kolev, K. 1973. A new pest of pear in this country. *Rastit. Zasht.* 21:28 (in Bulgarian)

113. Korobitsin, V. G. 1964. Whiteflies— pests of ornamental plants. *Zashch. Rast. Vred Bolez* 1964(pt. 8):46–47 (in Russian)

114. Las, A. 1979. Male courtship persistence in the greenhouse whitefly, *Trialeurodes vaporariorum* Westwood (Homoptera: Aleyrodidae). *Behaviour* 72:107–25

115. Latta, R. 1937. The Rhododendron whitefly and its control. *USDA Circ. No. 429.* pp. 1–8

116. Li, T., Maschwitz, U. 1983. Sexual pheromone in the green house whitefly *Trialeurodes vaporariorum* Westw. *Z. Angew. Entomol.* 95:439–46

117. Li, T., Vinson, S. B., Gerling, D. 1989. Courtship and mating behavior of *Bemisia tabaci* (Homoptera: Aleyrodidae). *Environ. Entomol.* 18:800–6

118. Liotta, G., Maniglia, G. 1974. Essais de lutte contre *Dialeurodes citri* (Ashmead) (Homoptera—Aleyrodidae) sur mandarinier en Sicile. XXVI Intl. Symp. Fytofarmacie Fytiatrie, 7 Mei 1974. I; II. *Meded. Fac. Landbouwwet. Rijksuniv. Gent.* 39:875–83

119. Lloyd, L. L. 1922. The control of the greenhouse white fly (*Asterochiton vaporariorum*) with notes on its biology. *Ann. Appl. Biol.* 9:1–34

120. Martin, J. H. 1987. An identification guide to common whitefly pest species of the world (Homoptera: Aleyrodidae). *Trop. Pest Manage.* 33:298–322

121. Maskell, W. M. 1895. Contributions towards a monograph of the Aleurodidae, a family of Hemiptera-Homoptera. *Trans. N. Z. Inst.* 28:411–449

122. McKenzie, H. L. 1967. *Mealybugs of California.* Los Angeles: Univ. Calif. Press. 525 pp.

123. Meyerdirk, D. E., Hart, W. G., Burnside, J. 1979. Marking and dispersal study of adults of the citrus blackfly,

Aleurocanthus woglumi. Southwest. Entomol. 4:325–29

124. Meyerdirk, D. E., Moreno, D. S. 1984. Flight behavior and color-trap preference of *Parabemisia myricae* (Kuwana) (Homoptera: Aleyrodidae) in a citrus orchard. *Environ. Entomol.* 13:167–70

125. Mittler, T. E. 1953. Amino-acids in phloem sap and their excretion by aphids. *Nature* 172:207

126. Mittler, T. E. 1971. Dietary amino acid requirements of the aphid *Myzus persicae* affected by antibiotic uptake. *J. Nutr.* 101:1023–28

127. Morrill, A. W., Back, E. A. 1911. Whiteflies injurious to citrus in Florida. *USDA Bur. Entomol. Bull. 92.* 109 pp.

128. Mound, L. A. 1962. A new genus and four new species of whitefly from ferns. *Rev. Zool. Bot. Afr.* 64:127–32

129. Mound, L. A. 1963. Host-correlated variation in *Bemisia tabaci* (Gennadius) (Homoptera: Aleyrodidae). *Proc. R. Entomol. Soc. London* 38:171–80

130. Mound, L. A. 1983. Biology and identity of whitefly vectors of plant pathogens. In *Plant Virus Epidemiology,* ed. R. T. Plumb, J. M. Thresh, pp. 305–13. Oxford, England: Blackwell Scientific

131. Mound, L. A., Halsey, S. H. 1978. *Whitefly of the World.* New York: Wiley. 340 pp.

132. Munir, M., Schneider, B., Schiweck, H. 1987. 1-0-α-D-Glucopyranosyl-D-fructose: Darstellung aus Saccharose und ihre Reduktion zu 1-0-α-D-Glucopyranosyl-D-glucitol. *Carbohydr. Res.* 164:477–85

133. Muniyappa, V. 1980. Whiteflies. In *Vectors of Plant Pathogens,* ed. K. F. Harris, K. Maramorosch, pp. 39–85. New York: Academic

134. Nechols, J. R., Tauber, M. J. 1977. Age-specific interaction between the greenhouse whitefly and *Encarcia formosa:* influence of the parasite on host development. *Environ. Entomol.* 6:207–10

135. Onillon, J. C. 1970. Premieres observation sur la biologie d'*Aleurothrixus floccosus* Mask. (Homopt., Aleurodidae) dans le Sud-Est de la France. *Al Awamia* 37:105–9

136. Onillon, J. C. 1975. Contribution a l'etude de la dynamique des populations d'Homopteres infeodes aux Agrumes. V.3. Evolution des populations d'*A. Flocosus* Mask. (Homopt. Aleurodidae) pendant les trois annees suivant l'introduction de *Cales noacki* How. (Hymenopt., Aphelinidae). *Fruits* 30: 237–45

137. Owen, D. F. 1978. Why do aphids synthesize melezitose? *Oikos* 31:264–67
138. Paulson, G. S., Beardsley, J. W. 1985. Whitefly (Hemiptera: Aleyrodidae) egg pedicel insertion into host plant stomata. *Ann. Entomol. Soc. Am.* 78:506–8
139. Paulson, G. S., Beardsley, J. W. 1986. Development, oviposition and longevity of *Aleurothrixus floccosus* (Maskell) (Homoptera: Aleyrodidae). *Proc. Hawaii. Entomol. Soc.* 26:97–99
140. Perez Ibanez, T., Alberti Maurici, J., Calderon Forns, E., Martiniex-Canales Murcia, G., Vinaches Gomis, P. 1973. Generaciones anuales de *Aleurothrixus howardi* Quant. mosca blance de los agrios. *Bol. Inf. Plagas* 109:17–19
141. Perkins, H. H. 1983. Identification and processing of honeydew-contaminated cottons. *Text. Res. J.* pp. 508–12
142. Perkins, H. H. 1987. Sticky cotton. *Proc. West. Cotton Prod. Conf.* pp. 53–55
143. Peterson, G. D. Jr. 1955. Biological control of the orange spiny whitefly in Guam. *J. Econ. Entomol.* 48(6):681–83
144. Poinar, G. O. 1965. Observations on the biology and ovipositional habits of *Aleurocybotus occiduus* (Homoptera: Aleyrodidae) attacking grasses and sedges. *Ann. Entomol. Soc. Am.* 58:618–20
145. Pollard, D. G. 1955. Feeding habits of the cotton whitefly, *Bemisia tabaci* Genn. (Homoptera: Aleyrodidae). *Ann. Appl. Biol.* 43:664–71
146. Prasad, V. G. 1954. *Neomaskellia bergii* Sign.—another whitefly pest of sugarcane in Bihar. *Indian J. Entomol.* 16:254–60
147. Priesner, H., Hosny, M. 1932. Contributions to the knowledge of the white flies (Aleurodidae) of Egypt (I). *Bull. Minist. Agric. Egypt 121.* 8 pp.
148. Priesner, H., Hosny, M. 1934. Contribution to the knowledge of the whiteflies (Aleurodidae) of Egypt (III). *Bul. Minist. Agric. Egypt 145.* 11 pp.
149. Quaintance, A. L. 1900. Contributions toward a monograph of the American Aleurodidae. *USDA Tech. Series No. 8,* pp. 9–64
150. Quaintance, A. L., Baker, A. C. 1913. Classification of the Aleyrodidae. Part I. *USDA Tech. Series No. 27.* pp. 1–93
151. Russell, L. M. 1948. The North American species of whiteflies of the genus *Trialeurodes. USDA Misc. Pub. No. 635.* 85 pp.
152. Sanzin, R. 1915. Citrus whitefly (*Aleurodes citri*) on lemons and oranges in the province of Mendoza, Argentina. *La Enologia Argentina, Mendoza* 1:42–43
153. Schrader, F. 1920. Sex determination in the white-fly (*Trialeurodes vaporariorum*). *J. Morph.* 34:267–305
154. Shah, A. H., Patel, M. B., Patel, G. M. 1986. Record of a coccinellid predator (*Serangium parcesetosum* Sci.) of sugarcane whitefly in south Gujarat. *Gujarat Agric. Univ. Res. J.* 12:63–64
155. Southwood, T. R. E., Reader, P. M. 1976. Population census data and key factor analysis for the Viburnum whitefly, *Aleurotrachelus jelinekii* (Frauenf.), on three bushes. *J. Anim. Ecol.* 45:313–25
156. Spiller, N. J., Llewellyn, M. 1987. Honeydew production and sap ingestion by the cereal aphids *Rhopalosiphum padi* and *Metopolophium dirhodum* on seedlings of resistant and susceptible wheat species. *Ann. Appl. Biol.* 110:585–90
157. Summy, K. R., Gilstrap, F. E., Hart, W. G., Caballero, J. M., Saenz, I. 1983. Biological control of citrus blackfly (Homoptera: Aleyrodidae) in Texas. *Environ. Entomol.* 12:782–86
158. Swirski, E., Izhar, Y., Wysoki, M., Blumberg, D. 1986. Overwintering of the Japanese bayberry whitefly, *Parabemisia myricae*, in Israel. *Phytoparasitica* 14:281–86
159. Targe, A., Deportes, L. 1953. L'aleurode des agrumes, *Dialeurodes citri* Ash. dans les Alpes-Maritimes. Premiers resultats d'experimentations de traitements. *Phytoma Déf. Cult.* 6:9–15
160. Taylor, R. A. J. 1985. Migratory behavior in the Auchenorrhyncha. In *The Leafhoppers and Planthoppers,* ed. L. R. Nault, J. G. Rodrigues, pp. 259–88. New York: Wiley
161. Thompson, C. R., Reinert, J. A. 1983. Annotated bibliography of the citrus blackfly, *Aleurocanthus woglumi* Ashby (Homoptera: Aleyrodidae). *Bibl. Entom. Soc. Am.* 1:11–41
162. Trehan, K. N. 1940. Studies on the British white-flies (Homoptera: Aleyrodidae). *Trans. R. Entomol. Soc. London* 90:575–616
163. Trehan, K. N., Butani, D. K. 1960. Bibliography of Aleyrodidae. *Beitr. Entomol.* 10:330–88
164. van Lenteren, J. C., Noldus, L. P. J. J. 1990. Whitefly-plant relationships: behavioural and ecological aspects. See Ref. 78, pp. 47–89
165. van Lenteren, J. C., Woets, J. 1977. Development and establishment of

biological control of some glasshouse pests in the Netherlands. In *Pest Management in Protected Culture Crops*, ed. F. F. Smith, R. E. Webb, pp. 81–87

166. van Lenteren, J. C., Woets, J. 1988. Biological and integrated pest control in greenhouses. *Annu. Rev. Entomol.* 33:239–69

167. Viggiani, G. 1984. Bionomics of the Aphelinidae. *Annu. Rev. Entomol.* 29: 257–76

168. Von Arx, R., Baumgartner, J., Delucci, V. 1983. Developmental biology of *Bemisia tabaci* (Genn.) (Sternorrhyncha, Aleyrodidae) on cotton at consistant temperatures. *Bull. Sor. Entomol. Suisse* 56:389–99

169. Walker, G. P. 1985. Stylet penetration by the bayberry whitefly, as affected by leaf age in lemon, *Citrus limon*. *Entomol. Exp. Appl.* 39:115–21

170. Walker, G. P. 1987. Probing and oviposition behavior of the bayberry whitefly (Homoptera: Aleyrodidae) on young and mature lemon leaves. *Ann. Entomol. Soc. Am.* 80:524–29

171. Walker, G. P., Aitken, D. C. G. 1985. Oviposition and survival of bayberry whitefly, *Parabemisia myricae* (Homoptera: Aleyrodidae) on lemons as a function of leaf age. *Environ. Entomol.* 14:254–57

172. Walker, G. P., Gordh, G. 1990. The occurrence of apical labial sensilla in the Aleyrodidae and evidence for a contact chemosensory function. *Entomol. Exp. Appl.* 51:215–24

173. Watts, W. S., Alam, M. 1973. Spray trials against the citrus blackfly (*Aleurocanthus woglumi*) on limes in the Oman. *Misc. Rep. Overseas Dev. Admin. For. Commonwealth Off. 8*. 7 pp.

174. Weber, H. 1931. Lebensweise und Umweltbeziehungen von *Trialeurodes vaporariorum* (Westwood) (Homoptera: Aleurodina). *Z. Morph. Okol. Tiere* 23:575–753

175. Wigglesworth, V. B. 1965. *The Principles of Insect Physiology*. England: Chapman and Hall. 827 pp. 6th ed.

176. Williams, C. B. 1917. Some problems of sex ratio and parthenogenesis. *J. Genet.* 6:255–67

177. Wilson, G. F. 1929. The rhododendron white fly. *J. R. Hortic. Soc.* 54:214–17

178. Woets, J., van Lenteren, J. C. 1976. The parasite-host relationship between *Sucaisia formosa* (Hymenoptera: Aphelinidae) and *Trialeurodes vap* (Homoptera: Aleyrodidaes) VI. The enfluence of the host plant on the greenhouse whitefly and its parasite *Encarsia formosa*. *Bull. OILB/SROP* 4:151–64

179. Wootton, R. J., Newman, D. J. S. 1979. Whitefly have the highest contractions frequencies yet recorded in non-fibrillar muscles. *Nature* 280:402–3

180. Yothers, W. W. 1913. Spraying for whiteflies in Florida. *USDA Cir. 168*. 8 pp.

181. Zalom, F. G., Natwick, E. T., Toscano, N. C. 1985. Temperature regulation of *Bemisia tabaci* (Homoptera: Aleyrodidae) populations in Imperial Valley cotton. *J. Econ. Entomol.* 78:61–64

Annu. Rev. Entomol. 1991. 36:459–84

AEDES ALBOPICTUS IN THE AMERICAS

Karamjit S. Rai

Department of Biological Sciences, University of Notre Dame, Notre Dame, Indiana 46556

KEY WORDS: mosquito population genetics, vector competence, dengue and dengue hemorrhagic fevers, ribosomal DNA, mitochondrial DNA

PERSPECTIVE AND OVERVIEW

The subgenus *Stegomyia* of the genus *Aedes* contains approximately 110 described species divided into 7 groups, including the *scutellaris* group, which is subdivided into the *scutellaris* and the *albopictus* subgroups consisting of 34 and 11 species, respectively (98).

Aedes albopictus has a wide distribution. It is thought to have originated in southeast Asia (115) and has spread as far west as Madagascar and the Seychelles and east through the Indomalayan and the Oriental regions— China, Japan, Marianas Islands, and New Guinea (30, 54, 57, 68, 123). It became established in Hawaii sometime between 1830 and 1896 (46, 61) and in Guam between 1944 and 1948 (2, 58). Within the past 25 years, the species has colonized the Solomon and Santa Cruz Islands in the South Pacific (28, 91).

In a grant application to the NIH in October 1983 entitled "Genetic Differentiation of the *Aedes albopictus* Subgroup," I predicted that "introduction and subsequent establishment of *A. albopictus* in the continental United States cannot be discounted in view of the well documented success of this species to move through commerce routes and to colonize new regions. . . . For the continental United States, the potential public health importance of this species becomes even greater when one realizes that this species is considerably more cold tolerant than *A. [Aedes] aegypti*. Thus, if it were

459

introduced to the United States, it might become established in more northern (Midwestern) states where *A. aegypti* does not exist. In Japan and China where its distribution reaches just about 40°N latitude and in the Oriental region (North India, West Pakistan, Nepal) it can withstand long periods of freezing."

Subsequent events have borne out both these predictions. Relatively large populations of *A. albopictus* were found throughout Harris County, Texas in August 1985 (116). From there, the species spread quickly so that, by the summer of 1989, it was widely distributed in some 18 states in the United States extending as far south as Brownsville, Texas, and Polk County, Florida, and north throughout the midwest from Kansas City, Missouri, and Chicago, Illinois, to Baltimore, Maryland, in the east. In the south, the species has even moved to Matamoros, Mexico, across the border from Brownsville (82, 131). Using data on the distribution of the species in north Asia, Nawrocki & Hawley (87) estimated that the 0°C isotherm is the northern limit of the overwintering range and the −5°C isotherm limits the maximum northward expansion during summers. In the United States, this is much farther north than the related *A. aegypti* can colonize and unquestionably puts most of the midwestern and eastern parts of the country in probable contact with *A. albopictus*. Furthermore, mosquito workers in Houston, New Orleans, and Memphis have observed that locations in and around these cities that yielded *A. aegypti* in previous years are now producing primarily *A. albopictus* (82; D. Sprenger, G. Carmichael, B. Kelly, personal communication).

In Brazil, an infestation of *A. albopictus* was first detected during a large dengue epidemic in 1986 (33). The populations have now spread to four states and the city of São Paulo (32).

The introduction of *A. albopictus* into the United States and Brazil has been regarded as the most singular medical entomological event of the past decade in the Americas (69).

Because colonizations usually involve a few individuals, the event may go undetected for many years. As a result, formal studies are usually carried out years to decades after the actual colonization. The detection of *A. albopictus* in the continental United States and Brazil, presumably very shortly after its introduction, provides a rare opportunity to study various aspects of the biology, population genetics, and vector competence of a species that is rapidly colonizing a new continent. This review highlights observations in these fields with particular emphasis on *A. albopictus* populations in the Americas.

Earlier reviews have appeared on the genetics (97) and the biology (48) of *A. albopictus*. This review by and large excludes information contained in these earlier reviews except in areas where much additional work has been done since they were written. For details concerning mate-choice tests, karyotypes, chromosome C-banding, meiosis and spermatogenesis, variation in abundance and distribution of families of repeated DNA sequences, and genome organization in this species, one should refer to the genetics review (97). Interest in research on *A. albopictus* has increased dramatically since its introduction into the Americas in mid 1980s. Therefore, work done by many laboratories and cited in this text as personal communications or as unpublished data will soon appear in print.

VECTOR POTENTIAL AND DISEASE EPIDEMICS

The introduction and establishment of *A. albopictus* in the United States has potentially serious implications. It is a primary vector of dengue and dengue hemorrhagic fevers (DHF), particularly in rural areas of Southeast Asia (110, 115). It may also play an important role in the epidemiology of other viruses such as Chickungunya (74, 125, 135), Japanese encephalitis (106, 134), San Angelo (126), La Crosse (124), and several other viruses (114). The species has adapted itself increasingly to urban habitats such as Singapore in Southeast Asia (19) and Chicago, Houston, New Orleans, and other cities in the US (82), enhancing the potential for urban epidemics. Among all the viruses that *A. albopictus* can transmit, dengue poses the most serious threat in the Americas, followed by the potential for La Crosse transmission in the midwestern United States. In addition, because the species can breed in urban, rural, and forest areas, it could facilitate urban cycles of yellow fever in South America by bridging the jungle and urban environments (89).

Dengue and Dengue Hemorrhagic Fevers

A. albopictus has been incriminated as the vector responsible for dengue and DHF epidemics in several locations including Hawaii (128), Japan (112), Indonesia (63), the Seychelles (15, 77), southern China (94), Thailand (37), Singapore (19), and Malaysia (29).

Epidemic dengue activity has increased dramatically in the tropics because increased air travel has provided an efficient mechanism for the dispersal of dengue viruses (39). As a result, dengue has become the most important arbovirus disease of humans during the past 20 years and potentially fatal DHF has shown a 300-fold increase (from 2,067 cases in 1967 to over 600,000 in 1987) in its incidence among children in Southeast Asia (40). Up to 100 million cases of dengue infection world-wide may occur each year (45). Over two billion people are at risk of infection (40).

Dengue and DHF continue to pose serious public health problems in the Americas as well. Dengue activity has remained at a relatively high level in recent years; epidemics of all four dengue serotypes have occurred in various parts of the region. *A. aegypti* has been hitherto implicated as the only vector (40). In Puerto Rico, dengue has persisted as an endemic disease and epidemics have occurred there repeatedly since 1963 (40). The importance of dengue as a serious public health problem in the western hemisphere was underscored in 1981 when a major epidemic of dengue-2 and DHF (over 340,000 cases and 158 deaths) occurred in Cuba (43). In 1982, widespread epidemics of dengue were reported from Mexico in the east to Surinam in the west. The total number of cases reported for the Americas in 1987 (128,430) more than quadrupled compared with 31,337 cases in 1984; major epidemics occurred in 1985 in Aruba, Mexico, Nicaragua, and Brazil (112a). A major dengue-1 epidemic swept seven states in Brazil with more than a million cases during 1986–1987 (112b, 112c). Moreover, many dengue endemic countries in the Americas have sporadic cases of DHF on a regular basis (40).

Many suspected cases of dengue have been imported into the continental United States over the past several years. However, the first indigenous dengue-1 transmission since 1945 (27 confirmed cases) occurred in southern Texas in 1980 and caused considerable concern regarding the possibility of more extensive future outbreaks (44). This concern has been considerably heightened by the presence of *A. albopictus* in the continental United States. Furthermore, in view of the demonstrated ability of *A. albopictus* to transovarially transmit dengue virus in laboratory experiments (107) and the sexual transmission of dengue by male to female mosquitoes (105), the species has the potential to effectively import the virus into the US and initiate an epidemic without an infected human reservoir population. This threat made it imperative to learn the origins of the United States populations and to genetically characterize them. This knowledge can aid in assessments of the likelihood that the introduced populations are carrying and/or transmitting dengue. Although the importance of transovarial transmission in nature is not yet clear (130), possible future introductions of *A. albopictus* from dengue infested areas, should they occur, increase the probability that the dengue virus will become established in the United States.

La Crosse Encephalitis

La Crosse virus, a member of the California Serogroup (Bunyaviridae) is endemic in several midwestern states in the US. In most years, this virus causes the most frequently diagnosed arthropod-borne encephalitis in the US (131), although the numbers involved in any given year are relatively small. In six states in the La Crosse Belt (Ohio, Indiana, Illinois, Missouri, Kentucky, and North Carolina), *A. albopictus* is well established. Although *Aedes triseriatus* is the primary vector of La Crosse, the presence of *A. albopictus* in these states puts the human population at an enhanced risk.

Vector Competence: Experimental Analysis

A. albopictus is more susceptible than *A. aegypti* to oral infection with each of the four dengue serotypes (106). Moreover, 13 geographic strains of *A. albopictus* from Asia and Hawaii (42) varied markedly in oral susceptibility to dengue 1–4 viruses. Therefore, the determination of the geographic origin(s) of the US populations is important in assessing the danger of the introduced populations to public health in the US.

Considerable work has been done to determine differences in the vector competence of *A. albopictus* populations recently introduced to the Americas for particular strains of dengue. Boromisa et al (12) compared vector competence for dengue-1 virus from Fiji in three strains of *A. albopictus* from Houston, Memphis, and New Orleans with that of three native strains from Malaysia and one each from Japan (Tokyo) and Hawaii (Oahu) and with a strain of *A. aegypti* from Houston. The US strains of *A. albopictus* were closer to the Japanese *A. albopictus* and the Houston *A. aegypti* than to the Asian *A. albopictus* strains in their ability to transmit the virus.

In laboratory experiments, strains of *A. albopictus* from Houston have been shown to be competent vectors for dengue, yellow fever and Ross River viruses (80), and Rift Valley fever (127). Similarly, a strain of *A. albopictus* from Brazil (Cariacica City, Espirito Santo State) was able to transmit all four dengue virus serotypes and a sylvan strain of yellow-fever virus (78). Furthermore, the transmission rates of an epizootic strain of Venezuelan equine encephalomyelitis virus were shown to be significantly higher (25%) in two South American (São Paulo and Santa Tereza) strains of *A. albopictus* than in two North American (Houston and Alsace) strains (5%) (J. R. Beaman & M. J. Turell, unpublished data).

Strains of *A. albopictus* from the US and Japan show more or less similar transmission rates for La Crosse virus in the laboratory, but differ significantly from those of tropical Asian strains (T. Streit, personal communication). Furthermore, a population of *A. albopictus* from Harris County, Texas, has been shown to transmit La Crosse virus at rates equal to or greater than seven *A. triseriatus* field populations from the La Crosse endemic region (38). However, an intensive survey by the Centers for Disease Control (CDC) personnel for virus isolates from field-caught *A. albopictus* females collected from several active breeding sites has been negative for the presence of La Crosse virus to date. A newly recognized virus has been isolated from mosquitoes collected in Potosi, Missouri, and is being characterized (B. Francy, personal communication). Serologically, this virus is a member of the Bunyaviridae and preliminary results suggest that it shows low virulence in suckling mice and hamsters (P. Heard, personal communication).

Studies on the genetic basis of vector competence are important. Vector competence is probably influenced by alleles at several loci representing a polygenic character or nature (41, 120). With such traits, environmental effects can play an important role in determining the phenotype. It is, therefore, important to determine what proportion of the phenotypic variation is attributable to genetic effects. Gubler & Rosen (42) undertook selection of *A. albopictus* strains for increased resistance and susceptibility to oral infection with dengue-2 virus. Attempts to increase susceptibility were unsuccessful, but a nearly refractory strain was obtained. Preliminary crossing experiments between susceptible and resistant strains resulted in intermediate susceptibility to infection in the offspring. These data suggested that susceptibility is genetically controlled. Boromisa et al (12) observed major differences in the midgut and disseminated infection rates among the various geographic strains of *A. albopictus*. For example, the Houston strain showed evidence of a significant midgut infection barrier to dengue-1 virus, while the strains from Memphis and New Orleans did not. The salivary gland barrier(s) were a major factor in limiting transmission, and the extent of their occurrence was characteristic of the geographic strains employed (12). An analysis of crosses among the appropriate strains, such as Houston or Oahu with New Orleans or Memphis, should help clarify the role of possible genetic factors for the observed differences (R. Thomas, personal communication).

Finally, the ability of a vector to cause a disease epidemic does not depend on its relative vector competence alone. As shown by studies of the sylvatic *A. aegypti*–borne yellow-fever epidemic in Nigeria in 1987, even a relatively incompetent mosquito vector can cause a disease epidemic if its population density is high (79). Field (82) and population-genetic evidence (64) suggest that the population densities of *A. albopictus* in the US have increased considerably over the last three to four years.

Nucleic acid probes have recently been developed with good sensitivity to detect *A. albopictus* mosquitoes infected with dengue-2 virus (52, 65, 66, 88). These should aid in improved virus surveillance in laboratory and field studies as well as in studies on genetics of vector competence.

PHOTOPERIODIC SENSITIVITY

Seventeen strains of *A. albopictus* were tested for photoperiodically induced embryonic diapause by exposing females, in the laboratory, to either long or short day conditions (50). The six strains tested from the US (Texas, Louisiana, Florida, Tennessee, Indiana) and three from northern Asia (Beijing, Tokyo, and Korea) showed a photoperiodically induced egg diapause, i.e. when exposed to the short day lengths, the females laid eggs that either failed to hatch or resulted in low hatches. The same was true for three other strains tested earlier from Shanghai, China (129), and Nagasaki and Kyoto, Japan (60, 83). However, none of eight strains tested from tropical Asia

(Mauritius, Madagascar, and northern Taiwan) (50) nor six from Brazil (22) showed any photoperiodic response.

All the *A. albopictus* strains from the US and Japan had a critical photoperiod (hours of light/24 h inducing 50% diapause) between 13 and 14 hours of light. Strains from the higher latitudes (northern strains) had higher critical photoperiods than those from the lower latitudes (southern strains). However, the regression slope of critical photoperiod versus latitude of origin was relatively shallow and less well defined in the US than in the Japanese strains (W. A. Hawley, unpublished data, cited in 93). In addition to the latitude, the critical photoperiod was shown to be influenced by the rearing temperature and the nutritional status of the larvae (93). It has also been demonstrated in laboratory experiments that photoperiodic response can be lost rapidly under selection. Selection for a nearly diapause-free strain was accomplished in just two generations from the photoperiodically sensitive Tokyo strain of *A. albopictus* by exposing the pupae and adults to short photoperiods (L:D 9:15) and rearing the larvae that hatched in each generation at 21°C, L:D 16:8 (93).

COLD-HARDINESS

Exposure of embryonated eggs laid by females of 6 North American and 2 Japanese strains that were subjected to long-day photoperiods to subfreezing temperatures in the laboratory (-10°C for a 24-hour period after they were cold-conditioned at 5°C for 2 weeks) caused mortality of 22% or less. Conversely, similar treatment of eggs of two *A. albopictus* strains from tropical Malaysia and of a laboratory strain of *A. aegypti* resulted in nearly 100% mortality, whereas a subtropical strain from Taiwan exhibited intermediate mortality (50). Thus, the degree of cold-hardiness of the North American *A. albopictus* strains is similar to those of the Japanese strains. Nevertheless, as in the case of the photoperiodic response, differences in cold-hardiness were observed among the North American strains.

Field-overwintering studies over a three-year period at three outdoor locations in Indiana have corroborated the results of the above-mentioned laboratory experiments. Eggs of North American and Japanese *A. albopictus* strains survive at higher rates than those of strains from tropical Asia, Hawaii, and Brazil. Within the US, strains from Indiana were more cold-tolerant than those from Texas, Louisiana, and Florida. Furthermore, prolonged cold-conditioning and photoperiodically induced diapause enhanced the cold-hardiness of *A. albopictus* eggs (49).

REPRODUCTIVE DIFFERENTIATION

Crosses were made among 8 US and 12 Asian strains of *A. albopictus* from various parts of its range to detect any reproductive isolation, as measured by

egg hatch rates contrasted with those of control crosses (K. S. Rai, un-
published data). The strains tested were: Brazoria, Houston, and Harris
County, Texas; New Orleans, Louisiana; Memphis, Tennessee; Evansville
and Indianapolis, Indiana; Oahu, Hawaii; Cariacica and Santa Tereza, Brazil;
Peradeniya, Sri Lanka; Kuala Lumpur, Malaysia; Kent Ridge, Singapore;
Tokyo and Zama City, Japan; Pontianak, Indonesia; Calcutta, India; Candos,
Mauritius; and Koh Samui, Thailand. Except in crosses involving Mauritius
females, the egg hatch rates from these interstrain crosses were not signifi-
cantly different from those of the control intrastrain crosses. Certain crosses,
such as those involving Koh Samui females and Calcutta males and Kent-
Ridge females and Brazoria males, produced somewhat reduced egg hatches.
However, these isolated observations need to be repeated.

Mauritius females produced infertile eggs when crossed with males from all
other strains tested. This infertility was postzygotic because virtually 100% of
the Mauritius females were inseminated by males of the other strains (K. S.
Rai, unpublished data). The reciprocal crosses (Mauritius males crossed with
females of all other strains) produced high egg-hatch rates ranging from
76–98%, i.e. not significantly different from controls. The cause of this
asymmetric egg hatch remains to be determined. Such unilateral incompatibility has also been observed in both intra- and interspecific crosses among
certain members of the *Aedes scutellaris* subgroup (96).

Crosses were also made between the Tananareve, Madagascar, strain of *A.
albopictus* and several other species in the *albopictus* subgroup. All these
crosses were incompatible and yielded infertile eggs (97).

GENETIC DIFFERENTIATION

An important objective of our work with *A. albopictus* has been to genetically
characterize geographic variation between, and within, various populations
throughout the range of the species, particularly emphasizing the US populations and employing a variety of parameters as follows.

Allozymes

Black et al (7) analyzed allelic and genotypic frequencies at seven enzymatic
loci in several US populations of *A. albopictus* from New Orleans and in and
around Houston and one population each from Memphis, Jacksonville,
Evansville, and Indianapolis. Unique alleles and relatively high levels of
heterozygosity were detected in New Orleans, Houston, and Indianapolis,
indicating relatively large, and possibly independent, introductions into these
cities. Low heterozygosities with no unique alleles were observed in other
populations, indicative of population bottlenecks ensuing from a relatively
small number of founding individuals or through insecticidal control efforts at

these locations (Memphis, Evansville, and Jacksonville). Partitioning of variance showed that approximately 77% of the total variance was attributable to the within-population component and 23% to the between-population component. This result indicated that the amount of differentiation among collections within a city was three to four times as large as the variance among cities, suggestive of extensive local differentiation. Evidently, much genetic drift accompanied the establishment of local populations in New Orleans and Houston and local gene flow was sufficiently restricted to maintain differentiation (7). In a subsequent study, similar results were obtained using 10 enzymatic loci in 11 native populations from peninsular Malaysia and Borneo (8). This study indicated that extensive local differentiation observed in US populations did not ensue from the recent colonization but was rather characteristic of the breeding structure of the species.

To further quantify the patterns of genetic variation among populations throughout the distribution of the species and to attempt to trace the geographic origin(s) of the US and Brazilian populations, S. Kambhampati, W. C. Black, and I (unpublished data) conducted a worldwide survey of genetic variation at eight polymorphic loci in 57 populations collected from 9 countries (US, 28; Brazil, 4; Malaysia, 9; Sri Lanka, 1; Borneo, 2; India, 2; China, 3; Japan, 7; and Mauritius, 1). In all cases, the samples were derived either from eggs collected by placing ovitraps in the field or by capturing adult females, which then oviposited in the laboratory. The P_1 or F_1 adults thus obtained were used in the allozyme analyses.

The Southeast Asian populations showed slightly more polymorphism than those from US, Brazil, and Japan. Unique alleles were also present most often in populations from southern Asia. Furthermore, as observed earlier by Black et al (7, 8), extensive differentiation occurred at the local level, i.e. among locations within cities and among cities within countries. Differentiation was also evident in samples collected from different habitats in the same region, e.g. from bamboo stumps versus waste boxes from Hunan, China. These results confirmed that genetic drift at the local level is integral to the breeding structure of *A. albopictus* throughout its distribution (S. Kambhampati, W. C. Black, & K. S. Rai, unpublished data). Furthermore, a stepwise multiple discriminant analysis of allele frequencies in each of the 57 populations showed that the populations were genetically distinct from one another and that the populations from a given country formed distinct clusters (Figure 1). The probability of assigning a population to its correct region was greater than 98%. Although they did not overlap, the populations from Japan, China, and the US clustered close together, corroborating the suggestion of Hawley et al (50) regarding the northern Asian origin of the US populations. In addition, our results indicated that the Brazilian *A. albopictus* may also have originated in Japan. The Brazilian populations clustered near the Japan-China-US group, and they were closest to the Japanese populations (Figure 1).

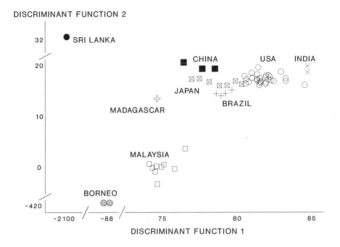

DISCRIMINANT FUNCTION 2

DISCRIMINANT FUNCTION 1

Figure 1   A plot of a multiple discriminant analysis of allele frequencies showing 9 broad groupings among 57 populations of *Aedes albopictus*. Populations from each country are represented by a different symbol accompanied by the name of the country. The populations from Japan, China, the US, and Brazil form one loose cluster, suggesting that the US and Brazilian populations originated in northern Asia (S. Kambhampati, W. C. Black, and K. S. Rai, unpublished data).

Similarly, a multivariate analysis of allele frequencies at 19–22 loci in *A. aegypti* established clustering of populations in a manner that reflected their geographic proximity (92, 121). Subdivided population structure resulting from restricted gene flow was also reported in *Anopheles gambiae* in western Kenya where significantly different distributions of restriction fragment arrays of rDNA intergenic spacers were obtained from field sites less than 10 km apart (75).

Genome Sizes

Through the use of quantitative cytophotometry of Feulgen-stained primary spermatocytes, haploid (1C) nuclear DNA amounts have been determined in 47 geographic populations of *A. albopictus* from 16 countries. Of these, 12 populations were assayed from 8 states in the continental United States and 3 from Hawaii (70, 99). Overall, the haploid C values of DNA ranged nearly threefold from 0.62 ± 0.02 (means ± SE) picograms (pg) in Koh Samui, Thailand, to 1.66 ± 0.08 pg in Harris County. Within continental US populations, the values ranged from 1.03 pg in Chambers County, Texas, to 1.66 pg in Harris County. The haploid DNA amounts in populations from various parts of Japan ranged from 0.76 ± 0.03 pg in Nagasaki to 1.16 ±

0.05 pg in Zama City. Statistical comparisons of populations revealed significant differences in the nuclear DNA amounts (ANOVA, all P < 0.05). However, a multiple range test did not result in strict geographical groupings. Thus, unlike allozyme analysis, no correlation was observed between geographic location and DNA content.

Although the populations from the US and Brazil possess a higher amount of nuclear DNA, no evidence suggests that increased genome size results from the recent range expansion.

Using DNA-reassociation kinetics, Black & Rai (10) showed that the intraspecific variation in DNA amount in strains of *A. albopictus* results mainly from highly repetitive DNA sequences. For example, the proportion of highly repetitive DNA in the Mauritius strain of *A. albopictus* possessing 1.32 pg DNA/haploid nuclear genome was 2.5 times greater than in the Calcutta strain possessing 0.86 pg DNA. The amounts of fold-back, highly repetitive, and middle-repetitive DNA increased linearly with genome size, though not at similar rates. Further, the frequency of eight highly repetitive DNA sequences isolated from a Hawaiian strain varied considerably among different strains of *A. albopictus* as a function of their geographic location and genome sizes (76). Like other culicine mosquitoes, *A. albopictus* possesses three pairs of chromosomes (95). As shown by in situ hybridization, the highly repetitive DNA sequences have different patterns of distribution on the three pairs of chromosomes (A. Kumar & K. S. Rai, unpublished data).

A lot of debate has concerned the organismic function of genome size variation. DNA content has been found to correlate with a variety of cellular and organismic attributes (17). Ferrari and Rai (31) found a significant correlation between genome size and development rate in different strains of *A. albopictus*. The larger the genome size, the slower is the rate of development from eggs to adults.

Ribosomal DNA

Molecular genetic surveys of multigene families have received considerable attention in recent years. Of the many multigene families, the family coding for ribosomal RNA (rDNA) has been used widely in population genetic analyses. This family is composed of 100–1500 copies of tandemly repeated genes in animals. These genes code for 5.8S, 18S, and 28S RNA components of the ribosome. Nontranscribed spacer regions (NTS), which connect adjacent copies, display extensive intraspecific variation, while the coding regions are conserved within a species (35). In situ hybridization with H^3 rDNA probes has shown that in *A. albopictus* a single locus of rDNA cistron per haploid karyotype is located on the smallest chromosome (A. Kumar & K. S. Rai, unpublished data). The larvae contain 300–400 copies (90).

We (9) conducted a worldwide nested spatial survey of variation in the rDNA gene family in 17 populations of *A. albopictus* from 9 countries at 4 levels (in individuals, in populations within a city, in city populations, and in countries) to determine the patterns of divergence in individual components of the rDNA cistron. We also compared these patterns in newly established US populations with those from the native range to determine if the observed patterns could be used to identify the origin of US populations (9).

DNA was isolated from larvae of the Mauritius strain and used in the construction of a genomic library. Three clones of ribosomal cistrons were isolated from the partial library using 18S and 28S rDNA probes from *Calliphora erythrocephala* kindly provided by K. Beckingham. These three clones were mapped using 10 restriction enzymes. The 18S and 28S coding regions were identical in each of the clones in their restriction sites, whereas the NTS varied among all three in that clones 2 and 3 had different-sized inserts.

To probe for variation in each region, the rDNA cistron was further subcloned into five regions corresponding to the 18S, the 28S (α), the 28S (β) subunits and the two structurally dissimilar halves of the NTS, designated Px1 and Px8. These clones were used to probe for intraspecific variation. The 18S and 28S (α and β) coding regions were similar in size among all 17 populations of *A. albopictus* examined, while continuous and extensive variation existed in the NTS among and within populations. The Px1 region contained multiple 190–base pair (bp) *Alu*I repeats nested within larger *Xho*I repeats of various sizes. The *Alu*I repeat region of the NTS contained many length variants. No repeats were found in the second region, which gave rise to relatively fewer variants (9). Individual mosquitoes in a given population carried a unique set of spacers. This observation differs markedly from that of *Anopheles albimanus* (4) and *Drosophila* species (21, 132), in which polymorphism in spacer diversity exists mainly among, rather than within, populations. However, spacer lengths in *A. albopictus* varied most at the population level, which indicates little conservation of spacers in a population. Of the 17 *A. albopictus* populations examined, only the Assam, India, population appeared to follow the *Anopheles/Drosophila* pattern, although here too, the frequency of the 4 spacer classes among the individuals examined was considerably different.

A partitioning of variation in rDNA in *A. albopictus* showed that the average diversity among individuals within samples accounted for 65% of total diversity. Differences among samples within cities accounted for 5%, differences among cities within a country for 4%, and differences between countries accounted for 25% of total diversity (9).

Although no correlation between the genetic (based on the entire lengths of the NTS) and geographic distances was observed, a positive correlation existed when either one or the other region (Px1 or Px8) of the NTS were used in the analyses. Nevertheless, populations from the US were not consistently similar to any one of the foreign populations examined, and diversity clearly arose independently in adjacent populations. Furthermore, as suggested by the diversity of bands found in the US populations, the diversity of the NTS may have changed rapidly during colonization (9).

Mitochondrial DNA

Mitochondrial DNA (mtDNA) provides a sensitive tool to determine lineages and evolutionary history of a species. It is maternally inherited, has a relatively simple organization and, in most organisms, evolves 5–10 times faster than single-copy nuclear DNA. Furthermore, genomes of different maternal clones do not recombine. As a consequence, several studies have used restriction enzyme analysis of mtDNA to determine population structure (1, 84).

We surveyed 17 populations of *A. albopictus* from 10 countries including 6 populations from the US for mtDNA restriction fragment length polymorphism (S. Kambhampati & K. S. Rai, unpublished data). These populations were derived from Candos, Mauritius; Cariacica, Brazil; Delhi, India; Ebina, Saga, and Zama City, Japan; Hong Kong; Kuala Lumpur, Malaysia; Peradeniya, Sri Lanka; Singapore City; Taipei, Taiwan; and, in the US, Galveston County, Texas; Indianapolis; Jacksonville; Memphis; New Orleans; and Manoa Valley, Hawaii. Total genomic DNA was extracted from each population and digested with 18 different restriction enzymes, separated on 1% agarose gels, Southern blotted, and probed with mtDNA of *Anopheles quadrimaculatus* cloned in its entirety into plasmids and kindly made available by S. Mitchell and A. Cockburn.

Six restriction enzymes did not reveal any cutting sites. The remaining 12 yielded 1,060 fragments for the 17 populations. Only three populations showed the presence of novel mtDNA haplotypes: Mauritius with *Hae*III and *Hpa*II, Hong Kong with *Eco*RI and *Hin*fI and Singapore with *Hae*III. The mean *F* value (i.e. proportion of shared sites) was 0.995. A value of 1 means complete identity.

*Eco*RV, *Hin*fI, *Hpa*II, and *Xho*I were used to survey within-population variation in individual mosquitoes in nine populations (Kabeshima and Sebruri from Japan and Brazoria, Galveston, Houston, East Saint Louis, Jacksonville, Memphis, and Savannah from the US). Except for one individual from Brazoria, which possessed a 1,810 bp mtDNA fragment when assayed with *Hin*fI rather than the 1,588 bp fragment observed in all other individuals, no variation in restriction sites was observed for any of the enzymes tested. The amount of mtDNA polymorphism observed between and within various populations of *A. albopictus* is considerably lower than animals in general, as well as other insect species (1).

The paucity of mtDNA polymorphism in *A. albopictus* markedly contrasts with the extensive variation observed among several species in the *A. scutellaris* group including three other species in the *A. albopictus* subgroup (S. Kambhampati & K. S. Rai, unpublished data). The results observed for *A. albopictus* probably reflect (*a*) a relatively recent introduction of the species into the US, (*b*) the expansion of its range to most other locations outside its native range, perhaps during the past (approximately) 100 years, and (*c*) extensive gene flow through human transport associated with large-scale movement of tires infested with *A. albopictus* eggs.

Temporal Changes in Gene Frequencies

A survey of temporal variation at 10 loci in 17 populations from locations in and around Houston between 1986 and 1988 revealed significant, but non-directional, changes (64). Overall variance in allele frequencies remained unchanged as did the levels of variation within and among populations, leading to the authors' conclusion that ". . . . the breeding structure of *A. albopictus* in the US did not differ substantially from that in a native habitat, either soon after colonization or after several generations in the new habitat" (64). Also, the fact that no significant decrease in genetic variation was observed (64) once again suggests that the introduction into the US was fairly large.

Preliminary results of studies on temporal variation in the rDNA NTS of several US populations indicated that the high level of variation observed by Black et al (9) is being maintained with seemingly random fluctuations in the frequencies of the various size repeats (S. Kambhampati & K. S. Rai, unpublished data).

ORIGIN AND SPREAD OF *AEDES ALBOPICTUS* IN THE AMERICAS

North America

Based on allozyme analysis (S. Kambhampati, W. C. Black, & K. S. Rai, unpublished data), and photoperiodic and cold-hardiness responses (50), it is reasonably certain that the North American populations originated in northern Asia. Observations on transmission patterns of dengue-1 virus (12) and of La Crosse virus (T. Streit, personal communication) among strains of different geographic origins generally corroborate the proposed derivation of North American strains in northern Asia, probably Japan. More specifically, an analysis of critical photoperiods suggests that the original founder population of *A. albopictus,* established in Houston, may have been introduced from a port city south of Tokyo, (W. A. Hawley, unpublished data, cited in 93).

The pattern of used-tire imports to the US also suggests a probable Japanese origin of North American *A. albopictus* (23, 50, 82). That *A. albopictus* entered North America as eggs and/or larvae in used tires is virtually certain. Craven et al (23) reported that *A. albopictus* larvae were imported into the United States at the substantial rate of 20 infested wet tires per 10,000 inspected at the port of entry in Seattle, Washington. Furthermore, of the 3.2 million tires imported into the US in 1985, 2.8 million came from Asian countries with indigenous *A. albopictus* populations. The pattern of used-tire imports to the US from among the Asian countries suggests a probable Japanese origin of at least one introduction of *A. albopictus* into North America. CDC monitoring of used tires originating in Japan, and offloaded in Seattle, indicated that 11 out of 2,613 tires inspected on October 6–7, 1986, contained *A. albopictus* larvae (82). This observation points to a substantial infestation rate for tires entering the US from Japan.

All the above lines of evidence indicate the probable Japanese origin of at least one North American introduction of *A. albopictus.* However, some traits, such as photoperiodic sensitivity, cold-hardiness, and vector competence to viruses, represent complex physiological characteristics whose expression is modulated by multiple genetic and environmental factors (13, 47, 53, 93) and is seemingly subject to rapid selection or adaptation (53, 93). Although conclusions concerning geographic origins based on such characteristics may not be definitive, in conjunction with genetic markers (such as allozymes) whose genetic basis is well-defined, they did provide useful information in the case of *A. albopictus.*

The movement of *A. albopictus,* particularly in recent times, appears to be associated with random movement of used tires around the globe on a massive scale. This is certainly the case in the US and probably in most other countries as well. During the period 1978–1985, the US imported 11,590,921 used tires from 58 countries (17 Asian, 12 American, 24 European, 2 African, 3 Pacific) and exported 6,343,856 used tires to 62 countries (15 Asian, 24 American, 14 European, 6 African, 3 Pacific countries) (102). Imported tires are shipped to numerous locations in the United States. For example, during 1985, 30 US states were specified to US customs as destinations for used tires (23). Of these 30 states, 15 are now known to be infested with *A. albopictus.* Such tire trade provides an efficient mechanism for world-wide dispersal of *A. albopictus.*

A. albopictus can introduce several arboviruses into the United States, particularly through transovarially infected eggs. Future importations of populations from other geographic regions can enhance the genetic variability of existing populations and/or introduce new genotypes with regard to their adaptability to colonize areas where the species hitherto was absent. Beginning January 1, 1988, under the Public Law 78-410, Public Health Service Act, Section 361, and 42 CFR 71-32 (c) (10), the CDC required that all used tires coming from Asia be certified as dry, clean, and free of insects. The CDC is monitoring compliance with this regulation (82). However, in the absence of any regulations concerning inter- and intrastate movement of tires within the US, or their export from the US to other countries and vice versa, further expansion of *A. albopictus* in the Americas and elsewhere is likely. The actual extent of such expansion will be determined by several factors, including, but not restricted to, the type(s) and amount(s) of various genotypes imported and the environmental conditions that such introduced genotypes can adapt to.

Several authors have suggested that because existing populations in the US exhibit photoperiodic sensitivity and cold-hardiness consistent with a temperate Asian origin, expansion of the species' range into south Florida, south Texas, or Mexico may not be possible because of their inability to adapt to subtropical climates (23, 82). Nevertheless, recent reports on the southward spread of the species in the continental US tend to belie the above speculation. *A. albopictus* populations were recently observed in Laredo, Brownsville, and Misson, Texas; Polk County, Florida; and Matamoros, Mexico (131; G. B. Craig, Jr., personal communication). This southern movement may have been associated with a relatively rapid adaptation to changed photoperiodic sensitivity (G. B. Craig, Jr., personal communication). Nevertheless, for reasons not entirely understood, the species has been unable, thus far, to colonize western parts of the US, and the continents of Australia and Africa.

Brazil

Three possibilities may be considered concerning the origin of *A. albopictus* populations in Brazil:

1. Because the six Brazilian populations studied are nonphotoperiodic (22) as are all the other tropical strains (50), *A. albopictus* may have entered Brazil, possibly in bamboo stumps, from Southeast Asia (89).
2. If *A. albopictus* entered Brazil also through commercial trading in used tires, Japan may be considered a probable source (S. Kambhampati, W. C. Black, & K. S. Rai, unpublished).
3. *A. albopictus* may have entered Brazil through North America (22).

Both options 2 and 3 would require selection for a nonphotoperiodic response following the introduction of the mosquitoes into Brazil from the photoperiodically sensitive populations in Japan and/or the US. In view of the demonstration that the photoperiodic response is lost rapidly under selection

(93), the possibility that a strain from North America has colonized Brazil, and then incorporated the nondiapause response, cannot be ruled out. As proposed by Smith (115), *A. albopictus* likely originated in treeholes in tropical Asia. Craig & Hawley (22) presented a plausible historical pattern of the distribution and range expansion of *A. albopictus* populations and the associated switching in the photoperiodic response patterns. They (22) suggested that in time "it colonized artificial containers, developed a closer association with man, and was likely transported northward by him. In temperate Asia, a photoperiodic response evolved. Now, after transport to a new, temperate climate, this offshoot of tropical *A. albopictus* is heading southward to the Neotropics."

INSECTICIDAL TOLERANCE AND RESISTANCE AND BIOLOGICAL CONTROL

Determination of insecticidal susceptibility and resistance status of *A. albopictus* introduced to the Americas is important in terms of the management of the species and for comparative purposes of putative source populations from Asia. Native populations from India, Malaysia, Southeast Asia, the Philippines, and Japan show resistance to DDT and dieldrin. Among the commonly used organophosphorus adulticides or larvicides, populations of *A. albopictus* from Singapore and Vietnam show resistance to malathion, from Malaysia to fenthion, and from Madagascar to fenitrothion (14). Little variability was observed in the susceptibility of the larvae of *A. albopictus* from Tokyo and Nagasaki to fenitrothion, fenthion, temephos, permethrin, and DDT. The highest level of resistance was observed for fenthion (122).

In the US, strains of *A. albopictus* from Houston and Liberty County, Texas, showed resistance to malathion and tolerance to bendiocarb and resmethrin among a limited number of insecticides tested to date in laboratory experiments (67, 104). Similarly, a strain from Oakhill, Ohio, showed resistance to malathion and four other organophosphorus compounds (D. Wesson, personal communication). In addition, several field populations in Harris County exhibited resistance to malathion following its field application (67). In preliminary experiments, however, larvae of the Kentucky strain of *A. albopictus* have been shown to be susceptible to malathion (20). These observations imply either (*a*) introductions of different genotypes and/or different frequencies of resistance and susceptibility alleles at the above mentioned locations or (*b*) development of rapid resistance following insecticide treatments after the mosquitoes' arrival in the US. At any rate, as emphasized by the Pan American Health Organization (89), repeated monitoring of *A. albopictus* populations in the US is essential. Finally, no significant difference was seen between the dengue-infected and uninfected control group of mosquitoes in their response to malathion (101).

Because malathion is the most widely used adulticide along the Gulf Coast and elsewhere in the US, the observed resistance of *A. albopictus* populations in states like Texas and Louisiana may present a serious problem in the event of a disease outbreak (104).

Limited work has also been done to evaluate the potential of various biological control agents and growth inhibitors. These include *Bacillus thuringiensis* var. *israelensis* and *Bacillus sphaericus* (3, 20, 24, 72, 103); the fungus pathogen, *Metarhizum anisopliae* (100); a mosquito virus, Yokoshoji (59); a growth inhibitor, Dimilin (56); and the protozoans, *Ascogregrina armigerei* (119) and *Ascogregrina taiwanensis* (D. Wesson, personal communication). *B. thuringiensis* var. *israelensis* gave good control of *A. albopictus* larvae in automobile tires for periods up to eight weeks in Malaysia (72).

COMPETITION WITH *A. AEGYPTI* AND *A. TRISERIATUS*

Extensive literature concerns competitive interactions between *A. albopictus* and *A. aegypti* in their native habitats in Southeast Asia (reviewed in 48, 54, 97). Surveys in several major cities in Southeast Asia, e.g. Bangkok (109), Calcutta (36), Saigon (111), Manila (62), and Jakarta, Java (54), have fairly well established that *A. aegypti* populations have spread throughout Southeast Asia since the turn of the century and increased in abundance in several of these locations. Simultaneously, the abundance of *A. albopictus* has declined from many of these same locations, e.g. Bangkok (108), Kuala Lumpur (117), Calcutta (36, 113), and probably in Jakarta (54) and Saigon (111) as well. The reasons for such competitive interaction between these two species are not clearly understood.

In the United States, the situation may be just the reverse. First, *A. albopictus* was introduced to Hawaii sometime during 1830–1896 (46, 61). *A. aegypti*, which was widespread in Hawaii in 1892, may have been introduced there at approximately the same time as *A. albopictus*. However, by 1912, the latter species was much more abundant than the former, and, by 1943–1944, 85% of the day-time mosquitoes were *A. albopictus* and 15% were *A. aegypti* (36). The continental United States (i.e. Houston and New Orleans) has also experienced a steady decline in populations of *A. aegypti* concomitant with a corresponding increase in *A. albopictus* densities following introduction of *A. albopictus* in the 1980s.

Experimental analyses of larval competition between the two species under laboratory and field conditions in Southeast Asia indicated that *A. aegypti* outcompetes *A. albopictus* (18, 73, 81, 118). Black et al (11) undertook a series of laboratory experiments on competition between the Houston and New Orleans strains of the two species to resolve the possible reasons for apparently contradictory observations from the native habitats in Southeast Asia and those from the US. Larval survival, size, and development rates under different densities of pure and mixed cultures, oviposition preferences, and ethological isolation of the two species were analyzed. The larval-competition results showed that, in mixed cultures, survival of the US *A. albopictus* strains was significantly lower than those of *A. aegypti* strains, particularly under optimal diet conditions. The developmental rate of the Houston strains of *A. albopictus* also was delayed compared with that of *A. aegypti* in mixed cultures. Ethological mate choice tests, in which females of one or the other species were placed with males of both, showed *A. albopictus* males to be less sexually aggressive than males of *A. aegypti*. These results did not suggest that the US *A. albopictus* populations were more competitive in the laboratory than those of *A. aegypti*.

Nasci et al (86), using males of one species (either *A. aegypti* or *A. albopictus*) and females of the other have shown that, in the laboratory, males of *A. albopictus* were very effective in inseminating *A. aegypti* females while *A. aegypti* males rarely inseminated *A. albopictus* females, especially if *A. aegypti* females were available. More importantly, interspecific matings between *A. albopictus* males and *A. aegypti* females were observed in the field using mark-release-recapture methods. The frequency of such matings is expected to increase as the population density of *A. albopictus* increases. Such interspecific matings, in time, could cause species displacement through pheromone-mediated sterilization of *A. aegypti* females (71).

Reasons other than those ensuing from competitive interaction may also be responsible for the recent decline of *A. aegypti* populations in the United States. As is well documented, *A. aegypti* populations had declined rather dramatically in the southern US in the 1960s (51, 85) before the presence of *A. albopictus* in these regions.

Ho et al (55) studied larval interaction in bispecific and trispecific mixed populations involving US strains of *A. albopictus, A. aegypti,* and *A. triseriatus*. Results involving *A. albopictus* and *A. aegypti* were generally similar to those of Black et al (11) and showed that intraspecific competition was greater than interspecific competition. However, their studies on larval competition involving *A. triseriatus* were rather instructive. Among the three *Aedes* species examined, *A. triseriatus* was the weakest competitor: its larvae were slower in development and showed greater mortality than either *A. albopictus* or *A. aegypti* (55). Furthermore, results of field studies of larval competition in tire habitats suggested that *A. albopictus* outcompeted *A. triseriatus* in mixed-breeding habitats. These observations led Ho et al (55) to conclude that "the new immigrant, *A. albopictus,* may spread continuously northward if no effective control measures are taken."

CONCLUDING REMARKS

In view of the ability of *A. albopictus* to colonize newer habitats and to thereby expand its range, and the fact that in the Americas it has already become associated with its native breeding habitat, treeholes, both in the US (34) and Brazil (25), one can safely conclude that the species is here to stay in the western hemisphere and that it has become a part of the local fauna. This colonization parallels the introduction of another formerly Asian mosquito species, *Aedes (Finlaya) togoi,* along the coasts of British Columbia and Washington (6, 133).

Much has been learned concerning the biology of *A. albopictus* in the Old and the New World. Various geographic populations have shown extensive variability in genetic and biological parameters: allozymes, components of the nuclear and mitochondrial genomes, vector competence, response to changes in day length, and overwintering survival. Much remains to be learned, however, concerning the biology of *A. albopictus* in the Americas, particularly, with passage of time, its potential involvement as a vector for both exotic and endemic viruses, and those features of its genotype that make it a successful colonizer. Determination of the blood-meal sources of this species in the Americas is important: does it feed on the eastern chipmunk and the grey squirrel reservoirs of La Cross virus? Long-range monitoring of temporal and spatial changes in population-genetic structure, the insecticidal susceptibility and resistance status, particularly of well established foci, and the future patterns of its spread will be equally important.

Evidence from several sources, particularly from allozymes, comparative studies of photoperiodic sensitivity and cold-hardiness of embryonated eggs, suggests a north-Asian origin, probably Japan, for North American populations of *A. albopictus,* through the importation of used tires. Allozyme and vector competence data indicate a relatively large, possibly more than one, introduction into the US. From the original port(s) of entry, probably in Texas and Louisiana, the populations spread through adult dispersal over short distances, resulting in extensive local differentiation through genetic drift, and over fairly long distances through random movement of tires, causing gene flow. The Japanese or the North American populations may be the source of the Brazilian strains rapidly followed by selection for photoperiodic insensitivity.

Rather surprisingly, the US *A. albopictus* has not exhibited characteristics of a species introduced to a new location. The populations in the US have not shown much, if any, evidence of founder effects, i.e. evident loss of genetic variability. The average genetic heterozygosities involving several enzyme loci have been generally similar in the US to those in the native habitats (7, 8, 64).

Ecologically, *A. albopictus* is a generalist and can adapt to diverse environmental conditions both in the tropics and in the temperate zone. Clearly, the species can exploit disturbed habitats very effectively. However, its curious absence in the western United States and the continents of Australia and Africa suggests that the species may not be able to adapt to certain, yet unknown, environmental conditions.

ACKNOWLEDGMENTS

I thank Drs. Srinivas Kambhampati, Arun Kumar, Rex Thomas, Paul Grimstad, and George Craig, Jr. for helpful comments on earlier drafts of the manuscript, Dawn Verleye and Chris Bosio for assistance in literature survey, and Tracy Frost for preparing the typescript. I wish to express special thanks to several former and current postdoctoral research associates/trainees and graduate students in my laboratory, in particular Drs. W. Black, J. Ferrari, R. Boromisa, K. McLain, D. Pashley, S. Kambhampati, A. Kumar, and N. Rao for their significant contributions to the research represented herein. I also thank several colleagues at the University of Notre Dame and elsewhere for access to unpublished manuscripts/data. Research in the author's laboratory has been supported by National Institutes of Health grants 5R01 AI-21443 and 5T32 AI-07030.

Literature Cited

1. Avise, J. C., Arnold, J., Ball, R. M., Bermingham, E., Lamb, T., et al. 1987. Intraspecific phylogeography: The mitochondrial DNA bridge between population genetics and systematics. *Annu. Rev. Ecol. Syst.* 18:489–522
2. Bailey, S. F., Bohart, R. M. 1952. A mosquito survey and control program in Guam. *J. Econ. Entomol.* 45:947–52
3. Barjac, H., Coz, J. 1979. Sensibilite comparee de six especes differentes de moustiques a *Bacillus thuringiensis* var. *israelensis*. *Bull. WHO* 57:139–41
4. Beach, R. F., Mills, D., Collins, F. H. 1989. Structure of ribosomal DNA in *Anopheles albimanus* (Diptera: Culicidae). *Ann. Entomol. Soc. Am.* 82:641–48
5. Belkin, J. N. 1962. *The mosquitoes of the South Pacific (Diptera: Culicidae),* Vol. 1. Berkeley: Univ. Calif. Press. 608 pp.
6. Belton, P. 1980. The first record of *Aedes togoi* (Theo.) in the United States—aboriginal or ferry passenger? *Mosq. News* 40:624–26
7. Black, W. C. IV, Ferrari, J. A., Rai, K. S., Sprenger D. 1988. Breeding structure of a colonizing species: *Aedes albo-*
   pictus (Skuse) in the United States. *Heredity* 60:173–81
8. Black, W. C. IV, Hawley, W. A., Rai, K. S., Craig, G. B. Jr. 1988. Breeding structure of a colonizing species: *Aedes albopictus* (Skuse) in Peninsular Malaysia and Borneo. *Heredity* 61:439–46
9. Black, W. C. IV, McLain, D. K., Rai, K. S. 1989. Patterns of variation in the rDNA cistron within and among world populations of a mosquito, *Aedes albopictus* (Skuse). *Genetics* 121:539–50
10. Black, W. C. IV, Rai, K. S. 1988. Genome evolution in mosquitoes: Intra- and interspecific variation in repetitive DNA amounts and organization. *Genet. Res. Camb.* 51:185–96
11. Black, W. C. IV, Rai, K. S., Turco, B. J., Arroyo, D. C. 1989. Laboratory study of competition between United States strains of *Aedes albopictus* and *Aedes aegypti* (Diptera: Culicidae). *J. Med. Entomol.* 26:260–71
12. Boromisa, R. D., Rai, K. S., Grimstad, P. R. 1987. Variation in the vector competence of geographic strains of *Aedes albopictus* for dengue-1 virus. *J. Am. Mosq. Control Assoc.* 3:378–86

13. Bradshaw, W. E. 1976. Geography of photoperiodic response in a diapausing mosquito. *Nature London* 262:384–86

14. Brown, A. W. A. 1986. Insecticide resistance in mosquitoes: A pragmatic review. *J. Am. Mosq. Control Assoc.* 2:123–40

15. Calisher, C. H., Nuti, M., Lazuick, J. S., Ferrari, J. D. M., Kappus, K. D. 1981. Dengue in the Seychelles. *Bull. WHO* 59:619–22

16. Deleted in proof

17. Cavalier-Smith, T. 1985. Introduction: The evolutionary significance of genome size. In *The Evolution of Genome Size,* ed. T. Cavalier-Smith, pp. 1–35. New York: Wiley

18. Chan, K. L., Chan, Y. C., Ho, B. C. 1971. *Aedes aegypti* (L.) and *Aedes albopictus* (Skuse) in Singapore City. 4. Competition between species. *Bull. WHO* 44:643–49

19. Chan, Y. C., Ho, B. C., Chan, K. L. 1971. *Aedes aegypti* (L.) and *Aedes albopictus* (Skuse) in Singapore City. 5 Observations in relation to dengue haemorrhagic fever. *Bull. WHO* 44:651–58

20. Cilek, J. E., Moorer, G. D., Delph, L. A., Knapp, F. W. 1989. The Asian tiger mosquito, *Aedes albopictus,* in Kentucky. *J. Am. Mosq. Control Assoc.* 5:267–68

21. Coen, E. S., Strachnan, T., Dover, G. 1982. Dynamics of concerted evolution of ribosomal DNA and histone gene families in the *melanogaster* species subgroup of *Drosophila. J. Mol. Biol.* 158:17–35

22. Craig, G. B. Jr., Hawley, W. A. 1990. The Asian tiger mosquito, *Aedes albopictus:* Whither, whence and why not in Virginia? In *Entomology in Virginia: New Problems and New Approaches,* ed. R. D. Fell, pp. 1–10. Blacksburg, VA: Va. Polytech. Inst. and State Univ. Press

23. Craven, R. B., Eliason, D. A., Francy, D. B., Reiter, P., Campos, E. G., et al. 1988. Importation of *Aedes albopictus* and other exotic mosquito species into the United States in used tires from Asia. *J. Mosq. Control Assoc.* 4:138–42

24. Dagnogo, M., Coz, J. 1982. Un insecticide biologique: *Bacillus sphaericus.* 1. Activate larvicide de *B. sphaericus* sur quelques especes et souches de moustiques. *Cah. ORSTOM Ser. Entomol. Med. Parasitol.* 22:133–38

25. Castro Gomez, A., Marques, G. R. A. 1988. Encontro de criadouro natural de *Aedes (Stegomyia) albopictus* (Skuse),

estado de Sao Paulo, Brasil. *Rev. Saude Publ. Sau Paulo.* 22:245

26. Deleted in proof

27. Deleted in proof

28. Elliott, S. A. 1980. *Aedes albopictus* in the Solomon, Santa Cruz Islands, South Pacific. *Trans. R. Soc. Trop. Med. Hyg.* 74:747–48

29. Fang, R., Lo, E., Lim, T. W. 1984. The 1982 dengue epidemic in Malaysia: Epidemiological, serological and virological aspects. *Southeast Asian J. Trop. Med. Public Health* 15:51–58

30. Feng, L. C. 1938. The geographical distribution of mosquitoes in China. *Proc. 7th Int. Congr. Entomol. Berlin.* pp. 1579–88. Weimar, Druck: Ushmann

31. Ferrari, J. A., Rai, K. S. 1989. Phenotypic correlates of genome size variation in *Aedes albopictus. Evolution* 43:895–99

32. Ferreira Neto, J. A., Lima, M. M., Aragao, M. B. 1987. First observations on *Aedes albopictus* in Brazil. *Cad. Saude Publ. Rio de Janiero* 3:56–61 (In Portuguese)

33. Forattini, O. P. 1986. Identificacao de *Aedes (Stegomyia) albopictus* no Brasil. *Rev. Saude Publ. Sau Paulo* 20:244–45

34. Foster, B. E. 1989. *Aedes albopictus* larvae collected from treeholes in southern Indiana. *J. Am. Mosq. Control Assoc.* 5:95

35. Gerbi, S. A. 1985. Evolution of ribosomal DNA. In *Molecular Evolutionary Genetics,* ed. R. J. MacIntyre, pp. 419–517. New York: Plenum

36. Gilotra, S. K., Rozeboom, L. E., Bhattacharya, N. C. 1967. Observations on possible competitive displacement between populations of *Aedes aegypti* Linnaeus and *Aedes albopictus* (Skuse) in Calcutta. *Bull. WHO* 37:437–46

37. Gould, D. J., Yuill, T. M., Moussa, M. A., Simasathien, P., Rutledge, L. C. 1968. An insular outbreak of dengue hemorrhagic fever. III. Identification of vectors and observations on vector ecology. *Am. J. Trop. Med. Hyg.* 17:609–18

38. Grimstad, P. R., Kobayashi, J. F., Zhang, M., Craig, G. B. Jr. 1989. Recently introduced *Aedes albopictus* in the United States: Potential vector of La Crosse virus (Bunyaviridae: California serogroup). *J. Am. Mosq. Control Assoc.* 5:422–27

39. Gubler, D. J. 1988. Dengue. In *The Arboviruses: Epidemiology and Ecology,* ed. T. P. Monath, 2:223–61. Boca Raton, FL: CRC

40. Gubler, D. J. 1989. *Aedes aegypti* and

Aedes aegypti–borne disease control in the 1990s: Top down or bottom up. *Am. J. Trop. Med. Hyg.* 40:571–78

41. Gubler, D. J., Novak, R., Mitchell, C. J. 1982. Arthropod vector competence—epidemiological, genetic, and biological considerations. See Ref. 117a, pp. 343–78

42. Gubler, D. J., Rosen, L. 1976. Variation among geographic strains of *Aedes albopictus* in susceptibility to infection with dengue viruses. *Am. J. Trop. Med. Hyg.* 25:318–25

43. Guzman, M. G., Kouri, G. P., Bravo, J., Soler, M., Vazquez, S., et al. 1984. Dengue haemorrhagic fever in Cuba. II. Clinical investigations. *Trans. R. Soc. Trop. Med. Hyg.* 78:239–41

44. Hafkin, B., Kaplan, J. E., Reed, C., Elliott, L. B., Fontaine, R., et al. 1982. Reintroduction of dengue fever into the continental United States. I. Dengue surveillance in Texas, 1980. *Am. J. Trop. Med. Hyg.* 31:1222–28

45. Halstead, S. B. 1988. Pathogenesis of dengue: Challenges to molecular biology. *Science* 239:476–81

46. Hardy, D. E. 1960. Culicidae. In *Insects of Hawaii, Diptera: Nematocera-Brachycera,* ed. E. C. Zimmermann, 10:18–22, 81–90. Honolulu: Univ. Hawaii Press

47. Harvey, G. T. 1957. The occurrence and nature of diapause-free development in the spruce budworm, *Choristoneura fumiferana* (Clem.) (Lepidoptera: Tortricidae). *Can. J. Zool.* 35:549–72

48. Hawley, W. A. 1988. The biology of *Aedes albopictus. J. Am. Mosq. Control Assoc.* 4(suppl. 1):1–39

49. Hawley, W. A., Pumpuni, C. B., Brady, R. H., Craig, G. B., Jr. 1989. Overwintering survival of *Aedes albopictus* (Diptera: Culicidae) eggs in Indiana. *J. Med. Entomol.* 26:122–29

50. Hawley, W. A., Reiter, P., Copeland, R. S., Pumpuni, C. B., Craig, G. B. Jr. 1987. *Aedes albopictus* in North America: Probable introduction in used tires from northern Asia. *Science* 236:1114–16

51. Hayes, G. R. Jr., Ritter, A. B. 1966. The diminution of *Aedes aegypti* infestations in Louisiana. *Mosq. News* 26:381–83

52. Henchal, E. A., Narupiti, S., Feighny, R., Padmanabhan, R., Vakharia, V. 1987. Detection of dengue virus RNA using nucleic acid hybridization. *J. Virol. Meth.* 15:187–200

53. Henrich, V. C., Denlinger, D. L. 1983. Genetic differences in pupal diapause incidence between two selected strains of the flesh fly. *J. Hered.* 74:371–74

54. Ho, B. C., Chan, K. L., Chan, Y. C. 1973. The biology and bionomics of *Aedes albopictus* (Skuse). In *Vector Control in Southeast Asia,* ed. Y. C. Chan, K. L. Chan, B. C. Ho, pp. 125–43. Singapore: Univ. Singapore Press

55. Ho, B. C., Ewert, A., Chew, L. M. 1989. Interspecific competition among *Aedes aegypti, Ae. albopictus* and *Ae. triseriatus* (Diptera: Culicidae): Larval development in mixed cultures. *J. Med. Entomol.* 26:615–23

56. Ho, C.-M., Hsu, T. R., Wu, J. Y., Wang, C. H. 1987. Effect of dimilin, a chitin synthesis inhibitor, on the growth and development of larvae of *Aedes albopictus* (Skuse). *Chin. J. Entomol.* 7:131–42

57. Huang, Y.-M. 1972. Contributions to the mosquito fauna of southeast Asia. XIV. The subgenus *Stegomyia* of *Aedes* in southeast Asia. I. The Scutellaris group of species. *Contrib. Am. Entomol. Inst. Ann Arbor* 9:1–109

58. Hull, W. B. 1952. Mosquito survey of Guam. *US Armed Forces Med. J.* 3:1287–95

59. Igarashi, A., Matsuo, S., Bundo-Morita, K., Oda, T., Mori, A., Fujita, K. 1986. Larvicidal effect of an insect virus of mosquitoes: An attempt to use as a potential agent of biological control on mosquitoes. *Trop. Med.* 28:191–208

60. Imai, C., Maeda, O. 1976. Several factors affecting on hatching of *Aedes albopictus* eggs. *Jpn. J. Sanit. Zool.* 27:367–72 (In Japanese with English Summary)

61. Joyce, C. R. 1961. Potentialities for accidental establishment of exotic mosquitoes in Hawaii. *Proc. Hawaii. Entomol. Soc.* 17:403–13

62. Jueco, N. L., Cabrera, B. D. 1974. Investigations on the ecology and biology of *Aedes aegypti* and *Ae. albopictus:* Seasonal incidence and larval breeding preference. *Kalikasan* 3:187–92

63. Jumali, S., Gubler, D. J., Nalim, S., Eram, S., Saroso, J. S. 1979. Epidemic dengue hemorrhagic fever in rural Indonesia. III. Entomological studies. *Am. J. Trop. Med. Hyg.* 28:717–24

64. Kambhampati, S., Black, W. C. IV, Rai, K. S., Sprenger, D. 1990. Temporal variation in genetic structure of a colonizing species: *Aedes albopictus* in the United States. *Heredity* 64:281–87

65. Kerschner, J. H., Vorndam, A. V., Monath, T. P., Trent, D. W. 1986. Genetic and epidemiological studies of dengue type 2 viruses by hybridization

using synthetic deoxyoligonucleotides as probes. *J. Gen. Virol.* 67:2645–61

66. Khan, A. M., Wright, P. J. 1987. Detection of flavivirus RNA in infected cells using photobiotin-labelled hybridization probes. *J. Virol. Meth.* 15:121–30

67. Khoo, B. K., Sutherland, D. J., Sprenger, D., Dickerson, D., Nguyen, H. 1988. Susceptibility status of *Aedes albopictus* to three topically applied adulticides. *J. Am. Mosq. Control Assoc.* 4:310–13

68. Knight, K. L., Stone, A. 1977. A catalog of the mosquitoes of the world (Diptera: Culicidae), Vol. 6. College Park, MD: Entomol. Soc. Am., Thomas Say Found. 610 pp. 2nd ed.

69. Knudsen, A. B. 1986. The significance of the introduction of *Aedes albopictus* into the southern United States with implications for the Caribbean, and perspectives of the Pan American Health Organization. *J. Am. Mosq. Control Assoc.* 2:420–23

70. Kumar, A., Rai, K. S. 1990. Intraspecific variation in nuclear DNA content among world populations of a mosquito, *Aedes albopictus* (Skuse). *Theor. Appl. Genet.* 79:748–52

71. Leahy, M. G., Craig, G. B. Jr. 1967. Barriers to hybridization between *Aedes aegypti* and *Aedes albopictus* (Diptera: Culicidae). *Evolution* 21:41–58

72. Lee, H. L., Cheong, W. H. 1987. Field evaluation of the efficacy of *Bacillus thuringiensis* H-14 for the control of *Aedes (Stegomyia) albopictus* (Skuse). *Mosq.-Borne Dis. Bull.* 3:57–63

73. Macdonald, W. W. 1956. *Aedes aegypti* in Malaya. II. Larval and adult biology. *Ann. Trop. Med. Parasitol.* 50:399–414

74. Mangiafico, J. A. 1971. Chikungunya virus infection and transmission in five species of mosquito. *Am. J. Trop. Med. Hyg.* 20:642–45

75. McLain, D. K., Collins, F. H., Brandling-Bennett, A. D., Were, J. B. O. 1989. Microgeographic variation in rDNA intergenic spacers of *Anopheles gambiae* in western Kenya. *Heredity* 62:257–64

76. McLain, D. K., Rai, K. S., Fraser, M. J. 1987. Intraspecific and interspecific variation in the sequence and abundance of highly repeated DNA among mosquitoes of the *Aedes albopictus* subgroup. *Heredity* 58:373–81

77. Metselaar, D., Grainger, C. R., Oei, K. G., Reynolds, D. G., Pudney, M., et al. 1980. An outbreak of type 2 dengue fever in the Seychelles, probably transmitted by *Aedes albopictus* (Skuse). *Bull. WHO* 58:937–43

78. Miller, B. R., Ballinger, M. E. 1988. *Aedes albopictus* mosquitoes introduced into Brazil: Vector competence for yellow fever and dengue viruses. *Trans. R. Soc. Trop. Med. Hyg.* 82:476–77

79. Miller, B. R., Monath, T. P., Tabachnick, W. J., Ezike, V. I. 1989. Epidemic yellow fever caused by an incompetent mosquito vector. *Trop. Med. Parasit.* 40:396–99

80. Mitchell, C. J., Miller, B. R., Gubler, D. J. 1987. Vector competence of *Aedes albopictus* from Houston, Texas, for dengue serotypes 1 to 4, yellow-fever and Ross River viruses. *J. Am. Mosq. Control Assoc.* 3:460–65

81. Moore, C. G., Fisher, B. R. 1969. Competition in mosquitoes: Density and species ratio effects on growth, mortality, fecundity, and production of growth retardant. *Ann. Entomol. Soc. Am.* 62:1325–31

82. Moore, C. G., Francy, D. B., Eliason, D. A., Monath, T. P. 1988. *Aedes albopictus* in the United States: rapid spread of a potential disease vector. *J. Am. Mosq. Control Assoc.* 4:356–61

83. Mori, A., Oda, T., Wada, Y. 1981. Studies on the egg diapause and overwintering of *Aedes albopictus* in Nagasaki, Japan. *Trop. Med.* 23:79–90

84. Mortiz, C., Dowling, T. E., Brown, W. M. 1987. Evolution of animal mitochondrial DNA: Relevance for population biology and systematics. *Ann. Rev. Ecol. Syst.* 18:269–92

85. Morlan, H. B., Tinker, M. E. 1965. Distribution of *Aedes aegypti* infestations in the United States. *Am. J. Trop. Med. Hyg.* 14:892–99

86. Nasci, R. S., Hare, S. G., Willis, F. S. 1989. Interspecific mating between Louisiana strains of *Aedes albopictus* and *Aedes aegypti* in the field and laboratory. *J. Am. Mosq. Control Assoc.* 5:416–21

87. Nawrocki, S. J., Hawley, W. A. 1987. Estimation of the northern limits of distribution of *Aedes albopictus* in North America. *J. Am. Mosq. Control Assoc.* 3:314–17

88. Olson, K., Blair, C., Padmanabhan, R., Beaty, B. 1988. Detection of dengue virus type 2 in *Aedes albopictus* by nucleic acid hybridization with strand-specific RNA probes. *J. Clin. Microbiol.* 26:579–81

89. Pan American Health Organization. 1987. Control of *Aedes albopictus* in the Americas. *PAHO Bull.* 21:314–24

90. Park, J., Fallon, A. M. 1990. Mosquito ribosomal RNA genes: Characterization of gene structure and evidence for changes in copy number during development. *Insect Biochem.* 20:1–11

91. Pashley, D. N., Pashley, D. P. 1983. Observations on *Aedes (Stegomyia)* mosquitoes in Micronesia and Melanesia. *Mosq. Systemat.* 15:41–49

92. Powell, J. R., Tabachnick, W. J., Arnold, J. 1980. Genetics and the origin of a vector population: *Aedes aegypti*, a case study. *Science* 208:1385–87

93. Pumpuni, C. B. 1989. *Factors influencing photoperiodic control of egg diapause in Aedes albopictus (Skuse).* PhD thesis. Notre Dame, IN: Univ. Notre Dame. 148 pp.

94. Qiu, F., Zhang, H., Shao, L., Li, X., Luo, H., Yu, Y. 1981. Studies on the rapid detection of dengue virus antigen by immunofluorescence and radioimmunoassay. *Chin. Med. J.* 94:653–58

95. Rai, K. S. 1963. A comparative study of mosquito karyotypes. *Ann. Entomol. Soc. Am.* 56:160–70

96. Rai, K. S. 1983. Genetic and chromosomal differentiation in the *Aedes (Stegomyia) scutellaris* group. In *Genetics: New Frontiers*, pp. 99–111, *Proc. 15th Int. Congr. Genet. New Delhi*. New Delhi: Oxford & IBH

97. Rai, K. S. 1986. Genetics of *Aedes albopictus. J. Am. Mosq. Control Assoc.* 2:429–36

98. Rai, K. S., Pashley, D. P., Munstermann, L. E. 1982. Genetics of speciation in aedine mosquitoes. See Ref. 117a, pp. 84–129

99. Rao, P. N., Rai, K. S. 1987. Inter and intraspecific variation in nuclear DNA content in *Aedes* mosquitoes. *Heredity* 59:253–58

100. Ravallec, M., Riba, G., Vey, A. 1989. Susceptibility of larvae of *Aedes albopictus* (Diptera: Culicidae) to the entomopathogenic hyphomycete *Metarhizium anisopliae* (Metsch) Sorokin. *Entomophaga* 34:209–17 (In French)

101. Rawlins, S. C., Eliason, D. A., Moore, C. G., Campos, E. G. 1988. Effects of dengue-1 infection in *Aedes albopictus* on its susceptibility to malathion. *J. Am. Mosq. Control Assoc.* 4:372–73

102. Reiter, P., Sprenger, D. 1987. The used tire trade: A mechanism for the worldwide dispersal of container breeding mosquitoes. *J. Am. Mosq. Control Assoc.* 3:494–501

103. Ren, G. X., Sun, G. H., Xu, R. M., Zhang, J. S., Lu, B. L. 1987. Toxicity assay of acetone powder from *Bacillus sphaericus* strain TS-1 on seven species of mosquito larvae. *Acta Entomol. Sin.* 30:21–25

104. Robert, L. L., Olson, J. K. 1989. Susceptibility of female *Aedes albopictus* from Texas to commonly used adulticides. *J. Am. Mosq. Control Assoc.* 5:251–53

105. Rosen, L. 1987. Sexual transmission of dengue viruses by *Aedes albopictus. Am. J. Trop. Med. Hyg.* 37:398–402

106. Rosen, L., Rozeboom, L. E., Gubler, D. J., Lien, J. C., Chaniotis, B. N. 1985. Comparative susceptibility of mosquito species and strains to oral and parenteral infection with dengue and Japanese encephalitis viruses. *Am. J. Trop. Med. Hyg.* 34:603–15

107. Rosen, L., Shroyer, D. A., Tesh, R. B., Freier, J. E., Lien, J. C. 1983. Transovarial transmission of dengue viruses by mosquitoes: *Aedes albopictus* and *Aedes aegypti. Am. J. Trop. Med. Hyg.* 32:1108–19

108. Rudnick, A. 1965. Studies on the ecology of dengue in Malaysia: a preliminary report. *J. Med. Entomol.* 2:203–8

109. Rudnick, A., Hammon, W. M. 1960. Newly recognized *Aedes aegypti* problems in Manila and Bangkok. *Mosq. News* 20:247–49

110. Rudnick, A., Marchette, N. J., Garcia, R. 1967. Possible jungle dengue-recent studies and hypothesis. *Jpn. J. Med. Sci. Biol.* 20(supp):69–74

111. Russell, P. K., Quy, D. V., Nisalak, A., Simasathien, P., Yuill, T. M., Gould, D. J. 1969. Mosquito vectors of dengue viruses in South Vietnam. *Am. J. Trop. Med. Hyg.* 18:455–59

112. Sabin, A. B. 1952. Research on dengue during World War II. *Am. J. Trop. Med. Hyg.* 1:30–500

112a. San Juan Laboratories, Centers for Disease Control. 1986. *Dengue Surveillance Summary*. No. 33. 5 pp.

112b. San Juan Laboratories, Centers for Disease Control. 1987. *Dengue Surveillance Summary*. No. 46. 6 pp.

112c. San Juan Laboratories, Centers for Disease Control. 1988. *Dengue Surveillance Summary*. No. 56. 5 pp.

113. Senior-White, R. 1934. Three years mosquito work in Calcutta. *Bull. Med. Res.* 25:551–96

114. Shroyer, D. A. 1986. *Aedes albopictus* and arboviruses: A concise review of the literature. *J. Am. Mosq. Control Assoc.* 2:424–28

115. Smith, C. E. G. 1956. The history of dengue in tropical Asia and its probable

relationship to the mosquito *Aedes aegypti*. *J. Trop. Med. Hyg.* 59:243–52

116. Sprenger, D., Wuithiranyagool, T. 1986. The discovery and distribution of *Aedes albopictus* in Harris County, Texas. *J. Am. Mosq. Control Assoc.* 2:217–19

117. Stanton, A. T. 1920. The mosquitoes of far eastern ports with special references to the prevalence of *Stegomyia fasciatus* F. *Bull. Entomol. Res.* 10:333–34

117a. Steiner, W. M., Tabachnick, W. J., Rai, K. S., Narang, S., eds. 1982. *Recent Developments in the Genetics of Insect Disease Vectors.* Champaign, IL: Stipes

118. Sucharit, S., Tumrasvin, W., Vutikes, S., Viraboonchai, S. 1978. Interactions between larvae of *Aedes aegypti* and *Aedes albopictus* in mixed experimental populations. *Southeast Asian J. Trop. Med. Public Health* 9:93–97

119. Sulaiman, I. 1987. Susceptibility of *Armigeres subalbatus* and three species of *Aedes* to *Ascogregarina armigerei*. *Trop. Biomed.* 4:145–49

120. Tabachnick, W. J., Aitken, T. H. G., Beaty, B. J., Miller, B. R., Powell, J. R., Wallis, G. P. 1982. Genetic approaches to the study of vector competence of *Aedes aegypti*. See Ref. 117a, pp. 413–32

121. Tabachnick, W. J., Powell, J. R. 1979. A world-wide survey of genetic variation in the yellow fever mosquito, *Aedes aegypti*. *Genet. Res.* 34:215–29

122. Takahashi, M., Chieko, S., Wada, Y., Ito, T. 1985. Insecticide susceptibility in *Aedes albopictus* (Skuse). *Jpn. J. Sanit. Zool.* 36:251–53

123. Tanaka, K., Mizusawa, K., Saugstad, E. S. 1979. A revision of the adult and larval mosquitoes of Japan (including the Ryukyu Archipelago and the Ogasawara Islands) and Korea (Diptera: Culicidae). *Contrib. Am. Entomol. Inst. Ann Arbor* 16:1–987

124. Tesh, R. B., Gubler, D. J. 1975. Laboratory studies of transovarial transmission of La Crosse and other arboviruses by *Aedes albopictus* and *Culex fatigans*. *Am. J. Trop. Med. Hyg.* 24:876–80

125. Tesh, R. B., Gubler, D. J., Rosen, L. 1976. Variation among geographic strains of *Aedes albopictus* in sus-

ceptibility to infection with Chickungunya virus. *Am. J. Trop. Med. Hyg.* 25:326–35

126. Tesh, R. B., Shroyer, D. A. 1980. The mechanism of arbovirus transovarial transmission in mosquitoes: San Angelo virus in *Aedes albopictus*. *Am. J. Trop. Med. Hyg.* 29:1294–1404

127. Turell, M. J., Bailey, C. C., Beaman, J. R. 1988. Vector competence of a Houston, Texas strain of *Aedes albopictus* for Rift Valley fever virus. *J. Am. Mosq. Control Assoc.* 4:94–96

128. Usinger, R. L. 1944. Entomological phases of the recent dengue epidemic in Honolulu. *US Public Health Rep.* 59:423–30

129. Wang, R. L. 1966. Observations on the influence of photoperiod on egg diapause in *Aedes albopictus* (Skuse). *Acta Entomol. Sin.* 15:75–77

130. Watts, D. M., Harrison, B. A., Pantuwatana, S., Klein, T. A., Burke, D. S. 1985. Failure to detect natural transovarial transmission of dengue viruses by *Aedes aegypti* and *Aedes albopictus* (Diptera: Culicidae). *J. Med. Entomol.* 22:261–65

131. Wesson, D., Hawley, W., Craig, G. B. Jr. 1990. Status of *Aedes albopictus* in the Midwest: La Crosse Belt distribution, 1988. *Proc. Ill. Mosq. Vect. Control Assoc.* 1:11–15

132. Williams, S. M., DeSalle, R., Strobeck, C. 1985. Homogenization of geographical variants at the non-transcribed spacer of rDNA in *Drosophila mercatorum*. *Mol. Biol. Evol.* 2:338–46

133. Wood, D. M., Dang, P. T., Ellis, R. A. 1979. The mosquitoes of Canada (Diptera: Culicidae). *The Insects and Arachnids of Canada,* part 6. Biosystemat. Res. Inst., Can. Dept. Agric. Publ. No. 1686. 390 pp.

134. Wu, C. J., Wu, S. Y. 1957. The species of mosquitoes transmitting Japanese B type encephalitis in Fukien. *Acta Microbiol. Sin.* 5:27–32

135. Yamanishi, H., Konishi, E., Sawayama, T., Matsumura, T. 1983. The susceptibility of some mosquitoes to Chikungunya virus. *Jpn. J. Sanit. Zool.* 34:229–33 (In Japanese with English summary)

Annu. Rev. Entomol. 1991. 36:485–509

ENVIRONMENTAL IMPACTS OF CLASSICAL BIOLOGICAL CONTROL[1]

Francis G. Howarth

J. Linsley Gressitt Center for Research in Entomology, Bernice Pauahi Bishop Museum, P.O. Box 19000-A, Honolulu, Hawaii 96817

KEY WORDS: extinctions, endangered species, genetically engineered organisms, nontarget organisms, alien species

---

PERSPECTIVES AND OVERVIEW

Biological control, the use of living organisms to control pest populations, dates from ancient times (117, 127). However, 100 years have passed since Albert Koebele intentionally introduced the Australian vedalia lady beetle via New Zealand (6, 86) into California orange groves in 1889, where it spectacularly controlled the cottony-cushion scale. This milestone marks the start of modern classical biological control—the importation and release of an organism outside its natural range for the purpose of controlling a pest species. Classical biological control has also included certain other introductions that enhance beneficial organisms, e.g. pollinators (8, 118, 127), scavengers (8, 11, 118), and competitors (11, 82).

Undertaking a review of the environmental impacts of this control method is timely as the second century of classical biological control begins and as the rapid development of biotechnology creates the specter of new risks from the purposeful release into the environment of an increasing number of artificially engineered organisms (98, 120). Furthermore, mounting criticism of chemi-

---

[1]This paper is dedicated to the late Wayne C. Gagné whose wit and committment to insect conservation provided many insights to this review.

0066-4170/91/0101-0485$02.00

cals in pest control, including increasing public fear and the banning of pesticides, and development of resistance to pesticides have inspired renewed enthusiasm for classical biological control introductions. The many successes in biological control and the available data on the history of introductions have been well reviewed (6, 11, 23, 70, 117, 129).

The introduction of biological control agents has often been declared to be environmentally safe and risk free (4, 7, 17, 23, 28, 59, 72, 82, 126). However, adequate data to defend this assertion on safety have not been systematically gathered. In 1899, David Sharp (114) wrote to L. O. Howard concerning Koebele's program of importing biological control agents into Hawaii:

> It is important that a permanent record shall be secured of what Mr. Koebele has done in matters that may affect the fauna, and we shall be much obliged if you will draw up a statement as full as you can on these points. Mr. Koebele is actually making a huge biological experiment, and the particulars should be fully recorded, though it must be very long before the results can be at all accurately estimated.

Unfortunately, Sharp's advice has not been heeded, as few workers have studied the effects of purposefully introduced species on nontarget organisms or other aspects of the environment (75, 76). The limited information on the environmental impacts of biological control is scattered within the control and ecological literature or remains unpublished or, worse, ungathered. Often, different names have been used for the same organism, making comparisons between the control and ecological literature difficult. Records of biological control importations and releases for many regions are also scanty (5). Koebele and many other early workers recorded only those species that they thought were successful (118), and for many regions no records exist for the vast majority of introductions that were tried (5, 118). Therefore, absence of evidence of negative environmental impacts is not evidence of absence of these impacts.

Documentation of significant environmental impacts is accumulating, and awareness of the problem is increasing (30, 57, 61, 75, 94, 96, 97, 125, 129). Much of the evidence of environmental damage has been gathered serendipitously during unrelated field studies (e.g., 54, 90, 106). Several recent workshops and symposia on biological control have included sessions on environmental impacts (e.g. 66, 135, 76, 134).

Biological control introductions are part of the much larger problem of the invasion of new areas by alien species, which are recognized as a major factor in species extinctions (63, 79). This form of biological pollution is one of the most critical problems facing managers of natural areas and nature reserves and has been the subject of important recent reviews (29, 49, 77, 79, 83). The main premise of classical biological control is based on the fact that alien organisms disrupt established populations.

This review calls attention to the wide variety of concerns and problems inherent in purposeful introductions. If the outcomes of purposeful introductions, including those of classical biological control and genetically engineered organisms, are to become predictable, we must ask the right questions and gather appropriate data to answer them (88, 94). A review of environmental conflicts resulting from past purposeful introductions should provide clues useful in foreseeing the environmental risks of future actions. This knowledge should also allow greater skill in preventing or mitigating environmental problems. Although this review mainly concerns insects and other arthropods, microorganisms, molluscs, and vertebrates are included because these groups have also been introduced by entomologists acting under the aegis of biological control. Impacts on both the human and natural environments are covered. Natural environments also have significant economic, cultural, aesthetic, scientific, and other human values. In reviewing the negative aspects of classical biological control, I do not intend to slight its positive aspects. Past and current workers have used the best methods and theories available in attempting to improve human welfare, in many instances with spectacular success (6).

BENEFICIAL ASPECTS OF BIOLOGICAL CONTROL

The many benefits to the environment of classical biological control have been well documented (e.g. 6, 11, 13, 23, 47, 47a, 64, 70, 100, 109, 117, 129, 134). Most classical biological control programs have been aimed at pests in agroecosystems and disease vectors with significant successes measured in long-term economic and public health benefits (e.g. 6, 11, 13, 23, 47, 100, 129). Biological control of pests usually has been safer to public health than has chemical control (7, 94, 95, 97).

Many earlier programs targeted alien weeds and pests that also incidentally had invaded native habitats. Undoubtedly, the control of many of these aliens [e.g. rabbits and cactus in Australia (49, 70), the shrub *Lantana* in Hawaii (37), and scale insects and cactus in South Africa (44, 109, 140)] benefitted native species. A dilemma presented by alien invasions into nature reserves is that any control attempt will have some negative impacts. However, an uncontrolled alien species may endanger more native species than would intrusive but successful efforts to control it (57, 62, 109). Biological control has special promise in some cases because it can target the pest and achieve control without many of the adverse impacts of mechanical or chemical methods (109). The level of control sometimes achieved makes it economically appealing in inaccessible terrains or over large tracts.

NEGATIVE IMPACTS OF CLASSICAL BIOLOGICAL CONTROL

Negative environmental impacts of biological control introductions have not been well documented in the literature (61, 104, 107, 125). The major environmental risks have been perceived as only those that affect the success of the control program (7, 43), and most studies of the risks and negative impacts have considered only the human environment [i.e. whether or not the pest was controlled, or whether the agent damaged crop plants, beneficial organisms, or human health (5, 7, 11, 75, 76, 96, 97)]. In fact, biological control agents affect the environment in a variety of ways. They have failed to control the pest, enhanced the targeted pest, synergistically interacted with other organisms to enhance pest problems, affected public health, and attacked nontarget organisms. In short, some have become pests themselves.

Endangered Species and Extinctions

In no other aspect is classical biological control strategy more in conflict with environmental protection than as the cause of species extinctions. Most extinction studies have been done in hindsight after biologists realize that a species has disappeared. A widely accepted tenet among ecologists holds that biological control agents cannot cause extinctions (7, 22, 109, 127). Yet extinctions of both target and nontarget species have been well documented, and local extirpation of pests by biological control agents may be a common phenomenon (88, 89, 96).

LEVUANA MOTH   The control of the coconut moth, *Levuana iridescens,* on Fiji by the purposeful introduction of the tachinid fly *Bessa remota* from Malaysia in 1925 was described in great detail (124) and is often cited as a classic example of successful biological control (23, 100, 117). The last authentic specimen of this endemic monotypic genus of Zygaenidae was collected in 1929 (106, 124), although the species may have survived until the 1940s (104, 106). The species, a widespread local pest, became endangered in less than two years, and its demise is probably the best documented study of extinction among the insects (124). Another unrelated zygaenid, *Heteropan dolens,* was extirpated from Fiji at the same time (106). The impacts of *B. remota* on other nonpestiferous native Fijian Lepidoptera were not recorded, even though the fly still occurs on Fiji, parasitizing nontarget species (107). Tothill et al (124) were unable to find the major alternate hosts in Fiji.

PACIFIC ISLAND LAND SNAILS   Beginning in the mid 1950s, three predatory land snails (*Gonaxis kibweziensis, G. quadrilateralis,* and *Euglandina rosea*) were introduced into the Hawaiian Islands to control the alien giant African snail, *Achatina fulica* (37). *E. rosea,* in particular, moved away from *Achatina*-infested areas and invaded native forests, where it has been strongly implicated in the extinction of several species of endemic tree snails. The complete extermination of a well-studied population of the endemic Oahu tree snail *Achatinella mustelina* corresponded with the arrival and increase of *E. rosea* in the study site (54). Achatinellines are poorly adapted to predation pressure and are unable to cope with *E. rosea* (52, 53). Disease or unknown mortality factors may also have been important, but the final blow was *E. rosea* (52). Most of the 41 recognized species of *Achatinella* are extinct, and the remaining populations are officially listed as endangered under the United States Endangered Species Act. *E. rosea* even enters shallow water to prey on aquatic snails and has been implicated in the decimation of the native Hawaiian lymnaeid snails (R. Kinzie, in preparation).

Predatory snails have been moved to other islands in spite of little evidence that they reduce pest *Achatina* populations (80, 121). *E. rosea* was introduced into Moorea, French Polynesia, in 1977, and the seven species of *Partula* endemic to Moorea disappeared with the advancing wave of *E. rosea* (10, 90, 130). *Partula suturalis, P. taeniata, P. tohiveana, P. mooreana, P. aurantia, P. mirabilis,* and *P. exigua* are extinct in the wild. All except *P. exigua* still exist in captivity in laboratories and zoos in Europe (90). The Moorean *Partula* had been the object of an intensive genetic research program, and their distributions and status were well mapped before the importation of *E. rosea.* Thus, the documentation of extinction is irrefutable. Had biologists not been in the field studying the ecology of *Partula* on Moorea when the *E. rosea* population was expanding, the extinction process probably would have been missed, and the evidence against *E. rosea* would have continued to be circumstantial. Clarke et al (10) stated "Their loss is not merely a tragedy for students of genetics, it is also a warning about the potentially devastating effects of some programs in 'biological control'. . . ."

A similar scenario is now happening on Tahiti, where the native lowland *Partula* still exist in areas not yet occupied by *E. rosea* (90). In contrast, native *Partula* species continue to thrive on the island of Huahine where *E. rosea* has not been introduced, even though two lowland species are exploited for jewelry. The ineffectiveness of *E. rosea* in controlling *Achatina* populations, as well as the commercial and scientific values of native land snails, should discourage the importation of *E. rosea* to Huahine or any other area (69, 121).

HAWAIIAN PENTATOMIDS   The native pentatomid genera *Coleotichus* and *Oechalia* declined sharply after the successful introduction in 1962 of the tachinid *Trichopoda pilipes* and the scelionid *Trissolcus basalis* for control of the alien southern green stink bug, *Nezara viridula* (39, 63). *Coleotichus blackburniae,* the largest and most conspicuous true bug in Hawaii, was locally abundant even on lowland alien acacias until the early 1970s. The harlequin stink bug, *Murgantia histrionica,* was inadvertantly introduced to Hawaii in 1924 and remained a local minor pest on Oahu and Kauai until at least 1965 (21, 37). It was extirpated from Hawaii along with its introduced parasite *Trissolcus murgantiae* in the late 1960s and early 1970s (37) at the same time that the *Nezara* parasites were building up and *Oechalia* and *Coleotichus* were declining. The evidence for the cause of the decline of *Oechalia* is circumstantial. However, *Murgantia* and *Coleotichus* were tried as alternate hosts for mass rearing these parasites in the laboratory, and *Coleotichus* was found to be susceptible to both the egg and adult parasites (21, 113).

THE LARGE BLUE   The British population of the large blue butterfly, *Maculina arion,* became extinct following the unsanctioned biological control of rabbits by the *Mixoma* virus (84). When the rabbit population crashed, the early successional grazed habitat changed and no longer supported the conspicuous butterfly (84, 130). Two lessons come from this event (84). First, many biological control introductions are made by unregulated private interests who rarely document their actions or show concern for environmental issues. Second, the effects of an alien species can be complex and difficult to predict (84).

OTHER ARTHROPODS   The damselfly *Megalagrion pacificum* was once the most abundant damselfly in lowland habitats in Hawaii. It disappeared from Oahu Island after 1910 following the introduction of *Gambusia* and other fish for mosquito control (85). Native Hawaiian shrimp have been extirpated from coastal pools by topminnows and tilapia introduced for mosquito and aquatic weed control (78).

Gagne & Howarth (42) believed that biological control introductions were the major factor in the extinction of at least 15 species of the larger native moths of Hawaii. Five of the 15 species (*Hedylepta euryprora, H. fullawayi, H. meyricki, H. musicola,* and *Agrotis crinigera*) were the direct targets of biological control introductions (93, 118, 139). A large number of generalist lepidopteran predators and parasites have been purposefully introduced to Hawaii for biological control (37), and many have the potential to attack some native species (37, 62, 139). Zimmerman (138) believed that the rarity and extinction of native Hawaiian predators, especially *Odynerus* wasps, resulted from the diminution in numbers of native prey. The renowned Hawaiian entomologist R. C. L. Perkins (93) recognized that many native insect species were becoming extinct from the predators and parasites introduced by Koebele, but he felt that the process was inevitable.

In New Zealand, imported Lepidoptera parasites, especially the tachinid *Trigonospila brevifacies* and ichneumonid *Glabrodorsum stokesii*, have spread far from their target habitats on farms and orchards and now occur in native forests where they are causing a decline of some endemic moths (104, 107). Coprophagous beetles imported to disperse dung, thereby controlling several veterinary pests, may be severely competing with native species in New Zealand (18) and North America (31).

The South American toad *Bufo marinus*, introduced to Australia in 1935, now occupies over 40% of Queensland, including undisturbed native habitats, and poses a serious threat to the native fauna (35, 36). The toad is also implicated in the extirpation of an introduced parasite of the suger cane beetle in Puerto Rico (117).

Commercial applications over wide areas of forests of self-perpetuating disease organisms capable of attacking both beneficial and pest species can affect community structure and cause collapse of regulating species with a consequent outbreak of pests (31, 96), as well as cause extinctions. Flexner et al (34) reviewed published studies on the effects of microbial pesticides on the natural enemies of target species. Although a few natural enemies appear to be sensitive to microbial pesticides, most of the negative effects were indirect and related to the death or reduction of their hosts. Some parasites may succumb indirectly from reduced fitness or abnormal behavior; e.g. *Nosema* infection in the braconid *Cotesia glomeratus* decreases the wasps' ability to enter diapause and therefore greatly increases winter mortality. These longer term effects are probably important but are not well studied (34).

The effects of insect pathogens on vulnerable nontarget species related to the target are not well known. *Bacillus thuringiensis israelensis* (Bti serotype H-14), which is widely used for mosquito control, has caused significant mortality of both mayfly and dragonfly larvae (137) and can moderately to severely affect the populations of chironomids, ceratopogonids, and dixids (34). The nematode *Steinernema feltiae* (= *Neoplectana carpocapsae*) can infect at least 250 insect species in several orders (43, 96). Timper et al (122), presented a scenario of dispersal by infected individuals, i.e. little "Typhoid Marys," that result in continual infection in many natural habitats. Yet introduction of the nematode is claimed to be environmentally safe (43). Tests on nontarget organisms in one region will not suffice for predicting an agent's impacts in a different region, as indicated by Hokkanen & Pimentel's new association model (59) and the empirical evidence.

VERTEBRATES   Honegger (60) listed seven species of reptiles on Caribbean islands that had been driven to extinction by the introduced mongoose, *Herpestes auropunctatus*. The disappearance of some native Pacific island birds and lizards has been attributed to the introduction of the harrier [*Circus approximans* (133)], myna birds [*Acridotheres tristis* (133)], and mongoose (123, 133). Mosquitofish, *Gambusia* spp., have caused extinctions of native fish (51, 78, 81, 88). The introduced viral disease mixomatosis extirpated introduced rabbits from Tierra del Fuego (26). The toad *B. marinus* may have eliminated a native frog, *Rana vittigera,* from the Philippines (100) and is considered a threat to several native vertebrates in Australia (36). The potential spread of the purposely introduced seed weevils, *Microlarinus* spp., from the main Hawaiian Islands to the Northwest Hawaiian Islands portends ill for the endangered Laysan finch, *Telespyza cantans* because the seeds of one of their hosts, *Tribulus cistoides,* comprise an important food source for the bird during drought (15). Reduction of native prey populations by both inadvertently and puposely introduced predators and parasites is believed to be important in the decline and extinction of native birds in Hawaii (3, 62).

PLANTS   In contrast to animals, no plant species appear to have been driven to extinction by biological control introductions (57, 70). This observation may reflect the greater care and stricter guidelines required for the introduction of herbivores (4, 16, 71, 127), the smaller number of programs and introductions to control plants (70), and the less likelihood of extinctions among lower–trophic level organisms (87). However, the potential for extinctions of plants is demonstrated by the disruptions to natural vegetation, including extinctions, caused by inadvertent alien species introductions (57, 125). Three native *Cirsium* species, which are attacked by agents introduced to the United States, are candidates for endangered species status (125). Biological control agents have extirpated weeds from local areas or specific habitats (57, 140). In Hawaii, the native *Tribulus cistoides* has become rare as a consequence of the control of the congeneric puncture vine *T. terrestris* by the seed weevils *Microlarinus* spp. (70). Some associated native herbivores may be eradicated if their host plant becomes rare (41, 84, 99).

Factors Affecting the Degree of Risk to Nontarget Organisms

Organisms introduced in biological control programs have the potential to feed on all available suitable hosts or affect, either directly or indirectly, associated species, including nontarget species. The outcomes of these encounters may range from negligible impacts to extinctions. Examples of significant population reductions of native species are found in many regions, and the problem may be widespread. The relative level of risk can be correlated with permanency of the agent in the environment, host range, habitat range, genetic plasticity, behavior, mutualistic relationships, and vulnerability of the target region (61, 62). Pimentel et al (98) and Teidje et al (120) list additional attributes relating to the risks of releasing genetically engineered organisms.

That only rare species are vulnerable to extinction or are of conservation concern (57, 109) is a myth. The demise of the levuana moth demonstrated that, if a species is vulnerable to a novel perturbation (62), it can become extinct no matter how abundant it is. Additionally, diseases can impact the most abundant species more than they impact rarer species (33, 96, 99).

PERMANENCE   Introductions for biological control are usually irreversible. This attribute is often said to be desirable (23, 38), but the chance that an agent will negatively affect a nontarget species, either directly or indirectly, increases with time. The more generations and longer time that an agent persists after introduction, the wider its potential geographic range and the greater the chance that host and habitat shifts can occur (84) or that other conflicts can develop between the agent and human interests. Generally, inundative releases of organisms unable to reproduce in the new environment pose less environmental risk than those establishing permanent populations (120). For example, in 1985 and 1986, sterile grass carp were released into waterways in California for effective control of the aquatic weed *Hydrilla* (103). However, the effects of innundative releases should be critically evaluated (1, 126).

HOST RANGE   Many workers have argued in favor of using polyphagous species as agents in biological control programs (e.g. 38, 43, 89); however, polyphagous agents have the greatest potential to destroy nontarget organisms (62, 96, 109). Hokkanen & Pimental (59) proposed using unnatural enemies, that is, agents that attack a related species but did not evolve with the pest. These new associations can have devastating effects. Most of the species extinctions recorded in this review resulted from such new associations. A relatively high percentage (21%) of the 33 established alien arthropods introduced to control weeds in North America have been reared from native nontarget hosts (125). In Hawaii, published incidental observations demonstrate that at least 14% (33/243) of alien biological control agents attack nontarget native or beneficial species (37), but no agent introduced to Hawaii since 1969 has been recorded to feed on a desirable species (37). However, with time and appropriate monitoring, these ratios will increase (e.g. 14).

A biological control agent feeding on both a common weed or pest and nontarget species can more severely impact one by maintaining its population on the other (125). The presence of alternate hosts was cited as a major factor in most of the documented extinctions cited above. Interspecific competition is often more important for native species growing in native communities, whereas weeds and pests by definition are dominant in the community partly because they have escaped or been released from competition (125). The alien agent is usually free from its own regulators (7), enabling it to put greater stress on its target and any vulnerable nontarget species. The combined impact of an agent and competition is often greater than the sum of each alone. Thus, changing cultural practices in conjunction with control efforts to increase interspecific competition will often achieve better control.

HABITAT RANGE   Exacerbating the impacts on nontarget species is the fact that alien organisms do not respect human boundaries but invade all available suitable habitats within their new range. A myth probably unintentionally promotes the assumption that agroecosystems can be treated in isolation from neighboring natural areas. The ability of biological control introductions to be pervasive presents special environmental problems (62). Some agents have been found far from their intended areas (35, 54, 62, 84, 90, 93, 102, 104), but few studies have examined their spread from the release site (76). Some control agents are habitat or niche specific rather than host specific and can severely affect susceptable nontarget species (104, 131). Stenotopic species involve less risk than eurytopic species. Vertebrates generally have wider tolerances than invertebrates, and a large proportion of introduced vertebrates have become disruptive invaders (33, 98). Use of vertebrates in biological control is discouraged (73, 94, 97, 127). Colonial and social insects rank with vertebrates in their ability to invade and disrupt natural ecosystems (62), and consequently should not be introduced into new areas.

GENETIC PLASTICITY   The risk that an organism will mutate and attack nontarget hosts is related to its genetic plasticity. Microorganisms have a greater propensity for change than higher taxa (34, 84, 96). Phytophagous and entomophagous insects often shift to new hosts in new areas (e.g. 9, 14, 31, 37, 59, 62, 97, 101, 102, 104, 109, 124, 125). The shift from ecological specialization to generalization or vice versa in some insects and other organisms may have a relatively simple genetic basis and may occur with ease (9, 98). Thus, we need to understand both the genetics of the proposed agent and the ecosystems into which it can spread (9). Genetically engineered organisms may have unstable genotypes, making mutations more likely (74, 98, 120). Also, microorganisms may exchange genetic material laterally between species, allowing some novel genes to escape their intended purpose (74, 98, 120).

BEHAVIOR   Behavioral attributes, such as dispersal-ability and host-searching and host-handling behaviors, can enhance an agent's ability to invade new habitats and attack nontarget species. Many parasites lay eggs on unsuitable hosts (21). Even though the parasite may not develop, its presence may reduce the fitness of the nontarget host.

Social and colonial arthropods historically have had far greater impacts on native ecosystems than have other invading species (62). The role of ants in extinctions is well recognized (62, 138). Ant populations have been manipulated locally for biological control since ancient times (96, 117, 136), but except for their movement within regions (47a) their use in classical biological control is apparently undocumented.

MUTUALISMS   Control agents with mutualistic associations can have a greater impact on both the pest and nontarget species than species acting alone (62). The wide host range and virulence of the entomogenous nematode *Steinernema feltiae* is aided by its mutualistic bacterium, *Xenorhabdus nematophilus* (43, 122). Disease vectors are well known as disruptive invaders, e.g. scolytid beetles with their ambrosia fungi (33). The *Mixoma* virus was so successful against rabbits in Australia because efficient vectors, i.e. mosquitoes, were present (84). Additional risks are involved in using a disease and/or vector in a biological control program because a more efficient vector, reservoir, or virulent disease may subsequently establish and upset the control program or impact nontarget species.

VULNERABILITY OF THE TARGET REGION   Most of the extinctions documented in this review occurred on islands or in freshwater habitats. In part, this may relate to the greater use of biological control on islands (11, 37), and to the better documentation of extinctions on islands and in aquatic habitats (29, 88). Local extinctions are frequent events in many freshwater habitats, in part because the simpler habitat has fewer refuges, allowing species interactions to run to conclusion (88). The aquatic milieu also moderates the physical environment, possibly making biotic interactions relatively more important in population control than abiotic factors (88). Island habitats are more confined, often have fewer refuges, and have more equitable climates than comparable continental habitats. The lack of refuge populations of *Levuana* and the equitable climate, which allowed continuous breeding of the control agent, were noted by Tothill et al (124) as important factors in the moth's vulnerability to the control agent. The impacts of broad spectrum microbial pesticides may be more severe among nontarget native insect faunas of islands that evolved without contact with many of these diseases (62, 63).

Biological control is being expanded into the tropics (127). Biotic interactions, both competition and host-prey, are often more important than abiotic parameters in the tropics compared to temperate regions (27). The assumption that native herbivores would competitively exclude the alien biological control agent (92, 108) is not true either empirically (62, 98, 125) or theoretically (62, 87). Thus, biological control agents may have greater effects on both target and vulnerable nontarget species in the tropics.

RISK OF EXTINCTIONS BY CHEMICAL PEST CONTROL   Chemical pest control is more widely used and is aimed at a far greater array of pests than is biological control (95, 96). Because of the perceived environmental problems, monitoring of impacts is more thorough for chemical than biological pest control (46, 84, 95). Although the environmental impacts have been profound and populations of native species severely reduced or extirpated by pesticides, no extinctions caused by agricultural chemicals have been well documented (46, 84, 99, 130). However, some extinctions may have been overlooked. Herbicides may have indirectly caused some herbivore extinctions by reducing the population of hosts and by allowing agricultural conversion of marginal lands (84), and some parasites of birds of prey may have died out during the lowest levels of their hosts' populations (99). However, the difference in risks of irreversible impacts between biological and chemical methods may be real, and may result from the different protocols used. Chemical pesticides do not mutate, persist, disperse, or reproduce to the extent that living organisms do, and the monitoring program, development of resistance, and changing economic climate have forced pesticides and their application to be changed over time, giving sensitive nontarget species a chance to recover (84).

Economics

INDIRECT COSTS   Although the direct costs of classical biological control are considered advantageous when compared to other methods, the indirect costs should be considered in any economic analysis. These include prerelease studies, postrelease monitoring for efficacy and impacts on nontarget organisms, and the delay in achieving control after release.

Until recently, biological control was considered harmless, and any available and promising species were tried (22, 126). Prerelease studies to improve the success rate were considered a waste of resources and time (126).

The attempt to control the Karoo caterpillar, *Loxostege frustalis,* in South Africa was one of the most expensive and prolonged biological control efforts, yet it failed to achieve any control. Subsequent research found that control could be accomplished by changing pasture management (1, 55). Over 60 parasites and predators were introduced into northeastern North America to control the gypsy moth, *Lymantria dispar,* with little effect on the pest population (5, 96). Similar scenarios have occurred in nearly every region (5, 11, 70, 100, 117, 118).

Since only 10–20% of introduced species become established, and of those only a small percentage affect control (5, 57, 96, 104), the greatest boon to biological control methodology would be to increase the success rate and thus reduce the indirect costs (5). Simberloff (115), in reviewing problems of establishment among aliens, stressed the necessity of including the details of failures in order to understand why some succeed. He lamented that most biological control programs do not include efforts to follow up on the reasons for failures, and thereby do not supply the data to improve success rates. Greater understanding of the ecology of both the pest system and the agents is urgently needed to improve methodology (4, 32, 96), especially now that the environmental impacts are recognized (31, 62).

PESTIFEROUS BIOLOGICAL CONTROL AGENTS   Most documented economic problems caused by classical biological control agents have been minor compared to the damage created by the original target pest (97, 117). However, nearly all regions have had exceptions, and some biological control agents have become pests themselves. Often such cases are conspicuous and are popularized to disparage pest control programs. Vertebrates are most frequently blamed (12, 26, 36, 67, 94, 97, 98, 109, 123, 127). For example, Pimentel et al (98) state without naming the vertebrate species that all five introduced to the United States for biological control have become pests. Weed control agents also have a real potential to become pests on desirable plants (57, 96, 97, 125). The most notorious example (55, 94, 125) is the apparent host switch by the bug *Teleonemia scrupulosa* from the weed *Lantana* to a sesame crop in Uganda, although the problem was temporary because *T. scrupulosa* did not reproduce on sesame (47).

At least two classical biological control agents became the targets of subsequent introductions. The lizard *Anolis grahami*, introduced into Bermuda for fruit fly control, became the target of the introduced Kiskadee bird, *Pitangus sulphuratus*, which has itself also become a pest there (13). Ironically, the original target, the Mediterranean fruit fly, *Ceratitis capitata*, no longer occurs on Bermuda (13). The Japanese introduced the monitor lizard, *Varanus indicus*, to some Micronesian islands to control rats. After the lizard became pestiferous they introduced *B. marinus* to Kosrae in hopes that the toads would poison the lizards. Both species are now well-established pests (112).

In 1959, cattle egrets were introduced to Hawaii to control pests of cattle. In two decades, they increased 120-fold to a minimum 13,000 birds to become a serious pest of prawn farms (91). They also compete with native endangered water birds, making control difficult without further endangering the protected species (91).

Like chemical control, biological control agents may disrupt existing controls, allowing secondary pest outbreaks, or previously established agents may interfere with subsequent programs (30, 32, 45, 47, 62, 101, 102, 117). For example, natural enemies may be extirpated following an epizootic, thereby allowing a resurgence of the pest (96). In New Zealand, the purposefully imported *Copidosoma floridanum* has replaced the native egg parasites, *Trichogrammatoidea* spp., as the principal parasite of the pest noctuid *Chrysodeixis eriosoma*. Since *C. floridanum* encourages its host caterpillars to consume more than unparasitized caterpillars and is poorly synchronized with the host, the amount of damage may have locally increased after the introduction (104, 105). The recent intentional establishment and spread of the lady beetle, *Coccinella septempunctata* (110), into North America is of some concern (30, 110). Because this lady beetle is the dominant aphidophagous coccinellid in Europe, it has the potential to displace native North American aphidophagous species as well as to disrupt some established biological control systems (30).

Several scarabs have been intentionally introduced to remove cow dung from pastures both to facilitate regeneration of grasses and to reduce the food supply of the horn fly (8, 11, 37). In Hawaii, these beetles act synergistically with the mongoose, *H. auropunctatus,* in upland pastures. The mongoose might not maintain its high population in pastures and neighboring forests, with the resultant negative impacts on native birds and invertebrates, if the dung beetles were not an abundant food resource (123). In Puerto Rico, the mongoose has enhanced population levels of the more pestiferous roof rat by reducing its ground-dwelling competitor, the Norway rat (96).

When a pest is targeted without understanding the true problem, the control program may be successful, but the pest may simply be replaced by another, sometimes worse, pest (30, 57, 96, 125). Weed control programs offer the best examples (40, 129), and often the problem results from land management practices (e.g. grazing schedule) rather than a weed problem (1, 55, 125).

Public Health

Biological control agents may even affect public health. The mongoose, *H. auropunctatus,* is considered the major wild reservoir of rabies, as well as a reservoir of leptospirosis, on Puerto Rico (96, 97). The carnivorous snail *E. rosea* is an efficient carrier of rat lung worm, which may infect humans and domestic animals (80). Frogs and toads that are poisonous to humans and domestic animals have been introduced and have caused public health problems (36, 37). In some systems, mosquito fish, *Gambusia* sp., may affect other predator populations and favor bilharziasis vectors (51). Some introduced wasps can sting (e.g. tiphiids, ampulicids, and sphecids) (11, 37, 96). Insect pathogens may infect humans under some conditions or cause toxic or allergic responses (76, 96). Human pathogens may contaminate mass rearing facilities for producing microbial pesticides and be disseminated with the formulation (96). In Hawaii, cattle egrets often roost and feed near airports where they pose serious risks to airline safety (91).

Conflicts over What Is a Pest or Weed

Changes in a region's economy and the irreversibility of biological control introductions can create acute conflicts of interest among different groups within society (30, 31, 61, 125). Some may see the target as beneficial or as having special values, e.g. native species, and some may feel the risks to nontarget organisms, both alien and native, are unacceptable (62, 84, 125). Some perceived conflicts can be outlandish (56). These conflicts are better recognized for plants than for animal pests (71, 104, 125), and guidelines for the importation of agents to control weeds are stronger than those for agents to control invertebrate species (16, 71). The most notorious conflict concerned the biological control of the invasive alien weed *Echium plantagineum* in Australia, which resulted in a long, bitter court battle and eventually led to Australia's Biological Control Act (25). The act requires that an environmental impact study be prepared for biological control introductions (25, 57). Anticipating a similar fight between ranchers and bee keepers concerning gorse (*Ulex europaes,* a spiny leguminous shrub) control, the New Zealand Department of Scientific and Industrial Research voluntarily submitted a detailed environmental impact statement (58).

In Madagascar, the scale *Dactylopius opuntiae* controlled both weedy *Opuntia* spp. and important cattle forage species, causing in some areas more harm than good (47). Similar conflicts between the classical biological control of weedy opuntias and forage crops have occurred in South Africa (2) and Hawaii (65). In South Africa, *Acacia mearnsi* and other Australian acacias are grown in timber plantations and are also important forage for honey bees; yet some are also major invasive weeds (44).

The CAB International Institute of Biological Control is currently attempting to resolve two conflicting projects in southeast Asia and the Pacific. In one, the introduction of a *Heteropsylla* sp. is being considered to control the weed *Mimosa invisa*. In the other, biological control agents are being sought for a recently adventive complex of other *Heteropsylla* spp., which attack the related tree *Leucaena leucocephala* (127, 129). The conflict is especially complex because, in many parts of the region, *L. leucocephala* is also regarded as a major invasive weed (116, 127).

Similar conflicts of interest occur with animal control programs, but insects and other invertebrates have fewer advocates in western society (24, 61, 62, 104). The increasing awareness of the biodiversity crisis (132), including loss of native habitats and the attendant rapid loss of species, means that conflicts between control technologies and conservation will intensify.

Pest control may affect land management, e.g. it may encourage attempts to exploit marginal land beyond its sustainable carrying capacity, even when the land's greater values lie in watershed, native wildlife conservation, and scenic vistas (84, 104).

Another dilemma concerns the control of false pests or nuisance species of minor importance. Pests and weeds are often defined by public policy on the basis of conspicuousness rather than on sound ecological or economic importance (55, 61, 62, 104). Marketing policies often require produce that is free of insect damage known to be of no consequence to quality or storage

(22). Insect feeding damage on ornamentals is often targeted even when the damage is cosmetic rather than economic (61, 62, 126). Public misconceptions concerning risks can hinder implementation of ecologically sound control programs (56, 57, 62).

SOLUTIONS

The first step in finding solutions is to recognize that classical biological control introductions have environmental risks and that adequate resources are needed to minimize them. Reliance on classical biological control without proper analysis abets the belief that aliens are not a serious problem. This belief undermines quarantine efforts and limits the chances of finding real solutions. For example, potentially more serious weeds than those currently targeted for biological control continue to be introduced into Hawaii (40, 116). Current policy naively dictates that, if any of the new arrivals become invasive, biological agents can be introduced to control them. However, the real problem is the continued introduction of the alien species (40, 62, 116).

Regulatory Solutions

Quarantine programs need to be strengthened. Quarantines are effective in two ways. They slow the rate of pest introductions and reinforce efforts to change the cultural bias that favors the introduction of alien species. Individuals and agencies who attempt to introduce an organism into a new land undertake a grave responsibility. Between 1/10 and 1/100 of all organisms intentionally introduced in North America, including those for biological control, have become serious pests (98). This risk is too high, and ways should be developed to reduce it (98).

The environmental risks associated with the introduction of alien organisms for biological control are sufficient to justify the creation of legal safeguards (56). The first biological control law in Australia provides a model for other nations. Currently, in the United States and many other nations, different state and federal agencies have separate and sometimes conflicting mandates concerning environmental protection, public health, and agriculture, which results in a lack of clear responsibility for overseeing the environmental impacts of biological control introductions (98).

Expanded review processes and protocols for the release of alien and newly engineered organisms are needed (56, 57, 61, 62, 75, 76, 97) and should allow input from the affected community, i.e. scientific disciplines, commercial interests, and the public, to minimize the conflicts of interest. To reduce the risks, all potential environmental impacts deserve to be addressed. Native species that may be at risk should have standing.

The review process should include criteria and mechanisms to separate true pests needing control from those causing only trivial or noneconomic damage (62). Often the problem is one of management rather than of the presence of a true pest; i.e. without changed management practices, the pest may not be controlled (1) or the controlled pest may be replaced by another noxious pest (41, 125). Releases should only be made in the public interest after the ecological consequences are considered (57).

The protocols for weed control need to be strengthened and applied to programs aimed at other pests (16, 61, 62, 104). Introductions to control native organisms or organisms that have close relatives either taxonomically or ecologically in the native biota should be attempted only in special circumstances in which their ecology is well known and in which the negative impacts will be less than positive ones. The effects on associated native species dependent on the native target should be considered (57). Organisms with known wide host or ecological ranges should not be introduced (62, 104, 109). The myth that agroecosystems exist in isolation from surrounding environments needs to be deflated, especially with regard to releases of novel or alien organisms.

Research

Prerelease studies of potential risks should include host specificity tests on an appropriate set of nontarget species, especially involving any that are closely related to the target or are likely to be either directly or indirectly affected (104, 125, 128). These studies should also include descriptions of the target area and all surrounding habitats likely to be invaded by the control agent. As much of the autecology of the target and candidate agents as possible should be worked out in the source area before the importation and quarantine stage. The purpose of the search should be to first understand the pest problem. If biological control proves desirable, the most promising candidate for effective control of the pest with minimal environmental risk should be sought (1, 62, 104, 109). For example, a target weed could be planted in a suitable environment in its presumed place of origin, and potential candidates evaluated as they colonized the plantation.

Importing agencies need to make commitments to support long-term postrelease monitoring to determine both the efficacy of the agent and the impacts on nontarget species. We need to monitor range expansions, host changes, and effects on target and nontarget species [e.g. the recent establishment of the catholic gypsy moth parasite *Coccygominus disparis* in North America (111)]. Determinations of the causes of failures are also important (115). Long-term ecological research is needed to improve the understanding of community ecology and the impacts of alien species. The goal is to make applied ecology a predictive science (31).

Often, if a phenomenon is not specifically searched for during a research program, it will be overlooked (88, 94). Many environmental impacts of biological control agents have been missed because the introduction was considered safe and no impact was sought (62, 104).

Basic research programs should be encouraged and supported to improve the science. Purposeful introductions provide superb opportunities for determining the genetics of colonizing species and understanding evolutionary questions, such as the effects of bottlenecks. This should lead to practical applications in selecting future control agents (9, 50, 119, 127). Effective biological control requires a good knowledge of the systematics both of the pest and control agents, as well as the suite of nontarget organisms (20, 61).

FUTURE DIRECTIONS

Environmental Risks in the Tropics

Most ecological research has been based on temperate systems, and many of the conclusions may be inappropriate for the tropics. We know so little about biodiversity, ecology, community structure, and host relationships in most tropical ecosystems that assessment of the impacts of biological control agents will be difficult. Theoretically, environmental impacts may be more severe in the tropics than in temperate zones. The state of systematics research and the survey of biodiversity are so incomplete that many species of insects and their relatives may become extinct without ever being recognized or cataloged, unless special efforts are made (132).

Classical biological control may be an inappropriate technology in some Third World countries. Insects are often held in higher esteem in many cultures than they are in the West, and in many societies they provide 10% or more of the protein source (24). Alien entomophages or entomopathogens could significantly reduce local food resources (24).

Environmental Risks of Clandestine Programs

A message the public is receiving from the biological control community and the media is that there are no risks associated with biological control, i.e. "using nature to fight nature." Well-meaning but misinformed individuals might introduce species that could severely impact the environment. Some dispersal of biological control agents to new areas are undocumented or are unauthorized introductions outside of normal channels [e.g. B. marinus into Australia (36) and myxomatosis into England (84)]. Some generalist entomophagous organisms appear to have recently entered Hawaii in this way (S. McKeown, personal communication). Over three dozen mail order houses will ship living "beneficial" invertebrates and microbial pesticides to customers worldwide (19). Unless regulated, such suppliers are potential sources for disruptive alien species.

Environmental Risks from Genetically Engineered Organisms

The risks to the environment resulting from releases of genetically engineered organisms will be analogous to the releases of classical biological control agents (74, 98, 120) because the same cultural, political, and economic pressures will act on the releases and also because many releases will target similar pest control problems (74, 98, 120). This review points out several lessons.

1. Negative environmental effects will result from the release of novel genetically engineered organisms.
2. To adequately document the negative impacts, appropriate questions need to be asked and research designed to answer them.
3. Self-dispersing organisms will find their way to all suitable available habitats and not stay within the prescribed area without management (74).
4. Self-reproducing organisms and certain long-lived ones may affect ecosystems at many levels and in direct and indirect ways far into the future (74). Therefore, engineered organisms should be created so that they cannot establish wild populations, or if establishment is desired, then extra precautions concerning the risks are necessary (31).
5. Extinctions of nontarget species are probable, but by recognizing vulnerable species early, many of these extinctions may be prevented. Therefore, candidate vulnerable species should be monitored (84).
6. Some individuals and agencies will try to use or disperse the artificial organisms for their own short-sighted benefit (98). Therefore, extra precautions, security, and enforcement will be needed.

On the other hand, biotechnology promises to control pests without some of the detrimental side effects of other control methods, unless we fall into the same political and economic pitfalls that have plagued both chemical and biological control. Use of new genetic tools may enable researchers to modify native species to control alien as well as native pests (64). Releasing modified native species will usually have more predictable results than releasing an alien species, although some risk will remain (74, 120). With a better understanding of the ecosystem, one can find native controls for alien and native pests (48).

CONCLUSIONS

Purposeful introductions under the aegis of classical biological control have been remarkably successful in controlling numerous weeds and pests, thereby benefitting agriculture, public health, and natural ecosystems. However, biological control agents can also damage the environment. In fact, they have been strongly implicated in the extinctions of nearly 100 species of animals world-wide. The clearest examples, and some of the best-documented extinctions known, are from islands. A few of these were the targets of the biological control agents, but most were desirable nontarget organisms. Most of the environmental impacts were either recognized circumstantially in hindsight or were discovered serendipitously by researchers studying the affected organism in the field at the critical time. Thus, the majority of the environmental impacts of biological control, including most species extinctions, undoubtedly have never been recognized nor recorded. The greater the number of introduced organisms in the environment is, the greater the potential for harm (31, 62, 98). We need an accurate predictive theory so that only the most promising and least risky alien organisms, including engineered ones, are introduced for a given purpose (31).

In reality, pest control can have no panaceas. Any action to limit or kill a species will affect other species and will pose some environmental risk (61, 94, 98). Given the high reproductive potential and genetic plasticity of insects, the development of resistance to artificial population controls is a natural phenomenon. Some pest species have evolved ways to cope with any single control method applied to drastically reduce their populations, whether the control is chemical, biological, cultural, or genetic. Human agroecosystems are young and maintained for high harvestable productivity. The large acreages planted with one or a few crops amount to millions of succulent, attractive bait stations. Eventually one or more species will break through our defenses and become a pest. The long-term goals should be to optimize yields on a sustainable basis rather than maximize short-term returns (68) by using a full range of control methods based on a firm knowledge of ecology and systematics. Human relationships to the environment, especially as it relates to land use and agriculture, are in a period of change. Environmental concerns can no longer be regarded as trivial, but are becoming paramount in the development of a sustainable economy (84).

ACKNOWLEDGMENTS

I thank especially my wife Nancy for editorial assistance and helpful suggestions on the manuscript. I thank the colleagues who offered encouragement and constructive advice and sent reprints and pertinent references, especially L. E. Ehler and D. A. Spiller, U. C. Davis; L. Knudsen, USDA; I. Macdonald, Univ. of Cape Town, South Africa; and D. Pimentel, Cornell Univ. A. Manning, Bishop Museum, Honolulu, discovered and kindly supplied the Sharp quote. I thank R. Cowie, N. L. Evenhuis, and S. L. Montgomery, Bishop Museum, for critically reviewing early drafts; and J. W. Beardsley, N. J. Reimer, and J. S. Strazanac, Univ. of Hawaii, Honolulu; L. V. Giddings and L. Knutson, USDA; D. J. Greathead and J. K. Waage, CAB International Institute for Biological Control, UK; and L. E. Ehler for their constructive comments on the manuscript.

Literature Cited

1. Annecke, D. P., Moran, V. C. 1977. Critical reviews of biological pest control in South Africa. 1. The Karoo caterpillar, *Loxostege frustalis* Zeller (Lepidoptera: Pyralidae). *J. Entomol. Soc. S. Afr.* 40:127–45

2. Annecke, D. P., Moran, V. C. 1978. Critical reviews of biological pest control in South Africa. 2. The prickly pear, *Opuntia ficus-indica* (L.) Miller. *J. Entomol. Soc. S. Afr.* 41:161–88

3. Banko, W. E. 1978. Some limiting factors and research needs of endangered Hawaiian forest birds. *Proc. 2nd Conf. Nat. Sci. Hawaii, Volcanoes Natl. Park,* pp. 17–25. Honolulu: C. P. S. U. Univ. Hawaii

4. Batra, S. W. T. 1982. Biological control in agroecosystems. *Science* 215:134–39

5. Beirne, B. P. 1985. Avoidable obstacles to colonization in classical biological control of insects. *Can. J. Zool.* 63:743–47

6. Caltagirone, L. E. 1981. Landmark examples in classical biological control. *Annu. Rev. Entomol.* 26:213–32

7. Caltagirone, L. E., Huffaker, C. B. 1980. See Ref. 76, pp. 103–9

8. Cameron, P. J., Hill, R. L., Valentine, E. W., Thomas, W. P. 1987. *Invertebrates Imported into New Zealand for Biological Control of Invertebrate Pests and Weeds, for Pollination, and for Dung Dispersal, from 1874 to 1985. DSIR Bull. 242.* 51 pp.

9. Carson, H. L., Ohta, A. T. 1981. Origin of the genetic basis of colonizing ability. In *Evolution Today, Proc. 2nd Int. Congr. Syst. Evol. Biol.,* ed G. G. E. Scudder, J. L. Reveal, pp. 365–70. Pittsburgh, PA: Hunt Inst. Bot. Doc. Carnegie-Mellon Univ.

10. Clarke, B., Murray, J., Johnson, M. S. 1984. The extinction of endemic species by a program of biological control. *Pac. Sci.* 38:97–104

11. Clausen, C. P., ed. 1978. *Introduced Parasites and Predators of Arthropod Pests and Weeds: A World Review, US Dep. Agric. Handbook 480.* Washington, DC: US Dep. Agric. 545 pp.

12. Coblentz, B. E., Coblentz, B. A. 1985. Control of the Indian mongoose *Herpestes auropunctatus* on St. John, US Virgin Islands. *Biol. Conserv.* 33:281–88

13. Cock, M. J. W., ed. 1985. *Review of Biological Control of Pests in the Commonwealth Caribbean and Bermuda up to 1982. CAB CIBC Tech. Commun. No. 9.* 218 pp.

14. Conant, P. 1991. Note on *Uroplata girardi* Pic (Coleoptera: Chrysomelidae) feeding on commercial basil. *Proc. Hawaii. Entomol. Soc. for 1989.* 31: In press

15. Conant, S. 1988. Saving endangered species by translocation. *BioScience* 38:254–57

16. Coulson, J. R., Soper, R. S. 1988. Protocols for the introduction of biological control agents in the United States. In *Plant Quarantine,* ed. R. Kahn, 3:1–35 Boca Raton, FL: CRC

17. Council of Entomology Department Administrators. Dec. 1988 *Entomology Research Initiatives,* pp. 1–11

18. Cumber, R. A. 1961. The interaction of native and introduced insect species in New Zealand. *Proc. NZ Ecol. Soc.* 8:55–60

19. Daar, S., Olkowski, H., Olkowski, W., 1989. Directory of producers of natural enemies of common pests. *The IPM Practitioner* 11(4):15–18

20. Danks, H. V. 1988. Systematics in support of entomology. *Annu. Rev. Entomol.* 33:271–96

21. Davis, C. J. 1964. The introduction, propagation, liberation, and establishment of parasites to control *Nezara viridula* variety *smaragdula* (Fabricius) in Hawaii (Heteroptera: Pentatomidae). *Proc. Hawaii. Entomol. Soc.* 18:369–75

22. Davis, D. W., Hoyt, S. C., McMurty, J. A., AliNiazee, M. T. 1979. *Biological Control and Insect Pest Management. Agric. Exp. Stn. Univ. CA Bull. 1911*

23. DeBach, P. 1974. *Biological Control By Natural Enemies.* London: Cambridge Univ. Press. 323 pp.

24. DeFoliart, G. R. 1989. The human use of insects as food and as animal feed. *Bull. Entomol. Soc. Am.* 35:22–35

25. Delfosse, E. S. 1988. *Echium* appeal won by CSIRO. *News Bull. Aust. Entomol. Soc.* 24:149–52

26. Dobson, A. P. 1988. Restoring island ecosystems: the potential of parasites to control introduced mammals. *Conserv. Biol.* 2:31–39

27. Dobzhansky, T. 1950. Evolution in the tropics. *Am. Sci.* 38:208–21

28. Doutt, R. L. 1972. Biological control: parasites and predators. In *Pest Control Strategies for the Future (National Academy of Sciences),* pp. 288–97. Washington, DC: Nat. Acad. Sci.

29. Drake, J. A., Mooney, H. A., diCastri, F., Groves, R. H., Kruger, F. J., et al., eds. 1989. *SCOPE 37 Biological In-*

vasions: A Global Perspective. Chichester: Wiley. 525 pp.

30. Ehler, L. E. 1990. Environmental impact of introduced biological-control agents: implications for agricultural biotechnology. In Risk Assessment in Agricultural Biotechnology, ed. J. J. Marois, G. Bruening. Oakland: Calif. Div. Agric. & Nat. Res. In press

31. Ehler, L. E. 1990. Planned introductions in biological control. In Assessing Ecological Risks of Biotechnology, ed. L. R. Ginzburg. Butterworths Biotechnol. Ser. In press

32. Ehler, L. E., Andres, L. A. 1983. Biological control: exotic natural enemies to control exotic pests. In Exotic Plant Pests and North American Agriculture, ed. C. L. Wilson, C. L. Graham, pp. 395–418. New York: Academic

33. Elton, C. S. 1958. The Ecology of Invasions by Animals and Plants. London: Methuen. 181 pp.

34. Flexner, J. L., Lighthart, B., Croft, B. A. 1986. The effects of microbial pesticides on non-target, beneficial arthropods. Agric. Ecosyst. Environ. 16:203–54

35. Floyd, R. B., Easteal, S. 1986. See Ref. 50, p. 151

36. Freeland, W. J. 1986. Invasion north successful conquest by the cane toad. Aust. Nat. Hist. 22:69–72

37. Funasaki, G. Y., Lai, P.-Y., Nakahara, L. M., Beardsley, J. W., Ota, A. K. 1988. A review of biological control introductions in Hawaii: 1890 to 1985. Proc. Hawaii. Entomol. Soc. 28:105–60

38. Fuxa, J. R. 1987. Ecological considerations for the use of entomopathogens in IPM. Annu. Rev. Entomol. 32:225–51

39. Gagné, W. C. 1983. New egg-laying record of Trichopoda. Notes and Exhibitions. Proc. Hawaii Entomol. Soc. 24(2,3):191

40. Gagné, W. C. 1986. Hawaii's botanic gardens: panacea or Pandora's box in the conservation of Hawaii's native flora? Newsl. Hawaii Bot. Soc. 25(1):7–10

41. Gagné, W. C. 1988. Conservation priorities in Hawaiian natural systems. BioScience 38:264–71

42. Gagné, W. C., Howarth, F. G. 1985. Conservation status of endemic Hawaiian Lepidoptera. Proc. 3rd Congr. Eur. Lepid., Cambridge, 1982, pp. 74–84. Karlsruhe: Soc. Eur. Lepidopterol. 211 pp.

43. Gaugler, R. 1988. Ecological considerations in the biological control of soil-inhabiting insects with entomopath-ogenic nematodes. Agric. Ecosyst. Environ. 24:351–60

44. Galdenhuys, C. J. 1986. See Ref. 77, pp. 275–83

45. Goeden, R. D., Louda, S. M. 1976. Biotic interference with insects imported for weed control. Annu. Rev. Entomol. 21:325–42

46. Grant, I. F. 1989. Monitoring insecticide side-effects in large-scale treatment programmes: tsetse spraying in Africa. In Pesticides and Non-Target Invertebrates, ed. P. C. Jepson, pp. 43–69. Wimborne, Dorset, UK: Intercept. 240 pp.

47. Greathead, D. J. 1971. A Review of Biological Control in the Ethiopian Region. CAB Tech. Bull. 5. Slough, England: Commonw. Agric. Bureau. 162 pp.

47a. Greathead, D. J., ed. 1976. A Review of Biological Control in Western and Southern Europe. Tech. Commun, No. 7. Slough, UK: Commonw. Agric. Bur. 182 pp.

48. Gross, H. R. 1987. Conservation and enhancement of entomophagous insects—a perspective. J. Entomol. Sci. 22:97–105

49. Groves, R. H. 1986. See Ref. 50, pp. 137–49

50. Groves, R. H., Burdon, J. J., eds. 1986. Ecology of Biological Invasions. Cambridge: Cambridge Univ. Press. 166 pp.

51. Haas, R., Pal, R. 1984. Mosquito larvivorous fishes. Bull. Entomol. Soc. Am. 30:17–25

52. Hadfield, M. G. 1986. Extinction in Hawaiian Achatinelline snails. Malacologia 27:67–81

53. Hadfield, M. G., Miller, S. E. 1989. Demographic studies of Hawaii's endangered tree snails: Partulina proxima. Pac. Sci. 43:1–16

54. Hadfield, M. G., Mountain, B. S. 1981. A field study of a vanishing species, Achatinella mustelina, (Gastropoda, Pulmonata), in the Waianae Mountains of Oahu. Pac. Sci. 34:345–58

55. Harris, P. 1980. Evaluating biocontrol of weeds projects. Proc. 5th Int. Symp. Biol. Contr. Weeds Brisbane, Australia. pp. 345–53. Melbourne: Commonw. Sci. Ind. Res. Org.

56. Harris, P. 1987. The need for biological control legislation in Canada. Biological Control Workshop, Winnipeg, 9–10 Oct. 1986, pp. 55–59. Ottawa: Agric. Can.

57. Harris, P. 1988. Environmental impact of weed-control insects. BioScience 38:542–48

58. Hill, R. 1987. The Biological Control of Gorse (Ulex europaeus L.) in New Zea-

land: An Environmental Impact Assessment. Christchurch: DSIR. 56 pp. + 7 appendicies.
59. Hokkanen, H. M. T., Pimentel, D. 1989. New associations in biological control: theory and practice. *Can. Entomol.* 121:829–40
60. Honegger, R. E. 1981. List of amphibians and reptiles either known or thought to have become extinct since 1600. *Biol. Conserv.* 19:141–58
61. Howarth, F. G. 1983. Biological control: panacea or Pandora's box? *Proc. Hawaii. Entomol. Soc. 1980* 24:239–44
62. Howarth, F. G. 1985. Impacts of alien land arthropods and mollusks on native plants and animals in Hawaii. In *Hawaii's Terrestrial Ecosystems: Preservation and Management,* ed. C. P. Stone, J. M. Scott, pp. 149–79. Honolulu: Univ. Hawaii
63. Howarth, F. G. 1990. Hawaiian terrestrial arthropods: an overview. *Occas. Pap. Bernice Pauahi Bishop Mus.* 30:4–26
64. Hoy, M. A., Cunningham, G. L., Knutson, L., eds. 1983. *Biological Control of Pests by Mites. Univ. Calif. Spec. Publ. 3304.* Berkeley: Univ. Calif. 185 pp.
65. Huang, S., Tamashiro, M. 1966. The susceptibility of *Cactoblastis cactorum* (Berg) to *Bacillus thuringiensis* var. *thuringiensis. Proc. Hawaii. Entomol. Soc.* 19:213–21
66. Huffaker, C. B. 1986. Introduction to symposium of biological control. *Agric. Ecosyst. Environ.* 15:85–93
67. Hurlbert, S. H., Zedler, J., Fairbanks, P. 1972. Ecosystem alteration by mosquito fish *(Gambusia)* predation. *Science* 175:639–41
68. International Union of Conservation of Nature and Natural Resources. 1980. *World Conservation Strategy.* Gland, Switzerland: Int. Union. Conserv. Nat. Nat. Resour. 40 pp.
69. Johnson, M. S., Murray, J., Clarke, B. 1986. Allozymic similarities among species of *Partula* on Moorea. *Heredity* 6:319–27
70. Julien, M. H., ed. 1987. *Biological Control of Weeds: A World Catalog of Agents and Their Target Weeds.* Slough, London: CAB CIB Contr. 108 pp. 2nd ed.
71. Klingman, D. L., Coulson, J. R. 1983. Guidelines for introducing foreign organisms into the United States for the biocontrol of weeds. *Bull. Entomol. Soc. Am.* 29:55–61
72. Lai, P. Y. 1988. Biological control: A

positive point of view. *Proc. Hawaii. Entomol. Soc. 1980* 28:179–90
73. Laird, M. 1985. Conclusion. In *Biological Control of Mosquitos,* ed. H. C. Chapman, A. R. Barr, M. Laird, D. C. Weidhaas, 23:216–18. Fresno, CA: Am. Mosq. Control Assoc.
74. Levin, S. A., Harwell, M. A. 1986. Potential ecological consequences of genetically engineered organisms. *Environ. Manage.* 10:495–513
75. Luck, R. F. 1986. Biological control of California red scale. In *Ecological Knowledge and Environmental Problem-Solving Concepts and Case Studies,* pp. 166–89. Washington, DC: Natl. Acad. Press. 388 pp.
76. Lundholm, B., Stackerud, M., eds. 1980. *Environmental Protection and Biological Forms of Control of Pest Organisms. Ecol. Bull. 31,* Stockholm. 171 pp.
77. Macdonald, I. A. W., Kruger, F. J., Ferrar, A. A. 1986. *The Ecology and Management of Biological Invasions in Southern Africa. Proc. Natl. Synth. Symp. Ecol. Biol. Invasions.* Cape Town: Oxford Univ. Press. 324 pp.
78. Maciolek, J. A. 1984. Exotic fishes in Hawaii and other islands of Oceania. In *Distribution, Biology and Management of Exotic Fishes,* ed. W. R. Courtenay Jr., J. R. Stauffer Jr., pp. 131–61. Baltimore: Johns Hopkins Univ. Press
79. Maynard Smith, J. 1989. The causes of extinction. *Philos. Trans. R. Soc. London Ser. B* 325:241–52
80. Mead, A. R. 1979. Economic malacology with particular reference to *Achatina fulica.* In *The Pulmonates,* ed. V. Fretter, J. Peake, 2B:1–150. New York: Academic
81. Minckley, W. L., Deacon, J. E. 1968. Southwestern fishes and the enigma of endangered species. *Science* 159:1424–31
82. Moon, R. D. 1980. Biological control through interspecific competition. *Environ. Entomol.* 9:723–28
83. Mooney, H. A., Drake, J. A., eds. 1986. *Ecology of Biological Invasions of North America and Hawaii.* New York: Springer-Verlag
84. Moore, N. W. 1989. *The Bird of Time.* Cambridge: Cambridge Univ. Press. 290 pp.
85. Moore, N. W., Gagné, W. C. 1982. *Magalagrion pacificum* (McLachlan): A preliminary study of the conservation requirements of an endangered species. *Rep. Odonata Specialist Group, Int. Union Conserv. Nat. 3 Utrecht.* 5 pp.

86. Morales, C. F., Hill, R. L. 1989. See Ref. 125a, p. 43
87. Mueller-Dombois, D., Howarth, F. G. 1981. In *Island Ecosystems: Biological Organization in Selected Hawaiian Communities*. US/IBP Synth. Ser. 25, ed. D. Mueller-Dombois, K. W. Bridges, H. L. Carson, pp. 337–54. Pennsylvania: Hutchinson Ross. 583 pp.
88. Murdoch, W. W., Bence, J. 1987. General predators and unstable prey populations. In *Predation Direct and Indirect Impacts on Aquatic Communities*, ed. W. C. Kerfoot, A. Sih, 2:17–30. Hanover/London: Univ. Press New England. 386 pp.
89. Murdoch, W. W., Chesson, J., Chesson, P. L. 1985. Biological control in theory and practice. *Am. Nat.* 125:344–66
90. Murray, J., Murray, E., Johnson, M. S., Clarke, B. 1988. The extinction of *Partula* on Moorea. *Pac. Sci.* 42:150–53
91. Paton, P. W. C., Fellows, D. P., Tomich, P. Q. 1986. Distribution of cattle egret roosts in Hawaii with notes on the problems egrets pose to airports. *Elepaio* 46:143–47
92. Peschken, D. P. 1984. Host range of *Lema cyanella* (Coleoptera: Chrysomelidae), a candidate for biocontrol of Canadian thistle and four stenophagous, foreign thistle insects in North America. *Can. Entomol.* 116:1377–84
93. Perkins, R. C. L. 1897. The introduction of beneficial insects into the Hawaiian Islands. *Nature* 55(1430):499–500
94. Pimentel, D. 1980. See Ref. 76, pp. 11–24
95. Pimentel, D., Andow, D. 1984. Pest management and pesticide impacts. *Insect Sci. Appl.* 5:141–49
96. Pimentel, D., Glenister, C., Fast, S., Gallahan, D. 1982. *Environmental risks associated with the use of biological and cultural pest controls*. NTIS Rep. No. 1 PB-83-168-716. Springfield, VA: Natl. Tech. Inf. Serv. 165 pp.
97. Pimentel, D., Glenister, C., Fast, S., Gallahan, D. 1984. Environmental risks of biological pest controls. *Oikos* 42:283–90
98. Pimentel, D., Hunter, M. S., LaGro, J. A., Efroymson, R. A., Landers, F. T., et al. 1989. Benefits and risks of genetic engineering in agriculture. *BioScience* 30:606–14
99. Pyle, R., Bentzien, M., Opler, P. 1981. Insect conservation. *Annu. Rev. Entomol.* 26:233–58
100. Rao, V. P., Ghani, M. A., Sankaran, T., Mathur, K. C. 1971. *A Review of the Biological Control of Insects and Other Pests in South-East Asia and the Pacific Regions*. London: CAB. 149 pp.
101. Reimer, N. J. 1988. Predation on *Liothrips urichi* Karny (Thysanoptera: Phlaeothripidae): a case of biotic interference. *Environ. Entomol.* 17:132–34
102. Reimer, N. J., Beardsley, J. W. 1986. Some notes of parasitization of *Blepharomastix ebulealis* (Guenee) (Lepidoptera: Pyralidae) in Oahu forests. *Proc. Hawaii. Entomol. Soc.* 27:91–93
103. Remington, M. D., Stocker, R. K. 1989. See Ref. 125a, p. 66
104. Roberts, L. I. N. 1986. The practice of biological control—implications for conservation, science and the community. *Weta News Bull. Entomol. Soc. NZ* 9:76–84
105. Roberts, L. I. N., Cameron, P. J. 1990. *Chrysodeixis eriosoma* (Doubleday), green looper (Lepidoptera: Noctuidae). In *A Review of Biological Control of Invertebrate Pests and Weeds in New Zealand 1874 to 1987*, ed. P. J. Cameron, R. C. Hill, J. Bain, W. D. Thomas. Wallingford, UK: Tech. Commun. CAB Int. In press
106. Robinson, G. S. 1975. *Macrolepidoptera of Fiji and Rotuma*. Oxon, UK: Classey. 362 pp. + 545 figures
107. Russell, D. A. 1986. The role of the entomological society in insect conservation in New Zealand. *Weta* 9:44–54
108. Sailer, R. I. 1978. Our immigrant insect fauna. *Entomol. Soc. Am. Bull.* 24:3–11
109. Samways, M. J. 1988. Classical biological control and insect conservation: are they compatible? *Environ. Conserv.* 15:349–54
110. Schaefer, P. W., Dysart, R. J., Specht, H. B. 1987. North American distribution of *Coccinella septempunctata* (Coleoptera: Coccinellidae) and its mass appearance in coastal Delaware. *Environ. Entomol.* 16:368–73
111. Schaefer, P. W., Fuester, R. W., Chianese, R. J., Rhoads, L. D., Tichenor, R. B. Jr. 1989. Introduction and North American establishment of *Coccygominus disparis* (Hymenoptera: Ichneumonidae), a polyapagous pupal parasite of Lepidoptera, including gypsy moth. *Environ. Entomol.* 18:1117–25
112. Schreiner, I. 1990. Biological control introductions in the Caroline and Marshall Islands. *Proc. Hawaii. Entomol. Soc.* 29:57–69
113. Shahjahan, M., Beardsley, J. W. Jr. 1975. Egg viability and larval penetration in *Trichopoda pennipes pilipes*

Fabricius (Diptera: Tachinidae). *Proc. Hawaii. Entomol. Soc.* 22:133–36

114. Sharp, D. 1899. *Letter to L. O. Howard dated 14 April 1899.* Ms. in National Archives Record Group 7. Bureau of Entomol. and Plant Quarantine Incoming Letters

115. Simberloff, D. 1986. Introduced insects: a biogeographic and systematic perspective. *Ecol. Stud.* 58:1–26

116. Smith, C. W. 1985. Alien plants: Impacts, research, and management needs. In *Hawaii's Terrestrial Ecosystems: Preservation and Management*, ed. C. P. Stone, J. M. Scott, pp. 180–250. Honolulu: Univ. Hawaii

117. Sweetman, H. L. 1958. *The Principles of Biological Control.* Dubuque, IA: Brown. 560 pp.

118. Swezey, O. H. 1931. Records of introduction of beneficial insects into the Hawaiian Islands. In *Handbook of Insects and Other Invertebrates of Hawaiian Sugar Cane Fields*, compiler F. X. Williams, pp. 368–89. Honolulu: Advertiser Publ. 400 pp.

119. Templeton, A. R. 1979. Genetics of colonization and establishment of exotic species. In *Genetics in Relation to Insect Management, Rockefeller Found. Conf. 31, March–5 April, 1978*, ed. M. A. Hoy, J. J. McKelvey Jr. pp. 41–49 Bellagio, Italy: Rockefeller Found.

120. Tiedje, J. M., Colwell, R. K., Grossman, Y. L., Hodson, R. E., Lenski, R. E., et al. 1989. The planned introduction of genetically engineered organisms: ecological considerations and recommendations. *Ecology* 70:298–315

121. Tillier, S., Clarke, B. C. 1983. Lutte biologique et destruction du patrimoine genetique: le cas des mollusques gasteropodes pulmones dans les territoires francais du Pacifique. *Genet. Sel. Evol.* 15:559–66

122. Timper, P., Kaya, H. K., Gaugler, R. 1988. Dispersal of the entomogenous nematode *Steinernema feltiae* (Rhabditida: Steinernematidae) by infected adult insects. *Environ. Entomol.* 17:546–50

123. Tomich, P. Q. 1986. *Mammals of Hawaii.* Honolulu: Bishop Museum Spec. Publ. 76. 375 pp.

124. Tothill, J. D., Taylor, T. H. C., Paine, R. W. 1930. *The Coconut Moth in Fiji (a History of Its Control by Means of Parasites).* London: Imp. Inst. Entomol. 269 pp., 34 plates

125. Turner, C. E. 1985. Conflicting interests in biological control of weeds. *Proc. 6th Int. Symp. Biol. Contr. Weeds, 1984*, pp. 203–25. Vancouver: Agric. Can.

125a. University of California, Riverside. 1989. Program and Abstracts. *Int. Vedalia Symp. Biol. Control: A Century of Success, Riverside, CA, 1989.* Riverside, CA: Univ. Calif. 104 pp.

126. van Lenteren, J. C. 1983. The potential of entomophagous parasites for pest control. *Agric. Ecosyst. Environ.* 10:143–58

127. Waage, J. K., Greathead, D. J. 1988. Biological control: challenges and opportunities. See Ref. 134, pp. 111–28

128. Wapshere, A. J. 1974. A strategy for evaluating the safety of organisms for biological weed control. *Ann. Appl. Biol.* 77:201–11

129. Waterhouse, D. F., Norris, K. R. 1987. *Biological Control: Pacific Prospects.* Melbourne: Inkata. 454 pp.

130. Wells, S. M., Pyle, R. M., Collins, N. M., compilers. 1983. *The IUCN Invertebrate Red Data Book.* Gland, Switzerland: IUCN. 632 pp.

131. Whitfield, J. B., Wagner, D. L. 1988. Patterns in host ranges within the nearctic species of the parasitoid genus *Pholetesor* Mason (Hymenoptera: Braconidae). *Environ. Entomol.* 17:608–15

132. Wilson, E. O. 1985. The biological diversity crisis: a challenge to science. *Issues Sci. Technol.* 2:20–29

133. Wodzicki, K. 1981. Some nature conservation problems in the South Pacific. *Biol. Conserv.* 21:5–18

134. Wood, R. K. S., Way, M. J., eds. 1988. Biological control of pests, pathogens, and weeds: developments and prospects. *Philos. Trans. R. Soc. London Ser. B* 318:109–376

135. Deleted in proof

136. Youngs, L. C. 1983. Predaceous ants in biological control of insect pests of North American forests. *Bull. Entomol Soc. Am.* 29:47–50

137. Zgomba, M., Petrovic, D., Srdic, Z. 1986. Mosquito larvicide impact on mayflies (Ephemeroptera) and dragonflies (Odonata) in aquatic biotopes. *Odonatologica* 16:221–22 (Abstr.)

138. Zimmerman, E. C. 1948. *Insects of Hawaii*, Vol. 1, *Introduction*. Honolulu: Univ. Hawaii Press. 206 pp.

139. Zimmerman, E. C. 1958. *Insects of Hawaii*, Vol. 7, *(Macrolepidoptera)*. Honolulu: Univ. Hawaii Press. 542 pp.

140. Zimmerman, H. G., Moran, V. C., Hoffmann, J. H. 1986. See Ref. 77, pp. 269–74

Annu. Rev. Entomol. 1991. 36:511–34

MATERNAL EFFECTS IN INSECT LIFE HISTORIES

Timothy A. Mousseau[1] and Hugh Dingle[2]

Department of Entomology, University of California, Davis, California 95616

KEY WORDS: evolution, diapause, development, polymorphisms and polyphenisms, insect endocrines

INTRODUCTION

Natural selection can occur when phenotypic variation in a trait related to fitness is heritable (26, 39). The trait can be simple (e.g. appendage length) or complex (e.g. shape, reaction norms), but as long as individuals vary in their expression, and such variance is genetically based, natural selection can act. If over several generations natural selection is consistent and unconstrained by other factors such as antagonistic pleiotropy (145) or physiological thresholds (170), then a population will evolve.

Although selection acts upon phenotypic variation, the resulting changes in genotypic frequencies affect evolution. Thus, the developmental mechanisms linking phenotype to genotype are of fundamental concern to an understanding of evolutionary change. Because development is an elaborative process and adult phenotype is a summation of activities accrued over the life span of the individual, developmental events occurring early in the life cycle can significantly influence the phenotype at later stages. In some cases, such developmental variation will be dampened as a result of homeostatic mechanisms [e.g. targeted growth of body size in mice (136), brain size allometry (135), or developmental programming (183)], while in others developmental

---

[1]Present address: Department of Biology, University of South Carolina, Columbia, South Carolina 29208.

[2]Order of authorship decided by coin toss.

511

variation will be amplified [e.g. antlers in deer or canines in carnivores (129)]. In both cases, however, genetic variation and covariation among early developmental events determine the ultimate phenotypic variation subject to the forces of natural selection.

Maternal effects represent developmental influences extended across life cycle stages in which genetic or environmental differences in the maternal generation are expressed as phenotypic differences in the offspring. Thus, natural selection acts upon phenotypic variation that results from events in the previous generation. In the past, maternal effects were often dismissed by plant and animal breeders as troublesome contributors to covariance among sibs (41). Only recently have the evolutionary consequences of such cross-generational transmission been rigorously explored and appreciated (84, 134, 137).

The mechanisms of cross-generational transmission are diverse. In mammals, large mothers may produce more milk than small mothers, which correlates to size of offspring (41), whereas turtle mothers can control sex determination by placing eggs at varying temperatures (17). In the guppy and the mosquito fish, offspring weight at birth is primarily a function of maternal genotype and environment, along with the important "grandfather" effects that contribute to maternal genotype (130, 131). Plant seeds are often under direct maternal control, and effects frequently carry over to the early seedling stages of development; seed size, dormancy, and germination are often influenced by maternal environment and genotype (138, 160).

In this review, we survey the growing literature on maternal effects in insects and address the following questions: 1. Which characters are most likely to be susceptible to maternal effects? 2. What are the environmental factors most likely to influence maternal control? 3. What stages in the maternal life cycle are most sensitive to environmental factors? 4. Which stages during the development of offspring are most sensitive to maternal control? 5. Which biochemical mechanisms have been captured for maternal control of development? And 6. what is the evidence for genetic variation for maternal effects within and among populations? This review explores maternal effects in insects in relation to the potential significance of these effects to life history.

MATERNAL EFFECTS ON DIAPAUSE

Diapause is an essential component of insect life histories. It provides a mechanism for tolerance of adverse climatic or food conditions and for synchronization of life cycles. Some insects, especially those inhabiting temperate climates, are obligate diapausers (52, 121, 180), while many species rely on environmental cues, usually photoperiod but sometimes temperature or host availability, as a trigger for the induction or avoidance of diapause (24, 25, 161, 171).

The diapausing stage itself, or the stages immediately prior to it, are usually most sensitive to environmental cues (25, 157). The decision to diapause (or not) results from individual genotype, behavior, and physiology (35, 67, 120). In many instances, however, diapause is significantly influenced by the environmental conditions experienced by the parental generation, and in some species the control of diapause may rest entirely with the mother (2, 52, 81, 148, 190). Here we review parental control of diapause, starting with Kogure's (85) seminal work with *Bombyx mori*.

Sensitive Stages

The maternal regulation of offspring diapause shows clear patterns (Table 1). First, although any stage of maternal development may be involved in regulating the diapause of offspring, more than half (60%) the studies surveyed

Table 1 Maternal effects on diapause[a]

Sensitive stage in maternal generation	Diapausing stage in offspring	Environmental cues		References
		photo	temp	
Egg	Egg	X		45, 65, 85, 190
Larva	Egg	X	X	2, 3, 7, 37, 38, 45, 47, 52, 65, 73, 79, 81, 82, 107, 126, 151, 177, 190
	Larva	X	X	166
	Adult	X		33
Pupa	Egg	X		2, 3, 7, 118, 126, 177
	Larva	X	X	4, 155
	Pupa	X	X	23
Adult	Egg	X	X	3, 13, 40, 52, 59, 66, 73, 107, 111, 112, 118, 126, 132, 148, 163, 177, 179
	Larva	X	X	4, 31, 32, 128, 153, 154, 156, 158, 159, 161, 178, 180, 182, 187, 190
	Prepupa	X	X	16
	Pupa	X	X	1, 10, 23, 31, 32, 187
	Adult	X		56, 57, 105, 139, 146

[a] Most species respond to either photoperiod (photo) or temperature (temp), or both, as the principle environmental cues used in maternal regulation of diapause.

found that adults are the primary stage for maternal regulation. Second, species that diapause as eggs are most likely to be susceptible to maternal effects (57% of reported studies). These patterns contrast with the observations of Saunders (157) that only 28% of 350 species surveyed exhibited diapause as adults, and only 17% of the species exhibited egg or embryonic diapause. Thus, the pattern suggested in Table 1 likely represents a real phenomenon and not just a sampling artifact.

In general, the decision to diapause is most influenced by environmental conditions contemporary to the stage at which diapause occurs or the stage(s) immediately prior to it. Of the 61 species exhibiting adult reproductive diapause reviewed by Danks (25), 25 were sensitive to environmental cues only in the adult stage, 20 were sensitive mainly as larvae, and 16 were sensitive as both larvae and adults. Danks reported no cases in which environmental conditions experienced by eggs influenced adult diapause. In pupal diapause, most species are sensitive as larvae (5, 14, 101–103, 184), while others are sensitive in both embryonic and larval stages (1, 23, 24, 27–29, 53, 181). In some cases, the pupal stage itself is also sensitive (23, 127). In most studies for which the sensitive stage for larval diapause has been examined, it occurs in the one or two instars immediately prior to the diapausing stage (25, 157); eggs are occasionally sensitive (25), [four species in one study (25)], and in only one case was the embryo sensitive and not the larva (24, 25). Embryonic diapause is most often influenced by the mother but is often also subject to environmental cues experienced by the embryos themselves (e.g. 3, 69, 74, 112, 126, 164, 172).

The above observations no doubt reflect the physiological mechanisms associated with developmental regulation in insects. Maternal control of diapause in offspring often involves cytoplasmic transmission of hormonal factors (see below) that are likely to be dampened or amplified by environmental and physiological conditions experienced by the developing offspring.

Environmental Cues

Because of the reliability of photoperiod as an indicator of season at most geographic locations (24), it is the environmental cue most often used by mothers to control diapause in offspring (Table 1). Diapause in most insects cues to absolute photoperiod (25), and the typical responses can be grouped into four major categories, labelled by Beck (8) as Type I, long-day response with diapause only at short photoperiods; Type II, short-day response with diapause only at long photoperiods; Type III, short-day, long-day response with diapause only within a narrow range of photoperiods; and Type IV, long-day, short-day response with diapause at all but a narrow range of photoperiods. Most insect species, including those with maternal transmission, exhibit Type I diapause responses (25, 157, 171). This observation is not surprising given the predominance of maternal control of egg diapause, which is most often associated with overwintering dormancy (25); adults that experience the short photoperiods of fall would be expected to produce diapausing embryos. Kogure's (85) classic work with the silkworm *Bombyx mori* provides a notable exception because overwintering eggs are induced to diapause if eggs and young larvae of the preceding generation are exposed to long photoperiods.

Fewer studies have tested for the influence of temperature, either alone or in combination with photoperiod. This no doubt reflects the bias of many researchers that photoperiod is the primary cue for environmental modulation of diapause induction. But many of the studies found that temperature plays a significant role (23, 139, 158). In the mosquito *Aedes togoi*, the proportion of diapausing eggs increased with shorter photoperiods (179). Reduced temperature increased the proportion of diapausing eggs, no matter what photoperiod the mothers experienced. In the parasitoid wasp *Trichogramma evanescens*, mothers that experienced short photoperiods produced a high proportion of diapausing larvae, which was reduced if mothers were reared at higher temperatures (190).

In a comparison of three geographically distant populations of the mosquito *Aedes atropalpus*, Beach (7) found little effect of temperature on maternal induction of diapause in a northern population (latitude 45° N), while in a southern population (latitude 14° N) high temperature could completely block the effect of photoperiod. An intermediate population (latitude 34° N) exhibited an intermediate effect of temperature on diapause.

Interaction between temperature and photoperiod may often result from the effect of temperature on development rather than a direct effect on diapause per se. For example, in the southern population of *A. atropalpus*, egg diapause is triggered when nine or more short-day photoperiod cycles occur during the sensitive fourth instar and pupal stages of the maternal generation (7). High temperatures cause individuals to complete the photosensitive period in less than nine days, thereby averting diapause. Thus, temperature effects on development time interact with the number of diel cycles during the photosensitive period as a means of controlling diapause response. In the northern populations of the mosquito, only four diel cycles are required to induce diapause; thus temperatures must be much higher to sufficiently accelerate development and reduce the incidence of diapause in eggs.

Occasionally, temperature has been found to be the primary cue in maternal induction of diapause. For example, larval diapause in the chalcid wasp *Mormoniella vitripennis* was induced only when adult mothers were chilled at 10°C; no other abiotic factor influenced diapause (161).

In some cases, photoperiod and temperature affect morphological development of the parental generation, which can be correlated with diapause incidence in the offspring. For example, in the lepidopteran *Orgyia thyellina,* long photoperiods result in long-winged mothers that tend to produce nondiapausing eggs, while mothers that experience short photoperiods develop into short-winged morphs that produce diapausing eggs (81). In the locust *Locusta pardalina,* egg diapause is also associated with the morph of the mother. Matthée (110) found that small, dark gregaria mothers produced on average 42% diapausing eggs, while large and light-colored solitaria mothers produced 100% diapausing eggs.

Maternal effects on diapause intensity have also been reported. In the blowfly *Calliphora vicina,* maternal photoperiod affects both the incidence of diapause in offspring and the duration of favorable environmental conditions necessary to break diapause (178). Mothers exposed to short photoperiods produced mainly diapausing larvae that require more than three months of cooling for reactivation. Females reared under longer photoperiods produced far fewer diapausing offspring, and these reactivated after only two months of cooling. Similar results have been obtained for egg diapause in the moth *Orgyia antiqua* (108), and for adult reproductive diapause of the mite *Neoseiulus fallacis* (138) and the milkweed bug *Oncopeltus fasciatus* (56, 57).

Maternal Age

In many insects, maternal age affects diapause in offspring. In most cases, the frequency of diapause increases with maternal age (52, 110, 113, 124, 133, 141, 152, 154, 156, 158, 165, 166), although the reverse situation has also been reported (16, 128). In the flesh fly, *Sarcophaga bullata,* pupal diapause increases from 0% at 10 days maternal age to 24.5% at 26 days for mothers selected to produce nondiapausing offspring under short-day conditions. Long-day mothers produce 37.4% diapausing offspring in their first brood and approximately 70% diapausing offspring at 26 days of age (14). In the parasitoid *Aphidus nigripes,* prepupal diapause generally declined significantly with maternal age, although individuals varied, and some females exhibited the reverse tendency (16).

Maternal age effects often result from extrinsic influences, both biotic and abiotic, on maternal development and physiology. In *M. vitripennis,* diapause in larvae could be changed from 0 to 97% simply by depriving mothers of hosts for five days (155, 161). Starvation can have similar effects, as observed in *A. atropalpus* in which food deprivation during the maternal fourth instar increased the incidence of diapausing eggs (7). Host deprivation, starvation, and temperature are similar in that all can prolong the developmental period during which females are sensitive to environmental cues; alternatively, they can delay the onset of reproduction to an age at which females switch from the production of nondiapausing to diapausing offspring (155).

Table 2 Geographic or genetic variation in maternal control of diapause[a]

Species	Photoperiod and temperature cues	Effects	References
Aedes atropalpus	photo	+	2, 7, 73
Aedes freeborni	photo	+	33
Aedes sollicitans	photo	+	3
Allonemobius fasciatus	photo	+	120, 150, M. Bradford & D. A. Roff[b]
Calliphora vicina	photo	+	178, 180, 182
Dianamobius fascipes	photo + temp	+	107
Dianamobius taprobanensis	photo	+	107
Haematobia irratans	photo + temp	+?	31, 32
Heliothis virescens	photo	+	9
Lucilia caesar	photo	+?	22, 43, 132, 133
Melanoplus sanguinipes	photo	+	H. Dingle & T. A. Mousseau[c]
Nasonia vitripennis	photo + temp	+	152, 154, 156, 161, 188
Neoseiulus fallacis	photo + temp	−	139
Oncopeltus fasciatus	photo	+?	56, 57
Orgyia antiqua	photo + temp	+?	37, 82, 83, 91, 108
Orgyia thyellina	photo	+	81, 151
Peripsocus quadrifasciatus	photo + temp	+	38
Pteronemobius fascipes	photo + temp	+	79

[a] Species designated by a + display significant variation among populations or selected lines for maternal control of diapause. Species designated with a +? are suspected of being genetically variable. There was no evidence (−) for geographic variation in maternal control in *Neoseiulus fallacis*.
[b] Unpublished data.
[c] In preparation.

Geographic Variation

Both the incidence and intensity of diapause can vary among populations of the same species in relation to the seasonal characteristics of the site of origin (24, 25, 157). Although relatively few studies have examined geographic variation of maternal effects on diapause, the evidence suggests that maternal sensitivity to environmental cues such as temperature and photoperiod also varies among populations, usually in a seemingly adaptive manner (Table 2). Among 15 populations of the psocid *Peripsocus quadrifasciatus* from temperate North America, critical photoperiod for maternally induced embryonic diapause increased approximately 15 minutes for every degree increase of a population's latitude (38). Altitude also significantly affected critical photoperiod. In addition, the shape of the photoperiodic response curve varied geographically such that northern mothers exhibited a sharper transition from the production of nondiapausing offspring to diapausing offspring.

Aedes atropalpus also displays geographic variation in maternal induction of embryonic diapause (2, 7, 73). The critical photoperiod for diapause induction varies from approximately L:D 15:9 in northern populations (> latitude 41° N) to L:D 12:12 for a population collected from El Salvador (latitude 14° N). Populations from Texas (latitude 30° N) and Georgia (latitude 35° N) exhibited an intermediate critical photoperiod of about L:D 13:11. The interaction between temperature and photoperiod also varied among populations.

Few studies have been made on genetic differences responsible for geographic variation in maternal control of diapause. The fact that populations differ significantly when reared under common garden conditions is strong evidence in support of genetic divergence. However, it is not known if this divergence is largely additive or the result of dominance or epistatic effects.

Some studies have investigated intrapopulation genetic variation for maternal effects on diapause (10, 56, 57). For three populations of the milkweed bug O. fasciatus, Groeters & Dingle (56, 57) found no evidence for significant heritability of reaction norms governing maternal effects on reproductive diapause. However, in this case, the statistical power was low, and a negative result only implies a low heritability, a result consistent with this trait's association with fitness (19, 144). In a study of two noctuids, Heliothis zea and H. virescens, Benschoter (10) found significant differences between lines selected for diapause and nondiapause, suggesting significant intrapopulation genetic variance for photoperiod response; unfortunately, insufficient information was given to assess realized heritability of maternal control of diapause. E. B. Vinogradova & I. P. Tsutskova (personal communication) have examined variation in maternal control of diapause in hybrids of sixteen variants of C. vicina. They found that F1 hybrids between diapausing and nondiapausing strains strongly resembled the mother, while the F2 were intermediate in their expression of diapause. Their results support a hypothesis of polygenic control of maternal effects on diapause, similar in mechanism to that observed for offspring weight in the guppy and the mosquito fish (130, 131).

POLYMORPHISMS AND POLYPHENISMS

Animal photoperiodism was first discovered through studies of a maternally induced effect, the production of sexual forms in aphids (104). Both because of the intrinsic interest of their complex life cycles and because of their importance as crop pests, aphids have continued to be the focus of intensive study of polymorphisms and polyphenisms. As did Hardie & Lees (63), we shall restrict the term polymorphism to genotypic differences among individuals and use polyphenism to indicate differences induced in the same genotype by extrinsic factors (see also 162).

Aphid life cycles can be approximately divided into two types, each in turn regulated by a complex set of signals (63, 93, 95). In most aphids of temperate climates, both sexual and parthenogenic forms occur (holocycly), while many tropical, and some temperate species, have lost the capacity to produce sexuals (anholocycly). In addition, species may remain on one host plant over successive generations (autoecy) or may alternate between a primary, usually woody, host and one or more species of summer herbaceous plants (heteroecy). In most holocyclic aphids, a series of viviparous parthenogenetic virginoparae develop under long-day conditions and give rise in autumnal short days to sexual females (oviparae) and males. In heteroecious species like *Aphis fabae,* an additional obligate alate gynopara that migrates to the winter host is interposed between virginopara and ovipara. If virginoparae are crowded from birth, they give rise to alate (winged) progeny in contrast to apterous individuals born when parents are uncrowded. The extensive telescoping of aphid life cycles in which the largest embryos already contain oocytes destined to be generation 3 (63, 95) creates a situation highly suited for the evolution of maternal effects and effects lasting for several generations (12).

Three species of aphid, *A. fabae, Megoura viciae,* and *Myzus persicae,* have been the most intensively studied with respect to the control of polyphenisms (12, 61–64, 93, 95–97). In all three species, control of the production of oviparae (gynoparae in *A. fabae* and *M. persicae*) and males is basically similar. Exposure of virginoparous females to short-day photoperiods while they are still nymphs results in production of sexual forms via a maternal effect acting through the endocrine system (see below). The sex ratio often departs widely from 1 : 1 (93), and each species has a characteristic birth order of males. *M. viciae* males are born in the middle of the parturition sequence, after 20 or so females, while *M. persicae* apterous viviparae bear females for 4 days and then switch to males (12). Likewise, in another well-studied aphid, *Acyrthosiphum pisum,* males are born last; after 50 or so females, 20–30 males are produced consecutively (76). Thus in addition to the photoperiod effect mediated through the mother, there is also a maternal-age effect. In all the above species, crowding of virginoparae results in the production of winged offspring that can then migrate to a new host plant. Crowding is apparently detected by direct physical contact with other aphids (94). Species may vary considerably in their response to crowding. For example, slight contact is all that is necessary in *A. pisum* (88), a species living on ephemeral host plants, while intense crowding is necessary to produce alates in the anholocyclic *Aphis nerii,* a tropical and subtropical species infesting long-lived perennial hosts (60). With few exceptions, alate mothers can produce only apterous offspring (63).

Genetic variance for the maternal control of morph production occurs among clones of those aphid species that have been examined for its presence.

Morph frequency differences among clones as well as between holocyclic and anholocyclic clones occur in *M. persicae* (11). In *A. pisum* (88) and *A. nerii* (58), clonal differences occur in sensitivity to crowding and hence in the production of alate females. Male alary polymorphism in *A. pisum* was exhibited by three types of clones: those with only apterous or alate males, and those with apterous and alate males in approximately equal proportions (168). Clonal frequencies varied geographically, suggesting some adaptation in morph production to local conditions.

The telescoped life histories of aphids also lead to multigenerational effects transmitted through reproducing females. The photoperiodic effect across three generations in *M. persicae* (12) was noted above. A striking case of a grandmaternal effect occurs in *A. pisum* (100). Mothers born to grandmothers one to two days after attaining adulthood produced no alate offspring when crowded, but those born two to three days later produced alates at maximum levels. Effects of maternal environments can thus span two or more generations and profoundly influence actual and potential rates of life history evolution (84).

Maternal influences on the production of different morphs are not confined to aphids. The cowpea seed beetle, *Callosobruchus maculatus,* has two morphs, an active dispersing form and a sedentary one, each with distinct morphologies. Eggs laid late in the oviposition period are more likely to develop into the active morph than eggs laid earlier (149), an apparent bet-hedging strategy on the part of the mother against future crowding (114). In the red linden bug, *Pyrrhocoris apterus,* offspring of mothers of a macropterous strain kept under long-day conditions were about 90% macropterous whereas offspring of mothers kept under short-day conditions became macropterous in long days and brachypterous in short days (70). A particularly interesting case of maternal determination of form occurs in the collembolan *Orchesella cincta* in Holland (72). This species produces a winter and a summer morph that differ in several life history traits. These differences are produced by strong negative maternal effects, so that individuals display traits opposite to those of their mothers but similar to those of their grandmothers. This represents a clear adjustment of generations to the seasonal cycle in this species because a given individual encounters the environment of its grandparents rather than its parents. Maternal effects in both *C. maculatus* and *O. cincta* evidently play a role in ecological fine tuning.

Fine tuning is also a characteristic of phase polyphenism in various species of migratory locusts (reviews in 63, 75, 86). All active stages of locust life cycles are sensitive to environmental inputs influencing phase, but maternal influences are superimposed on direct effects. In *Locusta migratoria,* offspring of crowded mothers are darker and march more vigorously than those of isolated mothers. In *Schistocerca gregaria* and *Nomadacris septenfasciata,* eggs from crowded mothers have higher lipid reserves and produce nymphs passing through one less instar than nymphs from eggs of isolated mothers. Thus, maternal effects have important ecological consequences.

MATERNAL INFLUENCE ON OFFSPRING QUALITY

Maternal effects resulting from age or environmental inputs can often profoundly influence offspring fitness. This has been forcefully argued by Wellington (185, 186), who has emphasized spatial and temporal variations in the interactions between insects and their habitat. Using classic studies of forest Lepidoptera, mainly *Malacosoma californiacum pluviale,* the tent caterpillar, he demonstrated how the history of both the individual and its ancestors could influence current responses. The "quality" of an individual affected its response to stressful environments.

Studies of other Lepidoptera and other insects have led to similar conclusions. In the gypsy moth, *Lymantria dispar,* the host species and phenolic level in the maternal diet affected pupal weights, dispersal period, and development time of offspring (M. Rossiter, unpublished manuscript). Egg size mediated these effects; individuals from larger eggs hatched earlier; females developed faster and males grew larger; and individuals from eggs yolked first showed twice the resistance to a microbial pesticide as those from later-yolked eggs. In *Drosophila melanogaster,* malathion resistance is subject to maternal influences (167). In the blood-feeding tse-tse, *Glossina morsitans,* host species can influence offspring size and presumably survival (90). In the tobacco budworm, *Heliothis virescens,* addition of the naturally occurring flavonoid quercetin to an artificial diet on which the insect was reared results in reduced growth of offspring larvae raised on diets including quercetin or rutin, whereas offspring maintained on regular diets or diets containing gramine grow normally (54). The response of larvae to stress is thus mediated by an interaction between the stress itself and the maternal diet. Similar effects were noted when parents were reared on tobacco and cotton, the usual host plants; time to pupation of their offspring on quercetin and rutin diets was longer than when parents were reared on an artificial diet lacking these components. The influence of maternal stress on offspring fitness thus depends on interactions between maternal and offspring environments, adding a further dimension to notions of population quality (54).

Maternal age also influences offspring quality and interacts with proximate factors in the offspring environment (125; W. R. Tschinkel, unpublished manuscript). In the tenebrionid guano beetle, *Zophobas atratus,* birth order influenced several larval characters (W. R. Tschinkel, unpublished manuscript); larvae born later grew more slowly and weighed less at metamorphosis, but larval density also influenced this birth-order effect. At high densities, early-born larvae grew faster and were smaller at pupation than late-born larvae. Early-born females also laid more eggs with a higher hatching success and lived longer so that the number of eggs/lifespan was greater. Again interactions with density and other factors were noted. Similar birth-order effects were noted earlier for *Tenebrio molitor* (98, 99). Birth-order effects in which offspring of young mothers display higher survival, more rapid development, or both occur in houseflies (142) and in *Drosophila* (6, 18, 42, 77). Females evidently invest more in early offspring to produce adults of higher quality, a life history strategy that makes sense in species with high population turnover (r-selection).

Much of the maternal influence on offspring quality may be mediated via egg size (reviewed in 86, 87). Older mothers, for example, may lay smaller eggs with subsequent consequences for development, survival, and eventual adult size. The latter may in turn influence fecundity or competitive ability. An interesting example occurs in the bruchid *C. maculatus* (114) in which variation in adult size largely results from nongenetic maternal effects. The large larvae of the S-strain typically produce only a lone survivor when two eggs are laid on a single azuki bean. In contrast, two larvae of the smaller I-strain can survive in a single bean. Interactions between maternally controlled size and resource availability thus influence the payoffs of various life-history strategies.

PARENTAL EFFECTS ON DEVELOPMENT TIME

Development time, the period between the hatching of a larva or nymph and its eclosion to the adult stage, can be of considerable importance to an insect. The duration of the juvenile stage has an important bearing on population growth parameters because of its relation to age at first reproduction (34), and it can also influence body size and hence, indirectly, fecundity (143). It also may be critical to the induction and timing of diapause (173).

Studies of the influence of maternal age on development time in the milkweed bugs *Oncopeltus cingulifer* and *O. fasciatus* suggest a possible adaptive function (P. C. Frumhoff, unpublished manuscript). The population of *O. fasciatus* is migratory in North America and must return south in the autumn because it cannot overwinter in cold climates. *O. cingulifer* is tropical or subtropical and breeds year round. *O. fasciatus* offspring born of young mothers (< 50 days old) take about 28 days to develop, while offspring of old mothers (> 65 days) take only 25 days but hatch from smaller eggs and suffer reduced survivorship. The accelerated development would, however, increase the chances for these late offspring to reach adulthood and migrate before frosts eliminate them. The cost of producing smaller eggs (reduced survivorship) may represent a trade-off. *O. cingulifer* has no maternal-age effect on development time or on survivorship even though later eggs are significantly smaller.

Maternal effects influencing development time are also known in *D. melanogaster* (115). A particularly interesting parental effect in *D. melanogaster*, which is apparently parental nuclear and not maternal cytoplasmic, has been reported by Giesel and his colleagues (49–51). Progeny of flies reared in a short-day environment displayed shorter development times than progeny of long-day flies. Both maternal and paternal photoperiods influenced offspring phenotype and crosses indicated that the effects were roughly equal (50). Selection was ruled out. Photoperiod and density also interact because the progeny of short-day parents are sensitive to culture density while progeny of long-day parents are not. The more rapid development, especially at low density, of short-day offspring is consistent with a spring bloom population, whereas the lower-density sensitivity of long-day offspring may be advantageous in more dense summer populations (50). The effect is mediated by metabolic rates that are higher in short-day progeny (51). The effect of parent on metabolic rate is, however, overwhelmed when offspring photoperiods substantially differ from those of the parents (89). The surprising nature of these apparently nuclear transgenerational effects needs further study in *Drosophila* and should also be carefully looked for in other insects.

ENDOCRINE MEDIATION OF MATERNAL EFFECTS

The importance of maternal effects in insect life cycles has led to a search for an agent or agents responsible for transgenerational transfer of information from mother to offspring. Interest has naturally focused on the endocrine system. The most extensive studies involve the commercial silkworm *B. mori* (reviews in 30, 190). Classic studies initiated by Fukuda (45) and Hasegawa (65) identified a diapause hormone (a neuropeptide) that causes the ovarioles to produce diapause eggs. In its absence, and under the influence of juvenile hormone, the eggs develop without diapause. The diapause hormone (DH) is released from storage cells in the suboesophageal ganglion after synthesis in the brain (46, 169). Control of release results from a complex of stimulatory and inhibitory mechanisms traceable to both protocerebrum and tritocerebrum (109). The action of DH in causing diapause is not clear, but it may result from reduced permeability of the eggshell to oxygen (80).

The presence of DH has now been detected in several species of moth (30). The closest parallel to *B. mori* is found in the control of diapause in eggs of the tussock moth *O. antiqua* (82). Several species in which DH has been found diapause as pupae, not eggs, so that the hormone cannot function in the same manner as in *Bombyx* or *Orgyia*. In *B. mori,* DH also functions in several other metabolic pathways (30, 190), and presumably it does so in other Lepidoptera. DH was probably captured secondarily for its role in maternally controlled egg diapause.

In flesh flies, *S. bullata,* diapause results from shutdown of the brain–prothoracic gland axis, and this shutdown is prevented by the diapause-averting maternal effect in *S. bullata* females exposed to short days (67, 140). The effect of short-day exposure occurs during embryonic and larval life, but the information preventing diapause is transferred to the ovary sometime between the end of larval life and the third day of adulthood. Some progeny of short-day mothers that would otherwise be expected to avert diapause as a consequence of maternal effects can be induced to diapause by injecting hemolymph from one- to two-day-old long-day adults or central nervous system extracts from long-day larvae. Attempts to repeat these results with known hormones or their analogues were unsuccessful, and the chemical nature of the active diapause-promoting agent remains unknown. Also unknown is the exact nature of the information transfer system of the diapause-suppressing maternal effect. Central nervous system extracts from short-day larvae did lower diapause incidence in long-day progeny, but the observed diapause incidence was within the range of long-day values in other experiments and, according to Rockey et al (140), cannot be accepted with confidence.

Unlike the situation with *S. bullata,* egg diapause in the Australian plague locust *Chortoicetes terminifera* appears to be influenced by levels of ecdysteroids in the egg (55). This is one of many examples of ecdysteroids of maternal origin in insect eggs (68). In *B. mori,* one of the actions of DH may be suppression of ecdysteroids because application of exogenous 20-hydroxyecdysone can promote resumption of development in diapause eggs (48).

The maternally mediated polyphenisms of aphids appear to be regulated by juvenile hormone (JH) and its analogues (61, 63, 64, 95, 116). Experiments with *A. fabae, M. viciae,* and *A. pisum* all demonstrate that short-day, ovipara-producing aphids will in contrast produce parthenogenetic viviparous progeny after topical application of JH or JH analogues. At least in *A. fabae,* JH is apparently not involved in the apterous-alate dimorphism of parthenogenetic virginoparae influenced by crowding, although it can prevent the production of winged gynoparae under short-day conditions (63). The experiments of Lees (92) demonstrating the photosensitivity of the protocerebrum in the vicinity of the so-called Group-I neurosecretory cells point to control of JH by this region of the brain. Cauterization experiments (61) support the view that Group-I cells synthesize a neurohormone [most likely a neuropeptide (78)] that regulates corpus allatum size and juvenile hormone synthesis. The photoperiodically induced maternal effect would thus be mediated by photosensitive Group-I cells regulating synthesis (and release?) of JH that in turn would influence ovipara (gynopara) production. Much of this model needs confirmation. However, JH, a hormone already present in insects and regulating metamorphosis and ovary maturation, has evidently been captured for a role in determining polyphenisms (63), a situation parallel to that of the capture of DH in *B. mori* cited above.

Juvenile hormone has also been captured to influence phase polyphenism in migratory locusts including maternal effects (19–21, 123). Reduction of JH in solitary females by hemi-allatectomy resulted in decreased ovary size and production of larger eggs characteristic of gregarious females. About 20% of the nymphs that hatched displayed at least partial gregarious color patterns. The reverse experiment of implanting corpora allata into gregarious females caused some offspring to display solitary phase coloration. The level of JH in females evidently influences the programming of secretion, the sensitivity to JH of the offspring, or both.

Although our general understanding of the mechanisms by which maternal control is exerted is still inchoate, it is easy, at least in principle, to see how a maternal effect could influence the first stage of the next generation. Thus, hormones produced by the mother evidently directly regulate prenatal physiological pathways of an egg or embryo (in aphids) to induce diapause or sexual differentiation. Our ignorance is profound in those cases in which the expression of the maternal effect occurs much later in the development of the progeny, e.g. in adult diapause or age at first reproduction. The events leading to later expression must be triggered in the egg or embryo, but the exact developmental programs that follow and their regulation remain a considerable challenge to insect physiologists.

CONCLUSIONS

The most important conclusion of this review is that maternal effects in insects are as diverse and complex as the species and traits that exhibit them. They influence incidence and intensity of diapause, production of sexual forms, wing polyphenism, dispersal behavior, development time, growth rate, resistance to chemicals or microbial infection, and survival. Environmental factors most often associated with maternal control are photoperiod and temperature, cues that permit prediction of season at most latitudes. Maternal age and diet have also been found to have important effects on offspring fitness.

The cumulative evidence supports the hypothesis that maternal control of development in offspring has evolved as an adaptive response to uncertainties associated with life in a temporally and spatially heterogeneous environment.

Maternal effects described here are a special case of variation, often adaptive, known as phenotypic plasticity, in which individual genotypes exhibit variable life histories in response to environmental cues (56, 57, 176). Maternal effects are distinct from common phenotypic plasticity in that environmental effects experienced by the mother are expressed as phenotypic variation in offspring. In almost all cases reviewed here, maternal effects were correlated with predictable patterns of seasonality and resulted in appropriate switches in life history. For example, short photoperiods and cold temperatures (indicators of impending winter) experienced by mothers often resulted in diapausing offspring, while long photoperiods and high temperatures often averted diapause in offspring.

The evolutionary consequences of this transgenerational interaction are numerous and complex. In a recent theoretical treatment, Kirkpatrick & Lande (84) found that because of lags in evolutionary response to selection on maternal traits and the fact that the rate and direction of evolution depend upon the inheritance of traits not directly under selection, populations could evolve for an indefinite number of generations after selection has ceased. More importantly, the response can be larger or smaller than expected for direct genetic effects, and even in the opposite direction, thus allowing maternal selection to result in maladaptive evolution. On the other hand, appropriate phenotypic plasticity for temporal heterogeneity may evolve by selection for environmentally cued maternal effects in spite of constraints on plasticity imposed by cross-environment genetic correlations (56). Our survey of the literature indicates that indeed maternal effects in insects need not be limited by evolutionary constraints (and may serve in part to circumvent them), although such contraints may explain why maternal effects are not ubiquitous.

Few studies have examined the genetics of maternal effects in natural insect populations. In studies on the milkweed bug *O. fasciatus,* Groeters & Dingle (56, 57) found no evidence for genetic variance within populations for maternal control of adult reproductive diapause, although their methods were not sensitive to small amounts of variance. Benschoter (10), on the other hand, found significant responses to selection for diapausing and nondiapausing strains in two noctuids. Given the fitness consequences of many maternal effects, additive genetic variance would probably be low as a result of intense selection (119, 144). Significant geographic variation among populations in maternal control, however, suggests that low genetic variance does not represent a fundamental constraint to the evolution of maternal effects in most cases.

A more likely constraint to the evolution of maternal effects results from the kinds of biochemical mechanisms involved in the inter-generational transmission of environmental cues. Maternal control of development is mediated through a diverse array of endocrine functions that interact and also control a wide variety of developmental activities in addition to those associated with maternal effects. The implication is that the ability of insects to evolve hormones specifically for maternal regulation of offspring is severely limited. Rather, the overwhelming evidence suggests that endocrine functions normally used in other areas of development have been captured during evolution for maternal regulation of growth and development of offspring. This capture of hormones evolved for one function by another seems to be a general principle of insect endocrinology (30, 63, 123) and may be an important limitation on correlated pathways of development.

A further generality to emerge from this review is that maternal effects on diapause are strongest between mother and embryo. Similar associations have been reported in fish (130, 131), plants (138, 160), and mammals (117). This observation may result from constraints imposed by biochemical mechanisms and developmental pathways involved in maternal effects; alternatively, natural selection may have favored short-term over long-term control. We have noted that maternal control in aphids can extend across as many as three generations (12), presumably as a result of their telescoped life histories. In most species reviewed, however, environmental conditions experienced late in maternal development were most likely to influence offspring, and it was the early developmental stages that were most likely to be influenced. This differential influence may be a simple consequence of growth and development; initially, egg cytoplasm is maternally constituted, and it is only following development that the embryo's own biochemical machinery takes over.

An important objective of a review such as this is to point out promising future areas of research. Two strike us as particularly interesting and important. First, we know far too little about the biochemical mechanisms by which mothers direct future offspring performance. Second, a large gap exists in our understanding of the genetic basis for variation in maternal control. Studies of the genetic and biochemical mechanisms determining maternal effects in a broad array of taxa are imperative for a full understanding of the evolution of insect life histories and the genetic regulation of biochemical processes crucial for insect development. Further investigations of mode of deployment and interactions among life-history traits should result in new insights concerning the implications of maternal effects for adaptive strategies of insects and the management of insect populations.

ACKNOWLEDGMENTS

We thank Larry Harshman, Tim Prout, and Bruce Riska for comments on the manuscript and the many entomologists who sent us reprints, galleys, and unpublished manuscripts and data. We extend special thanks to H. V. Danks whose thorough review of insect dormancy launched and inspired this paper. A NSERC (Canada) Postdoctoral Fellowship to T. A. M. and NSF grants to H. D. are acknowledged.

Literature Cited

1. Adkisson, P. L., Roach, S. H. 1971. A mechanism for seasonal discrimination in the photoperiodic induction of pupal diapause in the bollworm *Heliothis zea* (Boddie). In *Biochronometry*, ed. M. Menaker, 1:272–80. Washington DC: Natl. Acad. Sci.
2. Anderson, J. F. 1968. Influence of photoperiod and temperature on the induction of diapause in *Aedes atropalpus* (Diptera: Culicidae). *Entomol. Exp. Appl.* 11:321–30
3. Anderson, J. F. 1970. Induction and termination of embryonic diapause in the salt marsh mosquito, *Aedes solicitans* (Diptera: Culicidae). *Bull. Conn. Agric. Exp. Stn.* No. 711. 22 pp.
4. Anderson, J. F., Kaya, H. R. 1974. Diapause induction by photoperiod and temperature in the elm spanworm egg parasitoid, *Oencyetus* sp. *Ann. Entomol. Soc. Am.* 67:845–49
5. Barker, R. J., Mayer, A., Cohen, C. F. 1963. Photoperiod effects in *Pieris rapae*. *Ann. Entomol. Soc. Am.* 56:292–94
6. Barnes, P. T. 1984. A maternal effect influencing larval viability in *Drosophila melanogaster*. *J. Heredity* 75:288–92
7. Beach, R. 1978. The required day number and timely induction of diapause in geographic strains of the mosquito, *Aedes atropalpus*. *J. Insect Physiol.* 24:449–55
8. Beck, S. D. 1980. *Insect Photoperiodism*. New York: Academic. 387 pp. 2nd ed.
9. Benschoter, C. A. 1968. Influence of light manipulation on diapause of *Heliothis zea* and *H. virescens*. *Ann. Entomol. Soc. Am.* 61:1272–74
10. Benschoter, C. A. 1970. Culturing *Heliothis* species (Lepidoptera: Noctuidae) for investigation of photoperiod and diapause relationships. *Ann. Entomol. Soc. Am.* 63:699–701
11. Blackman, R. L. 1972. The inheritance of life-cycle differences in *Myzus persicae* (Sulz.) (Hem. Aphididae). *Bull. Entomol. Res.* 61:281–94
12. Blackman, R. L. 1975. Photoperiodic determination of the male and female sexual morphs of *Myzus persicae*. *J. Insect Physiol.* 21:435–53
13. Bocher, J. 1975. Notes on the reproductive biology and egg diapause in *Nysius groenlandicus* (Zett.) (Heteroptera: Lygaeidae) *Vidensk. Medd. Dan. Naturhist. Foren. Khobenhavn* 138:21–38
14. Bonnemaison, L. 1976. Action de la photopériod sur l'induction de la pyrale du mais (*Ostrinia nubilalis* Hbn., Lep., Noctuidae). *Ann. Soc. Entomol. Fr.* 11:767–81
15. Deleted in proof
16. Brodeur, J., McNeil, J. N. 1989. Biotic and abiotic factors involved in diapause induction of the parasitoid, *Aphidus nigripes* (Hymenoptera: Aphidiidae). *J. Insect Physiol.* 35:969–74
17. Bull, J. J., Vogt, R. C., Bulmer, M. G. 1982. Heritability of sex ratio in turtles with environmental sex determination. *Evolution* 36:333–41
18. Cadieu, N. 1983. Maternal influence and the effect of heterosis on the viability of *Drosophila melanogaster* during different pre-imaginal stages. *Can. J. Zool.* 61:1152–55
19. Cassier, P. 1966. Effects de l'ablation d'un corps allate sur la fecondite et la descendance des femelles isolees du criquet migrateur (*Locusta migratoria migratorioides* R. et F.). *Insectes Soc.* 13:17–28
20. Cassier, P. 1966. L'activitie des corps allates et la reproduction du criquet migrateur africain, *Locusta migratoria migratorioides* R. et F. *Bull. Sci. Zool. Fr.* 91:133–48
21. Cassier, P., Papillon, M. 1968. Effets des implantations de corps allates sur la reproduction des femelles groupees de *Schistocerca gregaria* (Forsk.) et sur le polymorphisme de leur descendance. *CR. Acad. Sci. Paris* 266 D:1048–51
22. Cragg, J. B., Cole, P. 1952. Diapause in *Lucilia sericata* (Mg.) (Diptera). *J. Exp. Biol.* 29:600–4
23. Cullen, J. M., Browning, T. O. 1978. The influence of photoperiod and temperature on the induction of diapause in pupae of *Heliothis punctigera*. *J. Insect Physiol.* 24:595–601
24. Danilevskii, A. S. 1965. *Photoperiodism and Seasonal Development of Insects*. Edinburgh, London: Oliver and Boyd

25. Danks, H. V. 1987. *Insect Dormancy: An Ecological Perspective.* Ottawa: Biological Survey of Canada Monograph Series No. 1
26. Darwin, C. 1859. *The Origin of Species.* London: Murray
27. Denlinger, D. L. 1970. Embryonic determination of pupal diapause induction in the flesh fly *Sarcophaga crassipalpus* Macquart. *Am. Zool.* 10:320–21
28. Denlinger, D. L. 1971. Embryonic determination of pupal diapause induction in the flesh fly *Sarcophaga crassipalpus.* *J. Insect Physiol.* 17:1815–22
29. Denlinger, D. L. 1972. Induction and termination of pupal diapause in *Sarcophaga* (Diptera: Sarcophagidae). *Biol. Bull.* 142:11–24
30. Denlinger, D. L. 1985. Hormonal control of diapause. See Ref. 76a, 8:353–411
31. Depner, K. R. 1961. The effect of temperature on development and diapause of the horn fly, *Siphona irritans* (L.) (Diptera: Muscidae). *Can. Entomol.* 93:855–59
32. Depner, K. R. 1962. The effects of photoperiod and ultraviolet radiation on the incidence of diapause in the horn fly, *Haematobia irritans* (L.) (Diptera: Muscidae). *Int. J. Biometeorol.* 5:68–71
33. Depner, K. R., Harwood, R. F. 1966. Photoperiodic responses of two latitudinally diverse groups of *Anopheles freeborni* (Diptera: Culicidae). *Ann. Entomol. Soc. Am.* 59:7–11
34. Dingle, H. 1990. The evolution of life histories. In *Population Biology,* ed. K. Wöhrmann, S. Jain, pp. 267–89 Berlin: Springer
35. Dingle, H., Hegmann, J. P. 1982. *Evolution and Genetics of Life Histories.* New York: Springer-Verlag
36. Deleted in proof
37. Doskoeil, J. 1957. Beitrag zur Kenntnis der Insektendiapause. 2. Einfluss der Beleuchtungslange auf die Entstehung der Diapause der Eier. *Vestn. Cesk. Spol. Zool.* 21:273–83
38. Eertmoed, G. E. 1978. Embryonic diapause in the psocid, *Peripsocus quadrifasciatus:* photoperiod, temperature, ontogeny and geographic variation. *Physiol. Entomol.* 3:197–206
39. Endler, J. 1986. *Natural Selection in the Wild.* Princeton: Princeton Univ. Press
40. Ewen, A. B. 1966. A possible endocrine mechanism for inducing diapause in eggs of *Adelphocoris lineolatus* (Goeze) (Hemiptera: Miridae). *Experentia* 22:470
41. Falconer, D. S. 1989. *Introduction to Quantitative Genetics.* New York: Longman 3rd ed.
42. Fleuriet, A., Vageille, M. C. 1982. On the maintenance of the polymorphism at the ref (2)P locus in populations of *Drosophila melanogaster. Genetica* 59:203–10
43. Fraser, A., Smith, W. F. 1963. Diapause in larvae of green blowflies (Diptera: Cyclorrhaphia: *Lucilia* spp.). *Proc. R. Entomol. Soc. London* (A) 38:90–97
44. Deleted in proof
45. Fukuda, S. 1951. Factors determining the production of non-diapause eggs in the silkworm. *Proc. Jpn. Acad.* 27:582–86
46. Fukuda, S., Takeuchi, S. 1967. Studies on the diapause factor-producing cells in the suboesophageal ganglion of the silkworm *Bombyx mori* L. *Embryologia* 9:333–53
47. Gaylor, M. T., Sterling, W. L. 1977. Photoperiodic induction and seasonal incidence of embryonic diapause in the cotton fleahopper, *Pseudatomoscelis seriatus. Ann. Entomol. Soc. Am.* 70: 893–97
48. Gharib, B., Legay, J. M., de Reggi, M. 1981. Potentiation of developmental abilities of diapausing eggs of *Bombyx mori* by 20-hydroxyecdysone. *J. Insect Physiol.* 27:711–13
49. Giesel, J. T. 1986. Genetic correlation structure of life history variables in outbred, wild *Drosophila melanogaster:* Effects of photoperiod regimen. *Am. Nat.* 128:593–603
50. Giesel, J. T. 1988. Effects of parental photoperiod on development time and density sensitivity of progeny of *Drosophila melanogaster. Evolution* 42: 1348–50
51. Giesel, J. T., Lanciani, C. A., Anderson, J. F. 1989. Effects of parental photoperiod on metabolic rate in *Drosophila melanogaster. Fla. Entomol.* 71:499–503
52. Glinyanaya, Y. I. 1975. The importance of daylength in the control of seasonal cycles and diapause in some Psocoptera. *Entomol. Rev.* 54(1):10–13
52a. Goryshin, N. I., ed. 1972. *Problems of Photoperiodism and Diapause in Insects.* Leningrad: Leningrad Univ. Press
53. Gnagey, A. L., Denlinger, D. L. 1984. Photoperiodic induction of pupal diapause in the flesh fly, *Sarcophaga crassipalpis:* embryonic sensitivity. *J. Comp. Physiol. B* 154:91–96
54. Gould, F. 1988. Stress specificity of maternal effects in *Heliothis virescens* (Boddie) (Lepidoptera: Noctuidae) lar-

vae. *Mem. Entomol. Soc. Can.* 146: 191–97

55. Gregg, P. C., Roberts, B., Wentworth, S. L. 1987. Levels of ecdysteroids in diapause and non-diapause eggs of the Australian plague locust, *Chortoicetes terminifera* (Walker). *J. Insect Physiol.* 33:237–42

56. Groeters, F. R., Dingle, H. 1987. Genetic and maternal influences on life history plasticity in response to photoperiod by milkweed bugs *(Oncopeltus fasciatus)*. *Am. Nat.* 129:332–46

57. Groeters, F. R., Dingle, H. 1988. Genetic and maternal influences on life history plasticity in milkweed bugs *(Oncopeltus fasciatus)*: response to temperature. *J. Evol. Biol.* 1:317–33

58. Groeters, F. R., Dingle, H. 1989. The cost of being able to fly in the milkweed-oleander aphid, *Aphis nerii* (Homoptera: Aphididae). *Evol. Ecol.* 3:313–26

59. Gustin, R. D. 1974. Termination of diapause in the painted leafhopper, *Endria inimica*. *Ann. Entomol. Soc. Am.* 67:607–9

60. Hall, R. W., Ehler, L. E. 1980. Population ecology of *Aphis nerii* on oleander. *Environ. Entomol.* 9:338–44

61. Hardie, J. 1987. The corpus allatum, neurosecretion and photoperiodically controlled polymorphism in an aphid. *J. Insect Physiol.* 33:201–5

62. Hardie, J. 1987. The photoperiodic control of wing development in the black bean aphid, *Aphis fabae*. *J. Insect Physiol.* 33:543–49

63. Hardie, J., Lees, A. D. 1985. Endocrine control of polymorphism and polyphenism. See Ref. 76a, 8:441–90

64. Hardie, J., Lees, A. D. 1985. The induction of normal and teratoid viviparae by a juvenile hormone and kinoprene in two species of aphids. *Physiol. Entomol.* 10:65–74

65. Hasegawa, K. 1951. Studies in voltinism in the silkworm *Bombyx mori* L., with special reference to the organs concerning determination of voltinism (a preliminary note). *Proc. Jpn. Acad.* 27: 667–71

66. Helfert, B. 1980. Die regulative Wirkung von Photoperiode und Temperatur auf den Lebszklus okologisch unterschiedlicher Tettigoniiden-Arten (Orthoptera, Saltatoria). 2. Teil: Embryogenese und Dormanz der Filialgenera. *Zool. Jahrb. Abt. Syst. Oekol. Geogr. Tiere* 107:449–500

67. Henrich, V. C., Denlinger, D. L. 1982. A maternal effect that eliminates pupal diapause in progeny of the flesh fly, *Sar-*

cophaga bullata. *J. Insect Physiol.* 28:881–84

68. Hoffmann, J. A., Lagueux, M. 1985. Endocrine aspects of embryonic development in insects. See Ref. 76a, 1:435–60

69. Hogan, T. W. 1960. The onset and duration of diapause in eggs of *Acheta commodus* (Walk.) (Orthoptera) *Aust. J. Biol. Sci.* 13:14–29

70. Honek, A. 1980. Maternal regulation of wing polymorphism in *Pyrrhocoris apterus:* Effect of cold activation. *Experientia* 36:418–19

71. Isayev, V. A. 1975. Photoperiodic induction of diapause in the egg phase in *Culicoides pulicaris punctatus* Mg. (Diptera, Ceratopogonidae). *Parazitologiya* 9:501–606

72. Janssen, G. M., de Jong, G., Joosse, E. N. G., Scharloo, W. 1988. A negative maternal effect in springtails. *Evolution* 42:828–34

73. Kalpage, K. S. P., Brust, R. A. 1974. Studies on diapause and female fecundity in *Aedes atropalpus*. *Environ. Entomol.* 3:139–45

74. Kappus, K. D., Venard, C. E. 1967. The effects of photoperiod and temperature on the induction of diapause in *Aedes triseriatus* (Say). *J. Insect Physiol.* 13:1007–19

75. Kennedy, J. S. 1961. Continuous polymorphism in locusts. *Symp. Entomol. Soc. London* 1:80–90

76. Kenten, J. 1955. The effect of photoperiod and temperature on reproduction in *Acyrthosiphon pisum* (Harris) and the forms produced. *Bull. Entomol. Res.* 46:599–624

76a. Kerkut, G. A., Gilbert, L. I., eds. 1985. *Comprehensive Insect Physiology, Biochemistry, and Pharmacology,* Vols. 1–9. Oxford: Pergamon

77. Kerver, W. J. M., Rotman, G. 1987. Development of ethanol tolerance in relation to the alcohol dehydrogenase locus in *Drosophila melanogaster*. II. The influence of phenotypic adaptation and maternal effect on alcohol supplemented media. *Heredity* 58:239–48

78. Khan, M. A. 1988. Brain-controlled synthesis of juvenile hormone in adult insects. *Entomol. Exp. Appl.* 46:3–17

79. Kidokoro, T., Masaki, S. 1978. Photoperiodic response in relation to variable voltinism in the ground cricket, *Pteronemobius fascipes* Walker (Orthoptera: Gryllidae). *Jpn. J. Ecol.* 28:291–98

80. Kim, S.-E. 1987. Changes in eggshell permeability to oxygen during the early

developmental stages of diapause eggs of *Bombyx mori. J. Insect Physiol.* 33:229–35

81. Kimura, M. T., Masaki, S. 1977. Brachypterism and seasonal adaptation in *Orgyia thyellina* Butler (Lepidoptera, Lymantriidae). *Kontyu* 45:97–106

82. Kind, T. V. 1965. Neurosecretion and voltinism in *Orgyia antiqua* L. (Lepidoptera, Lymantriidae). *Entomol. Rev.* 44(3):326–27

83. Kind, T. V. 1972. Endocrine determination of diapause in the moth *Orgyia antiqua* (Lepidoptera, Lymantriidae). See Ref. 52a, pp. 210–28

84. Kirkpatrick, M., Lande, R. 1989. The evolution of maternal characters. *Evolution* 43:485–503

85. Kogure, M. 1933. The influence of light and temperature on certain characteristics of the silkworm, *Bombyx mori. J. Dep. Agric. Kyushu Imp. Univ.* 4:1–93

86. Labeyrie, V. 1967. Physiologie de la mère et etat de la progeniture chez les insectes. *Bull. Biol.* 101:13–71

87. Labeyrie, V. 1988. Effets maternels et biologie des populations d'insectes. *Mem. Entomol. Soc. Can.* 146:153–69

88. Lamb, R. J., MacKay, P. A. 1979. Variability in migratory tendency within and among natural populations of the pea aphid, *Acyrthosiphon pisum. Oecologia* 39:289–99

89. Lanciani, C. A., Giesel, J. T., Anderson, J. F., Emerson, S. S. 1990. Photoperiod-induced changes in metabolic response to temperature in *Drosophila melanogaster* Meigen. *Funct. Ecol.* 4:41–46

90. Langley, P. A., Pimley, R. W., Mews, A. R., Flood, M. E. T. 1978. Effect of diet composition on feeding, digestion, and reproduction in *Glossina morsitans. J. Insect Physiol.* 24:233–38

91. Lees, A. D. 1955. *The Physiology of Diapause in Arthropods.* Cambridge: Cambridge Univ. Press

92. Lees, A. D. 1964. The location of the photoperiodic receptors in the aphid *Megoura viciae* Buckton. *J. Exp. Biol.* 41:119–33

93. Lees, A. D. 1966. The control of polymorphism in aphids. *Adv. Insect Physiol.* 3:207–77

94. Lees, A. D. 1967. The production of the apterous and alate forms in the aphid *Megoura viciae* Buckton, with special reference to the role of crowding. *J. Insect Physiol.* 13:289–318

95. Lees, A. D. 1983. The endocrine control of polymorphism in aphids. In *Endocrinology of Insects,* ed. R. G. H.

Downer, H. Laufer, pp. 369–77. New York: Liss

96. Lees, A. D. 1984. Parturition and alate morph determination in the aphid *Megoura viciae. Entomol. Exp. Appl.* 35:93–100

97. Lees, A. D. 1986. Some effects of temperature on the hour glass photoperiod timer in the aphid *Megoura viciae. J. Insect Physiol.* 32:79–89

98. Ludwig, D., Fiore, C. 1960. Further studies on the relationship between parental age and the life cycle of the mealworm, *Tenebrio molitor. Ann. Entomol. Soc. Am.* 53:595–600

99. Ludwig, D., Fiore, C. 1961. Effects of parental age on offspring from isolated pairs of the mealworm, *Tenebrio molitor. Ann. Entomol. Soc. Amer.* 54:463–64

100. MacKay, P. A., Wellington, W. G. 1977. Maternal age as a source of variation in the ability of an aphid to produce dispersing forms. *Res. Pop. Ecol.* 18:195–209

101. Mansingh, A., Smallman, B. N. 1966. Photoperiod control of an "obligatory" pupal diapause. *Can. Entomol.* 98:613–16

102. Mansingh, A., Smallman, B. N. 1967. Effect of photoperiod on the incidence and physiology of diapause in two saturniids. *J. Insect Physiol.* 13:1147–62

103. Mansingh, A., Smallman, B. N. 1971. The influence of temperature on the photoperiodic regulation of diapause in saturniids. *J. Insect Physiol.* 17:1735–39

104. Marcovitch, S. 1924. The migration of the Aphididae and the appearance of the sexual forms as affected by the relative length of daily light exposure. *J. Agric. Res.* 27:513–22

105. Masaki, S. 1973. Climatic adaptation and photoperiodic response in the band-legged ground cricket. *Evolution* 26:587–600

106. Deleted in proof

107. Masaki, S., Shirado, I., Nagase, A. 1987. Tropical, subtropical and temperate life cycles in ground crickets. *Insect Sci. Appl.* 8:475–81

108. Maslennikova, V. A. 1972. The effect of hormonal balance of diapausing insects on their reactivation. See Ref. 52a, pp. 229–41

109. Matsutani, K., Sonobe, H. 1987. Control of diapause-factor secretion from the suboesophageal ganglion in the silkworm *Bombyx mori:* The roles of the protocerebrum and tritocerebrum. *J. Insect Physiol.* 33:279–85

110. Matthee, J. J. 1951. The structure and physiology of the egg of *Locustana pardalina* (Walk.). *Sci. Bull. Dep. Agric. For. Un. S. Afr.* No. 316. 83 pp.
111. McGinnis, K. M., Brust, R. A. 1982. The effect of photoperiod and temperature on the induction of embryonic diapause in *Aedes togoi* (Theobald) (Diptera: Culicidae) and overwintering. *Proc. Entomol. Soc. Manitoba* 38:23
112. McHaffey, D. G., Harwood, R. F. 1970. Photoperiod and temperature influences on diapause in eggs of the floodwater mosquito, *Aedes dorsalis* (Meigen) (Diptera: Culicidae). *J. Med. Entomol.* 7:631–44
113. McNeil, J. N., Rabb, R. L. 1973. Physical and physiological factors in diapause initiation of two hyperparasites of the tobacco hornworm, *Manduca sexta. J. Insect Physiol.* 19:2107–18
114. Messina, F. J. 1990. Alternative life-histories in *Callosobruchus maculatus:* Environmental and genetic basis. In *Proc. 2nd Int. Symp. Bruchids and Legumes,* ed. K. Gujii. Amsterdam: Junk
115. Mills, A., Hartmann-Goldstein, I. 1985. Maternal age, development time, and position effect variegation in *Drosophila melanogaster. Genet. Sel. Evol.* 17: 171–78
116. Mittler, T. E., Nassar, S. G., Staal, G. B. 1976. Wing development and parthenogenesis induced in progenies of kinoprene-treated gynoparae of *Aphis fabae* and *Myzus persicae. J. Insect Physiol.* 22:1717–25
117. Monteiro, L. S., Falconer, D. S. 1966. Compensatory growth and sexual maturity in mice. *Anim. Prod.* 8:179–92
118. Mori, A., Oda, T., Wada, Y. 1981. Studies on the egg diapause and overwintering of *Aedes albopictus* in Nagasaki. *Trop. Med. Nagasaki* 23:79–90
119. Mousseau, T. A., Roff, D. A. 1987. Natural selection and the heritability of fitness components. *Heredity* 59:181–97
120. Mousseau, T. A., Roff, D. A. 1989. Adaptation to seasonality in a cricket: patterns of phenotypic and genotypic variation in body size and diapause expression along a cline in season length. *Evolution* 43:1483–96
121. Muller, H. J. 1970. Formen der dormanz bei insekten. *Nova Acta Leopoldiana* 35:7–27
122. Deleted in proof
123. Nijhout, H. F., Wheeler, D. E. 1982. Juvenile hormone and the physiological basis of insect polymorphisms. *Q. Rev. Biol.* 57:109–33

124. Parrish, D. S., Davis, D. W. 1978. Inhibition of diapause in *Bathyplectes curculionis,* a parasite of the alfalfa weevil. *Ann. Entomol. Soc. Am.* 71:103–7
125. Parsons, P. A. 1964. Parental age and the offspring. *Q. Rev. Biol.* 39:258–75
126. Pinger, R. R., Eldridge, B. F. 1977. The effect of photoperiod on diapause induction in *Aedes canadensis* and *Psorophora ferox* (Diptera: Culicidae). *Ann. Entomol. Soc. Am.* 70:437–41
127. Prokopy, R. J. 1968. Influence of photoperiod, temperature, and food on initiation of diapause in the apple maggot. *Can. Entomol.* 100:318–29
128. Raina, A. K., Bell, R. A. 1974. Influence of dryness of larval diet and parental age on diapause in the pink bollworm, *Pectinophora gossypiella* (Saunders). *Environ. Entomol.* 3:316–18
129. Rensch, B. 1959. *Evolution above the Species Level.* New York: Columbia Univ. Press
130. Reznick, D. 1981. "Grandfather effects": the genetics of offspring size in the mosquito fish *Gambusia affinis. Evolution* 35:941–53
131. Reznick, D. 1982. Genetic determination of offspring size in the guppy *(Poecilia reticulata). Am. Nat.* 120:181–88
132. Ring, R. A. 1967. Maternal induction of diapause in the larvae of *Lucilia caesar* L. (Diptera: Calliphoridae). *J. Exp. Biol.* 46:123–36
133. Ring, R. A. 1967. Photoperiodic control of diapause induction in the larvae of *Lucilia caesar* L. (Diptera: Calliphoridae). *J. Exp. Biol.* 46:117–22
134. Riska, B. 1989. Composite traits, selection response, and evolution. *Evolution* 43:1172–91
135. Riska, B., Atchley, W. R. 1985. Genetics of growth predict patterns of brain size evolution. *Science* 229:668–71
136. Riska, B., Atchley, W. R., Rutledge, J. J. 1984. A genetic analysis of targeted growth in mice. *Genetics* 107:79–101
137. Riska, B., Rutledge, J. J., Atchley, W. R. 1985. Covariance between direct and maternal effects in mice, with a model of persistant environmental influences. *Genet. Res.* 45:287–97
138. Roach, D. A., Wulff, R. D. 1987. Maternal effects in plants. *Annu. Rev. Ecol. Syst.* 18:209–35
139. Rock, G. C., Yeargan, D. R., Rabb, R. L. 1971. Diapause in the phytoseiid mite *Neoseiulus* (T.) *fallacis. J. Insect Physiol.* 17:1651–59
140. Rockey, S. J., Miller, B. B., Denlinger,

D. L. 1989. A diapause maternal effect in the flesh fly, *Sarcophaga bullata:* Transfer of information from mother to progeny. *J. Insect Physiol.* 35:553–58

141. Rockey, S. J., Denlinger, D. L. 1986. Influence of maternal age on incidence of pupal diapause in the flesh fly, *Sarcophaga bullata*. *Physiol. Entomol.* 11:199–204

142. Rockstein, M. 1957. Longevity of male and female houseflies. *J. Gerontol.* 12:253–56

143. Roff, D. A. 1986. Predicting body size with life history models. *Bioscience* 36:316–23

144. Roff, D. A., Mousseau, T. A. 1987. Quantitative genetics and fitness: lessons from *Drosophila*. *Heredity* 58:103–18

145. Rose, M. R. 1982. Antagonistic pleiotropy, dominance, and genetic variation. *Heredity* 48:63–78

146. Rosenthal, S. S. 1968. Photoperiod in relation to diapause in *Hypera postica* from California. *Ann. Entomol. Soc. Am.* 61:531–34

147. Deleted in proof

148. Ryan, R. B. 1965. Maternal influence on diapause in a parasitic insect, *Coeloides brunneri* Vier. (Hymenoptera: Braconidae). *J. Insect Physiol.* 11:1331–36

149. Sano-Fujii, I. 1979. The effect of parental age and development rate on the production of active form of *Callosobruchus maculatus* (F.) (Coleoptera: Bruchidae). *J. Stored Prod. Res.* 2:187–95

150. Sarai, D. S. 1967. Effects of temperature and photoperiod on embryonic diapause in *Nemobius fasciatus* (DeGeer) (Orthoptera, Gryllidae). *Quest. Entomol.* 3:107–35

151. Sato, T. 1977. Life history and diapause of the white-spotted tussock moth, *Orgyia thyellina* Butler (Lepidoptera: Lymantriidae). *Jpn. J. Appl. Entomol. Zool.* 21:6–14

152. Saunders, D. S. 1962. The effect of the age of female *Nasonia vitripennis* (Walker) (Hymenoptera, Pteromalidae) on the incidence of larval diapause. *J. Insect Physiol.* 8:309–18

153. Saunders, D. S. 1965. Larval diapause induced by a maternally operating photoperiod. *Nature* 206:739–40

154. Saunders, D. S. 1965. Larval diapause of maternal origin: induction of diapause in *Nasonia vitripennis* (Walk.) (Hymenoptera, Pteromalidae). *J. Exp. Biol.* 42:495–508

155. Saunders, D. S. 1966. Larval diapause of maternal origin. II. The effects of

photoperiod and temperature on *Nasonia vitripennis*. *J. Insect. Physiol.* 12:569–81

156. Saunders, D. S. 1966. Larval diapause of maternal origin. III. The effect of host shortage on *Nasonia vitripennis*. *J. Insect Physiol.* 12:899–908

157. Saunders, D. S. 1982. *Insect Clocks.* Oxford: Pergamon 2nd ed.

158. Saunders, D. S. 1987. Maternal influence on the incidence and duration of larval diapause in *Calliphora vicina*. *Physiol. Entomol.* 12:331–38

159. Saunders, D. S., Macpherson, J. N., Cairncross, K. D. 1986. Maternal and larval effects of photoperiod on the induction of larval diapause in two species of fly, *Calliphora vicina* and *Lucilia sericata*. *Exp. Biol.* 46:51–58

160. Schaal, B. A. 1984. Life history variation, natural selection, and maternal effects in plant populations. In *Perspectives in Plant Population Ecology*, ed. R. Dirzo, J. Sarakhan, pp. 188–206. Sunderland, MA: Sinauer

161. Schneiderman, H. A., Horwitz, J. 1958. The induction and termination of facultative diapause in the chalcid wasps *Mormoniella vitipennis* (Walker) and *Tritneptis klugii* (Ratzeburg). *J. Exp. Biol.* 35:520–51

162. Shapiro, A. M. 1976. Seasonal polyphenism. *Evol. Biol.* 9:259–333

163. Shinkaji, N. 1975. Seasonal occurence of the winter eggs and environmental factors controlling the evocation of diapause in the common conifer spider mite, *Oligonychus ununguis* (Jacobi), on chestnut (Acarina: Tetranychidae). *Jpn. J. Appl. Entomol. Zool.* 19:105–111

164. Shroyer, D. A., Craig, G. B. Jr. 1980. Egg hatchability and diapause in *Aedes triseriatus* (Diptera: Culicidae). *Ann. Entomol. Soc. Am.* 73:39–43

165. Simmonds, F. J. 1946. A factor affecting diapause in hymenopterous parasites. *Bull. Entomol. Res.* 37:95–97

166. Simmonds, F. J. 1948. The influence of maternal physiology on the incidence of diapause. *Phil. Trans. R. Soc. London Ser. B* 233:385–414

167. Singh, R. S., Morton, R. A. 1981. Selection for malathion resistance in *Drosophila melanogaster*. *Can. J. Genet. Cytol.* 23:355–69

168. Smith, M. A. H., MacKay, P. A. 1989. Genetic variation in male alary dimorphism in populations of pea aphid, *Acyrthosiphon pisum*. *Entomol. Exp. Appl.* 51:125–32

169. Sonobe, H., Keino, H. 1975. Diapause factor in the brains, subesophageal gan-

glia and prothoracic ganglia of the silk-worm. *Naturwissenschaften* 62:348–49
170. Tauber, C. A., Tauber, M. J., Nechols, J. R. 1987. Thermal requirements for development in *Chrysopa acuta:* a geographically stable trait. *Ecology* 68: 1479–88
171. Tauber, M. J., Tauber, C. A., Masaki, S. 1986. *Seasonal Adaptations of Insects.* New York: Oxford Univ. Press
172. Tauthong, P., Brust, R. A. 1977. The effect of photoperiod on diapause induction, and temperature on diapause termination in embryos of *Aedes campestris* Dyar and Knab (Diptera: Culicidae). *Can. J. Zool.* 55:129–34
173. Taylor, F. 1986. Toward a theory for the evolution of the timing of hibernal diapause. In *The Evolution of Insect Life Cycles,* ed. F. Taylor, R. Karban, pp. 236–57. New York: Springer
174. Deleted in proof
175. Vaz Nunes, M., Kenny, N. A. P., Saunders, D. S. 1990. The photoperiodic clock in the blowfly *Calliphora vicina. J. Insect Physiol.* 36:61–67
176. Via, S., Lande, R. 1987. Evolution of genetic variability in a spatially heterogeneous environment: effects of genotype-environment interaction. *Genet. Res. Camb.* 49:147–56
177. Vinogradova, E. B. 1965. An experimental study of the factors regulating induction of imaginal diapause in the mosquito *Aedes togoi* Theob. (Diptera, Culicidae) *Entomol. Rev.* 44(3):309–15
178. Vinogradova, E. B. 1974. The pattern of reactivation of diapausing larvae in the blowfly, *Calliphora vicina. J. Insect Physiol.* 20:2487–96
179. Vinogradova, E. B. 1975. The role of photoperiod reaction and temperature in the induction of diapause in the egg stage in *Aedes caspius caspius* Pall. (Dipt., Culicidae). *Parazitologia* 9:385–92
180. Vinogradova, E. B. 1975. Intraspecific variation in the reactions controlling larval diapause in *Calliphora vicina. Entomol. Rev.* 54(4):11–20

181. Vinogradova, E. B. 1976. Embryonic photoperiodic sensitivity in two species of fleshflies, *Parasarcophaga similis* and *Boettcherisca septenttrionalis. J. Insect Physiol.* 22:819–23
182. Vinogradova, E. B., Zinovjeva, K. B. 1972. Maternal induction of larval diapause in the blowfly, *Calliphora vicina. J. Insect Physiol.* 18:2401–9
183. Waddington, C. H. 1948. The concept of equilibrium in embryology. *Folia Biotheor.* 3:127–38
183. Way, M. J., Hopkins, B. A. 1950. The influence of photoperiod and temperature on the induction of diapause in *Diataraxia oleracea* L. (Lepidoptera). *J. Exp. Biol.* 27:365–76
184. Wellington, W. G. 1976. Applying behavioral studies in entomological problems. *Perspectives in Forest Entomology,* pp. 87–97. New York: Academic
185. Wellington, W. G. 1977. Returning the insect to insect ecology: Some consequences for pest management. *Environ. Entomol.* 6:1–8
186. Wellso, S. G., Adkisson, P. L. 1964. Photoperiod and moisture as factors involved in the termination of diapause in the pink bollworm, *Pectinophora gossypiella. Ann. Entomol. Soc. Am.* 57:170–73
187. Wilson, G. R., Horsfall, W. R. 1970. Eggs of floodwater mosquitoes. XII. Installment hatching of *Aedes vexans* (Diptera: Culicidae). *Ann. Entomol. Soc. Am.* 63:1644–47
188. Wylie, H. G. 1958. Factors that affect host-finding by *Nasonia vitripennis* (Walk.) (Hymenoptera, Pteromalidae). *Can. Entomol.* 90:597–608
189. Yamashita, O., Hasegawa, K. 1985. Embryonic diapause. See Ref. 76a, 1:407–34
190. Zaslavsky, V. A., Umarova, T. Y. 1982. Photoperiodic and temperature control of diapause in *Trichogramma evanescens* Westw. (Hymenoptera, Trichogrammatidae). *Entomol. Rev.* 60 (4):1–12

Annu. Rev. Entomol. 1991. 36:535–60

BIONOMICS OF LEAF-MINING INSECTS

H. A. Hespenheide

Department of Biology, University of California, Los Angeles, California 90024-1606

KEY WORDS: biogeography, herbivory, insect-plant interactions, life history, parasitoids

INTRODUCTION

In recent years, leaf-mining insects have taken a conspicuous place in the ecological literature that is probably disproportionate to the effect of these insects in natural communities. Although a small number of species significantly damage agricultural crops or forest trees, most affect their plant hosts relatively little, either as individuals or as populations. Whereas externally feeding herbivores typically remove the evidence of their damage by internalizing it, leaf-miners leave a more or less conspicuous record of their presence that may persist long after they depart (135). Some mines may even be considered attractive, providing the viewer has the proper aesthetic view. More importantly, the fact that these herbivores live out most of their life histories within the confines of a mine, as well as the persistence of the record of the mine, allows an ecologist to observe or to reconstruct those lives, to measure the effect on the host plant, and to observe associated fauna such as parasitoids.

The ease with which leaf mines can be studied has made them a popular system for ecological research. In North America, the first major descriptive review of leaf-mining insects was that of Needham et al (125). In Europe, the monumental and encyclopedic works of Hering (82, 83) and Fulmek (61) provided a firm base for subsequent ecological work. In the past two decades, interest in leaf-miners has increased, perhaps stimulated by Opler's classic papers on the ecology of leaf-mining Lepidoptera of California oaks [(136,

535

0066-4170/91/0101-0353$02.00

137) themselves outgrowths of studies of Powell (152)], and by Askew's & Shaw's (4) study of parasitoid communities in England. Other major workers on natural systems of leaf-miners include J. H. Lawton in England and D. Simberloff, T. L. Bultmann, and S. H. Faeth in the United States. In general, this research has focused on a small number of major systems: oaks in California, Arizona, Florida, and England; hollies in North America and England; and beech in Europe. Agricultural research other than that on *Liriomyza* (142) includes the classic study by Taylor (182) on the biological control of the coconut palm leaf-miner in Fiji and, more recently, studies on the alfalfa blotch leaf miner *Agromyza frontella* (44) and on *Lithocolletis* species on apple (151).

This review focuses on studies that are more conceptual than descriptive in approach, both for natural populations or communities and for economically important species. I have not attempted to include studies of the genus *Lyriomyza*, partly because these insects have been recently reviewed (142) and partly because their polyphagy is unusual among leaf-miners, nor have I considered the needle-miners of gymnosperms. Taxonomic treatments are included only when they include useful bibliographies or discussion of host-plant relationships. Most of the included studies were done on North American or western European Lepidoptera or Coleoptera, and the latter are somewhat emphasized because of my familiarity with them. This paper first surveys the taxa that exhibit this mode of feeding and then discusses the relationships of leaf-miners with their hosts, emphasizing the factors that determine the diversity of miner faunas and the components and consequences of leaf selection. A review of mortality factors emphasizes the roles of intraspecific competition, leaf abcission, and parasitoids, the last of which has become a separate object of research. The paper concludes with a brief review of economic studies.

TAXA OF LEAF-MINERS

Hering (82, 83) and Needham et al (125) have done general reviews of the taxa of leaf-mining insects. These give many details of host plant taxa and descriptions of mines and life histories, and will not be summarized here. Although all of the four major orders Diptera, Lepidoptera, Coleoptera, and Hymenoptera have leaf-mining taxa, the greatest diversity of forms and number of species are in the Lepidoptera, then in the fly family Agromyzidae, and relatively fewer in the Coleoptera and Hymenoptera: for Britain and a fauna of nearly 700 species (63) the proportions are 57% Diptera, 33% Lepidoptera, 8% Coleoptera, and 3% Hymenoptera. This overall pattern is probably maintained in most localities (e.g. 4, 63) but varies from habitat to habitat (3, 63) (see below) and shows some change with geography (see below), perhaps especially in the relative importance of the Coleoptera and the Agromyzidae.

Powell (152) recently reviewed the feeding methods and host plants of the Lepidoptera. Much of the recent ecological work on leaf-miners has been done on Lepidoptera, as will be seen below. Major families include the Cosmopterygidae (124, 127, 165), Gelechiidae (65, 66), Gracillariidae (5, 36, 115, 123, 126, 139, 151, 186), Heliozelidae (88, 110), Incurvariidae (120, 185), Lyonetiidae (60, 132, 133, 164), and Nepticulidae (144).

Beetle leaf-miners are concentrated in the three families Buprestidae, Chrysomelidae, and Curculionidae. In the Buprestidae, miners are placed in the subfamily Trachyinae (187) and are primarily tropical in distribution (see below) and poorly known (28, 38, 173, 184, 188). Most chrysomelid leaf-miners are also tropical or subtropical and concentrated in several tribes of the subfamily Hispinae (116, 168, 169) (another significant portion of the subfamily are external feeders on monocots). Weevils include the family Rhynchitidae, some members of which mine fallen leaves (69), as well as several small subfamilies of the Curculionidae that mine living leaves. The Camarotinae, Prionomerinae, and Tachygoninae (1, 98) are largely tropical; the Rhynchaeninae are north temperate in distribution (2), among which the weevil *Rhynchaenus fagi* is one of the most-studied leaf-mining insects (11–13, 128, 155).

Leaf-mining flies are concentrated in the Agromyzidae, but also occur in the Anthomyiidae (64), Drosophilidae (31–34), and Ephydridae (170, 171), among others. Agromyzids have been extensively studied, perhaps because of their temperate distribution, but also because of their economic importance (142). The alfalfa blotch leaf-miner, *A. frontella,* is one of the most-studied leaf-miners (67, 77–80, 122, 156–159), as are *Phytomyza* species on *Ilex* (76, 92, 95, 96, 141, 147–150) and *Ranunculus* species (177, 181). Hymenoptera that mine leaves are sawflies of the family Tenthredinidae and have been infrequently studied ecologically (183).

HOST PLANT INTERACTIONS

Host Specificity and Species Diversity

As endophytic herbivores, leaf-miners have generally narrow host preferences. Most are monophagous, a smaller proportion oligophagous, and only a few are polyphagous [*Liriomyza* (142)]. This pattern is shown by the agromyzids of British Umbelliferae (105), for all leaf-miners on British trees (29), and for all leaf-miners on California oaks (136), but probably holds for other faunas as well (H. A. Hespenheide, unpublished data). Species that use more than one species of host usually occur on other members of the same plant genus, and the few recorded shifts of plant hosts among leaf-miners involve congeneric (9) or confamilial plants (9, 146, 191). Species of *Cameraria* feeding on evergreen instead of deciduous *Quercus* differ in that species that feed on evergreens show greater host specificity than species that feed on deciduous hosts (138, 139).

In the Buprestidae, leaf-miners tend to be associated with more modern plant groups compared to wood-boring subfamilies and are therefore perhaps most recently evolved (187), although the genus *Neotrachys* occurs on ferns (85). In a comprehensive survey of taxon-taxon host relationships, Powell (152) showed that evolutionary radiations in microlepidoptera were based on larval feeding niches rather than on evolutionary relationships among plant hosts, despite a bias in data (only the Holarctic fauna is well known) that might tend to show specialization. Although some buprestid genera are apparently restricted to only a few plant families (*Taphrocerus* on sedges and palms, *Hylaeogena* on Bignoniaceae), others use several unrelated families in the manner of microlepidoptera (H. A. Hespenheide, unpublished data).

Southwood (167) first pointed out that the number of herbivores on British trees correlated with the frequency of fossil records of the plants, after which Strong (174) showed that present-day distributions were an even better predictor of fauna size, based on the analogy of host plants as islands (93). Similarly, Opler (136) showed that the number of species of leaf-mining insects on California oaks was very closely related to the geographic area occupied by the oaks. When just the leaf-mining component of the faunas of British trees was analyzed (29), only 19% of the variation could be explained by area, whereas an additional 23% could be accounted for if the number of species in each plant genus were considered. A subsequent analysis (62) showed that if plant height and the number of other genera in a plant's order were considered, 69% of the variation could be explained and area ceased to have a significant effect. A survey of agromyzids associated with British Umbelliferae (105) found that a combination of geographic area, plant size, leaf form, and occurrence in aquatic habitats explained about 50% of the variation in size of the fauna. When local abundance and the number of occupied habitats were considered (58), addition of the latter explained a total of 61% of variation. Lawton and others (35, 175) have studied the herbivore faunas of bracken *(Pteridium aquilinum)* at different geographic localities. These faunas, including leaf-miners, are of a size predicted by the plant's abundance, but differ in guild representation, so that several species of miners occur in England and Papua New Guinea but few in New Mexico and South Africa.

Biogeographic Patterns of Diversity

Most basic ecological studies have been carried out in temperate zone localities (England, California) or at the margins of the subtropics (Florida,

Arizona), although a few economically important tropical species have been studied (see below). Leaf-miners are common in the tropics and it is clear that these faunas differ from those in temperate zones. Powell (152) quotes Opler as having found leaf-miners on half of 102 tree species in Costa Rica. Four leaf-miners—two lepidopterans, one agromyzid, one buprestid—were among the herbivores of *Byttneria aculeata* at the La Selva Biological Station, Costa Rica (86), and a hispine beetle uses this plant in Panama (H. A. Hespenheide, unpublished data). Although beetles are a minor component of temperate zone faunas, they are much more numerous in the tropics. Only a few descriptive studies of tropical leaf-miners have been published (59, 81, 98–101). To date, I have collected or reared 130 species of leaf-mining beetles at La Selva (71 Buprestidae, 41 Hispinae, 18 Curculionidae), compared to 53 species known in all of Britain (63) and 98 from the United States (19). A survey of leaf-mining beetles associated with the plant genus *Cecropia* in central Panama (H. A. Hespenheide & N. Boren, unpublished data) showed seven buprestids in three genera, two hispines, and two weevils of the genus *Tachygonus*.

Figure 1 shows that the proportion of leaf-mining species of Buprestidae is higher in the Neotropics compared to both the Northern and Southern temperate zones. The figure is based only on the two subfamilies Agrilinae and Trachyinae; if all members of the family were included, the trend would be more striking because the other wood-boring subfamilies are disproportionately rare in the tropics. Leaf-miners are the dominant buprestids in Neotropical wet forests: at La Selva they comprise 80% of the subfamilies in Figure 1 and 73% of all buprestids. A geographic analysis of the leaf-mining buprestids at the generic level (87) shows finer scale changes in faunal composition. These trends include ecological characteristics; for example, although all members of the genus *Taphrocerus* from La Selva and northward use members of the Cyperaceae as hosts, several species in central Panama use palms (H. A. Hespenheide, unpublished data). Most buprestid species in the southwestern United States (90%) and Mexico (79%) have a naked pupa that remains in the leaf, whereas in Costa Rica and Panama, a greater proportion (30 and 43%, respectively) construct a pupal case, some of which are cut out of the leaf, perhaps because of higher rates of ant predation (H. A. Hespenheide, unpublished data). Parasitoid faunas also differ (see below). Much additional research is needed on leaf-miners in other parts of the world before a general picture is complete.

Host Plant Defenses

Pre-ovipositional host-plant defenses are those factors that affect a female miner's choice of plants on which to oviposit. Only a few of these have been studied. Leaf-color polymorphism has been shown to affect herbivory by

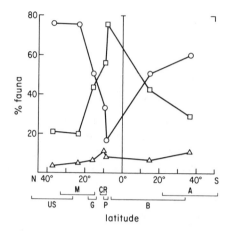

Figure 1 Latitudinal changes in the relative proportions of larval ecologies (*circle,* wood-borers; *square,* leaf-miners, *triangle,* soft stem borers) among members of the Buprestidae (Agrilinae, Coraebinae, and Trachyinae only). The reversal of the trend south of the Equator suggests an ecological rather than a historical process is involved. US = United States, *n* (number of species in fauna of each country) = 212; M = Mexico, *n* = 322; G = Guatemala, *n* = 113; CR - Costa Rica, *n* = 100; P = Panama, *n* = 139; B = Brazil, *n* = 949; A = Argentina, *n* = 139; data from Reference 19.

leaf-miners on *Byttneria* (166); mine density varies with frequencies of green and white-variegated leaf morphs and is lower on the variegated morph. The mechanism is unclear and possibly habitat-dependent because morph frequencies and rates of herbivory covaried. Leaf trichomes can interfere with access to the leaf surface; gracillarids were less common on pilose than glabrous individuals of *Arbutus xalapensis* (48). *Tildenia* on horsenettle *(Solanum carolinense)* is apparently forced to be endophytic and suffers higher parasitism because of dense trichomes, whereas another *Tildenia* on ground cherry *(Physalis heterophylla)* with sparser trichomes is more vagile and can escape parasitoids (66). Size (see below) and dissection (105) of leaves are also important.

Plant chemistry almost certainly provides cues for location of the host as well as poses problems for the larvae after oviposition and hatching. One method of dealing with chemical defenses might be to avoid tissues with defensive compounds and to mine cell layers without them (56); however, only one of 18 species studied on oak in Florida restricted mining to cell layers low in tannins (53). *Phytomyza ilicfolia* mined middle cell layers of palisade mesophyll of *Ilex opaca,* even though the layers were high in saponins, and avoided abaxial cells with crystals (96). Latex is a defense that requires specialized methods of response by herbivores; externally feeding species typically cut the major veins (45). Several Neotropical Buprestidae mine latex-producing plants, especially members of the genus *Leiopleura* on *Ficus* and on various Apocynaceae (H. A. Hespenheide, unpublished data). Species of *Pachyschelus*, including *P. aversus*, feed on species of *Sapium* and force the latex out onto the surface of the leaf above the mine (57; G. Vogt & H. A. Hespenheide, unpublished data).

Leaf venation can affect mining. Nielsen (128) described a "coincidence factor" of mortality in which early instar miners could not complete mining because of developmental increase in scleratization of crossveins of beech leaves. The development of the weevil must be in synchrony (i.e. coincide) with that of the beech leaf so that early-instar larvae can mine younger leaves; cool weather delays oviposition and subsequent larval development past a critical point in leaf hardening. Likewise, *Phytomyza* may time its emergence to coincide with newly flushed leaves of *Ilex* that are easier for adult flies to feed and oviposit in (149). Other leaf-miners that use younger leaves may also do so for reasons of leaf structure, and even under normal conditions miners may be constrained by patterns of venation.

Feeny (56) noted that 23 of 26 leaf-miners of *Quercus robur* feed in late summer when tannins are in high concentration. West (190) showed experimentally that survivorship and pupal weight are actually greater for *Phyllonorycter harrisella* on undamaged leaves of *Q. robur* in spring than on older leaves in summer. Larval survivorship decreases with increased leaf damage, apparently because of wound-induced responses of plants. Late-season mining is hypothesized to be an adaptation to avoid interactions with externally feeding caterpillars producing wounds; i.e. feeding patterns were not a response to leaf chemistry per se, but to avoid asymmetrical interspecific competition. Reciprocally, mined leaves of *Betula pendula* were consistently avoided by externally feeding caterpillars when given a choice of damaged or undamaged leaves (71), although significant mortality to *R. fagi* is caused by externally feeding larvae that open mines (128). Alternatively, competition between leaf-miners and externally feeding larvae may (50, 51) or may not (72) be mediated by parasitoids. Plants of *Cardamine cordifolia* that were experimentally stressed had increased herbivory by the leaf-miner *Scaptomyza nigrita* (S. M. Louda & S. K. Collinge, unpublished data). In two other situations, one experimental (150) and one natural (8), stressed trees experienced leaf-miner outbreaks after refoliation.

Plants also engage in active postovipositional defenses. Jarrah trees *(Eucalyptus marginata)* are polymorphic for resistence to the leaf-miner *Perthida glyphopa* (118, 185); although one individual was not oviposited in for chemical/anatomical reasons, most others extruded or killed eggs by cell growth in the vicinity. Leaf abcission is another common plant response to leaf-mining, discussed below as a major mortality factor.

LIFE HISTORY AND ECOLOGY

Adult Reproductive Behavior

Mating behavior of adult leaf-miners has not been extensively studied. Leaf-miner matings may well be single, at least for female Lepidoptera (151), but beetle matings are known to be multiple (97, 173). Protracted copulation is known for both groups (97, 151, 173), and postcopulatory escort behavior has been described for the hispine *Odontota dorsalis* (97).

Oviposition behavior has been studied primarily for the agromyzids (see also 120) *Phytomyza* on *Ranunculus* spp. in relation to egg and larval density (177) and *Agromyza* on alfalfa (159). Both agromyzids puncture the leaves with the ovipositor for both oviposition and for feeding; *Agromyza* females mark eggs with an oviposition pheromone (122) that can influence the outcome of larva-larva competition (158). In experimental choice situations, female *Agromyza* rank unexploited leaves above marked leaves or those with many nutrition holes above those with late-instar larvae (159). Members of the buprestid genus *Pachyschelus* place their eggs under small flaps of epidermis and then alter the leaf in the vicinity of the egg by cutting additional flaps without eggs, perhaps to affect leaf chemistry in the vicinity of the egg (H. A. Hespenheide, unpublished data). Egg placement is discussed below.

Fecundity has been measured for relatively few species (5, 36, 115, 123) and depends on food resources for the adults in those cases examined. *Phyllonorycter blancardella* females produced an average of 44–67 eggs when offered sugar solution and 20–29 when offered water only (151). *Liriomyza trifolii* produced 78 eggs when aphid honeydew was available on tomato leaflets and only 12 on leaflets without honeydew (195). *Phyllocnistis* is reported to feed on foliar nectaries of *Populus tremuloides* in British Columbia (36), and agromyzids have been seen feeding at extrafloral nectaries on *Byttneria* in Costa Rica (H. A. Hespenheide, unpublished data). Most leaf-miners lay eggs singly, but a few species lay eggs in a group from which the larvae will form a communal mine. The most common clutch size for the fly *Pegomya nigritarsis* (64) was also the most productive. In contrast, the most frequent clutch size for *Pachyschelus psychotriae* on *Connarus* in Panama was the least productive (H. A. Hespenheide & C. Kim, unpublished).

Adult longevity varies among species depending on life history; for example, buprestids of the genus *Taphrocerus* overwinter as adults (28) as does *R. fagi* (11, 12). Longevity also depends on access to food in some species that have been studied: *P. blancardella* females lived 8–14 days with access to sugar solution and 4–7 days with water only (151); *L. trifolii* females lived 2.4 days on glass, 4.3–5.1 days on tomato leaflets without honeydew, and 9.7 days when aphid honeydew was available (195). *R. fagi* uses a small set of adult hosts other than *Fagus* at different seasons (13). Feeding by adults can be extensive and damaging to plants—both hosts and nearby plants (146, 192). It has been estimated that *A. frontella* females make over 3,700 feeding perforations, equivalent to 1.1 cm^2 per female in damage to alfalfa (77).

Although studies of leaf-miner mortality have concentrated on larvae (see below), adults are also susceptible to predation and have evolved antipredator defenses. Hispine chrysomelids are often involved in mimicry complexes, usually with beetles in the Lycidae or Lampyridae (14), that are probably Mullerian in character. Leaf-mining Buprestidae of the genera *Taphrocerus* and *Leiopleura* have been hypothesized to mimic flies (84); some members of the genera *Hylaeogena* and *Pachyschelus* have a rotund form and patterns of setae resembling that of coccinellids (H. A. Hespenheide, unpublished data); and others share color patterns with Chrysomelidae of subfamilies other than the Hispinae. Aiello & Vogt (1) report an unusual double mimicry in which the leaf-mining weevil *Tachygonidius dasypus* resembles stingless bees in flight and assumes a fungus-like position on a leaf when disturbed. Some tropical microlepidoptera probably also participate in mimicry complexes.

Larval Ecology

DISTRIBUTION OF MINES AND MINERS   The types of mines have generally not been considered for their influence on the ecology of their inhabitants, except for that of tentiform mines on parasitoids (20). Extreme types are the blotch and serpentine mines, but many variations exist (82, 143). Whether variants are means of exploiting hosts or avoiding parasitoids remains to be investigated.

The distribution of leaf-miners has been discussed at several different spatial scales: between different habitats, among and within plants within a single habitat, and among and within leaves of a single plant. Over a successional gradient, major faunal components changed between annuals, which were dominated by agromyzids, and perennials, which were dominated by lepidopterans; a larger number of species were on perennials, and earlier plant stages had a greater density of miners (63). Differences in the fauna of *Pteridium* between different habitats were more complex (121). Within habitats, miners are unequally distributed among trees (22, 123, 185), in some cases preferring younger trees over older ones (115). Densities often decrease from edges to the interior of a habitat (53, 124), in one case because of parasite pressure (170). Miners on isolated plants may have higher survivorship, but show no increased densities and derive most of their population from immigration (39, 55).

Within a plant (usually a tree), miners may prefer upper portions of the canopy (131) or, more usually, lower portions (31, 90, 92, 115, 120, 121, 123, 131, 145, 185), or show no preference (145). Such preferences are usually presumed to be for sun (26, 31, 34, 124) or shade (26), but shaded plants may show more damage than preferred plants in the sun (32), and miners may prefer sun leaves but survive better in the shade (26). In *Brachys tessellatus,* beetles and mines are more common in the upper, sunnier quadrants of the canopy earlier in the season and in lower, more shaded portions later in the year (184). Peripheral leaves may be more heavily mined than interior ones (24, 165). The dispersion of mines among leaves is often aggregated (10, 21, 123, 147) rather than uniform (22, 36), but this varies among species (121). Leaves chosen for oviposition are usually undamaged (165) (see above) and may be either younger [4 of 18 species (53), some (63, 183, 185)] or older [14 of 18 species (53), some (63, 127)]. Leaf size may not influence preference (23, 50) but usually does; smaller leaves are preferred by smaller species (23), or larger leaves are simply preferred (88, 165). Larger leaves may increase probability of survival of single larvae (23), but may also increase density of mines (25, 124, 183); communal mines have been suggested to be an adaptation for exploiting larger leaves (25). Mines may be more frequent toward the base of the leaf (128) or toward the apex (10), or be randomly distributed (65, 123). When more than two mines occur on a leaf, they are often on opposite sides of the midvein (10, 36, 172). Mines may be superficial or full-depth (82), and the two sides of a leaf may serve as discrete habitats for superficial miners (36). Mines of different species may co-occur on leaves more frequently than expected by chance (21) without apparent competition, perhaps because of preferences for different portions of the leaf (130).

PATTERNS OF SURVIVORSHIP AND MORTALITY   Studies of survivorship in leaf-mining insects usually means larval survivorship. Mortality of eggs is infrequently estimated (41, 151, 184), is often underestimated in death-assemblage studies (38), and may result from parasitoids (below), predation (151), or unknown factors (155). Studies of survivorship and mortality have constructed life tables (115, 155, 162) or survivorship curves (5, 38) and often concern themselves with population regulation, either by key factor analysis (41) or with density dependence or independence. Many studies based on death assemblages document a large proportion of mortality due to unknown causes [12–22% (38), 9–33% (55), 17–58% (119), 30–43% (154), 19% (162), 11% (165) (see also 41)]. The major identifiable mortality factors are parasitoids, intraspecific competition, leaf abcission, predators, host plant defenses [plant chemical resistance (190), developmental inedibility (128, 155); see above], and abiotic factors. Over-winter mortality is usually a combination of terrestrial predators and abiotic factors (36, 37, 151). These will be considered in turn.

Parasitism   Parasitoids on leaf-miners have been studied both as a source of mortality to miners and for their own sake (see below). Parasitoids may kill leaf-miners by oviposition and subsequent feeding by the parasitoid larva, as well as by feeding of the adult parasitoid on the larval host. Host feeding by adults is difficult to separate from plant antibiosis (6, cf 190), but has been observed (5) or measured in the field (154) or in laboratory studies at times to be a more frequent cause of host death than oviposition (178, 181).

Rates of parasitism are frequently measured (115, 119, 134, 154, 165) and a variety of patterns have been observed that are in some cases contradictory. For example, rates of parasitism by *Holcothorax* and *Sympiesis* on *Phyllonorycter ringoniella* (162) and by pupal parasitoids on *Phytomyza* (76) were density independent, whereas rates of parasitism by parasitoids on *Phyllonorycter* sp. (123) and on *Stilbosis* (124) on oak and by *Chrysocharis* on *Phytomyza* larvae (76) were density dependent. Exclosure of most parasites and predators led to increased densities of *Cameraria* on oak, which implied that the parasites and predators probably regulate *Cameraria* populations (54). In some studies, rates of parasitism are higher later in the season (65, 123, 185), while, in others, they have been higher on the first than on the second host generation (5, 106), although either overall mortality was greater in the second generation (5) or parasitism was higher again in the third generation (106). Rates of parasitism have been shown to vary with leaf size (37). Only one study has discussed encapsulation of endoparasitoid eggs as a host defense mechanism (5).

Intraspecific competition   Intraspecific competition can either take the form of cannibalism, in which one larva kills a conspecific that occurs with it in a mine *(interference competition),* or of preempting the conspecifics' use of the leaf by mining it first *(exploitation competition).* More rarely, feeding by adults on a leaf may preclude mining by a larva (128, 157). On the other hand, because intraspecific competition depends on the presence of at least two mines in a leaf and then on either large numbers of mines (185) or large mines relative to the size of the leaf (67), it may not occur at all when densities of mines are low, or be relatively minor as a source of mortality (5, 123, 126). For example, as many as seven *R. fagi* emerged from a single *Fagus* leaf with little larva-larva competition (128). Larval survivorship did not decrease because of intraspecific competition when exclosure of parasites and predators led to increased densities of *Cameraria* (54). All three forms of competition occur in *A. frontella* (156–158), and the relative importance of the two forms of larva-larva competition can be predicted from larval density and food availability (43).

Exploitation competition depends on multiple mines in a leaf as well as their size and distribution on the leaf (see above). Confluent mines of *Labdia* occurred on 17% of multiply mined *Acacia* phyllodes and on less than 8% of all mined phyllodes (127). Up to 210 mines of *Perditha* can occur on Jarrah leaves; 64 insects matured in one leaf with 178 mines, but the usual maximum is 40–60 (185). Some leaves were mined so intensely that no larvae matured and the leaf was totally consumed. A number of studies report either lower survival (67) or lower pupal weight (25, 183), or both (147, 171), as the number of mines per leaf increases.

Cannibalism reportedly occurs in several species other than *A. frontella,* although it has been reported to not occur in other species (165). Overall mortality from cannibalism was 11% for *Phyllonorycter* on apple in Japan and increased with larval density to nearly 50% for approximately 25 larvae/leaf (162). Larval death from interference accounted for 3–12% of mortality and was density dependent for *Lithocolletis* on oak (10) and for *Agromyza* (156). Mortality from larval competition was 53% for *Lithocolletis salicifoliella* on *P. tremuloides* (115).

Leaf abcission   Owen (141) first suggested that holly trees *(Ilex aquifolia)* regulate the frequency of leaves mined by *Phytomyza ilicis* by shedding them earlier than normal and thereby regulate the abundance of the miner. Faeth et al (52, but cf. 94) showed that mined leaves abcised significantly early in three oak species and constituted one third of the mortality of two species of miners. Subsequent work has shown a variety of effects of miners on abcission and abcission on miner mortality, but most researchers feel abcission is a general response to damage and not an attempt to regulate miner numbers (153).

Rhynchaenus accelerated leaf abcission on beech, while *Phyllonorycter* spp. did not accelerate abcission on holm oak *(Quercus ilex)* (153), but abcission was affected by the within-tree distribution of mines. *Vaccinium macrocarpon* leaves mined by *Coptodisca negligens* were abcised earlier than undamaged leaves, those with two mines earlier than those with one (112). Abcission rate was about 10% higher for *Quercus geminata* leaves mined by *Stilbosis quadricustatella* (165). Several other studies report early abcission of mined leaves (36, 60, 88, 107, 185) or differential susceptibility of leaves to abcission based on position on tree (24, 92) or the pattern (172) or density of mining (10) on leaves.

Timing of abcission determines the effect on herbivore demography. Delayed (four to five months) post-hatch development of larval *P. ilicis* synchronizes adult emergence with leaf flush and postpones early leaf abcission of infested leaves (92). *Phytomyza ilicicola* (95, cf 147) and *A. frontella* (79) increased leaf abcission, but abcission did not always decrease miner survivorship because leaves dropped after the flies pupated and may even have reduced parasitism for *Phytomyza* since parasitoids did not search fallen leaves (95). Fourth-instar larvae of *Coptodisca* on *Vaccinium* had 90% survival from abcised leaves (112). Leaf abcission accounted for < 3% of mortality for *Phyllonorycter* spp. on beech and holm oak (154). On the other hand, early abcission by *Q. geminata* was the largest source of mortality for *Stilbosis* larvae that were not close to pupation (165), for *Brachys ovatus* on *Quercus virginiana* (38), and for *Lithocolletis quercus* on *Quercus calliprinos* (10).

Englebrecht et al (47) pointed out that leaf-miners could be found completing their life cycle in "green islands" of otherwise senescent leaves and that these areas of the leaf contained high levels of cytokinins, perhaps secreted by the larvae (46) to maintain their food source until pupation. Such green islands allowed some postdrop survival for *Phyllonorycter* spp. on beech and holm oak following leaf abcission (154). Green islands have also been reported for *Stilbosis juvantis* on oak (50) [but not for *S. quadricustatella* (165)], and for other species of miners and plants (for reviews, see 82, 94).

Predators and abiotic factors   Predators other than parasitoids have been given less attention, almost certainly because the hit-and-run nature of predation makes it difficult to observe. Invertebrate predators are probably primarily ants (119). Predation rates on *Eriocraniella* sp. mining younger leaves of *Quercus nigra* were only 4% when ants were exclosed, compared to rates of 25–42% on control trees (49). Although Hinkley (89) documented a drastic reduction in parasitoids following dieldrin application on sweet potatoes to control weevils, he attributed the subsequent outbreaks of two species of lepidopteran leaf-miners to elimination of ant predation. Krombein (103) observed the vespid wasp *Leptochilus tylocephalus* opening a gracillariid blotch mine on *Galactia volubilis*. The vespid *Symmorphus canadensis* apparently specializes on leaf-miners, with both Lepidoptera and Coleoptera as prey (104). Numbers of adult agromyzids have been taken from cells of the sphecid wasp *Ectemnius paucimaculatus* (102). Other invertebrate predators include anthocorid (128), nabid, and mirid bugs and mites (151) on trees, carabid beetles on overwintering individuals in fallen leaves (151) or estivating in litter (119), earwigs (119), and spiders feeding on adults (151).

Several studies have documented feeding from mines by birds and demonstrated significant levels of predation, although most have been forced to infer such feeding based on empty mines. Werner & Sherry (189) observed individual Cocos Finches specializing on lepidopteran leaf-miner larvae on *Hibiscus tiliaceus*. Owen (140) reported that mines of *P. ilicis* were significantly denser on leaves of *Ilex aquifolium* that had more prickles per leaf, but that *Parus caeruleus* took a significantly smaller proportion of larvae from these leaves than from leaves with fewer prickles and lower densities of mines. An inverse relationship between predation by *Parus* and density of *Phytomyza* might result from the difference in scale between the foraging of the larger bird and the dispersion of the smaller mines, rather than to numbers of prickles (76). Itamies & Ojanen (91), on the other hand, found that leaves of *Alnus* with greater numbers of mines of *Lithocolletis* spp. had higher predation rates by *Parus* spp. They also suggested that the birds preferred full size mines and thus avoided parasitized larvae that were smaller, and consequently had a greater influence on the populations of the moths. *Parus* species also prey on *Rhynchaenus* larvae (155). Larvae of *Perthida* were found in the stomachs of nine species of birds (119).

Abiotic mortality factors include the *coincidence factor*, described above, in which cool weather delays larval development as the leaf matures. The action of wind or rain weakens and opens mines [accounting for 5% of *Rhynchaenus* mortality (155)]; this may especially be a problem in the wet tropics (G. Vogt, unpublished data). Frost killed 24% of *P. blancardella* pupae overwintering in fallen leaves of apple (151).

Population Behavior

Number of generations per year, generation time, and presence of diapause are obviously interrelated. Temperate species are the only ones studied in detail to date, most of which must undergo a winter season diapause. Whether there is more than one generation per year depends on resource availability. One might expect that species feeding on annual or deciduous plants face temporally condensed resources and produce a single generation, whereas species feeding on perennial or evergreen plants face extended resources and thus produce more generations. However, Opler (138, 139) found that California leaf-miners feeding on evergreen oaks had fewer annual generations, longer larval periods, lower populations, greater host specificity, and were larger in size compared to those using deciduous oaks. He hypothesized that evergreen oaks are better protected chemically (30). These generalizations did not hold for three oaks in Florida that differed in leaf size and persistence: they shared the same pool of miners, had equal densities of miners when leaf sizes were scaled, and had miner densities that were inversely related to leaf nitrogen (53). On the other hand, length of development depends on foliar nitrogen for *Stilbosis* on two species of oak, and lower nitrogen levels lead to a longer larval period (124).

Diapause can occur at any stage with species overwintering as eggs, larvae, pupae, or adults. *Coptodisca* spp. on cranberry and *Chamaedaphne* overwinter as eggs up to nine months (88, 110). Larvae of several species of leaf-miners feeding on evergreen trees exhibit protracted development and continue to mine through the winter, pupating and emerging when new leaves are flushed in the spring (10, 147, 149). Pupae or prepupal larvae are probably the most common diapausing stage. Although several leaf-miners are known to overwinter as adults (28, 36, 115), diapause has been studied physiologically only in *R. fagi* (11). Of 11 leaf-miners on *Quercus agrifolia*, five underwent diapause as an egg and five or six as prepupae or pupae (137).

Within a season, bivoltine species of *Phyllonorycter* had second-generation population densities that were two to six times the first generation densities (5 and references therein, 123). When individual plants refoliate out of synchrony with the rest of the population, leaf-miners that specialize on young foliage have been observed to produce outbreak population levels (150), in one case perhaps by staggering emergence (8).

Few leaf-miner populations have been studied for more than three years. Significant between-year variability in miner densities has been attributed to large year-to-year fluctuations (53). Population cycles similar to those in temperate insects and mammals occur in tropical insects, including coffee leaf-miners (*Leucoptera* spp.) and their parasitoids studied over six years (16–18). Godfray & Hassell (64a), using these data, showed theoretically that host-parasite systems can cycle given appropriate parasite generation times relative to those of their host. No clear population cycles were detected in a 10-year study of *R. fagi* in Ireland (41), although some years saw sustained growth or decline.

PARASITOIDS OF LEAF-MINERS

Faunal Size

Askew & Shaw (7) have pointed out that the parasitoids of endophytic herbivores such as leaf-miners are much more easily studied and more completely known than those of exophytic hosts because researchers can reconstruct ecological events after the fact. Perhaps partly for this reason, the documented parasitoid faunas of British leaf-miners are larger than those for any other herbivore feeding niche, averaging about 10 species per host (3, 75). This relationship also largely holds on a global basis (73), perhaps because leaf-miners are relatively immobile compared to external feeders and are more conspicuous than other endophytic forms such as gall-makers.

Egg parasitoids of leaf-miners have been little studied, although a number of parasitoid taxa are involved and host mortality rates can be high. Mymarids have been reared from eggs of the buprestid genus *Taphrocerus* (173), and up to 70% of *Taphrocerus* eggs are parasitized toward the close of the season (28). Two species of eulophids were reared from *B. tessellatus* on oak and 13% of eggs were parasitized (184). Other egg parasitoids have been reared from *Pachyschelus* (but only 1% of eggs of *Pachyschelus psychotriae* were parasitized; H. A. Hespenheide & C. Kim, unpublished data) and emergence holes have been seen in eggs of hispines (H. A. Hespenheide, unpublished data).

In an analysis of the parasitoids of leaf-miners of British deciduous trees, Askew & Shaw (4) found that most belonged to three subfamilies of the Eulophidae, with a fourth eulophid subfamily parasitizing weevils and pteromalids parasitizing agromyzids. There are fewer ichneumonid than braconid parasites of leaf-miners, perhaps because leaf-miners are too small to be exploited by the relatively larger ichneumonids (163). A broader perspective both in terms of geography and in host taxa would lengthen the list of parasitoid taxa (113). Braconids are often important parasitoids (68, 108, 148), as are pteromalids and chalcidids on tropical Buprestidae and Hispinae, as well as tachinids on Hispinae (H. A. Hespenheide, unpublished data).

Ecological Specialization

Comparison of parasitoid faunas from a variety of miners on a variety of British plants showed that species were shared more frequently on the basis of plant-host taxon than miner taxon (3, 4). Parasites of *Phyllonorycter* spp. on birch (*Betula* sp.) and oak (*Q. robur*) could be divided in two ecological groups differing in life history and host specificity: endoparasites with high reproductive potential and narrow host range that parasitize early in development and allow the host to continue to develop (koinobionts); or parasites with low reproductive potential and broad host range that parasitize and kill the host late in development (idiobionts) (6, 7). The large faunas of leaf-miners on trees consist largely of idiobionts [82% (73, 74)] and are similar in size in all areas (73). The number of idiobionts is greater in shrubs and trees (more apparent plants) than in herbs (3), although all three groups have the same numbers of koinobionts, suggesting that interspecific competition is important (74). These large-scale generalizations require testing; experimentally moving a miner into the mine of a congener on a different host with a larger parasitoid fauna did not increase parasitism, suggesting parasitism is host- and not plant-based (66). Tropical leaf-mining buprestids and hispines appear to have largely separate faunas even though they share some plant hosts (H. A. Hespenheide, unpublished data). Generalized host preference allows other miner species to serve as alternate hosts for the biological control of economically important miners (15, 40, 109, 111). Several studies have shown that endoparasitic parasitoids develop and emerge in close synchrony with their hosts (68, 106, 108, 148, 181).

Foraging behavior of parasitoids has received some attention. In the field, mines are detected in flight, apparently visually, although the mode of distinguishing suitable plant hosts is unknown (27). Laboratory experiments suggest plant hosts are located chemically, mines by vision, and feeding larvae by sound (179 and references therein). Parasitoids might mediate competition between miners and externally feeding herbivores by being attracted to damaged leaves (50, 51), but experimental damage had no effect on parasitism (72). Oviposition in or feeding on the host is influenced by host density (176). Subsequent search is affected by a marking pheromone (180). Parasitoid longevity and rates of foraging, oviposition, and host feeding have been shown in the field (27) or experimentally to be temperature dependent (181).

Relatively little attention has been paid to the potential influence of mine morphology on susceptability to parasitism. The observation that eulophids parasitizing tentiform miners have longer ovipositers than those parasitizing other types of mines suggests tentiform mines may be a defense against parasitoids (20). Unusual larval refuges have been observed in several tropical Hispinae that may reduce parasitism (H. A. Hespenheide, unpublished data).

EFFECT OF LEAF-MINERS ON PLANTS

A number of studies have reported the proportion of a leaf mined by a single miner [*Agromyza*, 27% of alfalfa (68); *Coptodisca*, 7.6% of vegetative and 47% of floral leaves of *Chamaedaphne* (88)]. The amount of leaf area mined has an effect on photosynthesis that is complicated to analyze. Leaf-miners often prefer shaded (131) or older (88) leaves that are less productive photosynthetically (see above). On the other hand, photosynthesis may be affected in disproportion to the area mined if the leaf is abcised prematurely (see above). *Phaseolus* mined by *L. trifolii* replaced mined palisade mesophyll with other photosynthetically active cells such that when one fourth of the area was mined, photosynthesis decreased by < 10% (114). New (126, 127) has calculated the actual weight of leaf consumed by individual miners; Nielsen (129) has estimated annual energetic losses for a forest.

Photosynthesis was measured directly as a function of number of *Bucculatrix* miners on pear in Japan, and this analysis showed a < 10% decrease in photosynthesis with 15 or fewer miners and a > 30% decrease with more than 50 miners (60). Leaf-miners typically cause a relatively small amount of damage to an individual tree because of low population densities and minor damage from single mines. *Phyllonorycter* spp. on beech and holm oak mined < 3% of leaf area, although mines accounted for up to 55% of herbivory on oak (153). Reduction in photosynthetic area from mines and mining-caused abcission was about 2% for *Coptodisca* on *Chamaedaphne* (88). Densities of individual species or cumulative density of all leaf-miners were less than 1% on *Betula pendula* (63). At the other extreme, leaf-miners may lead to complete defoliation (36, 132, 133) of plants in outbreak situations.

ECONOMIC IMPORTANCE

Examples of damage to agriculture or trees from leaf-miners other than *Liriomyza* (142) are varied. Sweet potato *(Ipomoea batatas)* has reportedly been attacked by different species of *Bedellia* on Fiji and Hawaii (89) and in California (164), as well as by a less-damaging gracillariid on Fiji. Although outbreaks of *Nepticula* have been recorded on pecan *(Carya illinoensis)* in Georgia (144), a study of pecan-leaf life histories in Texas (161) found that leaf-miners were widespread but caused limited damage. Leaf-mining activity of *Bucculatix* causes premature leaf abcission on pear *(Pyrus communis)* in Japan (60), as does the activity of *Phyllonorycter crataegella* on apple *(Pyrus malus)* in Connecticut (107), *Agromyza* on alfalfa *(Medicago sativa)* in New Jersey (79), and *Coptodisca* on cranberry *(V. macrocarpon)* in Connecticut (112). In alfalfa, yield losses averaged 7.7% in 1981–82 in New Jersey and Pennsylvania; loss of digestible dry matter ranged between 7.5–19.9% in cited studies (78). At the extreme, *Leucoptera coffeella* and *L. caffeina* are reported to cause almost total defoliation of coffee *(Coffea arabica)* in Kenya (132, 133).

Leaf miners can produce other effects on plants. The locust leaf-miner *(O. dorsalis)* is destructive to its noneconomic host *Robinia pseudoacacia*, but has become of economic importance because of its use of such alternate larval hosts as soybean *(Glycine max)* (146) and ornamental plants (191). Adults also cause feeding damage to leaves of plants other than their larval hosts (13, 192 and references therein). Cyclic outbreaks of the hispine beetle *Coelaenomenodera elaeidis* resulted in severe damage to foliage and reduced production of oil palm *(Elaeis guineensis)* in Ghana (15). *Perthida* reduces girth increment of jarrah in Australia by 33–47% to 64–83% (117). Depending on level of infestation, *P. crataegella* caused premature fruit drop or reduced fruit size of apple, and it and *P. blancardella* reduced fruit set the following season in two cultivars of apple in New York (160).

Notable examples of biological control of leaf-mining herbivores have been achieved for the hispines *Promecotheca coeruleipennis* and *P. cumingi* on coconut *(Cocos nucifera)* on Fiji (182) and Sri Lanka (42), respectively, as well as for *A. frontella* in the United States (44). Other attempts have had mixed results or are in development (40). Conversely, leaf-miners have been considered for biocontrol of pest plants: of *Lantana camara* by *Octotoma* and *Uroplata* (70, 193, 194; references in 168) and of *Echium plantagineum* by *Dialectia* (186).

CONCLUSIONS

Biologists tend to treat certain organisms as model systems (e.g. *R. fagi*) because they are well known. This review points out that leaf-miners have great ecological diversity even in closely related groups, and only a few species have been studied in detail. Perhaps especially because the tropical faunas of leaf-miners are largely unstudied and almost certainly greater in size than temperate faunas, it is premature either to take global comparisons very seriously or to know how representative our current knowledge is.

ACKNOWLEDGMENTS

I owe much of my interest and enthusiasm for leaf-mining insects to George Vogt and thank him for sharing a small part of his knowledge of them. My own studies of leaf-mining beetles have been supported at times by NSF grant DEB 76-10109, by the UCLA Academic Senate, and by myself.

Literature Cited

1. Aiello, A., Vogt, G. 1986. *Tachygonus dasypus* (Coleoptera: Curculionidae) Observations on an unusual tropical weevil. *Proc. K. Ned. Akad. Wet. C.* 89:117–20

2. Anderson, R. S. 1989. Revision of the subfamily Rhynchaeninae in North America (Coleoptera: Curculionidae). *Trans. Am. Entomol. Soc.* 115:207–312

3. Askew, R. R. 1980. The diversity of insect communities in leafminers and plant galls. *J. Anim. Ecol.* 49:817–29

4. Askew, R. R., Shaw, M. R. 1974. An account of the Chalcidoidea (Hymenoptera) parasitising leaf-mining insects of deciduous trees in Britain. *Biol. J. Linn. Soc. London* 6:289–335

5. Askew, R. R., Shaw, M. R. 1979. Mortality factors affecting the leaf-mining stages of *Phyllonorycter* (Lepidoptera: Gracillariidae) in oak and birch. I. Analysis of the mortality factors. *Zool. J. Linn. Soc. London* 67:31–49

6. Askew, R. R., Shaw, M. R. 1979. Mortality factors affecting the leaf-mining stages of *Phyllonorycter* (Lepidoptera: Gracillariidae) in oak and birch. II. Biology of the parasite species. *Zool. J. Linn. Soc. London* 67:51–64

7. Askew, R. R., Shaw, M. R. 1986. Parasitoid communities: their size, structure and development. In *Insect Parasitoids*, ed. J. Waage, D. Greathead, pp. 225–64. London: Academic

8. Auerbach, M., Simberloff, D. 1984. Responses of leaf miners to atypical leaf production patterns. *Ecol. Entomol.* 9:361–67

9. Auerbach, M., Simberloff, D. 1988. Rapid leaf-miner colonization of introduced trees and shifts in sources of herbivore mortality. *Oikos* 52:41–50

10. Auerbach, M., Simberloff, D. 1989. Oviposition site preference and larval mortality in a leaf-mining moth. *Ecol. Entomol.* 14:131–40

11. Bale, J. S. 1979. The occurrence of an adult reproductive diapause in the univoltine life cycle of the beech leaf mining weevil, *Rhynchaenus fagi* L. *Int. J. Invert. Reprod.* 1:57–66

12. Bale, J. S. 1981. Seasonal distribution and migratory behavior of the beech leaf mining weevil, *Rhynchaenus fagi* L. *Ecol. Entomol.* 6:109–18

13. Bale, J. S., Luff, M. L. 1978. The food plants and feeding preferences of the beech leaf mining weevil, *Rhynchaenus fagi* L. *Ecol. Entomol.* 3:245–49

14. Balsbaugh, E. U. Jr. 1988. Mimicry and the Chrysomelidae. See Ref. 93a, pp. 261–84

15. Bernon, G., Graves, R. C. 1979. An outbreak of the oil palm leaf miner beetle in Ghana with reference to a new alternative host for its parasite complex. *Environ. Entomol.* 8:108–12

16. Bigger, M. 1973. An investigation by Fourier analysis into the interaction between coffee leaf-miners and their larval parasites. *J. Anim. Ecol.* 42:417–34

17. Bigger, M. 1976. Oscillations of tropical insect populations. *Nature* 259:207–9

18. Bigger, M., Tapley, R. G. 1969. Prediction of outbreaks of coffee leaf-miners on Kilamanjaro. *Bull. Entomol. Res.* 58:601–17

19. Blackwelder, R. E. 1944. Catalogue of the Coleoptera of the Americas south of Mexico. *US Natl. Mus. Bull.* 185:306–41

20. Brandl, R., Vidal, S. 1987. Ovipositor length in parasitoids and tentiform leaf mines: adaptations in eulophids (Hymenoptera: Chalcidoidea). *Biol. J. Linn. Soc. London* 32:351–55

21. Bultman, T. L., Faeth, S. H. 1985. Patterns of intra- and interspecific association in leaf-mining insects on three oak host species. *Ecol. Entomol.* 10:121–29

22. Bultman, T. L., Faeth, S. H. 1986. Experimental evidence for intraspecific competition in a lepidopteran leaf-miner. *Ecology* 67:442–48

23. Bultman, T. L., Faeth, S. H. 1986. Leaf size selection by leaf-mining insects on *Quercus emoryi* (Fagaceae). *Oikos* 46:311–16

24. Bultman, T. L., Faeth, S. H. 1986. Selective oviposition by a leaf miner in response to temporal variation in abcission. *Oecologia* 69:117–20

25. Bultman, T. L., Faeth, S. H. 1986. Effect of within-leaf density and leaf size on pupal weight of a leaf-miner, *Cameraria* (Lepidoptera: Gracillariidae). *Southwest. Nat.* 31:201–06

26. Bultman, T. L., Faeth, S. H. 1988. Abundance and mortality of leaf miners on artificially shaded emory oak. *Ecol. Entomol.* 13:131–42

27. Casas, J. 1989. Foraging behavior of a leafminer parasitoid in the field. *Ecol. Entomol.* 14:257–65

28. Chapman, R. N. 1923. Observations on the life history of *Taphrocerus gracilis* (Say). *Cornell Agric. Exp. Sta. Mem.* 671:1–13

29. Claridge, M. F., Wilson, M. R. 1982. Insect herbivore guilds and species-area relationships: leaf-miners on British trees. *Ecol. Entomol.* 7:19–30

30. Coley, P. D. 1988. Effects of plant growth rate and leaf lifetime on the amount and type of anti-herbivore defense. *Oecologia* 74:531–36

31. Collinge, S. K., Louda, S. M. 1988. Patterns of resource use by a drosophiolid (Diptera) leaf miner on a native crucifer. *Ann. Entomol. Soc. Am.* 81:733–41

32. Collinge, S. K., Louda, S. M. 1988. Herbivory by leaf miners in response to experimental shading of a native crucifer. *Oecologia* 75:559–66

33. Collinge, S. K., Louda, S. M. 1989. Influence of plant phenology on the insect herbivore/bittercress interaction. *Oecologia* 79:111–16

34. Collinge, S. K., Louda, S. M. 1989. *Scaptomyza nigrita* Wheeler (Diptera: Drosophilidae), a leaf miner of the native crucifer, *Caramine cordifolia* A. Gray (Bittercress). *J. Kans. Entomol. Soc.* 62:1–10

35. Compton, S. G., Lawton, J. H., Rashbrook, V. K. 1989. Regional diversity, local community structure and vacant niches: the herbivorous arthropods of bracken in South Africa. *Ecol. Entomol.* 14:365–73

36. Condrashoff, S. F. 1964. Bionomics of the aspen leaf miner *Phyllocnistis populiella* Cham. (Lepidoptera: Gracillariidae). *Can. Entomol.* 96:857–74

37. Connor, E. F. 1984. The causes of overwintering mortality on *Phyllonorycter* on *Quercus robur*. *Ecol. Entomol.* 9:23–28

38. Connor, E. F. 1988. Cohort and death assemblage estimates of survival rates and causes of mortality in *Brachys ovatus* (Weber)(Coleoptera, Buprestidae). *Am. Midl. Nat.* 120:150–55

39. Connor, E. F., Faeth, S. H., Simberloff, D. 1983. Leaf-miners on oaks: the role of immigration and in situ reproductive recruitment. *Ecology* 64:191–204

40. Cornelius, S. J., Godfray, H. J. C. 1984. Natural parasitism of the chrysanthemum leaf-miner *Chromatomyia syngenesiae* H. (Dipt.: Agromyzidae). *Entomophaga* 29:341–45

41. Day, K. R., Watt, A. D. 1989. Population studies of the beech leaf mining weevil *(Rhynchaenus fagi)* in Ireland and Scotland. *Ecol. Entomol.* 14:23–30

42. Dharmadhikari, P. R., Perera, P. A. C. R., Hassen, T. M. F. 1977. A short account of the biological control of *Promecotheca cumingi* (Col.:Hispidae) the coconut leafminer, in Sri Lanka. *Entomophaga* 22:3–18

43. Dohse, L. A., McNeil, J. N. 1988. An intraspecific competition model for the leafminer, *Agromyza frontella* (Rondani). *Can. Entomol.* 120:779–86

44. Drea, J. J. Jr., Hendrickson, R. M. Jr. 1986. Analysis of a successful classical biological control project: the alfalfa blotch leafminer (Diptera: Agromyzidae) in the northeastern United States. *Environ. Entomol.* 15:448–55

45. Dussourd, D. E., Eisner, T. 1987. Vein-cutting behavior: insect counterploy to the latex defense of plants. *Science* 237:898–901

46. Engelbrecht, L. 1971. Cytokinin activity in larval infected leaves. *Biochem. Physiol. Pflanz.* 162:8–27

47. Engelbrecht, L., Orban, U., Heese, W. 1969. Leaf-miner caterpillars and cytokinins in the "green islands" of autumn leaves. *Nature* 223:319–21

48. Ezcurra, E., Gomez, J. C., Becerra, J. 1987. Diverging patterns of host use by phytophagous insects in relation to leaf pubescence in *Arbutus xalapensis* (Ericaceae). *Oecologia* 72:479–80

49. Faeth, S. H. 1980. Invertebrate predation of leaf-miners at low densities. *Ecol. Entomol.* 5:111–14

50. Faeth, S. H. 1985. Host leaf selection by leaf miners: interactions among three trophic levels. *Ecology* 66:870–75

51. Faeth, S. H. 1986. Indirect interactions between temporally separated herbivores mediated by the host plant. *Ecology* 67:479–94

52. Faeth, S. H., Connor, E. F., Simberloff, D. 1980. Early leaf abcission: a neglected source of mortality for folivores. *Am. Nat.* 117:409–15

53. Faeth, S. H., Mopper, S., Simberloff,

D. 1981. Abundances and diversity of leaf-mining insects on three oak host species: effects of host-plant phenology and nitrogen content of leaves. *Oikos* 37:238–51

54. Faeth, S. H., Simberloff, D. 1981. Population regulation in a leaf-mining insect, *Cameraria* sp. nov., at increased field densities. *Ecology* 62:620–24

55. Faeth, S. H., Simberloff, D. 1981. Experimental isolation of oak host plants: effects on mortality, survivorship, and abundances of leaf-mining insects. *Ecology* 62:625–35

56. Feeny, P. 1970. Seasonal changes in oak leaf tannins and nutrients as a cause of spring feeding by winter moth caterpillars. *Ecology* 51:565–81

57. Fiebrig, K. 1908. Eine Schaum bildene Kaferlarve *Pachyschelus* spec. (Bupr. Sap.) Die Ausscheidung von Kautschuk aus der Nahrung und dessen Verwertung zu Schutzzwecken (auch bei Rhynchoten). *Z. Wiss. Insektenbiol.* 4:333–39, 353–63

58. Fowler, S. V., Lawton, J. H. 1982. The effects of host-plant distribution and local abundance on the species richness of agromyzid flies attacking British umbellifers. *Ecol. Entomol.* 7:257–65

59. Frost, S. W. 1931. The habits of leaf-mining Coleoptera on Barro Colorado Island, Panama. *Ann. Entomol. Soc. Am.* 24:396–404

60. Fujiie, A. 1982. Ecological studies on the population of the pear leaf miner, *Bucculatrix pyrivorella* Kuroko (Lepidotera: Lyonetiidae). VI. Effects of injury by the pear leaf miner on leaf fall and photosynthesis of the pear tree. *Appl. Entomol. Zool.* 17:188–93

61. Fulmek, L. 1962. *Parasitinsecten der Blattminierer Europas*. The Hague: Junk. 203 pp.

62. Godfray, H. C. J. 1984. Patterns in the distribution of leaf-miners on British trees. *Ecol. Entomol.* 9:163–68

63. Godfray, H. C. J. 1985. The absolute abundance of leaf miners on plants of different successional stages. *Oikos* 45:17–25

64. Godfray, H. C. J. 1986. Clutch size in a leaf-mining fly (*Pegomya nigritarsis:* Anthomyiidae). *Ecol. Entomol.* 11:75–81

64a. Godfray, H. C. J., Hassell, M. P. 1989. Discrete and continuous insect populations in tropical environments. *J. Anim. Ecol.* 58:153–74

65. Gross, P. 1988. Life histories and geographic distributions of two leafminers, *Tildenia georgei* and *T. inconspicuella*

(Lepidoptera: Gelechiidae), on solanaceous weeds. *Ann. Entomol. Soc. Am.* 79:48–55

66. Gross, P., Price, P. W. 1988. Plant influences on parasitism of two leafminers: a test of enemy-free space. *Ecology* 69:1506–16

67. Guppy, J. C. 1981. Bionomics of the alfalfa blotch leafminer, *Agromyza frontella* (Diptera: Agromyzidae), in eastern Ontario. *Can. Entomol.* 113:593–600

68. Guppy, J. C., Meloche, F., Harcourt, D. G. 1988. Seasonal development, behavior, and host synchrony of *Dacnusa dryas* (Nixon) (Hymenoptera: Braconidae), parasitizing the alfalfa blotch leafminer, *Agromyza frontella* (Rondani) (Diptera: Agromyzidae). *Can. Entomol.* 120:145–52

69. Hamilton, R. W. 1980. Notes on the biology of *Eugnamptus collaris* (Fabr.) (Coleoptera: Rhynchitidae), with descriptions of the larva and pupa. *Coleopt. Bull.* 34:227–36

70. Harley, K. L. S. 1969. The suitability of *Octotoma scabripennis* Guer. and *Uroplata girardi* Pic (Coleoptera: Chrysomelidae) for the control of *Lantana* (Verbenaceae) in Australia. *Bull. Entomol. Res.* 58:835–43

71. Hartley, S. E., Lawton, J. H. 1987. Effects of different types of damage on the chemistry of birch foliage, and the responses of birch feeding insects. *Oecologia* 74:432–37

72. Hawkins, B. A. 1988. Foliar damage, parasitoids and indirect competition: a test using herbivores of birch. *Ecol. Entomol.* 13:301–8

73. Hawkins, B. A. 1990. Global patterns of parasitoid assemblage size. *J. Anim. Ecol.* 59:57–72

74. Hawkins, B. A., Askew, R. R., Shaw, M. R. 1990. Influences of host feeding-niche and foodplant type on generalist and specialist parasitoids. *Ecol. Entomol.* 15:In press

75. Hawkins, B. A., Lawton, J. H. 1987. Species richness for parasitoids of British phytophagous insects. *Nature* 326:788–90

76. Heads, P. A., Lawton, J. H. 1983. Studies on the natural enemy complex of the holly leafminer: the effects of scale on the detection of aggregative responses and the implications for biological control. *Oikos* 40:267–76

77. Hendrickson, R. M. Jr., Barth, S. E. 1978. Biology of the alfalfa blotch leafminer (Diptera: Agromyzidae). *Ann. Entomol. Soc. Am.* 71:295–98

78. Hendrickson, R. M. Jr., Day, W. H. 1986. Yield losses caused by alfalfa

blotch leafminer (Diptera: Agromyzidae). *J. Econ. Entomol.* 79:988–92

79. Hendrickson, R. M. Jr., Dysart, R. J. 1983. Leaflet abcission caused by alfalfa blotch leafminer (Diptera: Agromyzidae). *J. Econ. Entomol.* 76:1075–79

80. Hendrickson, R. M. Jr., Plummer, J. A. 1983. Biological control of alfalfa blotch leafminer (Diptera: Agromyzidae) in Delaware. *J. Econ. Entomol.* 76:757–61

81. Hering, E. 1942. Neotropische Buprestidenminen. *Arb. Physiol. Angew. Entomol.* 9:241–49

82. Hering, E. M. 1951. *Biology of the Leaf Miners.* The Hague: Junk. 520 pp.

83. Hering, E. M. 1957. *Bestimungstabellen der Blattminen von Europa,* Bd. 1–3. The Hague: Junk. 1185 pp.

84. Hespenheide, H. A. 1973. A novel mimicry complex: Beetles and flies. *J. Entomol. London A* 48:49–56

85. Hespenheide, H. A. 1982. A revision of Central American species of *Neotrachys* (Coleoptera: Buprestidae). *Coleopt. Bull.* 36:328–49

86. Hespenheide, H. A. 1985. The visitor fauna of extrafloral nectaries of *Byttneria aculeata* (Sterculiaceae): Relative importance and roles. *Ecol. Entomol.* 10:191–204

87. Hespenheide, H. A. 1991. The fauna of La Selva. In *La Selva: Ecology and History of a Neotropical Rainforest,* ed. L. McDade, K. S. Bawa, G. S. Hartshorn, H. A. Hespenheide. Chicago: Univ. Chicago Press. In press

88. Hileman, D. R., Lieto, L. F. 1981. Mortality and area reduction in leaves of the bog shrub *Chamaedaphne calyculata* (Ericaceae) caused by the leaf miner *Coptodisca kalmiella* (Lepidoptera: Heliozelidae). *Am. Midl. Nat.* 106:180–88

89. Hinckley, A. D. 1963. Lepidopterous ieafminers on sweet potato in Fiji. *Bull. Entomol. Res.* 53:665–70

90. Hinckley, A. D. 1972. Comparative ecology of two leaf miners on white oak. *Environ. Entomol.* 1:358–61

91. Itamies, J., Ojanen, M. 1977. Autumn predation of *Parus major* and *P. montanus* on two species of *Lithocolletis* (Lepidoptera: Lithocolletidae). *Ann. Zool. Fennici* 14:235–41

92. James, R., Pritchard, I. M. 1988. Influence of the holly leaf miner, *Phytomyza ilicis* (Diptera: Agromyzidae), on leaf abcission. *J. Nat. Hist.* 22:395–402

93. Janzen, D. H. 1968. Host plants as islands in evolutionary and contemporary time. *Am. Nat.* 102:592–95

93a. Jolivet, P., Petitpierre, E., Hsiao, T. H., eds. 1988. *Biology of Chrysomelidae.* Dordrecht, Netherlands: Kluwer Academic

94. Kahn, D. M., Cornell, H. V. 1983. Early leaf abscission and folivores: comments and considerations. *Am. Nat.* 122:428–32

95. Kahn, D. M., Cornell, H. V. 1989. Leafminers, early leaf abscission, and parasitoids: a tritrophic interaction. *Ecology* 70:1219–26

96. Kimmerer, T. W., Potter, D. A. 1987. Nutritional quality of specific leaf tissues and selective feeding by a specialist leafminer. *Oecologia* 71:548–51

97. Kirkendall, L. R. 1984. Long copulations and post copulatory "escort" behaviour in the locust leaf miner, *Odontota dorsalis* (Coleoptera: Chrysomelidae). *J. Nat. Hist.* 18:905–19

98. Kogan, M. 1963. Uma nova especie do genero *Tachygonus* Schoenherr, 1833 e observacoes sobre seus habitos minadores (Coleoptera, Curculioinidae). *Rev. Brasil. Biol.* 23:85–94

99. Kogan, M. 1963. Contribuicao ao conhecimento da sistematica e biologia de buprestideos minadores do genero *Pachyschelus* Solier, 1833 (Coleoptera, Buprestidae). *Mem. Inst. Oswaldo Cruz* 61:429–57

100. Kogan, M. 1964. Notas biologicas e descricao de uma nova especie do genero *"Brachys"* Solier, 1833, minador de folhas de *"Inga sessilis"* (Coleoptera, Buprestidae). *Rev. Brasil. Biol.* 24:393–404

101. Kogan, M. 1964. Obsevacoes sobre a sistematica e a etologia de um buprestideo do genero *Leiopleura* Deyrolle, 1864, minador de folhas de Jaqueira (Coleoptera, Buprestidae). *Anais II Con. Latino Am. Zool. Sao Paulo, 1962* 1:197–206

102. Krombein, K. V. 1964. Natural history of Plummers Island, Maryland. XVIII. The hibiscus wasp, an abundant rarity, and its associates (Hymenoptera: Sphecidae). *Proc. Biol. Soc. Wash.* 77:73–112

103. Krombein, K. V. 1964. Results of the Archbold Expeditions. No. 87 Biological notes on some Floridian wasps (Hymenoptera, Aculeata). *Am. Mus. Novit.* 2201. 27 pp.

104. Krombein, K. V. 1967. *Trap-Nesting Wasps and Bees: Life Histories, Nests, and Associates.* Washington, DC: Smithsonian. 570 pp.

105. Lawton, J. H., Price, P. W. 1979. Species richness of parasites on hosts: agromyzid flies on the British Umbelliferae. *J. Anim. Ecol.* 48:618–37

106. Maier, C. T. 1982. Parasitism of the apple blotch leafminer, *Phyllonorycter crataegella*, on sprayed and unsprayed apple trees in Connecticut. *Environ. Entomol.* 11:603–10

107. Maier, C. T. 1983. Effect of the apple blotch leafminer (Lepidoptera: Gracillariidae) on apple leaf abcission. *J. Econ. Entomol.* 76:1265–68

108. Maier, C. T. 1984. Abundance and phenology of parisitoids of the spotted tentiform leafminer, *Lithocolletis blancardella* (Lepidoptera: Gracillariidae), in Connecticut. *Can. Entomol.* 116:443–49

109. Maier, C. T. 1988. Parisitoid fauna of two *Phyllonorycter* spp. (Lepidoptera: Gracillariidae) on wild cherries, and similarity to fauna of apple leafminers. *Ann. Entomol. Soc. Am.* 81:460–66

110. Maier, C. T. 1988. Life cycle of *Coptodisca negligens* (Lepidoptera: Heliozelidae) on cranberry. *J. Econ. Entomol.* 81:497–500

111. Maier, C. T. 1988. Gracillariid hosts of *Sympiesis marylandensis* (Hymenoptera: Eulophidae) in New England. *Ann. Entomol. Soc. Am.* 81:728–32

112. Maier, C. T. 1989. Accelerated abscission of cranberry leaves damaged by the leafminer, *Coptodisca negligens* (Lepidoptera: Heliozelidae). *Environ. Entomol.* 18:773–77

113. Mariau, D. 1988. The parasitoids of Hispinae. See Ref. 93a, pp. 449–61

114. Martens, B., Trumble, J. T. 1987. Structural and photosynthetic compensation for leafminer (Diptera: Agromyzidae) injury on lima beans. *Environ. Entomol.* 16:374–78

115. Martin, J. L. 1956. The bionomics of the aspen blotch-miner, *Lithocolletis salicifoliella* Cham. (Lepidoptera: Gracillariidae). *Can. Entomol.* 88:155–68

116. Maulik, M. A. 1937. Distributional correlation between hispine beetles and their host-plants. *Proc. Zool. Soc. London Ser. A* 1937:129–59

117. Mazanec, Z. 1974. Influence of Jarrah leaf miner on growth of Jarrah. *Aust. For.* 37:42–32

118. Mazanec, Z. 1985. Resistance of *Eucalyptus marginata* to *Perthida glyphopa* (Lepidoptera: Incurvariidae). *J. Aust. Entomol. Soc.* 24:209–21

119. Mazanec, Z. 1987. Natural enemies of *Perthida glyphopa* Common (Lepidoptera: Incurvariidae). *J. Aust. Entomol. Soc.* 26:303–8

120. Mazanec, Z., Justin, M. J. 1986. Oviposition behaviour and dispersal of *Perthida glyphopa* Common (Lepidoptera: Incurvariidae). *J. Aust. Entomol. Soc.* 26:149–59

121. McGavin, G. C., Brown, V. K. 1986. Variation in populations of mine- and gall-forming Diptera and the growth form of their host plant, bracken [*Pteridium aquilinum* (L.) Kuhn]. *J. Nat. Hist.* 20:799–816

122. McNeil, J. N., Quiring, D. T. 1983. Evidence of an oviposition-deterring pheromone in the alfalfa blotch leafminer, *Agromyza frontella* (Rond.) (Diptera: Agromyzidae). *Environ. Entomol.* 12:990–92

123. Miller, P. F. 1973. The biology of some *Phyllonorycter* species (Lepidoptera: Gracillariidae) mining leaves of oak and beech. *J. Nat. Hist.* 7:391–409

124. Mopper, S., Faeth, S. H., Boecklen, W. J., Simberloff, D. S. 1984. Host-specific variation in leaf miner population dynamics: effects on density, natural enemies and behavior of *Stilbosis quadricustatella* (Cham.) (Lepidoptera: Cosmopterigidae). *Ecol. Entomol.* 9:169–77

125. Needham, J. G., Frost, S. W., Tothill, B. H. 1928. *Leaf-Mining Insects.* Baltimore, MD: Williams & Wilkins. 351 pp.

126. New, T. R. 1976. Aspects of exploitation of *Acacia* phyllodes by a mining lepidopteran, *Acrocercops plebeia* (Gracillariidae). *J. Aust. Entomol. Soc.* 15:365–78

127. New, T. R. 1979. Biology of *Labdia* sp. (Lepidoptera: Cosmopterygidae), a miner in phyllodes of *Acacia. Aust. J. Zool.* 27:529–36

128. Nielsen, B. O. 1968. Studies on the fauna of beech foliage 2. Observations on the mortality and mortality factors of the beech weevil [*Rhynchaenus (Orchestes) fagi* L.] (Coleoptera: Curculionidae). *Nat. Jutl.* 14:99–125

129. Nielsen, B. O. 1978. Aspects of the population ecology and energetics of some beech leaf–feeding insects. *Nat. Jutl.* 20:259–72

130. Nielsen, B. O. 1978. Food resource partition in the beech leaf–feeding guild. *Ecol. Entomol.* 3:193–201

131. Nielsen, B. O., Ejlersen, A. 1978. The distribution pattern of herbivory in a beech canopy. *Ecol. Entomol.* 2:293–99

132. Notley, F. B. 1948. The *Leucoptera* leaf miners of coffee on Kilamanjaro. I. *Leucoptera coffeella* Guer. *Bull. Entomol. Res.* 39:399–416

133. Notley, F. B. 1956. The *Leucoptera* leaf miners of coffee on Kilamanjaro. II. *Leucoptera coffeella* Guer. *Bull. Entomol. Res.* 46:899–912

134. Oatman, E. R. 1985. Parasites associated with lepidopterous leaf miners on apple in northeastern Wisconsin. *J. Econ. Entomol.* 78:1063–66
135. Opler, P. A. 1973. Fossil lepidopterous leaf mines demonstrate the age of some insect-plant relationships. *Science* 179:1321–23
136. Opler, P. A. 1974. Oaks as evolutionary islands for leaf-mining insects. *Am. Sci.* 62:67–73
137. Opler, P. A. 1974. Biology, ecology, and host specificity of Microlepidoptera associated with *Quercus agrifolia* (Fagaceae). *Univ. Calif. Publ. Entomol.* No. 75. 83 pp.
138. Opler, P. A. 1978. Interaction of plant life history components as related to arboreal herbivory. In *The Ecology of Arboreal Folivores*, ed., G. G. Montgomery, pp. 23–31. Washington, DC: Smithsonian Inst.
139. Opler, P. A., Davis, D. R. 1981. The leafmining moths of the genus *Cameraria* associated with Fagaceae in California (Lepidoptera: Gracillariidae). *Smithson. Contrib. Zool. No. 333.* 58 pp.
140. Owen, D. F. 1975. The efficiency of blue tits, *Parus caerulus*, preying on the larvae of *Phytomyza ilicis. Ibis* 117:515–16
141. Owen, D. F. 1978. The effect of a consumer *Phytomyza ilicis* on seasonal leaf-fall in the holly *Ilex aquifolium. Oikos* 31:268–71
142. Parella, M. P. 1987. Biology of *Liriomyza. Annu. Rev. Entomol.* 32:201–24
143. Payne, J. A., Tedders, W. L., Cosgrove, G. E., Foard, D. 1972. Larval mine characteristics of four species of leaf-mining Lepidoptera in pecan. *Ann. Entomol. Soc. Am.* 65:74–81
144. Payne, J. A., Tedders, W. L., Gentry, C. R. 1971. Biology and control of a pecan serpentine leaf-miner, *Nepticula juglandifoliella. J. Econ. Entomol.* 64:92–93
145. Phillipson, J., Thompson, D. J. 1983. Phenology and intensity of phyllophage attack on *Fagus sylvatica* in Wytham Woods, Oxford. *Ecol. Entomol.* 8:315–30
146. Poos, F. W. 1940. The locust leaf-miner as a pest of soybean. *J. Econ. Entomol.* 33:742–45
147. Potter, D. A. 1985. Population regulation of the native holly leaf-miner, *Phytomyza ilicicola* Loew (Diptera: Agromyzidae), on American holly. *Oecologia* 66:499–505
148. Potter, D. A., Gordon, F. C. 1985. Parasites associated with the native holly leafminer, *Phytomyza ilicicola* Loew (Diptera: Agromyzidae), on American holly in Kentucky. *J. Kans. Entomol. Soc.* 58:727–30
149. Potter, D. A., Kimmerer, T. W. 1986. Seasonal allocation of defense investment in *Ilex opaca* Aiton and constraints on a specialist leafminer. *Oecologia* 69:217–24
150. Potter, D. A., Redmond, C. T. 1989. Early spring defoliation, secondary leaf flush, and leafminer outbreaks on American holly. *Oecologia* 81:192–97
151. Pottinger, R. P., LeRoux, E. J. 1971. The biology and dynamics of *Lithocolletis blancardella* (Lepidoptera: Gracillariidae) on apple in Quebec. *Mem. Entomol. Soc. Can. No. 77.* 437 pp.
152. Powell, J. A. 1980. Evolution of larval food preferences in microlepidoptera. *Annu. Rev. Entomol.* 25:133–59
153. Pritchard, I. M., James, R. 1984. Leaf mines: their effect on leaf longevity. *Oecologia* 64:132–39
154. Pritchard, I. M., James, R. 1984. Leaf fall, a source of leaf miner mortality. *Oecologia* 64:140–41
155. Pullin, A. S. 1985. A simple life table study based on development and mortality in the beech leaf mining weevil *Rhynchaenus fagi* L. *J. Biol. Educ.* 19:152–56
156. Quiring, D. T., McNeil, J. N. 1984. Exploitation and interference larval competition in the dipteran leaf miner *Agromyza frontella* (Rondani). *Can. J. Zool.* 62:421–27
157. Quiring, D. T., McNeil, J. N. 1984. Adult-larval intraspecific competition in *Agromyza frontella* (Diptera: Agromyzidae). *Can. Entomol.* 116:1385–91
158. Quiring, D. T., McNeil, J. N. 1984. Intraspecific competition between different aged larvae of *Agromyza frontella* (Rondani) (Diptera: Agromyzidae): advantages of an oviposition-deterring pheromone. *Can. J. Zool.* 62:2192–96
159. Quiring, D. T., McNeil, J. N. 1987. Foraging behavior of a dipteran leaf miner on exploited and unexploited hosts. *Oecologia* 73:7–15
160. Reissig, W. H., Weires, R. W., Forshey, C. G. 1982. Effects of gracillariid leafminers on apple tree growth and production. *Environ. Entomol.* 11:958–63
161. Ring, D. R., Harris, M. K., Olszak, R. 1985. Life tables for pecan leaves in Texas. *J. Econ. Entomol.* 78:888–94
162. Sekita, N., Yamada, M. 1979. Studies on the population of the apple leaf miner *Phyllonorycter ringoniella* Matsumura (Lepidoptera: Lithocolletidae). III.

Some analyses of the mortality operating upon the population. *Appl. Entomol. Zool.* 14:137–48

163. Shaw, M. R., Askew, R. R. 1976. Ichneumonoidea (Hymenoptera) parasitic upon leaf-mining insects of the orders Lepidoptera, Hymenoptera, and Coleoptera. *Ecol. Entomol.* 1:127–33

164. Shorey, H. H., Anderson, L. D. 1960. Biology and control of the morning glory leaf miner, *Bedellia somnulentella*, on sweet potatoes. *J. Econ. Entomol.* 53:1119–22

165. Simberloff, D., Stiling, P. 1987. Larval dispersion and survivorship in a leaf-mining moth. *Ecology* 68:1647–57

166. Smith, A. P. 1986. Ecology of a leaf color polymorphism in a tropical forest species: habitat selection and herbivory. *Oecologia* 69:283–87

167. Southwood, T. R. E. 1961. The number of species of insect associated with various trees. *J. Anim. Ecol.* 30:1–8

168. Staines, C. L. Jr. 1989. A revision of the genus *Octotoma* (Coleoptera: Chrysomelidae, Hispinae). *Insecta Mundi* 3:41–56

169. Staines, C. L. Jr., Staines, S. L. 1989. A bibliography of New World Hispinae (Coleoptera: Chrysomelidae). *Md. Entomol.* 3:83–122

170. Stiling, P. D., Brodbeck, B. V., Strong, D. R. 1982. Foliar nitrogen and larval parasitism as determinants of leaf-miner distribution patterns on *Spartina alterniflora*. *Ecol. Entomol.* 7:447–52

171. Stiling, P. D., Brodbeck, B. V., Strong, D. R. 1984. Intraspecific competition in *Hydrellia valida* (Diptera: Ephydridae), a leaf-miner of *Spartina alterniflora*. *Ecology* 65:660–62

172. Stiling, P. D., Simberloff, D., Anderson, L. C. 1987. Non-random distribution patterns of leaf-miners on oak trees. *Oecologia* 73:116–19

173. Story, R. N., Robinson, W. H., Pienkowski, R. L., Kok, L. T. 1979. The biology and immature stages of *Taphrocerus schaefferi*, a leaf-miner of yellow nutsedge. *Ann. Entomol. Soc. Am.* 72:93–98

174. Strong, D. R. 1974. Nonasymptotic species richness models and the insects of British trees. *Proc. Natl. Acad. Sci. USA* 71:2766–69

175. Strong, D. R., Lawton, J. H., Southwood, R. 1984. *Insects on Plants, Community Patterns and Mechanisms*. Oxford: Blackwell Sci. 313 pp.

176. Sugimoto, T. 1978. Host-attacking behavior of a eulophid parasite, *Kratochviliana* sp., to the leaf mining host, *Phytomyza ranunculi* (Diptera: Agromyzidae) during its stay on the leaf. *Res. Pop. Ecol.* 19:197–208

177. Sugimoto, T. 1980. Models of the spatial pattern of egg population of Ranunculus leaf mining fly, *Phytomyza ranunculi* (Diptera: Agromyzidae), in host leaves. *Res. Pop. Ecol.* 22:13–22

178. Sugimoto, T., Ishii, M. 1979. Mortality of larvae of a Ranunculus leaf mining fly, *Phytomyza ranunculi* (Diptera: Agromyzidae) due to parasitization and host-feeding by its eulophid parasite *Chrysocharis pentheus* (Hymenoptera: Eulophidae). *Appl. Entmol. Zool.* 14:410–18

179. Sugimoto, T., Shimono, Y., Hata, Y., Nakai, A., Yahara, M. 1988. Foraging for patchily-distributed leaf-miners by the parasitoid, *Dapsilarthra rufiventris* (Hymenoptera: Braconidae). III. Visual and acoustic cues to a close range patch-location. *Appl. Entmol. Zool.* 23:113–21

180. Sugimoto, T., Tsjimoto, S. 1988. Stopping rule of host search by the parasitoid *Chrysocharis pentheus* (Hymenoptera: Eulophidae), in host patches. *Res. Pop. Ecol.* 30:123–33

181. Sugimoto, T., Yasuda, I., Ono, M., Matsunaga, S. 1982. Occurrence of a Ranunculus leaf mining fly, *Phytomyza ranunculi* and its eulophid parasitoids from fall to summer in the low land. *Appl. Entmol. Zool.* 17:139–43

182. Taylor, T. H. C. 1937. *The Biological Control of an Insect in Fiji. An Account of the Coconut Leaf-Mining Beetle and Its Parasite Complex*. London: Imperial Inst. Entomol. 239 pp.

183. Toumi, J., Niemela, P., Manilla, R. 1981. Leaves as islands: interactions of *Scolioneura betuleti* (Hymenoptera) miners on birch leaves. *Oikos* 37:146–52

184. Turnbow, R. H. Jr., Franklin, R. T. 1981. Bionomics of *Brachys tessellatus* in coastal plain scrub oak communities. *Ann. Entomol. Soc. Am.* 74:351–58

185. Wallace, M. M. H. 1970. The biology of the jarrah leaf miner, *Perthida glyphopa* Common (Lepidoptera: Incurvariidae). *Austr. J. Zool.* 18:91–104

186. Wapshere, A. J., Kirk, A. A. 1977. biology and host specificity of the *Echium* leaf miner, *Dialectica scalariella* (Zeller) (Lepidoptera: Gracillariidae). *Bull. Entomol. Res.* 67:627–33

187. Weidlich, M. 1986. Zum rezenten Wirtspflanzenspektrum der Buprestidae unter Berucksichtigung phylogenetischer Aspekte. *Dtsch. Entomol. Z.* 33:83–93

188. Weiss, H. B., West, E. 1922. Notes on the *Desmodium* leaf miner, *Pachyschelus laevigatus* (Say) (Col.: Buprestidae). *Entomol. News* 33:180–83

189. Werner, T. K., Sherry, T. W. 1987. Behavioral feeding specialization in *Pinaroloxias inornata*, the "Darwin's Finch" of Cocos Island, Costa Rica. *Proc. Natl. Acad. Sci. USA* 84:5506–10

190. West, C. 1985. Factors underlying the late seasonal appearance of the lepidopterous leaf-mining guild on oak. *Ecol. Entomol.* 10:111–20

191. Wheeler, A. G. Jr. 1989. Japanese pagodatree: a host of locust leafminer, *Odontota dorsalis* (Thunberg)(Coleoptera: Chrysomelidae). *Coleopt. Bull.* 34:95–98

192. Williams, C. E. 1989. Damage to woody plants by the locust leafminer, *Odontota dorsalis* (Coleoptera: Chrysomelidae), during a local outbreak in an Appalachian oak forest. *Entomol. News* 100:183–87

193. Winder, J. A., Harley, K. L. S. 1982. The effects of natural enemies on the growth of *Lantana* in Brazil. *Bull. Entomol. Res.* 72:599–616

194. Winder, J. A., Harley, K. L. S., Kassulke, R. C. 1984. *Uruplata lantanae* Buzzi and Winder (Coleoptera: Chrysomelidae: Hispinae), a potential biological control agent of *Lantana camara* in Australia. *Bull. Entomol. Res.* 74:327–40

195. Zoebisch, T. G., Schuster, D. J. 1987. Longevity and fecundity of *Liriomyza trifolii* (Diptera: Agromyzidae) exposed to tomato foliage and honeydew in the laboratory. *Environ. Entomol.* 16:1001–3

Annu. Rev. Entomol. 1991. 36:561–86

VEGETATIONAL DIVERSITY AND ARTHROPOD POPULATION RESPONSE

D. A. Andow

Department of Entomology, University of Minnesota, St. Paul, Minnesota 55108

KEY WORDS: polyculture, plant diversity, intercropping, cultural controls, integrated pest management

INTRODUCTION

Studies of agroecosystems during the past 30 years have lead several agricultural scientists to question the commitment of modern industrial agriculture to high intensity monocultural production. Additionally, current research directions in integrated pest management emphasize biological interactions among insect pests, natural enemies, and other crop pests, such as weeds. These inquiries have led to a recent rebirth in interest and research activities on cultural and biological controls in entomology.

Vegetational diversity plays a central role in this research renaissance. If one considers it broadly, vegetational diversity involves mixing different kinds of plants in a plant community, but, to paraphrase Vilfredo Pareto (112), vegetational diversity appears like a bat; within it one can find both birds and mice. More specifically, vegetational diversity varies in three ways: the kinds, the spatial array, and the temporal overlap of the plants in the mixture. In most cases, the mixed plants are different plant species. These plants might be two crops, which is called intercropping; a crop and a weed, which is called weedy culture; or a crop and a beneficial noncrop, which is known by many names including nursery crops, living-mulches, cover-cropping, etc. In some cases, however, different plant genotypes are mixed (41), including polyvarietal mixtures of agronomically dissimilar genotypes

561

and multilines of agronomically similar genotypes. The spatial arrangement of the plants can vary widely from small scale, intimate mixtures to larger-scale mixtures. The former can include mixed plantings with no specific row arrangement, mixed rows with the plants mixed within rows, and row mixtures with the plant species in alternating rows. The larger scale mixtures can include strip mixtures with the plants mixed in alternating strips of rows including strip cropping and grassy waterways, and mixtures of fields of plants (35) including trap crops, hedgerows, irrigation ditches, conservation reserve program lands, and woodlots. The temporal overlap between plants can range from none, as in crop rotations, to intermediate as in relay cropping (3) to complete as in intercropping. The overlap can be asymmetrical; for example in a dry bean-banana intercropping system, dry beans are associated with bananas throughout their lives, but bananas are associated with dry bean for only a short period of their lives.

This tremendous variety supplies rich food for thought and experimentation, but at the same time it has generated a large collection of partially digested observations scattered throughout the literature. In this review, I concentrate on spatially intimate mixtures of different plant species with maximal temporal overlap. I refer to these mixtures as polycultures and the corresponding bare-ground sole-crop fields as monocultures. These polycultures are the most ecologically complex of all of the systems of vegetational diversity because inter- and intraspecific plant competition occurs simultaneously with herbivory and the plants are interspersed at spatial scales similar to the shorter movement scales of the arthropod herbivores and natural enemies. Polycultures have received considerably more research attention than other forms of vegetational diversity (but see 11, 14), and understanding the ecology of arthropod response to polycultures should lead to an understanding of the response of arthropods to other forms of vegetational diversity.

THE PROBLEM OF POLYCULTURE

The number of potential ecological interactions between the plants, arthropod herbivores, and arthropod natural enemies, and the possible evolutionary responses of each population to any of the others, creates a veritable Gordian knot of complexity. For example, a relatively simple ecosystem of 2 plant species, 6 herbivore species, and 6 natural enemy species has 91 potential two-way and 364 potential three-way ecological interactions and at least an equal number of possible evolutionary responses. Clearly, experimental analysis of each potential interaction would be unrealistic. Furthermore, if one considers the number of possible plant combinations and the concomitant shuffling of the herbivore and natural enemy fauna, the problem convolutes upon itself and rapidly increases in complexity. Yet the central problem remains: how do arthropods respond to polycultures compared to monocultures? Undoing this knot will transform our understanding of insect ecology and pest control.

Perhaps as a response to the complexity of the problem, two alternate approaches have evolved. The first seeks to unravel the problem by minutely examining each thread of the knot. This empirical approach is usually utilitarian, focusing on economically significant polycultures and associated herbivores, as reviewed by Altieri et al (12) and others (47, 108, 163). Frequently, no effort is expended toward developing a general understanding of the response of arthropods to polyculture (108). Indeed, while never explicitly stated, one detects an implied commitment to the idea that nature is inherently idiosyncratic and that there is no general understanding to discover. Species may evolve to have individualistic responses to their environment (84) because each species is subject to a unique set of selective forces. If this is true, then robust generalizations about arthropod response to polycultures may not exist. Certainly, in most of this work, a utilitarian ideology prevails, and the utility of general understanding is questioned when specific information is required for any particular situation (158).

The second approach seeks to cut the knot of complexity with simple but elegant theory (51, 52, 129). This theoretical approach clearly seeks general understanding of the response of arthropods to polyculture, and indeed, assumes that such general understanding exists, that regularities await discovery, and that nature is not inherently idiosyncratic. These generalizations might emerge from evolutionary processes if regular patterns in selective forces have acted on arthropods in relation to polycultures, or if phylogeny has created an evolutionary inertia that constrains evolution to certain paths.

The question I address in this review asks if generalities underlie arthropod response to polyculture or if these responses are idiosyncratic, depending on the particular combination of plants and arthropods. These generalities might relate to the actual responses of the arthropods or to the mechanisms acting to create those responses.

THEORY

The diversity-stability hypothesis states that the greater is the biological diversity of a community of organisms, the greater is the stability of that community (52, 71, 96, 99, 111, 116). It generated a broad interest in arthropod response to vegetational diversity, which was one of the most important sources of evidence supporting the hypothesis (63). As reviewed by Elton (52) and Pimentel (116), arthropod pest outbreaks were more likely to occur in monocultures than in polycultures. Subsequent historical (63), theoretical (95, 102), and empirical (63, 106, 158) investigations suggested that the diversity-stability hypothesis probably originated in the 18th-century political economics of Spencer (143), a belief that the wondrous variety of nature must have some purpose in an orderly world, and ageless folk wisdom about not putting all the eggs in one basket (63). These studies also indicated that the hypothesis had no logical force (95, 102) and that it was not unambiguously supported by the empirical data (63, 106, 158). Interest in arthropod response to polycultures, however, remained (106, 158).

These critiques served to shift the focus of research to goals related to pest control. Control of arthropod pests is not usually a question of stabilizing pest populations, it is usually a question of suppressing pest populations. If a large population density of pest arthropods can be tolerated, then one goal of pest control could be to reduce the magnitude of population fluctuations of the pest. Often, however, large pest populations are intolerable, so the goal of pest control is to lower pest population density (106, 145, 158). Indeed, Perrin (115) suggested that because polycultural cropping systems are so prevalent in many areas of the world, it behooves us to understand the ecology of arthropod response to polyculture in order to improve pest management in these systems.

The modern theory on the effects of polyculture on arthropods can be traced primarily to the writings of Elton (51, 52) and Root (129). Elton (52) originally suggested that arthropod herbivore populations would be more likely to surge to outbreaks in monocultures than polycultures. Pimentel (116) reasoned that if this hypothesis were true, then the time-averaged population density of arthropod herbivores would be higher in monocultures than poly-cultures. Furthermore, if the magnitude of population fluctuations were similar in both monocultures and polycultures, then population outbreaks would be more likely in monocultures than in polycultures. Tahvanainen & Root (147) called this phenomenon *associational resistance;* plants associated with taxonomically diverse plant species would suffer less herbivore attack than plants not so associated.

Another major prediction is that herbivore response to polycultures will be determined by natural enemies, resource concentration, or both. According to Root's *enemies hypothesis* (129), generalist and specialist natural enemies are expected to be more abundant in polycultures and therefore suppress herbi-vore population densities more in polycultures than in monocultures. General-ist predators and parasitoids should be more abundant in polycultures than monocultures because (*a*) they switch and feed on the greater variety of herbivores that become available in polycultures at different times during the growing season (51), (*b*) they maintain reproducing populations in polycul-tures while in monocultures only males of some parasitoids are produced (87), (*c*) they can utilize hosts in polycultures that they would normally not encounter and use in monocultures (32, 87), (*d*) they can exploit the greater variety of herbivores available in different microhabitats in the polyculture (129), and prey or hosts are (*e*) more abundant (23, 134) or (*f*) more available in polycultures. The amount of time available for predaceous carabid beetles to forage for prey was greater in polycultures than monocultures probably because polycultures had a moister, shadier soil surface microclimate, which enabled some of the beetles to forage during the day as well as at night (38). Detailed study on the movement of two carabid species, however, implied that monocultures can be more favorable habitats than polycultures at some plant densities (114).

Specialist predator and parasitoid populations are expected to be more abundant and effective in polycultures than monocultures because (*g*) prey or host refuges in polycultures enable the prey or host populations to persist, which stablizes predator-prey and parasitoid-host interactions, while in monocultures predators and parasitoids drive their prey or host populations to extinction and become extinct themselves shortly thereafter (129). Prey or host populations will recolonize these monocultures and rapidly increase. Sheehan (134) suggested that specialist parasitoids might be less abundant in polycultures than monocultures because (*h*) chemical cues used in host finding will be disrupted and the parasitoids will be less able to find hosts to parasitize and feed upon in polycultures and (*i*) the indistinct boundary at the edges of polycultures will be hard to recognize and they will be more likely to leave polycultural habitats than monocultures. In addition, Andow & Prokrym (22) showed that structural complexity, or the connectedness of the surface on which a parasitoid searches, can strongly influence parasitoid host-finding rates; an implication is (*j*) that structurally complex polycultures would have less parasitism than structurally simple monocultures.

Finally, both generalist and specialist natural enemies should be more abundant in polycultures than monocultures because (*k*) more pollen and nectar resources are available (39) at (*l*) more times during the season in polycultures than monocultures (129, 153).

I do not attempt a comprehensive review of the enemies hypothesis here, but both Sheehan (134) and Russell (131) have reviewed parts of this hypothesis. At this juncture, suffice it to say that only one published study has evaluated the relative influence of hypotheses *a-l* on a natural enemy population. Populations of the coccinellid *Coleomegilla maculata* were examined in maize monocultures and maize-bean-squash polycultures (23). For this generalist predator, only hypotheses *a*, phenology of alternate prey; *e*, abundance of total prey; *k*, abundance of pollen and nectar; and *l*, phenology of pollen and nectar, were compared, although *b*, *c*, and *g* were not applicable. Predator densities were greater in monocultures where prey and pollen abundance was greater, not in polycultures where the seasonal distribution of potential foods was more even. *C. maculata* had a higher foraging success rate where prey abundance (prey/plant surface area) was greater, and as a consequence, stayed in monocultures longer than polycultures. Further investigation of hypotheses *a-l* are needed before generalizations about natural enemies will be possible.

According to the *resource concentration hypothesis,* many herbivores, especially those with a narrow host range, are more likely to find and remain on host plants that are concentrated, i.e. that occur in large, dense, or pure stands (129). Other stand characteristics, such as the perimeter-to-area ratio of the stand, and stand quality can affect herbivore foraging behavior (129) and could be included in the resource concentration hypothesis, but here I focus on the effects of stand purity. Because some authors (10, 131) attempt to distinguish between resource concentration, associational resistance, and plant apparency (55), a note of clarification is in order. Associational resistance refers to the reduced herbivore attack that a plant experiences in association with genetically or taxonomically diverse plants and occurs because either the enemies hypothesis, the resource concentration hypothesis, or both hypotheses are occurring. Plant apparency refers to the ease of host finding by herbivores in relation to plant life-history characteristics (55). Thus, plants experience associational resistance, and arthropods can respond to plant-stand characteristics, i.e. resource concentration, or plant life-history characteristics, i.e. plant apparency.

TESTS OF ASSOCIATIONAL RESISTANCE

Associational resistance can be tested by comparing the effect of arthropod herbivores on plants that are associated with taxonomically (or genetically) different plants with plants that are associated with either other similar plants or no other plants. In experimental work, the first comparison is called a replacement series or substitutive design, in which individual plants are substituted for each other in experimental plots, and the second comparison is called an additive design, in which the taxonomically different plants are added to the original plant population. Combining both designs, an exhaustive review of the literature conducted for this article has produced 209 studies on 287 herbivorous arthropod species (for a complete list of species, see 16, supplemented by 4–7, 9, 24, 25, 31, 33, 34, 38, 40, 42, 45, 50, 61, 62, 67–70, 82, 85, 88, 89, 92–94, 97, 100, 107, 109, 114, 118, 119, 135, 139, 151, 156, 159). Approximately 51.9% of these herbivores (149 species) had lower population densities on plants in polycultures, while only 15.3% (44 species) had higher densities on plants in polycultures (Table 1). Thus, plants in polycultures are likely to have fewer individuals of a given species feeding on them than plants in monocultures.

Table 1 Numbers of arthropod species with particular responses
to additive and substitutive polycultures[a]

| | Population density of arthropod species in polyculture compared to monoculture | | | |
	Variable[b]	Higher	No change	Lower
Herbivores	58	44	36	149
	(20.2)	(15.3)	(12.5)	(51.9)
Monophagous	42	17	31	130
	(19.1)	(7.7)	(14.1)	(59.1)
Polyphagous	16	27	5	19
	(23.9)	(40.3)	(7.5)	(28.4)
Natural enemies	33	68	17	12
	(25.6)	(52.7)	(13.2)	(9.3)
Predators	27	38	14	11
	(30.3)	(42.7)	(15.7)	(12.4)
Parasitoids	6	30	3	1
	(15.0)	(75.0)	(7.5)	(2.5)

[a] Percentage of total number of species is in parentheses. Additional details
of the methods used are presented elsewhere (16).
[b] A variable response means that an arthropod species did not consistently
have a higher or lower population density in polycultures compared to
monocultures when the species response was studied several times.

Although individual arthropod species are likely to be less abundant, more
herbivore species may feed on plants in polycultures than monocultures.
Herbivore load is the total biomass (dry weight) of herbivores per unit dry
weight of consumable plant material (129) and is meant to estimate the impact
of the entire herbivore fauna on a plant in both polycultures and monocultures. Herbivore load should, however, be considered only a partial measure
because the same biomass of different herbivore species acting at different
times during plant growth is likely to have different effects on the plant. Root
(129) showed that collards grown next to meadow vegetation had lower
herbivore loads than collards grown next to collards, but herbivore loads have
rarely been measured in polycultures and monocultures.

Another way to assess the effect of the entire herbivore fauna on plants in
polycultures and monocultures is to eliminate the herbivores from both systems with insecticides (20). Factorial experimental designs, which compare
monocultures and polycultures with and without arthropod pests, allow comparison of absolute and proportional yield loss in monocultures and polycultures. Six published studies yield data amenable to this analysis (15, 33, 89,
109, 110, 159). Absolute (or proportional) yield loss was lower in polycultures than in monocultures in 14 (or 10) cases out of 20 in which arthropod
herbivore density was lower in polyculture than monoculture. In a few cases,
yield loss was higher in the polycultures than the monocultures. The studies

suggest that yield loss in polycultures should be less than or equal to yield loss in monocultures when the crop is not tolerant of herbivore injury, and when interspecific plant competition is not very severe in the polyculture.

The empirical data confirm the associational resistance hypothesis, that is, herbivorous arthropod species are less abundant on plants in polycultures on average, and plants in polycultures frequently experience lower yield loss from herbivores than plants in monocultures. Results vary in specific cases, however. Some (15.3%) herbivores are more abundant in polycultures than monocultures and many (20.2%) respond variably to polyculture. In addition, yield losses can be unaffected, or even higher in polycultures. While the associational resistance hypothesis is generally confirmed, it is not yet particularly useful for pest management and needs to be refined to account for the numerous exceptions that occur.

ENEMIES VERSUS RESOURCE CONCENTRATION

The enemies and resource concentration hypotheses are not mutually exclusive hypotheses, but how they might be related is unclear. Are they complementary, in the sense that their actions are mutually reinforcing in polycultures (131), or are they antagonistic, so that the effect of resource concentration acts in opposition to the effect of natural enemies, as implied by recent work on natural enemies in maize polycultures (23, 24)? Approximately 52.7% (68 species) of the natural enemy species had higher densities on plants in polycultures, but only 9.3% (12 species) had lower densities (Table 1). In the experimental studies reviewed by Russell (131), nine studies reported higher mortality rates from predators or parasitoids in polycultures, four reported no difference, and only two reported lower rates in polycultures. Thus, the empirical literature supports the conclusion that natural enemies and resource concentration are complementary mechanisms (131). The two studies in which lower rates of predation and parasitism were observed in polycultures, however, were conducted in maize-bean-squash and maize-clover polycultures (23, 24). Therefore, in some polycultures natural enemies and resource concentration may be antagonistic mechanisms. In addition, the influence of prey or host density and the spatial proximity of the experimental plots may be confounded in these results. A more robust theory that predicts when natural enemies and resource concentration are complementary, independent, and antagonistic mechanisms needs to be developed.

A different question that has generated considerable investigation asks which hypothesis better accounts for the observed responses of arthropod herbivores to polycultures. Given that both mechanisms occur together, does one explain the variation in herbivore response better than the other?

Before testing theoretical predictions with empirical data, several caveats about the empirical literature should be made. First, experimental comparisons between monocultures and substitutive polycultures confound vegetational diversity and plant density because the density of each component plant species is lower in the polyculture than the monoculture. This influences herbivore response. Examples are: the aphid *Aphis craccivora* was more abundant on peanuts in substitutive polycultures but less abundant in additive polycultures (54); the aphid *Brevicoryne brassicae* was less abundant on brussels sprouts in additive polycultures, but had similar abundance in substitutive polycultures (7); and the cicadellid *Dalbulus maidis* was less abundant on maize in additive polycultures, but had similar abundance in substitutive polycultures (118). Power's (118) observations are particularly instructive about the complications involved. A maize-bean polyculture and a low- and high-density maize monoculture were planted so that the low-density monoculture–polyculture comparison was an additive comparison and the high-density monoculture–polyculture comparison was a substitutive comparison. Individual *D. maidis* left both additive and substitutive monocultures slower than the polyculture, but remaining populations were diluted to a lower per-plant density in the substitutive (high-density) monoculture than the additive (low-density) monoculture. Because host density and vegetational diversity can be subtly confounded in substitutive designs, I report only herbivore response in additive polycultures.

Second, several arthropod species have been studied numerous times and have exhibited variable responses to polycultures. I used a 70%-consistency criterion for classifying species responses. If more than 70% of the studies showed a consistent response by that herbivore species, then it was classified as having either a higher, no change, or lower population density in polyculture. Otherwise it was classified as having a variable response. Further methodological details are available (16). Finally, the literature is of highly variable quality, ranging from casual observations to rigorous experimental evaluations. While casual observations by excellent observers probably have considerable validity, I evaluated herbivore response both in additive experiments in which investigators conducted statistical hypothesis testing (4, 7, 8, 13, 18, 21, 23, 34, 43, 48–50, 53, 54, 58, 61, 66, 70, 72–80, 82, 85, 88, 89, 91, 92, 94, 101, 104, 110, 113, 117, 118, 124–126, 132, 133, 135, 136, 138, 140, 148–151, 155, 159–161) and in all reported additive polycultures. When I restricted the data to statistically analyzed, additive experimental designs, a greater proportion of species are evaluated as having a variable response and no change in density compared with all additive polycultures (Table 2). This probably occurred because many of the casual observations are single observations, which makes variable responses impossible, and because researchers are more likely to report casual observations in which arthropod densities are different in monocultures and polycultures. Given

Table 2 Numbers of herbivore species with particular responses to additive polycultures for monophagous and polyphagous species[a]

	Population density of arthropod species in polyculture compared to monoculture			
	Variable	Higher	No change	Lower
All additive polycultures				
Herbivores	28	24	14	70
	(20.6)	(17.6)	(10.3)	(51.5)
Monophagous	21	8	11	62
	(20.6)	(7.8)	(10.8)	(60.8)
Polyphagous	7	16	3	8
	(20.6)	(47.1)	(8.8)	(23.5)
Additive polycultures with statistical tests				
Herbivores	24	10	14	32
	(30.0)	(12.5)	(17.5)	(40.0)
Monophagous	18	3	11	27
	(30.5)	(5.1)	(18.6)	(45.8)
Polyphagous	6	7	3	5
	(28.6)	(33.3)	(14.3)	(23.8)

[a] Percentage of total number of species is in parentheses.

these caveats, the natural enemies and resource concentration hypotheses can be compared.

Monophagous vs Polyphagous Herbivores

Aiyer (2) distinguished monophagous arthropod herbivores, which feed on only one of the plant species in a polyculture, and polyphagous herbivores, which feed on more than one plant species in the polyculture. The enemies and resource concentration hypotheses generate different predictions about the relative abundance of monophagous and polyphagous herbivores in monocultures and polycultures (15, 16, 128). A monophagous herbivore is predicted to have a lower abundance in polycultures regardless of the relative influence of the two hypotheses. But for a polyphagous species, if the enemies hypothesis better accounted for its response, its population density should be lower in polycultures than monocultures because predators and parasitoids are not known to prey and parasitize differentially on monophagous and polyphagous herbivores. On the other hand, if the resource concentration hypothesis better accounted for the polyphagous herbivore's response, it should frequently be more abundant in additive polycultures than monocultures.

The effect of arthropod-host range on arthropod response to polycultures is substantial (Table 2). More polyphagous species had higher density in polycultures (33.3%) than monophagous species (5.1%) and more monophagous species had lower density in polycultures (45.8%) than polyphagous species (23.8%) (X^2 = 12.43, 3 d.f., p < 0.05). In addition, particular arthropod herbivores respond to polycultures differently depending on the number of host plants in the polyculture. For example, the cicadellid *Scaphytopius acutus* had higher population density on peach trees associated with a ground cover of red clover or mixed rosaceous weeds, which are favored host plants, but lower population density on peach trees associated with a ground cover of a nonhost grass compared to monocultures (104). Similar results were obtained for the chrysomelids *Paranapiacaba waterhousei* and *Cerotoma ruficornis rogersi* (125, 126, 157). These data suggest that resource concentration has a greater influence on herbivore response to polyculture than natural enemies, but do not exclude the possibility that natural enemies act concurrently (15, 16, 128, 134).

The predicted effect of resource concentration on polyphagous herbivores is considerably more obscure than its effect on monophagous herbivores. Two types of polyphagous herbivores can be distinguished (17, 19): sequential polyphages that alternate hosts between generations in a temporal sequence and simultaneous polyphages that alternate hosts within a generation with individuals moving from host to host. If immigration into polycultures limits herbivore populations in polycultures because either host finding is more difficult or the number of potential immigrants is low, then sequential polyphages are expected to have higher populations on their second host in polycultures than in monocultures because they have already colonized the polycultures during their first generation. If populations are limited because herbivores leave polycultures much faster than monocultures, then sequential polyphages are expected to have lower populations on their second host in polycultures than in monocultures because they will leave the polycultures and accumulate in the monocultures.

The empirical data are inconclusive on this issue. Of the nine populations of sequential polyphages I reviewed (17), seven were more abundant and only one was less abundant in polycultures than in monocultures on the second host (one showed no response). In eastern Washington, the aphid *Myzus persicae* builds up populations on weeds in the spring and transfers to sugar beet in later generations. Wallis & Turner (162) burned out the weeds from drainage ditches in a 22-square-mile area, and, compared to an unburned area, populations were 51–91% lower, and incidence of beet western yellow disease, which is transmitted by *M. persicae,* was 77–84% lower. Similarly, Barnes (30) suggested that the lygalid *Nysius raphanus* became a pest on grape after building up populations on nearby weed patches, and Laster & Meridith (90) suggested that the mirid *Lygus lineolaris* was a greater pest on cotton after building up populations on an early host weed. These examples suggest that immigration to polycultures limits herbivore use of polycultures. Several factors complicate the analysis of sequential polyphages (17, 19): the initial host may be such a poor host that it has little effect on subsequent populations (29, 56, 59); arthropods might leave polycultures rather than transfer to the second host because the second host does not become suitable until later; and other factors such as weather might control populations on the second host.

The predictions of the resource concentration hypothesis do not extrapolate for simultaneous polyphages. Monophagous herbivores can be considered a special case of simultaneous polyphages, in which one plant is preferred and suitable as food while the other plant is not preferred and is unsuitable as food. Generalizing to simultaneous polyphages, the resource concentration hypothesis would predict that herbivore abundance will be lower on the more preferred, more suitable host in polycultures than the comparable monoculture. Contrary to this prediction, all possible responses have been observed (19). Simultaneous polyphages are both less abundant (36, 37, 82, 126) and more abundant (89, 126, 152) on the more-preferred and more suitable host in polycultures; conversely, they are both more abundant (81, 103, 104, 126) and less abundant (9, 61, 146) on the less-preferred and less suitable host in polycultures. Of 29 examples summarized (17), 13 populations were less abundant and 15 populations were more abundant in polycultures than monocultures. The details of herbivore movement and host preference will probably be essential for understanding these responses (17, 19). For example, movement by *D. maidis* across maize rows was inhibited by beans, which influenced the herbivore's population dynamics and its vectoring of corn stunt disease (118). Furthermore, host preference can change rapidly as host quality changes. The geometrid *Alsophila pometaria* prefers to feed on young leaf tissue of both *Quercus coccinea* and *Q. alba*. As host tissue ages, food quality rapidly changes. Unexpectedly, when either plant is rare in association with the other, it suffers greater herbivory from *A. pometeria* than when it is common (60). *Q. coccinea* leafs out about 10 days earlier than *Q. alba*, and when *Q. coccinea* is rare, larvae accumulate on it because it is the only food available. When *Q. alba* is rare, larvae accumulate on it because it is the best food available; all of the *Q. coccinea* leaves are older and less preferred at the time *Q. alba* leafs out.

Annual vs Perennial Polycultures

Several authors have suggested that herbivores have greater difficulty finding annual plants than perennial plants (55, 123) and that natural enemies have a greater effect on arthropod herbivores in perennial cropping systems than in annual cropping systems (64, 65, 120, 141). These ideas imply that differences between monocultures and polycultures are more likely related to the effects of resource concentration in annual systems, while in perennial systems the differences are more likely related to the action of natural enemies

Table 3 Numbers of herbivore species with particular responses to additive and substitutive annual and perennial polycultures for monophagous and polyphagous species[a]

| | Population density of arthropod species in polyculture compared to monoculture | | | |
	Variable	Higher	No change	Lower
Annual	51	24	31	100
	(24.8)	(11.7)	(15.0)	(48.5)
Monophagous	39	6	27	83
	(25.2)	(3.9)	(17.4)	(53.5)
Polyphagous	12	18	4	17
	(23.5)	(35.3)	(7.8)	(33.3)
Perennial	7	20	5	49
	(8.6)	(24.7)	(6.2)	(60.5)
Monophagous	3	11	4	47
	(4.6)	(16.9)	(6.2)	(72.3)
Polyphagous	4	9	1	2
	(25.0)	(56.3)	(6.2)	(12.5)

[a] Percentage of total number of species is in parentheses.

(15, 128). This hypothesis generates two predictions: if resource concentration better accounts for herbivore response to polycultures, then (*a*) more herbivore populations should be more abundant in perennial polycultures than in annual polycultures and (*b*) the difference between monophagous and polyphagous herbivores discussed above should be greater in annual systems than in perennial systems.

In using all reported responses of arthropods to substitutive and additive polycultures because the number of arthropod species examined in statistically analyzed, additive polycultures is too few, the empirical data suggest that (*a*) more herbivore species were more abundant in perennial polycultures (24.7%) than in annual polycultures (11.7%) and (*b*) the difference between monophagous and polyphagous herbivores was greater in annual systems than in perennial systems (Table 3). These data provide additional evidence that the response of arthropods to polycultures is more likely to be related to the effects of resource concentration than to the action of natural enemies.

Clear support for the effect of perenniality on herbivore response to polyculture is still lacking. No study analyzing arthropod response in comparable annual and perennial cropping systems has been published, and the current literature has not implicated perenniality as a significant factor in the response of an herbivore to polyculture. Specifically, the comparison between annual and perennial systems entails many factors that are confounded with perenniality, including the degree of soil disturbance and the year-to-year continuity of the crop. For example, eggs of the noctuid *Euxoa ochrogaster,* which feeds on asparagus, usually hatch in the spring before asparagus shoots start to grow. If they hatch near patches of spring weeds, the young larvae survive to feed on the asparagus when it sprouts—otherwise they die. Thus, the lack of soil disturbance and the availability of spring weeds, not the perenniality of asparagus, lead to increased damage by *E. ochrogaster* (148).

Case Studies

The structure of most experimental tests of the relative strength of resource concentration and natural enemies has been one of sufficiency; that is, the response of the population under study has been consistent with predictions from one of the hypotheses and a plausible mechanism has been proposed. For example, in support of the resource concentration hypothesis, Tahvanainen & Root (147) demonstrated that the chrysomelid *Phyllotreta cruciferae* colonized host plants in monocultures faster than polycultures (see also 50, 85). Similarly, Risch (127) showed that the chrysomelid *Acalymma themei* adults colonized monocultures faster than polycultures, and suggested that beetles stayed longer on host plants than nonhosts and therefore would stay longer in monocultures. Bach (27) clearly demonstrated that this mechanism occurred for *Acalymma vittatum,* and for *Acalymma innubrum* Bach (28) showed that both host finding and host leaving were involved. In support of the enemies hypothesis, Letourneau (93) showed that parasitoid abundance and parasitism rates on a squash herbivore were higher in polycultures and that predation by the anthocorid *Orius tristicolor* on thrips was higher in polycultures (94). Also, Speight & Lawton (142) showed that the carabid *Pterostichus melanarius* was more abundant and predation rates were higher in polycultures.

Several studies provided convincing evidence that natural enemies had no effect on the herbivores, but this evidence has been based on a dearth of natural enemies in the experimental systems. For example, parasitism of *P. cruciferae* was rare during peak beetle densities and the researchers argued that predation was uncommon and consequently natural enemies could not have contributed to beetle response to polyculture (147). Similarly, natural enemies probably had little effect on *A. themei* and *A. vittatum* because parasitoids were not found (27) or were extremely rare (127). Predators were also rare and no more abundant in polycultures than monocultures (27, 127), and prey disappearance was low and similar in each habitat (127). Although the absence of natural enemies might be characteristic of these agricultural systems, experimental tests in which natural enemies exerted significant mortality would be more convincing tests of the two hypotheses.

Tests of the relative influence of resource concentration and natural enemies have been structured as if they were independent. The enemies hypothesis has been assessed by examining the expected mechanism (predation and parasitism by natural enemies), while the resource concentration hypothesis has been assessed by inferring the mechanism from patterns of herbivore movement. Unfortunately, these two different approaches do not share the same units of measure and therefore the relative significance of the two hypotheses generally cannot be critically evaluated.

One way of solving this problem is to use demographic analysis. Arthropod response can be partitioned into its demographic components, i.e. colonization, per capita oviposition, and mortality of the immatures and adults. The enemies hypothesis represents one mortality factor and the resource concentration hypothesis is one mechanism influencing colonization. Demographic analysis (98) measures the relative influence of the demographic components on the population response of the herbivore.

This analysis was used to examine the response of the coccinellid *Epilachna varivestis,* an herbivore of beans, *Phaseolus vulgaris,* to three different polycultures (18). Adult colonization was lower and egg mortality from predators was higher in polycultures, so both the resource concentration and enemies hypotheses were confirmed. Demographic analysis, however, showed that adult colonization, which accounted for 58% of the variation in beetle population density, was the most important demographic component to respond to polyculture. Larval and pupal mortality was of secondary importance, accounting for 30% of the variation, and egg mortality was insignificant, accounting for less than 2% of the variation. Moreover, larval and pupal mortality was highest in the monoculture, which directly contradicts the enemies hypothesis. In this case, resource concentration, not natural enemies, caused beetle populations to be lower in polycultures.

Resource Concentration

Although still incomplete, current evidence suggests that the resource concentration hypothesis better accounts for the observed patterns of herbivore response to polycultures. This generalization, however, is of rather limited use for making predictions about specific herbivores in specific polycultures. About 30% of all herbivores have a variable response to polyculture (Table 2), and many exceptions to the generalization occur. Some monophagous herbivores are more abundant in polycultures and the response of polyphagous herbivores cannot be predicted in advance. Furthermore, natural enemies can strongly influence herbivores in polycultures. Indeed, a theory that predicts when natural enemies will exert significant influence on herbivores in polycultures awaits development.

MECHANISMS UNDERLYING RESOURCE CONCENTRATION

Resource concentration is predicted to influence the number of herbivores per plant by modifying the balance between immigration to and emigration from the plant (129). Plants in polycultures may sustain lower herbivore populations because herbivores have difficulty finding them, leave them more quickly, or have difficulty relocating them after leaving. Behavioral observation can demonstrate that an herbivore has difficulty finding its host, although this demonstration can be complicated. For example, the anthomyiid *Delia brassicae* laid fewer eggs near host plants in polycultures than monocultures (132). Laboratory flight experiments, however, showed that flies landed on trays containing hosts in polycultures at the same rate as on trays containing hosts in monocultures (44). Host plants were not hidden by nonhosts; instead the nonhosts caused flies to move more, spend less time laying eggs, and leave host plants faster (44).

In addition, properly designed population studies can evaluate the effects of host finding and host leaving. Both Kareiva (85) and Elmstrom et al (50) showed that data from frequent observations during the colonization period into polycultures and monocultures can be used to estimate immigration and emigration parameters using simple population dynamic models. Using population accumulation curves and observations on the disappearance rate of extremely high densities of *P. cruciferae,* Elmstrom et al (50) showed that polycultures reduced host-finding and increased host-leaving rates compared to monocultures. The causes of these responses remain obscure, and the relative effects of the following mechanisms have not been critically examined, compared, or distinguished in any published study.

Chemical and Visual Interference

The diversity of olfactory stimuli emanating from polycultures might mask the olfactory cues used by monophagous herbivores to find their host plants or otherwise confuse or repel these herbivores. In a choice test between host plants with tomato or ragweed odors versus host plants alone, Tahvanainen & Root (147) showed that *P. cruciferae* was more likely to move to host plants alone than host plants associated with nonhost odors. Although their experimental design may not have controlled for equal concentrations of host odors, their results suggested that nonhost plant chemicals reduced the probability of host use by the beetle. The precise mechanism, however, is not known. While some nonhost plant odors are repellent to some herbivores (1), the sensory basis of chemical masking is obscure. For a nonhost plant chemical to mask a host odor, the nonhost odor must interfere with the neural output from the host-odor receptor or affect central-nervous-system processing of the host-odor stimulus.

Stanton (144) proposed a simple model of host-plant finding by herbivores using long distance olfactory stimuli. Herbivores respond to their olfactory stimuli upon random encounter with a part of the odor plume in which odor concentration is greater than their receptor sensitivity. Concentration gradients in the odor plumes are determined by the size (larger size equals wider plume) and density (greater density equals greater odor concentration and plume length) of the host stand. Stanton (144) predicted that an herbivore with relatively insensitive olfactory receptors will be unable to detect subtle quantitative differences in these concentration gradients because diffusive and convective processes rapidly dissipate host olfactory stimuli to low, undetectable concentrations short distances away from the host stand. An herbivore with highly sensitive receptors will be able to respond to subtle quantitative differences in concentration gradients of host odors because it can detect the very low concentrations far from the host stand. One can generalize this model to polycultural systems: if nonhost plant odors interfere in some way with herbivore olfaction, then host finding by herbivores with low olfactory sensitivity is unlikely to be affected by polycultures, whereas herbivores with high olfactory sensitivity should be much less common in polycultures. These ideas have not yet been critically tested.

Polycultures might also interfere with visual host finding cues. Smith (138) found that aphid colonization of brussels sprouts was less in polycultures than monocultures and was less when green burlap was placed between host plants than when brown burlap was so placed. The nonhost plants and the green burlap may have reduced the contrast between green plants and brown soil and made the host plants less attractive to colonizing aphids. Rauscher (121) showed that the papilionid *Battus philenor* recognized host plants by leaf shape and oviposited less on host plants surrounded by nonhost vegetation compared to host plants around which all nonhost vegetation had been clipped (122).

Plant Quality

Modification of host-plant quality is concomitant with polycultural systems. It is highly improbable that a different plant species can be added or substituted into a stand of another species without altering intra- and interspecific plant competition and some aspect of host-plant quality. Plant quality can influence herbivore host finding because different quality plants can release different concentrations of chemicals used as host-finding stimuli by herbivores (57), but this effect has not been documented in polycultures. More commonly, the herbivore encounters plants of variable quality and departs from poorer-quality plants more rapidly (83). Few investigators have attempted to evaluate the effects of host quality on the response of herbivores in polycultures, but none have completely succeeded.

One approach is to measure some aspect of host-plant quality in monocultures and polycultures. If no differences are observed, then the inference is that host-plant quality exerts little effect (18, 105); if differences are observed, then the effects are removed by covariance analysis (21). Although this approach can be a reasonable approximation, it is inadequate because unobserved or unmeasured qualitative factors could be important to the herbivores. Another approach is to grow host plants in pots and expose the potted plants in monocultures and polycultures to foraging herbivores (46, 97). Although this approach controls for variation in plant quality, potted plants are invariably more light, nutrient, or water stressed than naturally growing plants, and this stress could obscure some of the effects of polyculture. The final approach has been to deliberately alter host-plant quality, such as by using fertilizers (118). This allows quantitative comparison of the effects of polyculture and one particular aspect of plant quality, but other untested, qualitative factors might be important in polycultures. Risch (127) was fortunate to find that the preferred, higher-quality host plants of A. themei were in polycultures, so plant quality could not have caused beetle populations to be lower in polycultures.

Plant quality can exert subtle effects on herbivores. For example, aphids on squash plants were less abundant in maize-bean-squash polycultures than in squash monocultures (23). The plants in both systems had the same number of leaves, but the leaves were larger and older in squash monocultures because the shaded squash leaves in polycultures senesced more rapidly. Aphids were invariably found on the older leaves in both systems and reached very high densities on the oldest leaves in the monocultures. No very old leaves were in the polycultures, so aphids did not have the opportunity to reach the population densities that occurred in the monocultures.

The effect of plant quality can be influenced by other factors. P. cruciferae responded to small host-plant size by rapidly leaving the plants (83). Its ability to respond to variation in host size, however, was influenced by the proximity of nearby hosts. If larger hosts were nearby, the beetle discriminated among hosts and accumulated on the larger hosts. If these hosts were farther away, beetles could not locate the larger hosts and appeared to discriminate poorly.

Nonhost Surface Area

Herbivores encounter nonhost plant surfaces as they forage in polycultures and these encounters can influence movement patterns and population density. For example, Empoasca fabae (86) and Epilachna varivestis (18) population densities were negatively correlated with nonhost-plant biomass in polycultures. The most exacting potential mechanism has been called the "fly-paper" effect (154); upon encounter with nonhost plants, individuals are "lost" to the population. For example, Farrell (54) suggested that aphid densities were lower in certain peanut-bean mixtures because aphids were caught on the hooked trichomes on bean leaves. Two more moderate hypotheses can be distinguished. If herbivores do not discriminate effectively between host and nonhost tissue, they can be "diluted" on the nonhost surface and fewer would be associated with their host plant in polycultures. More frequently, herbivores can effectively distinguish hosts and nonhosts, and typically they will leave nonhosts more rapidly than hosts (18, 26, 27, 28, 127). In this searching process, herbivores waste time searching on nonhost plants, and, as a consequence, leave polycultures more rapidly than monocultures.

Other Factors

Polycultures could provide more shade, protection from desiccation by wind, lower mid-day temperatures, and other modifications of microhabitat. These modifications can affect herbivore movement and the activity of natural enemies. For example, shade caused A. *themei* (127) and A. *vittatum* (17) to leave host squash plants rapidly, and population densities of both beetles were lower in shady maize-bean-squash polycultures than in the sunny monocultures. Additional factors include edge effects (144) and the structural complexity or connectedness of surfaces on which herbivores search (22).

TOWARD AN ELABORATION OF THEORY

While some of the major twists in the Gordian knot of vegetational diversity can be perceived, we are a long way from unraveling its complexity. Broad predictions of the theory of herbivore response to polycultures have been confirmed by the empirical evidence, yet these generalizations appear to be of limited use in pest management. The associational resistance hypothesis has many exceptions, and these cannot yet be accounted for. The resource concentration hypothesis largely accounts for many of the observed responses by herbivores to polyculture, but the response of polyphages cannot yet be predicted. Moreover, a theory that predicts when natural enemies will exert significant mortality in polycultures is entirely lacking.

Perhaps we have been too bold to suggest that a single ecological theory would account for the variety of responses in a taxonomic group as species rich as are the Arthropoda. The evolutionary history of particular arthropodan lineages could require multiple ecological theories. For example, *Pieris rapae* (Lepidoptera: Pieridae) is sometimes more abundant in polycultures than monocultures in the eastern United States (130, but see 21). The reasons for its unusual response may be related to its evolutionary history (130). A host plant of *P. rapae* is a highly unpredictable environment for the developing larva whether it is in a polyculture or a monoculture. Apparently to minimize the variance in generation-to-generation replacement rate, ovipositing butterflies scatter their eggs widely among potential hosts by tending to fly

580   ANDOW

linear routes and passing over many potential hosts (130). Aphids, with weak flight abilities and parthenogenic reproductive biology, may respond to ecological and evolutionary forces very differently than pierid butterflies. Evolutionary theory might provide a thread by which to further unravel the complexity of vegetational diversity.

Literature Cited

1. Ahmad, S. 1983. *Herbivorous Insects: Host Seeking Behavior and Mechanisms.* New York: Academic
2. Aiyer, A. K. Y. N. 1949. Mixed cropping in India. *Indian J. Agric. Sci.* 19:439–543
3. Akhanda, A. M., Mauco, J. R., Green, V. E., Prine, G. M. 1978. Relay intercropping peanut, soybean, sweet potato and pigeon pea in corn. *Proc. Soil Crop Sci. Fla.* 37:95–101
4. Ali, A. D., Reagan, T. E. 1985. Vegetation manipulation impact on predator and prey populations in Louisiana sugarcane ecosystems. *J. Econ. Entomol.* 78:1409–14
5. Ali, A. D., Reagan, T. E. 1986. Influence of selected weed control practices on araneida faunal composition and abundance in sugarcane. *Environ. Entomol.* 15:527–31
6. Ali, A. D., Reagan, T. E., Flynn, J. L. 1984. Influence of selected weedy and weed-free sugarcane habitats on diet composition and foraging activity of the imported fire ant (Hymenoptera:Formicidae). *Environ. Entomol.* 13:1037–41
7. Altieri, M. A. 1984. Patterns of insect diversity in monocultures and polycultures of brussels sprouts. *Prot. Ecol.* 6:227–32
8. Altieri, M. A., Francis, C. A., van Schoonhoven, A., Doll, J. A. 1978. A review of insect prevalence in maize (*Zea mays* L.) and bean (*Phaseolus vulgaris* L.) polycultural systems. *Field Crops Res.* 1:33–44
9. Altieri, M. A., Gliessman, S. R. 1983. Effects of plant diversity on the density and herbivory of the flea beetle, *Phyllotreta cruciferae* Goeze, in California cropping systems. *Crop Prot.* 2:497
10. Altieri, M. A., Letourneau, D. K. 1982. Vegetation management and biological control in agroecosystems. *Crop Prot.* 1:405–30
11. Altieri, M. A., Letourneau, D. K. 1984. Vegetation diversity and insect pest outbreaks. *CRC Crit. Rev. Plant Sci.* 2(2):131–69
12. Altieri, M. A., van Schoonhoven, A., Doll, J. A. 1977. The ecological role of weeds in insect pest management systems: A review illustrated by bean (*Phaseolus vulgaris*) cropping systems. *PANS* 23:195–205
13. Altieri, M. A., Whitcomb, W. H. 1980. Weed manipulation for insect pest management in corn. *Environ. Manage.* 4:483–89
14. Andow, D. A. 1983. The extent of monoculture and its effects on insect pest populations with particular reference to wheat and cotton. *Agric. Ecosys. Environ.* 9:25–35
15. Andow, D. A. 1983. Effect of agricultural diversity on insect populations. In *Environmentally Sound Agriculture,* ed. W. Lockeretz, pp. 91–115. New York: Praeger
16. Andow, D. A. 1986. Plant diversification and insect population control in agroecosystems. In *Some Aspects of Pest Management,* ed. D. Pimentel, pp. 277–348. Ithaca, New York: Dep. Entomol. Cornell Univ.
17. Andow, D. A. 1988. Management of weeds for insect manipulation in agroecosystems. In *Weed Management in Agroecosystems: Ecological Approaches,* ed. M. A. Altieri, M. Z. Liebman, pp. 265–301. Boca Raton, FL: CRC Press
18. Andow, D. A. 1990. Population dynamics of an insect herbivore in simple and diverse habitats. *Ecology* 71:1006–17
19. Andow, D. A. 1990. Control of arthropods using crop diversity. In *CRC Handbook on Pest Management,* ed. D. Pimentel, Boca Raton, FL: CRC Press In press. 2nd ed.
20. Andow, D. A. 1991. Yield loss in vegetationally diverse habitats. *Environ. Entomol.* Submitted
21. Andow, D. A., Nicholson, A. G., Wien, H. C., Willson, H. R. 1986. Insect populations on cabbage grown with living mulches. *Environ. Entomol.* 15:293–99
22. Andow, D. A., Prokrym, D. R. 1990. Plant structural complexity and host-

finding by a parasitoid. *Oecologia* 82: 162–65

23. Andow, D. A., Risch, S. J. 1985. Predation in diversified agroecosystems: relations between a coccinellid predator and its food. *J. Appl. Ecol.* 22:357–72
24. Andow, D. A., Risch, S. J. 1987. Parasitism in diversified agroecosystems: Phenology of *Trichogramma minutum* (Hymenoptera:Trichogrammatidae). *Entomophaga* 32:255–60
25. Aveling, C. 1981. The role of *Anthocoris* species (Hemiptera:Anthocoridae) in the integrated control of the damson-hop aphid (*Phorodon humuli*). *Ann. Appl. Biol.* 97:143
26. Bach, C. E. 1980. Effects of plant diversity and time of colonization on an herbivore-plant interaction. *Oecologia* 44:319–26
27. Bach, C. E. 1980. Effects of plant density and diversity on the population dynamics of a specialist herbivore, the striped cucumber beetle, *Acalymma vittata* (Fab.). *Ecology* 61:1515–30
28. Bach, C. E. 1984. Plant pattern and herbivore population dynamics: plant factors affecting the abundance of a tropical cucurbit specialist (*Acalymma innubrum*). *Ecology* 65:175–90
29. Banham, F. L. 1971. Native hosts of western cherry fruit fly (Diptera: Tephritidae) in the Okanagan Valley of British Columbia. *J. Entomol. Soc. B.C.* 69:29
30. Barnes, M. M. 1970. Genesis of a pest: *Nysius raphanus* and *Sisymbrium irio* in vineyards. *J. Econ. Entomol.* 63:1462
31. Barney, R. J., Lamp. W. O., Armbrust, E. J., Kapusta, G. 1984. Insect predator community and its response to weed management in spring-planted alfalfa. *Prot. Ecology* 6:23–33
32. Beard, R. L. 1964. Parasites of muscoid flies. *Bull. WHO* 31:491–93
33. Berberet, R. C., Stritzke, J. F., Dowdy, A. K. 1987. Interactions of alfalfa weevil (Coleoptera:Curculionidae) and weeds in reducing yield and stand of alfalfa. *J. Econ. Entomol.* 80:1306–13
34. Boiteau, G. 1984. Effect of planting date, plant spacing, and weed cover on populations of insects, arachnids, and entomophthoran fungi in potato fields. *Environ. Entomol.* 13:751–56
35. Brandenburg, R. L., Kennedy, G. G. 1982. Intercrop relationships and spider mite dispersal in a corn/peanut agroecosystem. *Ent. Exp. Appl.* 32:269–76
36. Branson, T. F., Ortman, E. E. 1967. Host range of larvae of the northern corn rootworm (Coleoptera:Chrysomelidae). *J. Kans. Entomol. Soc.* 40:412
37. Branson, T. F., Ortman, E. E. 1970.

The host range of larvae of the western corn rootworm: further studies. *J. Econ. Entomol.* 63:800
38. Brust, G. E., Stinner, B. R., McCartney, D. A. 1986. Predation by soil inhabiting arthropods in intercropped and monoculture agroecosystems. *Agric. Ecosyst. Environ.* 18:145–54
39. Bugg, R. L., Ehler, L. E., Wilson, L. T. 1987. Effect of common knotweed (*Polygonum viculare*) on abundance and efficiency of insect predators of crop pests. *Hilgardia* 55(7):1–53
40. Bugg, R. L., Wilson, L. T. 1989. *Ammi visnaga* (L.) Lamarck (Apiaceae): Associated beneficial insects and implications for biological control, with emphasis on the bell-pepper agroecosystem. *Biol. Agric. and Hort.* 6:241–68
41. Cantelo, W. W., Sanford, L. L. 1984. Insect population response to mixed and uniform plantings of resistant and susceptible plant material. *Environ. Entomol.* 13:1443–45
42. Capinera, J. L., Weissling, T. J., Schweizer, E. E. 1985. Compatibility of intercropping with mechanized agriculture: Effects of strip intercropping of pinto beans and sweet corn on insect abundance in Colorado. *J. Econ. Entomol.* 78:354–57
43. Centro Internacional de Agricultura Tropical. 1979. *Bean Program 1978 Annual Report.* Cali, Columbia: CIAT
44. Coaker, T. H. 1980. Insect pest management in *Brassica* crops by intercropping. Integr. Control Brassica Crops. *Sect. Reg. Ouest Palearetique, West. Palearctic Reg. Sect.* 3:117–25
45. Croft, B. A. 1975. Tree fruit pest management. In *Introduction to Insect Pest Management*, ed. L. Metcalf, W. Luckman, p. 471. New York: Wiley
46. Cromartie, W. J. Jr. 1975. The effect of stand size and vegatational background on the colonization of cruciferous plants by herbivorous insects. *J. Appl. Ecol.* 12:517–33
47. Cromartie, W. J. Jr. 1981. The environmental control of insects using crop diversity. In *Handbook of Pest Management in Agriculture*, ed. D. Pimentel, 1:223–51. Boca Raton, FL: CRC Press
48. Dempster, J. P. 1969. Some effects of weed control on the numbers of the small cabbage white (*Pieris rapae* L.) on Brussels sprouts. *J. Appl. Ecol.* 6:339–46
49. Dempster, J. P., Coaker, T. H. 1974. Diversification of crop ecosystems as a means of controlling pests. In *Biology in Pest and Disease Control*, ed. D. Price-

582   ANDOW

Jones, M. E. Solomon, pp. 106–14. Oxford: Blackwell

50. Elmstrom, K. M., Andow, D. A., Barclay, W. W. 1988. Flea beetle movement in a broccoli monoculture and diculture. *Environ. Entomol.* 17:299–305

51. Elton, C. S. 1927. *Animal Ecology.* London: Sidgwick and Jackson

52. Elton, C. S. 1958. *The Ecology of Invasions by Animals and Plants.* London: Methuen

53. Ezueh, M. I. 1977. The implications of interplanting cowpea through maize on the major pests of cowpea. In *11th Annu. Meeting Nigerian Entomol. Soc. Ife, 19–22 December 1977.* Ibadan: Entomol. Soc. Nigeria. 12 pp.

54. Farrell, J. A. K. 1976. Effects of intersowing with beans on the spread of groundnut rosette virus by *Aphis craccivora* Koch (Hemiptera, Aphididae) in Malawi. *Bull. Entomol. Res.* 66:331–33

55. Feeny, P. P. 1976. Plant apparency and chemical defense. *Rec. Adv. Phytochem.* 10:1–40

56. Finch, S., Ackley, C. M. 1977. Cultivated and wild host plants supporting populations of the cabbage root fly (*Erioischia brassicae:* Dipt., Anthomyiidae). *Ann. Appl. Biol.* 85:13

57. Finch, S., Skinner, G. 1982. Upwind flight by the cabbage root fly, *Delia radicum. Physiol. Entomol.* 7:387–99

58. Flaherty, D. 1969. Ecosystem trophic complexity and willamette mite *Eotetranychus willamettei* Ewing (Acarina:Tetranychidae). *Ecology* 50:911–16

59. Foott, W. H. 1968. The importance of *Solanum carolinense* L. as a host of the pepper maggot *Zonosemata electa* (Say) (Diptera:Tephritidae) in southwestern Ontario. *Proc. Entomol. Soc. Ont.* 98:16

60. Futuyma, D. J., Wasserman, S. S. 1980. Resource concentration and herbivory in oak forests. *Science* 210:920–22

61. Gliessman, S. R., Altieri, M. A. 1982. Polyculture cropping has advantages. *Calif. Agric.* 36(7):15–16

62. Goldburg, R. J. 1987. Sequential flowering of neighboring goldenrods and the movements of the flower predator *Epicauta pennsylvanica. Oecologia Berlin* 74:247–52

63. Goodman, D. 1975. The theory of diversity-stability relationships in Ecology. *Q. Rev. Biol.* 50:238–66

64. Hall, R. W., Ehler, L. E. 1979. Rate of establishment of natural enemies in classical biological control. *Bull. Entomol. Soc. Am.* 25:280–82

65. Hall, R. W., Ehler, L. E., Bisabri-Ershadi, B. 1980. Rate of success in classical biological control of arthropods. *Bull. Entomol. Soc. Am.* 26:111–14

66. Hassanein, M. H., Khalil, F. M., Eisa, M. A. 1971. Contribution to the study of *Thrips tabaci* Lind. in Upper Egypt. *Bull. Soc. Entomol. Egypte* 54:133–40

67. Helenius, J. 1989. The influence of mixed intercropping of oats with field beans on the abundance and spatial distribution of cereal aphids (Homoptera: Aphididae). *Agric. Ecosyst. Environ.* 25:53–73

68. Helenius, J., Ronni, P. 1989. Yield, its components and pest incidence in mixed intercropping of oats (Avena sativa) and field beans (Vicia faba). *J. Agric. Sci. Finland* 61:15–31

69. Hodek, I. 1973. *Biology of Coccinellidae.* The Hague: Junk Academia

70. Horn, D. J. 1988. Parasitism of cabbage aphid and green peach aphid (Homoptera:Aphididae) on collards in relation to weed management. *Environ. Entomol.* 17(2):354–58

71. Hutchison, G. E. 1959. Homage to Santa Rosalia or why are there so many kinds of animals? *Am. Nat.* 93:145–59

72. International Crops Research Institute for the Semi Arid Tropics. 1976. Cropping Entomology. *Annual Report, 1975–76,* pp. 178–87. Hyderabad, India: ICRISAT

73. International Crops Research Institute for the Semi Arid Tropics. 1977. Cropping Systems Entomology. *Annual Report, 1976–1977,* pp. 164–66. Hyderabad, India: ICRISAT

74. International Crops Research Institute for the Semi Arid Tropics. 1978. Cropping Entomology. *Annual Report, 1977–78,* pp. 204–10. Hyderabad, India: ICRISAT

75. International Crops Research Institute for the Semi Arid Tropics. 1979. Cropping Entomology. *Annual Report, 1978–79,* pp. 209–13. Hyderabad, India: ICRISAT

76. International Crops Research Institute for the Semi Arid Tropics. 1980. Cropping Entomology. *Report of Work, 1979–80.* Hyderabad, India: ICRISAT

77. International Rice Research Institute. 1973. Multiple cropping, lessons from traditional technology. *Annual Report, 1972,* pp. 16–25. Los Banos, Philippines: IRRI

78. International Rice Research Institute. 1974. *Annual Report for 1973.* Los Banos, The Philippines: IRRI

79. International Rice Research Institute.

1978. *Annual Report for 1975.* Los Banos, The Philippines: IRRI
80. International Rice Research Institute. 1983. *Annual Report for 1981.* Los Banos, The Philippines: IRRI
81. Jacques, R. L. Jr., Peters, D. C. 1971. Biology of *Systena frontalis,* with special reference to corn. *J. Econ. Entomol.* 64:135
82. Johnson, T. B., Turpin, F. T., Bergman, M. K. 1984. Effect of foxtail infestation on corn rootworm larvae (Coleoptera:Chrysomelidae) under two corn-planting dates. *Environ. Entomol.* 13:1245
83. Kareiva, P. M. 1982. Experimental and mathematical analyses of herbivore movement: quantifying the influence of plant spacing and quality on foraging discrimination. *Ecol. Monogr.* 52:261–82
84. Kareiva, P. M. 1983. Influence of vegetation texture on herbivore populations: resource concentration and herbivore movement. In *Variable Plants and Herbivores in Natural and Managed Systems,* ed. R. F. Denno, M. S. McClure, pp. 259–89. New York: Academic
85. Kareiva, P. M. 1985. Finding and losing host plants by *Phyllotreta:* patch size and surrounding habitat. *Ecology* 66:1809–16
86. Kingsley, P. C., Scriber, J. M., Harvey, R. G. 1986. Relationship of weed density with leafhopper populations in alfalfa, Wisconsin. *Agric. Ecosyst. Environ.* 17:281–86
87. Kulman, H. M. 1970. Community effect of biological control in mixed stands. *Proc. 6th World Forestry Congress, Madrid,* 2:1991–95. Barcelona: Comercial y Artes Gráficas
88. Lambert, J. D. H., Arnason, J. T., Serratos, A., Philogene, B. J. R., Faris, M. A. 1987. Role of intercropped red clover in inhibiting European corn borer (Lepidoptera:Pyralidae) damage to corn in eastern Ontario. *J. Econ. Entomol.* 80:1192–96
89. Lamp, W. O., Morris, M. J., Armbrust, E. J., Kapusta, G. 1984. Selective weed control in spring-planted alfalfa: effect on leafhoppers and planthoppers *Homoptera: Auchenorrhyncha,* with emphasis on potato leafhopper. *Environ. Entomol.* 13:207–13
90. Laster, M. L., Meridith, W. R. Jr. 1974. Evaluating the response of cotton cultures to tarnished plant bug injury. *J. Econ. Entomol.* 67:686
91. Latheef, M. A., Irwin, R. D. 1980. Effects of companionate planting on snap bean insects, *Epilachna varivestis* and *Heliothis zea. Environ. Entomol.* 9:195–98
92. Letourneau, D. K. 1986. Associational resistance in squash monocultures and polycultures in tropical Mexico. *Environ. Entomol.* 15:285–92
93. Letourneau, D. K. 1987. The enemies hypothesis: tritrophic interaction and vegetational diversity in tropical agroecosystems. *Ecology* 68:1616–22
94. Letourneau, D. K., Altieri, M. A. 1983. Abundance patterns of a predator, *Orius tristicolor* (Hemiptera:Anthocoridae), and its prey, *Frankliniella occidentalis* (Thysanoptera:Thripidae): Habitat attraction in polycultures versus monocultures. *Environ. Entomol.* 12:1464–69
95. Levins, R. 1970. Complex Systems. In *Towards a Theoretical Biology: 3 Drafts,* ed. C. H. Waddington, pp. 73–88. Edinburgh: Edinburgh Univ. Press
96. MacArthur, R. H. 1955. Fluctuations of animal populations and a measure of community stability. *Ecology* 36:533–36
97. Maguire, L. A. 1984. Influence of surrounding plants on densities of *Pieris rapae* (L.) eggs and larvae (Lepidoptera:Pieridae) on collards. *Environ. Entomol.* 13:464–68
98. Manly, B. F. J. 1977. The determination of key factors from life table data. *Oecologia* 13:111–17
99. Margalef, R. 1968. *Perspectives in Ecological Theory.* Chicago: Univ. of Chicago Press
100. Martin, R. C., Arnason, J. T., Lambert, J. D. H., Isabelle, P., Voldeng, H. D., Smith, D. L. 1989. Reduction of European corn borer (Lepidoptera:Pyralidae) damage by intercropping corn with soybean. *J. Econ. Entomol.* 82:1455–59
101. Matteson, P. C. 1982. The effects of intercropping with cereals and minimal permethrin applications on insect pests of cowpea and their natural enemies in Nigeria. *Trop. Pest Manage.* 28:372–80
102. May, R. M. 1973. *Stability and Complexity in Model Ecosystems.* Princeton, NJ: Princeton Univ. Press 265 pp.
103. McClure, M. S. 1980. Role of wild host plants in the feeding, oviposition, and dispersal of *Scaphytopius acutus* (Homoptera:Cicadellidae), a vector of peach X-disease, *Environ. Entomol.* 9:265
104. McClure, M. S., Andreadis, T. G., Lacy, G. H. 1982. Manipulating orchard cover to reduce invasion by leafhopper vectors of peach X-disease. *J. Econ. Entomol.* 75:64–68
105. Muma, M. H. 1961. The influence of cover crop cultivation on populations of

injurious insects and mites in Florida citrus groves. *Fla. Entomol.* 44:61–68

106. Murdoch, W. W. 1975. Diversity, complexity, stability and pest control. *J. Appl. Ecol.* 12:795–807

107. Nordlund, D. A., Chalfant, R. B., Lewis, W. J. 1984. Arthropod populations, yield and damage in monocultures and polycultures of corn, beans and tomatoes. *Agric. Ecosyst. Environ.* 11:353–67

108. Norris, R. F. 1982. Interactions between weeds and other pests in the agroecosystem. *Biometeorology in Integrated Pest Management, Conf. Biometeorol. Integrated Pest Management 1980*, ed. J. L. Hatfield, I. J. Thompson. New York: Academic

109. Norris, R. F., Cothran, W. R., Burton, V. E. 1984. Interactions between winter annual weeds and Egyptian alfalfa weevil (Coleoptera:Curculionidae) in alfalfa. *J. Econ. Entomol.* 77:43–52

110. O'Donnell, M. S., Coaker, T. H. 1975. Potential of intra-crop diversity for the control of brassica pests. *Proc. 8th Brit. Insect. Fung. Conf.* 1:101–7. Brighton: Br. Insecticide and Fungicide Counc.

111. Odum, E. P. 1953. *Fundamentals of Ecology.* Philadelphia: Saunders

112. Ollman, B. 1976. *Alienation.* Cambridge: Cambridge Univ. Press. 2nd ed.

113. Perfect, T. J., Cook, A. G., Critchley, B. R. 1979. *Effects of intercropping maize and cowpea on pest incidence.* Ibadan, Nigeria: IITA Internal report. 38 pp.

114. Perfecto, I., Horwith, B., Vandermeer, J., Schultz, B., McGuinness, H. Dos Santos, A. 1986. Effects of plant diversity and density on the emigration rate of two ground beetles, *Harpalus pennsylvanicus* and *Evarthrus sodalis* (Coleoptera:Carabidae), in a system of tomatoes and beans. *Environ. Entomol.* 15:1028–31

115. Perrin, R. M. 1977. Pest management in multiple cropping systems. *Agro-Ecosystems* 3:93–118

116. Pimentel, D. 1961. Species diversity and insect population outbreaks. *Ann. Entomol. Soc. Am.* 54:76–86

117. Pitre, H. N., Boyd, F. J. 1970. A study of the role of weeds in corn fields in the epidemiology of corn stunt disease. *J. Econ. Entomol.* 63:195–97

118. Power, A. G. 1987. Plant community diversity, herbivore movement, and insect-transmitted disease of maize. *Ecology* 68:1658–69

119. Power, A. G., Rosset, P. M., Ambrose, R. J., Hruska, A. J. 1987. Population response of bean insect herbivores to

inter- and intraspecific plant community diversity: Experiments in a tomato and bean agroecosystem in Costa Rica. *Turrialba* 37:219–26

120. Price, P. W., Bouton, C. E., Gross, P., McPheron, B. A., Thompson, J. N., Weis, A. E. 1980. Interactions among three trophic levels. *Ann. Rev. Ecol. Syst.* 11:41–65

121. Rauscher, M. D. 1978. Search image for leaf shape in a butterfly. *Science* 200:1071–73

122. Rauscher, M. 1981. The effect of native vegetation on the susceptibility of *Aristolochia reticulata* (Aristolochiaceae) to herbivore attack. *Ecology* 62:1187–95

123. Rhoades, D. F., Cates, R. G. 1976. Toward a general theory of plant antiherbivore chemistry. *Rec. Adv. Phytochem.* 10:168–213

124. Risch, S. J. 1979. A comparison, by sweep sampling, of the insect fauna from corn and sweet potato monocultures and dicultures in Costa Rica. *Oecologia* 42:195–211

125. Risch, S. J. 1980. Fewer beetle pests on beans and cowpeas interplanted with banana in Costa Rica. *Turrialba* 30:229–30

126. Risch, S. J. 1980. The population dynamics of several herbivorous beetles in a tropical agroecosystem: The effect of intercropping corn, beans, and squash in Costa Rica. *J. Appl. Ecol.* 17:593–612

127. Risch, S. J. 1981. Insect herbivore abundance in tropical monocultures and polycultures: An experimental test of two hypotheses. *Ecology* 62:1325–40

128. Risch, S. J., Andow, D. A., Altieri, M. A. 1983. Agroecosystem diversity and pest control: data, tentative conclusions, and new research directions. *Environ. Entomol.* 12:625–29

129. Root, R. B. 1973. Organization of a plant-arthropod association in simple and diverse habitats: The fauna of collards (*Brassica oleracea*). *Ecol. Monogr.* 43:95–124

130. Root, R. B., Kareiva, P. M. 1984. The search for resources by cabbage butterflies (*Pieris rapae*): Ecological consequences and adaptive significance of markovian movements in a patchy environment. *Ecology* 65:147–65

131. Russell, E. P. 1989. Enemies hypothesis: A review of the effect of vegetational diversity on predatory insects and parasitoids. *Environ. Entomol.* 18:590–99

132. Ryan, J., Ryan, M. F., McNaeidhe, F. 1980. The effect of interrow plant cover on populations of the cabbage root fly,

Delia brassicae (Wiedemann). *J. Appl. Ecol.* 17:31–40

133. Sawyer, A. J. 1976. *The effect on the cereal leaf beetle of planting oats with a companion crop.* MS thesis. Lansing, MI: Mich. State Univ. 145 pp.

134. Sheehan, W. 1986. Response by specialist and generalist natural enemies to agroecosystem diversification: a selective review. *Environ. Entomol.* 15:456–61

135. Shelton, M. D., Edwards, C. R. 1983. Effects of weeds on the diversity and abundance of insects in soybeans. *Environ. Entomol.* 12:296

136. Singh, R. N., Singh, K. M. 1978. Influence of intercropping on succession and population build up of insect pests in early variety of red gram, *Cajanus cajan* (L.) Millsp. *Indian J. Entomol.* 40:361–75

137. Sluss, R. R. 1967. Population dynamics of the walnut aphid, *Chromaphis juglandicola* (Kalt.) in northern California. *Ecology* 48:41–58

138. Smith, J. G. 1976. Influence of crop background on aphids and other phytophagous insects on Brussels sprouts. *Ann. Appl. Biol.* 83:1–13

139. Solomon, B. P. 1981. Windbreaks as a source of orchard pests and predators. In *Pests, Pathogens and Vegetation,* ed. J. M. Thresh, p. 273. Boston: Pitman

140. Solomon, B. P. 1981. Response of a host-specific herbivore to resource density, relative abundance, and phenology. *Ecology* 62:1205–14

141. Southwood, T. R. E. 1977. The relevance of population dynamic theory to pest status. In *Origins of Pest, Parasite, Disease and Weed Problems,* ed. J. M. Cherrett, G. R. Sagar, pp. 35–54. Oxford: Blackwell

142. Speight, M. R., Lawton, J. H. 1976. The influence of weed-cover on the mortality imposed on artificial prey by predatory ground beetles in cereal fields. *Oecologia* 23:211–23

143. Spencer, H. 1866. *Principles of Biology.* New York: Appleton

144. Stanton, M. L. 1983. Spatial patterns in the plant community and their effects upon insect search. In *Herbivorous Insects: Host-Seeking Behavior and Mechanisms,* ed. S. Ahmad, pp. 125–56. New York: Academic

145. Stern, V. M., Smith, R. F., Van den Bosch, R., Hagen, K. S. 1959. The integration of chemical and biological control of the spotted alfalfa aphid: the integrated control concept. *Hilgardia* 29:81–101

146. Stride, G. O. 1969. Investigations into

the use of a trap crop to protect cotton from attack by *Lygus vosseleri. J. Entomol. Soc. South Afr.* 32:469–77

147. Tahvanainen, J. O., Root, R. B. 1972. The influence of vegetational diversity on the population ecology of a specialized herbivore, *Phyllotreta cruciferae* (Coleoptera: Chrysomelidae). *Oecologia* 10:321–46

148. Tamaki, G., Moffitt, H. R., Turner, J. E. 1975. The influence of perennial weeds on the abundance of redbacked cutworm on asparagus. *Environ. Entomol.* 4:274–76

149. Taylor, T. A. 1977. Mixed cropping as a input in the management of crop pests in tropical Africa. *Afr. Environ.* 2(4)/3(1):111–26

150. Theunissen, J., den Ouden, H. 1980. Effects of intercropping with *Spergula arvensis* on pests of Brussels sprouts. *Entomol. Exp. Appl.* 27:260–68

151. Tingey, W. M., Lamont, W. J. Jr. 1988. Insect abundance in field beans altered by intercropping. *Bull. Entomol. Res.* 78:527–35

152. Toba, H. H., Kishaba, A. N., Bohn, G. W., Hield, H. 1977. Protecting muskmelons against aphid-borne viruses. *Phytopathology* 67:1418–23

153. Topham, M., Beardsley, J. W. Jr. 1975. Influence of nectar source plants on the New Guinea sugarcane weevil parasite, *Lixophaga sphenophori* (Villeneuve). *Proc. Hawaii. Entomol. Soc.* 22:145–54

154. Trenbath, B. R. 1977. Interactions among diverse hosts and diverse parasites. *Ann. N.Y. Acad. Sci.* 287:124–50

155. Tukahirwa, E. M., Coaker, T. H. 1982. Effect of mixed cropping on some insect pests of brassicas; reduced *Brevicoryne brassicae* infestations and influences on epigeal predators and the disturbance of oviposition behavior in *Delia brassicae. Entomol. Exp. Appl.* 32:129–40

156. Uvah, I. I. I., Coaker, T. H. 1984. Effect of mixed cropping on some insect pests of carrots and onions. *Entomol. Exp. Appl.* 36:159–67

157. Valverde, R. A., Moreno, R., Gamez, R. 1982. Incidence and some ecological aspects of cowpea severe mosaic virus in two cropping systems in Costa Rica. *Turrialba* 32:29–32

158. van Emden, H. F., Williams, G. F. 1974. Insect stability and diversity in agro-ecosystems. *Annu. Rev. Entomol.* 19:455–75

159. van Schoonhoven, A., Cardona, C., Garcia, J., Garzón. 1981. Effect of weed covers on *Empoasca kraemeri*

Ross and Moore populations and dry bean yields. *Environ. Entomol.* 10:901–7

160. Venugopal, M. S., Palanippan, S. P. 1976. Influence of intercropping in sorghum on the incidence of sorghum shootfly. *Madras Agric. J.* 63:572–73

161. Villamajor, F. G. Jr. 1976. Relations of cotton yield to insect infestation, plant diversification, fertilization, and pesticide application. *Kalikasan Philipp. J. Biol.* 5:175–86

162. Wallis, R. L., Turner, J. E. 1969. Burning weeds in drainage ditches to suppress populations of green peach aphids and incidence of beet western yellows disease in sugarbeets. *J. Econ. Entomol.* 62:307

163. Zandstra, B. H., Motooka, P. S. 1978. Beneficial effects of weeds in pest management—a review. *PANS* 24:333–38

Annu. Rev. Entomol. 1991. 36:587–609

LYME BORRELIOSIS: Relation of Its Causative Agent to Its Vectors and Hosts in North America and Europe[1]

R. S. Lane

Department of Entomological Sciences, University of California, Berkeley, California 94720

J. Piesman

Center for Infectious Diseases, Division of Vector-Borne Infectious Diseases, Fort Collins, Colorado 80522

W. Burgdorfer

Laboratory of Vectors and Pathogens, Rocky Mountain Laboratories, Hamilton, Montana 59840

KEY WORDS: ticks, *Borrelia burgdorferi*, vertebrates, ecology, epizootiology

INTRODUCTION

In 1981, the spirochete now known to cause Lyme disease was isolated from the midgut diverticula of *Ixodes dammini* ticks from Shelter Island, New York (26). This organism, which was subsequently named *Borrelia burgdorferi* (54), is only the second species of spirochete to be associated with ixodid (hard-bodied) ticks (30). Since then, several additional species of ixodid ticks from five genera (*Amblyomma, Dermacentor, Haemaphysalis, Ixodes, Rhipicephalus*) have been found infected naturally with *B. burgdorferi*, but only members of the genus *Ixodes* and, more specifically, those in the *I. (Ixodes) ricinus* complex, appear to be primary vectors in North America, Europe, and Asia (5). The principal species involved are *I. dammini* in the northeastern and upper midwestern United States, *I. pacificus* in the far-western United States, and *I. ricinus* and *I. persulcatus* in Eurasia. *I. scapularis* in the southcentral and southeastern United States has been found infected naturally with and to be an efficient experimental vector of *B. burgdorferi*, but it has not been implicated as a vector to humans (28, 29, 74, 89, 96). Except for *Borrelia theileri* and *B. recurrentis*, which are transmitted to domestic livestock and humans by ixodid ticks and lice, respectively, all other described species of *Borrelia* (n = 16) have been associated with argasid (soft-bodied) ticks, particularly those in the genus *Ornithodoros* (30).

The uniqueness of the *B. burgdorferi*–ixodid tick associations and the global importance of Lyme disease as a public health and veterinary problem stimulated numerous investigations of the relationship of this parasite with some of its tick vectors and vertebrate hosts during the 1980s. Several recent reviews (4, 5, 25, 30, 31, 90, 100, 110) have examined the findings stemming from these studies. The reader is also referred to Spielman et al (112) for an earlier review of the ecology of Lyme disease in the United States, to Barbour & Hayes (18) for an overview of the biology of borreliae, and to Steere (114) for a comprehensive review of the causation, vectors, epidemiology, clinical manifestations, pathogenesis, diagnosis, and treatment of Lyme disease.

This review evaluates the literature relevant to the relation of *B. burgdorferi* to its arthropod vectors and vertebrate hosts in Europe and particularly in North America. Space limitations preclude a more comprehensive review of the biology of the chief European vector, *I. ricinus*, which is one of the most intensively studied ticks in the world. Instead, what little is known about the relation of this tick to *B. burgdorferi* and a few of the more significant European vertebrate hosts of the Lyme borreliosis spirochete are summarized briefly. We also had to be somewhat selective as to which references from North America we included. Furthermore, we divided the rapidly expanding North American literature on this subject along geographic lines (i.e. eastern and western) to facilitate the ensuing discussion.

EASTERN NORTH AMERICA

Ixodes dammini

Since the seminal discovery of the Lyme disease spirochete, *B. burgdorferi*, in the midgut of *I. dammini* (26), this tick has been demonstrated to be the primary vector of Lyme disease in the northeastern and upper midwestern United States.

LIMITS ON HABITAT AND DISTRIBUTION   Initial studies on the ecological distribution of *I. dammini* took place principally in coastal areas of New England (37, 97). The proximity of *I. dammini* populations to coastal areas led one investigator to suggest that mean winter temperature restricted this tick's distribution (84). The existence of active foci of *I. dammini* in Minnesota and Wisconsin (42, 44, 53), however, demonstrates that *I. dammini* is not restricted to coastal environments with mild mean winter temperatures. Even in the northeastern United States, thriving inland populations of *I. dammini* have become established in Westchester (46) and Dutchess (14) counties in New York. The macrogeographical factors limiting the distribution of *I. dammini* are not well understood. It seems safe to predict, however, that the requirement for high humidity levels by members of the *I. ricinus* complex restricts *I. dammini* from dispersing to arid regions of North America. High altitude (106) may also be restrictive because of the extreme desiccation suffered by ticks in sparsely covered windy areas.

The microecological factors that limit the distribution of *I. dammini* are complex. Island habitats containing dense brush and characterized by vegetation such as bayberry, wild rose, poison ivy, green-brier, and scrub oak appear to be ideal for populations of *I. dammini* (51, 97). Inland populations of *I. dammini,* by contrast, may be extremely abundant in heavily forested areas with high canopy. Areas of Westchester County, where Lyme disease is endemic, have been characterized as eastern deciduous woodland with a mix of oak, sugar maple, and tulip (41); the area of greatest *I. dammini* abundance in Wisconsin was described as the tension zone between the northern coniferous-hardwood forest and the southern deciduous forest (53). In general, *I. dammini* adults and immatures are found in heavily forested or brushy areas and edge habitat, but not in open spaces. An exception to this rule has been reported in Westchester County where spirochete-infected *I. dammini* were found on well-maintained lawns in residential areas (46). Thus, spillover from forested areas to open spaces may occur in areas of dense tick populations.

The history of land use patterns in Lyme disease–endemic areas provides clues as to the factors that limit the distribution of *I. dammini*. During the second half of the twentieth century, much of the northeastern United States was converted from intense agricultural use to suburban residential development. Where forests were once cleared and fields maintained, woodlots have reemerged especially where zoning patterns encourage large (≥ 1 acre) properties. The result is a mosaic of woods in various stages of growth, substantial edge vegetation, ornamental plantings, and maintained lawn areas. This mosaic appears to provide the ideal habitat for populations of animal hosts and *I. dammini* ticks to become established and thrive. Interestingly, areas of the midwest that are not heavily forested and remain under intense agricultural pressure have not become heavily infested with *I. dammini* (99).

A thorough understanding of the ecological constraints on *I. dammini* populations may lead to environmentally sound means of controlling Lyme disease.

MEASURING *IXODES DAMMINI* POPULATIONS   Diverse means of surveying tick populations have been utilized to measure the abundance of *I. dammini*. By far the most commonly used method has been to collect mammalian and avian hosts and examine them for attached ticks (11, 37, 42, 48, 53, 83, 97, 98, 128). In areas where *I. dammini* is sparse, host examination may be the most cost-efficient method for determining whether *B. burgdorferi*–infected ticks are present. In areas where *I. dammini* is abundant, direct sampling of questing tick populations by dragging or flagging a piece of cloth over vegetation may give a more accurate picture. *I. dammini* and other ixodid tick populations can be quantified as the number of ticks collected per person-hour of sampling (64, 93), the number of ticks per square meter of transect (47), the number of ticks per 100 drag samples (59, 69), or the number of ticks removed from a person's clothing (52). Carbon dioxide– (dry ice) baited traps have also been tried (46, 51) with limited success because *I. dammini* appears to respond over a much shorter distance than other tick species such as *Amblyomma americanum*.

Models for predicting the behavior of *I. dammini* populations (41, 50, 83) are presently in their infancy compared with extensive models available for other tick species such as *Dermacentor variabilis* (86). Once time-tested computer models are available for *I. dammini,* the predictive value of such models may aid in planning Lyme disease prevention and control strategies.

SEASONALITY OF *IXODES DAMMINI*   Lyme disease is clearly a late spring, summer illness in the northeastern and upper midwestern United States, where approximately two-thirds of human victims suffer onset from May through August (38, 81, 116). The peak month is July. The seasonal questing (host-seeking) pattern of nymphal *I. dammini* precedes the peak of human Lyme disease cases. Nymphs quest from April to September, reaching maximum activity in May and June (37, 79, 93, 97, 127). Questing nymphal populations are still high during early July, but fall off toward mid to late July. Recreational and residential outdoor human activity picks up substantially during early July.

The peak of larval *I. dammini* questing follows that of nymphs by two to three months. The maximum activity is in August and September (79, 97, 127). The fact that larval feeding follows nymphal feeding increases the efficiency of the enzootic cycle of *B. burgdorferi* because it ensures that juvenile rodents become infected with spirochetes (May, June) before serving as hosts to larvae (August, September), thus perpetuating the infection from year to year (112). The Lyme disease spirochete may overwinter in fed larvae or in flat nymphs after replete larvae have molted in summer or fall.

Adult *I. dammini* quest during the cooler months from October to May (37, 41, 98). The feeding pattern of adult *I. dammini* on white-tailed deer and dogs is bimodal, with a large fall peak from October to December and a smaller spring peak from March to May. The question arises as to whether adults that feed in fall and spring are of the same or different cohorts. Experimental evidence suggests that spring-feeding adults are part of the fall-questing populations that did not find hosts. Thus, fall-questing adult *I. dammini* that were marked with fluorescent powder were recaptured while questing in spring (41). Replete females, fed in fall or spring, probably do not oviposit until the warmer summer months. In Massachusetts, Yuval & Spielman (128a) determined the longevity of all three parasitic stages of *I. dammini* under field conditions and the duration of the developmental cycle of this tick. They concluded that the life cycle of *I. dammini* may range from two to four years and appears to be regulated by host abundance and physiological mechanisms. The life history of *I. dammini* between questing periods in other regions and the issue of diapause deserve research attention.

HOST ASSOCIATIONS OF *IXODES DAMMINI*   *I. dammini,* particularly the immature stages, has a broad host range. Larvae and nymphs reportedly infest 31 mammalian and 49 avian species (4–6); the adults are somewhat more selective, and are limited to 13 species of medium or large-sized mammals. Two hosts, however, stand out as central to the survival of *I. dammini* and of the Lyme disease spirochete: the white-footed mouse, *Peromyscus leucopus,* and the white-tailed deer, *Odocoileus virginianus.*

Many surveys have shown the white-footed mouse to be the most commonly captured and most frequently parasitized host for immature *I. dammini* (12, 37, 53, 70, 79, 83, 97, 127). Studies in coastal Massachusetts have emphasized the importance of the white-footed mouse as a host for immature *I. dammini* (70, 83, 97, 110, 127); the hypothesis that the white-footed mouse is the only important host for immature *I. dammini,* however, is arguable. Small mammals such as gray squirrels, *Sciurus carolinensis;* chipmunks, *Tamias striatus;* and short-tailed shrews, *Blarina brevicauda,* may be important in other locations (53, 79). Medium-sized mammals such as raccoons, *Procyon lotor;* opossums, *Didelphis virginiana;* and striped skunks, *Mephitis mephitis,* may serve as secondary hosts to immature *I. dammini* (12, 48); large mammals such as white-tailed deer also feed large numbers of immature *I. dammini* (98, 119). Finally, the importance of avian hosts for immature *I. dammini* (8, 11, 15, 19) requires an exhaustive species-by-species evaluation. The rich and diverse vertebrate fauna in Lyme disease–endemic regions make a comprehensive evaluation of the relative role of each species as host

for immature *I. dammini* (and other members of the *I. ricinus* complex) a
challenging task.

Since adult *I. dammini* attach preferentially to large vertebrates, issues
surrounding the importance of such hosts are more clearcut. Adult *I. dammini*
can be found on medium-sized mammals like raccoon, opossum, and skunk
(48), but the numbers that attach to these hosts are usually low. Larger species
of wildlife such as bear, moose, bobcat, lynx, coyote, and especially fox
(111) may be heavily infested with adult *I. dammini,* but these hosts are not
particularly abundant in many of the Lyme disease–endemic regions, e.g.
suburban New York. Livestock, such as sheep, horses, and cattle, may also
be infested with *I. dammini,* but their role as hosts for adult *I. dammini* has
not been evaluated carefully. Domestic pets, particularly dogs, which have
been used to monitor adult *I. dammini* populations, may support populations
of the adults (41, 44). Attention has been directed to the potential role of cats
as phoretic hosts of *I. dammini* (39), but systematic studies of the number of
adult *I. dammini* infesting cats are unavailable.

Evidence that the white-tailed deer is the primary host for adult *I. dammini*
is compelling. Infestations of white-tailed deer with adult *I. dammini* may be
intense (up to 500 adults per deer), and a high proportion of the deer
population may be infested (10, 80, 98, 106). Islands inhabited by deer were
found to contain active populations of *I. dammini* and spirochete-infected
ticks, whereas islands without deer lacked *I. dammini* (9, 126). A natural
experiment was conducted on Great Island, Massachusetts, in which a defined
deer herd was virtually eradicated over a two-year period (128). As a result,
immature *I. dammini* populations infesting rodents were reduced significantly
during the three years following removal of deer, presumably due to deprivation of the principal hosts for adults. The population of immatures on the
nonintervention island (Nantucket) remained stable. This experiment suggested that *I. dammini* populations can be impacted by eradication of deer.
However, many questions remained unanswered. Why were ticks still present
on the island three years after deer removal? Had dogs, foxes, or other
medium-sized mammals been able to support the *I. dammini* population at a
reduced level? Had the immature ticks been reduced to a level where enzootic
transmission of spirochetes ceased? A continuation of the Great Island experiment or similar experiments in other locations should provide answers to these
questions. Finally, many authors have noted the apparent disappearance of
deer from much of the northeastern United States in the early part of the
twentieth century, the population explosion of deer in recent decades, and the
seemingly simultaneous occurrence of the Lyme disease epidemic and unprecedented population levels of white-tailed deer (6, 9, 98, 109, 112, 126).
Thus, the Lyme disease epidemic appears to be intimately linked in time
and space with burgeoning populations of white-tailed deer; but the fact that

some areas with high deer populations have not yet experienced Lyme disease suggests caution in attributing direct cause and effect to this relationship.

RESERVOIR COMPETENCE   To serve as a competent reservoir of the Lyme disease spirochete, most of the host population should be infected with *B. burgdorferi*, should serve as host to large numbers of larval *I. dammini*, and should be extremely infectious to these ticks. The white-footed mouse certainly fits this description. Of 22 white-footed mice collected in Connecticut, 19 (86%) were infected with *B. burgdorferi* (7). As discussed above, the white-footed mouse serves as host to a large proportion of the larval *I. dammini* population, and these mice are quite infectious to feeding ticks; most larvae removed from naturally infected white-footed mice had become infected with the Lyme disease spirochete (43, 70, 83).

Other small mammals like chipmunks (7, 83), squirrels, and shrews may play important roles as secondary reservoirs, but their reservoir competence or potential has not been thoroughly evaluated in diverse habitats. The cottontail rabbit, *Sylvilagus floridanus*, may play a significant role in a parallel cycle of *B. burgdorferi* (4, 13, 120, 121). Lyme disease spirochetes have been isolated from cottontail rabbits (13), and these lagomorphs serve as hosts for both *I. dammini* and *Ixodes dentatus;* both these tick species are excellent vectors of *B. burgdorferi* (121). The rabbit–*I. dentatus* parallel cycle may be of particular importance in areas where the enzootic white-footed mouse–*I. dammini* cycle is inefficient or does not occur.

Medium-sized mammals like raccoons, opossums, and skunks, serve as hosts to immature *I. dammini* (48) and may be infected with *B. burgdorferi* (12). Interestingly, larvae derived from medium-sized mammals often fail to molt to the nymphal stage when brought into the laboratory. This may result from immune-intolerance on the part of these hosts (110). The role of domestic animals and larger species of wildlife such as bear, coyote, lynx, bobcat, moose, and fox as reservoirs of *B. burgdorferi* has not been evaluated carefully so far.

Birds have been found to be infected with *B. burgdorferi*, and infected ticks have been recovered from them (8, 11, 108, 123). In a study on Naushon Island, Massachusetts, however, infected larvae were not recovered from the predominant avian species at the time larval *I. dammini* were feeding (grey catbird, *Dumetella carolinensis*) (82). In contrast, studies in Connecticut revealed that nymphs that had fed as larvae on American robins and house wrens were infected with spirochetes (15). Defining the role of birds as reservoirs in endemic regions and as hosts for spreading spirochetes to new locations is another challenging field for more intensive research.

The issue of whether white-tailed deer serve as reservoirs of *B. burgdorferi* is controversial. Antibodies to *B. burgdorferi* and viable spirochetes have been detected in the blood of white-tailed deer (21, 22, 76). On the other hand, 19 deer from Massachusetts were negative for spirochetes; only 2 of 185 larvae derived from these deer produced infected nymphs, a rate compatible with transovarial transmission and not with a high reservoir competence for deer (119). Additional studies utilizing serology, spirochete isolation, and xenodiagnosis are required to determine more precisely the reservoir competence of white-tailed deer for *B. burgdorferi*.

NATURAL INFECTION RATES   The enzootic cycle of *B. burgdorferi* involving *I. dammini* is extremely efficient. Adult *I. dammini* have two chances to acquire infection transstadially, i.e. during the larval and nymphal feeding periods. In general, the infection rate of questing adult *I. dammini* for *B. burgdorferi* in enzootic areas is approximately 50% (26, 46, 94, 107, 108); in highly enzootic areas, however, up to 100% of adult *I. dammini* have been found to be infected (24). Questing nymphal *I. dammini*, having had one opportunity to acquire infection, display an infection rate of approximately 25% (12, 21, 46, 73, 93, 94, 107). Questing *I. dammini* larvae that are infected presumably acquired spirochetes via transovarial passage. Surveys of flat larvae revealed that 0 of 148 were infected in Connecticut (115) and 2 of 274 (< 1%) were infected in Massachusetts (91). In the laboratory, 44 of 2,297 (1.9%) larvae were infected (77). The efficient process by which *I. dammini* obtains spirochetes from reservoir hosts and passes them transstadially probably renders the inefficient transovarial passage superfluous.

TRANSMISSION OF SPIROCHETES   The ability of transovarially infected larval *I. dammini* to transmit *B. burgdorferi* is unknown. Transmission of *B. burgdorferi* by adult *I. dammini* to laboratory rabbits has been documented (24, 26). Onset of human disease, however, only rarely occurs during the activity period of adult *I. dammini* (38, 105), perhaps because the adults, once attached to humans, are easily detected and removed before they have time to transmit an infectious dose of spirochetes.

Nymphal *I. dammini* are efficient vectors of *B. burgdorferi*. Vector competence experiments demonstrated that ≥ 2/3 of individual nymphal *I. dammini* successfully transmitted *B. burgdorferi* to white-footed mice and hamsters (43, 92). The efficiency of transmission by nymphal *I. dammini* is directly related to the duration of feeding; if nymphs are removed prior to 48 h of attachment, the risk of transmission is slight (95). Unlike the larger adults, the diminutive nymphs (< 2 mm long) are often overlooked and remain attached long enough to transmit *B. burgdorferi*. Therefore, daily tick checks and prompt removal are an important component of our effort to prevent Lyme disease.

Another member of the *I. ricinus* complex, *I. scapularis,* is also an efficient experimental vector of *B. burgdorferi* (28, 89, 96). Nevertheless, surveys in the southern United States have shown that Lyme disease spirochetes are rare or absent in natural populations of *I. scapularis* (74, 96). Besides *I. dentatus* (see above), other ticks reported to be infected naturally with *B. burgdorferi* in the eastern United States include *A. americanum* (74, 104, 107) and *Dermacentor variabilis* (7). Initial studies with a northern isolate of *B. burgdorferi* demonstrated that *A. americanum* and *D. variabilis* did not retain spirochetes transstadially (96). These experiments should be repeated with Lyme disease spirochetes isolated from the southern United States. Similarly, reports that insects (i.e. mosquitoes, horseflies, and deerflies) may be involved in the transmission of *B. burgdorferi* (73, 75, 78) deserve close scrutiny as do reports that contact transmission between vertebrates may occur (33, 34).

The dynamics of the pathogen-vector relationship between *B. burgdorferi* and *I. dammini* have been reviewed recently by Burgdorfer et al (31), and a recent ultrastructural study by Zung et al (129) has shed further light on this subject.

WESTERN NORTH AMERICA

The Western Black-Legged Tick, Ixodes pacificus

DISTRIBUTION, SEASONALITY, AND QUESTING BEHAVIOR   In the far-western United States, the western black-legged tick, *I. pacificus,* has been implicated as the primary vector of *B. burgdorferi* to humans and other animals (32, 60, 64, 87). *I. pacificus* occurs along the Pacific Coast from California north to British Columbia and in Nevada, Utah, and Idaho. In California, this tick has been studied intensively since the mid 1970s. It has been collected in 53 of 58 counties at elevations ranging from sea level to more than 2,150 m (49). Adult ticks are active mainly from November through May, whereas the larvae and nymphs are most active from March through June (49, 57, 65). Approximately 80 species of vertebrates have been recorded as hosts for this tick (16, 49). Hosts of the immatures include alligator lizards (*Gerrhonotus* spp.), spiny lizards (*Sceloporus* spp.), rodents (e.g. *Peromyscus* spp.), and birds. The adults infest medium to large-sized mammals such as bears, deer, dogs, horses, and humans.

In northern California, environmental factors governing the questing behavior of adult ticks were investigated at coastal and inland sites. Diel (24-h) questing by both sexes was correlated positively with relative humidity and negatively with ambient temperature (and less so with light and soil surface temperature) in experimental tick arenas (72), but efforts to confirm these findings under natural conditions by flagging adult ticks from chaparral or grassland diurnally (sunrise to sunset only) yielded erratic results (69). Multiple regression analyses revealed that tick abundance was not consistently associated with either ambient temperature or relative humidity. The lack of significant associations between tick abundance and meteorological factors may have resulted from the reduced number of sample points per date when natural vegetation was sampled during daylight hours only. This in turn would reduce the power of this test statistic to detect any but the strongest associations (69). Similar behavioral studies with greater replication need to be done in various habitats to clarify further the diurnal host-seeking behavior of immature and adult *I. pacificus* and of other ixodid ticks that harbor *B. burgdorferi;* such information is valuable in assessing the risk of exposure to vector ticks diurnally and seasonally within different habitat types.

NATURAL INFECTION RATES   Naversen & Gardner (87) were the first to report a confirmed case of Lyme disease from western North America, and they were also the first to implicate adult *I. pacifus* as a vector to humans. A 32-year-old man developed erythema chronicum migrans after being bitten by a tick tentatively identified as *I. pacificus* while hiking in Sonoma County, California, in June 1975. Four more cases of Lyme disease were reported among residents of California (one patient, however, apparently contracted her infection in Reno, Nevada) in 1983 (36), and since then hundreds of cases have been reported by California state health authorities (e.g. 301 cases for 1986 and 1987 combined).

In 1982, an initial survey of adult *I. pacificus* for Lyme disease spirochetes was undertaken in northern California and southwestern Oregon (32). In striking contrast to the remarkably high spirochete-infection rate of 61% of adult *I. dammini* from Shelter Island, New York (26), only 25 (1.5%) of 1,687 adult *I. pacificus* collected from vegetation were infected with spirochetes that resembled *B. burgdorferi* morphologically and immunologically. Spirochetes isolated from a single male tick reacted with monoclonal antibodies (H5332) specific for *B. burgdorferi,* and the protein profile of whole-cell lysates of this isolate was nearly identical to the profiles of two isolates of *B. burgdorferi* from the eastern United States. Thus, the presence of *B. burgdorferi* in western North America was established, and, despite major differences in the vectors and vertebrate hosts of this spirochete in the western compared to the eastern United States, the organism was found to be virtually identical from coast to coast.

Significantly, 8 (32%) of the 25 infected ticks from California and Oregon had generalized infections in their tissues, whereas spirochetes in the remaining 17 ticks were restricted to the midgut. Spirochetes in six of the eight ticks that had generalized infections exhibited reduced immunofluorescent staining reactivity in tissues besides the midgut. More recently, spirochetes in up to 46% of infected *I. pacificus* adults from vegetation in California fluoresced weakly to moderately in the midgut or other tissues, and ticks containing such spirochetes typically had disseminated infections (20, 64, 67). Generalized spirochetal infections occur much less often (i.e. in ≤ 5% of adult ticks) in unfed *I. dammini* and the related European vector, *I. ricinus* (26, 27). The potential epidemiologic and epizootiologic implications of the higher prevalence of generalized infections in adult *I. pacificus* are twofold: first, spirochetes in unfed female ticks with disseminated infections are more likely to be passed transovarially than are spirochetes in females with restricted (midgut) infections, and second, ticks with disseminated infections may transmit the infection to vertebrates more rapidly while feeding.

Transovarial transmission of *B. burgdorferi* in *I. pacificus* has been demonstrated (61). Transstadial passage of *B. burgdorferi* in *I. pacificus* was proven by Burgdorfer (24), who infected up to 17.5% of nymphal ticks by feeding them as larvae on experimentally infected domestic rabbits and allowing replete ticks to molt to nymphs prior to examining them. In a separate study, one of three wild-caught *I. pacificus* females whose ovaries contained spirochetes yielded 100% infected progeny that maintained the spirochetes transstadially and passed them ovarially to 90–97% of F2 filial ticks in four of five cases (61). Filial ticks infected with spirochetes via eggs and by subsequent transstadial passage invariably had generalized tissue infections characterized by reduced immunofluorescent staining reactivity. This finding offered a plausible explanation as to why 32–46% of infected, unfed, wild-caught adult ticks possessed disseminated tissue infections characterized by weakly to moderately fluorescing spirochetes, i.e. these ticks may have been the progeny of females that had passed spirochetes via eggs.

The infectivity of ovarially passed spirochetes for vertebrates warrants additional investigation. Experimental efforts to infect 12 domestic rabbits by feeding ovarially infected *I. pacificus* filial ticks on them produced inconsistent results, and all attempts to culture spirochetes passed ovarially by this tick have been unsuccessful (56, 61). On the other hand, three of five New Zealand white rabbits exposed to two to four systemically infected *I. pacificus* males seroconverted within 30 to 61 days (R. S. Lane, unpublished data). In these experiments, males were paired with equal numbers of uninfected female ticks inside feeding capsules fastened to the shaven backs of the rabbits. These findings suggest that some spirochetes passed via eggs may be infective for vertebrates even after prolonged passage in tick tissues. Also, this is the first experimental evidence demonstrating that *I. pacificus* males do feed and can transmit *B. burgdorferi*.

In northern California, spirochete-infected *I. pacificus* adults have been collected from vegetation and/or animals in 10 coastal or inland counties, and spirochetes similar to if not identical with *B. burgdorferi* have been recovered from ticks in seven of these counties (20, 32, 60, 61, 64, 65, 69). In southwestern Oregon, infected ticks have been obtained only from Douglas County (32). In both states, infection rates in *I. pacificus* adults from disjunct populations have ranged from 0 to 5.9%, but mostly between 1 and 2%; infection rates in *I. pacificus* immatures removed from western fence lizards or rodents in northern California have been comparable (65; R. S. Lane & J. E. Loye, unpublished data). Unfortunately, similar tick-spirochete surveys have not been reported for other states or provinces in western North America that lie within the known geographic distribution of *I. pacificus* (i.e. Nevada, Utah, Idaho, Washington, British Columbia). Notably, cases of Lyme borreliosis have been reported from all of these territories (81, 122).

Several factors have been proposed to account for the low spirochetal infection rates in populations of *I. pacificus:* (*a*) the apparent reservoir incompetence of the western fence lizard (a major host of *I. pacificus* immatures in many areas), (*b*) the seasonal peak of feeding by larval ticks precedes that of nymphs, and (*c*) the lack of frequent feeding by nymphs on reservoir-competent rodents such as the deer mouse, *Peromyscus maniculatus* (58, 65). The vector competence of *I. pacificus,* currently under investigation, may also be lower than that of other members of the *I. ricinus* complex (24).

Other Potential Western Vectors of B. burgdorferi

B. burgdorferi or related borreliae have been detected by direct immunofluorescence, isolation, or both in 11 tick species in six genera from the western United States and Texas; it also has been isolated from the cat flea (*Ctenocephalides felis*) in Texas (101). The ticks include *Amblyomma americanum* and *A. maculatum* from Texas (101, J. A. Rawlings & G. Teltow, personal communication), *Dermacentor occidentalis* from California (64), *Dermacentor parumapertus* and *D. variabilis* from Texas (101), *Haemaphysalis leporispalustris* and *Ixodes neotomae* from California (62), *I. pacificus* from California and Oregon (20, 32, 60, 61, 64, 65, 67, 69), *I. scapularis* from Texas (J. A. Rawlings & G. Teltow, personal communication), tentatively *Ornithodoros coriaceus* from California (66), and *Rhipicephalus sanguineus* from Texas (101). One or more stages of eight of these tick species attach to humans occasionally, but only *I. pacificus* and *A. americanum* have been implicated as vectors to humans, the latter in New Jersey (104). The three tick species that rarely or never attach to humans are *D. parumapertus,* *H. leporispalustris,* and *I. neotomae.* Also, both *Amblyomma* spp., all three *Dermacentor* spp., *H. leporispalustris, I. neotomae,* and *I. scapularis* reportedly feed in part on lagomorphs, an observation that supports recent field and experimental evidence implicating jack rabbits (*Lepus californicus*) and cottontails (*Sylvilagus* spp.) as potential enzootic reservoirs in California, Texas, and the northeastern United States (4, 13, 35, 62, 68, 101, 120, 121).

Furthermore, *Ixodes* (*Ixodiopsis*) *angustus* has been implicated as a vector of *B. burgdorferi* to humans in Washington State even though spirochetes have so far not been detected in this tick (40), which is widely distributed in North America including the Pacific region from California to Alaska (55). Although it feeds principally on rodents, *I. angustus* does attach to people occasionally (49, 55).

The vector competencies of these ticks and of the cat flea for *B. burgdorferi* have not been reported, though *I. pacificus* and the other four tick species found infected in California are under investigation. However, transovarial passage of borreliae has been demonstrated for *H. leporispalustris;* 67% of the F1 (larval) progeny of a female obtained from a jack rabbit were infected with spirochetes that reacted to *Borrelia-* (genus) specific monoclonal antibodies (62). Isolation of *B. burgdorferi* from the cat flea and detection of antibodies to this spirochete in 36% of 85 serum specimens from domestic cats in Texas (101) are exciting findings that deserve further evaluation. Although direct contact transmission among rodents may be responsible for maintenance of Lyme disease in areas outside the known distribution of its tick vectors (33), the possibility that wild rodent fleas or other bloodsucking arthropods besides ticks (e.g. mites) might be involved should be explored (58).

Few insects besides fleas have been examined for spirochetes in western North America or Texas. An exception is the western tree-hole mosquito (*Aedes sierrensis*) in northern California. This mosquito was selected for study because it feeds often on deer and because spirochetes (of undetermined species) have been detected in the blood of 27–56% of native and exotic deer in northwestern California (60). *A. sierrensis* females collected in spring and summer from an area in Mendocino County in which Lyme borreliosis is endemic (kindly provided by Drs. J. R. Anderson and J. O. Washburn) were examined for spirochetes. Tissue smears prepared from the squashed heads and foreguts of 341 parous individuals were tested by direct immunofluorescence, but none was found to contain spirochetes (R. S. Lane, unpublished data). In contrast, *B. burgdorferi* has been detected in 7–8% of three species of *Aedes* mosquitoes in Connecticut (75).

Vertebrate Hosts and Potential Reservoirs

Recent studies of the bionomics of *I. pacificus* have shown that western fence lizards (*Sceloporus occidentalis*) and Columbian black-tailed deer (*Odocoileus hemionus columbianus*) are among the most important hosts for immature and adult *I. pacificus,* respectively (63, 65, 124). As many as 35 *I. pacificus* larvae and nymphs have been collected, on average, from western fence lizards in some localities during peak tick abundance in spring (S. A. Manweiler & R. S. Lane, unpublished data), and a similar number of predominantly adult ticks has been obtained from deer in fall (124). In comparison, rodents, lagomorphs, and birds have yielded much lower burdens of immature or adult *I. pacificus* (58, 62, 80a).

To determine the relative importance of birds versus lizards as hosts for *I. pacificus* immatures, a bird-lizard-tick survey was conducted in a focus of Lyme borreliosis in the Sierra Nevada foothills of northcentral California (80a). *I. pacificus* was not found on 71 birds captured with mist nets in an oak woodland between December 1988 and March 1989, and only 5 ticks were removed from 3 of 67 (4%) birds caught in spring 1989. However, 15 of 16 (94%) western fence lizards collected in the same habitat in spring yielded 369 *I. pacificus* (172 larvae, 197 nymphs) for an average of 23 ticks per lizard. Although 138 birds in 24 species were examined during this initial survey, many more species of birds comprising larger samples from diverse habitats need to be examined to clarify the association of *I. pacificus* to birds. The foregoing observations suggest that many species of birds inhabiting oak woodlands in the Sierra Nevada foothills are considerably less important than lizards as hosts of *I. pacificus*. If so, then birds would be relatively unimportant in the epizootiology of *B. burgdorferi* in California in contrast to their significant role as hosts of spirochetes and *I. dammini* immatures in the upper midwestern and northeastern United States (see above).

A similar comparative study of *I. pacificus* burdens on *Peromyscus* mice and western fence lizards was performed in chaparral and woodland-grass habitats in northern California (65; R. S. Lane & J. E. Loye, unpublished data). The deer mouse (*P. maniculatus*) and the piñon mouse (*P. truei*) were the two most abundant species of small rodents captured with Sherman traps, whereas the western fence lizard was the most numerous reptile observed in both habitats. When populations of *I. pacificus* immatures were peaking in spring, the prevalence and abundance of this tick were statistically significantly greater on lizards than on both species of mice in almost every pairwise comparison (65). Further studies will be conducted to estimate the abundance of various species of vertebrates and their associated ticks to determine which vertebrate species supports the greatest biomass of *I. pacificus* immatures in each of these habitats (65).

Because western fence lizards were found to be a major host of immature *I. pacificus* in some endemic foci for Lyme borreliosis, the susceptibility (= reservoir competence) of this reptile for *B. burgdorferi* was studied (56). Repeated efforts to infect juvenile and adult western fence lizards with spirochetes by inoculation with low-passage, tick-derived isolates or by feeding infected ticks on them were largely unsuccessful, as were attempts to isolate spirochetes from the blood or tissues of 20 wild-caught individuals.

However, recent serosurveys of western fence lizards ($n = 163$) conducted in three widely spaced localities in northern California revealed that low percentages of lizards do seroconvert after exposure to *I. pacificus* ticks, especially in spring (S. A. Manweiler, R. S. Lane, & C. H. Tempelis, unpublished data). It is not known if the occurrence of plasma antibodies to *B. burgdorferi* in lizards is merely indicative of antigenic exposure to such spirochetes following their inoculation by infective ticks or if spirochetal replication actually occurs in the bodily fluids or tissues of such lizards. In the same investigation, however, 0 of 366 larval and only 2 (0.4%) of 521 nymphal *I. pacificus* removed from these lizards and examined for spirochetes were infected. By contrast, approximately 1–3% of the adult *I. pacificus* collected from vegetation in all three localities were found to contain *B. burgdorferi*. These findings strongly support the conclusion that western fence lizards contribute little to the transmission cycle of *B. burgdorferi* (56).

Serum or plasma antibodies to *B. burgdorferi* or other borreliae have also been detected in 15 mammalian species in the western United States or Texas. These include axis deer (*Cervus axis*), Barbary sheep (*Ammotragus lervia*), black-tailed jack rabbit (*Lepus californicus*), bovine (*Bos* sp.), brush rabbit (*Sylvilagus bachmani*), cat (*Felis catus*), Columbian black-tailed deer (*O. hemionus columbianus*), coyote (*Canis latrans*), deer mouse (*P. maniculatus*), desert cottontail (*Sylvilagus audubonii*), dog (*Canis familiaris*), fallow deer (*Cervus dama*), horse (*Equus caballus*), piñon mouse (*P. truei*), and white-tailed deer (*O. virginianus*) (35, 58, 60, 62, 68, 101, 118; J. E. Madigan & R. S. Lane, unpublished data). The highest seroprevalences have been reported for brush rabbits ($\leq 100\%$), jack rabbits ($\leq 90\%$), and Columbian black-tailed deer (38%) in California, and for Barbary sheep (91%) and coyotes ($\leq 53\%$) in Texas. The high seropositivity rates in brush rabbits and jack rabbits suggest that these species (and perhaps other lagomorphs) would be useful as sentinel animals for surveillance of borreliosis in wildlands or rural areas of the West in the same fashion that domestic dogs have been proposed as sentinels for Lyme borreliosis in human environments (45). Caveats in assuming that borrelial antibodies in populations of wildlife (or humans) are directed against *B. burgdorferi* have been discussed and need not be repeated, except to reiterate that the results of spirochetal serosurveys should be interpreted cautiously, particularly in California, Texas, or other states where additional tick-borne borreliae (*B. coriaceae, B. hermsii, B. parkeri, B. turicatae*) may coexist (60, 62, 102).

In California, spirochetemia has been detected in three species of deer, two species of *Peromyscus* mice, and in the black-tailed jack rabbit, but attempts to isolate spirochetes from the blood or tissues of these animals have been unsuccessful. Consequently, experiments are being conducted with several species of rodents and Columbian black-tailed deer to determine their reservoir-competencies for *B. burgdorferi*. Spirochetes identified as *B. burgdorferi* with species-specific monoclonal antibodies have been isolated from a bushy-tailed woodrat (*Neotoma cinerea*) in northern California (J. Theis, personal communication) and from coyotes in southern Texas (35).

EUROPE

Ixodes ricinus

The castor bean or sheep tick, *I. ricinus,* is the nominal species for the *I. ricinus* complex. As early as 1909, it was suspected to be associated with erythema migrans (3). However, not until Burgdorfer et al (26) discovered a spirochete in the midgut of *I. dammini* to be the causative agent of Lyme disease did researchers detect similar spirochetes in *I. ricinus* midgut smears that had been prepared during 1978 in connection with tick-rickettsial surveys in Switzerland (23). The association of spirochetes with and the isolation of these microorganisms from live *I. ricinus* were accomplished in the spring of 1982 by examination of ticks collected from the Seewald Forest on the Swiss Plateau (17, 27).

I. ricinus is the most common tick in Europe (1). It occurs in woodlands and pastures from Ireland, Britain, and southern Scandinavia to Spain, Portugal, most of southern Europe, Central Europe, USSR to the Caspian Sea, and northern Iran. It has also been collected in northern Africa whereto it is apparently dispersed by migrating birds.

I. ricinus feeds indiscriminately, i.e. on numerous species of reptiles, birds, and small-, medium-, and large-sized mammals. Rodents, insectivores, reptiles, and birds are fed upon by larvae and nymphs, whereas medium- to large-sized wild and domestic animals serve as hosts for adults.

Successful infestation by *I. ricinus* and by all other members of the *I. ricinus* complex depends on fulfilling at least three biological and microclimatic requirements: (*a*) availability of suitable hosts, (*b*) temperature fluctuations between -10°C and 35°C in which extreme values last for short periods of time only, and (*c*), most importantly, a constant relative humidity of at least 80% in the air and of near saturation in the soil. The ideal biotope of these ticks, therefore, is a deciduous forest with damp soil covered by a dense and thick layer of undergrowth.

I. ricinus has been considered to be the only tick vector of *B. burgdorferi* in Europe. However, recent tick-spirochete surveys in West Germany have led to the detection of spirochetes in 13 (11.7%) of 111 *Ixodes hexagonus,* a tick that commonly parasitizes hedgehogs (*Erinaceus europaeus*), Mustelidae, and occasionally humans (71).

The prevalence of spirochete-infected *I. ricinus* collected in nature varies widely. Thus, in the initial survey conducted in western Switzerland, 112 (29.4%) of 381 adult ticks were infected (23). Subsequent surveys in that country revealed infection rates from 5 to 34% in nymphal and adult ticks (2, 88). In Austria, infection rates, ranging from 4 to 40% were recorded (113), and of 2,403 ticks from seven regions in southern Germany, 328 (13.6%) were infected with spirochetes (125). Similar studies conducted in Sweden yielded infection rates ranging from 6 to 21% in adult ticks (117).

Small rodents, particularly species of *Apodemus* and *Clethrionomys,* serve as important hosts for the immature stages of *I. ricinus* and as sources for infecting ticks with *B. burgdorferi.* Like the white-footed mouse (*P. leucopus*) in the northeastern and midwestern United States, these rodents are highly susceptible to the Lyme disease spirochete and, once infected, remain so for life (A. Aeschlimann, personal communication). All stages of *I. ricinus* occur almost simultaneously throughout the season of activity, although nymphal and adult ticks start feeding about four weeks before larval ticks. Spirochetal infection, therefore, is acquired by larvae parasitizing rodents that had become infected by nymphs.

The development of *B. burgdorferi* in *I. ricinus* has been the subject of intensive investigations at the Zoological Institute of the University of Neuchâtel in Switzerland (85; L. Gern, unpublished information). Observations reported so far tend to be similar to those obtained with *I. dammini* in the United States (103, 129). Accordingly, spirochetal distribution in the majority of infected unfed ticks from nature is limited to the midgut. Once such ticks begin to feed, the spirochetes penetrate the midgut epithelium and induce a more or less heavy systemic infection that includes the tissues of the salivary glands and ovary. The presence of spirochetes in salivary glands and the absence of these organisms in the pharyngeal cavity strongly suggest that transmission occurs via saliva. Transovarial transmission, when it does occur, results in low filial infection rates inasmuch as less than 5% of wild-caught larvae are infected (2, 125).

CONCLUSIONS

Changing human land-use patterns during the second half of the twentieth century appear to have provided an environmental mosaic suitable for wildlife and *I. dammini* to thrive in parts of the northeastern and midwestern United States. Critical hosts for the enzootic cycle of *I. dammini*–transmitted *B. burgdorferi* include the white-footed mouse as a host for immature ticks and as a reservoir of the spirochete and white-tailed deer as a host for adult *I. dammini.* The nymphal stage of *I. dammini* seems to be primarily responsible for transmission of the Lyme disease spirochete.

In western North America and Texas, 10 species of ixodid ticks and, provisionally, a single argasid tick contain *B. burgdorferi* or closely related spirochetes, but the relations of only two of these ticks (*A. americanum* and *I. pacificus*) to their associated spirochetes have been studied. In the Far West, *I. pacificus* appears to be the primary tick vector of *B. burgdorferi* to humans. Although research is being conducted in California to elucidate the vector competencies of *I. pacificus*, *D. occidentalis*, and several other ticks (including two nonhuman-biting species that may perpetuate *B. burgdorferi* in enzootic foci), similar studies are needed in other regions including the Pacific Northwest and western Canadian provinces. Likewise, the relations of these ticks to their vertebrate hosts and the reservoir competencies of such hosts for *B. burgdorferi* have received little or no attention in most western states or provinces except for northern California. Recent field and laboratory evidence suggest that the heavy utilization of western fence lizards as hosts by *I. pacificus* in parts of the Far West may be one of several biologic factors contributing to the low spirochete-infection rates reported for this tick.

The sheep tick, *I. ricinus*, the most common tick in Europe, is the primary vector of *B. burgdorferi* on that continent. Like the related American *Ixodes* ticks, *I. ricinus* has a broad host range and most unfed, infected individuals of this tick harbor spirochetes in their midguts only. Various species of *Apodemus* and *Clethrionomys* rodents appear to serve as long-term sources of *B. burgdorferi* infection for noninfected *I. ricinus* larvae that feed on them. Although *I. ricinus* is unquestionably one of the best-studied ticks worldwide, much still remains to be learned about its role in the epizootiology of Lyme borreliosis in Europe and elsewhere.

ACKNOWLEDGMENTS

We thank R. N. Brown for checking the literature citations and J. A. Rawlings and J. Theis for permission to cite previously unpublished findings. Some of the research reviewed here was supported in part by US Public Health Service Grant AI22501 from the National Institutes of Health to R. S. Lane.

Literature Cited

1. Aeschlimann, A. 1981. The role of hosts and environment in the natural dissemination of ticks. Studies on a Swiss population of *Ixodes ricinus* L., 1758. In *Review of Advances in Parasitology*, pp. 859–69. Warszawa

2. Aeschlimann, A., Chamot, E., Gigon, F., Jeanneret, J.-P., Kessler, D., et al. 1986. *B. burgdorferi* in Switzerland. *Zentralbl. Bakteriol. Mikrobiol. Hyg. A* 263:450–58

3. Afzelius, A. 1921. Erythema chronicum migrans. *Acta Derm. Venereol.* 2:120–25

4. Anderson, J. F. 1988. Mammalian and avian reservoirs for *Borrelia burgdorferi. Ann. N. Y. Acad. Sci.* 539:180–91

5. Anderson, J. F. 1989. Epizootiology of *Borrelia* in *Ixodes* tick vectors and reservoir hosts. *Rev. Infect. Dis.* 11:S1451–59

6. Anderson, J. F. 1989. Ecology of Lyme disease. *Conn. Med.* 53:343–46

7. Anderson, J. F., Johnson, R. C., Magnarelli, L. A., Hyde, F. W. 1985. Identification of endemic foci of Lyme disease: isolation of *Borrelia burgdorferi* from feral rodents and ticks (*Der-*

macentor variabilis). *J. Clin. Microbiol.* 22:36–38

8. Anderson, J. F., Johnson, R. C., Magnarelli, L. A., Hyde, F. W. 1986. Involvement of birds in the epidemiology of the Lyme disease agent *Borrelia burgdorferi*. *Infect. Immun.* 51:394–96

9. Anderson, J. F., Johnson, R. C., Magnarelli, L. A., Hyde, F. W., Myers, J. E. 1987. Prevalence of *Borrelia burgdorferi* and *Babesia microti* in mice on islands inhabited by white-tailed deer. *Appl. Environ. Microbiol.* 53:892–94

10. Anderson, J. F., Magnarelli, L. A. 1980. Vertebrate host relationships and distribution of ixodid ticks (Acari: Ixodidae) in Connecticut, USA. *J. Med. Entomol.* 17:314–23

11. Anderson, J. F., Magnarelli, L. A. 1984. Avian and mammalian hosts for spirochete-infected ticks and insects in a Lyme disease focus in Connecticut. *Yale J. Biol. Med.* 57:627–41

12. Anderson, J. F., Magnarelli, L. A., Burgdorfer, W., Barbour, A. G. 1983. Spirochetes in *Ixodes dammini* and mammals from Connecticut. *Am. J. Trop. Med. Hyg.* 32:818–24

13. Anderson, J. F., Magnarelli, L. A., LeFebvre, R. B., Andreadis, T. G., McAninch, J. B., et al. 1989. Antigenically variable *Borrelia burgdorferi* isolated from cottontail rabbits and *Ixodes dentatus* in rural and urban areas. *J. Clin. Microbiol.* 27:13–20

14. Anderson, J. F., Magnarelli, L. A., McAninch, J. B. 1987. *Ixodes dammini* and *Borrelia burgdorferi* in Northern New England and upstate New York. *J. Parasitol.* 73:419–21

15. Anderson, J. F., Magnarelli, L. A., Stafford, K. C. III. 1990. Bird-feeding ticks transstadially transmit *Borrelia burgdorferi* that infect Syrian hamsters. *J. Wildl. Dis.* 26:1–10

16. Arthur, D. R., Snow, K. R. 1968. *Ixodes pacificus* Cooley and Kohls, 1943: its life-history and occurrence. *Parasitology* 58:893–906

17. Barbour, A. G., Burgdorfer, W., Hayes, S. F., Péter, O., Aeschlimann, A. 1983. Isolation of a cultivable spirochete from *Ixodes ricinus* ticks of Switzerland. *Curr. Microbiol.* 8:123–26

18. Barbour, A. G., Hayes, S. F. 1986. Biology of *Borrelia* species. *Microbiol. Rev.* 50:381–400

19. Battaly, G. R., Fish, D., Dowler, R. C. 1987. The seasonal occurrence of *Ixodes dammini* and *Ixodes dentatus* (Acari: Ixodidae) on birds in a Lyme disease endemic area of southeastern New York State. *J. N. Y. Entomol. Soc.* 95:461–68

20. Bissett, M. L., Hill, W. 1987.

Characterization of *Borrelia burgdorferi* strains isolated from *Ixodes pacificus* ticks in California. *J. Clin. Microbiol.* 25:2296–2301

21. Bosler, E. M., Coleman, J. L., Benach, J. L., Massey, D. A., Hanrahan, J. P., et al. 1983. Natural distribution of the *Ixodes dammini* spirochete. *Science* 220:321–22

22. Bosler, E. M., Ormiston, B. G., Coleman, J. L., Hanrahan, J. P., Benach, J. L. 1984. Prevalence of the Lyme disease spirochete in populations of white-tailed deer and white-footed mice. *Yale J. Biol. Med.* 57:651–59

23. Burgdorfer, W. 1984. Discovery of the Lyme disease spirochete and its relation to tick vectors. *Yale J. Biol. Med.* 57:515–20

24. Burgdorfer, W. 1984. The New Zealand white rabbit: an experimental host for infecting ticks with Lyme disease spirochetes. *Yale J. Biol. Med.* 57:609–12

25. Burgdorfer, W. 1989. Vector/host relationships of the Lyme disease spirochete, *Borrelia burgdorferi*. In *Rheumatic Disease Clinics of North America*, ed. R. C. Johnson, 15:775–87. Philadelphia: Saunders

26. Burgdorfer, W., Barbour, A. G., Hayes, S. F., Benach, J. L., Grunwaldt, E., Davis, J. P. 1982. Lyme disease—a tick-borne spirochetosis? *Science* 216:1317–19

27. Burgdorfer, W., Barbour, A. G., Hayes, S. F., Peter, O., Aeschlimann, A. 1983. Erythema chronicum migrans—a tick-borne spirochetosis. *Acta Trop.* 40:79–83

28. Burgdorfer, W., Gage, K. L. 1986. Susceptibility of the black-legged tick, *Ixodes scapularis*, to the Lyme disease spirochete, *Borrelia burgdorferi*. *Zentralbl. Bakteriol. Mikrobiol. Hyg. A* 263:15–20

29. Burgdorfer, W., Gage, K. L. 1987. Susceptibility of the hispid cotton rat *(Sigmodon hispidus)* to the Lyme disease spirochete *(Borrelia burgdorferi)*. *Am. J. Trop. Med. Hyg.* 37:624–28

30. Burgdorfer, W., Hayes, S. F. 1989. Vector/spirochete relationships in louse-borne and tick-borne borrelioses with emphasis on Lyme disease. In *Advances in Disease Vector Research*, 6:127–50. New York: Springer-Verlag

31. Burgdorfer, W., Hayes, S. F., Corwin, D. 1989. Pathophysiology of the Lyme disease spirochete, *Borrelia burgdorferi*, in ixodid ticks. *Rev. Infect. Dis.* 11:S1442–50

32. Burgdorfer, W., Lane, R. S., Barbour, A. G., Gresbrink, R. A., Anderson, J. R. 1985. The western black-legged tick,

Ixodes pacificus: a vector of *Borrelia burgdorferi. Am. J. Trop. Med. Hyg.* 34:925–30

33. Burgess, E. C., Amundson, T. E., Davis, J. P., Kaslow, R. A., Edelman, R. 1986. Experimental inoculation of *Peromyscus* spp. with *Borrelia burgdorferi:* evidence of contact transmission. *Am. J. Trop. Med. Hyg.* 35:355–59

34. Burgess, E. C., Patrican, L. A. 1987. Oral infection of *Peromyscus maniculatus* with *Borrelia burgdorferi* and subsequent transmission by *Ixodes dammini. Am. J. Trop. Med. Hyg.* 36:402–7

35. Burgess, E. C., Windberg, L. A. 1989. *Borrelia* sp. infection in coyotes, black-tailed jack rabbits and desert cottontails in southern Texas. *J. Wildl. Dis.* 25:47–51

36. Campagna, J., Lavoie, P. E., Birnbaum, N. S., Furman, D. P. 1983. Lyme disease in northern California. *West. J. Med.* 139:319–23

37. Carey, A. B., Krinsky, W. L., Main, A. J. 1980. *Ixodes dammini* (Acari: Ixodidae) and associated ixodid ticks in south-central Connecticut, USA. *J. Med. Entomol.* 17:89–99

38. Ciesielski, C. A., Markowitz, L. E., Horsley, R., Hightower, A. W., Russell, H., et al. 1989. Lyme disease surveillance in the United States, 1983–86. *Rev. Infect. Dis.* 11:S1435–41

39. Curran, K. L., Fish, D. 1989. Increased risk of Lyme disease for cat owners. *N. Engl. J. Med.* 320:183

40. Damrow, T., Freedman, H., Lane, R. S., Preston, K. L. 1989. Is *Ixodes (Ixodiopsis) angustus* a vector of Lyme disease in Washington State? *West. J. Med.* 150:580–82

41. Daniels, T. J., Fish, D., Falco, R. C. 1989. Seasonal activity and survival of adult *Ixodes dammini* (Acari: Ixodidae) in southern New York State. *J. Med. Entomol.* 26:610–14

42. Davis, J. P., Schell, W. L., Amundson, T. E., Godsey, M. S., Spielman, A., et al. 1984. Lyme disease in Wisconsin; epidemiologic, clinical, serologic, and entomologic findings. *Yale J. Biol. Med.* 57:685–96

43. Donahue, J. G., Piesman, J., Spielman, A. 1987. Reservoir competence of white-footed mice for Lyme disease spirochetes. *Am. J. Trop. Med. Hyg.* 36:92–96

44. Drew, M. L., Loken, K. I., Bey, R. F., Swiggum, R. D. 1988. *Ixodes dammini:* occurrence and prevalence of infection with *Borrelia* spp. in Minnesota. *J. Wildl. Dis.* 24:708–10

45. Eng, T. R., Wilson, M. L., Spielman,

A., Lastavica, C. C. 1988. Greater risk of *Borrelia burgdorferi* infection in dogs than in people. *J. Infect. Dis.* 158:1410–11

46. Falco, R. C., Fish, D. 1988. Prevalence of *Ixodes dammini* near the homes of Lyme disease patients in Westchester County, New York. *Am. J. Epidemiol.* 127:826–30

47. Falco, R. C., Fish, D. 1989. Potential for exposure to tick bites in recreational parks in a Lyme disease endemic area. *Am. J. Public Health* 79:12–15

48. Fish, D., Dowler, R. C. 1989. Host associations of ticks (Acari: Ixodidae) parasitizing medium-sized animals in a Lyme disease endemic area of southern New York. *J. Med. Entomol.* 26:200–9

49. Furman, D. P., Loomis, E. C. 1984. The ticks of California (Acari: Ixodida). *Bull. Calif. Insect Surv.* 25:1–239

50. Ginsberg, H. S. 1988. A model of the spread of Lyme disease in natural populations. *Ann. N. Y. Acad. Sci.* 539: 379–80

51. Ginsberg, H. S., Ewing, C. P. 1989. Habitat distribution of *Ixodes dammini* (Acari: Ixodidae) and Lyme disease spirochetes on Fire Island, New York. *J. Med. Entomol.* 26:183–89

52. Ginsberg, H. S., Ewing, C. P. 1989. Comparison of flagging, walking, trapping, and collecting from hosts as sampling methods for northern deer ticks, *Ixodes dammini,* and lone-star ticks, *Amblyomma americanum* (Acari:Ixodidae). *Exp. Appl. Acarol.* 7:313–22

53. Godsey, M. S. Jr., Amundson, T. E., Burgess, E. C., Schell, W., Davis, J. P., et al. 1987. Lyme disease ecology in Wisconsin: Distribution and host preferences of *Ixodes dammini,* and prevalence of antibody to *Borrelia burgdorferi* in small mammals. *Am. J. Trop. Med. Hyg.* 37:180–87

54. Johnson, R. C., Schmid, G. P., Hyde, F. W., Steigerwalt, A. G., Brenner, D. J. 1984. *Borrelia burgdorferi* sp. nov.: etiologic agent of Lyme disease. *Int. J. Syst. Bacteriol.* 34:496–97

55. Keirans, J. E., Clifford, C. M. 1978. The genus *Ixodes* in the United States: A scanning electron microscope study and key to the adults. *J. Med. Entomol. Suppl.* 2:1–149

56. Lane, R. S. 1990. Susceptibility of the western fence lizard *(Sceloporus occidentalis)* to the Lyme borreliosis spirochete *(Borrelia burgdorferi). Am. J. Trop. Med. Hyg.* 42:75–82

57. Lane, R. S. 1990. Seasonal activity of two human-biting ticks. *Calif. Agric.* 44:23–25

58. Lane, R. S. 1990. Infection of deer mice and piñon mice *(Peromyscus* spp.) with spirochetes at a focus of Lyme borreliosis in northern California, USA. *Bull. Soc. Vector Ecol.* 15:25–32

59. Lane, R. S., Anderson, J. R., Yaninek, J. S., Burgdorfer, W. 1985. Diurnal host seeking of adult Pacific Coast ticks, *Dermacentor occidentalis* (Acari: Ixodidae), in relation to vegetational type, meteorological factors, and rickettsial infection rates in California, USA. *J. Med. Entomol.* 22:558–71

60. Lane, R. S., Burgdorfer, W. 1986. Potential role of native and exotic deer and their associated ticks (Acari: Ixodidae) in the ecology of Lyme disease in California, USA. *Zentralbl. Bakteriol. Mikrobiol. Hyg. A* 263:55–64

61. Lane, R. S., Burgdorfer, W. 1987. Transovarial and transstadial passage of *Borrelia burgdorferi* in the western black-legged tick, *Ixodes pacificus* (Acari: Ixodidae). *Am. J. Trop. Med. Hyg.* 37:188–92

62. Lane, R. S., Burgdorfer, W. 1988. Spirochetes in mammals and ticks (Acari: Ixodidae) from a focus of Lyme borreliosis in California. *J. Wildl. Dis.* 24:1–9

63. Lane, R. S., Emmons, R. W., Dondero, D. V., Nelson, B. C. 1981. Ecology of tick-borne agents in California. I. Spotted fever group rickettsiae. *Am. J. Trop. Med. Hyg.* 30:239–52

64. Lane, R. S., Lavoie, P. E. 1988. Lyme borreliosis in California: acarological, clinical, and epidemiological studies. *Ann. N. Y. Acad. Sci.* 539:192–203

65. Lane, R. S., Loye, J. E. 1989. Lyme disease in California: interrelationship of *Ixodes pacificus* (Acari: Ixodidae), the western fence lizard *(Sceloporus occidentalis),* and *Borrelia burgdorferi.* *J. Med. Entomol.* 26:272–78

66. Lane, R. S., Manweiler, S. A. 1988. *Borrelia coriaceae* in its tick vector, *Ornithodoros coriaceus* (Acari: Argasidae), with emphasis on transstadial and transovarial infection. *J. Med. Entomol.* 25:172–77

67. Lane, R. S., Pascocello, J. A. 1989. Antigenic characteristics of *Borrelia burgdorferi* isolates from ixodid ticks in California. *J. Clin. Microbiol.* 27:2344–49

68. Lane, R. S., Regnery, D. C. 1989. Lagomorphs as sentinels for surveillance of borreliosis in the far western United States. *J. Wildl. Dis.* 25:189–93

69. Lane, R. S., Stubbs, H. A. 1990. Host-seeking behavior of adult *Ixodes pacificus* (Acari: Ixodidae) as determined by flagging vegetation. *J. Med. Entomol.* 27:282–87

70. Levine, J. F., Wilson, M. L., Spielman, A. 1985. Mice as reservoirs of the Lyme disease spirochete. *Am. J. Trop. Med. Hyg.* 34:355–60

71. Liebisch, A., Olbrich, S., Brand, A., Liebisch, G., Mourettou-Kunitz, M. 1989. Naturliche Infektionen der Zeckenart *Ixodes hexagonus* mit Borrelien *(Borrelia burgdorferi).* *Tierarztl. Umsch.* 44:809–10

72. Loye, J. E., Lane, R. S. 1988. Questing behavior of *Ixodes pacificus* (Acari: Ixodidae) in relation to meteorological and seasonal factors. *J. Med. Entomol.* 25:391–98

73. Magnarelli, L. A., Anderson, J. F. 1988. Ticks and biting insects infected with the etiologic agent of Lyme disease, *Borrelia burgdorferi.* *J. Clin. Microbiol.* 26:1482–86

74. Magnarelli, L. A., Anderson, J. F., Apperson, C. S., Fish, D., Johnson, R. C., et al. 1986. Spirochetes in ticks and antibodies to *Borrelia burgdorferi* in white-tailed deer from Connecticut, New York State, and North Carolina. *J. Wildl. Dis.* 22:178–88

75. Magnarelli, L. A., Anderson, J. F., Barbour, A. G. 1986. The etiologic agent of Lyme disease in deer flies, horse flies, and mosquitoes. *J. Infect. Dis.* 154:355–58

76. Magnarelli, L. A., Anderson, J. F., Chappell, W. A. 1984. Antibodies to spirochetes in white-tailed deer and prevalence of infected ticks from foci of Lyme disease in Connecticut. *J. Wildl. Dis.* 20:21–26

77. Magnarelli, L. A., Anderson, J. F., Fish, D. 1987. Transovarial transmission of *Borrelia burgdorferi* in *Ixodes dammini* (Acari:Ixodidae). *J. Infect. Dis.* 156:234–36

78. Magnarelli, L. A., Freier, J. E., Anderson, J. F. 1987. Experimental infections of mosquitoes with *Borrelia burgdorferi,* the etiologic agent of Lyme disease. *J. Infect. Dis.* 156:694–95

79. Main, A. J., Carey, A. B., Carey, M. G., Goodwin, R. H. 1982. Immature *Ixodes dammini* (Acari: Ixodidae) on small animals in Connecticut, USA. *J. Med. Entomol.* 19:655–64

80. Main, A. J., Sprance, H. E., Kloter, K. O., Brown, S. E. 1981. *Ixodes dammini* (Acari: Ixodidae) on white-tailed deer *(Odocoileus virginianus)* in Connecticut. *J. Med. Entomol.* 18: 487–92

80a. Manweiler, S. A., Lane, R. S., Block, W. M., Morrison, M. L. 1990. Survey of birds and lizards for ixodid ticks and spirochetal infection in northern California. *J. Med. Entomol.* In press

81. Massachusetts Medical Society. 1989. Lyme disease—United States, 1987 and 1988. *Morb. Mortal. Wkly. Rep.* 38:668–72

82. Mather, T. N., Telford, S. R. III, MacLachlan, A. B., Spielman, A. 1989. Incompetence of catbirds as reservoirs for the Lyme disease spirochete *(Borrelia burgdorferi)*. *J. Parasitol.* 75:66–69

83. Mather, T. N., Wilson, M. L., Moore, S. I., Ribeiro, J. M. C., Spielman, A. 1989. Comparing the relative potential of rodents as reservoirs of the Lyme disease spirochete *(Borrelia burgdorferi)*. *Am. J. Epidemiol.* 130:143–50

84. McEnroe, W. D. 1977. The restriction of the species range of *Ixodes scapularis*, Say, in Massachusetts by fall and winter temperature. *Acarologia* 18:618–25

85. Monin, R. 1987. *Contribution à l'étude des modes de transmission de* Borrelia burgdorferi, *agent etiologique de la maladie de Lyme, chex* Ixodes ricinus. PhD thesis. Switzerland: Univ. Neuchâtel

86. Mount, G. A., Haile, D. G. 1989. Computer simulation of population dynamics of the American Dog Tick (Acari: Ixodidae). *J. Med. Entomol.* 26:60–76

87. Naversen, D. N., Gardner, L. W. 1978. Erythema chronicum migrans in America. *Arch. Dermatol.* 114:253–54

88. Péter, O. 1990. Lyme borreliosis in the state of Valais, Switzerland. *J. Int. Fed. Clin. Chem.* In press

89. Piesman, J. 1988. Vector competence of ticks in the southeastern United States for *Borrelia burgdorferi*. *Ann. N. Y. Acad. Sci.* 539:417–18

90. Piesman, J. 1989. Transmission of Lyme disease spirochetes *(Borrelia burgdorferi)*. *Exp. Appl. Acarol.* 7:71–80

91. Piesman, J., Donahue, J. G., Mather, T. N., Spielman, A. 1986. Transovarially acquired Lyme disease spirochetes *(Borrelia burgdorferi)* in field-collected larval *Ixodes dammini* (Acari: Ixodidae). *J. Med. Entomol.* 23:219

92. Piesman, J., Hicks, T. C., Sinsky, R. J., Obiri, G. 1987. Simultaneous transmission of *Borrelia burgdorferi* and *Babesia microti* by individual nymphal *Ixodes dammini* ticks. *J. Clin. Microbiol.* 25:2012–13

93. Piesman, J., Mather, T. N., Dammin, G. J., Telford, S. R. III, Lastavica, C. C., et al. 1987. Seasonal variation of transmission risk of Lyme disease and human babesiosis. *Am. J. Epidemiol.* 126:1187–89

94. Piesman, J., Mather, T. N., Donahue, J. G., Levine, J., Campbell, J. D., et al. 1986. Comparative prevalence of *Babesia microti* and *Borrelia burgdorferi* in four populations of *Ixodes dammini* in eastern Massachusetts. *Acta Trop.* 43: 263–70

95. Piesman, J., Mather, T. N., Sinsky, R. J., Spielman, A. 1987. Duration of tick attachment and *Borrelia burgdorferi* transmission. *J. Clin. Microbiol.* 25: 557–58

96. Piesman, J., Sinsky, R. J. 1988. Ability of *Ixodes scapularis, Dermacentor variabilis,* and *Amblyomma americanum* (Acari: Ixodidae) to acquire, maintain, and transmit Lyme disease spirochetes *(Borrelia burgdorferi)*. *J. Med. Entomol.* 25:336–39

97. Piesman, J., Spielman, A. 1979. Host-associations and seasonal abundance of immature *Ixodes dammini* in southeastern Massachusetts. *Ann. Entomol. Soc. Am.* 72:829–32

98. Piesman, J., Spielman, A., Etkind, P., Ruebush, T. K. II, Juranek, D. D. 1979. Role of deer in the epizootiology of *Babesia microti* in Massachusetts, USA. *J. Med. Entomol.* 15:537–40

99. Pinger, R. R., Glancy, T. 1989. *Ixodes dammini* (Acari: Ixodidae) in Indiana. *J. Med. Entomol.* 26:130–31

100. Pokorny, P. 1989. Incidence of spirochaeta *Borrelia burgdorferi* in arthropods and of antibodies in vertebrates. *Cesk. Epidemiol. Mikrobiol. Imunol.* 38:52–60

101. Rawlings, J. A. 1986. Lyme disease in Texas. *Zentrabl. Bakteriol. Mikrobiol. Hyg. A* 263:483–87

102. Rawlings, J. A., Lane, R. S. 1988. The prevalence of antibody to *Borrelia burgdorferi* among patients with undiagnosed central nervous system disease in California. *Ann. N. Y. Acad. Sci.* 539:489–91

103. Ribeiro, J. M. C., Mather, T. N., Piesman, J., Spielman, A. 1987. Dissemination and salivary delivery of Lyme disease spirochetes in vector ticks (Acari: Ixodidae). *J. Med. Entomol.* 24:201–5

104. Schulze, T. L., Bowen, G. S., Bosler, E. M., Lakat, M. F., Parkin, W. E., et al. 1984. *Amblyomma americanum:* a potential vector of Lyme disease in New Jersey. *Science* 224:601–3

105. Schulze, T. L., Bowen, G. S., Lakat, M. F., Parkin, W. E., Shisler, J. K. 1985. The role of adult *Ixodes dammini* (Acari: Ixodidae) in the transmission of Lyme disease in New Jersey, USA. *J. Med. Entomol.* 22:88–93

106. Schulze, T. L., Lakat, M. F., Bowen, G. S., Parkin, W. E., Shisler, J. K.

1984. *Ixodes dammini* (Acari: Ixodidae) and other ixodid ticks collected from white-tailed deer in New Jersey, USA. I. Geographical distribution and its relation to selected environmental and physical factors. *J. Med. Entomol.* 21:741–49

107. Schulze, T. L., Lakat, M. F., Parkin, W. E., Shisler, J. K., Charette, D. J., et al. 1986. Comparison of rates of infection by the Lyme disease spirochete in selected populations of *Ixodes dammini* and *Amblyomma americanum* (Acari: Ixodidae). *Zentralbl. Bakteriol. Mikrobiol. Hyg. A* 263:72–78

108. Schulze, T. L., Shisler, J. K., Bosler, E. M., Lakat, M. F., Parkin, W. E. 1986. Evolution of a focus of Lyme disease. *Zentralbl. Bakteriol. Mikrobiol. Hyg. A* 263:65–71

109. Spielman, A. 1976. Human babesiosis on Nantucket Island: transmission by nymphal *Ixodes* ticks. *Am. J. Trop. Med. Hyg.* 25:784–87

110. Spielman, A. 1988. Lyme disease and human babesiosis: Evidence incriminating vector and reservoir hosts. In *The Biology of Parasitism. A Molecular and Immunological Approach,* ed. P. T. Englund, A. Sher, pp. 147–65. New York: Liss

111. Spielman, A., Clifford, C. M., Piesman, J., Corwin, D. 1979. Human babesiosis on Nantucket Island, USA: description of the vector, *Ixodes (Ixodes) dammini,* n.sp. (Acarina: Ixodidae). *J. Med. Entomol.* 15:218–34

112. Spielman, A., Wilson, M. L., Levine, J. F., Piesman, J. 1985. Ecology of *Ixodes dammini*-borne human babesiosis and Lyme disease. *Annu. Rev. Entomol.* 30:439–60

113. Stanek, G., Pletschette, M., Flamm, H., Hirschl, A. M., Aberer, E., et al. 1988. European Lyme borreliosis. *Ann. N. Y. Acad. Sci.* 539:274–82

114. Steere, A. C. 1989. Lyme disease. *New Engl. J. Med.* 321:586–96

115. Steere, A. C., Grodzicki, R. L., Kornblatt, A. N., Craft, J. E., Barbour, A. G., et al. 1983. The spirochetal etiology of Lyme disease. *New Engl. J. Med.* 308:733–40

116. Steere, A. C., Sardinas, A. V., Lavoie, P. E., Birnbaum, N. J., Caputo, R. V., et al. 1981. Lyme disease—United States, 1980. *Morb. Mortal. Wkly. Rep.* 30:489–97

117. Stiernstedt, G. T. 1985. Tick-borne *Borrelia* infection in Sweden. *Scand. J. Infect. Dis.* (Suppl 45):1–70

118. Teitler, J., Madigan, J., DeRock, E., Pedersen, N., Carpenter, T., Franti, C. 1988. Prevalence of *Borrelia burgdorferi* antibodies in dogs in northern California: risk factors and zoonotic implications. *Ann. N. Y. Acad. Sci.* 539: 500–3

119. Telford, S. R. III, Mather, T. N., Moore, S. I., Wilson, M. L., Spielman, A. 1988. Incompetence of deer as reservoirs of the Lyme disease spirochete. *Am. J. Trop. Med. Hyg.* 39:105–9

120. Telford, S. R. III, Spielman, A. 1989. Competence of a rabbit-feeding *Ixodes* (Acari: Ixodidae) as a vector of the Lyme disease spirochete. *J. Med. Entomol.* 26:118–21

121. Telford, S. R. III, Spielman, A. 1989. Enzootic transmission of the agent of Lyme disease in rabbits. *Am. J. Trop. Med. Hyg.* 41:482–90

122. Todd, M. J., Carter, A. O., Galloway, T. D. 1989. Lyme disease—Canada. *Can. Dis. Wkly. Rep.* 15:135–37, 185

123. Weisbrod, A. R., Johnson, R. C. 1989. Lyme disease and migrating birds in the Saint Croix River Valley. *Appl. Environ. Microbiol.* 55:1921–24

124. Westrom, D. R., Lane, R. S., Anderson, J. R. 1985. *Ixodes pacificus* (Acari: Ixodidae): population dynamics and distribution on Columbian black-tailed deer *(Odocoileus hemionus columbianus). J. Med. Entomol.* 22:507–11

125. Wilske, B., Steinhuber, R., Bergmeister, H., Fingerle, V., Schierz, G., et al. 1987. Lyme-borreliose in Suddeutschland. Epidemiologische daten zum auftreten von erkrankungsfallen sowie zur durchseuchung von zecken *(Ixodes ricinus)* mit *Borrelia burgdorferi. Dtsch. Med. Wochenschr.* 112:1730–36

126. Wilson, M. L., Adler, G. H., Spielman, A. 1985. Correlation between abundance of deer and that of the deer tick, *Ixodes dammini* (Acari: Ixodidae). *Ann. Entomol. Soc. Am.* 78:172–76

127. Wilson, M. L., Spielman, A. 1985. Seasonal activity of immature *Ixodes dammini* (Acari: Ixodidae). *J. Med. Entomol.* 22:408–14

128. Wilson, M. L., Telford S. R. III, Piesman, J., Spielman, A. 1988. Reduced abundance of immature *Ixodes dammini* (Acari: Ixodidae) following elimination of deer. *J. Med. Entomol.* 25:224–28

128a. Yuval, B., Spielman, A. 1990. Duration and regulation of the developmental cycle of *Ixodes dammini* (Acari: Ixodidae). *J. Med. Entomol.* 27:196–201

129. Zung, J. L., Lewengrub, S., Rudzinska, M. A., Spielman, A., Telford, S. R. III, et al. 1989. Fine structural evidence for the penetration of the Lyme disease spirochete *Borrelia burgdorferi* through the gut and salivary tissues of *Ixodes dammini. Can. J. Zool.* 67:1737–48

Annu. Rev. Entomol. 1991. 36:611–36

ECOLOGICAL AND EVOLUTIONARY SIGNIFICANCE OF PHORESY IN THE ASTIGMATA

M. A. Houck

Department of Ecology and Evolutionary Biology, University of Arizona, Tucson, Arizona 85721

B. M. OConnor

Museum of Zoology and Department of Biology, University of Michigan, Ann Arbor, Michigan 48109-1079

KEY WORDS: Hemisarcoptes, Chilocorus, dispersal, hypopus, mites

PERSPECTIVES AND OVERVIEW

Phoresy is a common form of commensalism among animals in nature and one of the least explored. The term phoresy is applied to interspecific relationships in which one organism (the phoretic) attaches to another (the host) for the implied purpose of dispersal. Etymologically, the derivation from the Greek emphasizes the role of the host in the interaction (*phoras,* bearing) and at the same time the influence on the host (*phor,* thief). The review that follows is from the perspective of the phoretic, specifically phoretic mites of the Astigmata, which are masters of phoretic association and have often taken the concept to its extreme.

History of the Definition of Phoresy

Phoresy was recognized as a biological entity in the Acari in the mid 1700s but lacked explicit definition until 1896, when Lesne defined it as "those cases in which the transport host serves its passenger as a vehicle" (44). A

611

0066-4170/91/0101-0611$02.00

separate term, *symphorium* (Gr: *symphoros,* accompanying), was coined in 1917 (17) for commensal relationships in which an animal attaches to another "without becoming a parasite, on the outer surface of another animal species, without the occurrence of a mutualistic (reciprocal) relationship between the carried and the carrier." An entanglement of terms and applications quickly arose in the literature on phoresy and generated two obstacles: a proliferation in terminology and inadequate discrimination of time scales. Frequently, the discovery of a variant in the expression of phoresy provoked a new term in celebration of the observed novelty. Terminology became context-oriented and profuse. Also, early definitions contained no relative statement of time that would discriminate between associations that lasted only briefly and the colonization of the host during transport (such as barnacles attaching to whales).

Farish & Axtell (44) addressed these problems and contributed an explicit formalism to the definition. They restricted the definition to include any "phenomenon in which one animal actively seeks out and attaches to the outer surface of another animal for a limited time during which the attached animal (termed the phoretic) ceases both feeding and ontogenesis, such attachment presumably resulting in dispersal from areas unsuited for further development, either of the individual or its progeny." This perspective provides for the development of falsifiable hypotheses concerning trophic and ontogenetic interactions and restricts the range of included relationships.

Application of the Definition to the Astigmata

The definition of Farish & Axtell (44), however, does not entirely reflect phoretic associations in the Astigmata. It emphasizes proximate components, but phoresy is likewise the product of co-association through evolutionary time. Without that association, a temporal interaction must be considered as trivial migration (124) or accidental. Also, the implication that phoretics "actively seek" hosts is somewhat vague. Phoresy requires active discrimination among potential hosts, but astigmatid mites are often relatively quiescent while awaiting hosts and during transport. This behavior is necessary for the conservation of stored energy that is not, by definition, replaced during transit. The phoretic interval may be in terms of days, weeks, or even months (13).

We further tailor the definition by describing the movement itself as migration. Dispersal (the act of scattering) is a likely outcome of the movement of individuals (124), but it is not a necessary component of phoresy. When members of a population move in synchrony on hosts that respond similarly to environmental factors or are themselves aggregated, individuals may experience little dispersal. On the contrary, phoresy may provide a benefit to the phoretic that would be inhibited by dispersal (e.g. congregate colonization). We choose the term *superficially attached* over *external,* as in the original definition, because astigmatid mites can be transported within natural orifices of the host or under host structures that protect the phoretic from the external habitat. Finally, we characterize the emigrant habitat as *natal* to avoid a qualitative judgement concerning the emigrant habitat. In the Astigmata, migration is not necessarily directly correlated with habitat quality, but may be correlated with the life cycle of the host, and only indirectly correlated with habitat degradation.

Our operational definition of phoresy, therefore, is a phenomenon in which one organism (the phoretic) receives an ecological or evolutionary advantage by migrating from the natal habitat while superficially attached to a selected interspecific host for some portion of the individual phoretic's lifetime. Benefit is not conferred as a nutritional or developmental influence on the phoretic stage.

ECOLOGICAL BENEFITS OF PHORESY

Phoresy in the Acari (mites) encompasses a range of interspecific interactions from generalized, as in *Histiogaster arborsignis,* which uses over 40 host species belonging to three insect orders (105), or *Carpoglyphus lactis,* which can be collected from the head and tongues of at least nine species of nymphalid butterflies (128, 131), to complex co-evolved interactions (4, 51, 52, 75). The phoretic benefits from the host's capacity to move more effectively among habitats than could the unaided phoretic. Phoresy compensates for the disadvantages of small size in long-distance migration, the lack of morphological adaptation (e.g. wings) for independent migration, and the vulnerability to predation during migration. Phoresy requires that an organism possess structures enabling it to board a host and mechanisms for securing itself during transport. Additional adaptations that allow the phoretic to better distinguish hosts and quality habitats improve the probability of survival. Phoresy is a way of locating discrete or temporary microhabitats with a high degree of predictability (91).

Migration from spatially and temporally ephemeral habitats is often positively correlated with the degree of impermanence of the habitat (124). Phoresy provides a potential for colonization elsewhere when a temporary habitat degrades or disappears. Ephemeral habitats as defined here include carrion, dung, fungal mycelia or fruiting bodies, vascular plants of seral communities, nests of arthropods and vertebrates, phytotelmata, and temporary accumulations of decaying materials such as tree-holes and beach wrack. Rapid crowding is likely in these small, nutrient-rich habitats that are exploited by organisms with a short life cycle. Phoresy is an advantage under conditions of crowding in taxa with little propensity for diapause who live in an environment that is unpredictably patchy and where the patches are frequently renewed (e.g. dung pats) or have prohibitive distances between them (e.g. carrion).

The effectiveness and consequences of phoresy differ substantially because of differences in the effectiveness of attachment organs, specificity of host selection, mean distance of host migration, frequency and predictability of host availability, stage of the life cycle that is phoretic, durability of the phoretic stage exposed to environmental vagaries, colonization qualities of the phoretic, and the numbers of individuals participating in phoresy at any point in time.

DISPERSAL MODES IN THE ACARI

The most spectacular radiation of phoretic associations among animals occurs in the Acari (mites). With this radiation comes a diversification of adaptations for phoresy. A particular host species may attract a wide diversity of phoretics that occupy it simultaneously. For example, 11 species of Acari from 9 different families were collected from 28 specimens of the hermit flower beetle, *Osmoderma eremicola* (Coleoptera: Scarabaeidae) (95). Of the total collection (4,826 mites), 98% were astigmatid mites. Spatial distribution of mites on the host is often nonrandom; particular species are found on one specific part of the host. Location may vary by species and sex of the host (13).

Only one developmental stage is usually specialized for dispersal in mites, and the degree of specialization of that stage can be classified into unspecialized homeomorphs, specialized homeomorphs, and facultative heteromorphs (Figure 1). Unspecialized homeomorphs are relatively consistent in morphology with all other stages. Dispersal is accomplished by grasping the host with structures primarily adapted for other purposes, such as the chelicerae (e.g. female Macrochelidae) (Figure 1A) or pretarsal claws (e.g. deutonymphal Parasitidae). Transport may also result from the phoretic's exploitation of some peculiarity in host morphology. Adult laelapids, for example, are transported on carpenter bees by riding in an abdominal integumental pouch termed an *acarinarium* (84).

Specialized homeomorphs have modest morphological embellishments specifically adapted for attachment, such as corporeal or appendicular suckers. More marked adaptation occurs among polymorphic species of the Heterostigmata (e.g. *Pyemotes* and *Siteroptes*) (15, 90, 91) (Figure 1B), in which only a specialized female (phoretomorph) has the first pair of legs greatly enlarged for attachment. Another curious embellishment occurs in deutonymphs of the Uropodidae. Phoretic attachment to a host is possible because of the production of a hyaline anal pedicel (Figure 1C), a fluid strand

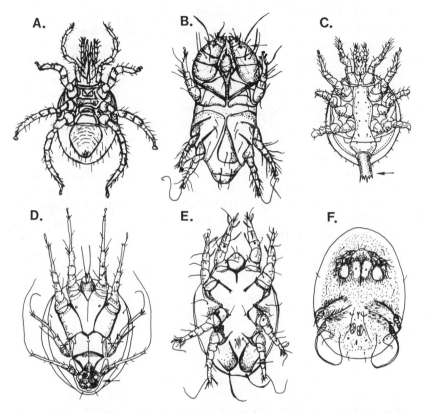

Figure 1   Diversity of morphological adaptations for phoresy in the Acari. Venter of: (*A*) a female *Macrocheles glaber*, redrawn from Evans et al (23); (*B*) a female *Pyemotes giganticus*, redrawn from Cross et al (15); (*C*) a uropodid mite, redrawn from Evans et al (23); (*D*) a deutonymph of *Hormosianoetus aeschlimanni*, redrawn from Fain (33); (*E*) a deutonymph of *Neoxenoryctes reticulatus*, redrawn from Fain & Philips (42); (*F*) a deutonymph of *Muridectes heterocephali*, redrawn from Fain (32). *D–F* are astigmatid mites.

secreted from the anus, which hardens into an umbilicus of attachment (1, 50, 70, 72). The anal pedicel has contributed significantly to a diversification of habitat utilization in the uropodid mites (1).

Facultative heteromorphs (optionally invoked) have a morphology completely disjunct from the general body plan of their particular taxon. Facultative heteromorphs (Figure 1D) are a common feature of the life history of the Astigmata and are the emphasis of the remainder of this review. These heteromorphs are characterized by a rudimentary gnathosoma, the absence of a mouth, the absence of a hollow gut, extensive sclerotization, and a caudoventral attachment organ (67, 68, 74, 77, 96, 112, 126, 130, 133). Such a morph is frequently referred to in the literature as a hypopus.

What is a hypopus?

The term hypopus stirs almost as much debate among acarologists as the concept of phoresy itself. The arguments are not polemic or semantic but focus on two critical issues: (*a*) the appropriate application of the term and (*b*) the interpretation of the homology of the hypopus within an ontogenetic context. The issues become coupled, as the following discussion reveals.

The term hypopus is employed in two ways, reflecting a basic dichotomy in evolutionary interpretation by various authors. First, it is used to delimit any heteromorphic stage of development, regardless of its specific morphological expression. Thus, *inert hypopus* is not an oxymoron but a qualification of type. Second, it delimits a particular morphological specialization (listed previously as a rudimentary gnathosoma, no mouth, etc), irrespective of placement in the ontogenetic sequence. A third universally recognized conflict in terminology exists in that facultative deutonymphs are so morphologically conservative that Dugès (20) originally applied the term *Hypopus* to designate a genus that was a polyphyletic assemblage of all known heteromorphic (hypopial) deutonymphs. He did not realize that the morph represented an ontogenetic step, and not a phylogenetic one.

The second important issue concerns the parsimony of hypothesized ontogenetic sequences. The accepted ancestral sequence in acariform mites proceeds as follows: egg → prelarva → larva → protonymph → deutonymph → tritonymph → adult. The problem of homology in the Astigmata rests in the interpretation of which particular developmental stage the hypopus represents. Homology with the deutonymphal stage is established by the fact that it is invariably positioned between the protonymphal stage and the tritonymphal stage and that it expresses ontogenetic transformations typical of deutonymphs (e.g. addition of a second pair of genital papillae, setae present on the trochanters, and accessory setae present on the fourth pair of legs). The hypothesized ontogenetic sequence with this interpretation is: egg → prelarva → larva → protonymph → facultative deutonymph (= hypopus) → tritonymph → adult. This hypothesis requires only an evolutionary modification of characters in one stage (hypopus).

The competing hypothesis assumes an addition of a unique instar (hypopus) intercalated into the ontogeny (9, 133) and requires a concomitant loss of another (either the tritonymph or the adult). While a variable number of nymphal instars occurs in argasid ticks, this phenomenon is unknown in the Astigmata. When individuals express the facultative heteromorph, in this scenario, two sequences of development are possible: (*a*) egg → prelarva → larva → protonymph → intercalated hypopus → deutonymph → adult; or (*b*) egg → prelarva → larva → protonymph → intercalated hypopus → deutonymph → tritonymph. The first (*a*) requires one addition (the hypopus) and one deletion (the tritonymph). The second (*b*) requires one addition (the hypopus) and the truncation of the ontogeny, which creates a paedomorphic sequence. When the facultative heteromorph is not invoked, two other possible sequences result: (*a*) egg → prelarva → larva → protonymph → deutonymph → adult, or (*b*) egg → prelarva → larva → protonymph → deutonymph → tritonymph. The first (*a*) requires one deletion (the tritonymph). The second (*b*) requires terminal truncation (paedomorphosis).

The hypothesis with the fewest ontogenetic assumptions is that the hypopus is a deutonymph that is facultative. In this review, we use the terms heteromorphic deutonymph or deutonymph to refer to what has previously been called a hypopus. The word hypopus has become confusedly entrenched in the literature over the last 100 years (87), and it may well continue to be, but if applied it should be defined in terms of explicit evolutionary assumptions.

THE ASTIGMATA

The Astigmata is a diverse and widely distributed monophyletic group of mites within the suborder Sarcoptiformes (98, 101). It is the most successful group of mites in establishing symbiotic relationships with both vertebrates and invertebrates. Few astigmatid mites are completely free-living throughout their ontogeny, and co-evolved interactions have enhanced the ecological and evolutionary radiation of the Astigmata, which comprises over 69 families and 785 genera (100).

Ancestrally comprising free-living fungivores (97), the group has diversified into many habitats and includes radiations of saprophages, nidicoles, synanthropic pests of stored foods, and permanent parasites. They are rarely hyperparasites of insects (81) or other mites (7). Relatively few species are truly phytophagous or predaceous (100). *Czenspinskia transversostriata* (Winterschmidtiidae) is an exception, but it is primarily a fungivore that consumes plant tissue. *C. transversostriata* is parthenogenetic, a characteristic of many phytophagous Acari.

The most-derived lineage in the Astigmata comprises a clade of more than 30 families, the species of which are permanent parasites of birds and mammals. This clade, termed Psoroptidia, includes species inhabiting the skin, hair and feather shafts, hair and feather follicles, intraquill spaces, subdermal tissues, and respiratory passages of vertebrate hosts. The heteromorphic deutonymph was apparently lost in the ancestor of this lineage. A few taxa exhibit some phoretic dispersal (e.g. *Strelkoviacarus* on hippoboscid flies) (74) but these utilize a nonspecialized homeomorphic stage, such as the adult female.

Several species of psoroptidid mites are of considerable medical and veterinary importance. *Sarcoptes scabiei,* the agent of human scabies and sarcoptic mange in animals (27), is ancestrally a human parasite that has colonized many species of domestic and wild mammals (101). Species of *Dermatophagoides* and related taxa are notorious as the house dust mites, causative agents of respiratory and dermal allergies in humans (6, 24, 38, 67, 127, 134). Secondary irritation to humans, from nonhuman associates, do occur as in asthmatic sensitivity to aspirated free-living species (6) or dermatitis (grocer's itch) from stored grains (67).

PHORESY IN THE ASTIGMATA

Phoresy clearly represents an ancestral characteristic of the Astigmata. The only known fossil astigmatid mite, *Amphicalvolia hurdi* (129), is encased in Oligocene or Miocene amber and is likely a congener of *Afrocalvolia* (B. M. OConnor, personal observation), a modern genus phoretically associated with subcortical beetles. The earliest extant derivative, *Schizoglyphus biroi*, exhibits all of the phoretic modifications of the heteromorphic deutonymph and is phoretic on a tenebrionid beetle (86). The ancient age of the group is also suggested by coevolutionary studies indicating sister-group relationships among mite associates of marsupial and placental mammals. As the hosts diverged in the late Jurassic, the Astigmata as a clade must have already been well established (100).

Phoresy is a common phenomenon in the life cycle of free-living Astigmata and has long been controversial with regard to origin and significance (128). Astigmatid mite taxonomy and systematics are hampered by the dimorphic life cycle. Because the free-living stages of many astigmatid mites are sequestered in unaccessable habitats, many species and higher taxa are known only from the phoretic stage that is more readily collected. For example, 36 of 56 nominal genera of Histiostomatidae are known only from this stage. In some cases, deutonymphal morphology is conservative to the extent that species recognition requires the complete ontogeny (125) or application of molecular techniques (61).

The frequency of the induction of the heteromorphic deutonymph varies among taxa. It may be suppressed and infrequent among individuals within a species (54, 63), a majority of individuals may invoke the dispersal stage (e.g. Histiostomatidae), or all individuals may do so (e.g. species of the genus *Chaetodactylus*) (25).

Morphology of the Heteromorphic Deutonymph

Heteromorphic deutonymphs have been variously categorized according to their functional morphology and mode of attachment (29, 31, 132, 140). Entomophilous forms (Figure 1D) retain the presumed ancestral deutonymphal morphology and attach to arthropod hosts via a caudoventral sucker. This type is found in all superfamilies in which species retain the deutonymph. Three different suites of modifications of this morphology occur, each having evolved independently in more than one lineage. Pilicolous forms (Figure 1E) generally have reduced suckers but possess modified ventral setae that attach to mammal hairs or insect setae. This form is found in the Glycyphagoidea and Histiostomatoidea (30, 36). In endofollicular and subdermal types (Figure 1F) the attachment organs are reduced or absent. They retain well-developed legs often used for holding on to mammalian hair follicles or subdermal tissues of birds and mammals. Many of these deutonymphs can be considered true parasites rather than phoretics in that they increase in size through neosomatic growth at the expense of host tissue (34, 82). This type is found in the Glycyphagoidea and Hypoderatoidea (26, 30). In inert or regressive deutonymphs, attachment structures are vestigial or absent and often the legs are also rudimentary or lacking (40). These deutonymphs do not form phoretic associations and either do not disperse or disperse passively via air currents. Inert forms are the only deutonymphal morphology in some species of *Acarus* and *Glycyphagus* (57, 71), but occur together with an entomophilous form in *Chaetodactylus* spp. (25) or an endofollicular form as in *Alabidopus* spp. (83).

Host Associations

Astigmatid deutonymphs most commonly occur in association with coleopterans and hymenopterans in arboreal or soil habitats, but many are nidicolous (nest inhabiting) associates of mammals and birds. The ancestral nidicolous condition has frequently been expanded to include synanthropic nests created by agroeconomic practices and edifices of humans, such as commercial granaries, warehouses (e.g. roots, bulbs, potatoes, meats, and fruit), barns, and mushroom houses (66, 97).

ARTHROPOD ASSOCIATES   Astigmatid deutonymphs are associated with a diversity of arthropods, but occur most frequently on Coleoptera, Hymenoptera, Diptera, and on other insect orders to a lesser extent. They also have minor association with the Myriapoda and terrestrial Crustacea. Several major radiations of astigmatid mites are almost exclusively composed of arthropod associates, for example the Histiostomatoidea, Hemisarcoptoidea, Canestrinioidea, and Acaroidea. Of these, the hemisarcoptid-host associations are the least well known (107, 108).

Some major insect groups do not normally form associations with astigmatid mites. Orthopteroids (except Dermaptera), hemipteroids, Trichoptera, most lepidopterans, and most other generally phytophagous or predaceous groups typically lack astigmatid mite associates. This nonassociation is primarily because the habitat exploited by these groups does not encourage the free-living stages of astigmatid mites.

COLEOPTERAN ASSOCIATES   Associations between astigmatid mites and beetles are likely very ancient (100) and most associations of histiostomatoid species are with Coleoptera. A great diversity of beetle families serve as phoretic hosts to the Astigmata and these associations are too numerous to present in detail.

Attachment organs of deutonymphs associated with beetles are usually entomophilous (32). However, deutonymphs of *Fibulanoetus* associated with African Scarabaeidae are pilicolous, attaching to host setae (36). Although the attachment organ is similar in structure to that of the pilicolous Glycyphagidae, convergent evolution is certainly responsible.

Most coleopteran associations involve habitat-associated mite species that move freely among habitats while forming nonspecific relationships with various hosts. Of 4,826 mites collected in a study of carabids in Canada, 9 families and 13 genera of mites were found on 554 carabids (of 8 genera and 11 species) (110). These mites are usually not site specific while being transported by a host.

Nonspecificity is likewise the rule for the Histiostomatidae and Acaridae that disperse among habitat patches of fungi and decaying wood. These taxa typically use various species of Carabidae, Staphylinidae, Tenebrionidae, and numerous taxa of subcortical beetles (79, 89, 92, 110, 130). Such mites typically attach externally rather than in the subelytral space, though subelytral transport does occur with some ipine bark beetles (80).

Host-specific relationships are most common where beetle species exploit rich temporary habitats such as carrion (e.g. *Pelzneria* and other *Nicrophorus*-associates) (99, 119) and vertebrate dung (e.g. *Rhopalanoetus* and other scarab associates) (100) or species that manufacture habitats through construction of brood galleries as in the Scolytidae (14, 89–91) and the somewhat social Passalidae.

The superfamily Canestrinioidea comprises species that live as external commensals or parasites on many beetle groups, notably the Carabidae, Scarabaeidae, Lucanidae, Passalidae, Tenebrionidae, and Chrysomelidae. Only the most primitive lineage, the genera *Megacanestrinia* and *Coleoglyphus* (117), retain the heteromorphic deutonymph. In these taxa associated with African Carabidae, feeding stages do occur in the subelytral space while deutonymphs congregate on the thoracic venter (B. M. OConnor, personal observation). These taxa provide inference for an evolutionary transition from phoresy to parasitism, with the concomitant loss of the phoretic stage in derived parasitic taxa.

HYMENOPTERAN ASSOCIATES   Astigmatid mites are typically the most abundant mites found in nests of cavity-dwelling and soil-dwelling bees and wasps. Ten described (and many undescribed) (106) species are mainly associated with solitary and social halictines (21). The most prevalent genus associated with soil-dwelling bees is *Anoetus,* which filter-feeds on microorganisms by skimming the surfaces of bee provisions and larvae in capped brood cells (21, 106). Phoretic deutonymphs of *Anoetus* are often site specific and reside in an acarinarium on the first gastral tergite of halictid bees. Host specificity, however, is not common. This nonspecificity enhances interspecific radiation among nesting bees, and phoretic association secondarily includes cleptoparasitic bees [e.g. *Sphecodes,* Halictidae, and *Nomada* (Anthophoridae)] and velvet wasps (Hymenoptera: Mutillidae) (103), which also visit bee nests.

Histiostomatids are common associates of bumble bees, honey bees (22), ants, and termites. They appear to not impede the brood development of these insects. The histiostomatids associated with nesting bees synchronize the production of deutonymphs with the emergence of adult hosts, upon which they are phoretic (21). Such life-history synchrony is well documented in the enslinielline winterschmidtiids (10, 75). Enslinielline deutonymphs are associated with a diversity of megachilid and colletid bees, as well as sphecid and vespid wasps (10, 11, 73, 75, 106).

The evolutionary interaction between the enslinielline *Kennethiella trisetosa* and its wasp host, *Ancistrocerus antilope* (Hymenoptera: Eumeninae), is especially curious (11). Females of *A. antilope* nest in tubular cavities in wood and deposit eggs in linear chains. Each egg is individually provisioned and sealed with mud. If ovipositing female wasps contain phoretic deutonymphs, the deutonymphs disembark in an egg chamber and are confined with the wasp egg at the time of cell-sealing. The life cycle of *K. trisetosa* is synchronized with that of the wasp and the next generation of deutonymphs occurs as the developing wasp reaches maturity. Neonate adult wasps chew out of the brood cell when mature and emerge with mites attached to pouches (acarinaria) on the propodeum. Curiously, female wasp larvae have evolved a behavior of killing mites in brood cells whereas male wasps have not, and female wasps emerge from the brood cell mite-free. When wasps copulate, deutonymphs migrate from the acarinarium on the male propodeum to a genital acarinarium present in both sexes. This strategy of venereal transmission and sex-biased phoresy is known to occur only in *K. trisetosa* associated with *A. antilope.*

The genus *Vidia* (Ensliniellinae) is a large group of cosmopolitan mites (most of which remain unnamed) that are phoretically associated with megachilid bees of the genera *Megachile, Creightonella,* and *Anthidium. Vidia* is not known to form mutualistic relationships with its host or to feed on the host during any stage of development. Free-living mites feed on fungus on leaves comprising the bee's nest, and heteromorphic deutonymphs attach to pupae prior to eclosion. Species-specific host associations are not common in *Vidia* (106).

Tyrophagus is a vagrant genus that inhabits a broad range of niches: animal nests, mushroom houses, vegetable crops, commercial cheeses, flower bulbs, granaries, grasses, and cereals (74). Until recently, researchers thought that the genus *Tyrophagus* did not produce deutonymphs (132). The first report of a deutonymphal form, from any species of *Tyrophagus,* was *Tyrophagus formicetorum* from the nest of the ant *Formica rufa* in Poland (37). It has a similar morphology to that of deutonymphs of *Forcellinia,* a genus of ant associates closely related to *Tyrophagus* (19, 139).

For a more thorough treatment of bee-associates, refer to the most recent summaries on the topic (21, 22, 103).

DIPTERA ASSOCIATES   The Diptera are common hosts of phoretic Astigmata because they are frequent visitors to temporary habitats such as dung, sap-fluxes, and phytotelmata. The histiostomatid genera *Myianoetus, Copronomia, Ameranoetus,* and *Xenanoetus* are restricted to dipteran associations (35, 85, 118). The stored product pest *Histiostoma feroniarum* is phoretic on sciarids and phorids (69).

Phoretics of *Naiadacarus arboricola* are attracted only to females of their syrphid hosts that visit water-filled treeholes. Phoretics do not attach to the males of syrphids because only female flies return to treeholes to oviposit (45). Other *Naiadacarus* species sometimes attach to male syrphids, but only in species that defend territories adjacent to female oviposition sites (104).

Hericia hericia are found in sap-flux from various deciduous trees (115). The cuticle of deutonymphs is of a hydrofuge type, and deutonymphs float on the flux surface. When phoretic hosts are attracted to the sap, deutonymphs swiftly attach to the host. Foliate setae on the first two pairs of legs function as small suction cups and are used for boarding the host (115). A typical caudoventral sucker plate is used for more tenacious attachment during transport.

MAMMAL ASSOCIATES   Astigmatid mites have phoretic associations with a wide diversity of mammal groups including rats, mice, hamsters, squirrels, chipmunks, marmots, bats, primates, badgers, armadillos, shrews, other insectivores, and opossums (135, 140). Essentially the entire radiation of the superfamily Glycyphagoidea (99) is associated with mammals, and most of these retain the heteromorphic deutonymph. Exceptions include the families Aeroglyphidae, Rosensteiniidae, and many taxa in the glycyphagid subfamily Fusacarinae. Many species of glycyphagids are common nest associates of wild mammals throughout the world (30). Most mammal associates attach to hair shafts of the host by claspers, which are modified conoidal setae of the ancestral attachment organ. This setal transformation occurred in the ancestor of the Glycyphagidae.

In general, heteromorphic deutonymphs are known to exploit mammals using three morphological variations (30) of the preglycyphagid morphology: (*a*) retention of ancestral claspers (pilicolous deutonymphs) as in the subfamilies Marsupialichinae, Glycyphaginae, Labidophorinae, Metalabidophorinae, and Fusacarinae); (*b*) regressed deutonymphs that are parasitic and transported within hair follicles (endofollicular deutonymphs) as in the families Chortoglyphidae, Pedetopodidae, Echimyopodidae, and the subfamilies of Glycyphagidae, Metalabidophorinae plus a clade comprising the Lophuromyopodinae and Ctenoglyphinae; and (*c*) within subcutaneous layers of the host tissue (hypodermic deutonymphs) of one known species from Echimyopodidae (41). Most hypodermic species belong to the Hypoderatoidea and are primarily associated with birds (26, 28) and not mammals.

One unusual species, *Xenocastor fedjushini*, occurs in Russia on the fur of beavers *(Castor fiber vistulanus)* (140). All homeomorphic stages of this mite occur in the fur of the host as well as in the nests (123). Phoresy by homeomorphic stages compensates for the apparent loss of the deutonymph in this species. A similar phenomenon is observed in the Rosensteiniidae, inhabitants of bat roosts, that do not form deutonymphs and in which homeomorphs are also phoretic on the fur of bats.

The potential for interspecific host-transfer to occur between taxa such as mammals, birds, and fleas (109) is present because of trophic or habitat preferences. For example, deutonymphs of two glycyphagids were collected from bird nests, *Dermacarus pilitarsus* from the nest of an American kestrel and *Neoxenoryctes reticulatus* from the nest of a screech owl. These mites are typically pilicolous rodent associates but Fain (42) believes that these deutonymphs came into association with the birds when rodent prey were brought back to the nest for consumption and presumed that they were probably accidental associates. These associations suggest common use of a nesting cavity by the raptors and squirrels (141). Such contacts over evolutionary time could provide opportunity for cross-host invasion.

STORED PRODUCT ASSOCIATIONS   Rodinov's theory (140) proposes that stored-product invasion resulted from opportunistic expansion of habitat utilization by nest or field species. Since bulk food storage is relatively recent, existing for only about 10,000 years (97), sufficient time has not passed for the evolution of a strictly human association to have become so widespread.

The largest number of mite species in stored products are astigmatid mites (97), and many genera (34 genera, 10 families) have independently radiated into stored products and house dust. Habitat expansion occurred where stored products were comparable in kind and quality to the aboriginal habitat. Heteromorphic deutonymphs of stored product mites occur in all four mor-

phological types, although species forming pilicolous or endofollicular deutonymphs are rare (8) and some taxa form no deutonymphs at all (e.g. Aeroglyphidae, Pyroglyphidae).

Sancassania berlesei, phoretic on chafer beetles (Scarabaeidae) (140), is
common in mushroom houses. *Rhizoglyphus echinopus,* which is phoretic on
the Narscissus fly and various other Diptera (e.g. *Scatopse;* scavenger flies)
(130, 140) is a destructive pest of onions, potatoes, and decorative bulbs.
Lardoglyphus ancestrally inhabited animal carcasses and now occurs in meat
warehouses through transport by dermestid and trogid beetles attracted to
these products. *Carpoglyphus lactis* (Gr: *carpos,* fruit) infests stored fruit,
rotten potatoes, flour, and has been collected being phoretic on floating cork
chips in a wine bottle (96). The phoretic association of *C. lactis* with nine
species of noctuid moths reflects an ancestral habit of nectar feeding, and the
lure of sweetness led both moth and mite into agroeconomic storehouses
(128).

Behavior of the Phoretic

INDUCTION OF THE FACULTATIVE DEUTONYMPAL STAGE   Heteromorphic
deutonymphs were first recognized by Michael (87), who initiated a discussion of what factors were responsible for the appearance of an occasionally
inserted heteromorphic stage. Since this seminal paper, acarologists have
discovered that deutonymphs may occur under a variety of endogenous or
ecological conditions (47, 53, 55, 56, 130), which differ considerably in
different species. Factors such as temperature, relative humidity, food quality, and overcrowding may be responsible for induction, and any environmental cue that signals habitat degradation is a potential trigger. Only infrequently
have investigators associated induction with genetically inherited traits (e.g.
16, 55, 60). Such explanations have been ad hoc and unsubstantiated, and
genetic influences on deutonymphal formation are rarely studied.

Limiting food has been shown to result in induction of deutonymphs in
Acarus immobilis (56), *Glycyphagus destructor* (114), and *Hemisarcoptes*
coccophagus (47). The content of food, as well as food quality, may influence
phoretic deutonymphs. *Rhizoglyphus echinopus* produced more deutonymphs
on mealworms than on mushrooms in both moist and dry environments, while
Histiostoma bakeri developed more heteromorphic deutonymphs on mushrooms (136). *H. bakeri* and *Histiostoma laboratorium* formed deutonymphs
only when the culture medium dried out. Growth of fungi and bacteria always
caused deutonymphs to be produced in great numbers. Woodring (136)
subjected *R. echinopus* to high CO_2 tensions; extreme over crowding (with
excess food); high concentrations of excretory products, bacteria, and fungi;
and variations in relative humidity and temperature. All failed to increase the
number of deutonymphs over controls. In most cases, the full complement of

potential environmental induction cues is yet to be determined, as is the relative importance of each.

Relative timing of developmental events preceding heteromorphic deutonymphal formation can signal the onset of production of phoretics. In *Sancassania boharti*, the onset of protonymphal quiescence was later in mites destined to become deutonymphs than in those destined to become tritonymphs (13). This pattern was also found in *Acarus immobilis* and *Naiadacarus arboricola*. In *N. arboricola*, the tritonymphal period was also protracted, possibly to recover energy losses incurred by the deutonymph (46).

Deutonymphal production may also be correlated with host phenology. The deutonymphs of *N. arboricola* occur seasonally, mainly in May and June, in aquatic habitats of water-filled treeholes, while the other developmental stages are present year round. Induction in this species appears to be a function of host emergence and not physical stress (45, 46, 104). A biological clock also appears to be the induction cue for *Hericia hericia*, which inhabits sap flux. Induction of heteromorphic deutonymphs occurs when maximum sap is flowing freely and is rich in carbohydrates. This time corresponds to the oviposition period for many potential host species that visit the sap flux (115).

Some astigmatid mites may utilize host secretions, as do some nonphoretic stages (75, 93, 94), to maintain synchrony with the host phenology. The developmental cycles of *Ensliniella kostylevi* and their hosts are closely correlated. This correlation is probably maintained by absorption of growth and differentiation hormones of the hosts by the mites (73, 75, 93, 94). Because some deutonymphs are known to die if restricted from contact with the host (e.g. *Hemisarcoptes*) (62, 63, 76, 136), host secretions may not only act as an environmental monitor but may be vital for survival and development.

ATTACHMENT BEHAVIOR   Facultative deutonymphs may respond to a variety of cues for host recognition, and these cues may be context dependent. Some are attracted to both genders of the host, as is *Histiostoma* associated with species of *Ips* (Coleoptera: Scolytidae) (78). Some phoretics respond only to female hosts, as does *Naiadacarus arboricola*. Some respond mainly to males, as does *Rhizoglyphus echinopus* collected from the scarab beetle *Osmoderma eremicola* (95) and *Kennethiella trisetosa*, which only matures on male larvae of the wasp *Ancistrocerus antilope* and is transported from the wasp nest in propodeal acarinaria of emergent males. Phoretics of *Sancassania boharti* are most common on the ventral surface of male alkali bees (Hymenoptera: Halictidae), but on female bees they may occur either dorsally or ventrally (13).

Questing (or Reiterstellung), the act of seeking hosts, in *Histiostoma* can occur spontaneously (59) but can also be provoked by tactile stimulation of gnathosomal setae or solenidia (e.g. *Carpoglyphus lactis*) (58, 122). The questing position consists of the first two pairs of legs raised in a vertical position, with the body anchored to the substrate by the caudoventral suckers. The third and fourth pair of legs buttress the angle of attachment. Jumping, to a height of 1–2 inches (2.5–5.0 cm), allows the mite to spring onto a passing drosophilid host (58). Phoretics may also be stimulated to embark onto a host by the stimulation of the setae of the first pair of legs (e.g. *Histiostoma laboratorium*) (122). Some less discriminating mites (e.g. *Sancassania berlesei*) have been known to attach to dead as well as live insects (121), and deutonymphs commonly persist following host mortality. Deutonymphs of *Sancassania,* often phoretic on scarab beetles, remain on the host when it dies and subsequent stages exploit the host as saprophages of necrotic host tissues (74, 116).

When attachment occurs, many deutonymphs may attach to a single individual host. Over 1,000 deutonymphs of *Rhizoglyphus* were found attached to a scarab beetle, *O. eremicola* (95), and 700 phoretics of *S. berlesei* (= *Caloglyphus rodionovi*) were attached to a single specimen of *Tenebrio molitor* (120), where they were systematically arranged for maximum compacting. Even when mites occur in acarinaria, numbers may be high. The greatest phoretic load on *A. antilope* was 407 *K. trisetosa* in the propodeal acarinarium of a male wasp and 238 in the genital acarinarium of a male (10).

Bilateral equality of deutonymphal distribution on a given host is common (13). Such distribution may minimize interference with host flight, and may be maintained by the mites. For example, if the bilateral symmetry of deutonympal attachment of *K. trisetosa* is disturbed, deutonymphs will redistribute themselves within 24 hours into roughly equal proportions (10). A similar pattern was observed with *Glyphanoetus nomiensis* and *S. boharti* on the alkali bee *Nomia melanderi* (Hymenoptera: Halictidae) (13, 14, 75, 137). Sometimes the choice of attachment site is important not only to the mobility of the host, but to prevent being groomed from the host. *Anoetus halictonidus* attach to the ventral surface of the forewings of *Halictus rubicunidus* (Hymenoptera: Halictidae). Neonate bees groom extensively before emerging from the brood chamber, and, while the upper surfaces of the wings can be efficiently cleaned, the bees cannot reach the lower surfaces (137).

Troxocoptes minutus is phoretic on *Trox costatus* (Coleoptera: Scarabaeidae), on which it rides in individual passenger compartments (dermal pits) on the external surface of the beetle elytra (43). A similar morphological congruence occurs between the mite *Boletoglyphus ornatus* and the darkling beetle *Bolitotherus cornutus* (39). Mobile stages of the mite are inhabitants of polypore fungi (102), and, when deutonymphs are produced, they attach to *B. cornutus* and position themselves into the deep puncta of the elytra.

DETACHMENT BEHAVIOR   Less is known concerning the detachment stimuli in the Astigmata. Stimulus for detachment has been inferred to correlate with oviposition of the host in nonastigmatid mites (75, 79), and it may also play a role in the Astigmata. Deutonymphs of *Histiostoma polypori,* which detach and then reattach to the same host with successive host molts, may respond to chemical changes in the host's cuticle (2).

Evolutionary Trade-Offs Associated with Phoresy

An opportunity exists for the development of predictive models designed to address the ecological trade-offs related to the facultative deutonymph and its role in the evolution of phoresy. Little has been done, and an exciting exploration waits in an area of evolutionary ecology that has virtually been ignored.

A simple descriptive model (88) has been proposed to conceptually explore phoresy in the Astigmata using two parameters: the number of migrators (N) with the associated probability of becoming a founder (F) and the probability of reaching a supportive habitat (P). In the model as applied to the Astigmata, the number of founders is a product of the number of phoretics and the likelihood of finding a new resource: $F = N P$. If the sexes are separate and migrate independently then: $F = (N_{males}P_{males}) + (N_{females}P_{females})$. The model predicts that when unmated males and females migrate together, only four mites are needed, on average, to assure that each sex will be represented in the founding population 87.5% of the time (88). If phoretics are inseminated or parthenogenetic, selection pressure for group migration is less.

If only a proportion of the population is committed to dispersal (P_d), as with facultative deutonymphs, d can be expressed as the biomass of deutonymphs (b_n) relative to the total biomass (B): $d = b_n/B$, where d varies from 0 to 1. If the phoretic is parthenogenetic, $d \leq 1.0$; if the phoretic is bisexual, with a sex ratio near $1:1$, and sexes are equivalent in mass, $d \leq 0.5$ when only females disperse. The number of founders (F) is a function of: (a) N, which can be increased through increased rates of reproduction or by increasing the proportion of phoretics; (b) generalization of food preferences by the free-living mite stages and the selection of generalist hosts; (c) increased survival during migration, resulting in an increase in P_d; (d) sensory discrimination of "quality" hosts that are predictable, frequent, and that have habitat requirements similar to the free-living mite stages; (e) production of small males, relative to females, that impact on limited resources less than full-sized males. Some female astigmatid mites can control adult male size through egg-retention and contraction of the developmental sequence. The mechanism of small-male production is well documented in *K. trisetosa* and occurs because some females retain eggs that incubate and hatch in her body. The resultant protonymphs leave the mother's body and are no larger than an egg. Such protonymphs molt directly into tritonymphs and then into small males. Other eggs, destined to become males, are not retained. They feed after hatching, go through the deutonymphal and tritonymphal stages, increasing in size with each molt, and result in normal-sized males. Thus, the main evolutionary adjustment (production of small males) is controlled by the female (12, 73).

Binns (3) suggests that deutonymphal production may be under the control of gravid females of *Sancassania moniezi,* which produces two female morphs. One morph is of normal size (1050–1250 μm) and lays eggs, the other is viviparous and very large (2000–2262 μm). The viviparous morph incubates and retains live larvae, protonymphs, and deutonymphs (140), which gives the female significant control over all aspects of development. To date, however, no experimental evidence supports Binns's claim.

HEMISARCOPTES: A BIOLOGICAL MODEL

Proximate Significance of Phoresy: Transition among Habitats

A phoretic association being investigated by the authors (48, 61–65) can serve as an example of the kinds of evolutionary patterns revealed through examination of phoresy in the Astigmata. This interaction concerns the astigmatid mite *Hemisarcoptes cooremani* and its host *Chilocorus cacti* (Coleoptera: Coccinellidae). Free-living stages of *Hemisarcoptes* are generalist predators of diaspidid scale insects, a speciose family parasitic on perennial vascular plants world-wide.

The entomophilous deutonymph is transported subelytrally on *C. cacti.* No evidence indicates habitat induction of the deutonymph in this species, though it may occur in other species (49). Only the phoretic heteromorph of *Hemisarcoptes* occurs on the beetle. The deutonymph is facultative and represents only 6% of the population (63). Phoretics that are produced and do not encounter a host do not complete ontogenesis, and they die. Since *H. cooremani* and *C. cacti* both prey on scale insects, postdeutonymphal development of the mite occurs within the habitat of an appropriate prey.

Phoresy in *Hemisarcoptes* can be considered in terms of the relative fitness trade-offs associated with invoking the facultative deutonymph. In high-quality habitats, the maximum fitness loss due to the phoretic stage is approximately 6% of all viable offspring. This fitness loss occurs when beetles are not present and the potential fitness gain of colonization, due to phoresy, is zero. The fitness gain from colonizing phoretics is positively correlated with the threat of local extinction in the natal habitat. The phoretic stage of *Hemisarcoptes* represents a bet-hedging strategy in ephemeral habitats of low predictability. Because only a small portion of reproductive investment is allotted to deutonymphs, its fitness cost is relatively low as compared to the potential benefits during ecological bottlenecks.

Following the examination of specimens of *Chilocorus* (> 10,000 specimens from among the 70 species) from holdings in major museums, we concluded that the *Hemisarcoptes-Chilocorus* association is genus specific rather than species specific. We also discovered that congregate phoresy (up to 800 mites/beetle) is common. The sex ratio of females to males is approximately 2:1 (63), which equals the ratio among deutonymphs following maturity. Thus, the probability of a substantial number of phoretics of each sex occurring on any particular beetle is high and little selection for the evolution of compensatory sexual strategies occurs (e.g. inseminated propagules or shifted sex ratios).

In summary, in *Hemisarcoptes,* the generalized prey preference expressed by free-living stages, the stenoxenic host-preference for a generalist predator by phoretics, the environmentally resistant migrators with well-developed attachment organs, and the congregate phoretic behavior all contribute to the increased probability of survival in a new habitat, in densities sufficient for successful mating.

Ultimate Evolutionary Significance of Phoresy: Transition to Parasitism

Extensive laboratory studies of *H. cooremani* and *C. cacti* indicated that: (*a*) ontogenesis is interrupted and deutonymphs do not survive without contact with *C. cacti;* (*b*) deutonymphs undergo a visible change (amber to cream, flattened to swollen) prior to disembarking from *C. cacti*; and (*c*) deutonymphs maintain a long transit time on *C. cacti* (5–21 days). This extended contact cannot be completely explained by the need to disperse because it also occurs when resources are plentiful. These observations led to the experimental testing of the hypothesis that some components essential to survival are acquired while *Hemisarcoptes* deutonymphs are associated with beetles (62).

The greatest source of available nutrients for phoretics while on *C. cacti* is reflexed beetle hemolymph. Adult *C. cacti* defend themselves from predators by reflex bleeding (111). Reflexed hemolymph of *C. cacti* coagulates upon deposition, gluing a predator's mouthparts together, and contains toxic alkaloids (111). Reflexed hemolymph frequently bathes the subelytral zone of *C. cacti,* but appears not to cause mechanical or chemical damage to attached phoretics.

Radio-labeling (HTO) studies revealed that phoretics of *H. cooremani* acquire materials directly from the beetle (62). The acquisition appears to be accomplished through the action of the discoidal suckers of the attachment organ, and an atrium has been observed in sagittal sections of the hindgut of other entomophilous deutonymphs (5, 68, 96, 133, 138). The hindgut appears to have a valve between the atrium and the vestigial anal opening on the sucker-plate.

Swelling in deutonymphs of *Hypodectes propus* (size increase from 150–1,500 μm) has previously been observed but was speculated to result from osmosis: "parvenu sous la peau, l'hypope augmente considérablement en taille, probablement à la suite de l'absorption de substances nutritives par osmose" (35). The method and context of uptake has not been explored, however. Likewise histological sections of hosts invaded by deutonymphs of *Echimyopus dasypus* illustrated hyperkeratosis and hypertrophy around embedded mites (41). However, the potential means of absorption of lysed tissues by endofollicular species is a matter of conjecture and remains unexplored.

The following hypothesis of evolutionary events has been proposed (62), which is compatible with observed phylogenetic patterns in the Astigmata, as well as with the experimental data: (*a*) the heteromorphic deutonymph primitively arose in the Astigmata for dispersal with low energetic costs to the host, and some modern species remain phoretic; (*b*) extended interspecific contact, over evolutionary time, resulted in chemical acclimatization to the host (for example to hemolymphal alkaloids) in some species; (*c*) adaptation to hemolymph led to hemolymphal utilization; (*d*) the interaction advanced from utlization of reflexed hemolymph (controlled by the host) to extracting hemolymph as required (host exploitation). The concomitant behavioral changes would have required that deutonymphs remain on the host to molt and that subsequent stages remain on the host as parasites. Little morphological adjustment would be required for parasitism by postdeutonymphal stages because chelicerae of nondeutonymphal stages are exapted to feeding on abdominal tissues of beetles. Chemical adaptation to hemolymphal toxins by the deutonymph conferred toxin resistance to subsequent feeding stages since the stages represent, genotypically, a single individual. This hypothesis suggests that the ultimate evolutionary significance of the deutonymphal morph is to serve as a transition from a free-living existence in temporary environments to a more stable parasitic existence.

Phylogenetic patterns of deutonymphal occurrence among the Astigmata are consistent with these conclusions. Host dependence varies in the Astigmata from free-living forms (with deutonymphs) to obligate parasitic forms. The deutonymph, however, has been found as part of the life cycle of only one parasite. Species of *Coleolyphus,* the most primitive canestrinioid taxon, have a deutonymph, but the other stages remain on the host as ectoparasites (117). All other canestriniids have lost the deutonymph. In general, heteromorphic deutonymphs and parasitic life cycles appear to be incompatible and have been completely eliminated in all astigmatid parasites of vertebrates (99). Evidence of parasitism by mites (family undetermined) is known from amber and has existed for at least 25–35 million years (113). The evolution of parasitism by mites is an ancient event, but it is also a process. Evidence that deutonymphs of *Hemisarcoptes* spp. acquire materials from their hosts while retaining the phoretic morphology implies a transitional compromise between complete parasitism and phoresy. The work on *H. cooremani* is the only current attempt to experimentally define such a relationship. Similar experimental studies on the behavioral and physiological ecology of astigmatid mites could lead to a greater understanding of co-evolution of mite-host associations and the mechanisms of the evolution of parasitism. While endoparasitic insects are rare and ectoparasitic grades do not exist in endoparasitic worms (31), the Astigmata contains all three (ectoparasites, endoparasites, and free-living) forms. Thus, this group offers a unique potential for uncovering extant transitional grades.

PROSPECTUS OF FUTURE DIRECTIONS

An essential part of any biological study is the accurate determination of a study organism as a unique entity. Currently, the most serious liability in the understanding of the evolution of phoretic interactions in the Astigmata is the need for taxonomic distinctions and phylogenetic interpretations of relationships. The number of scientists working on these problems is dwindling at a time when interest in co-evolution and migration is accelerating. We identify acarine taxonomy and systematics as a high priority for future concentration. We encourage the pursuit of the issue of the evolution of parasitism from the phoretic form, within a cladistic and an experimental context.

ACKNOWLEDGMENTS

We thank G. Eickwort, R. Strauss, and D. Wrensch for added useful comments on the manuscript. This work was supported by National Science Foundation grant BSR #83-07711 (B. M. OConnor & M. A. Houck) and Binational Agricultural Research and Development grant #IS-1397-87 (M. A. Houck).

Literature Cited

1. Athias-Binche, F. 1984. La phorésie chez les acariens uropodides (Anactinotriches), une stratégie écologique originale. *Acta Oecologica Oecol. Gen.* 5:119–33
2. Behura, B. K. 1956. The relationships of the Tyroglyphoid mite, *Histiostoma polypori* (Oud.) with the earwig *Forficula auricularia* Linn. *J. N. Y. Entomol. Soc.* 64:85–94
3. Binns, E. S. 1982. Phoresy as migra-

tion, some functional aspects of phoresy in mites. *Biol. Rev.* 57:571–620
4. Blackwell, M., Bridges, J. R., Moser, J. C., Perry, T. J. 1986. Hyperphoretic dispersal of *Pyxidiophora* anamorph. *Science* 232:993–95
5. Boczek, J., Jura, C., Krzysztofowicz, A. 1969. The comparison of the structure of the internal organs of postembryonic stages of *Acarus farris* (Oud.) with special reference to the

hypopus. In *Proc. 2nd Int. Congr. Acarol*, ed. G. O. Evans, pp. 265–71, Sutton Bonington, England, 1967. Budapest: Akadémiai Kiadó

6. Carter, H. F., Webb, G., D'Abrera, V. St. E. 1944. The occurrence of mites (Acarina) in human sputum and their possible significance. *Ind. Med. Gaz.* 79:4

7. Chant, D. A. 1960. An unusual instance of phoresy in the Acarina. *Entomol. News* 71:270–71

8. Chmielewski, W. 1975. Nowo stwierdzone stadium rozwojowe roztoczy z rodzaju *Ctenoglyphus*-hypopus *C. plumiger* (Koch, 1835) (Acarina: Glycyphagidae). *Zesz. Probl. Postepow Nauk Roln.* 171:261–68

9. Chmielewski, W. 1977. Powstawanie i znaczenie stadium hypopus w zyciu roztoczy z nadrodziny Acaroidea. *Pr. Nauk Inst. Ochr. Rosl.* 19:5–94

10. Cooper, K. W. 1955. Venereal transmission of mites by wasps, and some evolutionary problems arising from the remarkable association of *Ensliniella trisetosa* with the wasp *Ancistrocerus antilope*. Biology of Eumenine wasps II. *Trans. Am. Entomol. Soc.* 80:119–74

11. Cowen, D. P. 1984. Life history and male dimorphism in the mite *Kennethiella trisetosa* (Acarina: Winterschmidtiidae), and its symbiotic relationship with the wasp *Ancistrocerus antilope* (Hymenoptera: Eumenidae). *Ann. Entomol. Soc. Am.* 77:725–32

12. Cross, E. A. 1968. Descriptions of three new species of mites phoretic upon native bees. *Southwest Nat.* 13:325–34

13. Cross, E. A., Bohart, G. E. 1969. Phoretic behavior of four species of alkali bee mites as influenced by season and host sex. *J. Kans. Entomol. Soc.* 42:195–219

14. Cross, E. A., Moser, J. C. 1975. A new, dimorphic species of *Pyemotes* (Acarina: Tarsonemoidea) and a key to previously described forms. *Ann. Entomol. Soc. Am.* 68:723–32

15. Cross, E. A., Moser, J. C., Rack, G. 1981. Some new forms of *Pyemotes* (Acarina: Pyemotidae) from forest insects, with remarks on polymorphism. *Int. J. Acarol.* 7:179–96

16. Cutcher, J. J., Woodring, J. P. 1969. Environmental regulation of hypopial apolysis of the mite, *Caloglyphus boharti*. *J. Insect Physiol.* 15:2045–57

17. Deegener, P. 1917. Versuch zu einem System der Assoziations- und Sozietatsformen in Tierreiche. *Zool. Anz.* 49:1–16

18. Delfinado-Baker, M. D., Baker, E. W. 1983. New mites (*Sennertia*: Chaetodactylidae) phoretic on honey bees (*Apis mellifera* L.) in Guatemala. *Int. J. Acarol.* 9:117–21

19. Delfinado-Baker, M., Baker, E. W. 1989. A new mite in beehives: *Forcellinia faini*, n. sp. (Acari: Acaridae). *Am. Bee J.* Feb. 1989:127–28

20. Dugès, A. 1834. Recherches sur l'ordre des Acariens en général et la famille des Trombidies en particulier. *Ann. Sci. Nat. Zool.* 1:5–46

21. Eickwort, G. C. 1979. Mites associated with sweat bees (Halictidae). See Ref. 115a, 1:575–81

22. Eickwort, G. C. 1990. Associations of mites with social insects. *Annu. Rev. Entomol.* 35:469–88

23. Evans, G. O., Sheals, J. G., MacFarlane, D. 1961. *The Terrestrial Acari of the British Isles*, Vol. 1. London: British Museum. 219 pp.

24. Fain, A. 1966a. Nouvelle description de *Dermatophagoides pteronyssinus* (Trouessart, 1897). Importance de cet acarien en pathologie humaine (Psoroptidae: Sarcoptiformes). *Acarologia* 8:302–27

25. Fain, A. 1966b. Notes sur la biologie des acariens du genre *Chaetodactylus* et en particulier de *C. osmiae*, parasite des abeilles solitaires *Osmia rufa* et *O. cornuta* en Belgique. *Bull. Ann. Soc. R. Belge Entomol.* 102:249–261

26. Fain, A. 1967. Les hypopes parasites des tissus cellulaires des oiseaux (Hypodectidae: Sarcoptiformes). *Bull. Inst. R. Sci. Nat. Belge* 43:1–139

27. Fain, A. 1968a. Etude de la variabilité de *Sarcoptes scabiei* avec une révision des Sarcoptidae. *Acta Zool. Pathol. Antverp.* 47:1–196

28. Fain, A. 1968b. A new heteromorphic deutonymph (hypopus) of a sarcoptiform mite parasitic under the skin of a toucan. *J. Nat. Hist.* 2:459–61

29. Fain, A. 1968c. Notes on the heteromorphic deutonymphs (hypopi) of the Acaridiae (Acarina: Sarcoptiformes). *Parazitologiya Leningr.* 2:395–406

30. Fain, A. 1969a. Les deutonymphes hypopiales vivant en association phorétique sur les mammifères (Acarina: Sarcoptiformes). *Bull. Inst. Roy. Sci. Nat. Belge* 45:1–262

31. Fain, A. 1969b. Adaptation to parasitism in mites. *Acarologia* 11:429–49

32. Fain, A. 1971. Évolution de certains groupes d'hypopes en fonction du parasitisme (Acarina: Sarcoptiformes) *Acarologia* 13:171–75

33. Fain, A. 1980. *Hormosianoetus aeschli-*

manni n.g., n.sp. (Acari, Anoetidae) phorétique sur des Drosophiles d'élevage en Suisse. *Rev. Suisse Zool.* 87:753–56

34. Fain, A., Bafort, J. 1967. Cycle évolutif et morphologie de *Hypodectes* (Hypodectoides) *propus* (Nitzsch) acarien nidicole à deutonymphe parasite tissulaire des pigeons. *Bull. Acad. R. Belge Classe Sci.* 53:501–33

35. Fain, A., Britt, D. P., Molyneux, D. H. 1980. *Myianoetus copromyzae* sp. nov. (Acari, Astigmata, Anoetidae) phoretic on *Copromyza atra* (Meigen 1830) in Scotland. *J. Nat. Hist.* 14:401–3

36. Fain, A., Camerik, A. M., Lukoschus, F. S., Kniest, F. M. 1980. Notes on the hypopi of *Fibulanoetus* Mahunka, 1973, an anoetid genus with a pilicolous clasping organ. *Int. J. Acarol.* 6:39–44

37. Fain, A., Chmielewski, W. 1987. The phoretic hypopi of two acarid mites described from ants' nest: *Tyrophagus formicetorum* Volgin, 1948 and *Lasioacarus nidicolus* Kadzhaja and Sevastianov, 1967. *Acarologia* 28:53–61

38. Fain, A., Guerin, B., Hart, B. J. 1988. *Acariens et Allergies.* Allerbio, Cepharm: Varrennes en Argonne. 179 pp.

39. Fain, A., Ide, G. S. 1976. *Ellipsopus ornatus,* a new genus and species of Acaridae (Acari) phoretic on the beetle *Bolitotherus cornutus* (Panzer, 1794). *Entomol. News* 87:233–36

40. Fain, A., Lukoschus, F. S. 1977. New endofollicular or subcutaneous hypopi from mammals (Acarina: Astigmata). *Acarologia* 19:484–93

41. Fain, A., Lukoschus, F. S., Louppen, J. M. W., Mendez, E. 1973. *Echimyopus dasypus,* n. sp. a hypopus from *Dasypus novemcinctus* in Panama (Glycyphagidae, Echimyopinae: Sarcoptiformes). *J. Med. Entomol.* 10:552–55

42. Fain, A., Philips, J. R. 1977. Astigmatic mites from nests of birds of prey in U.S.A. I. Description of four new species of Glycyphagidae. *Int. J. Acarol.* 3:105–14

43. Fain, A., Philips, J. R. 1983. *Troxocoptes minutus* gen. n., sp. n. (Acari, Acaridae), a new hypopus phoretic on a beetle *Trox costatus. Bull. Ann. Soc. R. Belge Entomol.* 119:95–98

44. Farish, D. J., Axtell, R. C. 1971. Phoresy redefined and examined in *Macrocheles muscaedomesticae* (Acarina: Macrochelidae). *Acarologia* 13:16–29

45. Fashing, N. J. 1976a. The evolutionary modification of dispersal in *Naiadacarus arboricola* Fashing, a mite restricted to water-filled treeholes (Acarina: Acaridae). *Am. Midl. Nat.* 95:337–46

46. Fashing, N. J. 1976b. The resistant tritonymphal instar and its implications in the population dynamics of *Naiadacarus arboricola* Fashing (Acarina: Acaridae). *Acarologia* 18:704–14

47. Gerson, U. 1967. Observation on *Hemisarcoptes coccophagus* Meyer (Astigmata: Hemisarcoptidae), with a new synonym. *Acarologia* 9:632–38

48. Gerson, U., OConnor, B. M., Houck, M. A. 1990. Acari. In *Armored Scale Insects, Their Biology, Natural Enemies and Control,* ed. D. Rosen. Amsterdam: Elsevier. In press

49. Gerson, U., Schneider, R. 1982. The hypopus of *Hemisarcoptes coccophagus* Meyer (Acari: Astigmata: Hemisarcoptidae). *Acarologia* 23:171–76

50. Gordh, G. 1985. *Uropoda* sp. phoretic on *Elater lecontei* Horn. *Pan-Pac. Entomol.* 61:154

51. Greenberg, B. 1961. Mite orientation and survival on flies. *Nature,* 190:107–8

52. Greenberg, B., Carpenter, P. D. 1960. Factors in phoretic association of a mite and fly. *Science* 132:738–39

53. Griffiths, D. A. 1964a. Experimental studies on the systematics of the genus *Acarus* Linnaeus, 1758 (Sarcoptiformes, Acarina). In *Proc. 1st Int. Congr. Acarol. Acarologia* 6 (suppl.):101–16, Fort Collins, CO. Budapest: Akadémiai Kiadó

54. Griffiths, D. A. 1964b. A revision of the genus *Acarus* L., 1758 (Acaridae, Acarina). *Bull. Br. Mus. Nat. Hist. Zool.* 11:413–64

55. Griffiths, D. A. 1966. Nutrition as a factor influencing hypopus formation in the *Acarus siro* complex. *J. Stored Prod. Res.* 1:325–40

56. Griffiths, D. A. 1969. The influence of dietary factors on hypopus formation in *Acarus immobilis* Griffiths (Acari: Acaridae). In *Proc. 2nd Int. Congr. Acarol.* pp. 419–32. Budapest: Akadémiai Kiadó

57. Griffiths, D. A. 1970. A further systematic study of the genus *Acarus* L. 1758. (Acaridae, Acarina) with a key to species. *Bull. Br. Mus. Nat. Hist. Zool.* 19:85–118

58. Hall, C. C. Jr. 1959. A dispersal mechanism in mites. *J. Kans. Entomol. Soc.* 32:45–46

59. Hill, A., Deahl, K. L. 1978. Description and life cycle of a new species of *Histiostoma* (Acari: Histiostomidae) associated with commercial mushroom production. *Proc. Entomol. Soc. Wash.* 80:317–29

60. Hora, A. M. 1934. On the biology of the mite *Glycyphagus domesticus* de Geer (Tyroglyphidae, Acarina). *Ann. Appl. Biol.* 21:483–94

61. Houck, M. A. 1989. Isozyme analysis of *Hemisarcoptes* and its beetle associate *Chilocorus*. *Exp. Appl. Entomol.* 52:167–72

62. Houck, M. A., Cohen, A. 1990. Transition from a free-living life form to parasitism via phoresy: Experimental evidence from the mite *Hemisarcoptes*. *Exp. Appl. Acarol.* Submitted

63. Houck, M. A., OConnor, B. M. 1990. Ontogeny and life history of *Hemisarcoptes cooremani* (Acari: Hemisarcoptidae). *Ann. Entomol. Soc. Am.* 83:869–86

64. Houck, M. A., OConnor, B. M. 1990. Morphometric variation in the genus *Hemisarcoptes* (Acari: Acariformes): application of multivariate morphometrics. *Exp. Appl. Acarol.* Submitted

65. Houck, M. A., OConnor, B. M. 1990. Temperature and resource effects on character variance in *Hemisarcoptes* (Acari: Hemisarcoptidae). *Exp. Appl. Acarol.* Submitted

66. Hughes, A. M. 1976. *The Mites of Stored Food and Houses. Min. Agric. Fisheries Food, Tech. Bull. #9*, London: H. M. S. O. 400 pp.

67. Hughes, T. E. 1959. *Mites or the Acari.* London: Athlone. 225 pp.

68. Hughes, T. E., Hughes, A. M. 1939. The internal anatomy and postembryonic development of *Glycyphagus domesticus*. *Proc. Zool. Soc. London* 108:715–33

69. Hussey, N. W., Read, W. H., Hesling, J. J. 1969. *The Pests of Protected Cultivation.* London: Arnold

70. Johnston, D. E. 1961. A review of the lower upopodoid mites (former Thinozerconoidea, Protodinychoidea and Trachytoidea) with notes on the classification of the Uropodina (Acarina). *Acarologia* 3:522–45

71. Joyeaux, C., Baer, G. 1945. Morphologie, évolution et position systématique de *Catenotaenia pusilla* (Goeze, 1782), Cestode parasite de Rongeurs. *Rev. Suisse Zool.* 52:13–51

72. Kethley, J. 1983. Modifications of the deutonymph of *Uropodella laciniata* Berlese, 1888, for phoretic dispersal (Acari: Parasitiformes) *J. Ga. Entomol. Soc.* 18:151–55

73. Klompen, J. S. H., Lukoschus, F. S., OConnor, B. M. 1987. Ontogeny, life history and sex ratio evolution in *Ensliniella kostylevi* (Acaris: Winter-

schmidtiidae) *J. Zool. London* 213:591–607

74. Krantz, G. W. 1978. *A Manual of Acarology.* Corvallis: Oreg. State Univ. 509 pp. 2nd ed.

75. Krombein, K. V. 1961. Some symbiotic relationships between saproglyphid mites and solitary vespid wasps (Acarina, Saproglyphidae and Hymenoptera, Vespidae). *J. Wash. Acad. Sci.* 51:89–93

76. Kuo, J. S., Nesbitt, H. H. J. 1970. Termination of the hypopal stage in *Caloglyphus mycophagus* (Mégnin) (Acarina: Acaridae). *Can. J. Zool.* 48:529–37

77. Kuo, J. S., Nesbitt, H. H. J. 1971. Internal morphology of the hypopus of *Caloglyphus mycophagus. Acarologia* 13:156–70

78. Lieutier, F. 1978. Les acariens accociés a *Ips typographus* et *Ips sexdentatus* (Coleoptera: Scolytidae) en région Parasienne et les variations de leurs populations au course du cycle annuel. *Bull. Ecol.* 9:307–21

79. Lindquist, E. E. 1969. Mites and the regulation of bark beetle populations. In *Proc. 2nd Int. Congr. Acarol.* pp. 389–99. Budapest: Akadémia: Kiadó

80. Lindquist, E. E. 1975. Association between mites and other arthropods in forest floor habitats. *Can. Entomol.* 107:425–37

81. Lindquist, E. E., OConnor, B. M., Clulow, F. V., Nesbitt, H. H. J. 1979. Acaridae. Canada and its insect fauna, ed. H. V. Danks, *Mem. Entomol. Soc. Can.* 108:277–84

82. Lukoschus, F. S., Fain, A., Driessen, F. M. 1972. Life cycle of *Apodemus apodemi* (Fain, 1965) (Glycyphagidae: Sarcoptiformes). *Tijdschr. Entomol.* 115:325–39

83. Lukoschus, F. S., Scheperboer, G., Fain, A., Nadchatram, M. 1981. Life cycle of *Alabidopus asiaticus* sp. nov. (Acarina: Astigmata: Glycyphagidae) and hypopus of *Alabidopus malaysiensis* sp. nov. ex. *Rattus* spp. from Malaysia. *Int. J. Acarol.* 7:161–77

84. Madel, G. 1975. Vergesellschaftung der Milbenart *Dinogamasus villosior* mit der ostafrikanischen holzbiene *Xylocopa flavorufa* (Acarina:Laelapidae/Hymenoptera:Xylocopidae). *Entomol. Ger.* 1:144–50

85. Mahunka, S. 1972. Untersuchungen Über taxonomische und systematische Probleme bei der Gattung *Myianoetus* Oudemans, 1913, und der Unterfamilie Myianoetinae (Acari, Anoetidae). *Ann. Hist. Nat. Mus. Nat. Hung.* 64:359–72

86. Mahunka, S. 1978. Schizoglyphidae fam. n. and new taxa of Acaridae and Anoetidae (Acari: Acarida). *Acta Zool. Acad. Sci. Hung.* 24:107–31
87. Michael, A. D. 1884. The hypopus question, or the life history of certain Acarina. *J. Linn. Soc. London Zool.* 17:371–94
88. Mitchell, R. 1970. An analysis of dispersal in mites. *Am. Nat.* 104:425–31
89. Moser, J. C. 1976. Surveying mites (Acarina) phoretic on the southern pine beetle (Coleoptera: Scolytidae) with sticky traps. *Can. Entomol.* 108:809–13
90. Moser, J. C. 1981. Transfer of a *Pyemotes* egg parasite phoretic on western pine bark beetles to the southern pine beetle. *Int. J. Acarol.* 7:197–202
91. Moser, J. C., Cross, E. A. 1975. Phoretomorph: a new phoretic phase unique to the Pyemotidae (Acarina: Tarsonemoidea). *Ann. Entomol. Soc. Am.* 68:820–22
92. Moser, J. C., Roton, L. M. 1971. Mites associated with southern pine bark beetles in Allen Parish, Louisiana. *Can. Entomol.* 103:1775–98
93. Mostafa, A. R. I. 1970a. Saproglyphid hypopi (Acarina: Saproglyphidae) associated with wasps of the genus *Zethus* Fabricius. *Acarologia* 12:168–92
94. Mostafa, A. R. I. 1970b. Saproglyphid hypopi (Acarina: Saproglyphidae) associated with wasps of the genus *Zethus* Fabricius. *Acarologia* 12:383–401
95. Norton, R. A. 1973. Phoretic mites associated with the hermit flower beetle *Osmoderma eremicola* Knoch (Coleoptera: Scarabaeidae). *Am. Midl. Nat.* 90:447–49
96. Oboussier, H. 1939. Beiträge zur Biologie und Anatomie der Wohnungsmilben. *Z. Angew. Entomol.* 26:253–96
97. OConnor, B. M. 1979. Evolutionary origins of astigmatid mites inhabiting stored products. See Ref. 115a, 1:273–78
98. OConnor, B. M. 1981. *A systematic revision of the family-group taxa in the non-Psoroptidid Astigmata (Acari: Acariformes)*. PhD thesis. Ithaca, New York: Cornell Univ. 594 pp.
99. OConnor, B. M. 1982a. Acari: Astigmata. In *Synopsis and Classification of Living Organisms,* ed. S. B. Parker, 2:146–69. New York: McGraw Hill
100. OConnor, B. M. 1982b. Evolutionary ecology of astigmatid mites. *Annu. Rev. Entomol.* 27:385–409
101. OConnor, B. M. 1984a. Phylogenetic relationships among higher taxa in the Acariformes, with particular reference to the Astigmata. *Proc. 6th Int. Congr. Acarol,* 1:19–27
102. OConnor, B. M. 1984b. Acarine-fungal relationships: the evolution of symbiotic associations. In *Fungus-Insect Relationships: Perspectives in Ecology and Evolution,* ed. Q. Wheeler, M. Blackwell, pp. 354–81. New York: Columbia
103. OConnor, B. M. 1988. Coevolution in astigmatid mite-bee association. In *Africanized Honey Bees and Bee Mites,* ed. G. R. Needham, R. E. Paige, Jr., M. Delfinado-Baker, C. E. Bowman, pp. 339–46. Chichester: Harwood. 572 pp.
104. OConnor, B. M. 1989. Systematics, ecology and host associations of *Naiadacarus* (Acari: Acaridae) in the Great Lakes Region. *Great Lakes Entomol.* 22:79–94
105. OConnor, B. M. 1990. Ecology and host association of *Histiogaster arborsignis* (Acari: Acaridae) in the Huron Mountains of Northern Michigan. *Great Lakes Entomol.* In press
106. OConnor, B. M., Eickwort, G. C. 1988. Morphology, ontogeny, biology and systematics of the genus *Vidia* (Acari: Winterschmidtiidae). *Acarologia* 29: 147–74
107. OConnor, B. M., Houck, M. A. 1989a. A new genus and species of Hemisarcoptidae from Malaysia (Acari: Astigmata). *Int. J. Acarol.* 15:17–20
108. OConnor, B. M., Houck, M. A. 1989b. Two new genera of Hemisarcoptidae (Acari: Astigmata) from the Huron Mountains of Northern Michigan. *Great Lakes Entomol.* 22:1–10
109. OConnor, B. M., Pfaffenberger, G. S. 1987. Systematics and evolution of the genus *Paraceroglyphus* and related taxa (Acari: Acaridae) associated with fleas (Insects: Siphonaptera). *J. Parasit.* 73:1189–97
110. Olynyk, J. E., Freitag, R. 1979. Some phoretic association of ground beetles (Coleoptera: Carabidae) and mites (Acarina). *Can. Entomol.* 111:333–35
111. Pasteels, J. M., Deroe, C., Tursch, B., Braekman, J. C., Daloze, D., Hootele, C. 1973. Distribution et activités des alcaloides défensifs des Coccinellidae. *J. Insect Physiol.* 19:1771–84
112. Perron, R. 1954. Untersuchungen über Bau, Entwicklung und Physiologie der Milbe *Histiostoma laboratorium* Hughes. *Acta Zool. Stockholm* 35:71–106
113. Poinar, G. O. Jr. 1985. Fossil evidence

of insect parasitism by mites. *Int. J. Acarol.* 11:37–38

114. Polezhaev, V. G. 1938. Influence of food shortage on the formation of hypopi in *Glycyphagus destructor. Zool. Zh.* 17:617–21

115. Robinson, I. 1953. The hypopus of *Hericia hericia* (Kramer), Acarina Tyroglyphidae. *Proc. Zool. Soc. London* 123:267–272

115a. Rodriguez, J. G., ed. 1979. *Recent Advances in Acarology*, Vol. 1. New York: Academic. 631 pp.

116. Samsinák, K. 1970. Zwei neue Arten der Gattung *Sancassania* Oudemans, 1916 (Acari: Acaridae). *Zool. Anz.* 184:403–12

117. Samsinák, K. 1971. Die auf *Carabus*-Arten (Coleoptera, Adephaga) der palaearktischen Region lebenden Milben der Unterordnung Acariformes (Acari); ihre Taxonomie und Bedeutung für die Lösung zoogeographischer, entwicklungsgeschichtlicher und parasitophyletischer Fragen. *Entomol. Abh. St. Mus. Tierk. Dresden* 38:145–234

118. Samsinák, K. 1989. Mites on flies of the family Sphaeroceridae. II. *Acarologia* 30:85–105

119. Scheucher, R. 1957. Systematik und Oekologie der deutschen Anoetinen. See Ref. 125a. 1:233–384

120. Schulze, H. 1922. Beiträge zur Biologie von *Tyroglyphus mycophagus* (Mégnin). (Zerstörung einer Mehlwurmzucht durch diese Milbe). *Arb. Biol. Bundesanst. Landu Forstwirt. Berlin* 11:169–77

121. Schulze, H. 1924a. Über die Biologie von *Tyroglyphus mycophagus* Mégn., zugleich ein Beiträge zur Hypopusfrage. *Z. Morphol. Oekol. Tiere* 2:1–57

122. Schulze, H. 1924b. Zur Kenntnis der Dauerformen (Hypopi) der Mehlmilbe *Tyroglyphus farinae* (L.). *Zentralbl. Bakteriol. Parasitenk. Infektionskr.* 60:536–44

123. Shchur, L. E. 1970. Akaroidnie (Tiroglifoidnie) kleshchi, svyazannie s bobrom (*Castor fiber* L.) i mestami ego obitaniya. *Tezisi Dokladov, Vtoroe Akarologischeskoe Soveshchanie, Kiev, 1970* 2:242–43

124. Southwood, T. R. E. 1962. Migration of terrestrial arthropods in relation to habitat. *Biol. Rev.* 37:171–214

125. Spicka, E. J., OConnor, B. M. 1979. Description and life-cycle of *Glycyphagus (Myacarus) microti* sp. n. (Acarina: Sarcoptiformes: Glycyphagidae) from *Microtus pinetorum* from New York, U.S.A. *Acarologia* 21:451–76

125a. Stammer, H. J., ed. 1957. *Beiträge zur Systematik Oekologie Mitteleuropäis*

chen *Acarina*, Vol. 1. Leipzig: Acad. Verlagsges

126. Stolpe, S. G. 1938. The life cycle of the tyroglypid mites infesting cultures of *Drosophila melanogaster. Anat. Rec.* 72 (Suppl.):133–34

127. Traver, J. R. 1951. Unusual scalp dermatitis in humans caused by the mite *Dermatophagoides. Proc. Entomol. Soc. Wash.* 53:1–25

128. Treat, A. E. 1975. *Mites of Moths and Butterflies*. New York: Cornell Univ. Press. 362 pp.

129. Türk, E. 1963. A new tyroglyphid deutonymph in amber from Chiapas, México. *Univ. Calif. Publ. Entomol.* 31:49–51

130. Türk, E., Türk, F. 1957. Systematik und Oekologie der Tyroglyphiden Mitteleuropas. See Ref. 125a, 1:3–231

131. Vitzthum, H. G. 1940. Die Deutonympha von *Carpoglyphus lactis* (L. 1763) (Acari: Tyroglyphidae). *Zool. Anz.* 129:197–201

132. Volgin, V. I. 1971. The hypopus and its main types. In *Proc. 3rd Int. Congr. Acarol. Prague* pp. 381–83

133. Wallace, D. R. J. 1960. Observations on hypopus development in the Acarina. *J. Insect Physiol.* 5:216–29

134. Wharton, G. W. 1976. House dust mites. *J. Med. Entomol.* 12:577–621

135. Whitaker, J. O., Wilson, N. 1974. Host and distribution lists of mites (Acari), parasitic and phoretic, in the hair of wild mammals of North America, north of Mexico. *Am. Midl. Nat.* 91:1–67

136. Woodring, J. P. 1963. The nutrition and biology of saprophytic Sarcoptiformes. *Adv. Acarol.* 1:89–111

137. Woodring, J. P. 1973. Four new anoetid mites associated with halictid bees. *J. Kans. Entomol. Soc.* 46:310–27

138. Woodring, J. P., Carter, S. C. 1974. Internal and external morphology of the deutonymph of *Caloglyphus boharti* (Arachnida: Acari). *J. Morphol.* 144:275–95

139. Womersley, H. 1963. A new species of *Forcellinia* (Acarina: Tyroglyphidae) from hives in Western Australia. *Trans. R. Soc. South Aust.* 86:155–57

140. Zachvatkin, A. A. 1941. *Fauna of U.S.S.R., Arachnoidea.* Transl. A. Ratcliff, A. M. Hughes, 1959. Washington, DC: Am. Inst. Biol. Sci. 573 pp. (From Russian)

141. Philips, J. R., Dindal, D. L. 1990. Invertebrate populations in the nests of a screech owl (*Otus asis*) and an American kestrel (*Falco sparvarius*) in central New York. *Entomol. News* 101:170–92

Annu. Rev. Entomol. 1991. 36:637–57

INSECT HERBIVORY ON EUCALYPTUS

C. P. Ohmart

Scientific Methods, Inc., PO Box 599, Durham, California, 95938

P. B. Edwards

CSIRO Division of Entomology, PO Box 1700, Canberra, ACT Australia 2601

KEY WORDS: *Eucalyptus,* herbivory, insect-plant interactions, forest insects

INTRODUCTION

A great deal of research and discussion over the past 10 years has focused on herbivory in plant systems (35, 39, 74, 87, 141). Three recent reviews concentrated on herbivory in forests (6, 87, 141). The forests and woodlands of Australia are unique in that they are dominated almost completely by one tree genus, *Eucalyptus.* Not only does *Eucalyptus* play a dominant ecological role, it is economically important for timber, fiber, shelter for livestock, soil conservation, and watershed management.

The first publications devoted to insect herbivores associated with eucalypts in Australia were written by French (61) and Froggatt (62, 63). Both authors concentrated on economically important insect species. Their publications are the only major compilations of information on insect herbivores of eucalypts in Australia and, as a result, are still used by many researchers as the initial source of information on eucalypt insect pests. Only four brief synopses of important insect pests of eucalypts have subsequently been published (27, 64, 116, 119), while other research has focused on the ecological role of insect herbivores in eucalypt forests (59, 60, 109, 117, 121, 147).

0066-4170/91/0101-0637$2.00

Studies of eucalypt herbivory in agricultural regions have concentrated on rural dieback, a syndrome in which canopies of eucalypts progressively decline until ultimately the tree dies (95, 127). The underlying causes of rural dieback are complex and poorly understood (88), but in many cases chronic, high-level insect defoliation is a major attribute (73, 88, 95, 127).

Studies of herbivory of eucalypts that have been planted as exotics have been restricted to insects of economic importance; only limited summaries of this information have been published to date (12, 144, 158). In temperate regions, few indigenous species of insects have so far become critical pests (53, 114). The main problems arise from the accidental introduction of Australian insects into these regions. In tropical eucalypt plantations, the most economically important herbivores have been indigenous insects that have adapted to eucalypts, rather than those accidentally introduced from Australia (77).

EUCALYPTS AND THEIR INSECTS

Eucalyptus spp. are native to Australia, Indonesia, Papua New Guinea, and the Philippines and all but 2 of the more than 500 species are found in Australia (154). Eucalypts grow as forest trees 30–50 m high and with 30–70% projected ground cover in regions with more than 900 mm annual rainfall, and as woodland trees 10–25 m high and with 10–30% ground cover in regions with 400–900 mm rainfall. Australia has 41 million ha of forests, of which 86% are dominated by eucalypts, and 65 million ha of woodlands, which are almost entirely dominated by eucalypts (55). Eucalypts are now grown in plantations in an attempt to satisfy demand for pulp and timber and to reduce the rate of destruction of native forests, but recent estimates put the area under plantations in Australia at only 57,010 ha (52).

Eucalypts evolved under unique and harsh environmental conditions. Australia is largely an arid continent with a high frequency of fires. The vast majority of soils are very infertile, and particularly low in phosphate availability. Although the mature foliage of eucalypts is strikingly uniform, species vary considerably in many other characteristics. For example, they range in form from the world's tallest angiosperm, *E. regnans,* to mallee species under 6 m high that have several stems growing from one underground base. Furthermore, they grow from coastal to montane habitats and from very moist regions, excluding rainforests, to semi-arid areas. The wood of eucalypt species shows remarkable variation in color, weight, hardness, toughness, strength, elasticity, and durability and can be used for a wide range of products.

Eucalypts have many other distinctive features, several of which are pertinent to an understanding of the interactions between the trees and their herbivorous insects. First, eucalypts are evergreen (136). Leaves may live for three years or more, but on average most remain on the tree for about 18 months (79), a relatively short time for an evergreen species. Eucalypt leaves may be shed at any time of the year, often in response to flowering, fruiting, bursts of growth, fire, and insect attack (79, 130). In suitable environments, growth can occur continuously throughout the year, but elsewhere growth may be restricted to one or two periods a year. This difference has considerable implications for herbivores that feed on new growth.

One of the most characteristic features of eucalypt leaves is the presence of oil glands. Essential oils can comprise up to 5% of the fresh weight of leaves (130). Although the function of these oils is unknown (130), they probably provide some protection against insect and fungus attack. Eucalypt leaves also usually have high levels of other secondary compounds such as tannins, phenols, and surface waxes, which are characteristics often considered to be defences against phytophagous insects. Eucalypt leaves are sclerophyllous, and their toughness can be an effective barrier against leaf-eating insects (45, 90, 126).

A striking feature of eucalypts is the different leaf types produced during development of the tree. Juvenile leaves, produced by young eucalypts, differ markedly from the adult leaves produced by mature trees. The strongest degree of heterophylly occurs in the blue-gums (series Globulares), in which juvenile leaves are opposite, sessile, ovate, and glaucous, whereas adult leaves are alternate, petiolate, lanceolate, and dark green (130, 136). Such differences in leaf surface characteristics, shape, texture, and arrangement can have considerable implications for the ability of insects to colonize and utilize trees in different stages of growth (e.g. 45).

Eucalypts have a remarkable capacity to recover from defoliation by insects, fire, drought, or mechanical damage, which is a consequence of the trees' system of bud production. Eucalypts produce leafy shoots in four ways (79). First, shoots can be produced from stalked, naked buds found in the axil of every leaf. Second, shoots can grow from accessory buds, which are found at the base of naked buds, but which develop only if the naked buds and their leaves are eaten by insects or otherwise damaged. Additional accessory buds may grow if the replacement shoot is itself eaten, and this process may be repeated several times in a season. Third, new shoots can be produced from epicormic buds. When a leaf is shed, a shaft of tissue continues to grow out radially from the old leaf axil as the branch or trunk grows, and this tissue can produce one or more buds if the upper crown suffers severe damage. Thus, the trunk and branches have at least one potential epicormic bud for every leaf that has developed. The foliage produced by epicormic buds is juvenile in form, even if the tree has long since been producing only adult foliage. Finally, leafy shoots can grow from lignotubers, which occur in most species

and consist of a mass of vegetative buds, vascular tissue, and food reserves
(136). Lignotubers are woody swellings that form at the base of the stem in a
position corresponding to the axils of the cotyledons and first seedling leaves,
and they can produce new stems following severe or total destruction of the
main stem.

The major leaf eaters of eucalypts are adult and larval beetles (Scarabaeidae, Chrysomelidae, Curculionidae), stick insects (Phasmatidae), sawflies (Pergidae), and lepidopteran larvae (Limacodidae, Geometridae, Nolidae, Anthelidae, Lasiocampidae, Saturniidae). Leaves are mined by lepidopteran larvae from several families (Cosmopterygidae, Gracilariidae, Incurvariidae, and Nepticulidae) (104, 105) and by some sawflies (Hymenoptera: Pergidae). Sap-feeding insects of eucalypts include psyllids (Psyllidae),
leafhoppers (Eurymelidae), coreid bugs (Coreidae), and scale insects
(Eriococcidae). A great variety of galls are very common on eucalypt foliage
and can be formed by psyllids, coccids, wasps, and flies (51, 69, 130). Wood
and bark feeders include beetle and moth larvae (Cerambycidae, Hepialidae,
and Cossidae) and termites. As with many other angiosperms, no bark beetles
(Scolytidae) mass attack or kill eucalypts (118).

Eucalyptus seed forms within a woody fruit, or *gum nut* (130). Although
they are quite small, the seeds are fed upon within the fruit by beetles
(especially *Dryophilodes* spp.) and wasps (especially *Megastigmus* spp.) (3,
9) and by ants and lygaeid bugs after the tree has shed its fruit (1, 36, 51).
Ants may remove over 60% of the annual seedfall (4).

The suite of herbivorous insects associated with *Eucalyptus* differs in
several respects from the groups of insects found in northern hemisphere
forests. In particular, the dominance of leaf-eating beetles on eucalypts is
striking. Among the Lepidoptera, several moth families (including Pyralidae
and Noctuidae) and nearly all butterflies are almost entirely absent from
eucalypts (34). The great diversity of psyllids but almost total absence of
aphids from eucalypts differs sharply from the situation in tree genera elsewhere (42). Furthermore, eucalypts are unusual in that termites attack living,
mature trees (131). Finally, a great diversity of lepidopteran larvae utilize the
abundant and regular supply of leaf litter, which contrasts with northern
hemisphere forests where microorganisms predominate in the breakdown of
litter. In Australia, up to 3,000 species of oecophorid moth larvae may feed
on dead eucalypt leaves (34).

INSECT-PLANT INTERACTIONS

Eucalyptus foliage contains high concentrations of secondary compounds
such as tannins, other phenols, and essential oils. Work with other insect-
plant systems indicates that these compounds should have a significant influence on insect-eucalypt interactions. However, for the insects studied to

date, secondary compounds in eucalypt foliage seem to influence patterns and amounts of herbivory and larval performance very little. For example, a wide range of tannin and phenol concentrations do not affect the growth rate and nitrogen-use efficiency of *Paropsis atomaria* (Coleoptera: Chrysomelidae) larvae (57). Furthermore, variation in essential oil concentrations had no influence on patterns of herbivory on three species of eucalypts growing in monospecific stands, although two other species with high oil content showed a significant negative correlation between oil concentration and insect damage (111). Insects feeding on eucalypt foliage have evolved a variety of mechanisms for dealing with the compounds in eucalypt foliage. Some species avoid the oil glands by feeding around them (P. B. Edwards, unpublished observation); other species appear to tolerate oils, passing them through the gut unchanged (111) or sequestering them in special body structures (112), while other species detoxify oils (120). Tannins do not appear to interfere with protein digestion (57), and surfactants in the guts of some eucalypt defoliators may prevent tannins from precipitating ingested protein (125).

Nitrogen is the leaf component that appears to influence the growth of eucalypt defoliators most (57, 123, 124). *P. atomaria* has a threshold of nitrogen, below which larval growth is directly related to foliar nitrogen concentrations and above which larval growth and development proceed at an optimum rate regardless of the level of nitrogen ingested (57, 102, 123). *Paropsis charybdis,* like most other leaf-eating insects, is more efficient at converting nitrogen than energy (references in 46). When fed on *E. viminalis* with a nitrogen content of 2.33%, *P. charybdis* larvae assimilated 59% of the nitrogen but only 27% of the energy in the food, and adults assimilated 55% of the nitrogen but 39% of the energy (46). Because eucalypt foliage is low in nitrogen, eucalypt defoliators may need to consume large amounts of foliage to obtain sufficient nitrogen for development (57). In support of this hypothesis, foliage consumption by *P. atomaria* larvae is inversely related to foliar nitrogen levels, although in the field this species prefers newly flushed foliage that usually contains nitrogen levels above the threshold at which nitrogen is limiting to growth (126). Much of the research on the effect of eucalypt secondary compounds and nitrogen on insect performance has been done using *P. atomaria* as the test insect, and when more species are investigated, our understanding of insect-eucalypt interactions will be considerably broadened.

While eucalypts are known to have high levels of constitutive defences, the presence of induced defenses has not yet been demonstrated. Some eucalypt herbivores, after feeding on a leaf, sever the leaf from the tree at the petiole, possibly indicating an attempt to sabotage transmission of rapid induced defenses (44). No evidence has been found for delayed induced defenses; indeed, *P. atomaria* larval performance is similar on undamaged plants and plants with a history of damage (E. Haukioja & C. P. Ohmart, unpublished data). In some cases, performance on and/or attractiveness of eucalypt foliage seems to be enhanced by earlier damage, perhaps in response to higher nitrogen levels in regrowth foliage (84, 90).

Physical attributes of eucalypt leaves can also have an effect on herbivory. Eucalypts have tough, leathery leaves, which may influence the population dynamics of specific eucalypt defoliators. Many species, such as *P. atomaria*, are attracted to and feed preferentially on newly flushed eucalypt foliage, which is tender and high in nitrogen (126). As the foliage ages, leaf toughness increases rapidly (45, 90) and nitrogen declines. Soon, the leaves become too tough for *P. atomaria* larvae and they search for foliage that is more suitable. The threshold of leaf toughness is exceeded before nitrogen declines to the point at which it becomes limiting for growth (90, 126). Therefore, the phenology of the insect must be tuned to that of the leaves to ensure the presence of adequate food (126).

Sclerophylly may also be a significant factor in determining the host range of eucalypt defoliators. For instance, snow gums (*E. pauciflora*) have exceptionally tough leaves, and, in a field study on the distribution of 11 species of adult chrysomelids on 8 species of eucalypts, only 2% of all individuals were found on *E. pauciflora* (45). The leaf toughness of *E. pauciflora* in this study was nearly 50% higher than that considered detrimental to the survival of *P. atomaria* larvae on *E. blakelyi* (123). The host range of insects may vary depending on the suite of available hosts (58), and, in a region where only three eucalypt species occurred, several chrysomelids in fact showed a preference for *E. pauciflora* (110).

Leaf waxes are a feature of many eucalypt species, and a glaucous bloom is well developed on the juvenile foliage of many blue gum species. The observation that adult *P. charybdis* cannot grip onto waxy juvenile foliage leads to the suggestion that leaf waxes may confer resistance against adult leaf beetles (43). The almost total inability of several chrysomelid species to cling to leaves with a thick waxy bloom has since been quantified experimentally, and these data in large part correlate with the field observation of very low numbers of chrysomelids on eucalypts with glaucous leaves (45). Interestingly, chrysomelids are significant defoliators of the adult foliage of blue gums, which does not have a waxy bloom. One chrysomelid (*Chrysophtharta m-fuscum*) has overcome the difficulties of colonizing waxy-leaved species, partly by behavioral modifications, and is in fact a specialist on such trees (45).

HERBIVOROUS INSECTS IN FORESTS AND WOODLANDS

Despite the enormous variety of phytophagous insects associated with eucalypts in Australia, only a few of these have become serious forest pests

(27). Populations of some species can erupt into outbreaks over large areas, causing severe damage or tree death, but this phenomenon is rare. In all cases, these eruptive species are foliage feeders, mostly phasmatids, Coleoptera, or Lepidoptera.

Three species of stick insects (Phasmatodea: Phasmatidae), *Didymuria violescens, Ctenomorphodes tessulata,* and *Podocanthus wilkinsoni,* occasionally occur at outbreak levels in southeastern Australia and can completely defoliate forests (12, 17, 71, 96, 115, 137). Total defoliation causes severe stem growth reduction (> 80%) (96) and, if it occurs more than once, can cause more than 80% mortality of the forest stand (115). These species feed on a wide range of eucalypts but appear to favor particular species associations. Their life cycles last for two years. The egg overwinters twice before hatching. Therefore, heavy defoliation occurs every other year during an outbreak. Birds and several species of egg parasites are thought to be important controlling populations when phasmatids are at low density, but when outbreaks occur birds become ineffective because of their limited numerical response, and ultimately populations are limited by intraspecific competition for food (137).

The jarrah leafminer, *Perthida glyphopa* (Lepidoptera: Incurvariidae), has occurred at chronically high levels in some jarrah forests (*E. marginata*) in Western Australia for several decades (97–99). Tree density appears to greatly influence the abundance of *P. glyphopa.* High populations are usually associated with less-densely forested areas or artificial clearings caused by forestry operations (156). Heavy defoliation year after year can cause over 50% reduction in stem growth of trees, but the trees are rarely killed because defoliation occurs during the winter when the foliage is least photosynthetically active. In the spring, a new cohort of foliage is produced and is not damaged until the following winter (97). Three different levels of tree susceptibility to *P. glyphopa* have been identified in populations of *E. marginata* (98); one individual tree has been found on which oviposition has never been observed and therefore seems to be totally resistant; about 35% of the population of trees are attacked by females but the eggs either never hatch or larvae die when they are young; the remainder of the population of the trees are attacked and are susceptible to *P. glyphopa.*

The gum leaf skeletonizer, *Uraba lugens* (Lepidoptera: Nolidae) has been recorded at outbreak levels on at least eight species of eucalypts in many parts of Australia (11, 18, 72). Outbreaks are particularly prevalent on *E. camaldulensis* forests in the Murray River valley (18). Very large areas of eucalypt forest, sometimes exceeding 45,000 ha, can be severely defoliated (18). The gum leaf skeletonizer is found in every part of Australia except the Northern Territory and can feed on a wide range of eucalypt species. Survival among hosts is variable, however (108).

The autumn gum moth, *Mnesampela privata* (Lepidoptera: Geometridae), is a eucalypt herbivore in which ovipositing females prefer the glaucous juvenile-form foliage of the blue gum group of eucalypts (e.g. *E. bicostata*), but larvae can feed on a wide range of other eucalypts as long as they have leaves that can be tied into larval shelters (48). Eggs are laid mainly on small trees in the autumn and most defoliation occurs in the autumn and winter. Another geometrid, *Stathomorrhopa aphotista,* has occasionally occurred at outbreak levels, causing complete defoliation of several different species of eucalypts. For example, it completely defoliated over 1,000 ha of *E. amygdalina* and *E. viminalis* in Tasmania (49, 50).

Cup moth larvae, *Doratifera* spp. (Limacodidae), can cause considerable defoliation of woodland and urban eucalypts. Although the larvae are quite well known on account of the painful stings they can inflict (146), only limited ecological information is available on the genus (70, 146).

Leaf beetles of the family Chrysomelidae are an ecologically and economically important group of eucalypt herbivores, particularly on young trees. *Chrysophtharta bimaculata,* for example, is considered to be one of the most important forest insect pests in Tasmania. It can cause heavy defoliation on young as well as mature stands of trees (38, 65, 81). Both larvae and adults feed on young, expanding foliage of several commercially important eucalypts. Foliage of *E. nitens* is not attacked until the nonwaxy adult-form foliage is produced at about age five or after. Many species of chrysomelid beetles are associated with eucalypt woodlands (e.g. 45). *P. atomaria* is a common species in southeastern Australia, with a wide host range (22). Although it rarely causes high levels of herbivory, *P. atomaria* has been used extensively in studies on insect-eucalypt interactions (22, 57, 90, 120, 123, 124, 126, 149).

Gonipterus scutellatus (Coleoptera: Curculionidae) is better known as an introduced pest of eucalypts overseas than in Australian eucalypt forests and woodlands. It is indigenous to southeastern Australia and can cause considerable levels of herbivory, removing all of the current season's foliage in areas where populations are high (47; C. P. Ohmart, personal observation).

Adult scarabs, particularly *Anoplognathus* spp., feed on eucalypt foliage while the larvae develop in the soil, feeding on organic matter. The majority of the 40 species of *Anoplognathus,* commonly known as Christmas beetles, occur in coastal and subcoastal east Australia; only one species is recorded from Western Australia (2, 20, 24). Old-field environments provide an excellent habitat for larvae. When young eucalypts are planted on these sites, emerging adults feed on the foliage in large numbers and can cause considerable damage (25; F. R. Wylie, personal communication). Once canopy closure is reached, populations of scarabs decline and the level of herbivory caused by them becomes minimal (26). Christmas beetles can also substantially damage agricultural regions and are an important component in some rural diebacks (see below).

Many species of sawflies in the family Pergidae are prevalent in woodlands and young plantations. Detailed ecological information exists for only one species, *Perga affinis affinis* (21). *P. affinis affinis* larvae are gregarious, and many colonies on one tree can coalesce to form huge aggregations of thousands of individuals. Larvae spread out to feed at night, maintaining contact by tapping the foliage or branch with their abdomens. After feeding, they regroup and spend the day clustered on a branch. A heavily infested tree can be completely defoliated, at which time the colony moves down the trunk and across the ground to a nearby tree. Trees can withstand repeated defoliation by *P. affinis affinis* because it occurs during the winter and early spring when foliage removal least impacts tree growth and health (23).

Another pergid sawfly, *Lophyrotoma analis,* commonly causes heavy defoliation of *E. melanophora* in semi-arid woodlands throughout much of Queensland (62, 139). A related species, *Lophyrotoma interupta,* can defoliate isolated groups of trees in Tasmania (51).

Many species from the superfamily Psylloidea are common in eucalypt woodlands. The nymphs of many psyllids form a protective covering of starch called a *lerp,* which can be quite ornate and may aid in identification (150). The taxonomy of Australian psyllids has received considerable attention over the past 30 years, and the fact that many species that feed on eucalypts are species-specific has assisted in resolving the taxonomy and ecological relationships of the host plants (106). Even greater attention has been directed towards the ecology of psyllids to the extent that they are one of the most intensively studied groups of eucalypt-feeding insects in Australia. This emphasis partly reflects their importance as defoliators of eucalypts during outbreak periods, but mostly their suitability as model insects for developing and testing theories of population dynamics.

A 25-year study of *Cardiaspina albitextura* in *E. blakelyi* woodlands of eastern Australia, and a 7-year study of 4 other species on the same host, culminated in comprehensive and detailed life system analyses for each species (complete reference list in 31–33). A long-term study of *Cardiaspina densitexta* on the pink gum, *E. fasciculosa,* commenced in South Australia in 1959 (107). Part of this study was designed to investigate the causes of outbreaks in *E. fasciculosa* woodlands (161). The studies on *C. albitextura* and *C. densitexta* both revealed that food quality and quantity were major determinants of psyllid abundance. However, the former study indicated that high soil-moisture levels were required to initiate an outbreak (32), whereas the latter study led to the conclusion that drought was favorable for an outbreak because water stress increases the availability of nitrogen in the food and hence improves the survival of first-instar larvae (160). Subsequent studies of *C. albitextura* on *E. camaldulensis* indicated that the levels of leaf phenolics may be more important for survival of nymphs than nitrogen levels, which often are not limiting for psyllids (107). Other psyllids that can cause substantial damage include *Glycaspis* species in New South Wales (103) and *Creiis periculosa* on flooded gum (*E. rudis*) and wandoo (*E. wandoo*) in Western Australia (83).

Australian eucalypt forests are unusual in that mature trees can suffer considerable termite attack (131). It is quite common for the heartwood of living trees to be hollowed out by termites to form feeding galleries, and up to 15 trees can be attacked from one central colony (66). Up to 80% of timber losses in old-growth forests may be associated with termite activity (68), resulting in a reduction by half of the millable timber (67). Such "piping" of forest trees is not in itself detrimental to the living trees and is certainly beneficial for the many animals that use tree holes for shelter and nesting (131). Termite attack on heartwood is nearly always secondary and follows invasion by fungus, which in turn usually only occurs after physical damage, particularly that caused by fire (131). Termite damage in forests has increased since the arrival of Europeans in Australia, perhaps due to the higher incidence of fires hot enough to kill the cambium (131). The prospects for control therefore include protection from fire (131) or carefully planned control-burning (131).

Many wood-boring insects, both Coleoptera and Lepidoptera, attack living eucalypts, but most problems have so far occurred on stressed, dying, or dead trees (27, 134). Trees infected with hepialid or cossid moth larvae may be severely weakened by the activity of yellow-tailed black cockatoos, which remove large segments of wood to extract the larvae (100, 151). Such trees may subsequently be blown over, and losses of up to 25% of plantation trees have occurred (100).

ECOLOGICAL IMPACT OF INSECT HERBIVORY

Insect Defoliation in Eucalypt Forests

Until recently, a general feeling has prevailed that eucalypts in Australia sustain chronic, high levels of insect defoliation in the order of 20–50% leaf area loss annually (109, 147). Foresters have attributed part of the commercial success of eucalypt plantations overseas to the absence of insect and disease pests in these exotic plantations (136). Furthermore, high levels of defoliation have been recorded in various eucalypt forests and woodlands (13, 14, 18, 25, 32, 59, 65, 80, 81, 97, 109, 113, 137, 138, 147, 156). It has been suggested that this level of herbivory is important in regulating primary productivity of eucalypt forests and in short-term nutrient cycling (147).

More recently, some studies indicated that average defoliation rates in temperate eucalypt forests may not be any different than that recorded from other temperate forests in other parts of the world (117, 121) and is too low to affect nutrient cycling (121). Furthermore, abundance of insect herbivores in these forests is very similar to abundance of insect herbivores in northern temperate forests and much less abundant than expected based on claims of "chronically high" insect defoliation (122, 163). The most recent compilation of data indicates that defoliation recorded from eucalypt forests falls well within that recorded in other forest types in other parts of the world (87).

Rural Dieback of Woodland Eucalypts

Dieback in trees has been defined as a progressive dying from the extremity of any part of a plant (145). In eucalypts, this dying back is almost invariably preceded by the production of secondary shoots from epicormic buds on the affected branch (133), giving the tree a distinctive stag-headed appearance (73). The term *eucalypt dieback* has often been applied rather generally to include any disease that results in widespread damage or mortality of trees (132), including many cases of massive or repeated defoliation in eucalypt forests, discussed in earlier sections. However, a specific type of dieback, one nearly always associated with disturbed eucalypt woodlands in agricultural regions, has become known as *rural dieback* (82). The best-known case of rural dieback is in the northern tablelands area of New England, New South Wales, perhaps because it has reached the most advanced stage of any dieback in Australia. Dieback also occurs in other areas of the eastern states, and patches are found in western and southern Australia and Tasmania (references in 82, 88). Rural dieback is particularly severe in regions of intensive livestock production (165).

Many factors contribute to rural dieback, such as insect defoliation, pathogens, drought, salinity, soil compaction, and climatic changes (82, 132). The role of insects in rural dieback is equivocal (41, 73), even though repeated and heavy defoliation by insects is nearly always a feature (41, 82). Rarely do trees die with their foliage intact. What is not certain, however, is whether insect defoliation is a primary cause of dieback, stressing trees and rendering them more susceptible to other mortality factors (93, 94), or whether trees stressed by other factors become more susceptible to insect attack (88, 162).

Once defoliation of rural eucalypts has occurred, it can produce a positive-feedback loop of repeated cycles of defoliation and refoliation. The mechanisms of the feedback loop have been well demonstrated in *E. blakelyi*, a species susceptible to severe rural dieback. Foliage of mature *E. blakelyi* trees suffering from dieback has been found to be more heavily grazed than that of neighboring healthy trees, and this difference is correlated with superior nutritional qualities of the dieback foliage (85). This nutritional improvement occurs partly because the leaves on dieback trees are younger, but even when corrected for leaf age, the nitrogen content of the leaves is higher and their specific weight is lower. The latter attribute presumably renders the leaves less tough. That the superior leaf quality of dieback trees results from, as well as causes, defoliation, has been demonstrated experimentally (84). Foliage of healthy *E. blakelyi* trees was artificially clipped. The regrowth foliage on these trees was nutritionally superior to the foliage it replaced and was similar in quality to foliage on dieback trees (84). Thus, once a cycle of defoliation-refoliation is initiated, it can become self-perpetuating and may continue for many years. Repeated defoliations deplete the starch reserves of a tree, ultimately leading to its death (5). However, death is not inevitable. The crown can recover if the stress of defoliation is removed, either by protection with insecticide or from a natural decline in insect numbers (24).

Trees enter the defoliation-refoliation cycle either because the tree becomes nutritionally more suitable for insects or because of increased numbers of defoliating insects. The nutrient content of foliage may increase because of increased levels of soluble nitrogen in stressed trees (160, 162), the accumulation of nutrients in the soil under trees where stock animals congregate (89, 159), or the generally benign conditions where trees are growing (86). Numbers of herbivorous insects can increase for a variety of reasons. The loss of undergrowth in pastoral situations may reduce the shelter and nectar available to parasitoids and predators (37) and to insectivorous birds (56, 88), while land clearance and application of fertilizer may favor the soil-dwelling stages of defoliating scarabs (93).

Repeated defoliation by Christmas beetles (*Anoplognathus* spp.) is an important feature of rural dieback in the northern and southern tablelands of New South Wales (26, 28, 41, 75, 93, 94, 129). *E. blakelyi*, *E. nova-anglica* and *E. rubida* are particularly susceptible (19, 73, 93). The increased areas of improved pastures in rural regions provide an ideal breeding ground for Christmas beetle larvae and, as long as soil moisture levels are suitable (27), huge numbers of adults emerge in summer after spending one or two years in the soil as larvae (25). Adults seek out the few remaining trees in pastoral areas, and these can be stripped of all leaves within a week (73). Trees may be defoliated more than once in the same growing season by successive waves of Christmas beetles.

The decline and death of trees in rural Australia has become a matter of considerable public and scientific concern. Although species other than eucalypts are involved (82, 164), eucalypts have attracted the most attention because of their dominance in the Australian landscape. During the past 20 years, several national conferences have addressed all aspects of tree decline, including rural dieback. The proceedings of these conferences, as well as other review articles and books, provide information on the nonentomological components of rural dieback, as well as detailed situation reports and accounts of research activities (e.g. 73, 95, 127).

HERBIVORY IN *EUCALYPTUS* PLANTED AS AN EXOTIC

Because of eucalypts' ability to grow in a wide variety of habitats, particularly those poor in nutrients and water availability, they have become one of the most extensively planted tree genera in the world. More than 20 species are grown commercially in about 50 countries (114); Brazil alone has more than 4 million ha of eucalypt plantations. They are also planted in many countries for land stabilization, fuel, shelter, and as ornamentals. Observations made on insect herbivory in these environments have mainly related to economic injury. The general consensus is that only a few major economic problems have been caused by indigenous insect species adapting to eucalypts (10, 16, 29, 76, 114, 155); however, herbivorous insects that have been accidentally introduced from Australia have in some cases caused heavy defoliation and economic damage in overseas plantations.

Indigenous Insects Adapting to Eucalypts

Many indigenous insect herbivores have adapted to feeding on eucalypts planted in various parts of the world. For example, a total of 96 indigenous insect herbivore species has been recorded feeding on eucalypts in China (10), 94 in India (10), 223 in Brazil (54), 31 in New Zealand (167), 105 in Papua New Guinea, and 62 in Sumatra (R. Wylie, personal communication).

Termites have probably most successfully adapted to eucalypts in tropical and subtropical regions (10, 53, 54, 76, 128, 140, 148, 157). Most herbivory is caused by foraging workers consuming the lateral and tap roots of seedlings during the first year of planting (10, 128, 157). In some areas, 50–80% of the seedlings are killed (10, 157) and, if control measures are not taken, successful establishment of plantations is at risk. Herbivory is greater in areas that have a previous history of agricultural cultivation than in areas cleared of native vegetation (148, 157). A depletion of organic matter in the soil as a result of agricultural cultivation and a removal of termite habitats probably makes newly planted eucalypts very susceptible to attack. Termite herbivory declines after canopy closure reduces the accumulation of ground litter.

In Brazil, leaf-cutting ants, *Atta* and *Acromyrmex* spp., often completely defoliate eucalypts planted on sites with ant colonies. Twelve ant colonies per ha is sufficient to remove all eucalypt foliage on the site (91). If eucalypt plantations are abandoned in areas where leaf-cutting ants occur, stands would disappear because the ants would kill any regeneration (53).

Indigenous defoliators from several orders and families have caused locally heavy defoliation in many countries. Lepidopteran larvae in the family Geometridae and Noctuidae have defoliated eucalypts in most countries where they are planted; arctiids (Lepidoptera) have caused significant defoliation in South America; and in Africa lasiocampids, psychids, and saturniids have heavily defoliated eucalypts (76, 140). Tortricidae and Geometridae are considered the most important indigenous defoliator groups in India (114). *Hyomeces squamosus* (Coleoptera: Curculionidae) defoliates *E. camaldulensis* in Thailand (76).

Generalist herbivores such as scarabaeid larvae (Coleoptera), aphids, and tetranychid mites have caused noticeable herbivory in nurseries (7, 29, 53). Grasshoppers and crickets have occasionally defoliated newly established plantations in Italy, California, Africa, and Turkey (29, 53, 144). Hemipterans have caused significant shoot and tree dieback in some Pacific Islands (8, 53). Indigenous coleopteran and lepidopteran woodborers in several families have caused substantial herbivory in local areas (8, 29, 166). Ambrosia beetles have caused economic problems in Argentina, Uruguay, Figi, Western Samoa, and South Africa (53). A complex of 6 to 10 species of scolytids, platypodids, and bostrychids was associated with damage of eucalypt saplings in Brazil (30). Two of the most economically significant herbivores in Papua New Guinea are the buprestid *Agrilus opulentus* and the cossid moth *Zeuzera coffeae*. In some plantations, over 80% of trees can be attacked by *A. opulentus* and 20% killed (R. Wylie, personal communication).

Australian Insects Accidentally Introduced to Other Countries

Approximately 21 species of Australian insect herbivores have become established in countries where eucalypts are grown as exotics (16, 167). Eighteen of these species have become established only in New Zealand. *Phoracantha semipunctata* (Coleoptera: Cerambycidae) and *G. scutellatus* (Coleoptera: Curculionidae) have spread to many countries and caused substantial economic damage. Other species are of minor importance, except in New Zealand where *P. charybdis* is the most important species.

G. scutellatus was discovered in South Africa in 1916 and subsequently spread through the eucalypt-growing regions of Africa, France, and Italy (16, 29). It caused severe defoliation and threatened to make eucalypt plantations commercially nonviable (153). After unsuccessful attempts to control it with chemicals, an egg parasite, *Patasson nitens* (Hymenoptera: Mymaridae), was collected in Australia and subsequently released in South Africa (153). It proved to be a remarkably successful control agent and quickly reduced *G. scutellatus* populations to noneconomic levels.

P. semipunctata is by far the most widely spread eucalypt herbivore of Australian origin. It has become established in all major eucalypt growing regions of the world except India (16, 53, 54, 78, 92, 101, 135, 140, 142). It was first noticed in South Africa in 1906 (152) and has subsequently spread throughout Africa, the mediterranean region, the middle east, South America, and California (15, 40, 142). *P. semipunctata* has been studied little in Australia where, under normal conditions, it is restricted to debilitated trees and only causes noticeable tree mortality during severe droughts (134). In general, *P. semipunctata* has had its biggest economic and ecological impact in countries where droughts are relatively common, such as Spain, Portugal (16), and the mediterranean region (29).

CONCLUDING REMARKS

Socioeconomic pressures in Australia are encouraging the implementation of intensive management of eucalypt plantations for the production of wood fiber (119). This process will most likely bring about new insect pest-problems and a renewed need for more research on insect-eucalypt interactions (119). As more and more countries introduce eucalypts into new environments and establish ever larger areas of plantations, indigenous insects will become adapted to feeding on eucalypts and Australian species will be accidentally introduced into these plantings. These invasions will also increase pressure for research into insect-eucalypt interactions. The research reviewed in this article should provide a broad base on which to build our knowledge of insect herbivory on *Eucalyptus*.

ACKNOWLEDGMENTS

We thank Wolfgang Wanjura for considerable help with assembling the literature. Constructive comments on the manuscript were provided by Drs. L. B. Barton Browne and R. A. Farrow.

Literature Cited

1. Abbott, I., van Heurck, P. 1985. Comparison of insects and vertebrates as removers of seed and fruit in a Western Australian forest. *Aust. J. Ecol.* 10:165–68
2. Allsopp, P. G., Carne, P. B. 1986. *Anoplognathus vietor* sp. n. (Coleoptera: Scarabaeidae: Rutelinae) from west Queensland. *J. Aust. Entomol. Soc.* 25:99–101
3. Anderson, A. N., New, T. R. 1987. Insect inhabitants of fruits of *Leptospermum*, *Eucalyptus* and *Casuarina* in southeastern Australia. *Aust. J. Zool.* 35:327–36
4. Ashton, D. H. 1979. Seed harvesting by ants in forests of *Eucalyptus regnans* F.

Muell. in central Victoria. *Aust. J. Ecol.* 4:265–77
5. Bamber, R. K., Humphreys, F. R. 1965. Variation in sapwood starch levels in some Australian forest species. *Aust. For.* 29:15–23
6. Barbosa, P., Schultz, J. C., eds. 1987. *Insect Outbreaks*. New York: Academic
7. Barrett, R. L. 1978. *Forest nursery practice for the wattle regions in the Republic of South Africa*. Pietermaritzburg: Wattle Res. Inst.
8. Bigger, M. 1985. The insect pests of *Eucalyptus deglupta* in the Solomon Islands. See Ref. 77, pp. 142–46
9. Boland, D. J., Martensz, P. N. 1981. Seed losses in fruits on trees of *Eu-*

calyptus delegatensis. Aust. For. 44:64–67

10. Bolland, L., Wylie, F. R. 1987. *China-Australia Afforestation Proj. at Dongmen St. For. Farm Tech. Comm.* No. 28. 31 pp.

11. Brimblecombe, A. R. 1962. Outbreaks of the eucalypt leaf skeletonizer. *Queensl. J. Agric. Sci.* 19:209–17

12. Browne, F. G. 1968. *Pests and Diseases of Forest Plantation Trees.* Oxford: Clarendon

13. Burdon, J. J., Chilvers, G. A. 1974. Fungal and insect parasites contributing to niche differentiation in mixed species stands of eucalypt saplings. *Aust. J. Bot.* 22:103–14

14. Burdon, J. J., Chilvers, G. A. 1974. Leaf parasites on altitudinal populations of *Eucalyptus pauciflora* Sieb. ex Spreng. *Aust. J. Bot.* 22:265–69

15. Bytinski-Salz, H., Neumark, S. 1952. The eucalyptus borer (*Phoracantha semipunctata*) in Israel. *Trans. 9th Int. Congr. Entomol.* 1:696–99

16. Cadahia, D. 1986. Importance des insectes ravageurs de l'eucalyptus en region mediterraneenne. *EPPO Bull.* 16:265–83

17. Campbell, K. G. 1960. Preliminary studies in population estimation of two species of stick insects (Phasmatodea: Phasmatidae) occurring in plague numbers in highland forest areas of southeastern Australia. *Proc. Linn. Soc. NSW* 85:121–41

18. Campbell, K. G. 1962. The biology of *Roselia lugens* (Walk.) the gum-leaf skeletonizer moth, with particular references to the *Eucalyptus camaldulensis* Dehn. (River red gum) forests of the Murray Valley region. *Proc. Linn. Soc. NSW* 87:316–38

19. Campbell, K. G. 1964. Aspects of insect-tree relationships in forests of eastern Australia. In *Breeding Pest-Resistant Trees,* ed. H. G. Gerhold, R. E. McDermott, E. J. Schreiner, J. A. Winieski, pp. 239–50. Oxford: Pergamon. 505 pp.

20. Carne, P. B. 1957. A revision of the Ruteline genus *Anoplognathus* Leach (Coleoptera: Scarabaeidae). *Aust. J. Zool.* 5:88–143

21. Carne, P. B. 1962. The characteristics and behaviour of the sawfly *Perga affinis affinis* (Hymenoptera). *Aust. J. Zool.* 10:1–34

22. Carne, P. B. 1966. Ecological characteristics of the eucalypt defoliating chrysomelid *Paropsis atomaria. Aust. J. Zool.* 14:647–72

23. Carne, P. B. 1969. On the population dynamics of the eucalypt-defoliating sawfly *Perga affinis affinis* Kirby (Hymenoptera). *Aust. J. Zool.* 17:113–41

24. Carne, P. B. 1981. Three new species of *Anoplognathus* Leach, and new distribution records for poorly known species (Coleoptera: Scarabaeidae: Rutelinae). *J. Aust. Entomol. Soc.* 20:289–94

25. Carne, P. B., Greaves, R. T. G., McInnes, R. S. 1974. Insect damage to plantation-grown eucalypts in north coastal New South Wales, with particular reference to Christmas beetles (Coleoptera: Scarabaeidae). *J. Aust. Entomol. Soc.* 13:189–206

26. Carne, P. B., McInnes, R. S., Green, J. P. 1981. Seasonal fluctuations in the abundance of two leaf-eating insects. See Ref. 127, pp. 121–26

27. Carne, P. B., Taylor, K. L. 1978. Insect pests. In *Eucalypts for Wood Production,* ed. W. E. Hillis, A. G. Brown, pp. 155–68. Melbourne: CSIRO

28. Carter, J. J., Edwards, D. W., Humphreys, F. R. 1981. Eucalypt diebacks in New South Wales. See Ref. 127, pp. 27–30

29. Cavalcaselle, B. 1986. Les insectes nuisibles aux eucalyptus en Italie: importance des degats et methodes de lutte. *EPPO Bull.* 16:293–97

30. Clark, E. W. 1973. Report of the consultant in forest entomology, Brazil. *Proj. DP/BRA/71/545, Working Doc. No. 7*

31. Clark, L. R. 1964. The population dynamics of *Cardiaspina albitextura* (Psyllidae). *Aust. J. Zool.* 12:362–80

32. Clark, L. R., Dallwitz, M. J. 1974. On the relative abundance of some Australian Psyllidae that coexist on *Eucalyptus blakelyi. Aust. J. Zool.* 22:387–415

33. Clark, L. R., Dallwitz, M. J. 1975. The life system of *Cardiaspina albitextura* (Psyllidae), 1950–1974. *Aust. J. Zool.* 23:523–61

34. Common, I. F. B. 1980. Some factors responsible for imbalances in the Australian fauna of Lepidoptera. *J. Lepid. Soc.* 34:286–94

35. Crawley, M. J. 1983. *Herbivory: The Dynamics of Animal-Plant Interactions.* Berkeley: Univ. Calif. Press. 437 pp.

36. Cremer, K. W. 1965. Emergence of *Eucalyptus regnans* from buried seed. *Aust. For.* 29:119–24

37. Davidson, R. L. 1980. Local management. In *Focus on Farm Trees,* ed. N. M. Oates, P. J. Greig, D. G. Hill, P. A. Langely, A. J. Reid, pp. 89–98. Box Hill: Capital. 149 pp.

38. de Little, D. W. 1983. Life-cycle and aspects of the biology of Tasmanian Eucalyptus leaf beetle, *Chrysophtharta bimaculata* (Olivier) (Coleoptera: Chrysomelidae). *J. Aust. Entomol. Soc.* 22:15–18

39. Denno, R. F., McClure, M. S., eds. 1983. *Variable Plants and Herbivores in Natural and Planted Systems.* New York: Academic. 717 pp.

40. Duffy, E. A. J. 1960. *A monography of the immature stages of Australian atimber beetles (Cerambycidae)*, pp. 126–29 London: Brit. Mus.

41. Duggin, J. A. 1981. The use of ecological provinces and land systems in the study of Eucalyptus dieback in the New England tablelands. See Ref. 127, pp. 245–60

42. Eastop, V. F. 1978. Diversity of the Sternorrhyncha within major climatic zones. *Symp. R. Entomol. Soc. Lond.* 9:71–88

43. Edwards, P. B. 1982. Do waxes on juvenile Eucalyptus leaves provide protection from grazing insects? *Aust. J. Ecol.* 7:347–52

44. Edwards, P. B., Wanjura, W. J. 1989. Eucalypt-feeding insects bite off more than they can chew: sabotage of induced defenses? *Oikos* 54:246–48

45. Edwards, P. B., Wanjura, W. 1990. Physical attributes of eucalypt leaves and the host range of chrysomelid beetles. In *Proc. 7th Int. Symp. Insect-Plant Relationships. Symp. Biol. Hung.* 39: In press

46. Edwards, P. B., Whiteman, J. A. 1984. Energy and nitrogen budgets for larval and adult *Paropsis charybdis* feeding on *Eucalyptus viminalis. Oecologia Berlin* 61:302–10

47. Edwards, H. J. 1981. Eucalyptus weevil. *Tas. For. Comm. For. Pests Dis Leafl.* No. 8

48. Elliott, H. J., Bashford, R. 1978. The life history of *Mnesampela privata* (Guen.) (Lepidoptera: Geometridae), a defoliator of young eucalypts. *J. Aust. Entomol. Soc.* 17:201–4

49. Elliott, H. J., Bashford, R. 1983. Peppermint looper. *Tas. For. Comm. For. Pests Dis. Leafl.* No. 3

50. Elliott, H. J., Bashford, R., Palzer, C. 1980. Biology of *Stathomorrhopa aphotista* Turner (Lepidoptera: Geometridae), a defoliator of *Eucalyptus* spp. in southern Tasmania. *Aust. For.* 43:81–86

51. Elliot, H. J., de Little, D. 1985. *Insect Pests of Trees and Timber in Tasmania.* Hobart: Tas. For. Comm. 90 pp.

52. Elliott, H. J., Kile, G. A., Cameron, J. N. 1990. Biological threats to Australian Plantations: Implications for research and management. In *Prospects for Australian Plantations* ed. J. Dargavel, N. Semple. Canberra: CRES. In press

53. Food and Agriculture Organization. 1979. Diseases, pests and disorders. In *Eucalypts for Planting. Forestry Ser. No. 11* pp. 215–46. Rome: FAO. 677 pp.

54. Filho, E. B. 1985. Insects associated to eucalypt plantations in Brazil. See Ref. 77, pp. 162–78

55. Florence, R. G. 1985. Eucalypt forests and woodlands. In *Think Trees Grow Trees.* Dept. Arts, Heritage and Environ, pp. 29–49. Canberra: Aust. Gov. Publ. Ser. 210 pp.

56. Ford, H. A., Bell, H. L. 1982. Density of birds in eucalypt woodland affected to varying degrees by dieback. *Emu* 81: 202–8

57. Fox, L. R., Macauley, B. J. 1977. Insect grazing on *Eucalyptus* in response to variation in leaf tannins and nitrogen. *Oecologia Berlin* 29:145–62

58. Fox, L. R., Morrow, P. A. 1981. Specialization: Species property or local phenomenon? *Science* 211:887–93

59. Fox, L. R., Morrow, P. A. 1983. Estimates of damage by insect grazing on *Eucalyptus* trees. *Aust. J. Ecol.* 8:139–47

60. Fox, L. R., Morrow, P. A. 1986. On comparing herbivore damage in Australian and north temperate systems. *Aust. J. Ecol.* 11:387–93

61. French, C. 1900. *Handbook of the Destructive Insects of Victoria.* Melbourne: Govt. Printer

62. Froggatt, W. G. 1923. *Forest Insects of Australia.* Sydney: Govt. Printer

63. Froggatt, W. G. 1927. *Forest and Timber Borers.* Sydney: Govt. Printer. 107 pp.

64. Greaves, R. 1961. Insect Pests of Eucalyptus in Australia. Presented at 2nd World Eucalyptus Conf., Sao Paulo, Brazil. 9 pp. Canberra: Forestry and Timber Bureau

65. Greaves, R. 1966. Insect defoliation of eucalypt regrowth in the Florentine Valley, Tasmania. *APPITA* 19:119–26

66. Greaves, T. 1962. Studies of foraging galleries and the invasion of living trees by *Coptotermes acinaciformis* and *C. brunneus* (Isoptera). *Aust. J. Zool.* 10:630–51

67. Greaves, T., Armstrong, G. J., McInnes, R. S., Dowse, J. E. 1967. Timber losses caused by termites, decay, and fire in two coastal forests in New South Wales. In *Termites of Australian Forest Trees, Div. Entomol. Tech. Pap. No. 7*, pp. 4–33. Melbourne: CSIRO

68. Greaves, T., McInnes, R. S., Dowse, J. E. 1965. Timber losses caused by termites, decay and fire in an alpine forest in New South Wales. *Aust. For.* 29:161–74

69. Gullan, P. 1984. A revision of the gall-forming coccoid genus *Apiomorpha* (Homoptera: Eriococcidae). *Aust. J. Zool.* Suppl. 97. 203 pp.

70. Hadlington, P. 1966. Gum tree defoliation by cup moth caterpillers. *For. Timb.* 4(2):10–11

71. Hadlington, P., Hoschke, F. 1959. Observations on the ecology of the phasmatid *Ctenomorphodes tessulata* (Greg). *Proc. Linn. Soc. NSW* 84:146–59

72. Harris, J. A. 1974. The gum leaf skeletonizer *Uraba lugens* in Victoria. *For. Comm. Vict. For. Tech. Pap.* 21:12–18

73. Heatwole, H., Lowman, M. 1986. *Dieback, Death of an Australian Landscape*. Reed: Frenchs Forest. 150 pp.

74. Hedin, P. A., ed. 1983. *Plant Resistance to Insects*. Washington DC: Am. Chem. Soc. Symp. Ser. 208. 375 pp.

75. Hopkins, E. R. 1974. Crown dieback in eucalypt forests. See Ref. 95, pp. 1–16

76. Hutachasern, C., Sahasri, B. 1985. Insect pests of *Eucalyptus camaldulensis* Dehn. in the community woodlot in Thailand. *Z. Angew. Entomol.* 99:170–77

77. International Union of Forestry Research Organizations. 1985. *Noxiuous Insects to Pine and Eucalypt Plantations in the Tropics*. Curitiba, Parana, Brazil: IUFRO Working Party S2.07.07. 24–30 Nov. 213 pp.

78. Ivory, M. H. 1977. Preliminary investigations of the pests of exotic forest trees in Zambia. *Commonw. For. Rev.* 56(1):47–56

79. Jacobs, M. R. 1955. *Growth Habits of the Eucalypts*. Canberra: Govt. Printer

80. Journet, A. R. P. 1981. Insect herbivory on the Australian woodland eucalypt, *Eucalyptus blakelyi* M. *Aust. J. Ecol.* 6:135–38

81. Kile, G. A. 1974. Insect defoliation in eucalypt regrowth forests of southern Tasmania. *Aust. For. Res.* 6(3):9–18

82. Kile, G. A. 1981. An overview of eucalypt dieback in rural Australia. See Ref. 127, pp. 13–26

83. Kimber, P. C. 1981. Eucalypt diebacks in the southwest of Western Australia. See Ref. 127, pp. 37–43

84. Landsberg, J. 1990a. Dieback of rural eucalypts: Response of foliar dietary quality and herbivory to defoliation. *Aust. J. Ecol.* In press

85. Landsberg, J. 1990b. Dieback of rural eucalypts: Does insect herbivory relate to dietary quality of tree foliage? *Aust. J. Ecol.* In press

86. Landsberg, J. 1990c. Dieback of rural eucalypts: The effect of stress on the nutritional quality of foliage. *Aust. J. Ecol.* In press

87. Landsberg, J., Ohmart, C. P. 1989. Levels of insect defoliation in forests: Patterns and concepts. *Trends Ecol. Evol.* 4:96–100

88. Landsberg, J., Wylie, F. R. 1988. Dieback of rural trees in Australia. *Geojournal* 17:231–37

89. Landsberg, J., Morse, J., Khanna, P. 1990. Tree dieback and insect dynamics in remnants of native woodlands on farms. In *Australian Ecosystems: 200 years of Utilization, Degradation and Reconstruction. Proc. Ecol. Soc. Aust.* 16:149–65

90. Larsson, S., Ohmart, C. P. 1988. Leaf age and larval performance of the leaf beetle *Paropsis atomaria*. *Ecol. Entomol.* 13:19–24

91. Lima, P. P. S., Filho, E. B. 1985. The leaf cutting ants: Serious pests of forests in Brazil. See Ref. 77, pp. 147–48

92. Loyttyniemi, K. 1980. Control of *Phoracantha* beetles. *For. Dep. Div. For. Res. Zambia Res. Note No. 24*. 14 pp.

93. Mackay, S. M. 1978. Dying eucalypts of the New England tablelands. *For. Timb.* 14:18–20

94. Mackay, S. M., Humphreys, R. V., Clark, R. V., Nicholson, D. W., Lind, P. R. 1984. Native tree dieback and mortality on the New England tablelands of New South Wales. *For. Comm. NSW Res. Pap. No. 3*. 23 pp.

95. Marks, G. C., Idczak, R. M., eds. 1973. Eucalypt dieback in Australia. *Proc. Lakes Entrance Seminar, 1973*. Melbourne: Forest Commission of Victoria. 73 pp.

96. Mazanec, Z. 1967. Mortality and diameter growth in mountain ash defoliated by Phasmatids. *Aust. For.* 31:221–23

97. Mazanec, Z. 1974. Influence of jarrah leaf miner on the growth of jarrah. *Aust. For.* 37:32–42

98. Mazanec, Z. 1985. Resistance of *Eucalyptus marginata* to *Perthida glyphopa* (Lepidoptera: Incurvariidae). *J. Aust. Entomol. Soc.* 24:209–21

99. Mazanec, Z. 1987. Natural enemies of *Perthida glyphopa* Common (Lepidoptera Incurvariidae). *J. Aust. Ent. Soc.* 26:303–8

100. McInnes, R. S., Carne, P. B. 1978. Predation of cossid moth larvae by yellow-tailed black cockatoos causing losses in plantations of *Eucalyptus grandis* in

north coastal New South Wales. *Aust. Wild. Res.* 5:101–21

101. Mendel, Z., Golan, Y., Mador, Z. 1984. Natural control of the eucalyptus borer, *Phoracantha semipunctata* (F.) (Coleoptera: Cerambycidae), by the Syrian woodpecker. *Bull. Entomol. Res.* 74:121–27

102. Miles, P. W., Aspinall, D., Correll, A. T. 1982. The performance of two chewing insects on water-stressed food plants in relation to changes in their chemical composition. *Aust. J. Zool.* 30:347–55

103. Moore, K. M. 1961. Observations on some Australian forest insects. 7. The significance of the *Glycaspis* spp. (Hemiptera: Homoptera, Psyllidae) associations with their *Eucalyptus* spp. hosts. Erection of a new subgenus and descriptions of thirty-eight new species of *Glycaspis. Proc. Linn. Soc. NSW* 86:128–67

104. Moore, K. M. 1963. Observations on some Australian forest insects. 11. Two species of Lepidopterous leaf-miners attacking *Eucalyptus pilularis* Smith. *Aust. Zool.* 13:46–53

105. Moore, K. M. 1966. Observations on some Australian forest insects. 22. Notes on some Australian leaf-miners. *Aust. Zool.* 13:303–49

106. Moore, K. M. 1988. Associations of some *Glycaspis* species (Homoptera: Spondyliaspididae) with their *Eucalyptus* species hosts. *Proc. Linn. Soc. NSW* 110:19–24

107. Morgan, F. D. 1984. *Psylloidae of South Australia.* Adelaide: South Aust. Govt. Printer. 136 pp.

108. Morgan, F. D., Cobbinah, J. R. 1977. Oviposition and establishment of *Uraba lugens* (Walker), the gum leaf skeletonizer. *Aust. For.* 40:44–55

109. Morrow, P. A. 1977. The significance of phytophagous insects in the *Eucalyptus* forests of Australia. In *The Role of Arthropods in Forest Ecosystems,* ed. W. J. Mattson, pp. 19–29. New York: Springer-Verlag

110. Morrow, P. A. 1977. Host specificity of insects in a community of three codominant *Eucalyptus* species. *Aust. J. Ecol.* 2:89–96

111. Morrow, P. A., Fox, L. R. 1980. Effects of variation of *Eucalyptus* essential oil yield on insect growth and grazing damage. *Oecologia (Berl.)* 45:209–19

112. Morrow, P. A., Bellas, T. E., Eisner, T. 1976. Eucalyptus oils in the defensive oral discharge of Australian sawfly larvae (Hymenoptera: Pergidae). *Oecologia Berlin* 24:193–206

113. Morrow, P. A., LaMarche, V. C. 1978. Tree ring evidence for chronic insect suppression of productivity in subalpine *Eucalyptus. Science* 201:1244–46

114. Nair, K. S. S., Mathew, G., Varma, R. V., Sudheendrakumar, V. V. 1986. Insect pests of eucalypts in India. See Ref. 143, pp. 325–33

115. Neumann, F. G., Harris, J. A., Wood, C. H. 1977. The phasmatid problem in mountain ash forests of the central highlands of Victoria. *For. Comm. Vict. Bull.* 25. 43 pp.

116. Neumann, F. G., Marks, G. C. 1976. A synopsis of important pests and diseases in Australian forests and forest nurseries. *Aust. For.* 39:83–102

117. Ohmart, C. P. 1984. Is insect defoliation in eucalypt forests greater than that in other temperate forests? *Aust. J. Ecol.* 9:413–18

118. Ohmart, C. P. 1989. Why are there so few tree-killing bark beetles associated with angiosperms? *Oikos* 54:242–45

119. Ohmart, C. P. 1990. Insect pests in intensively-managed eucalypt plantations in Australia: Some thoughts on this challenge to a new era in forest management. *Aust. For.* In press

120. Ohmart, C. P., Larsson, S. 1989. Evidence for absorption of eucalypt essential oils by *Paropsis atomaria* Olivier (Coleoptera: Chrysomelidae). *J. Aust. Entomol. Soc.* 28:201–6

121. Ohmart, C. P., Stewart, L. G., Thomas, J. R. 1983. Leaf consumption by insects in three *Eucalyptus* forest types in southeastern Australia and their role in short-term nutrient cycling. *Oecologia Berlin* 59:322–30

122. Ohmart, C. P., Stewart, L. G., Thomas, J. R. 1983. Phytophagous insect communities in the canopies of three *Eucalyptus* forest types in south-eastern Australia. *Aust. J. Ecol.* 8:395–403

123. Ohmart, C. P., Stewart, L. G., Thomas, J. R. 1985. Effects of food quality, particularly nitrogen concentrations, of *Eucalyptus blakelyi* foliage on the growth of *Paropsis atomaria* (Coleoptera: Chrysomelidae). *Oecologia Berlin* 65:543–49

124. Ohmart, C. P., Stewart, L. G., Thomas, J. R. 1985. Effects of nitrogen concentrations of *Eucalyptus blakelyi* foliage on the fecundity of *Paropsis atomaria* (Coleoptera: Chrysomelidae). *Oecologia Berlin* 68:41–44

125. Ohmart, C. P., Thomas, J. R. 1988. Surfactant-producing microorganisms isolated from the gut of a *Eucalyptus*-feeding sawfly, *Perga affinis affinis. Oecologia Berlin* 77:140–42

126. Ohmart, C. P., Thomas, J. R., Stewart,

L. G. 1987. Nitrogen, leaf toughness and the population dynamics of *Paropsis atomaria* Oliver (Coleoptera: Chrysomelidae)—A hypothesis. *J. Aust. Entomol. Soc.* 26:203–7

127. Old, K. M., Kile, G. A., Ohmart, C. P., eds. 1981. Eucalypt dieback in forests and woodlands. *Proc. Conf. Eucalypt Dieback.* Canberra: CSIRO Div. For. Res. 282 pp.

128. Parihar, D. R. 1981. Termites affecting *Eucalyptus* plantations and their control in the arid region of India. *Z. Angew. Entomol.* 92:106–11

129. Parliment of New South Wales. 1979. *Final Report to the Minister for Conservation and Water Resources on the New England Dieback.* New South Wales: Govt. Printer. 7 pp.

130. Penfold, A. R., Willis, J. L. 1961. *The Eucalypts.* London: Hill

131. Perry, D. H., Lenz, M., Watson, J. A. L. 1985. Relationships between fire, fungal rots and termite damage in Australian forest trees. *Aust. For.* 48:46–53

132. Podger, F. D. 1973. The causes of eucalypt crown dieback: A review. See Ref. 95, pp. 17–35

133. Podger, F. D. 1981. Definition and diagnosis of diebacks. See Ref. 127, pp. 1–8

134. Pook, E. W., Forrester, R. K. 1984. Factors influencing dieback of drought-affected dry sclerophyll forest tree species. *Aust. For. Res.* 14:201–17

135. Powell, W. 1982. Age-specific life-table data for the *Eucalyptus* boring beetle, *Phoracantha semipunctata* (F.) (Coleoptera: Cerambycidae), in Malawi. *Bull. Entomol. Res.* 72:645–53

136. Pryor, L. D. 1976. *The Biology of the Eucalypts.* Southampton: Camelot

137. Readshaw, J. L. 1965. A theory of phasmatid outbreak release. *Aust. J. Zool.* 13:475–90

138. Readshaw, J. L., Mazanec, Z. 1969. Use of growth rings to determine past phasmatid defoliations of alpine ash forests. *Aust. For.* 33:29–36

139. Roberts, F. H. S. 1932. The cattle-poisoning sawfly. *Q. Agric. J.* 37:41–52

140. Schmutzenhofer, H. 1985. Status of forest insects in pine and eucalypt plantations in the tropics. See Ref. 77, pp. 12–19

141. Schowalter, T. D., Hargrove, W. W., Crossley, D. A. Jr. 1986. Herbivory in forested ecosystems. *Annu. Rev. Entomol.* 31:177–96

142. Scriven, G. T., Reeves, E. L., Luck, R. F. 1986. Beetle from Australia threatens eucalyptus. *Calif. Agric.* July–Aug. 1986:4–6

143. Sharma, J. K., Nair, C. T. S., Kedharnath, S., Kondas, S., eds. 1986. *Eucalypts in India. Past, Present, and Future. Proc. Nat. Seminar on Eucalypts in Indian Forestry, Kerala For. Res. Inst., Peechi, Kerala, India. 30–31 Jan.* Kerala: Kerala For. Res. Inst. 514 pp.

144. Singh, S., Singh, P. 1975. *Eucalyptus* diseases and insect pests in developing countries. *Document, Sec. World Tech. Const. For. Dis. Inst., New Delhi, India, 7–12 Apr.*

145. Society of American Foresters. 1950. *Forestry Terminology.* Washington: Munns. 93 pp.

146. Southcott, R. V. 1978. Lepidopterism in the Australian region. *Rec. Adelaide Child. Hosp.* 2:67–73

147. Springett, B. P. 1978. On the ecological role of insects in Australian eucalypt forests. *Aust. J. Ecol.* 3:129–39

148. Sudeendrakumar, V. V., Chacko, K. C. 1986. Effects of site preparation on incidence of termites in *Eucalyptus* plantations. See Ref. 143, pp. 364–66

149. Tanton, M. T., Epila, J. S. O. 1984. Parasitization of larvae of *Paropsis atomaria* in the Australian Capital Territory. *Aust. J. Zool.* 32:251–59

150. Taylor, K. L. 1962. The Australian genera *Cardiaspina* Crawford and *Hyalinaspis* Taylor (Homoptera: Psyllidae). *Aust. J. Zool.* 10:307–48

151. Tindale, N. B. 1953. On a new species of *Aneteus* (Lepidoptera, family Hepialidae) damaging eucalyptus saplings Tasmania. *Trans. R. Soc. South Aust.* 76:77–79

152. Tooke, F. G. C. 1928. A borer pest of *Eucalyptus.* The destructive *Phoracantha* beetle and its control. *Farming in South Africa* No. 79

153. Tooke, F. G. C. 1953. The eucalyptus snout-beetle, *Gonipterus scutellatus* Gyll. A study of its ecology and control by biological means. *Union South Afr. Dept. Agr. Entomol. Mem. 3.* 283 pp.

154. Turnbull, J. W. 1981. Eucalypts in China. *Aust. For.* 44:222–34

155. Vaccaro, N. C. 1985. Insectos registrados en plantaciones de *Eucalyptus* spp. y *Pinus* spp. en la region do Concondia, Entre Rios, Republica Argentina. See Ref. 77, pp. 112–15

156. Wallace, M. M. H. 1970. The biology of the jarrah leaf miner, *Perthida glyphopa* Common (Lepidoptera: Incurvariidae). *Aust. J. Zool.* 18:91–104

157. Wardell, D. A. 1987. Control of termites in nurseries and young plantations in Africa: Established practices and alternative courses of action. *Commonw. For. Rev.* 66(1):77–89

158. Wattle Research Institute. 1972. *Handbook on Eucalypt Growing. Notes on the Management of Eucalypt Plantations Grown for Timber in the Wattle-Growing Regions of South Africa.* Pietermaritzburg: Wattle Res. Inst.

159. Whalley, R. D. B., Robinson, G. G., Taylor, J. A. 1978. General effects of management and grazing by domestic livestock on the rangelands of the northern tablelands of New South Wales. *Aust. Rangeland J.* 1:174–90

160. White, T. C. R. 1969. An index to measure weather-induced stress of trees associated with outbreaks of psyllids in Australia. *Ecology* 50:905–9

161. White, T. C. R. 1970. Some aspects of the life history, host selection, dispersal, and oviposition of adult *Cardiaspina densitexta* (Homoptera: Psyllidae). *Aust. J. Zool.* 18:105–117

162. White, T. C. R. 1986. Weather, Eucalyptus dieback in New England, and a general hypothesis of the cause of dieback. *Pac. Sci.* 40:58–78

163. Woinarski, J. C. Z., Cullen, J. M. 1984. Distribution of invertebrates on foliage in forests of south-eastern Australia. *Aust. J. Ecol.* 9:207–32

164. Wylie, F. R. 1983. *Native tree dieback in southern Queensland: Its occurrence, severity and aetiology.* PhD thesis. St. Lucia: University of Queensland

165. Wylie, F. R., Johnston, P. J. M. 1984. Rural tree dieback. *Q. Agric. J.* 110:3–7

166. Yasuda, T., Abe, K. 1986. *Endoclita hosei* Tindale (Lepidoptera: Hepialidae) attacking eucalypts in Sabah, with descriptions of the immature and imaginal stages. *Appl. Entomol. Zool.* 21:417–23

167. Zondag, R. 1979. A check-list of insects attacking eucalypts in New Zealand. *N. Z. J. For.* 24:85–89

Annu. Rev. Entomol. 1991. 36:659–81

OFF-HOST PHYSIOLOGICAL ECOLOGY OF IXODID TICKS[1]

Glen R. Needham

Acarology Laboratory, Department of Entomology, Colleges of Agriculture and Biological Sciences, The Ohio State University, Columbus, Ohio 43210-1292

Pete D. Teel

Department of Entomology, Texas A&M University, College Station, Texas 77843-2475

KEY WORDS: water-balance physiology, tick survival, water-vapor sorption, Ixodidae, zoogeographic interpretations

INTRODUCTION

Ixodid ticks (Acari: Ixodidae) have notable abilities to imbibe and to concentrate a large volume of vertebrate blood. Through their feeding, they may act as vectors of disease or debilitate by exsanguination or injection of salivary toxin. The brief on-host interval is characterized by rapid metabolism and development while eliminating a water and electrolyte load several times the body weight of the tick as it increases in body size. In contrast, survival between blood meals comprises >90% of the tick's life and is highly conservative physiologically. Between blood meals, nutrient reserves must be economically used and body-water content maintained or desiccation and ultimately death results. During off-host periods, ticks experience the rigors of environmental stresses associated with the climate of the immediate habitat. Susceptibility to injury from temperature extremes and to desiccation varies with tick species, stage, sex, age, and physiological condition. Body-

[1]We dedicate this review in the memory of deceased Professors of Acarology Manning A. Price and George W. Wharton.

659

0066-4170/91/0101-0659$02.00

water homeostasis is one of the most important processes that influences off-host survival. Researchers have made considerable progress in understanding water balance in the Acari (3, 45, 46, 66, 78, 95, 96), but many questions remain unanswered. Some difficulties also exist because investigators variously interpret observations of water flow, water potentials, and the influence of humidity and temperature (95).

A complete understanding of the impact of environmental stressors on tick water balance, survivorship, and ecological interpretations has not emerged, yet these factors are essential to certain modelling objectives and interpretations (13). Computers and modelling techniques are tools that have been used to examine complex zoogeographic interrelationships (hosts, vegetation, meteorology, and pathogens) of a few well-known species, including *Boophilus microplus* (88), *Amblyomma americanum* (29, 63), *Dermacentor variabilis* (64), and *Rhipicephalus appendiculatus* (57, 71a). The objectives and interpretations of these models are limited by the degree to which fundamental relationships such as development, population growth, and survival are understood and/or by the extent of available data bases. Prior reviews of water balance by ticks between bloodmeals (45, 46, 66, 79) suggested that integumental permeability to water flux and the amount of water in a tick may be of greater value in interpreting survival potential and habitat associations of ixodid ticks than critical equilibrium activity (CEA).

Here we address off-host physiological ecology in light of recent findings on active water-vapor uptake and comparative studies of three- and one-host tick survival as a function of whole-body permeability and water content (39, 41, 83, 85; M. D. Sigal & G. R. Needham; P. D. Teel & Needham; Teel, O. F. Strey, Needham, & M. T. Longnecker, in preparation). We begin with brief overviews on tick biology, water balance, and integument as a preamble to the presentation on water-balance parameters and their correlations with off-host survival. We attempt this assessment to facilitate the application of whole-organism water-balance perspectives to arthropod (tick) survival at the suggestion of Eric Edney, who 15 years ago published his monograph on terrestrial-arthropod water balance (16).

For information on other factors that influence survival, the reader is referred to reports on diapause and drop-off rhythms (7), cold hardiness (50), and behavior (8, 30, 52a, 54, 55). Much of the literature concerning water balance of arthropods (including ticks) is found in reviews by Arlian & Veselica (3), Edney (16), Knülle (45), Machin (59), O'Donnell & Machin (71) and Wharton (95).

TICKS AS GORGING-FASTING ORGANISMS

An alternate off-host, on-host life-cycle strategy requires ticks to deal with very different natural selection factors in the two situations. Wharton (94) refers to animals using this lifestyle as gorging-fasting organisms. Ticks as a group survive longer than any other arthropod without food or drinking water. For example, a typical univoltine ixodid species spends a total of 12–21 days on the host (larvae, 3–5 days; nymphs, 3–5 days, adult females, 6–11 days) for an off-host annual percentage of 94–97. An even higher off-host percentage is possible when adults live for more than one season. This impressive capacity is the product of many adaptive features that conserve energy and water so that life may be extended for months or years (38, 54, 97). When exposed to optimal abiotic conditions in the absence of a host, ticks typically outlive the animals they parasitize (e.g. rodents). Those concerned with animal and human health must be cognizant of this factor when surveying hosts and ticks for the presence of disease organisms or if contemplating a tick or host reduction program.

The transformation from a fasting state, adapted for off-host existence in which dramatic abiotic (extrinsic) fluctuations may occur, to a gorging state, in which biotic (intrinsic) and more stable abiotic factors are offered by the host, challenges the tick while it attaches, feeds, and mates (4, 33, 35). The gorging interval is characterized by rapid metabolism and development accompanied by the elimination of excess ions and water back into the host. This intimate association results in the exchange of body fluids between parasite and host, which facilitates infection of the tick by microorganisms or transmission of disease agents to the host (5, 42). Ticks transmit a greater variety of infectious agents than any other group of blood-feeding arthropods (34). Tick saliva maintains the feeding lesion by injecting anti-edema components (76), and the mouth parts of some ixodid ticks are secured in place by an attachment cement that also serves as a gasket (62). The soft integument of the body of immatures and adult females grows, unfolds, and stretches to accommodate the high-volume fluid diet (2, 17, 24, 25). Male metastriate ticks, although attached for extensive periods between matings, take in little host fluid because a nonexpandable sclerotized scutum covers most of the dorsum. Mating generally occurs on the host after some interval of feeding by both sexes, and after engorgement the fed immature or adult female detaches and falls to the ground where development is completed or a preoviposition period is followed by egg laying. Most ixodids follow this pattern except for one- and two-host species in which one or both immature stages (larva and nymph) remain on the same animal to continue feeding and development, thus limiting intrastadial exposure to the off-host environment.

Off-host fasting is characterized by slow metabolism with lengthy intervals of inactivity, punctuated by movement within the microhabitat to increase water uptake, or to seek a position for detection of a passing blood-meal source (8, 54, 55). Long off-host survival increases the chance of locating a suitable animal. If microenvironmental temperature and relative humidity remain within upper and lower thresholds (without considering numerous negative biotic factors), then survival becomes a function of maintaining the delicate balance between judicious energy use and maintenance of body water content (38, 54, 97). Over the short term, especially when environmental stress is high because of seasonal factors and/or if the habitat type is less than optimal, water balance is probably the most critical factor. At the risk of oversimplification, energy depletion that could result in the inability to maintain water balance may lead to death after long-term survival.

WATER-BALANCE PHYSIOLOGY

The water-balance physiology of all terrestrial arthropods serves as the foundation for this review. Maintenance of body water is critical to these animals that have large surface-to-volume ratios. All terrestrial arthropods have a superficial layer of lipid that minimizes water loss from the animal and enables it to survive an otherwise desiccating environment (26, 27). The impediment to water loss offered by this barrier can be illustrated simply by placing a droplet of water about the size of a tick on a nonabsorptive surface and observing how long it takes to evaporate. This remarkable barrier is one of the chief reasons this assemblage of animals dominates the terrestrial environment, yet the integument is one of the least studied organs.

Water conservation is the product of many mechanisms, including an internal respiratory system that is guarded from the external environment by a valve and sometimes other specialized structures, excretion of nitrogenous waste in a water-conserving form (uric acid or guanine), and reabsorption of water from fecal material by the rectum. Water is obtained from drinking, eating hydrated food, vapor absorption (from air), and through metabolic processes. Water turnover between the arthropod and the surrounding environment may be dramatic, or it may involve the exchange of but a few water molecules per hour; the critical net result is that water balance must be maintained or the animal will perish.

The quantity of water in an insect or acarine may be expressed in different ways (3, 95). It is best expressed as the ratio of water weight to total weight times 100, or water as a percentage of total weight, and is estimated by subtracting the dry weight from total weight. Hemolymph serves as the main water reserve in ticks (37). Sigal (85) found that adult male and female *A. americanum* (lone star ticks) can lose as much as 70% of the water mass before losing locomotor ability. This percentage represents one of the highest tolerance levels reported for any arthropod (16), and adult lone star ticks are not considered particularly hardy compared to adults of other ixodid species.

Exchange of body water with the environment has been described using a combination of units (e.g. activity, osmotic pressure in millimeters of mercury, osmolality, freezing point depression, or NaCl equivalents; see 3, 16).

Because water moves, it seems logical to quantify water using the same measurement both inside and outside the animal, that is, as the ratio of the solution (solvent) to the total number of ions and molecules present. This method has been adopted by Arlian & Veselica (3), Edney (16), and Wharton & Richards (96), and these authors give further rationale for using activity in their studies. The ratio corresponds to the activity a_w of the solution (ideal). Activity of liquid water is the same as the activity of water (a_v) in air in equilibrium with a plane surface of an aqueous solution at the same temperature. The concentration of water in air is generally given in units of percentage RH or relative humidity, which equals the a_v times 100. For ticks, the a_w (of hemolymph or cytosol) is around 0.99 and that of the surrounding atmospheric a_v is generally less. As water tends to move to sites of lesser activity, ticks generally find themselves in a desiccating circumstance and must rely on other sources of water to make up the deficit.

Transpiration

The rate at which water leaves (transpiration) is determined by the chemical potential of the body fluids (a_w), the total water pool mass, and the permeability of the waterproofing barrier. Diffusion and bulk flow thoroughly mix water pools such that each molecule has the same chance of escaping as every other. This is in fact the definition of a single well-mixed compartment, and includes any separate compartment that operates in parallel (95). Exchangeable water then is unbound and is available for participation in transpiration. Sources of water for transpiration (e.g. hemolymph, gut, etc) and the internal exchange of water are discussed elsewhere (37, 66). Transpiration rates can be expressed as a constant percentage loss ($\% \ h^{-1}$) of this exchangeable water mass. Causes for changes in the rate are discussed below. An instanteous rate equals the product of the rate constant (permeability) and water mass (e.g. mg h^{-1}). The driving force for water movement is the concentration h of water molecules in the compartment, and though the water pool is finite, the driving force remains virtually the same, at least while the organism is alive (> 0.98 a_w). A doubling of hemolymph osmotic pressure has little impact on the a_w. Transpiration is responsible for most water loss, unless the respiration rate is significant (80).

Wharton (95) describes the simplification of Crank's (12) equations for solving the problem of evaporation from a surface and justifies not using Fick's law for describing water movement. The conditions that make simplification possible are: the water pool is well mixed; transpiration rate is slow compared to the rate of mixing and the rate of diffusion away from the surface after a water molecule escapes; and the geometry of one individual is essentially the same as every other. This latter condition allows for the comparison of ticks as whole organisms without expressing flux rates as a function of surface area.

Of concern also is the permeability with respect to symmetry of the integument, that is, does water move out at a different rate than it moves in? Water passes through the same integumental layers during sorption and transpiration except that the order is reversed, and the integument is, for practical purposes, symmetric in its resistance to water movement. Differences in rate of movement can be accounted for primarily in the driving forces contributed by the a_v outside or a_w inside, although Wharton (95) explains that water flow into dry air is somewhat slower than its movement into moist air. His rationale for mentioning this difference is given in his review (95). The integument was thought to be the site of active vapor sorption, and this belief misled some to think that the integument was asymmetric. Specific sites account for all known active uptake mechanisms in arthropods (59, 71).

Sorption

Sorption is driven by the free energy of water vapor and is proportional to the a_v because the source of water is virtually infinite. Any study of sorption must include an analysis of contributions by metabolic water during the experiment; thus, an estimate of respiratory rate can avoid confusion. When sorption rate (passive and active) equals transpiration rate, the water mass is in a state of equilibrium. Periods when net water loss occur can be compensated for by active uptake (drinking water, vapor absorption) and doing so greatly extends the life of the organism from days to weeks or months. The critical equilibrium activity is the a_v at which equilibrium weight is maintained and is related to whole-animal water balance (47).

Water-Vapor Sorption

Water-vapor absorption (WVA) is an energy-dependent process used by arthropods to obtain water directly from unsaturated air (71). WVA provides a source of water separate from all other avenues, including intake of hydrated food (blood), drinking liquid water, or metabolism. This ability is particularly important for organisms that experience long periods of starvation, have slow metabolic rates, have limited abilities to disperse, and drink liquid water infrequently, if at all. Most ticks (immatures and adults) extract water vapor from unsaturated air (39, 44, 46, 48, 51; M. D. Sigal, J. H. Machin & G. R. Needham, in preparation). Uptake intervals are intermittant and desiccation thresholds that might initiate the process have not been determined. Adult *Amblyomma variegatum* weights fluctuate $\pm 2\%$ of their original weight when held at 0.93 a_v; no specific lower value initiates uptake (79). One measure of WVA ability is the gravimetric determination of a CEA. If a test population of standardized, fasting specimens maintains body weight in an a_v below 0.99, then a WVA capability is most likely present. CEA values are the net result of many factors, including transpiration, the "pump" threshold a_v for active uptake (see below), and loss of water via avenues other than the integument. Each specimen has its own equilibrium a_v and weight at that moment in time in this laboratory determination. Published CEA values range from 0.75 to 0.94 a_v (46). Variability in reported CEA values is not surprising when one factors in age, stage, inherited variability, prior exposure to extrinsic factors, and what host(s) was fed upon (46). For adult *A. americanum*, values range from 0.80 to > 0.88 a_v (31, 81; M. D. Sigal & G. R. Needham, in preparation). These are lower than estimates for adult *Amblyomma maculatum*, ranging from > 0.86 (M. D. Sigal & G. R. Needham, in preparation) to 0.88–0.90 (51) and 0.92–0.93 (31). The *Amblyomma cajennense* CEA range was 0.90–0.92 (51). Adult one-host ticks that would normally not experience off-host conditions cannot maintain consistent equilibrium weights below saturation and apparently lack WVA capability (46; M. D. Sigal & G. R. Needham, in preparation). CEA values are useful especially when selecting a humidity for maintaining a laboratory colony or for choosing experimental conditions, but using them as indicators of habitat preference can be misleading as we will discuss later.

Machin (59) and O'Donnell & Machin (71) have encouraged investigators to determine the pump threshold because it provides more information about WVA. They describe a graphic procedure for estimating the pump threshold from the gain and loss rates above and below the CEA. Use of the term *pump* means energy was required to transport water against a concentration gradient. Pump threshold denotes the true physiological point (a_v) when uptake commences. The threshold a_v may be near the CEA if the integument of the tick is highly waterproofed; however, if the tick is particularly permeable, then the threshold will be lower than the CEA. Theoretically, a tick with a high pump threshold (near saturation) and leaky cuticle might not appear capable of WVA. This possibility is examined below. Pump thresholds for several tick species have been estimated by Machin (59), and values ranged from 74–89% relative humidity for *A. americanum* adults according to data from Sauer & Hair (81) and Hair et al (31). Factors that characterize a pump site are absorption capacity and absorption conductance (11). These factors have not been evaluated for ticks.

Examination of WVA capabilities reveals the following for different stages and physiological conditions. The egg stage of ticks is not believed to be capable of active uptake, but this condition has not been adequately studied. The unfed stages of most ticks examined thus far have WVA capability, unless that stage spends its unfed period on the host, as do nymphs of two-host species or nymphs and adults of one-host species. Fed larvae and nymphs of three-host *Ixodes dammini, Ixodes ricinus,* and *Haemaphysalis punctata* as well as engorged larvae and first-instar nymphs of the multi-host

Argas reflexus (Argasidae) are capable of active uptake (39, 41). Active uptake can commence within 1–3 days after feeding. Investigators (39, 41) found that larval *Dermacentor marginatus* (three-host) and nymphal *Hyalomma anatolicum excavatum* (three-host) do not display the capability of water-vapor absorption. Active uptake masked by loss via the integument and/or tracheal system (nymphs) cannot be ruled out at this time (39). In engorged *D. marginatus* nymphs, however, active uptake was present but did not lead to substantial net gain at 95% RH/15°C (39). Molting ticks lose uptake capability during the early pharate phase, just after apolysis (separation of cuticle from hypodermal cells of the integument), but *I. dammini* and *I. ricinus* regain uptake competence within hours postecdysis (39, 41). Each stage (unfed and engorged) must be examined to determine its WVA capability. Some specimens of adult *Dermacentor albipictus* intermittantly absorbed water vapor based on weighings, but most individuals were WVA incompetent (M. D. Sigal & G. R. Needham, in preparation).

Investigations of salivary gland structure at the light microscope level (40) suggest that engorged *I. ricinus* females may be water-vapor absorption competent until near the end of oviposition. Previously, Kahl (39) observed temporary weight gain in partially engorged preovipositing females but not in fully engorged females. He speculates that water vapor absorption should be impossible during oviposition because the gnathosoma is covered by a viscous secretion beginning a few days before oviposition commences. Partially engorged females, once having entered this phase, never gained weight. Fully engorged females usually lose water in the preoviposition interval, and Kahl notes that active uptake does not occur; however, loss of water via respiration and/or transpiration could mask water vapor absorption. To sort out some of this contradictory information, mouth-blocking experiments have been performed. Engorged preovipositing *I. ricinus* females with wax-blocked mouthparts did not alter the rate of water loss (39). M. Fahmy, G. R. Needham, & M. D. Sigal (in preparation) obtained the same result with *A. americanum*. Salivary glands from these females are being examined ultrastructurally to confirm or refute the hypothesis that uptake occurs prior to or during oviposition (M. Fahmy, G. R. Needham, & L. B. Coons, in preparation). At least for these two species, WVA apparently does not occur in the preoviposition interval.

Rudolph's & Knülle's (46, 78, 79) pioneering work demonstrated that the mouth was the site of uptake in ixodid and argasid ticks. They suggested that the salivary glands were involved in uptake by generating a concentrated hygroscopic salt solution. We (66) also proposed that uptake via hydrophilic cuticle might be responsible, as in the desert burrowing cockroach *Arenivaga investigata* (70, 71), which has hydrophilic setae. We readdressed the uptake hypothesis because oral fluid from rehydrating ticks had not been examined, and the cheliceral-sheath denticles offer a potentially large surface area for sorption. Oral fluid in ticks either changes the hydrophilic nature of the mouth-part cuticle, as with *A. investigata,* or serves as a hygroscopic substance itself that would support the solute-driven mechanism proposed at the outset. Oral secretions from two rehydrating *A. americanum* females (as determined using a recording electrobalance) have recently been analyzed as frozen thick sections of the mouth parts (M. D. Sigal, J. H. Machin, & G. R. Needham, in preparation). Fluid in the lateral salivarium compartment, near where the hypostome and chelicerae converge to form the basis capituli, thawed at between -10 and $-12°C$ (~6.5–5.4 osmoles). Such a solution would be in equilibrium with a humidity near the CEA range for this tick ($\sim90\%$). This observation confirms earlier suggestions that ticks use a solute-driven vapor-uptake mechanism rather than hydrophilic cuticle.

The salivary glands are the apparent source of the hygroscopic fluid responsible for uptake (40, 46, 58, 79). Comparative ultrastructural (65, 66) and light-level (40) descriptions of the type I (type A, argasids) salivary gland acini are given elsewhere.

Concentrated salt on the mouth parts of severely desiccated ticks may serve as temporary storage in water homeostasis. The salt is probably secreted there by the salivary glands (66, 79). As ticks desiccate and body fluids become concentrated, the salivary gland type I acini probably excrete the salt onto the mouth parts until rehydration is possible (40, 65). These salts delequesce at subsaturated humidities near the CEA for the tick (79), and may initiate active vapor uptake. M. D. Sigal, J. H. Machin, & G. R. Needham (in preparation) stimulated unfed adult *A. americanum* to produce concentrated oral solutions (0.43 to 5.11 osmoles, $n = 8$) within minutes of exposure to incandescent light and to heat. The residue dried immediately at a relatively low humidity.

Behavior-Associated Hydration State and Vapor Absorption

Lees (54) and Camin (8) presented some of the most stimulating thoughts on behavior associated with hydration state. Both recognized the balance between hydration state and host seeking as determinants for where a tick was positioned within the vegetation. It is beyond the scope of this review to discuss off-host behavior, but the reader is referred to reports on *Ixodes ricinus* (54, 56), *I. persulcatus* (16a), *Rhipicephalus appendiculatus* (72–74, 84), *A. maculatum* (18, 19), *Dermacentor occidentalis* (49), and *D. variabilis* (31a).

IXODID INTEGUMENT

Physical Properties of the Cuticle

Ticks have a large surface-to-volume ratio and are prone to desiccation. One of the most important factors that has enabled arthropods in general to be the dominant terrestrial animal form is the ability of the integument to restrict water loss. This factor was essential for the transition of arthropods from an aquatic to a terrestrial existence (27). Lipids associated with the epicuticle are thought to form the primary barrier to water loss by most terrestrial arthropods and plants (26). Other specialized integumental features restrict water loss—the spiracles of ixodid nymphs and adults (larvae have no spiracles) are covered by a spiracular plate and ventilation is regulated by a closing mechanism (80). The oral cavity is formed dorsally by closely opposed paired chelicerae, ventrally by the hypostome, and laterally by palpi that for many species partially enclose the dorso-ventral structures. This arrangement probably limits water loss from the mouth unless the palps are splayed. The anus is secured by valves. Hygroreceptors (setae) enable the tick to sense humidity gradients (32). Only a few investigators beginning with Lees (52) have studied tick integument. It is quite similar to insect integument, except the cement layer is apparently absent from the epicuticle of ixodids, making the unprotected lipid layer particularly susceptible to damage (2, 17, 24, 25).

Whole-Body Permeability

Integumental permeability, as shown by Lees (51), determines tick water-loss rates. Ticks held below the CEA die from desiccation very quickly if they are particularly permeable, while those that lose water slowly live quite some time by comparison (Table 1). Water-loss rates for ticks measured at 0 a_v are actually whole-body permeability estimations that include water loss via the respiratory system, from oral and anal excretions, and any other physiological process that involves the loss of water (66, 95; M. D. Sigal & G. R. Needham, in preparation). Freda & Needham (20) suggest that these additional avenues contribute proportionally much less than does integumentary loss, although an increase in locomotor activity can dramatically increase respiratory water loss (79).

An understanding of certain principles is critical for studies that estimate the permeability of an organism. First, gravimetric values of permeability must be determined at near 0 a_v. This determination eliminates any contribution of passive water to the water pool—otherwise permeability will be underestimated. That is not to say that experiments done in the presence of atmospheric water are of no value. Hair et al (31) report that *A. americanum* and *A. maculatum* at 32% RH showed a difference in permeability, and they suggest that this played a key role in habitats each species could occupy. Recent studies demonstrated that *A. americanum* water-loss rates from the whole-body water pool increased with age (M. D. Sigal & G. R. Needham, in preparation), which supports previous observations by Lees (51, 52) that ixodid CEAs increased with age. The greater rates of water loss probably raise water balance equilibrium thresholds in older ticks, resulting in CEAs at correspondingly higher relative a_vs.

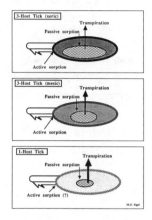

Figure 1 Whole-body permeability (passive sorption and transpiration) and active-water vapor uptake capability compared for three hypothetical adult ixodid tick species. Size of arrows indicates relative rate (large ~ fast, small ~ slow) of water movement. Limiting barrier represents integument: top figure is a tick that loses water slowly; the bottom figure is a leaky tick; and the middle panel represents a tick of intermediate integumental permeability. Internal eliptical shapes represent internal pools of different sizes from which water may be lost to the atmosphere. The top panel represents a tick that would survive in quite dry microhabitats and would move only occasionally to seek a higher humidity for rehydration; while the bottom panel represents a tick that is leaky, has a small water pool, may not take water from the air actively (via mouth), and would survive only a few days at most in a dry microhabitat. The middle panel diagrams a tick that may have to move daily between microhabitats for rehydration and those where a host may be accessed. Figure provided by M. D. Sigal.

A fundamental question concerning the water-balance physiology of arthropods including ticks is the amount of control they have over responding to changing ambient humidity conditions by altering integumental permeability (60, 61, 67–69). Integumental permeability control in ticks has not been adequately addressed. Short-term exposure (~ 2 weeks) to low humidity did not appear to influence transpiration rates (20). Ticks respond to superficial abrasion by repairing damage (28, 52). As already mentioned, permeability increases with age as indicated by whole-body permeability measurements, but the reason for this increase is unknown. Deposition and removal of lipids also occur during engorgement and molting (14, 15, 28). Evidence for the transition in cuticle permeability from an engorging on-host tick to an engorged off-host physiological condition emerges from several different studies (9, 81). The amount of extractable cuticular lipid per engorged female *Boophilus microplus* increased threefold over some nine days postdetachment (9), suggesting that the off-host individual is more protected from desiccation than one on the host. Engorged *A. americanum* females also seem more resistant to desiccation than unfed ticks (81). Alloscutal integument grows

Table 1 Comparison of water-balance and survival parameters which characterize adults of five ixodid species[a]

Three-host species	Whole-body[b] permeability (% h⁻¹)	Water mass newly emerged (mg)	Water mass 4-week old (mg)	50% Mortality[c] (days)	90% Mortality[c] (days)
Amblyomma americanum					
female	0.2827±.0340	4.30±0.33	2.58±0.15	41.3±1.3	49.2±1.8
(n)	(10)	(16)	(25)	(120)	(120)
male	0.3156±.0340	2.35±0.15	1.34±0.05	35.8±0.9	46.3±2.1
(n)	(10)	(14)	(24)	(120)	(120)
Amblyomma maculatum					
female	0.1067±.0003	6.89±0.54	2.963±0.275	37.7±2.0	63.7±2.2
(n)	(10)	(10)	(10)	(120)	(120)
male	0.1064±.0004	3.70±0.30	1.936±0.147	58.8±4.9	81.3±7.1
(n)	(10)	(10)	(10)	(120)	(120)
Amblyomma cajennense					
female	0.0638±.0044	Not Determined	5.11±0.22	269.5±48.8	Not reached as day 457
(n)	(20)		(20)	(120)	
male	0.0734±.0044		3.80±0.37	221.0±47.2	361.5
(n)	(20)		(20)	(100)[e]	(40)[e]
One-host species					
Dermacentor albipictus					
female	0.2449±.0021	1.68±0.08	Not Determined	23.4±0.7	35.6±0.6
(n)	(20)	(50)		(100)	(100)
male	0.3230±.0028	1.19±0.08		19.4±2.1	31.5±2.7
(n)	(20)	(49)		(150)	(150)
Boophilus annulatus					
female	1.5415±.0141	2.20±0.07	Not Determined	5.5[d]	>13.0
(n)	(22)	(22)		(25)	(25)
male	0.6908±.0066	1.01±0.03			
(n)	(20)	(20)			

[a] Values are means ± SE, parenthetic values are sample sizes (*n*); all experiments were conducted at 22–23°C; 14 : 10 h (L:D); ticks were ~4 weeks old when the experiment was begun, except for one-host whole-body permeability and survival studies done on newly emerged ticks (most would not live to 1 month). Data are from manuscripts in preparation by M. D. Sigal & G. R. Needham; P. D. Teel & Needham; Teel, O. F. Strey, Needham, & M. T. Longnecker.

[b] Water-loss rates determined in desiccator containing Drierite®; estimates water lost via all avenues. Values for species except *A. cajennense* were derived from regression analyses. *A. cajennense* values were based on arithmetic calculations from daily weighings. Loss rates represent constant percent loss of water available (exchangeable water mass). Permeability and water mass values for all species except *A. cajennense* are from Ref. 85.

[c] Survival experiments were performed at 0.75 a_v (= 75% relative humidity), which is below the critical equilibrium activities, so active water-vapor uptake was not likely. One-host ticks were newly emerged.

[d] Combined male and female survival of newly emerged ticks. Time to reach a nonambulatory or critical water mass for newly emerged cattle ticks was 65.5 ± 4.4 h (*n* = 20) for males and 27.1 ± 1.9 h (*n* = 22) for females (in desiccator containing Drierite®, 22–23°C) (85).

[e] Mortality level was not reached by all 6 replicates of 20 ticks each on day 457 of the study.

during feeding, a fact discovered by A. D. Lees in 1952 (53); therefore, unfolding of the cuticle to accommodate the blood meal could offer a greater surface area for transpiration. Eliminating excess water from the blood meal by transpiration could be an energy-saving avenue compared to exclusive secretion of fluid via the salivary glands. Thus, integumental permeability seems to change dramatically when the tick makes transitions from being unfed (off-host) to feeding (on-host) to being engorged (off-host). The regulation of these changes is unstudied in ticks.

COMPARISON OF WATER-BALANCE PARAMETERS FOR THREE- AND ONE-HOST IXODIDS

Standardized parameters for survival and water balance have been developed for adults of five ixodid species to compare suitability to off-host survival between three- and one-host adapted species (M. D. Sigal & G. R. Needham; P. D. Teel & Needham; Teel, O. F. Strey, Needham & M. T. Longnecker, in preparation). The three-host ticks, *Amblyomma americanum*, *A. maculatum* and *A. cajennense,* were selected for study based on their different hydrophilic and vegetation associations within overlapping geographic distributions, and on an inverse correlation of estimated CEA values with these ecological relationships. The predominate habitat association for *A. americanum* has been defined as woodlands associated with oak-hickory, post-oak, blackjack oak, and persimmon-sassafras-winged elm communities (30). In comparison, *A. maculatum* is predominately associated with more xeric habitats of grass and scrub communities on upland prairies in Oklahoma (83) and coastal prairies and mesquite-acacia–dominated thorn shrublands in Texas (19, 91). *A. cajennense* is similarly associated with the coastal prairie and thorn shrublands of southern Texas. Southern Texas is the northern limit of distribution for *A. cajennense* in the western hemisphere (6, 93). The distributions of all three species overlap in southern Texas, where *A. maculatum* and *A. cajennense* are generally more abundant. In central and eastern Texas, *A. americanum* is most abundant.

The one-host species *Boophilus annulatus* and *Dermacentor albipictus* are collected from medium and large herbivores in southern Texas. While feeding, ticks are firmly attached; however, pharate nymphs of one-host species are easily dislodged from attachment sites. The ability of subsequent unfed adults to survive in off-host environments for a time sufficient to encounter other hosts is of epidemiological significance. Ransom (75) observed that newly molted *B. annulatus* adults survive up to two weeks under undescribed conditions and still attach to a second host. Rudolph & Knülle (80) noted 14 days as the maximum longevity of unfed *B. annulatus* adults and nymphs at 98% RH and 25°C. Small numbers of nymphal and adult *D. albipictus* have been collected by cloth drag (36) and by CO_2 tick traps (21, 83), indicating a degree of successful off-host survival.

Comparisons of estimates for water mass, whole-body permeability, and survival at 0.75 a_v and 23°C (Table 1) among the one- and three-host ticks reflect adaptations to lifestyle and hygrophilic association described above. Note that under these experimental conditions, movement to optimal sites to prevent desiccation or to rehydrate were precluded, and the 0.75 a_v (for survival experiments) should be below the CEA for these ticks. Thus, WVA was not likely and passive sorption was the major contributor (along with metabolic water) to the water mass. Total water mass and whole-body permeability are estimates of how much water the body contains and how rapidly the body gives up water at 0 a_v, respectively. The two one-host species had smaller total water masses compared to the three-host species. Whole-body permeabilities were ranked as follows (Table 1) from least to most leaky: *A. cajennense* females $\leq$ *A. cajennense* males $<$ *A. maculatum* males $=$ *A. maculatum* females $<$ *D. albipictus* females $=$ *A. americanum* females $<$ *A. americanum* males $=$ *D. albipictus* males $<$ *B. annulatus* males $<$ *B. annulatus* females. Both whole-body permeability and water mass were found to be closely associated with the mortality profiles developed under constant environmental conditions of 0.33, 0.75, and 0.85 a_v at 23 and 33°C. The largest species, *A. cajennense,* had the least whole-body permeability and as a consequence lived the longest (269.5 $\pm$ 48.8 days for females, 50% mortality). The small *B. annulatus* lived the shortest time (5.5 days for adults, 50% mortality) and had the greatest whole-body permeability.

In general, the larger ticks lost water at a slower rate than smaller ones, with several exceptions. Compare the males and females for *B. annulatus* and *A. maculatum* (Table 1) in which this trend was not observed. Edney (16) showed that smaller organisms of the same shape have greater surface area–to–volume ratios than larger organisms. If size is a dominant factor in determining passive exchange rates across the integument, one would expect proportionately greater loss rates in the smaller organisms. For Psocoptera, disadvantageous surface area–to–volume ratios in smaller species are compensated for in some way, probably by reduced integumental conductances (71, 77). This is the case for male *B. annulatus* ($\sim$0.69% h^{-1}), which is much smaller than the female ($\sim$1.54% h^{-1}) but was much less permeable. Placing the ticks at 0 a_v showed that this permeability difference resulted in a significantly longer time for males (65 $\pm$ 4 h) to lose the ability to walk than for females (27 $\pm$ 2 h). Male cattle ticks remain on the host for weeks feeding and mating intermittantly, and possession of a less-leaky integument would protect them from desiccation as they move about the surface of a host. If they are groomed off, this protection (compared to females) would allow some time to find another host.

Hoogstraal & Kim (35) concluded that a reduction in tick size was very important to the success of some ixodids, determined in some cases by the relatively small size of the preferred host group and in other cases, such as in *Boophilus*, *Margaropus*, and *Haemaphysalis*, by "dominating evolutionary trend regardless of host size" (35). These researchers (33, 35) postulated that the one-host ixodid life style evolved in response to the wandering behavior of their restricted mammalian hosts or to accommodate seasonal host availability. With the exception of *Hyalomma scupense*, which may act either as a one- or two-host tick, all the described one-host species are in the Rhipicephalinae. The price of size reduction would be a larger surface-to-volume ratio and smaller total water pool. If this reduction was accompanied by increased whole-body permeability to water of stages that tended to remain on the host and an inability to absorb water from subsaturated air, the tick species would have progressively less flexibility in surviving off-host environments. *B. annulatus* and *D. albipictus* are relatively small ticks that apparently cannot absorb water from subsaturated air; however, they differ in whole-body permeability.

One-host species were short-lived (Table 1), but *D. albipictus* exhibited a rather remarkable ability to survive at 0.75 a_v, 23°C, which was more than 3–5 times the longevity of *B. annulatus* at 50% mortality and more than twice the longevity at 90% mortality. Although *D. albipictus* has a one-host lifestyle, adults retain the ability to survive off the host much more so than *B. annulatus*. These survival studies may explain winter-spring interhost transmission of *Anaplasma marginale* to cattle in areas such as central Texas when other potential vectors are not active (1). Some adult *D. albipictus* (10%) lived more than 30 days (P. D. Teel & G. R. Needham, in preparation), most likely because of the protection of a less-leaky integument. Neither *B. annulatus* nor most *D. albipictus* adults appear able to absorb water from subsaturated air. Some individual *D. albipictus* adults gained weight at subsaturated a_v, indicating that a vapor uptake capability may exist (85). CEA experiments for these one-host adults show that as a whole, subpopulations could not maintain equilibrium at 0.98 a_v. This was also the finding for *B. annulatus* by Rudolph & Knülle (80).

Comparisons of water balance and mortality characteristics among the adult three-host species show that survival under these experimental conditions was a function of water mass and permeability to water (Table 1). *Amblyomma americanum* was the smallest of the three species and correspondingly had the smallest total water mass. Whole-body permeability to water was the greatest in *A. americanum*. Thus, in a constant 0.75 a_v environment, this species was comparatively short-lived. In contrast, the considerably larger water masses and substantially lower whole-body permeabilities of *A. maculatum* and *A. cajennense* provided for prolonged survival. The survival of *A. maculatum* males was longer than that of females (under conditions for data in Table 1) when whole-body permeabilities were essentially equal, despite the smaller water masses of the males. In addition, survival for cohorts of this species in drier environments of 0.75 a_v, 33°C, and 0.33 a_v, 23°C, show equal to slightly greater longevity for males compared to females. While this difference is as yet unexplained, it has ecological significance because male *A. maculatum* become active by as much as 2.5 months earlier than females (18).

A comparison of estimates between *D. albipictus* and *A. americanum* is particularly interesting. Whole-body permeabilities were nearly equal for the same sex; females were slightly less leaky than males. The mean total water masses of these ticks when exposed to the survival test conditions of 0.75 a_v, 23°C, were larger for females than males in both species, but largest for *A. americanum*. Between species, *A. americanum* lived longer, and between sexes the females were longer lived. The species difference in mean longevity at 50% mortality (Table 1) was 17.9 and 16.4 days for females and males, respectively. When these species are placed in an environment of 0.33 a_v, 23°C, the higher desiccating potential results in shorter longevity (23–26 days at 90% mortality) for both species and a separation of mortality profiles by 2 days or less. Assuming that the exchangeable water pool is correlated with the total water mass, relative longevity under these conditions can be explained in terms of whole-body permeability and the exchangeable water mass.

Survival studies were conducted on newly emerged individuals (study began on day of emergence) of one-host species and the same treatment protocol was performed on four-week-old, three-host ticks because of the relative life expectancies and postmolt preparedness of each species. Estimates of water permeability show that newly molted adult ixodid integument is as impermeable to water as it will ever be (M. D. Sigal & G. R. Needham, in preparation) and suggests that postmolt tanning of the cuticle is not accompanied by greater protection against water loss. Adjustments in weight and water content were made in the molting and postmolting periods through defecation; however, the strategy for timing of defecation appears to vary between tick species. Adult *B. annulatus* were observed to defecate in the nymphal exuvium before ecdysis, whereas *D. albipictus* began defecating after ecdysis. All three species of three-host adults retained excretory products for days after ecdysis before voiding substantial quantities of moisture-laden fecal material (P. D. Teel, personal observation). Under field conditions, *A. maculatum* adults (18) have been observed to undergo postmolt quiescent periods lasting days to weeks. The difference in mean water masses for four-week-old *A. americanum* (Table 1) was 59 and 57% of the mean water mass of newly emerged female and male ticks, respectively. This is not to suggest that there has been equivalent change in the exchangeable water mass between newly molted and four-week old ticks. The resulting postmolt adjustment in water mass for *A. americanum* by week four puts this species on near-equal status with newly molted *D. albipictus*.

One criterion for selecting the three-host species for study was the inverse correlation between published CEA values and the hygrophilic associations of these species. New CEA estimations were made for the cohort of each of the three-host species included in our studies using the protocol outlined by Knülle & Wharton (47). Results were a_vs greater than or equal to 0.88 for *A. americanum* males and females; greater than or equal to 0.85 and 0.89 for male and female *A. maculatum*, respectively; 0.80–0.82 and 0.82–0.85 for male and female *A. cajennense*, respectively. Note that these values are quite different from those previously cited (46). An interpretation of these new values suggests that *A. cajennense* are more tolerant of xeric environments than both *A. americanum* and *A. maculatum*. CEA values have helped determine that certain species can absorb water from subsaturated air, and the relative CEA values have been used to provide generalized comparisons of hygrophilic associations. The variation of CEA values within and among tick species and the lack of a completely satisfactory association of the CEAs with environments for off-host adapted ticks has been discussed (46, 66).

Beyond the loss-rate and water-mass statistics, another consideration when studying survival is the apparent sensitivity of a tick to desiccation. Some ticks tolerate the loss of a greater percentage of body water and are seemingly more able to recover from desiccation resulting in an immobile or nonambulatory state. M. D. Sigal & G. R. Needham (in preparation) found *A. americanum* adults could not recover after desiccation at 0 a_v and subsequent return to 0.98 a_v. *Amblyomma cajennense* was far more resilient when sensitivity to desiccation was tested (P. D. Teel, O. F. Strey, G. R. Needham, & M. T. Longnecker, in preparation). Ticks were alternated between dehydration at 0 a_v until the nonambulatory status was reached and 0.96 a_v for 24 h. Females survived a maximum of 12 cycles over a total period of 61 days with a mean of 4.4 (SE ± 1.28) cycles per tick. Males tolerated a maximum of 11 cycles over a total period of 68 days with a mean of 4.0 (SE ± 0.87) cycles per tick. Female and male *A. cajennense* took an average of 21.6 and 19.8 days to reach the initial nonambulatory point, respectively. Therefore, the internal tolerance of body organ systems to water loss and concentration of solutes may vary greatly, and one cannot rely just on the rate loss and water mass to fully characterize water-balance physiology.

A question arises as to why the distribution of *A. cajennense* is limited in the US, assuming all stages are similarly as resistant to desiccation as adults. The southern Texas area in which *A. cajennense* is found is subtropical to semiarid and the lower annual rainfall limits are near 40.6 cm. *Amblyomma cajennense* adults are more resistant to desiccation than *A. maculatum*, yet populations of the latter species are found at more northern latitudes and at near similar rainfall limits. The distribution of *A. cajennense* in the Western Hemisphere extends through Mexico, Central, and South America east of the Andes Mountains (6, 93). The limits of this distribution approach 30° latitude north and south, suggesting that an inability to tolerate cold is a limiting factor in off-host survival of this species. Also, temperatures below a developmental threshold could limit its growth such that it is out of sequence with critical biotic and abiotic conditions. Sonenshine (86) concluded that the northern limit of established populations of *Dermacentor variabilis* was associated with the duration of temperature below 0°C, while extensions of populations west of the Eastern deciduous forest biotope were limited by susceptibility of all life stages to desiccation. Outside this biotope, foci of sustaining populations of *D. variabilis* are believed successful in part because of localized environmental conditions within the tolerance of this species.

SUMMARY

As gorging-fasting organisms, ticks are really two very different animals: one is adapted for on-host existence as a blood feeder, the other for conservative off-host existence with life-extending strategies to balance water and energy resources and to increase the chance of obtaining a bloodmeal. Whole-body permeability, water mass, capability of water-vapor uptake, and survivorship show strong associations with life-style and off-host adaptations among adult one- and three-host ticks. This whole-organism perspective provides a better assessment of hydrophilic associations among off-host–adapted three-host adults than CEA-based interpretations. Furthermore, this perspective supports the concept that the evolution toward a one-host lifestyle resulted in reduced capability to survive off-host environments. Examination of these associations for each developmental stage, including diapausing and nondiapausing, engorged and unengorged forms, would provide considerably deeper insight into zoogeographic and epidemiologic interpretations. Stages that are comparatively more susceptible to desiccation may be more sensitive indicators of geographic and hydrophilic associations (46, 86).

We have not examined in detail the influence of behavior on survival, but ticks will clearly survive longer if given the opportunity to seek out more optimal abiotic conditions. Differences in daily and seasonal behavior influence physiological aging and the balance of energy and water resources. This subject is in need of much greater study and the development of standardized techniques and analytical procedures.

Modelling provides an avenue to integrate knowledge of ticks as whole organisms, based upon physiological processes, with field ecology (16), and to develop decision-making tools for tick and tick-borne disease management (87). While growth-rate models rely on environmental temperatures, models of survivorship depend on our understanding of environmental stresses associated with temperature and moisture (22, 23, 43, 92). Saturation deficiency as a stressor in arthropod survival has been long recognized (16) and is directly applicable to off-host tick survivorship because it provides, within limits, a means of estimating the sink for transpiration from the exchangeable water pool.

Computer-based geographic information systems (GIS) developed for management of landscape resources (10) have provided data management and analytical tools necessary to assess tick-related zoogeographic and epidemiological issues (43, 57, 89, 90). Results provide decision-making information for policy development at governmental levels to management-strategy development at the landowner level. The foundation of each of these GISs is one or more models describing tick development and survival in terms of meteorological or climatic stressors and/or estimates of the suitability of existing or perturbed vegetation habitat types for tick survival. Development of these systems and interpretations of their results emphasize that a thorough understanding of off-host tick survival is essential at both species and population levels.

ACKNOWLEDGEMENTS

The authors thank Drs. Marvin D. Sigal and Olaf Kahl in particular for sharing unpublished information from their doctoral work and for comments on the manuscript. Our appreciation is extended to Dr. Willi Knülle, Dr. Donald E. Johnston, Kathleen Curran, and Deborah Jaworski for their comments on the manuscript, and to Otto F. Strey for his assistance with the survival study.

Literature Cited

1. Alderink, F. J., Dietrich, R. A. 1983. Economic and epidemiological implications of anaplasmosis in Texas beef cattle herds. *Texas Agricultural Experiment Station Bulletin, B-1426.* College Station, TX. 16 pp.
2. Amisova, L. I. 1983. The integument. In *An Atlas of Ixodid Tick Ultrastructure,* ed. A. Raikhel, H. Hoogstraal, 2:23–58. Washington DC: Entomol. Soc. Am. Special Publication. 289 pp.
3. Arlian, L. G., Veselica, M. M. 1979. Water balance in insects and mites. *Comp. Biochem. Physiol.* 64A:191–200
4. Balashov, Y. S. 1972. Bloodsucking ticks (Ixodoidea)—Vectors of diseases of man and animals. *Misc. Publ. Entomol. Soc. Am.* 8:161–376
5. Balashov, Y. S. 1984. Interaction between blood-sucking arthropods and their hosts, and its influence on vector potential. *Annu. Rev. Entomol.* 29:137–56
6. Barré, N., Uilenberg, G., Morel, P. C., Camus, E. 1987. Danger of introducing heartwater onto the *American mainland:* Potential role of indigenous and exotic *Amblyomma* ticks. *Onderstepoort J. Vet. Res.* 54:405–17
7. Belozerov, V. N. 1982. Diapause and biological rhythms in ticks. See Ref. 69a, pp. 469–500
8. Camin, J. H. 1963. Relations between host-finding behavior and life histories in ectoparasitic acarina. In *Advances in Acarology,* ed. J. H. Naegele, 1:411–24. New York: Cornell Univ. Press. 480 pp.
9. Cherry, L. M. 1969. The production of

cuticle by engorged females of the cattle tick, *Boophilus microplus* (Canestrini). *J. Exp. Biol.* 50:705–9

10. Coulson, R. N., Lovelady, C. N., Flamm, R. O., Sprandling, S. L., Saunders, M. C. 1990. Intelligent geographic information system for natural resource management. In *Quantitative Methods in Landscape Ecology*, ed. M. G. Turner, R. H. Gardner. Berlin: Springer-Verlag. In press

11. Coutchie, P. A., Machin, J. 1984. Allometry of water vapor absorption in two species of tenebrionid beetle larvae. *Am. J. Physiol.* 247:R230–36

12. Crank, J. 1975. *The Mathematics of Diffusion.* Oxford: Claredon Press. 2nd ed.

13. Curry, G. L., Feldman, R. M. 1987. *Mathematical Foundations of Population Dynamics. Texas Engineering Experiment Station Monograph Series No. Three*, ed. A. R. McFarland. College Station, TX: Texas A&M Univ. Press. 246 pp.

14. Davis, M.-T. B. 1974. Changes in critical temperature during nymphal and adult development in the rabbit tick, *Haemaphysalis leporispalustris* (Acari: Ixodides: Ixodidae). *J. Exp. Biol.* 60:85–94

15. Davis, M.-T. B. 1974. Critical temperature and changes in cuticular lipids in the rabbit tick, *Haemaphysalis leporispalustris*. *J. Insect Physiol.* 20:1087–1100

16. Edney, E. B. 1977. Water balance in land arthropods. In *Zoophysiology and Ecology*, Vol. 9, ed. D. S. Farner. New York: Springer-Verlag. 282 pp.

16a. Filippova, N. A. 1985. *Taiga Tick Ixodes persulcatus Schulze (Acarina, Ixodidae): Morphology, Systematics, Ecology, and Medical Importance.* Nauka. 416 pp.

17. Filshe, B. K. 1976. The structure and deposition of the epicuticle of the adult female cattle tick (*Boophilus microplus*). In *The Insect Integument*, ed. H. R. Hepburn, pp. 193–206. Amsterdam: Elsevier. 571 pp.

18. Fleetwood, S. C. 1985. *The environmental influences in selected vegetation microhabitats on the various life stages of* Amblyomma maculatum *Koch (Acari: Ixodidae).* PhD Dissertation. College Station, TX: Texas A&M Univ. 194 pp.

19. Fleetwood, S. C., Teel, P. D. 1983. Variation in activity of aging *Amblyomma maculatum* Koch (Acarina: Ixodidae) larvae in relation to vapor pressure deficits in pasture vegetation complexes. *Prot. Ecol.* 5:343–52

20. Freda, T. J., Needham, G. R. 1984.

Water exchange kinetics of the lone star tick *Amblyomma americanum*. In *Acarology VI*, ed. D. A. Griffiths and C. E. Bowman. 1:358–64. Chichester: Ellis Horwood Ltd. 645 pp.

21. Garcia, R. 1969. Reaction of the winter tick, *Dermacentor albipictus* (Packard) to CO_2. *J. Med. Entomol.* 6:286

22. Gardiner, W. P., Gettinby, G. 1983. A weather-based prediction model for the life-cycle of the sheep tick, *Ixodes ricinus* L. *Vet. Parasitol.* 13:77–84

23. Gardiner, W. P., Gray, J. S. 1986. A computer simulation of the effects of specific environmental factors on the development of the sheep tick, *Ixodes ricinus* L. *Vet. Parasitol.* 19:133–44

24. Hackman, R. H. 1982. Structure and function in the tick cuticle. *Annu. Rev. Entomol.* 27:75–95

25. Hackman, R. H., Filshe, B. K. 1982. The tick cuticle. See Ref. 69a, pp. 1–42

26. Hadley, N. F. 1981. Cuticular lipids of terrestrial plants and arthropods: a comparison of their structure, composition, and waterproofing function. *Biol. Rev.* 56:23–47

27. Hadley, N. F. 1984. Cuticle: ecological significance. In *Biology of the Integument*, Vol. 1, *Invertebrates*, ed. J. Bereiter-Hahn, A. G. Matoltsy, K. S. Richard, pp. 686–702. New York: Springer-Verlag

28. Hafez, M., El-Ziady, S., Hefnawy, T. 1970. Biochemical and physiological studies of certain ticks (Ixodoidea). Cuticular permeability of *Hyalomma (H.) dromedarii* Koch (Ixodidae) and *Ornithodoros (O.) savignyi* (Audouin) (Argasidae). *J. Parasitol.* 56:154–68

29. Haile, D. G., Mount, G. A. 1987. Computer simulation of population dynamics of the lone star tick, *Amblyomma americanum* (Acari: Ixodidae). *J. Med. Entomol.* 24:356–69

30. Hair, J. A., Bowman, J. L. 1986. Behavioral ecology of *Amblyomma americanum* (L.). See Ref. 81a, pp. 406–27

31. Hair, J. A., Sauer, J. R., Durham, K. A. 1975. Water balance and humidity preference in three species of ticks. *J. Med. Entomol.* 12:37–47

31a. Harlan, H. J., Foster, W. A. 1990. Micrometeorologic factors affecting field host-seeking activity of adult *Dermacentor variabilis* (Acari:Ixodidae). *J. Med. Entomol.* 27:471–79

32. Hess, E., Vlimant, M. 1986. Leg sense organs. See Ref. 81a, pp. 361–90

33. Hoogstraal, H. 1978. Biology of ticks. In *Tick-Borne Diseases and Their Vec-*

tors. *Proc. Int. Conf.*, ed. J. K. H. de Wilde, pp. 3–14. Edinburgh: Edinburgh Univ. Press. 567 pp.

34. Hoogstraal, H. 1985. Argasid and nutalliellid ticks as parasites and vectors. *Adv. Parasitol.* 24:135–238

35. Hoogstraal, H., Kim, K. C. 1985. Tick and mammal coevolution, with emphasis on *Haemaphysalis*. In *Coevolution of Parasitic Arthropods and Mammals*, ed. K. C. Kim, 10:505–68. New York: Wiley. 800 pp.

36. Howell, D. E. 1939. The ecology of *Dermacentor albipictus* (Packard). In *Proc. Sixth Pacific Science Congress,* pp. 439–58. Berkeley, CA: Pacific Sci. Assoc.

37. Hsu, M. H., Sauer, J. R. 1975. Ion and water balance in the feeding lone star tick. *Comp. Biochem. Physiol.* 52A:269–76

38. Jaworski, D. C., Sauer, J. R., Williams, J. P., McNew, R. W., Hair, J. A. 1984. Age-related effects on water, lipid, hemoglobin and critical equilibrium humidity in unfed adult lone star ticks (Acari: Ixodidae). *J. Med. Entomol.* 21:100–4

39. Kahl, O. 1989. *Untersuchungen zum Wasserhaushalt von Zecken (Acari: Ixodoidea) im Laufe ihrer postembryonalen Entwicklung unter besonderer Berücksichtigung der aktiven Wasserdampfsorption bei gesogenen Stadien.* PhD dissertation. Berlin: Freie Universität. 356 pp. (In German)

40. Kahl, O., Hoff, R., Knülle, W. 1990. Gross morphological changes of the salivary glands of *Ixodes ricinus* (Acari, Ixodidae) between bloodmeals in relation to active uptake of atmospheric water vapour. *Exp. Applied Acarol.* 9:In press

41. Kahl, O., Knülle, W. 1988. Water vapour uptake from subsaturated atmospheres by engorged immature ixodid ticks. *Exp. Applied Acarol.* 4:73–83

42. Kaufman, W. R. 1989. Tick-host interaction: a synthesis of current concepts. *Parasitol. Today* 5:50–59

43. King, D., Gettinby, G., Newson, R. M. 1988. A climate-based model for the development of the ixodid tick, *Rhipicephalus appendiculatus* in East Coast fever zones. *Vet. Parasit.* 29:41–51

44. Knülle, W. 1965. Equilibrium humidities and survival of some tick larvae. *J. Med. Entomol.* 2:335–38

45. Knülle, W. 1984. Water vapor uptake in mites and insects: an ecophysiological and evolutionary perspective. *Acarology* 6:71–82

46. Knülle, W., Rudolph, D. 1982. Humid-

47. ity relationships and water balance in ticks. See Ref. 69a, pp. 43–70

47. Knülle, W., Wharton, G. W. 1964. Equilibrium humidities in arthropods and their ecological significance. *Acarologia* 6:299–306

48. Kraiss-Gothe, A., Kalvelage, H., Gothe, R. 1989. Investigation into the critical equilibrium humidity, active atmospheric water absorption and water content of *Rhipicephalus evertsi mimeticus*. *Exp. Appl. Acarol.* 7:131–41

49. Lane, R. S., Anderson, J. R., Yaninek, J. S., Burgdorfer, W. 1985. Diurnal host seeking of adult Pacific Coast ticks, *Dermacentor occidentalis* (Acari: Ixodidae), in relation to vegetational type, meteorological conditions, and rickettsial infection rates in California, USA. *J. Med. Entomol.* 5:558–71

50. Lee, R. E., Baust, J. G. 1987. Cold-hardiness in the Antarctic tick, *Ixodes uriae. Physiol. Zool.* 60:499–506

51. Lees, A. D. 1946. The water balance in *Ixodes ricinus* L. and certain other species of ticks. *Parasitology* 37:1–20

52. Lees, A. D. 1947. Transpiration and the structure of the epicuticle in ticks. *J. Exp. Biol.* 23:379–410

52a. Lees, A. D. 1948. The sensory physiology of the sheep tick, *Ixodes ricinus* L. *J. Exp. Biol.* 25:145–207

53. Lees, A. D. 1952. The role of cuticle growth in the feeding process of ticks. *Proc. Zool. Soc. London* 121:759–72

54. Lees, A. D. 1964. The effect of aging and locomotor activity on the water transport mechanism of ticks. In *Proc. First Int. Congr. Acarology, Acarologia* 6:315–23

55. Lees, A. D. 1969. The behavior and physiology of ticks. *Acarologia* 11:397–410

56. Lees, A. D., Milne, A. 1951. The seasonal and diurnal activities of individual sheep ticks (*Ixodes ricinus* L.). *Parasitology* 41:189–208

57. Lessard, P., L'Eplattenier, R., Norval, R. A. I., Kundert, K., Dolan, T. T., et al. 1990. Geographic information systems for studying the epidemiology of cattle diseases caused by *Theileria parva. Vet. Rec.* 126:255–62

58. McMullen, H. L., Sauer, J. R., Burton, R. L. 1976. Possible role of uptake of water vapour by ixodid tick salivary glands. *J. Insect Physiol.* 22:1281–85

59. Machin, J. H. 1979. Compartmental osmotic pressures in the rectal complex of *Tenebrio* larvae: evidence for a single tubular pumping site. *J. Exp. Biol.* 82:123–37

60. Machin, J. H., Lambert, G. J., O'Don-

nell, M. J. 1985. Component permeabilities and water content in *Periplaneta* integument: role of the epidermis is re-examined. *J. Exp. Biol.* 117:155–69

61. Machin, J. H., O'Donnell, M. J., Kestler, P. 1986. Evidence against hormonal control of integumentary water loss in *Periplaneta americana*. *J. Exp. Biol.* 121:339–48

62. Moorhouse, D. E. 1969. The attachment of some ixodid ticks to their natural hosts. In *Proc. 2nd Int. Congr. Acarology*. ed. G. O. Evans. pp. 319–27. Budapest: Akademiai Nyomda. 652 pp.

63. Mount, G. A., Haile, D. G. 1987. Computer simulation of area-wide management strategies for the lone star tick, *Amblyomma americanum* (Acari: Ixodidae). *J. Med. Entomol.* 24:523–31

64. Mount, G. A., Haile, D. G. 1989. Computer simulation of population dynamics of the American dog tick, *Dermacentor variabilis* (Acari: Ixodidae). *J. Med. Entomol.* 26:60–76

65. Needham, G. R., Rosell, R., Greenwald, L., Coons, L. B. 1990. Ultrastructure of type I salivary gland acini in four species of ticks and the influence of hydration states on the type I acini of *Amblyomma americanum*. *J. Exp. Appl. Acarol.* 10: In press

66. Needham, G. R., Teel, P. D. 1986. Water balance by ticks between bloodmeals. See Ref. 81a, pp. 100–51

67. Noble-Nesbitt, J., Al-Shukur, M. 1987. Effects of desiccation, water-stress and decapitation on integumentary water loss in the cockroach, *Periplaneta americana*. *J. Exp. Biol.* 131:289–300

68. Noble-Nesbitt, J., Al-Shukur, M. 1988. Cephalic neuroendocrine regulation of integumentary water loss in the cockroach *Periplaneta americana* L. *J. Exp. Biol.* 136:451–59

69. Noble-Nesbitt, J., Al-Shukur, M. 1988. Involvement of the terminal abdominal ganglion in neuroendocrine regulation of integumentary water loss in the cockroach *Periplaneta americana* L. *J. Exp. Biol.* 137:107–17

69a. Obenchain, F. D., Galun, R., eds. 1982. *Physiology of Ticks*. New York: Pergamon. 509 pp.

70. O'Donnell, M. J. 1982. Hydrophilic cuticle—the basis for water vapour absorption by the desert burrowing cockroach, *Arenivaga investigata*. *J. Exp. Biol.* 99:43–60

71. O'Donnell, M. J., Machin, J. 1988. Water vapor absorption by terrestrial organisms. In *Advances in Comparative and Environmental Physiology*, 2:47–90. Berlin, Heidelberg: Springer-Verlag

71a. Perry, B. D., Lessard, P., Norval, R. A. I., Kundert, K., Kruska, R. 1990. Climate, vegetation and the distribution of *Rhipicephalus appendiculatus* in Africa. *Parasitol. Today* 6:100–4

72. Punyua, D. K., Newson, R. M. 1979. Diurnal activity behavior of *Rhipicephalus appendiculatus* in the field. See Ref. 76a, pp. 441–45

73. Punyua, D. K., Newson, R. M., Mutinga, M. J. 1985. Diurnal and seasonal activity of unfed adult *Rhipicephalus appendiculatus* (Acarina: Ixodidae) in relation to some intrinsic and extrinsic factors. I. Factors regulating activity. *Insect Sci. Applic.* 6:63–70

74. Punyua, D. K., Newson, R. M., Mutinga, M. J. 1985. Diurnal and seasonal activity of unfed adult *Rhipicephalus appendiculatus* (Acarina: Ixodidae) in relation to some intrinsic and extrinsic factors. III. Daily changes in water content. *Insect Sci. Applic.* 6:71–73

75. Ransom, B. H. 1906. Some unusual host relations of the Texas fever tick. USDA *Bur. Anim. Ind. Cir. No. 98.* 8 pp.

76. Ribeiro, J. 1989. Role of saliva in tick/host interactions. *Exp. Appl. Acarol.* 7:15–20

76a. Rodriguez, J. G., ed. 1979. *Recent Advances in Acarology*, Vol. 1. New York: Academic. 631 pp.

77. Rudolph, D. 1982. Site, process, and mechanism of active uptake of water vapour from the atmosphere in the Psocoptera. *J. Insect. Physiol.* 28:205–12

78. Rudolph, D., Knülle, W. 1974. Site and mechanism of water vapour uptake from the atmosphere of ixodid ticks. *Nature* 249:84–85

79. Rudolph, D., Knülle, W. 1978. Uptake of water vapor from the air: process, site and mechanism in ticks. See Ref. 82a, pp. 97–113

80. Rudolph, D., Knülle, W. 1979. Mechanisms contributing to water balance in non-feeding ticks and their ecological implications. See Ref. 76a, pp. 375–83

81. Sauer, J. R., Hair, J. A. 1971. Water balance in the lone star tick (Acarina: Ixodidae): the effects of relative humidity and temperature on weight changes and total water content. *J. Med. Entomol.* 8:479–85

81a. Sauer, J. R., Hair, J. A., eds. 1986. *Morphology, Physiology, and Behavioral Biology of Ticks*. Chichester, UK: Horwood. 510 pp.

82. Schmidt-Nielsen, K., Bolis, L., Maddrell, S. H. P., eds. 1978. *Comparative Physiology: Water, Ions, and Fluid Mechanics.* Cambridge: Cambridge Univ. Press. 360 pp.

83. Semtner, P. J., Hair, J. A. 1975. Evaluation of CO_2-baited traps for survey of *Amblyomma maculatum* Koch and *Dermacentor variabilis* Say (Acarina: Ixodidae). *J. Med. Entomol.* 12:137–38

84. Short, N. J., Floyd, R. B., Norvall, R. A. I., Sutherst, R. W. 1989. Survival and behavior of unfed stages of the ticks *Rhipicephalus appendiculatus, Boophilus decolratus,* and *B. microplus* under field conditions in Zimbabwe. *Exp. Appl. Acarol.* 16:215–36

85. Sigal, M. D. 1990. *The water balance physiology of the lone star tick,* Amblyomma americanum *(Acari: Ixodoidea), with ecophysiological comparisons to other ixodid species.* PhD dissertation. Columbus: Ohio State Univ. 180 pp.

86. Sonenshine, D. E. 1979. Zoogeography of the American dog tick, *Dermacentor variabilis.* In *Recent Advances in Acarology,* ed. J. G. Rodriguez, 2:123–34. New York: Academic. 569 pp.

87. Sutherst, R. W. 1987. The role of models in tick control. *Proc. Int. Conf. Vet. Prev. Med. Anim. Prod.,* ed. K. L. Hughes, pp. 32–37. Melbourne: Aust. Vet. Assoc.

88. Sutherst, R. W., Dallwitz, M. J., Utech, K. B. W., Kerr, J. D. 1977. Aspects of host finding by the cattle tick, *Boophilus microplus. Aust. J. Zool.* 25:159–74

89. Sutherst, R. W., Maywald, G. F. 1985. A computerised system for matching climates in ecology. *Agric. Ecosys. Environ.* 13:281–99

90. Teel, P. D. 1991. Application of modelling to the ecology of *Boophilus annulatus* (Acari:Ixodidae). Applications of computer models to medical/veterinary entomological problems. *J. Agric. Entomol.* In Press

91. Teel, P. D., Fleetweed, S. C., Huebner, G. L. Jr. 1983. An integrated sensing and data acquisition system designed for unattended, continuous monitoring of microclimate relative humidity and its use to determine the influence of vapor pressure deficits on tick (Acari: Ixodoidea) activity. *Agric. Meteor.* 27:145–54

92. Utech, K. B. W., Sutherst, R. W., Dallwitz, M. J., Wharton, R. H., Maywald, G. F., Sutherland, I. D. 1983. A model of the survival of larvae of the cattle tick, *Boophilus microplus,* on pasture. *Aust. J. Agric. Res.* 34:63–72

93. Walker, J. B., Olwage, A. 1987. The tick vectors of *Cowdria ruminantium* (Ixodoidea, Ixodidae, Genus *Amblyomma*) and their distribution. *Onderstepoort J. Vet. Res.* 54:353–79

94. Wharton, G. W. 1978. Uptake of water vapour by mites and mechanisms utilized by the acaridei. See Ref. 82a, pp. 79–95

95. Wharton, G. W. 1985. Water balance of insects. In *Comprehensive Insect Physiology, Biochemistry and Pharmacology,* ed. G. A. Kerkut, L. I. Gilbert, 14:565–601. New York: Pergamon

96. Wharton, G. W., Richards, A. G. 1978. Water vapor exchange kinetics in insects and acarines. *Annu. Rev. Entomol.* 23:309–28

97. Williams, J. P., Sauer, J. R., McNew, R. W., Hair, J. A. 1986. Physiological and biochemical changes in unfed lone star ticks, *Amblyomma americanum* (Acari: Ixodidae), with increasing age. *J. Med. Entomol.* 23:230–35

SUBJECT INDEX

A

Abamectins, 91-117
 degredation of, 96-98
Acalymna,
 innubrum, 574
 themei, 574, 578-79
 vittatum, 574, 579
Acarinaria, 621, 626
Acarus, 619
 immobilis, 624-25
Acoustical traps, 385
Acromyrmex, 649
Acrosterum hilare, 122
Aculops lycopersici, 102
Acyrthosiphum pisum, 519-20,
 524
Aedes
 aegypti, 143-44, 147-48, 150-
 51, 168, 365, 369, 459-
 60, 462-65, 468, 476-
 477
 albopictus, 459-84
 atropalpus, 515-517
 freeborni, 517
 gambiae, 320
 scutellaris, 466, 472
 sierrensis, 599
 sollicitans, 365, 517
 taeniorhynchus, 366
 togoi, 478, 515
 triseriatus, 168, 374, 462-63,
 477
Afrocalvolia, 618
Agathis thompsoni, 72-73
Age and maternal effects, 516,
 521-22
Agrilus opulentus, 650
Agrites, 123
Agromyza, 542, 546
 frontella, 536-57, 543, 45-46,
 552
Agrotis
 crinigera, 490
 ipsilon, 388
AIDS transmission, 367-68
Alabidopus, 619
Aleurocanthus
 husaini, 446
 spiniferus, 446, 449
 spinosus, 446
 woglumi, 437-38, 443-44,
 446, 449

Aleurocybotus occiduus, 436-38,
 446
Aleurodicus, 435
 cocois, 443, 446
 dispersus, 434
 pimentae, 446
Aleurolobus barodensis, 446
Aleuroplatus coronata, 434
Aleurothrixus, 432
 floccosus, 434, 438-39, 446,
 450
Aleurotrachelus jelinekii, 434,
 446, 449-50
Aleyrodes
 brassicae, 440
 fragariae, 446
 proletella, 432, 439, 443,
 446, 449
Alighting behavior, 73-75
Alkaloids, 70
Allatectomy and host-seeking
 behavior, 147-48
Allonemobius, 517
Allotropa, 268
Allozymes of mosquitoes, 466-
 68, 472-73
Alsophila pometaria, 82, 572
Amblyomma
 americanum, 590, 595, 598,
 604, 660, 665-75
 cajennense, 665, 670-71,
 673, 675-76
 maculatum, 598, 665, 668,
 670-71, 673-75
 variegatum, 664
Amblyseius, 267
 aerialis, 268
 fallacis, 313-14
 gossipi, 109-10
Ameranoetus, 622
Amino acids and storage hexamers, 207-23
Anagyrus, 266
 niombae, 264
Anastrepha suspensa, 168, 171
Ancistrocerus antilope, 621,
 625-26
Anemotaxis, 143
Anoetus, 621
 halictonidus, 626
Anomala orientalis, 386
Anopheles
 albimanus, 470

 freeborni, 150, 365
 gambiae, 168, 468
 labranchiae, 150
 quadrimaculatus, 374, 471
Anoplognathus, 644, 648
Antheraea polyphemus, 421
Anthidium, 621
Anthocharis cardamines, 76-77
Anthonomus grandis, 120, 122
Anthrenus flavipes, 108
Antonina graminis, 390
Ants
 and abamectins, 98-99
 and biological control, 495
 and *Eucalyptus*, 640, 649
 and leaf-miners, 538
 and mealybugs, 270
 and mites, 621
 and oviposition behavior, 72
 and rootworms, 235
 and turfgrass, 391
Aphids
 and abamectins, 102
 and *Eucalyptus*, 650
 and gene amplification, 1-23
 and insecticide resistance, 1-
 23
 and maternal effects, 518
 and polycultures, 569, 571,
 577-78, 580
Aphidus nigripes, 516
Aphis
 craccivora, 569
 fabae, 102, 111, 519, 524
 nerii, 519-20
Aphytis melinus, 109
Apis mellifera, 50, 208, 212
Apolysis
 and sensilla, 341
 and whiteflies, 435
Aporia crataegi, 80
Arboviruses and mosquitoes,
 473
Arenivaga investigata, 666
Argas reflexus, 666
Arylphorin, 207, 212-22
Associational resistance, 566-68
Asterobemisia carpini, 446, 449
Astigmata and phoresy, 611-32
Ataenius spretulus, 386-387,
 394
Athalia rosae, 124
Atherigona, 123

Atta, 649
Attagenus unicolor, 108
Attractants, 317-18
Avermectins in arthropod pest
    control, 91-117
Azadirachtin, 311, 316

B

Bacillus
    popilliae, 394-95
    thuringiensis, 315, 324, 395,
        476, 491
Baits for rootworms, 242-44
Battus philenor, 74-75, 81, 577
Bedbugs and HIV, 367-70
Bedellia, 551
Bees
    and abamectins, 97
    and dietary self-selection, 50-
        51
    and gene transformation, 175
    and mites, 621
Behavior
    and biological control, 494
    and plant protectants, 305-30
    and pheromones, 408
    of ticks, 660
Behavioral adaptations, 319-20
Bemisia tabaci, 123, 432, 434-
    45, 447, 450
Bessa remota, 488
Biodegradation of insecticides,
    239, 241, 392
Biological control
    of cassava pests, 257-83
    impact of, 485-509
    of leaf-miners, 550-52
    of mosquitoes, 475-76
    of rootworms, 234-36
    of turfgrass pests, 394-95
Biosystematics of chewing lice,
    185-203
Biting lice and retroviruses, 364
Blatta orientalis, 208, 216, 218
Blattella germanica, 99, 208,
    213
Blissus
    insularis, 389
    leucopterus, 389
Blood feeding behavior, 139-
    158
Boletoglyphus ornatus, 626
Bolitotherus cornutus, 626
Boll weevil, 122-24
Bombyx, 167
    mori, 209, 214, 218, 221-22,
        513, 515, 523-25
Boophilus
    annulatus, 670-71, 673-74
    microplus, 372, 660, 669
Borrelia burgdorferi, 587-604
Bovine retroviruses, 370-73

Brachus ovatus, 547
Brachyptery and oviposition be-
    havior, 81-82
Brevicoryne brassicae, 569
Bryobia praetiosa, 390
Bucculatrix, 551-52

C

Cactoblastis cactorum, 76
Calliphora, 208
    erythrocephala, 208, 211,
        214, 218-19, 470
    stygia, 208
    vicina, 214, 216-17, 220-21,
        516-18
Calliphorin, 211-212, 216-17,
    219-220
Callosobruchus maculatus, 318,
    520, 522
Calpodes ethlius, 209, 214, 216
Cameraria, 538, 545
Cannibalism, 545-46
Carbamates and rootworms, 236
Carbofuran, 237, 239-40
Carbohydrates and dietary self-
    selection, 47-49
Carbon dioxide
    and host-seeking behavior,
        140-142, 144-46
    and sensilla, 346
Carcinogens, 4
Cardenolides, 69
Cardiaspina
    albitextura, 645
    densitexta, 645
Carpoglyphus lactis, 613, 624,
    626
Cassava
    mealybugs, 257-83
    pests, 257-83
Celatoria diabroticae, 234
Ceratitis capitata, 168, 208,
    497
Cerotoma
    ruficornis, 571
    trifurcata, 122
Chaetodactylus, 618-19
Chamaedaphne, 548, 551
Chartocerus, 266
Chauliognathus, 235
Chemosensilla, 334-37
Chewing lice of pocket gophers,
    185-203
Chilo partellus, 123
Chilocorus cacti, 628
Chlorinated hydrocarbons and
    rootworms, 236
Cholinesterase, 4-5
Chorion, 2
Choristoneura
    fumiferana, 78
    occidentalis, 106, 412

Chortoicetes terminifera, 524
Chromosomes
    and gene amplification, 13-18
    and gene transformation, 160
    and mosquitoes, 461, 469
    and oviposition behavior, 68,
        72
    of pocket gophers, 186, 196-
        99
    and retroviruses, 356-57
    and storage hexamers, 221
    and translocation, 3, 7-8
Chrysocharis, 105, 545
Chrysodeixis eriosoma, 497
Chrysonotomyia, 105
Chrysophtharta
    bimaculata, 644
Chrysops flavidus, 366
Chrysoteuchia topiaria, 388
Chymomyza procnemis, 170-71
Cimex
    hemipterus, 369-70
    lectularius, 369-70
Circadian rhythms and host-
    seeking behavior, 152
Cladistic relationships for chew-
    ing lice, 199
Classical biological control,
    485-509
Cleptoparasities, 621
Coccinella septempunctata, 498
Coccygominus disparis, 501
Cockroaches
    and abamectins, 99-100
    and behavior, 312
    and dietary self-selection, 48-
        50, 53-54
    and HIV, 368
    and juvenile hormone, 153
    and storage hexamers, 213
    and water balance, 666
Coelaenomenodera elaeidis, 552
Coevolution, 65-66
Cold-hardiness
    of mosquitoes, 465
    of ticks, 660
Coleoglyphus, 620
Coleomegilla maculata, 565-66
Coleotichus blackburniae, 489-
    90
Colias, 318-19
    eurytheme, 67, 70, 75
Color and whiteflies, 443
Colotis, 80
Competition
    and leaf-miners, 545
    and mosquitoes, 476-477
Conogethes punctiferalis, 69
Copidosoma floridanum, 497
Copronomia, 622
Coptodisca, 548, 551-52
    negligens, 546
Coptotermes formosanus, 312

Corn earworm, 44
Corpora allata, 147, 151
Cospeciation in chewing lice,
    198-99
Cotesia glomeratus, 491
Cotinus nitida, 387
Crambus
    praefectellus, 388
    sperryellus, 388
Cratosomus flavofasciatus, 123
Creightonella, 621
Creiis periculosa, 646
Crop rotation and rootworms,
    230-32
Cross resistance, 4, 10
    and abamectins, 110
Cryptoloemus montrouzieri, 109
Ctenomorphodes tessulata, 643
Cucurbitacins, 317
    and rootworms, 242
    as plant protectants, 308
Culex, 14, 148-52, 369
    pipiens, 9-11, 144, 147, 149,
        374
    pipiens quinquefasciatus, 1-
        23, 364-65, 374
    tarsalis, 9, 11, 365
Culicoides variipennis, 365
Culiseta inornata, 365
Cuterebra horripilum, 342
Cuticle
    and sensilla, 333, 341
    and storage hexamers, 217
    of ticks, 667-71
Cuticulin and sensilla, 333
Cyclocephala, 394
    borealis, 386
    lurida, 386
Cyclodienes and rootworms, 239
Cydia pomonella, 69, 107
Czenspinskia transversostriata,
    617

D

Dactylopius opuntiae, 499
Dacus
    cucurbitae, 123
    frontalis, 123
Dalbulus maidis, 569, 572
Danaus plexippus, 78
DDT, 312, 314, 316, 320, 475
Defenses in trees, 25-42
Defoliation
    and Eucalyptus pests, 646
    and induced resistance, 28
Degradation of abamectins, 106
Delia brassicae, 576
Dendrites and sensilla, 333-342
Dengue, 461-64, 472, 476
Density dependence and popula-
    tion analysis, 296-97

Depressaria
    daucella, 121
    depressella, 121
    pastinacella, 76
Dermacarus pilitarsus, 623
Dermacentor
    albipictus, 666, 670-72, 674-
        75
    marginatus, 666
    occidentalis, 598, 604, 667
    parumapertus, 598
    variabilis, 590, 595, 598,
        660, 667, 676
Dermatophagoides, 618
Deterrency and plant pro-
    tectants, 313, 316-17
Diabrotica, 123, 308, 323
    barberi, 229
    undecimpunctata, 95, 230,
        309
    virgifera, 229-30
Diabroticite rootworms in corn,
    229-55
Dialectia, 552
Dialeurodes
    chittendeni, 433, 436, 447,
        449
    citri, 432, 435-36, 447, 449
    citrifolii, 432
Dialeurolonga elongata, 447
Dianamobius
    fasciatus, 517
    taprobanensis, 517
Diapause, 245
    and biological control, 491
    hormone, 523
    and host-seeking behavior,
        146, 149-151
    of leaf-miners, 548
    and maternal effects, 512-18
    of mosquitoes, 464-65
    and rootworm eggs, 231-32,
        245-46
    of ticks, 591, 660
Diatraea
    grandiosella, 214
    saccharalis, 81
Dicrodiplosis manihoti, 264
Didymuria violescens, 643
Diet switching, 44
Dietary self-selection, 43-63
Diglyphus intermedius, 105
Diomus, 263, 269
Dispersal and insect population
    analysis, 297-98
Distribution of sensilla, 342-47
Distributions of populations,
    286-87
DNA, 2-4, 8, 10-16, 18, 164,
    469-72
    and gene transformation, 159-
        83
    and HIV, 368

and mosquitoes, 461, 468-69
and retroviruses, 355-57, 359
sequencing, 207
and storage hexamers, 220
Dociostaurus maroccanus, 44
Doratifera, 644
Drosophila, 13-16, 159-83, 470,
    522
    crucigera, 208
    hawaiiensis, 166
    mauritiana, 166, 173
    melanogaster, 11-12, 160-77,
        200-21, 334, 341-42,
        345, 347, 521, 523
    mimica, 208
    mulleri, 208
    simulans, 166
    willistoni, 166
Dryophilodes, 640
Dust mites, 618

E

Ecdysis and sensilla, 341
Ecdysone and storage hexamers,
    216, 219
Ecdysteroids and diapause, 524
Echimyopus dasypus, 630
Ecology of ixodid ticks, 659
Economic thresholds for root-
    worms, 244-46
Economics of biological control,
    496-98
Ectelmnius paucimaculatus, 547
Eldana saccharina, 123, 131
EMA, 111
Empoasca fabae, 578
Endangered species, 488-92
Endocrines and maternal effects,
    523-25
Endocrinology and host-seeking
    behavior, 147-151
Ensliniella kostylevi, 625
Ephestia, 75, 167
Epidinocarsis
    diversicornis, 266, 268
    lopezi, 257-60, 263-66, 268-
        73
Epilachna
    borealis, 309
    varivestis, 111, 122, 575, 578
Epitrimerus pyri, 104
Equine infectious anemia virus,
    358-67
Eriocraniella, 547
Eriophyes
    cynodoniensis, 389
    discoridis, 101
Essential oils and Eucalyptus
    pests, 639, 641
Esterases and gene amplifica-
    tion, 6-9
Estragole, 242, 244

Eucalyptus, pests of, 637-57
Eugenol, 242, 244
Euphydryas, 83
  anicia, 76-77
  chalcedona, 77, 308
  editha, 67, 72-73, 75, 77
Eurytides marcellus, 74
Euschistus, 122
Euseius
  concordis, 268
  stipulatus, 109
Euthystira brachyptera, 52
Euxoa, 120
  messoria, 124
  ochrogaster, 573-74
  tessellata, 124
Evolution
  of behavioral resistance, 332-33
  of chewing lice, 198-99
  of oviposition behavior, 65-89
  of storage hexamers, 205-28
Excretion by whiteflies, 440
Exochomus, 263-64
Extinction and biological control, 488-92

F

Farnesene, 69
Fat body and storage hexamers, 205-23
Feeding behavior
  and dietary self-selection, 43-63
  and plant protectants, 306, 308
  and sensilla, 332
Feeding deterrents and sensilla, 334, 347
Feline retroviruses, 373
Fibulanoetus, 620
Fissicrambus mutabilis, 388
Flavoproteins, 213, 218-19, 220
Foraging, models for, 45
Formica rufa, 622
Fungi
  and cassava green mites, 267
  and cassava mealybugs, 266
  and rootworms, 235
  and turfgrass pests, 395-97

G

GABA and avermectins, 94
Galendromus
  annectens, 268
  helveolus, 268
Galleria mellonella, 209, 214
Ganespidium hunteri, 105
Gene
  amplification, 1-23
  flow, 199, 467-68, 472

vectors, 160, 164-5, 169, 172-78
Genetic
  changes and behavior, 318-22
  drift in mosquitoes, 467
Genetics
  of *Aedes albopictus*, 466-68
  and biological control, 494
  and diapause, 518
  of maternal effects, 526
  and oviposition behavior, 67-68
Geographic variation and diapause, 517
Geomydoecus, 185-203
  alcorni, 192-93
  alleni, 190
  betleyae, 194
  biagiae, 194
  bulleri, 190, 192
  californicus, 190, 194-95
  chapini minor, 188-84
  cherriei, 192
  chiapensis, 190-91
  copei, 190, 192
  coronadoi, 190
  costaricensis, 192
  dalgleishi, 190-91
  davidhafneri, 192
  expansus, 190
  geomydis, 190
  hoffmanni, 192
  mcgregori, 190, 193
  mexicanus, 190
  nebrathkensis, 197
  neotruncatus, 191
  oklahomensis, 197
  oregonus, 190, 195
  panamensis, 190, 192
  polydentatus, 193
  pygacanthi, 191
  quadridentatus, 190-91
  scleritus, 190
  setzeri, 192
  subcalifornicus, 190
  telli, 190, 193
  temaulipensis, 193
  texanus, 190-91
  thomomyus, 190, 194, 197
  tolucae, 190, 195
  trichopi, 190, 195
  truncatus, 190-91
  umbrini, 190
Glabrodorsum stokesii, 491
Glossina morsitans, 521
Glucosinolates, 69
Glutathione S-transferases, 5
Glycaspis, 646
Glycyphagus, 619
  destructor, 624
Glyphanoetus nomiensis, 626
Gonipterus scutellatus, 644, 650
Gonotrophic dissociation, 149

Grasshoppers
  and dietary self-selection, 44, 50, 52-53
  and *Eucalyptus*, 650
Greya subalba, 72-73, 79
Gustation and sensilla, 332, 335-36, 347
Gypsy moths, 44, 67-68

H

Habitat range and biological control, 494
Hadena, 75
  bicruris, 68, 74
Haemaphysalis
  leporispalustris, 598-99
  punctata, 665
Haematobia irritans, 342, 517
Halictus rubicunidus, 626
Hedylepta
  euryprora, 490
  fullawayi, 490
  meyricki, 490
  musicola, 490
Heliconius, 50, 75-76, 78
Heliothis, 11, 133, 411
  armigera, 106
  subflexa, 68
  virescens, 68, 106, 111, 315, 517-18, 521
  zea, 48-50, 52-58, 106, 110-11, 123, 125, 206, 209, 411, 518
Hemisarcoptes, 625
  coccophagus, 624
  cooremani, 628, 631
Hemocyanins, 206-10, 212, 220-22
Hemolymph proteins, 205-23
Hepatitis B virus, 367
Herbicides and gene amplification, 4
Herbivory on *Eucalyptus*, 637-57
Hericia hericia, 622, 625
Herpetogramma
  licarsisalis, 388
  phaeopteralis, 388
Heteropan dolens, 488
Heteropsylla, 499
Hexamerins, 205-28
Histiogaster arborsignis, 613
Histiostoma, 625
  bakeri, 624
  feroniarum, 622
  laboratorium, 624, 626
  polypori, 627
HIV
  and insects, 369
  transmission, 367-70
Hobo vectors, 166-67
Holcothorax, 545

Holobus, 267
Holomelina
  immaculata, 415-16
  lamae, 409
Homeopronematus anconai, 102
Homoeosoma electellum, 411
Honey bees and dietary self-
  selection, 50-51
Honeydew
  and dietary self-selection, 55
  and leaf-miner fecundity, 542
  of whiteflies, 431, 434, 440
Hormones and maternal effects,
  527
Hormosianoetus aeschlimanni,
  615
Host
  parasite relationships, 185-203
  plant resistance and turfgrass
    pests, 396-97
  preference in Lepidoptera, 65-
    89
  range and biological control,
    493
  seeking behavior in mos-
    quitoes, 139-158
  specificity of leaf-miners, 537
Houseflies, 6, 12-13
Human retroviruses, 368
Humidity
  and sensilla, 334
  and whiteflies, 445
Hyalomma
  anatolicum, 666
  scupense, 673
Hyalophora cecropia, 209, 213-
  14, 218-19, 220
Hybomitra lasiophthalma, 367
Hybrid dysgenesis, 165
Hylaeogena, 538, 543
Hylephila phyleus, 389
Hyomeces squamosus, 650
Hyperaspis
  jucunda, 263, 268
  notata, 268-69
  raynevali, 263, 269
  senegalensis, 263
Hyperecteina aldrichi, 394
Hyperodes, 387
Hyperparasitism, 266
Hyphantria cunea, 214
Hypochlora alba, 345
Hypodectes propus, 630
Hypopus, 615-17

I

Immunodeficiency viruses, 367-
  70
Indole, 243, 247
Induced resistance, 25-42
Induction of defenses in trees,
  25-42

Insecticide resistance
  and gene amplification, 1-23
  of mosquitoes, 475-76
Insecticides
  and behavior, 305
  and rootworms, 236
  and turfgrass pests, 391-92
Integrated pest management, see
  IPM
Intercropping, 561-62
Inversions and gene amplification, 15
Ionone, 243, 247
IPM
  and dietary self-selection, 43-
    61
  and pheromones, 408
  and turfgrass pests, 397-98
  and vegetational diversity,
    561
Ips, 130, 625
  typographus, 121-22
Ivermectin, 95
Ixodes
  angustus, 599
  dammini, 588-97, 600,602-3,
    665-66
  dentatus, 593, 595
  hemagonus, 602
  neotomae, 598
  pacificus, 588, 592,595, 598,
    604
  persulcatus, 588, 667
  ricinus, 588, 595, 597-98,
    602, 604, 665-67
  scapularis, 588, 595, 598

J

Japanese beetle, 386-87, 393-
  94, 396
Junonia coenia, 68
Juvenile hormone
  and host-seeking behavior,
    147-48, 151, 153
  and photoperiod, 523-25
  and sensilla, 341
  and storage hexamers, 213

K

Kairomones and trap cropping,
  128
Keiferia lycopersicella, 106
Kennethiella trisetosa, 621, 625-
  26
Key factor analysis, 295-99

L

Labdia, 545
La Crosse encephalitis, 462-63

Lactic acid and host-seeking be-
  havior, 140-49, 152
Lardoglyphus, 624
Larra bicolor, 394
Lasius neoiger, 235
Leafminers
  and abamectins, 104-05
  bionomics of 535-60
Learning
  and oviposition behavior, 75
  and plant protectants, 312,
    315
Leiopleura, 541
Lepidoptera and abamectins,
  106-108, 111
Leptidea sinapis, 74, 76-77
Leptinotarsa decemlineata, 111,
  122
Leucophaea maderae, 344
Leucoptera, 549
  caffeina, 552
  coffeella, 552
Levuana, 495
  iridescens, 488
Lice, 185-203
  and *Borrelia*, 588
  population model, 188-89
Life-tables
  and population analysis, 295
  for whiteflies, 445
Light and whiteflies, 443
Lipids and storage hexamers,
  219
Lipophorin, 213, 219
Liposomes, 175-76
Liriomyza, 536-37
  trifolii, 104-05, 123, 313-14,
    551
Listronotus maculicollis, 387
Lithocolletis, 536, 548
  salicifoliella, 546
  quercus, 547
Lobesia botrana, 107
Locusta
  migratoria, 54, 58, 168-69,
    208, 213, 217, 219, 310,
    345, 520
  pardalina, 516
Locusts
  and dietary self-selection, 52-
    54, 58
  and maternal effects, 520,
    525
Lophyrotoma
  analis, 645
  interupta, 645
Loxostege frustalis, 496
Lucilia
  caesar, 335, 517
  cuprina, 177, 208, 220-21
Lycia, 81
Lygus, 133
  elisus, 122

hesperus, 122
lineolaris, 571
rugulipennis, 123
voessleri, 123, 125
Lymantria dispar, 79, 81-82,
    106, 209, 212, 218, 220,
    496, 521
Lyme disease, 587-609
Lysosomes and storage hexa-
    mers, 219

M

Macrocheles glaber, 615
Maculina arion, 490
Malacasoma
    americana, 214, 346
    californiacum, 521
    neustria, 346
Maladera castanea, 386
Malathion and mosquitoes, 475
Mallophaga, 185-203
Mamestra
    brassicae, 339, 417
    configurata, 410
Manduca sexta, 50, 54, 106,
    209, 212, 215-217, 221-22,
    335, 338, 340
Margarodes, 390
Maruca testulalis, 124, 130
Maternal effects, 511-34
Mating behavior
    of leaf-miners, 542
    of whiteflies, 440-41
Mealybugs, 257-83
Mechanosensilla, 333-34
Megacanestrinia, 620
Megachile, 621
Megalagrion pacificum, 490
Megastigmus, 640
Megoura viciae, 519, 524
Melanoplus
    bilituratus, 52
    bivittatus, 53
    differentialis, 52-53
    femurrumbrum, 53
    sanguinipes, 53, 517
Meligethes, 126
    aeneus, 122
Metaseiulus occidentalis, 104,
    109
Microfilaria and avermectins, 95
Microlarinus, 492
Microtubules and sensilla, 333-
    34, 337-38
Migration
    and insect population analy-
        sis, 297-98
    of whiteflies, 444-45
Milbemycins, 93
Mimicry and leaf-miners, 543
Mites
    and abamectins, 95, 100-104,
        108-110

behavior of, 313
and cassava, 257
and diapause, 516
and leaf-miners, 547
and phoresy, 611
and plant protectants, 307
and pyrethroids, 321
and rootworms, 235-36, 242
and trap cropping, 125
and turf, 389-90
and whiteflies, 450
Mixed-function oxidases, 5-6,
    12
Mixoma virus, 490, 495
Mnesampela privata, 643
Models
    for foraging 45
    for turfgrass pests, 393
Molting and sensilla, 341
Mononychellus tanajoa, 257-83
Mormoniella vitripennis, 515-16
Mosquitoes, 459-84
    and biological control, 491,
        495
    and blood meals, 361
    and DDT, 320
    and gene amplification, 1-23
    and gene transformation, 159,
        168
    and genetic control, 163
    and hepatitis B virus, 367-69
    and HIV, 368
    and host-seeking behavior,
        139-158
    and insecticide resistance, 1-
        23
    and lyme disease, 599
    and maternal effects, 515
    and olfaction, 138-158
    and retroviruses, 361, 364,
        365, 368, 370, 374
Murgantia histrionica, 490
Muridectes heterocephali, 615
Musca domestica, 110, 208
Mutation rates, 165
Myrmica, 72
Mythimna separata, 415
Myxomatosis, 502
Myzus persicae, 1-23, 519-20,
    571

N

Naiadacarus arboricola, 622,
    625
Nasonia vitripennis, 517
Natural enemies
    of cassava pests, 258, 263
    and polycultures, 562, 564,
        568, 572, 574
    of rootworms, 234
    and trap cropping, 120, 125,
        128, 132, 134
    of whiteflies, 449-50

Nectar
    and dietary self-selection, 50-
        51, 55
    and mites, 624
    and natural enemies, 565
    and oviposition behavior, 77
Nematodes
    and avermectins, 92, 94-96
    and biological control, 491,
        495
    and gene transformation, 164,
        173
    and rootworms, 235
    and turfgrass pests, 395
Neoaplectana carpocapsae, 235
Neodiprion
    rugifrons, 308
    swainei, 308
Neomaskellia
    andropogonis, 447
    bergii, 447
Neoseiulus
    anonymus, 268
    californicus, 268
    fallacis, 517
    idaeus, 267-68
Neotrachys, 538
Neoxenoryctes reticulatus, 615,
    623
Nephelodes minians, 388
Nephotettix virescens, 123
Nepticula, 552
Nezara viridula, 122, 489
Nicrophorus, 620
Nilaparvata lugens, 123
Nomadacris septemfasciata, 521
Nomia melanderi, 626
Nosema and diapause, 491
Nutrients and dietary self-
    selection, 45-48
Nutrition
    and host-seeking behavior,
        151
    of whiteflies, 439-40
Nysius raphanus, 571

O

Octotoma, 552
Ocyptamus, 261
Odonaspis ruthae, 390
Odontota dorsalis, 542, 552
Odors and host-seeking be-
    havior, 143
Oechalia, 490
Ogyris amaryllis, 72
Olfaction
    in mosquitoes, 139-58
    and oviposition behavior, 74
    and sensilla, 144-45, 331,
        336-37, 346
Oligonychus ilicis, 307
Olizonychus pratensis, 390

Oncogenes, 2, 356-59
Oncopeltus
  fasciatus, 516-18, 522, 526
  cingulifer, 522-23
Operophthera, 81
Optimal foraging models, 45
Orchesella cincta, 520
Organophosphorus insecticides
  and gene amplification, 4-6
  and rootworms, 237, 239
Orgyia, 82
  antiqua, 516-17, 523
  pseudotsugata, 82
  thyellina, 516-17
Orius tristicolor, 574
Ornithodoros, 588
  coriaceus, 598
Oscinella frit, 389
Osmoderma eremicola, 614,
  625-26
Osmosensilla, 337
Ostrinia nubilalis, 124
Oviposition
  behavior, 65-89
  by leaf-miners, 542
  and plant protectants, 315

P

P-element system, 160-66
Pachyschelus, 542-43
  aversus, 541
  psychotriae, 542, 549
Panolis flammea, 69
Panonychus
  citri, 102
  ulmi, 104, 109
Papilio, 83
  machaon, 70
  oregonius, 67-68, 72, 75
  polyxenes, 68-69, 209
  protenor, 69
  xuthus, 68
  zelicaon, 67-68, 72, 75, 124
Parabemisia myricae, 437-38,
  441, 444-45, 447
Paraleyrodes, 438
Paralimna decipiens, 170
Paranapiacaba waterhousei, 571
Parapyrus manihoti, 260
Parasite-host associations, 185-
  203
Parasites and abamectins, 105
Parasitism by mites, 630-31
Parasitoids
  and leaf-miners, 540-41, 543-
    45, 549-51
  and oviposition behavior, 73
  and polycultures, 564, 568
  and trap cropping, 131
Paropsis
  atomaria, 641-42, 644
  charybdis, 641-42, 650

Parthenogenesis
  in lice, 193-94
  in whiteflies, 441
Patasson nitens, 650
Pathogens and rootworms, 235
Pectinophora gossypiella, 106
Pediasia trisecta, 388
Pegomya nigritarsis, 542
Pelzneria, 620
Penthaleus major, 390
Perditha, 545, 548
Perga affinis, 645
Peridroma saucia, 388
Periplaneta americana, 217,
  337, 344
Perisocus quadrifasciatus, 517
Permethrin and mosquitoes, 475
Perthida glyphopa, 78, 541,
  643
Pesticide resistance and trap
  cropping, 134
Pesticides and behavior, 311
Pests and vegetational diversity,
  561-86
Phaedon cochleariae, 123
Phagostimulants, 339, 347
Phenacoccus
  herreni, 260-61, 263
  madeirensis, 264
  manihoti, 257-83
Phenols and induced resistance,
  31-33
Pheromones, 407-30
  and age, 409-10
  and communication in moths,
    407-30
  and DDT, 316
  emossion of, 409
  and humidity, 417-18
  and oviposition behavior, 75
  as plant protectants, 317
  receptivity, 409
  and rootworms, 242
  and sensilla, 346
  and trap cropping, 124, 130
  and traps, 121, 419-22
  and turfgrass pests, 393
  and pathogens, 412-13, 422
  and photoperiod, 409-10, 413
  and whiteflies, 441
  and wind speed, 418-19
Phidippus audax, 110
Phigalia, 81
Phoracantha semipunctata, 650
Phoresy in the Astigmata, 611-
  36
Phormia regina, 55
Photoperiod
  and diapause, 514-20, 523
  and light intensity, 416
  and maternal effects, 512
  and mosquitoes, 464-65, 472,
    474
  and pheromones, 409-10, 413

  and temperature, 413
  and whiteflies
Phyllocnistis, 542
Phyllocoptruta oleivora, 101
Phyllonorycter, 546-47, 549-51
  crataegella, 552
  blancardella, 542, 548, 552
  harrisella, 541
  ringoniella, 545
Phyllophaga
  crenata, 387
  latifrons, 387
Phyllotreta cruciferae, 574,
  576, 578
Phytomyza, 537, 541-42, 545
  ilicicola, 308, 540, 546
Phytoseiid mites, 13, 101, 261,
  267-69, 313-14
Phytoseiulus persimilis, 13
Phytoseius finitimus, 101
Pieris, 70, 83
  brassicae, 69, 74-76, 80,
    214, 339
  napi, 73, 77, 318
  rapae, 69, 74-76, 80, 579
Piezodorus guildinii, 122
Plasmids
  and gene transformation, 166-
    74
  and retroviruses, 375
Platynota stultana, 409, 413
Plodia interpunctella, 174, 209,
  212, 410, 417
Plutella xylostella, 107-08
Podocanthus wilkinsoni, 643
Pollen
  and dietary self-selection, 50-
    51, 55
  and natural enemies, 565
  and pheromones, 411
Pollination
  and oviposition behavior, 78
  and rootworms, 246
Polycultures and pest pop-
  ulations, 561-86
Polymorphism and maternal
  effects, 518-25
Polyphagotarsonemus latus, 102
Polyphenism and maternal
  effects, 518-25
Polyploidy and gene amplifica-
  tion, 15
Pontia daplidice, 77
Popillia japonica, 386
Population
  analysis, 285-304
  dynamics of whiteflies, 445-
    50
Prisiphora erichsonii, 308
Prochiloneurus insolitus, 266
Promecotheca
  coeruleipennis, 552
  cumingi, 552
Prosapia bicincta, 390

Proteins and dietary self-
    selection, 47-49
Pseudaletia unipuncta, 388,
    410-11, 415
Pseudoplusia includens, 107
Psila rosae, 123, 131
Psorophora columbiae, 365-66
Psylla pyricola, 105
Psyllids and abamectins, 105
Pteronemobius fascipes, 517
Pterostichus melanarius, 575
Pyemotes, 614
    giganticus, 615
Pyrethroids and behavior, 313-
    15
Pyrrhocoris apterus, 520

R

Receptor specificity, 339
Reflex bleeding and mites, 629
Reproductive isolation, 465
Resistance
    to insecticides, 239, 241
    of plants, 234
Resource concentration and pest
    populations, 575-79
Retrotransposons, 356
Retroviruses, 355-81
Reverse transcriptase, 356-57,
    359, 369
Rhipicephalus
    appendiculatus, 660, 667
    sanguineus, 598
Rhizoglyphus, 626
    echinopus, 624-25
Rhizotrogus majalis, 386
Rhopalonoetus, 620
Rhynchaenus, 546, 548
    fagi, 437, 541-42, 545, 548-
        49, 552
RNA, 3-4, 8, 10, 12, 357
    and storage hexamers, 217,
        220
    and gene transformation, 178
    genomes, 355
    splicing, 171
    virions, 356
Rootworms, 229-55

S

Saliva and retroviruses, 362
Salivary glands and water bal-
    ance, 666-67
Sampling
    and insect population,s 285-
        304
    techniques, 292-95
Sancassania
    berlesei, 624, 626
    boharti, 625-26
    moniezi, 628

Sarcophaga
    bullata, 220, 516, 524
    peregrina, 215
Sarcoptes scabiei, 617
Scabies, 617
Scaphytopius acutus, 571
Scapteriscus
    acletus, 385
    vicinus, 385
Scaptomyza nigrita, 541
Schistocerca gregaria, 52, 310,
    521
Schizaphis graminum, 390
Schizoglyphus biroi, 618
Sclerotization and storage
    hexamers, 207, 217
Scolopalial sensilla, 333-34
Scolothrips, 267
    sexmaculatus, 104
Secondary compounds and *Eu-
    calyptus* pests, 639-40
Self-selection of optimal diets,
    43-63
Semiochemicals
    and behavioral ecology, 407-
        30
    and rootworms, 241-42, 248
Sensilla
    and host-seeking behavior,
        144-45
    of immature insects, 331-
        54
Sensory neurons, 337
Sequential
    estimations of populations,
        292-93
    sampling, 393
Sequestration of plant com-
    pounds, 73
Serotonin and dietary self-
    selection, 58
Sex ratios
    in mites, 629
    of parasitoids, 265
    and rootworms, 243
    in whiteflies, 442
Simulium, 95
Sinigrin, 69, 75
Siphoninus, 439
    phillyreae, 445, 447
Siteroptes, 614
Sitotroga cerealella, 412
Soil insecticides and rootworms,
    236-42
Solenopsis invicta, 98
Sophophora, 166
Spalgis lemolea, 264
Spatial patterns of populations,
    286-91
Sphenophorus
    parvulus, 387
    venatus, 387
Spiders
    and abamectins, 110

and dietary self-selection, 45,
    52
and leaf-miners, 547
Spilosoma congrua, 418
Spirochetes and ticks, 588
Spodoptera, 133
    eridania, 106
    exigua, 106, 111
    frugiperda, 5, 111, 124, 388
    littoralis, 54, 106
    litura,124, 131, 212, 214,
        220
Stable flies and retroviruses,
    366
Stathomorrhopa aphotista, 644
Stegomyia, 459
Steinernema feltiae, 235
Sterile-male release programs,
    163
Stethorus, 267
Stilbosis, 545, 548
    quadricustatella, 546-47
Stomoxys calcitrans, 374
Storage hexamers, 205-28
Strelkoviacarus, 617
Strip cropping, 562
Supella longipalpa, 53-54
Symmorphus canadensis, 547
Sympherobius maculipennis, 268
Sympiesis, 545
Synanthedon
    exitiosa, 107
    pictipes, 107

T

Tabanids and retroviruses, 366
Tabanus
    fuscicostatus, 362, 364, 366-
        67, 372
    lineola, 360
    nigrovittatus, 372
    nipponicus, 371-72
    quinquevittatus, 364, 366-67
    sulcifrons, 367
    trigeminus, 372
Tachygonidius dasypus, 543
Tachygonus,539
Tannins and *Eucalyptus* pests,
    639, 641
Taphrocerus, 538-39, 542-43,
    549
Tegeticula, 75, 78-79
Tehama bonifetella, 388
Telenomus dignus, 131
Teleonemia scrupulosa, 497
Temperature
    and diapause, 515-18
    and maternal effects, 512-13
    and sensilla, 334
Tenebrio molitor, 522, 626
Termites
    behavior of, 312

and *Eucalyptus*, 640, 646,
   649
and mites, 621
and storage hexamers, 213
Tetraleurodes
   acaciae, 446-47
   semilunaris, 447
   stanfordi, 437
Tetranychus
   pacificus, 104, 110
   urticae, 102-04, 109-11, 268,
      309, 313-14, 321
Thermo-hygrosensilla, 334
Thermoreceptors and host-
   seeking behavior, 145-46
Thomomydoecus, 185-203
   byersi, 194
   minor, 190, 195
   neocopei, 190, 195
   wardi, 190, 194-95
   zacatecae, 197
Thrips tabaci, 123
Ticks
   and HIV, 368
   and lyme disese, 587-609
   off host ecology, 659-81
   permeability of, 668-71
   and retroviruses, 372*Tildenia*,
      540
Tillage and rootworms, 232-33,
   240
Tineola bisselliella, 108
Tomicus piniperda, 122
Toxins amd behavior, 318
Toxorhynchites amboinensis,
   369
Translocation and gene
   amplification, 15-16, 18
Transmission
   of lyme disease, 544-95
   of retroviruses, 359-81
Transovarial infection, 594, 597
Transposable elements, 174
Transposons, 160-61, 164-67,
   169, 172-74

Trap cropping, 562
   in pest management, 199-138
   for rootworms, 233
Traps
   and pheromones
   and sound, 385-86
   for rootworms, 243-44
Trialeurodes
   abutiloneus, 448
   lauri, 448
   vaporariorum, 432-39, 441-
      43, 445, 448
Triatoma infestans, 213, 219
Tribolium, 339
   confusum, 50-51, 95
Trichodectus, 188
Trichogramma evanescens, 515
Trichogrammatoidea eldanae,
   130
Trichoplusia, 75, 213
   ni, 111, 220, 222, 409-10,
      414, 418-19
Trichopoda pilipes, 489
Trigonospila brevifacies, 491
Trissolcus
   basalis, 489
   murgantiae, 490
Trox costatus, 626
Troxocoptes minutus, 626
Turfgrass pest insects, 383-406
Typhlodromalus limonicus, 268
Typhlodromus occidentalis, 314
Tyria jacobaeae, 70
Tyrophagus formicetorum, 622

U

Uraba lugens, 643
Uroplata, 552
Utetheisa ornatrix, 419

V

Vasiform Orifice of whiteflies,
   434-36

Vector
   competence of mosquitoes,
      463-64
   DNA, 164
   potential of mosquitoes, 461
Vedalia beetle, 485
Vegetational diversity and pests,
   561-86
Vidia, 621
Virions, 356-58
Viruses and mosquitoes, 461
Vision and oviposition behavior,
   74
Vitamins and dietary self-
   selection, 48, 57
Vitellogenesis and host-seeking
   behavior, 146, 151
Vitellogenin and storage hexa-
   mers, 213, 218

W

Water balance of ticks, 660-76
Waxes of whiteflies, 434
Whitefly biologoy, 431-57
Wyeomyia
   aporonema, 153
   smithii, 153

X

Xenanoetus, 622
Xenocastor fedjushini, 623

Y

Yellow fever, 461, 463-64
Yolk deposition, 215

Z

Zeuzera coffeae, 650
Zophobas atratus, 521

CUMULATIVE INDEXES

CONTRIBUTING AUTHORS, VOLUMES 27–36

A

Aldrich, J. R., 33:211–38
Allan, S. A., 32:297–316
Allen, W. A., 35:379–97
Alstad, D. N., 27:369–84
Altieri, M. A., 29:383–402
Altner, H., 30:273–95
Ammar, E. D., 34:503–29
Andow, D. A., 36:561–86
Arends, J. J., 35:101–26
Arlian, L. G., 34:139–21
Axtell, R. C., 35:101–26
Azad, A. F., 35:553–69

B

Baker, H. G., 28:407–53
Baker, R. R., 28:65–89
Baker, T. C., 35:25–58
Balashov, Yu. S., 29:137–56
Barfield, C. S., 28:319–35
Beck, S. D., 28:91–108
Beckage, N. E., 30:371–413
Beeman, R. W., 27:253–81
Bell, W. J., 35:447–67
Bellows, T. S. Jr., 36:431–57
Bentley, M. D., 34:401–21
Berenbaum, M. R., 35:319–43
Berlocher, S. H., 29:403–33
Berry, S. J., 27:205–27
Billingsley, P. F., 35:219–48
Birch, M. C., 35:25–58
Blissard, G. W., 35:127–55
Blomquist, G. J., 27:149–72
Bloomquist, J. R., 34:77–96
Blum, M. S., 32:381–413
Bowen, M. F., 36:139–58
Bownes, M., 31:507–31
Bradley, T. J., 32:439–62
Braman, S. K., 36:383–406
Brittain, J. E., 27:119–47
Brogdon, W. G., 32:145–62
Brown, H. P., 32:253–73
Brown, T. M., 32:145–62
Burgdorfer, W., 36:587–609
Burk, T., 33:319–35
Burkholder, W. E., 30:257–72
Bush, G. L., 29:471–504
Byers, G. W., 28:203–28
Byrne, D. N., 36:431–57

C

Caltagirone, L. E., 34:1–16
Carlson, S. D., 35:597–621
Carruthers, R. I., 35:399–419
Catts, E. P., 27:313–38

Chalfant, R. B., 35:157–80
Chapman, R. F., 31:479–505
Chen, P. S., 29:233–55
Cheng, L., 30:111–35
Christensen, T. A., 34:477–501
Claridge, M. F., 30:297–317
Cochran, D. G., 30:29–49
Cohen, E., 32:71–93
Coleman, R. J., 34:53–75
Coulson, R. N., 32:415–37
Crawley, M. J., 34:531–64
Crego, C. L., 33:467–86
Croft, B. A., 29:435–70
Crossley, D. A. Jr., 31:177–94

D

Daly, H. V., 30:415–38
Danks, H. V., 33:271–96
Daoust, R. A., 31:95–119
Day, J. F., 32:297–316;
    34:401–21
De Jong, D., 27:229–52
DeFoliart, G. R., 32:479–505
Delcomyn, F., 30:239–56
Denlinger, D. L., 31:239–64
Denno, R. F., 35:489–520
Dettner, K., 32:17–48
Devonshire, A. L., 36:1–23
Diehl, S. R., 29:471–504
Dingle, H., 36:511–34
Dixon, A. F. G., 30:155–74
Dohse, L., 28:319–35
Doutt, R. L., 34:1–16
Drake, V. A., 33:183–210
Druk, A. Ya., 31:533–45
Dunn, P. E., 31:321–39
Dybas, R. A., 36:91–117

E

Edman, J. D., 32:297–316
Edmunds, G. F. Jr., 27:369–84;
    33:509–29
Edwards, J. S., 32:163–79
Edwards, P. B., 36:637–57
Eickwort, G. C., 27:229–52;
    35:469–88
Elkinton, J. S., 35:571–96

F

Fahmy, M. A. H., 31:221–37
Farrow, R. A., 33:183–210
Felsot, A. S., 34:453–76
Field, L. M., 36:1–23
Finch, S., 34:117–37
Fitt, G. P., 34:17–52

Fletcher, B. S., 32:115–44
Foil, L. D., 36:355–81
French, A. S., 33:39–58
Friedman, S., 36:43–63
Friend, J. A., 31:25–48
Fujita, S. C., 33:1–15
Futuyma, D. J., 30:217–38
Fuxa, J. R., 32:225–51

G

Gagné, W. C., 29:383–402
Galione, A., 35:345–77
Gamboa, G. J., 31:431–54
Gerling, D., 34:163–90
Getz, W. M., 27:447–66
Gould, F., 36:305–30
Grégoire, J.-C., 28:263–89
Grimstad, P. R., 32:479–505
Gut, L. J., 31:455–78
Gutierrez, A. P., 27:447–66

H

Hackman, R. H., 27:75–95
Halffter, G., 32:95–114
Hamilton, M. R. L., 35:521–51
Handler, A. M., 36:159–83
Hardy, J. L., 28:229–62
Hare, J. D., 35:81–100
Hargrove, W. W., 31:177–96
Harpaz, I., 29:1–23
Harris, M. K., 28:291–318
Hassell, M. P., 29:89–114
Haukioja, E., 36:25–42
Hawkins, C. P., 34:423–51
Haynes, K. F., 33:149–68
Hefetz, A., 34:163–90
Hellenthal, R. A., 36:185–203
Herren, H. R., 36:257–83
Hespenheide, H. A., 36:535–60
Higley, L. G., 31:341–68
Hildebrand, J. G., 34:477–501
Hogue, C. L., 32:181–99
Hokkanen, H. M. T., 36:119–38
Holman, G. M., 35:201–17
Homberg, U., 34:477–501
Houck, M. A., 36:611–36
Houk, E. J., 161–87; 28:229–62
House, G. J., 35:299–318
Howard, R. W., 27:149–72
Howarth, F. G., 28:365–89
Howarth, F. G., 36:485–509
Hoy, M. A., 30:345–70
Huddleston, E. W., 27:283–311
Hunt, H. W., 33:419–39

CONTRIBUTING AUTHORS 693

Hunter, P. E., 33:393–417
Hutchins, S. H., 31:341–68

I

Ikeda, T., 29:115–35
Illies, J., 28:391–406
Issel, C. J., 36:355–81

J

Jackai, L. E. N., 31:95–119
Jansson, R. K., 35:157–80
Jay, S. C., 31:49–65

K

Kaneshiro, K. Y., 28:161–78
Keh, B., 30:137–54
Kenmore, P. E., 33:367–91
Kevan, P. G., 28:407–53
King, E. G., 34:53–75
Kirschbaum, J. B., 30:51–70
Knight, A. L., 34:293–313
Kobayashi, F., 29:115–35
Kogan, M., 32:507–38
Kramer, L. D., 28:229–62
Krivolutsky, D. A., 31:533–45
Kuenen, L. P. S., 33:83–101
Kunkel, J. G., 36:205–28
Kuno, E., 36:285–304

L

Lacey, L. A., 31:265–96
Lamb, R. J., 34:211–29
Lane, R. S., 36:587–609
Lange, W. H., 32:341–60
Larsen-Rapport, E. W., 31:145–75
Lasota, J. A., 36:91–117
Lattin, J. D., 34:383–400
Laverty, T. M., 29:175–99
Law, J. H., 33:297–318
Lawton, J. H., 28:23–39
Levine, E., 36:229–55
Levine, J. F., 30:439–60
Liebhold, A. M., 35:571–96
Liss, W. J., 31:455–78
Lloyd, J. E., 28:131–60
Lockley, T., 29:299–320
Loftus, R., 30:273–95
Luck, R. F., 33:367–91
Lummis, S. C. R., 35:345–77

M

Ma, M., 30:257–72
MacMahon, J. A., 34:423–51
Maeda, S., 34:351–72
Matteson, P. C., 29:383–402
McCafferty, W. P., 33:509–29
McCaffery, A. R., 31:479–505

McKenzie, J. A., 32:361–80
McNeil, J. N., 36:407–30
Meeusen, R. L., 34:373–81
Moore, J. C., 33:419–39
Morse, R. A., 27:229–52
Mousseau, T. A., 36:511–34
Mumford, J. D., 29:157–74
Murdoch, W. W., 33:441–66

N

Nachman, R. J., 35:201–17
Nault, L. R., 34:503–29
Needham, G. R., 36:659–81
Neuenschwander, P., 36:257–83
Nicolas, G., 34:97–116
Norton, G. A., 29:157–74
Norton, G. W., 34:293–313

O

O'Brochta, D. A., 36:159–83
OConnor, B. M., 27:385–409
OConnor, B. M., 36:611–36
Ohmart, C. P., 36:637–57
Oloumi-Sadeghi, H., 36:229–55
Onstad, D. W., 35:399–419
Owens, E. D., 28:337–64
Owens, J. C., 27:283–311

P

Page, R. E. Jr., 31:297–320
Page, W. W., 31:479–505
Papaj, D. R., 34:315–50
Parrella, M. P., 32:201–24
Pasteels, J. M., 28:263–89
Pearson, D. L., 33:123–47
Pedigo, L. P., 31:341–68
Pellmyr, O., 36:65–89
Petersen, C. E., 28:455–86
Peterson, S. C., 30:217–38
Pfennig, D. W., 31:431–54
Piesman, J., 30:439–60
Piesman, J., 36:587–609
Pinder, L. C. V., 31:1–23
Plowright, R. C., 29:175–99
Popov, G. B., 35:1–24
Poppy, G. M., 35:25–58
Porter, A. H., 34:231–45
Potter, D. A., 36:383–406
Prestwich, G. D., 29:201–32
Price, R. D., 36:185–203
Pritchard, G., 28:1–22
Prokopy, R. J., 28:337–64; 34:315–50

R

Radcliffe, E. B., 27:173–204
Rai, K. S., 36:459–84
Rajotte, E. G., 35:379–97
Randolph, S. E., 30:197–216

Reeve, H. K., 31:431–54
Reeves, W. C., 28:229–62
Ribeiro, J. M. C., 32:463–78
Richardson, A. M. M., 31:25–48
Riechert, S. E., 29:299–320
Riley, J. R., 34:247–71
Robinson, M. H., 27:1–20
Roderick, G. K., 35:489–520
Rogers, D. J., 30:197–216
Rohrmann, G. F., 35:127–55
Rosario, R. M. T., 33:393–417
Ross, K. G., 30:319–43
Roush, R. T., 32:361–80
Rowell-Rahier, M., 28:263–89

S

Saint Marie, R. L., 35:597–621
Saunders, M. C., 32:415–37
Schal, C., 35:521–51
Schalk, J. M., 35:157–80
Scharrer, B., 32:1–16
Schmidt, J. O., 27:339–68
Schmutterer, H., 35:271–97
Schowalter, T. D., 31:177–96
Schuh, R. T., 31:67–93
Seal, M. D. R., 35:157–80
Seastedt, T. R., 29:25–46
Sehnal, F., 30:89–109
Shapiro, A. M., 34:231–45
Shapiro, J. P., 33:297–318
Shelley, A. J., 33:337–66
Shepard, B. M., 33:367–91
Shields, V. D., 36:331–54
Silk, P. J., 33:83–101
Sillans, D., 34:97–116
Smith, B. P., 33:487–507
Sögawa, K., 27:49–73
Soderlund, D. M., 34:77–96
Sonenshine, D. E., 30:1–28
Spangler, H., 33:59–81
Spielman, A., 30:439–60
Staal, G. B., 31:391–429
Stanford, J. A., 27:97–117
Stark, R. W., 27:479–509
Stimac, J. L., 28:319–35
Stinner, B. R., 35:299–318
Sullivan, D. J., 32:49–70
Sylvester, E. S., 30:71–88

T

Tallamy, D. W., 31:369–90
Taylor, C. W., 35:345–77
Taylor, L. R., 29:321–57
Teel, P. D., 36:659–81
Telfer, W. H., 36:205–28
Tempelis, C. H., 28:179–201
Terra, W. R., 35:181–200
Terriere, L. C., 29:71–88
Tesh, R. B., 33:169–81
Thompson, J. N., 36:65–89

Thompson. S. N.. 31:197–219
Thornhill. R.. 28:203–28
Thornton. I. W. B.. 30:175–96
Todd. J. W.. 34:273–92
Traniello. J. F. A.. 34:191–210
Turnipseed. S. G.. 32:507–38

U

Undeen. A. H.. 31:265–96

V

van Alphen. J. J. M.. 35:59–79
van Lenteren. J. C.. 33:239–69
Velthuis. H. H. W.. 34:163–90
Via. S.. 35:421–46
Viggiani. G.. 29:257–76
Villani. M. G.. 35:249–69
Visser. J. H.. 31:121–44
Visser. M. E.. 35:59–79

W

Waage. J. K.. 29:89–114
Waldbauer. G. P.. 36:43–63
Walde. S. J.. 33:441–66
Wallner. W. E.. 32:317–40
Wallwork. J. A.. 28:109–30
Waloff. N.. 35:1–24
Walter. D. E.. 33:419–39
Walton. R.. 33:467–86
Ward. J. V.. 27:97–117
Warren. C. E.. 31:455–78
Warren. G.. 34:373–81
Washino. R. K.. 28:179–201
Watts. D. M.. 32:479–505
Watts. J. G.. 27:283–311
Wearing. C. H.. 33:17–38
Wehner. R.. 29:277–98
Weinstein. L. H.. 27:369–84
Weis. A. E.. 33:467–86
Welch. S. M.. 29:359–81

Wells. M. A.. 33:297–318
Westigard. P. H.. 31:455–78
Whalon. M. E.. 29:435–70
Wiegert. R. G.. 28:455–86
Wikel. S. K.. 27:21–48
Wille. A.. 28:41–64
Williams. S. C.. 32:275–95
Wilson. M. L.. 30:439–60
Wirtz. R. A.. 29:47–69
Woets. J.. 33:239–69
Wood. D. L.. 27:411–46
Wood. T. K.. 31:369–90
Wright. M. S.. 35:201–17
Wright. R. J.. 35:249–69

Y

Yamane. A.. 29:115–35

Z

Zacharuk. R. Y.. 36:331–54

CHAPTER TITLES, VOLUMES 27–36

ACARINES, ARACHNIDS, AND OTHER NONINSECT ARTHROPODS
Courtship and Mating Behavior in Spiders   M. H. Robinson   27:1–20
Mite Pests of Honey Bees   D. De Jong, R. A. Morse, G. C.
     Eickwort   27:229–52
Evolutionary Ecology of Astigmatid Mites   B. M. OConnor   27:385–409
Oribatids in Forest Ecosystems   J. A. Wallwork   28:109–30
Pheromones and Other Semiochemicals of the
   Acari   D. E. Sonenshine   30:1–28
Recent Advances in Genetics and Genetic
   Improvement of the Phytoseiidae   M. A. Hoy   30:345–70
Biology of Terrestrial Amphipods   J. A. Friend, A. M. M. Richardson   31:25–48
Scorpion Bionomics   S. C. Williams   32:275–95
Associations of Mesostigmata with Other
   Arthropods   P. E. Hunter, R. M. T. Rosario   33:393–417
Host-Parasite Interaction and Impact of Larval
   Water Mites on Insects   B. P. Smith   33:487–507
Biology, Host Relations, and Epidemiology of
   Sarcoptes scabiei   L. G. Arlian   34:139–61
Associations of Mites With Social Insects   G. C. Eickwort   35:469–88

AGRICULTURAL ENTOMOLOGY
Insect Pests of Potato   E. B. Radcliffe   27:173–204
Rangeland Entomology   J. G. Watts, E. W. Huddleston,
     J. C. Owens   27:283–311
Integrated Pest Management of Pecans   M. K. Harris   28:291–318
Economics of Decision Making in Pest
   Management   J. D. Mumford, G. A. Norton   29:157–74
Developments in Computer-Based IPM
   Extension Delivery Systems   S. M. Welch   29:359–81
Modification of Small Farmer Practices for
   Better Pest Management   P. C. Matteson, M. A. Altieri,
     W. C. Gagné   29:383–402
Apple IPM Implementation in North America   M. E. Whalon, B. A. Croft   29:435–70
Insect Pests of Cowpeas   L. E. N. Jackai, R. A. Daoust   31:95–119
Economic Injury Levels in Theory and
   Practice   L. P. Pedigo, S. H. Hutchins,
     L. G. Higley   31:341–68
Perspectives on Arthropod Community
   Structure, Organization, and Development
   in Agricultural Crops   W. J. Liss, L. J. Gut, P. H.
     Westigard, C. E. Warren   31:455–78
Improved Detection of Insecticide Resistance
   Through Conventional and Molecular
   Techniques   T. M. Brown, W. G. Brogdon   32:145–62
Insect Pests of Sugar Beet   W. H. Lange   32:341–60
Computer-Assisted Decision-Making as
   Applied to Entomology   R. N. Coulson, M. C. Saunders   32:415–37
Ecology and Management of Soybean
   Arthropods   M. Kogan, S. G. Turnipseed   32:507–38
Evaluating the IPM Implementation Process   C. H. Wearing   33:17–38
Biological and Integrated Pest Control in
   Greenhouses   J. C. van Lenteren, J. Woets   33:239–69

695

Experimental Methods for Evaluating
Arthropod Natural Enemies   R. F. Luck. B. M. Shepard,
    P. E. Kenmore   33:367–91

The Ecology of *Heliothis* Species in Relation
to Agroecosystems   G. P. Fitt   34:17–52

Potential for Biological Control of *Heliothis*
Species   E. G. King. R. J. Coleman   34:53–75

Ecological Considerations in the Management
of *Delia* Pest Species in Vegetable Crops   S. Finch   34:117–37

Entomology of Oilseed *Brassica* Crops   R. J. Lamb   34:211–29

Economics of Agricultural Pesticide
Resistance in Arthropods   A. L. Knight. G. W. Norton   34:293–313

Insect Control With Genetically Engineered
Crops   R. L. Meeusen. G. Warren   34:373–81

Enhanced Biodegradation of Insecticides in
Soil: Implications for Agroecosystems   A. S. Felsot   34:453–76

Arthropods and Other Invertebrates in
Conservation Tillage Agriculture   B. R. Stinner. G. J. House   35:299–318

Ecology and Management of Arthropod Pests
of Poultry   R. C. Axtell. J. J. Arends   35:101–26

Ecology and Management of Sweet Potato
Insects   R. B. Chalfant. R. K. Jansson,
    M. D. R. Seal. J. M. Schalk   35:157–80

Integrated Suppression of Synanthropic
Cockroaches   C. Schal. M. R. L. Hamilton   35:521–51

The Changing Role of Extension Entomology
in the IPM Era   W. A. Allen, E. G. Rajotte   35:379–97

Trap Cropping in Pest Management   H. M. T. Hokkanen   36:119–38

Management of Diabroticite Rootworms in
Corn   E. Levine. H. Oloumi-Sadeghi   36:229–55

Ecology and Management of Turfgrass Insects   D. A. Potter. S. K. Braman   36:383–406

APICULTURE AND POLLINATION
Mite Pests of Honey Bees   D. De Jong, R. A. Morse,
    G. C. Eickwort   27:229–52

Insects As Flower Visitors and Pollinators   P. G. Kevan, H. G. Baker   28:407–53

Spatial Management of Honey Bees on Crops   S. C. Jay   31:49–65

BEHAVIOR
Courtship and Mating Behavior in Spiders   M. H. Robinson   27:1–20

Bioluminescence and Communication in
Insects   J. E. Lloyd   28:131–60

Visual Detection of Plants by Herbivorous
Insects   R. J. Prokopy, E. D. Owens   28:337–64

Defense Mechanisms of Termites   G. D. Prestwich   29:201–32

Astronavigation in Insects   R. Wehner   29:277–98

Pheromones and Other Semiochemicals of the
Acari   D. E. Sonenshine   30:1–28

Factors Regulating Insect Walking   F. Delcomyn   30:239–56

Pheromones for Monitoring and Control of
Stored-Product Insects   W. E. Burkholder, M. Ma   30:257–72

Acoustic Signals in the Homoptera: Behavior,
Taxonomy, and Evolution   M. F. Claridge   30:297–317

Host Odor Perception in Phytophagous Insects   J. H. Visser   31:121–44

Convergence Patterns in Subsocial Insects   D. W. Tallamy, T. K. Wood   31:369–90

The Evolution and Ontogeny of Nestmate
Recognition in Social Wasps   G. J. Gamboa, H. K. Reeve,
    D. W. Pfennig   31:431–54

Insect Hyperparasitism   D. J. Sullivan   32:49–70

Visual Ecology of Biting Flies   S. A. Allan, J. F. Day, J. D.
    Edman   32:297–316

Moth Hearing, Defense, and Communication   H. G. Spangler   33:59–81

Sex Pheromones and Behavioral Biology of the Coniferophagous *Choristoneura*	P. J. Silk, L. P. S. Kuenen	33:83–101
Sublethal Effects of Neurotoxic Insecticides on Insect Behavior	K. F. Haynes	33:149–68
Chemical Ecology of the Heteroptera	J. R. Aldrich	33:211–38
Insect Behavioral Ecology: Some Future Paths	T. Burk	33:319–35
Ecology and Behavior of *Nezara viridula*	J. W. Todd	34:273–92
Ecological and Evolutionary Aspects of Learning in Phytophagous Insects	D. R. Papaj, R. J. Prokopy	34:315–50
Chemical Ecology and Behavioral Aspects of Mosquito Oviposition	M. D. Bentley, J. F. Day	34:401–21
Environmental Influences on Soil Macroarthropod Behavior in Agricultural Systems	M. G. Villani, R. J. Wright	35:249–69
Searching Behavior Patterns in Insects	W. J. Bell	35:447–67
Self-Selection of Optimal Diets by Insects	G. P. Waldbauer, S. Friedman	36:43–63
Evolution of Oviposition Behavior and Host Preference in Lepidoptera	J. N. Thompson, O. Pellmyr	36:65–89
The Sensory Physiology of Host-Seeking Behavior in Mosquitoes	M. F. Bowen	36:139–58
Arthropod Behavior and the Efficacy of Plant Protectants	F. Gould	36:305–30
Behavioral Ecology of Pheromone-Mediated Communication in Moths and Its Importance in the Use of Pheromone Traps	J. N. McNeil	36:407–30

BIOCHEMISTRY
See PHYSIOLOGY AND BIOCHEMISTRY

BIOGEOGRAPHY
See SYSTEMATICS, EVOLUTION, AND BIOGEOGRAPHY

BIOLOGICAL CONTROL

The Chemical Ecology of Defense in Arthropods	J. M. Pasteels, J.-C. Grégoire, M. Rowell-Rahier	28:263–89
Spiders as Biological Control Agents	S. E. Riechert, T. Lockley	29:299–320
Nutrition and In Vitro Culture of Insect Parasitoids	S. N. Thompson	31:197–219
Insect Hyperparasitism	D. J. Sullivan	32:49–70
Biological and Integrated Pest Control in Greenhouses	J. C. van Lenteren, J. Woets	33:239–69
Experimental Methods for Evaluating Arthropod Natural Enemies	R. F. Luck, B. M. Shepard, P. E. Kenmore	33:367–91
The History of the Vedalia Beetle Importation to California and Its Impact on the Development of Biological Control	L. E. Caltagirone, R. L. Doutt	34:1–16
Potential for Biological Control of *Heliothis* Species	E. G. King, R. J. Coleman	34:53–75
Epizootiological Models of Insect Diseases	D. W. Onstad, R. I. Carruthers	35:399–419
Superparasitism as an Adaptive Strategy for Insect Parasitoids	M. E. Visser, J. J. M. van Alphen	35:59–79
Biological Control of Cassava Pests in Africa	H. R. Herren, P. Neuenschwander	36:257–83
Environmental Impacts of Classical Biological Control	F. G. Howarth	36:485–509

BIONOMICS
See also ECOLOGY

The Rice Brown Planthopper: Feeding Physiology and Host Plant Interactions	K. Sōgawa	27:49–73
Biology of Mayflies	J. E. Brittain	27:119–47

Biology of New World Bot Flies:
Cuterebridae   E. P. Catts   27:313–38
Biology of Tipulidae   G. Pritchard   28:1–22
Biology of the Stingless Bees   A. Wille   28:41–64
Biology of the Mecoptera   G. W. Byers, R. Thornhill   28:203–28
The Ecology and Sociobiology of Bumble
Bees   R. C. Plowright, T. M. Laverty   29:175–99
Bionomics of the Aphelinidae   G. Viggiani   29:257–76
Population Ecology of Tsetse   D. J. Rogers, S. E. Randolph   30:197–216
Bionomics of the Variegated Grasshopper
(Zonocerus variegatus) in West and Central
Africa   R. F. Chapman, W. W. Page   31:479–505
The Biology of Dacine Fruit Flies   B. S. Fletcher   32:115–44
Biology of Liriomyza   M. P. Parrella   32:201–24
Biology of Tiger Beetles   D. L. Pearson   33:123–47
Bionomics of the Large Carpenter Bees of the
Genus Xylocopa   D. Gerling, H. H. W. Velthuis,
                                                  A. Hefetz   34:163–90
Ecology and Behavior of Nezara viridula   J. W. Todd   34:273–92
Bionomics of the Nabidae   J. D. Lattin   34:383–400
Ecology and Management of the Colorado
Potato Beetle   J. D. Hare   35:81–100
Population Biology of Planthoppers   R. F. Denno, G. K. Roderick   35:489–520
Bionomics of Leaf-Mining Insects   H. A. Hespenheide   36:535–60

ECOLOGY
See also BIONOMICS; BEHAVIOR
Thermal Responses in the Evolutionary
Ecology of Aquatic Insects   J. V. Ward, J. A. Stanford   27:97–117
Effects of Air Pollutants on Insect Populations   D. N. Alstad, G. F. Edmunds, Jr.,
                                                  L. H. Weinstein   27:369–84
A Perspective on Systems Analysis in Crop
Production and Insect Pest Management   W. M. Getz, A. P. Gutierrez ·   27:447–66
Plant Architecture and the Diversity of
Phytophagous Insects   J. H. Lawton   28:23–39
Insect Territoriality   R. R. Baker   28:65–89
Dispersal and Movement of Insect Pests   R. E. Stinner, C. S. Barfield,
                                                  J. L. Stimac, L. Dohse   28:319–35
Ecology of Cave Arthropods   F. G. Howarth   28:365–89
Energy Transfer In Insects   R. G. Wiegert, C. E. Petersen   28:455–86
The Role of Microarthropods in
Decomposition and Mineralization
Processes   T. R. Seastedt   29:25–46
Host-Parasitoid Population Interactions   M. P. Hassell, J. K. Waage   29:89–114
Biology of Halobates (Heteroptera: Gerridae)   L. Cheng   30:111–35
Structure of Aphid Populations   A. F. G. Dixon   30:155–74
Genetic Variation in the Use of Resources by
Insects   D. J. Futuyma, S. C. Peterson   30:217–38
Pheromones for Monitoring and Control of
Stored-Product Insects   W. E. Burkholder, M. Ma   30:257–72
Biology of Freshwater Chironomidae   L. C. V. Pinder   31:1–23
Herbivory in Forested Ecosystems   T. D. Schowalter, W. W. Hargrove,
                                                  D. A. Crossley, Jr.   31:177–96
Dormancy in Tropical Insects   D. L. Denlinger   31:239–64
Insect Hyperparasitism   D. J. Sullivan   32:49–70
The Biology of Dacine Fruit Flies   B. S. Fletcher   32:115–44
Arthropods of Alpine Aeolian Ecosystems   J. S. Edwards   32:163–79
Biology of Riffle Beetles   H. P. Brown   32:253–73
Factors Affecting Insect Population Dynamics:
Differences Between Outbreak and
Non-Outbreak Species   W. E. Wallner   32:317–40
Evolutionary and Ecological Relationships of
the Insect Fauna of Thistles   H. Zwölfer   33:103–22

The Influence of Atmospheric Structure and
  Motions on Insect Migration   V. A. Drake, R. A. Farrow   33:183–210
Insect Behavioral Ecology: Some Future Paths T. Burk   33:319–35
Arthropod Regulation of Micro- and
  Mesobiota in Below-Ground Detrital Food
  Webs   J. C. Moore, D. E. Walter,
                                              H. W. Hunt   33:419–39
Spatial Density Dependence in Parasitoids   S. J. Walde, W. W. Murdoch   33:441–66
Reactive Plant Tissue Sites and the Population
  Biology of Gall Makers   A. E. Weis, R. Walton,
                                              C. L. Crego   33:467–86
The Ecology of *Heliothis* Species in Relation
  to Agroecosystems   G. P. Fitt   34:17–52
Foraging Strategies of Ants   J. F. A. Traniello   34:191–210
Remote Sensing in Entomology   J. R. Riley   34:247–71
Ecological and Evolutionary Aspects of
  Learning in Phytophagous Insects   D. R. Papaj, R. J. Prokopy   34:315–50
Chemical Ecology and Behavioral Aspects of
  Mosquito Oviposition   M. D. Bentley, J. F. Day   34:401–21
Guilds: The Multiple Meanings of a Concept   C. P. Hawkins, J. A. MacMahon   34:423–51
Insect Herbivores and Plant Population
  Dynamics   M. J. Crawley   34:531–64
Behavioral Ecology of Pheromone-Mediated
  Communication in Moths and Its
  Importance in the Use of Pheromone Traps   J. N. McNeil   36:407–30
Whitefly Biology   D. N. Byrne, T. S. Bellows, Jr. 36:431–57
Aedes albopictus in the Americas   K. S. Rai   36:459–84
Vegetational Diversity and Arthropod
  Population Response   D. A. Andow   36:561–86

EVOLUTION
See SYSTEMATICS, EVOLUTION, AND BIOGEOGRAPHY

FOREST ENTOMOLOGY
The Role of Pheromones, Kairomones, and
  Allomones in the Host Selection and
  Colonization Behavior of Bark Beetles   D. L. Wood   27:411–46
The Japanese Pine Sawyer Beetle as the
  Vector of Pine Wilt Disease   F. Kobayashi, A. Yamane, T. Ikeda 29:115–35
Population Dynamics of Gypsy Moth in North
  America   J. S. Elkinton, A. M. Liebhold   35:571–96
Induction of Defenses in Trees   E. Haukioja   36:25–42
Insect Herbivory on Eucalyptus   C. P. Ohmart, P. B. Edwards   36:637–57

GENETICS
Sexual Selection and Direction of Evolution
  in the Biosystematics of Hawaiian
  Drosophilidae   K. Y. Kaneshiro   28:161–78
Potential Implication of Genetic Engineering
  and Other Biotechnologies to Insect Control J. B. Kirschbaum   30:51–70
Recent Advances in Genetics and Genetic
  Improvement of the Phytoseiidae   M. A. Hoy   30:345–70
Imaginal Disc Determination: Molecular and
  Cellular Correlates   E. W. Larsen-Rapport   31:145–75
Expression of the Genes Coding for
  Vitellogenin (Yolk Protein)   M. Bownes   31:507–31
Ecological Genetics of Insecticide and
  Acaricide Resistance   R. T. Roush, J. A. McKenzie   32:361–80
Use of Hybridoma Libraries in the Study of
  the Genetics and Development of
  Drosophila   S. C. Fujita   33:1–15
Baculovirus Diversity and Molecular Biology   G. W. Blissard, G. F. Rohrmann   35:127–55

Ecological Genetics and Host Adaptation in Herbivorous Insects: The Experimental Study of Evolution in Nat. and Agric. Systems — S. Via — 35:421–46
Gene Amplification and Insecticide Resistance — A. L. Devonshire, L. M. Field — 36:1–23
Prospects for Gene Transformation in Insects — A. M. Handler, D. A. O'Brochta — 36:159–83

HISTORICAL
Frederick Simon Bodenheimer (1897–1959): Idealist, Scholar, Scientist — I. Harpaz — 29:1–23
Cultural Entomology — C. L. Hogue — 32:181–99
The History of the Vedalia Beetle Importation to California and Its Impact on the Development of Biological Control — L. E. Caltagirone, R. L. Doutt — 34:1–16
Sir Boris Uvarov (1889-1970): The Father of Acridology — N. Waloff, G. B. Popov — 35:1–24

INSECTICIDES AND TOXICOLOGY
Recent Advances in Mode of Action of Insecticides — R. W. Beeman — 27:253–81
Induction of Detoxication Enzymes in Insects — L. C. Terriere — 29:71–88
Derivatization Techniques in the Development and Utilization of Pesticides — M. A. H. Fahmy — 31:221–37
Chitin Biochemistry: Synthesis and Inhibition — E. Cohen — 32:71–93
Improved Detection of Insecticide Resistance Through Conventional and Molecular Techniques — T. M. Brown, W. G. Brogdon — 32:145–62
Ecological Genetics of Insecticide and Acaricide Resistance — R. T. Roush, J. A. McKenzie — 32:361–80
Sublethal Effects of Neurotoxic Insecticides on Insect Behavior — K. F. Haynes — 33:149–68
Neurotoxic Actions of Pyrethroid Insecticides — D. M. Soderlund, J. R. Bloomquist — 34:77–96
Enhanced Biodegradation of Insecticides in Soil: Implications for Agroecosystems — A. S. Felsot — 34:453–76
Properties and Potential of Natural Pesticides From the Neem Tree, Azadirachta indica — H. Schmutterer — 35:271–97
Gene Amplification and Insecticide Resistance — A. L. Devonshire, L. M. Field — 36:1–23
Avermectins, a Novel Class of Compounds: Implications for Use in Arthropod Pest Control — J. A. Lasota, R. A. Dybas — 36:91–117

MEDICAL AND VETERINARY ENTOMOLOGY
Immune Responses to Arthropods and Their Products — S. K. Wikel — 27:21–48
Biology of New World Bot Flies: Cuterebridae — E. P. Catts — 27:313–38
Mosquito Host Bloodmeal Identification: Methodology and Data Analysis — R. K. Washino, C. H. Tempelis — 28:179–201
Intrinsic Factors Affecting Vector Competence of Mosquitoes for Arboviruses — J. L. Hardy, E. J. Houk, L. D. Kramer, W. C. Reeves — 28:229–62
Allergic and Toxic Reactions to Non-Stinging Arthropods — R. A. Wirtz — 29:47–69
Interaction Between Blood-Sucking Arthropods and Their Hosts, and its Influence on Vector Potential — Yu. S. Balashov — 29:137–56
Scope and Applications of Forensic Entomology — B. Keh — 30:137–54
Ecology of Ixodes dammini–borne Human Babesiosis and Lyme Disease — A. Spielman, M. L. Wilson, J. F. Levine, J. Piesman — 30:439–60
Microbial Control of Black Flies and Mosquitoes — L. A. Lacey, A. H. Undeen — 31:265–96

Role of Saliva in Blood-Feeding by
  Arthropods   J. M. C. Ribeiro   32:463–78
Advances in Mosquito-Borne
  Arbovirus/Vector Research   G. R. DeFoliart, P. R. Grimstad,
                                                D. M. Watts   32:479–505
The Genus *Phlebovirus* and its Vectors   R. B. Tesh   33:169–81
Vector Aspects of the Epidemiology of
  Onchocerciasis in Latin America   A. J. Shelley   33:337–66
Biology, Host Relations, and Epidemiology of
  Sarcoptes scabiei   L. G. Arlian   34:139–61
Epidemiology of Murine Typhus   A. F. Azad   35:553–69
Transmission of Retroviruses by Arthropods   L. D. Foil, C. J. Issel   36:355–81
Aedes albopictus in the Americas   K. S. Rai   36:459–84
Lyme Borreliosis: Relation of Its Causative
  Agent to Its Vectors and Hosts in North
  America and Europe   R. S. Lane, J. Piesman,
                                                W. Burgdorfer   36:587–609

MORPHOLOGY
Structure and Function in Tick Cuticle   R. H. Hackman   27:75–95
The Functional Morphology and Biochemistry
  of Insect Male Accessory Glands and Their
  Secretions   P. S. Chen   29:233–55
Morphology of Insect Development   F. Sehnal   30:89–109
Ultrastructure and Function of Insect Thermo-
  and Hygroreceptors   H. Altner, R. Loftus   30:273–95
Transduction Mechanisms of Mechanosensilla   A. S. French   33:39–58
Structure and Function of the Deutocerebrum
  in Insects   U. Homberg, T. A. Christensen,
                                                J. G. Hildebrand   34:477–501
Scents and Eversible Scent Structures of Male
  Moths   M. C. Birch, G. M. Poppy,
                                                T. C. Baker   35:25–58
The Midgut Ultrastructure of Hematophagous
  Insects   P. F. Billingsley   35:219–48
Structure and Function of Insect Glia   S. D. Carlson, R. L. Saint Marie   35:597–621
The Function and Evolution of Insect Storage
  Hexamers   W. H. Telfer, J. G. Kunkel   36:205–28
Sensilla of Immature Insects   R. Y. Zacharuk, V. D. Shields   36:331–54

PATHOLOGY
Potential Implication of Genetic Engineering
  and Other Biotechnologies to Insect Control   J. B. Kirschbaum   30:51–70
Microbial Control of Black Flies and
  Mosquitoes   L. A. Lacey, A. H. Undeen   31:265–96
Ecological Considerations for the Use of
  Entomopathogens in IPM   J. R. Fuxa   32:225–51
Scents and Eversible Scent Structures of Male
  Moths   M. C. Birch, G. M. Poppy,
                                                T. C. Baker   35:25–58
Structure and Function of Insect Glia   S. D. Carlson, R. L. S. Marie   35:597–621
The Midgut Ultrastructure of Hematophagous
  Insects   P. F. Billingsley   35:219–48

PHYSIOLOGY AND BIOCHEMISTRY
The Rice Brown Planthopper: Feeding
  Physiology and Host Plant Interactions   K. Sōgawa   27:49–73
Structure and Function in Tick Cuticle   R. H. Hackman   27:75–95
Chemical Ecology and Biochemistry of Insect
  Hydrocarbons   R. W. Howard, G. J. Bloomquist   27:149–72
Maternal Direction of Oogenesis and Early
  Embryogenesis in Insects   S. J. Berry   27:205–27
Biochemistry of Insect Venoms   J. O. Schmidt   27:339–68

Insect Thermoperiodism ... S. D. Beck ... 28:91–108
Nitrogen Excretion in Cockroaches ... D. G. Cochran ... 30:29–49
Regulation of Reproduction in Eusocial
  Hymenoptera ... D. J. C. Fletcher, K. G. Ross ... 30:319–43
Endocrine Interactions Between Endoparasitic
  Insects and heir Hosts ... N. E. Beckage ... 30:371–413
Imaginal Disc Determination: Molecular and
  Cellular Correlates ... E. W. Larsen-Rapport ... 31:145–75
Nutrition and In Vitro Culture of Parasitoids ... S. N. Thompson ... 31:197–219
Sperm Utilization in Social Insects ... R. E. Page, Jr. ... 31:297–320
Biochemical Aspects of Insect Immunology ... P. E. Dunn ... 31:321–39
Anti Juvenile Hormone Agents ... G. B. Staal ... 31:391–429
Expression of the Genes Coding for
  Vitellogenin (Yolk Protein) ... M. Bownes ... 31:507–31
Insects as Models in Neuroendocrine Research ... B. Scharrer ... 32:1–16
Chitin Biochemistry: Synthesis and Inhibition ... E. Cohen ... 32:71–93
Biosynthesis of Arthropod Exocrine
  Compounds ... M. S. Blum ... 32:381–413
Physiology of Osmoregulation in Mosquitoes ... T. J. Bradley ... 32:439–62
Transduction Mechanisms of Mechanosensilla ... A. S. French ... 33:39–58
Sex Pheromones and Behavioral Biology of
  the Coniferophagous *Choristoneura* ... P. J. Silk, L. P. S. Kuenen ... 33:83–101
Chemical Ecology of the Heteroptera ... J. R. Aldrich ... 33:211–38
Lipid Transport in Insects ... J. P. Shapiro, J. H. Law,
  ... M. A. Wells ... 33:297–318
Immediate and Latent Effects of Carbon
  Dioxide on Insects ... G. Nicolas, D. Sillans ... 34:97–116
Expression of Foreign Genes in Insects Using
  Baculovirus Vectors ... S. Maeda ... 34:351–72
Evolution of Digestive Systems of Insects ... W. R. Terra ... 35:181–200
Insect Neuropeptides ... G. M. Holman, R. J. Nachman,
  ... M. S. Wright ... 35:201–17
Transmembrane Signaling in Insects ... S. C. R. Lummis, A. Galione,
  ... C. W. Taylor ... 35:345–77
The Sensory Physiology of Host-Seeking
  Behavior in Mosquitoes ... M. F. Bowen ... 36:139–58
Off-Host Physiological Ecology of Ixodid
  Ticks ... G. R. Needham, P. D. Teel ... 36:659–81

POPULATION ECOLOGY
Assessing and Interpreting the Spatial
  Distributions of Insect Populations ... L. R. Taylor ... 29:321–57
Sampling and Analysis of Insect Populations ... E. Kuno ... 36:285–304

SERICULTURE

SYSTEMATICS, EVOLUTION, AND BIOGEOGRAPHY
Changing Concepts in Biogeography ... J. Illies ... 28:391–406
Insect Molecular Systematics ... S. H. Berlocher ... 29:403–33
An Evolutionary and Applied Perspective of
  Insect Biotypes ... S. R. Diehl, G. L. Bush ... 29:471–504
The Geographical and Ecological Distribution
  of Arboreal Psocoptera ... I. W. B. Thornton ... 30:175–96
Insect Morphometrics ... H. V. Daly ... 30:415–38
The Influence of Cladistics on Heteropteran
  Classification ... R. T. Schuh ... 31:67–93
The Evolution and Ontogeny of Nestmate
  Recognition in Social Wasps ... G. J. Gamboa, H. K. Reeve,
  ... D. W. Pfennig ... 31:431–54
Fossil Oribatid Mites ... D. A. Krivolutsky, A. Ya. Druk ... 31:533–45
Chemosystematics and Evolution of Beetle
  Chemical Defenses ... K. Dettner ... 32:17–48

Biogeography of the Montane Entomofauna of
  Mexico and Central America   G. Halffter   32:95–114
Use of Hybridoma Libraries in the Study of
  the Genetics and Development of
  Drosophila   S. C. Fujita   33:1–15
Systematics in Support of Entomology   H. V. Danks   33:271–96
The Mayfly Subimago   G. F. Edmunds, Jr.,
      W. P. McCafferty   33:509–29
The Lock-and-Key Hypothesis: Evolutionary
  and Biosystematic Interpretation of Insect
  Genitalia   A. M. Shapiro, A. H. Porter   34:231–45
Bionomics of the Nabidae   J. D. Lattin   34:383–400
Evolution of Specialization in
  Insect-Umbellifer Associations   M. R. Berenbaum   35:319–43
Evolution of Oviposition Behavior and Host
  Preference in Lepidoptera   J. N. Thompson, O. Pellmyr   36:65–89
Biosystematics of the Chewing Lice of Pocket
  Gophers   R. A. Hellenthal, R. D. Price   36:185–203
Maternal Effects in Insect Life Histories   T. A. Mousseau, H. Dingle   36:511–34
Ecological and Evolutionary Significance of
  Phoresy in the Astigmata   M. A. Houck, B. M. OConnor   36:611–36

VECTORS OF PLANT PATHOGENS
Multiple Acquisition of Viruses and
  Vector-Dependent Prokaryotes:
  Consequences on Transmission   E. S. Sylvester   30:71–88
Leafhopper and Planthopper Transmission of
  Plant Viruses   L. R. Nault, E. D. Ammar   34:503–29